Organic Chemistry Through Solved Problems

Organic Chemistry Through Solved Problems

Goutam Brahmachari

Alpha Science International Ltd.
Oxford, U.K.

Goutam Brahmachari
Department of Chemistry
Visva-Bharati University
P.O. Santiniketan
West Bengal, India

Alpha Science International Ltd.
7200 The Quorum, Oxford Business Park North
Garsington Road, Oxford OX4 2JZ, U.K.

www.alphasci.com

Printed from the camera-ready copy provided by the Author.

ISBN 978-1-84265-398-2

Printed in India

To

My mother, my wife, and
little Asanjan (my son)

FOREWORD

I feel it a pleasure to write a foreword to the book, *Organic Chemistry Through Solved Problems*, written by Dr. Goutam Brahmachari.

This interactive book is designed to provide fundamental aspects of organic chemistry in a flexible way rather than a traditional approach. The book caters the requisite underpinning knowledge in the subject through solved problems and provides supplementary problems in each chapter enabling self-assessment as a tool to understanding the subject.

The contents of the book give a general impression about the way the author wants to deal with. The stepwise development of comprehensive knowledge in the areas of fundamental concepts, stereochemistry and reactive intermediate chemistry paved the way to travel through the rocky area of reaction mechanism. A suitable balance in approaching organic synthesis, pericyclic reactions and organic photochemistry, and complimentary spectroscopic techniques, the key to chemical analysis, along with heterocycles and carbohydrates follows. The addition of separate chapter each on heterocycles and carbohydrate chemistry has definitely enriched the contents of the book. Some of the problems are at a rather more advanced level, the aim being to extend the reader's knowledge and understanding.

While teaching, we felt the need of such a handy and concise book which can serve the undergraduate and to some extent the post-graduate students besides teachers and examinees to professional examinations. I must thank Dr. Brahmachari for his arduous effort in attempting to fulfill the long standing requirement in this area.

I am very much happy in recommending this book as a self-guide material for students and professionals.

Swapnadip Thakur
Former Professor of Chemistry
The University of Burdwan
Burdwan India

PREFACE

Modern organic chemistry is an intimate outcome of the combined knowledge of electronic theory, reaction mechanism, stereochemistry, spectroscopic techniques, and the interpretation of physical data, as well as the older areas of synthetic and degradative reactions. The students, particularly the beginners in organic chemistry are often thus overwhelmed by facts, concepts and quantity of subject matter dealt in enormous number of text and reference books. To meet the requirements of their syllabi, students have to consult various books — no one by itself is sufficient enough to meet all the requirements.

A fruitful way for the students, to make their knowledge sound and also to acquire command over the subject, is to deal with well-designed problems, basic as well as applied, covering various aspects of organic chemistry. To develop an all round conception one must need to build up a skill to tackle a problem and to solve it with perfection. Conception on a particular topic becomes enriched and nourished by handling well-documented problems. The present book — *Organic Chemistry Through Solved Problems* — has been designed with this particular goal, covering a vast area of organic chemistry. The main object is for students to learn by thinking and solving problems rather than by merely being told. Screening of problems on each topic is organized likewise, and the solutions are presented in such a manner so that the subject matter becomes clear, understandable and readily assimilated.

The book covers a wide area including fundamental concepts providing the students a solid understanding of the basics of organic chemistry. Chapters are successively arranged as Fundamental Concepts, Stereochemitsry, Reaction Intermediates, Organic Reaction Mechanism, Organic Syntheses, Pericyclic Reactions and Organic Photochemistry, Organic Spectroscopy, Heterocycles, and Carbohydrates. Besides the vast number of worked-out problems ranging from the very basic to the inner depth, each chapter provides supplementary problems so as to enable the students for self-assessment; working-out these problems independently will provide them the opportunity to brush up their knowledge and understanding of the subject. A special topic on the applications of some useful reagents in organic synthesis has been discussed in ***appendix-1*** that will also be very helpful.

During preparation of this book, a good number of problems have been adapted from the previous year question papers of a number of Indian Universities as well as previous year question papers of different competitive examinations involving Chemistry as a subject. For exact and specific clues to the problems, I have consulted numerous original literature sources (a very few of which have been referred in the text in respective places) in combination with various standard reference books. I am wholeheartedly acknowledging the help from all these sources, and express my deep gratitude to the respective authors and publishers concerned. A list of standard reference books that have been deeply consulted by me is cited in '***appendix-4***' as a recommendation for further studies.

This book can be used in support of a standard text, and also as a self-guide to the undergraduate and postgraduate students as well as to teachers and professionals; besides, examinees to various professional examinations including Joint UGC-CSIR Eligibility Test (NET), SLET, GATE, etc. will find the book much useful and essential.

I would like to express my deep sense of gratitude to Prof. S. Thakur (The University of Burdwan, Burdwan) for his keen interest in my manuscript and for writing a foreword to the book. I am also thankful to all of my research fellows — most specifically names of Arindam, Dilip, Sadhan and Shyamal are to be mentioned — for their constant inspiration during the period of writing and also for going through the proof. Thanks are due to my colleagues and students also.

Special thanks are due to Mr. N. K. Mehra, Managing Director and Publisher, Narosa Publishing House Pvt. Ltd., New Delhi, India for his earnest interest in publishing the work. I am also thankful to Mrs. Anupama Jauhry, the Production Manager, and other Staff-members of the Publishing House for sparing no pains to ensure that the highest standards of publication are maintained in bringing out the volume.

My effort will be successful only when it is found helpful to the students and teachers. Every step has been taken to make the manuscript errorless; in spite of that, some errors might have crept in. Any remaining error is, of course, of my own. Constructive comments on the content and approach of the book from the readers will be highly appreciated (my e-mail address is brahmg2001@yahoo.co.in).

Last but not the least, I wish to thank all of my family members, very particularly to my wife, Piyasi, for their well understanding and support throughout the entire period of writing. Without their support, this work would not have been possible.

Goutam Brahmachari

CONTENTS

CHAPTER 1

FUNDAMENTAL CONCEPTS

Introduction

The beginners in organic chemistry primarily need to conceive a clear idea on some fundamental concepts before going into the depth of the subject. Fundamental concepts mainly include IUPAC systems of nomenclatures, the concept of hybridization (bonding and structure), electronegativity, dipole moment, hydrogen bonding, hyperconjugation, inductive effects, resonance, tautomerism, steric effect, aromaticity, and acidity and basicity of organic acids and bases. Hence, this chapter has been designed to provide a clear view of these concepts through the detailed solution of ample of illustrative problems screened very carefully. Students are advised to follow them critically for better understanding of the subject.

Solved Problems

Problem 1.1 Give IUPAC nomenclature to each of the following molecules:

(i)
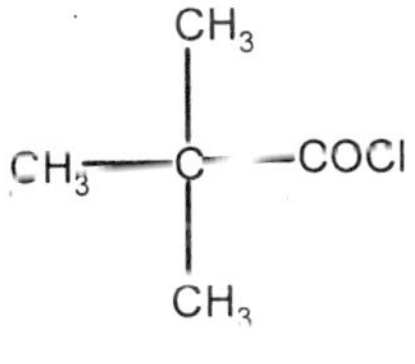

(ii)
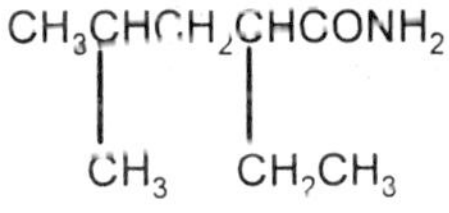

(iii)
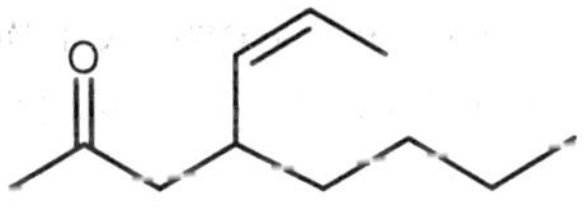

(iv)
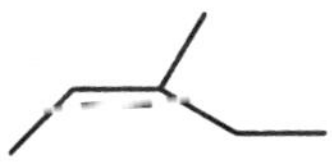

(v)
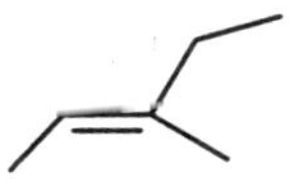

(vi)
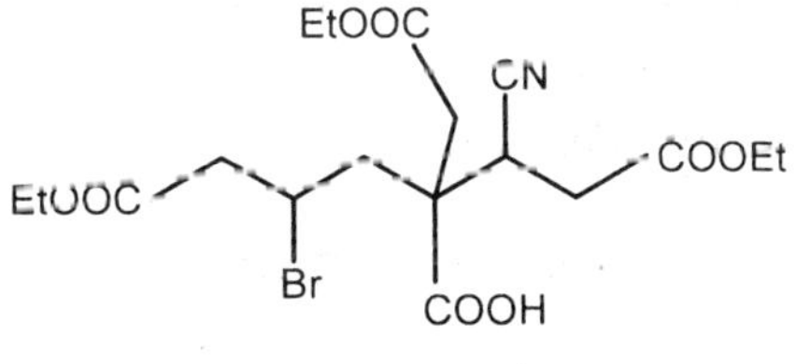

(vii) 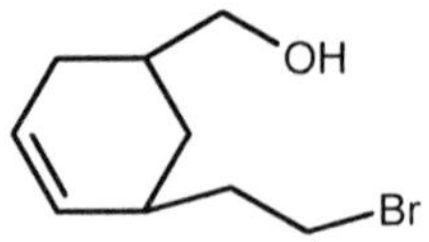 (viii) 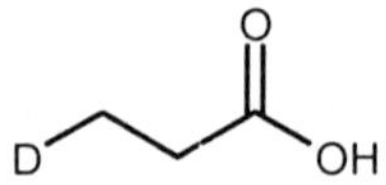 (ix) 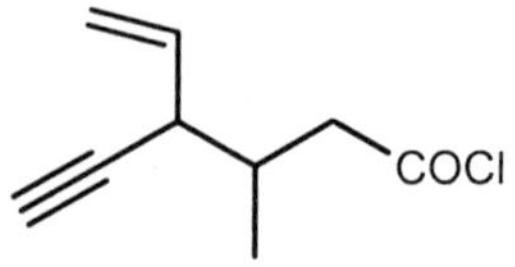

(x) 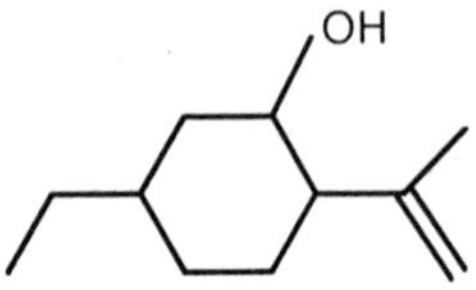 (xi) 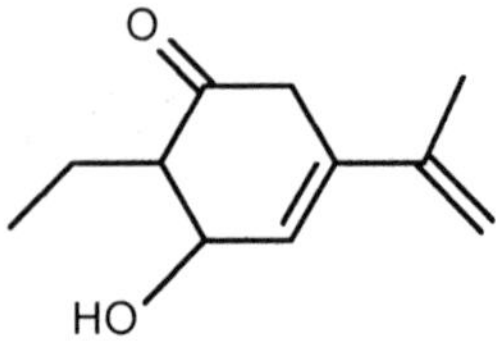 (xii) 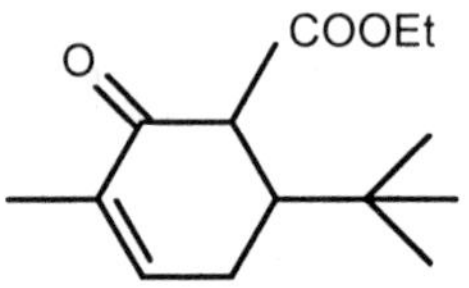

(xiii) 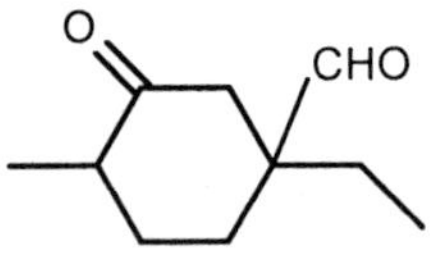 (xiv) 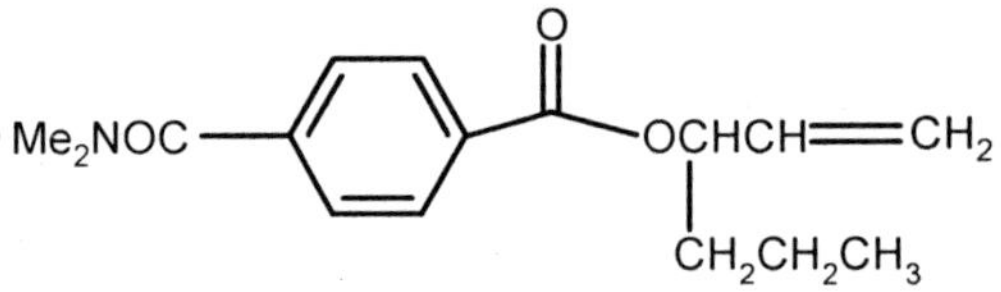 (xv) 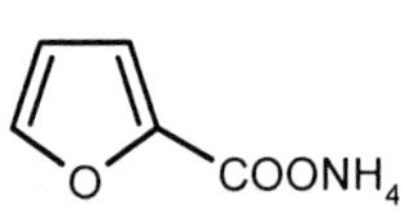

(xvi) 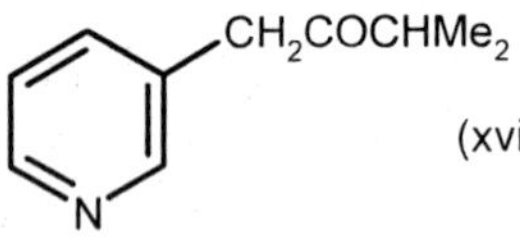 (xvii) 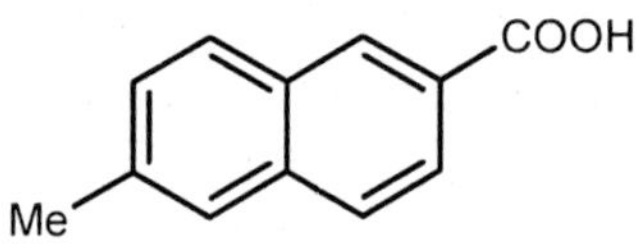 (xviii) 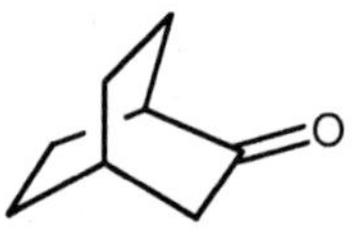

Solutions. (i) 2,2-Dimethylpropanoyl chloride; (ii) 2-Ethyl-4-methylpentanamide; (iii) 4-Butylhept-5-en-2-one; (iv) (*Z*)-3-Methyl-2-pentene; (v) (*E*)-3-Methyl-2-pentene; (vi) 4-Bromo-2-(2-carbethoxy-1-cyanoethyl)-2-(1-carbethoxymethyl)-5-carbethoxypentan-oic acid;(vii) [5-(2-Bromoethyl)cyclohex-2-en-1-yl]ethan-2-ol; (viii) 3-Deuterio-propanoic acid; (ix) 2-Chloroformyl]-3-ethynyl-pent-4-en-1-oic acid; (x) 2-[1-Methyl-1-ethenyl]-5-ethyl cyclohexanol; (xi) 3-[1-Methyl-1-ethenyl]-5-hydroxy-6-ethylcyclo- hex-3-en-1-one; (xii) Ethyl 6-[1,1-Dimethylethyl]-3-methyl-2-oxocyclohex-3-en-1-carboxylate; (xiii) 1-Ethyl-4-methyl-3-oxocyclohexane-1-carbaldehyde; (xiv) 1'-Propylprop-2'-enyl-4-*N*,*N*-dimethylcarbamoylbenzoate; (xv) Ammonium 2-furan carboxylate; (xvi) 3-Methyl-1-(3-pyridyl)-2-butanone; (xvii) 6-Methyl-2-naphthoic acid; (xviii) Bicylo[2.2.2]octan-2-one.

Problem 1.2 (a) What is 'hybrid orbital number' method used to determine the hybridized state of an atom? Use this method to determine the hybridized state of the underlined atoms.

(i) $:\underline{C}Cl_2$, (ii) $H_2\underline{C}{=}CH_2$, (iii) $H\underline{C}\equiv CH$, (iv) $CH_3C\equiv\underline{\ddot{N}}$, (v) $H_2\underline{\ddot{O}}:$

(b) Designate the state of hybridization of all the atoms involved in the following molecule:

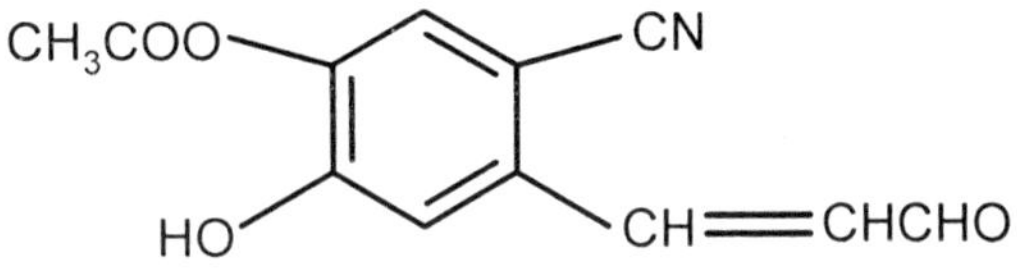

(c) Draw the orbital picture for the following molecules:
(i) ethylene, (ii) acetylene, (iii) prop-1,2-diene, (iv) acraldehyde, (v) ketene.

Solutions. (a) From the general concept of hybridization it is revealed that *each unshared and σ-bonded pair of electrons requires a hybrid orbital, but π–bonds do not*. The 'hybrid orbital number' needed by an atom within a molecule may be obtained as:

Hybrid orbital number = (number of σ-bonds) +(number of unshared pairs of electrons)

This method can be applied for a general use to determine hybridization state of an atom within a molecule. Let us determine the hybridization state for the atoms as concerned:

Entry	Atom (underlined) within the molecule concerned	Number of σ-bonds (X)	Number of unshared electron pairs (Y)	Hybrid orbital number (= X + Y)	Hybridiz-ation state
i	$:\underline{C}Cl_2$	2	1	3	sp^2
ii	$H_2\underline{C}{=}CH_2$	3	0	3	sp^2
iii	$\underline{C}_2H_2$	2	0	2	sp
iv	$CH_3C\underline{N}$	1	1	2	sp
v	$H_2\underline{O}$	2	2	4	sp^3

(b) Different atoms within the molecule are designated as a, b, & c according to their hybridization status: a = sp^3; b = sp^2; and c = sp.

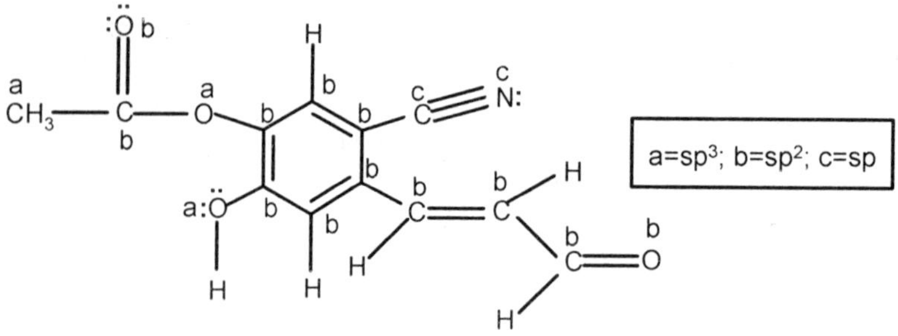

(c)

(i) Ethylene: $H_2C{=}CH_2$

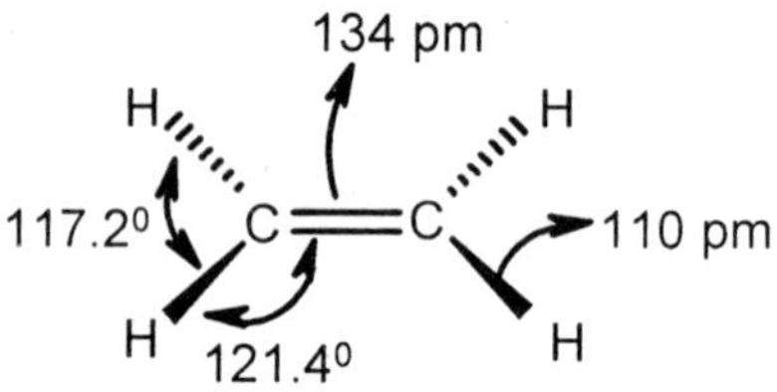

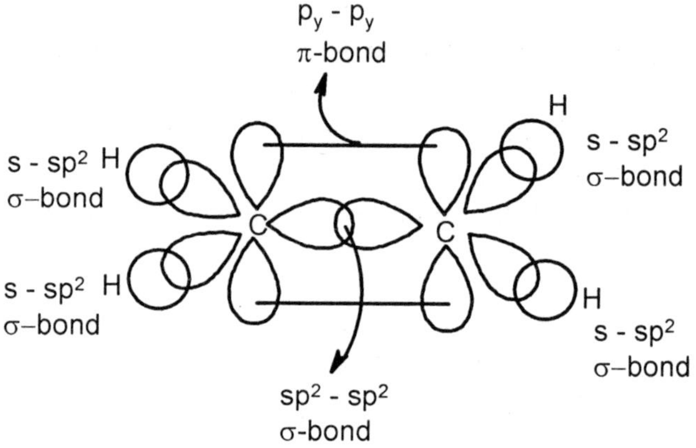

(ii) Acetylene: C_2H_2

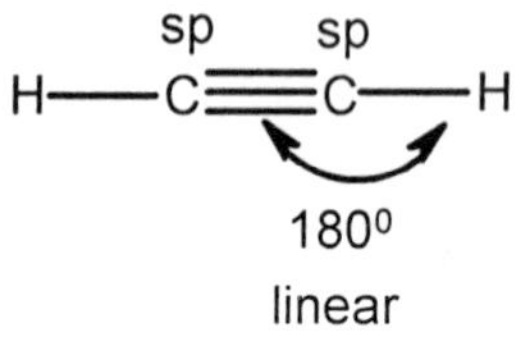

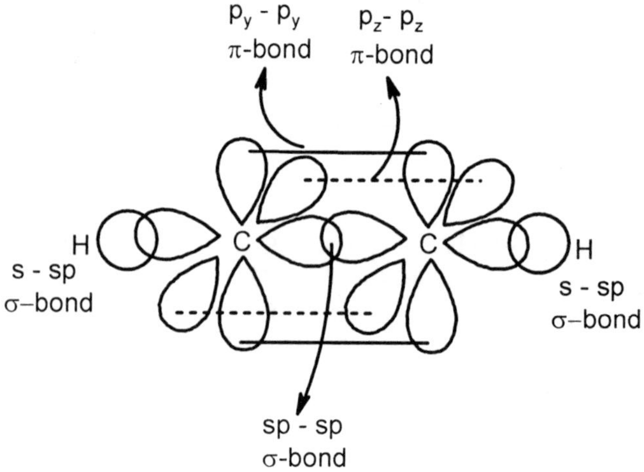

(iii) Prop-1,2-diene: It is an allene. The central carbon (C-2) is sp-hybridized, and the two end carbons are sp^2-hybridized. The two terminal carbon atoms lie in perpendicular planes. The orbital diagram for the molecule is given below:

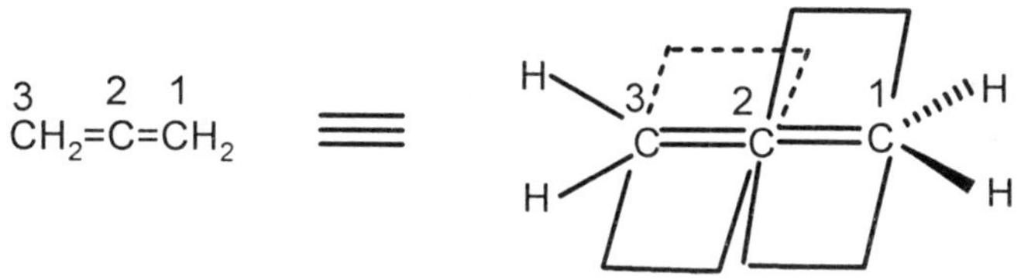

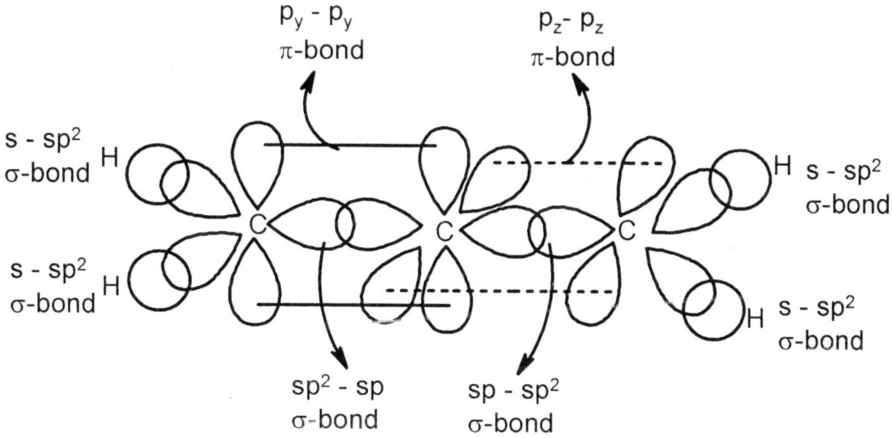

(iv) Acraldehyde: CH_2=CH-CHO
All the three carbons are sp^2 hybridized. The oxygen atom is probably sp^2 hybridized because it is σ-bonded to one carbon atom and has two lone pairs, requiring a total of three hybrid orbitals. However, it is impossible to verify whether the oxygen atom is really sp^2 hybridized because there is no experimental proof (bond angles of the lone pairs on oxygen cannot be measured experimentally) in support of this prediction.

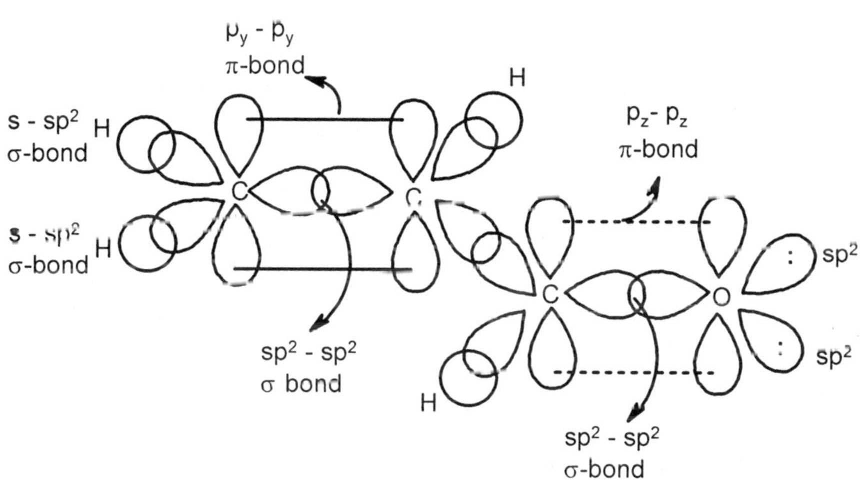

(v) Ketene: $CH_2=C=O$

Terminal carbon and oxygen atoms are sp^2 hybridized, whereas the central carbon is sp hybridized. However, prediction of sp^2 hybridized state for the oxygen has no experimental verification.

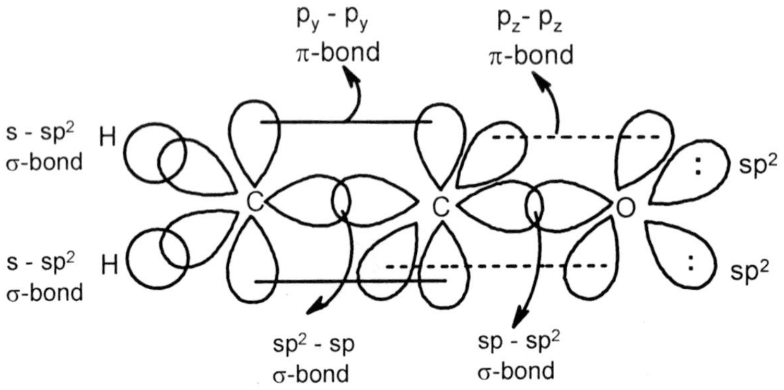

Problem 1.3 Compare the dipole moment of *cis*-1,2-dichloroethylene to that of the *trans*-isomer.

Solution.

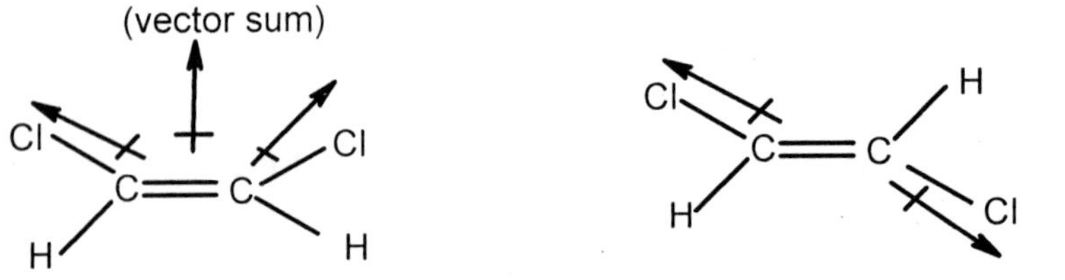

cis-1,2-dichloroethylene (μ =1.89 D)

trans-1,2-dichloroethylene (μ = 0 D)

Cis-1,2-dichloroethylene possesses considerable amount of dipole moment while the *trans*-isomer is of zero value. Both the molecules are planar — all the atoms in both of them lie in the same plane. In *cis*-isomer, the chlorine atoms are on the same side of the C=C double bond and hence, the individual C–Cl bond moments result to a net dipole moment (a vectorial sum) for the molecule. But in the *trans*-isomer, the chlorine atoms are on the opposite sides of the C=C double bond; the two individual C–Cl bond moments (which are equal in magnitude) become opposite in direction (anti-parallel), and thereby mutually cancel each other resulting a net dipole moment of zero for the molecule as a whole.

Problem 1.4 Comment on the dipole moment of *para*-dihydroxybenzene.

Solution.

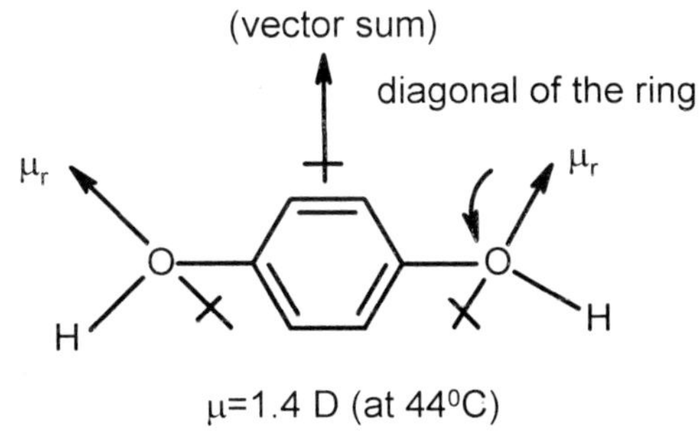

p-Dihydroxybenzene is found to possess appreciable amount of dipole moment (1.4 D at 44^{0}C). From the molecular diagram of the compound, it can easily be understood that the two O–H bonds are not along the diagonal of the ring, and hence the resultant group moments (μ_r) (one for each –OH group) are not in opposite direction. They result a net dipole moment for the molecule (μ 1.4D).

Problem 1.5 Between ethyl chloride and vinyl chloride which one is more polar?
Solution.

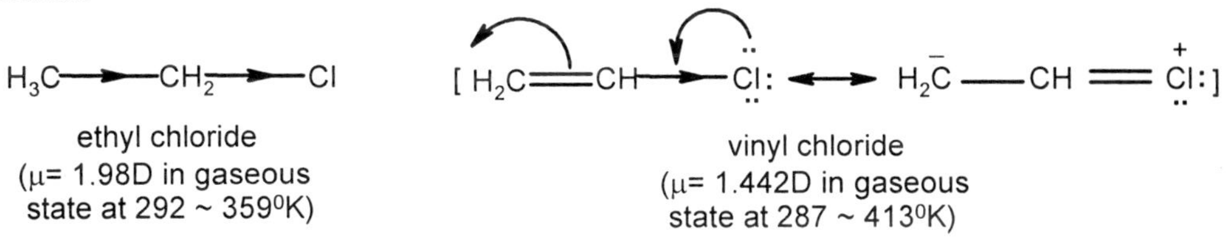

Chlorine is more electronegative than carbon; so the C–Cl bonding electron pair is attracted towards the chlorine atom making the bond polar. Displacement of the bonding electron pair towards chlorine is more pronounced in ethyl chloride in comparison to that in vinyl chloride. This is because of the fact that involvement of lone pair electrons on chlorine in vinyl chloride into resonance interaction with the C=C π-electrons opposes the usual displacement of C–Cl bonding electron pair towards chlorine to some extent. Moreover, C_1 carbon (sp^2 hybridized) in CH_2=CHCl, being more electronegative than that (sp^3 hybridized) in ethyl chloride, pushes the C–Cl bonding electrons towards the chlorine atom less effectively. But as a whole, there is still a net displacement of electrons towards chlorine in vinyl chloride, though is less pronounced. That's why ethyl chloride becomes more polar than vinyl chloride, and the former possesses greater dipole moment (μ) as mentioned above.

Problem 1.6 The dipole moment of 1,2-dichloroethane increases as the temperature is increased. Account for the fact.

Solution. Predominant conformations for 1,2-dichloroethane are the *anti*- and gauche (*P*-gauche and *M*-gauche) conformations.

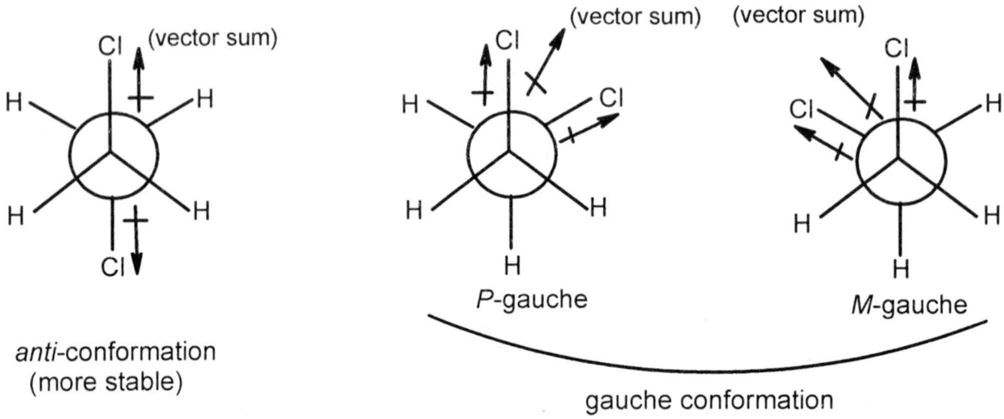

The dipole moment of *anti*-form is assumed to be zero since the two C–Cl dipoles are in opposite direction, but gauche conformations of the molecule possess a sort of dipole moment (μ~3.2D as calculated with reference to ethyl chloride). *Anti*-conformation is more stable than the gauche — gauche form is more energetic due to the presence of dipolar and steric repulsions within it. Thus in the gaseous state at 22^0C, 1,2-dichloroethane exists predominantly in the *anti*-form by 73%. Due to such conformational distribution the compound as a whole shows a lower value for its dipole moment at ordinary temperature. But as the temperature increases, the population of comparatively less stable and more polar gauche conformation increases — as a result of which the overall dipole moment of the compound is enhanced (μ 1.43 ~ 1.51D at 308 ~ 372^0K).

Problem 1.7 Comment on the dipole moment of *p*-nitroaniline.

Solution.

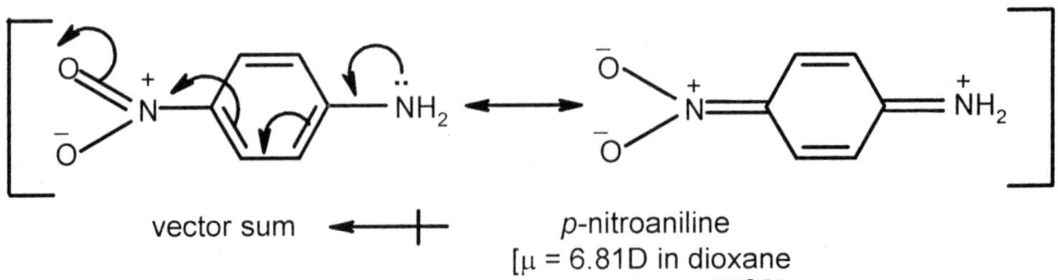

p-nitroaniline
[μ = 6.81D in dioxane solution at 25⁰C]

In *p*-nitroaniline, the group moments of $-NH_2$ (μ_{gas} = 1.48D) and $-NO_2$ (μ_{gas} = 4.19D) act in the same direction; so superficially it may be expected that the dipole moment of *p*-nitroaniline will be the sum of the two group moments. But in practice *p*-nitroaniline exhibits a greater value than that expected for its dipole moment. This is a consequence of charge separation to a greater extent. The strong electron-withdrawing $-NO_2$ group withdraws electrons by its –R effect, while the $-NH_2$ function donates electrons by its +R effect. The functionalities are oriented in 1,4-relation (*para*-orientation), and hence there occurs an effective conjugation resulting significant polarization. That's why dipole moment of *p*-nitroaniline is unusually high.

Problem 1.8 In each pair of compounds which one has the higher dipole moment?
(i) pyrrole; furan; (ii) butan-2,3-dione; cyclohex-3,5-dien-1,2-dione;
(iii) allyl bromide; vinyl bromide; (iv) *p*-nitroaniline; *p*-aminobenzoic acid;
(v) *p*-nitroaniline; 2,3,5,6-tetramethyl-4-nitroaniline.

Solutions. (i) pyrrole; furan

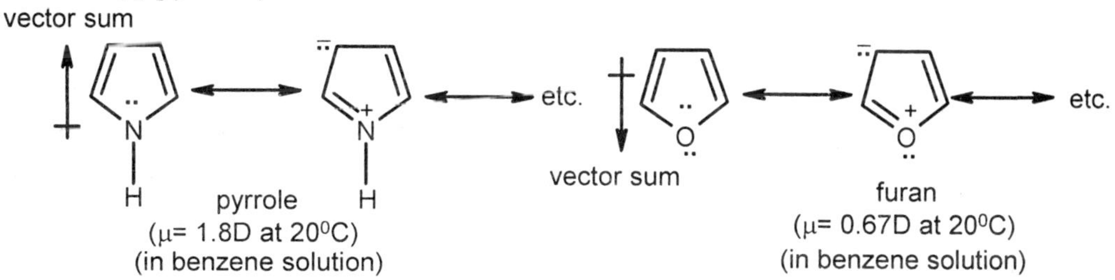

In pyrrole, the lone pair of electrons on nitrogen is extensively delocalised with the π-electrons of the ring so as to maintain aromaticity. Thus, resonance leads to the establishment of partial negative charges on the carbons and a partial positive charge on the nitrogen. At the same time, –I effect of nitrogen also plays its role in dragging the π-electrons towards itself from carbons — thus the electronic distribution in pyrrole is a balance of two opposing effects, of which the mesomeric effect is probably the more significant. This electronic drift away from nitrogen and towards the ring carbons offers a considerable dipole moment to the molecule, and the dipole directs towards the ring.

Analogous canonical forms, as drawn for pyrrole, may also be written for furan; but the higher electronegativity of oxygen implies that the polarized structures, with positive charges on oxygen, have little contribution. This is in accordance to the fact that the unshared electron pair on oxygen in furan is not well delocalized into the ring. The decreased mesomeric electronic drift away from oxygen is insufficient to overcome the inductive polarization (–I effect) towards the heteroatom; the net effect is to give a dipole of smaller moment directed towards the oxygen. A similar situation is observed in case of thiophene also ($\mu = 0.52D$).

▲ Another question-form: The dipole moment of furan is smaller than and opposite in direction from that of pyrrole. Account for the fact.

(ii) butan-2,3-dione; cyclohex-3,5-dien-1,2-dione

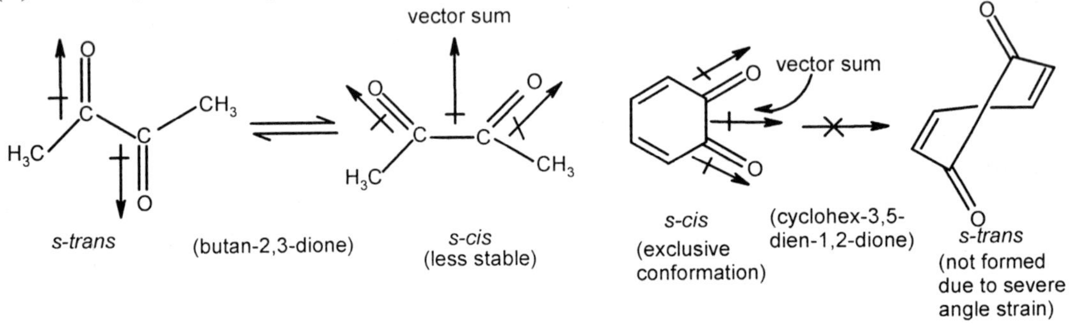

Butan-2,3-dione may exist in two conformations — *s-cis* and *s-trans* forms. The *s-cis* form is less stable, because it suffers from severe repulsive interactions between negatively polarized oxygen atoms of the two C=O dipoles. This form possesses considerable amount of dipole moment. But the predominant conformation of the compound is the *s-trans* form, since it suffers from minimum electronic repulsive interactions as the two C=O dipoles are oriented in *anti*-periplanar fashion; however its contribution towards the overall dipole moment of the compound is negligible. Thus net dipole moment of butan-2,3-dione is very small indeed.

On the contrary, cyclohex-3,5-dien-1,2-dione exits exclusively in *s-cis* conformation; the carbonyl groups are at about 60^0 angle with each other. Hence, the resultant of these two C=O group moments offers the molecule a large value for its dipole moment.

(iii) allyl bromide; vinyl bromide

The unshared pair of electrons on bromine in vinyl bromide takes part in conjugation with the π-electrons of C=C double bond; though the resonance is not so much effective, still it opposes the usual displacement of C–Br bonding electron pair towards the halogen to some extent. As a result the net effect of inductive polarization is thereby reduced. Such resonance interaction is not observed in allyl bromide. Consequently, the net displacement of σ-electrons of C_1–Br bond leading to the induct ive polarization is more effective in allyl bromide. Thus, allyl bromide should possess more dipole moment, which is the fact also.

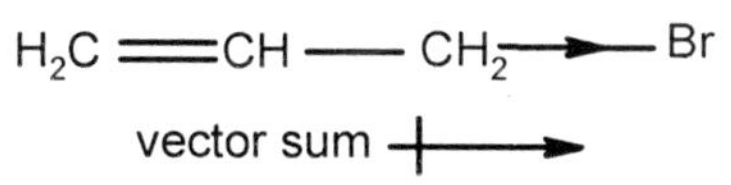

allyl bromide
(μ= 1.79D in benzene
solution at 20⁰C)

vinyl bromide
(μ= 1.407D in gaseous
state at 295 ~ 413⁰K)

(iv) *p*-nitroaniline; *p*-aminobenzoic acid

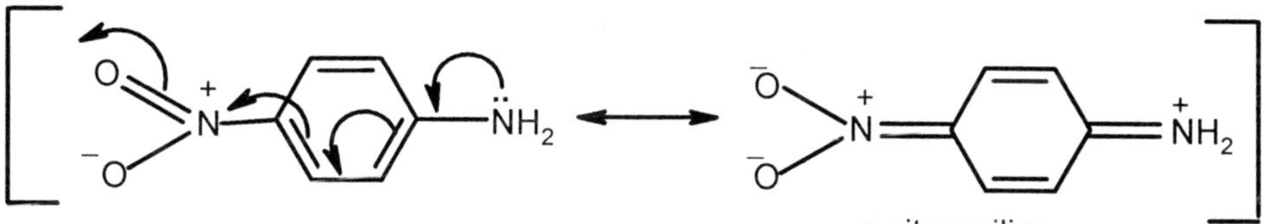

vector sum

p-nitroaniline
[μ = 6.81D in dioxane
solution at 25⁰C]

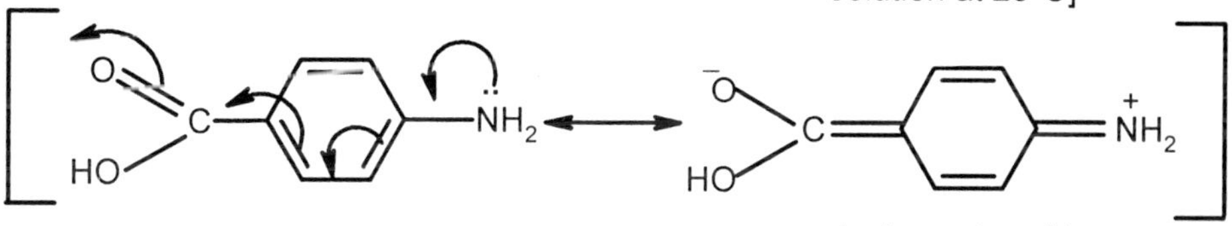

vector sum

p-aminobenzoic acid
[μ = 3.51D in dioxane
solution]

The nitro group has greater –R effect than carboxylic acid function. The former, thus, withdraws the π–electrons from the aromatic ring more strongly, and as a result of which the unshared pair of electrons on nitrogen of $-NH_2$ group (oriented at *para*-position) is delocalized more effectively in *p*-nitroaniline. Extent of polarization is more pronounced in this compound than that in *p*-aminobenzoic acid — the phenomenon is responsible for the greater dipole moment of *p*-nitroaniline.

(v) *p*-nitroaniline; 2,3,5,6-tetramethyl-4-nitroaniline

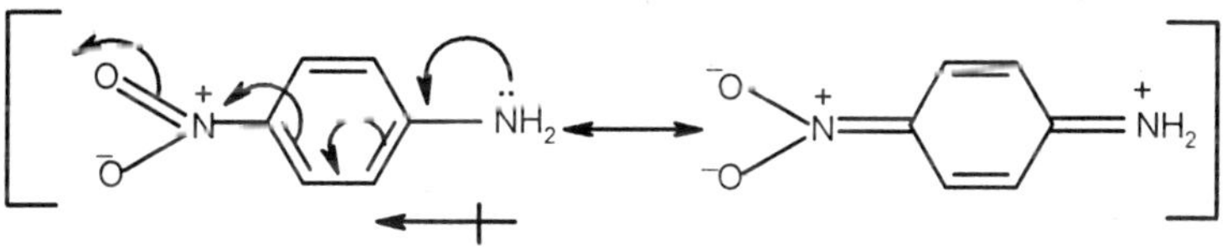

p-nitroaniline
[μ = 6.81D in dioxane
solution at 25⁰C]

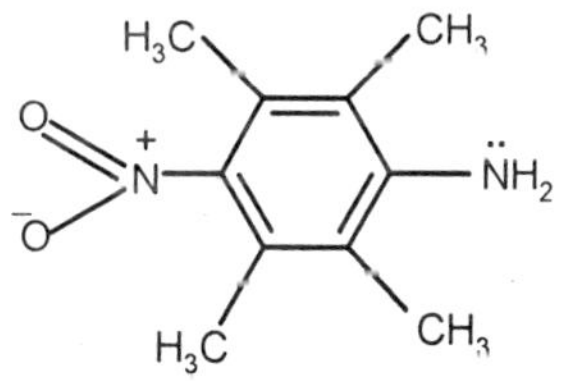

2,3,5,6-tetramethyl-4-nitroaniline

In 2,3,5,6-tetramethyl-*p*-nitroaniline, the bulky methyl groups present at both the *ortho*-positions with respect to $-NO_2$ as well as $-NH_2$ create considerable amount of steric repulsive interaction into the molecule. The large nitro group suffers more interaction and to get rid of such steric repulsive strain it goes out of the plane of the aromatic ring. Therefore, it can't take part in conjugation, and thereby extended conjugation is inhibited. This situation does not arise in the case of *p*-nitroaniline. Extent of polarization in *p*-nitroaniline is much more pronounced reflecting its greater dipole moment.

Problem 1.9 The dipole moment of chlorobenzene is larger than that of fluorobenzene, even though fluorine is more electronegative than chlorine. Justify the statement.

Solution.

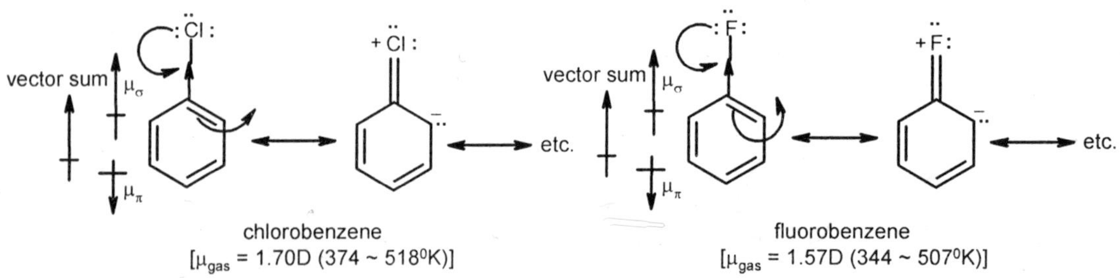

Both fluorine and chlorine show –I effect as well as +R effect. The above resonating structures involving the π-electrons of the aromatic ring and the unshared electron pairs of the halogen reflect their small contribution as they bear positive charge on electronegative halogen atoms. Thus, dipole moment for both the compounds is a consequence of inductive polarization towards the halogen atom. Chlorine and fluorine both are strongly electronegative — fluorine is more electronegative, but smaller in size than chlorine. Being smaller in size, fluorine with its unshared pairs of electrons undergoes resonance with the π-electrons of the ring more effectively relative to chlorine. Thus, +R effect is comparatively more pronounced in fluorobenzene. The directions of inductive polarization (μ_σ) and resonance polarization (μ_π) are in opposition. Hence the moment caused due to inductive electron withdrawal is much more counterbalanced by the moment caused due to resonance electron donation in fluorobenzene in comparison to that in chlorobenzene. Consequently, net dipole moment observed for chlorobenzene is greater than the value for fluorobernzene.

Problem 1.10 The following hydrocarbon has an unusually high dipole moment. Explain how a large dipole moment might arise?

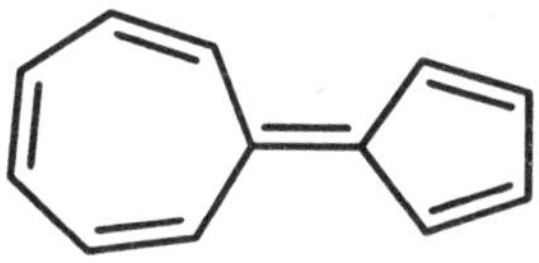

Solution. Sesquifulvalene, a twelve-π-electron system, is usually looked upon as a resonance hybrid of a non-polar form (I) and a dipolar form (II) that, in fact, is a representation of a number of charge-separated canonical structures.

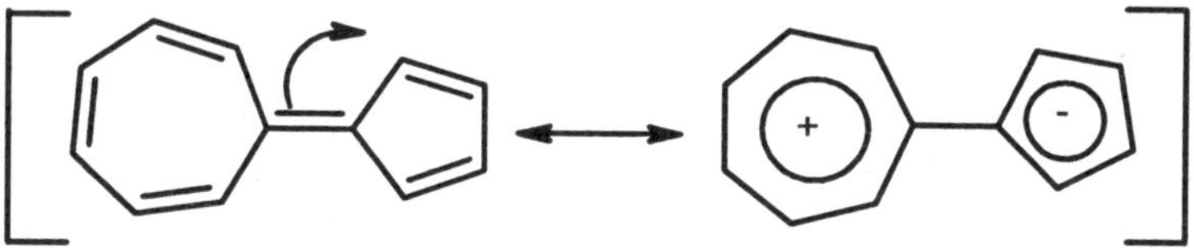

non-polar form (I) dipolar form (II)

The dipolar form (II) bears two units — one is cycloheptatrienyl cation and the other is cyclopentadienyl anion, both of which are aromatic in character and are stable enough. So this dipolar form contributes a lot to the hybrid. That's why sesquifulvalene becomes polar and possesses significant dipole moment.

Problem 1.11 Compound (I) has a stronger dipole moment than (II). Why?

(I) (II)

Solution.

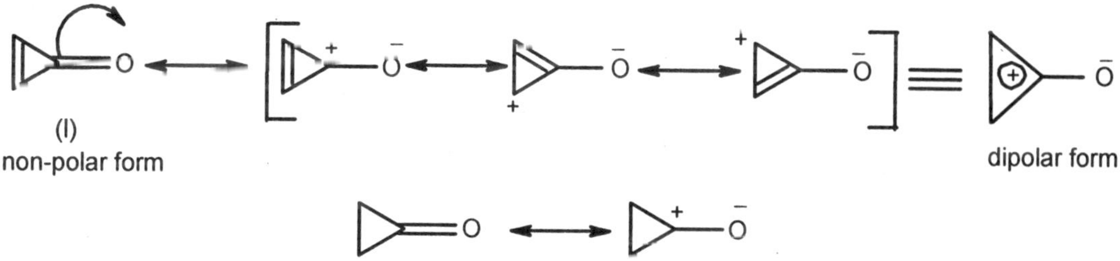

(I) non-polar form dipolar form

(II)

Dipolar form of cyclopentenone (I) attains aromaticity, and is stable. Hence this form contributes much to the hybrid of the molecule. On the other hand, such type of extra stability for the dipolar resonating structure of compound (II) does not arise. Polarity of compound (II) is developed only due to charge separation between carbonyl carbon and oxygen — extended charge separation as like that in cyclopropenone is not possible here. That's why (I) has a stronger dipole moment than (II).

Problem 1.12 Optically active isomer of 1,2-dichloro-1,2-diphenylethane has dipole moment greater than that of the *meso*-isomer. Account for the fact.

Solution. Optically active form [(+) or (–)] of 1,2-dichloro-1,2-diphenylethane exists in the following conformations:

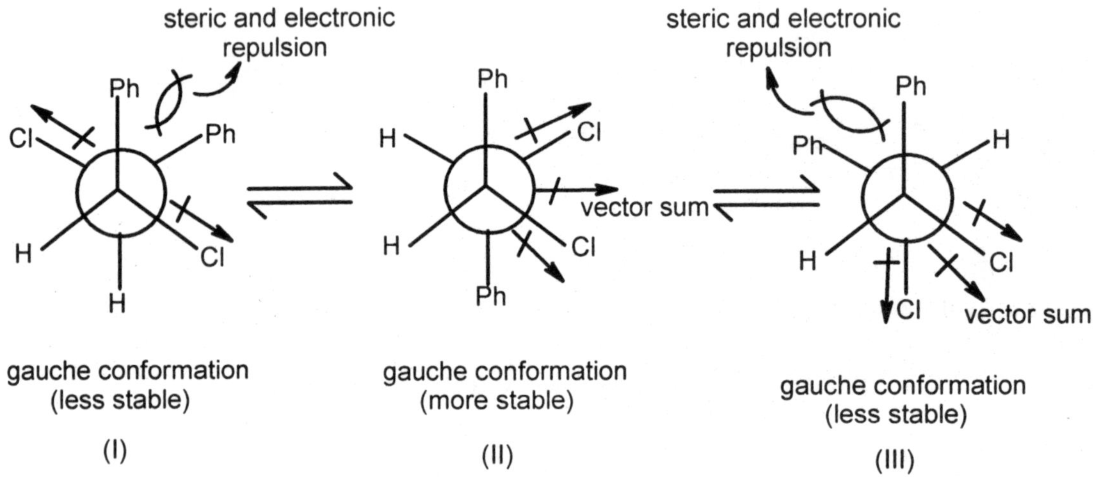

Conformations (I) and (III) are comparatively less stable than (II), because in both of them strong steric and electronic repulsive interactions between two bulky and electron-dense phenyl moieties are present. Out of these less stable conformations, the form (I) bears the two C–Cl bond dipoles in opposition to each other resulting its dipole moment nearly zero. But the C–Cl dipoles in the other less stable conformation (III) as well as in the more stable conformation (II) are in gauche orientation; thus over all dipole moment for the active isomer attains an appreciable value.

Meso-isomer of the compound also exists in three conformations (IV – VI as shown below). The more stable conformation (V) for the *meso*-isomer possesses no significant dipole moment since the two C–Cl bond dipoles are oppositely directed. A little amount of dipole moment for the *meso*-isomer is attributed to the gauche conformations (IV) and (VI), which are less predominant.

That's why optically active isomer of 1,2-dichloro-1,2-diphenylethane exhibits greater dipole moment than the *meso*-isomer.

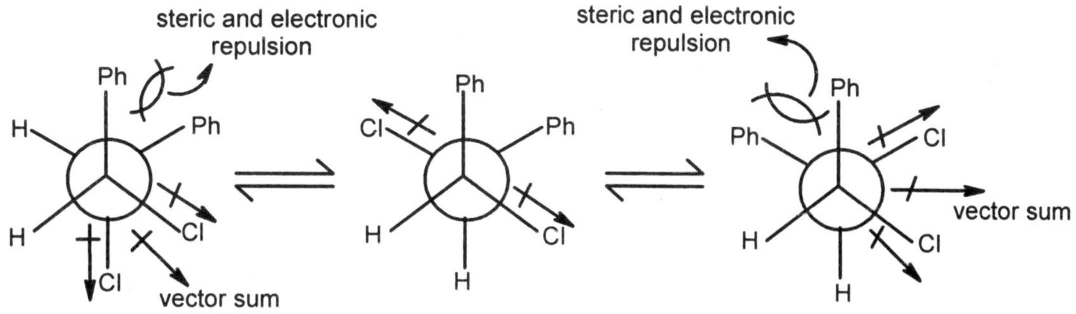

gauche conformation (less stable) (IV) — *anti*-conformation (more stable) (V) — gauche conformation (less stable) (VI)

Problem 1.13 *p*-Xylene boils at a lower temperature than *o*-xylene, but melts at a higher temperature than the other.

Solution. *Ortho*-xylene is weakly polar (μ=0.62D) whereas the *para*-isomer is practically non-polar (μ−0.1D). So dipole-dipole attractions, though very weak, are present among the *o*-xylene molecules — as a result of which the *o*-isomer boils relatively at higher temperature than the *p*-isomer (the difference in boiling points is very small).

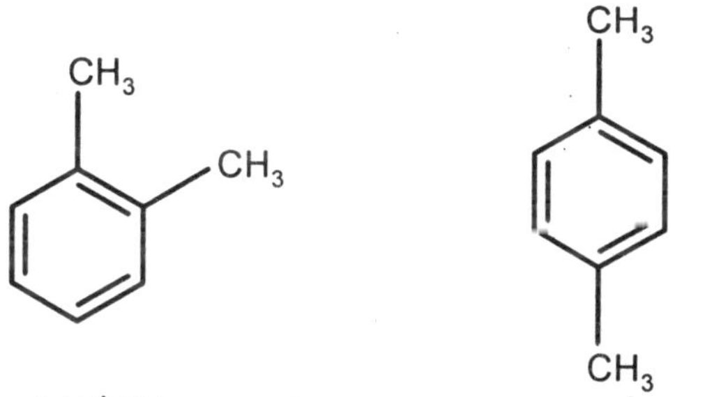

o-xylene
b.p. 144°C
m.p. −25°C

p-xylene
b.p. 138°C
m.p. 13°C

But the *p*-isomer melts at relatively higher temperature — the difference between the melting points is ~38°C. Melting point of a compound depends largely on its symmetry. The more symmetrical *p*-isomer fits better into the corresponding crystal lattice than the less symmetrical *o*-xylene. Consequently, *p*-xylene experiencing stronger intra-crystalline forces melts at higher temperature than the *o*-isomer. This phenomenon is utilized in the separation by fractional crystallization of *p*-xylene from the mixture of xylene isomers.

Problem 1.14 Offer an explanation for each of the following facts:

(a) *Ortho*-hydroxybenzaldehyde may be separated from a mixture containing both the *ortho*- and *para*-isomers of hydroxybenzaldehyde by steam distillation.

(b) *Ortho*-nitrophenol has a lower boiling point and lower water solubility than the *para*-isomer.

(c) The solubility of alcohols in water decreases with increase of molecular weight, and among isomeric alcohols the solubility increases with increase in branching of carbon chain.

(d) Hydrogen fluoride and ethyl fluoride are of comparable dipole moments, but the former has boiling point much greater than that of the latter, although ethyl fluoride is of larger molecular weight.

(e) *Syn*-(benzoyl)-benzil monoxime has lower melting point than the *trans*-isomer.

(f) Boiling point of 2-nitroresorcinol is lower than that of resorcinol.

Solutions. (a) In *o*-hydroxybezaldehyde (salicylaldehyde) the –OH and –CHO functionalities are in close proximity due to which intramolecular hydrogen bonding (chelation) is formed between the two groups giving rise to a stable six-membered ring. So the –OH group is no longer available to undergo molecular association through intermolecular hydrogen bonding, and the molecules exist in monomeric form. On the other hand, intramolecular hydrogen bonding is not possible in the *para*-isomer because the two functionalities in this molecule are held farther apart; hence the *para*-isomer undergoes molecular association through intermolecular hydrogen bonding between the *para*-isomeric molecules of hydroxybenzaldehyde.

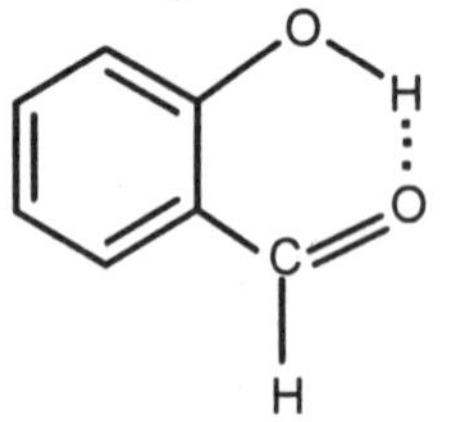

o-hydroxybenzaldehyde
(intramolecular hydrogen bonding; chelation)
m.p. -7^{0}C; b.p. 197^{0}C

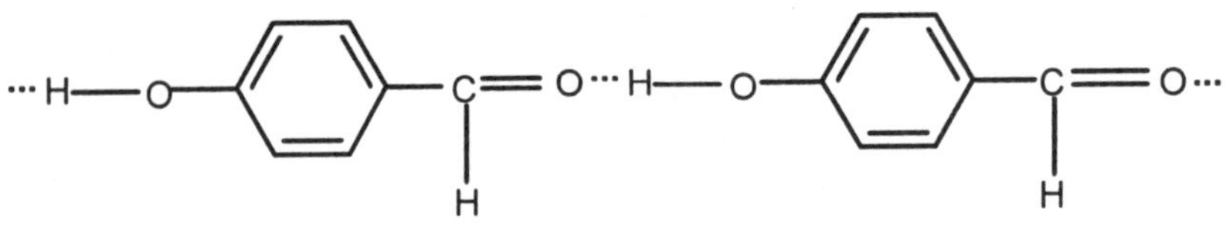

p-hydroxybenzaldehyde
(intermolecular hydrogen bonding; molecular association)
m.p. 115^{0}C

Difference in the hydrogen-bonding pattern (i.e. intra- and inter-molecular) is responsible for the observed difference in boiling points of the *o*- and *p*-hydroxybenzaldehydes. Since considerable amount of energy is required to separate the *p*-isomeric molecules by breaking intermolecular hydrogen bonding, the *p*-hydroxybenzaldehyde boils at much higher temperature than the *o*-isomer.

A compound having an appreciable vapour pressure at the boiling temperature of water can be steam-distilled. As the *o*-isomer is low-boiling and has appreciable vapour pressure at 100^0C, it can easily be separated from the mixture of *o*-and *p*-hydroxybenzaldehydes by steam-distillation.

(b)

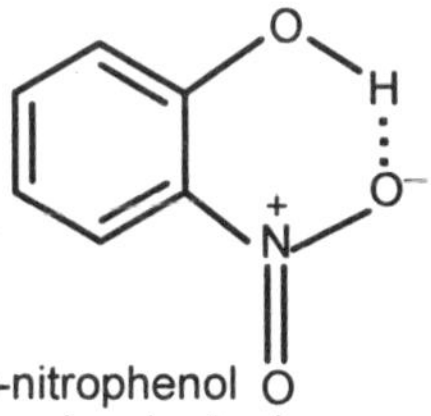

o-nitrophenol
(intramolecular hydrogen bonding; chelation)

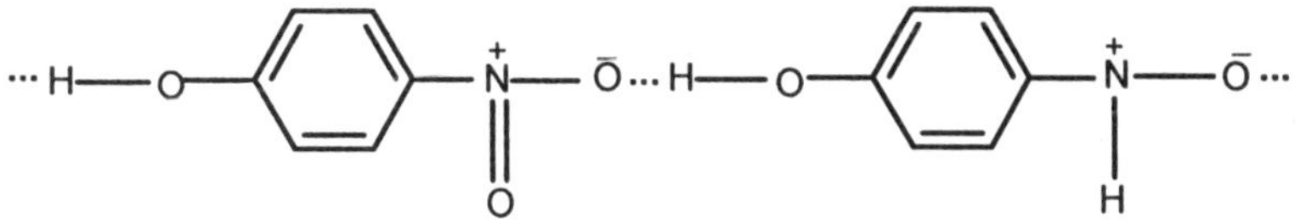

p-nitrophenol
(intermolecular hydrogen bonding; molecular association)

Due to presence of intramolecular hydrogen bonding in *ortho*-nitrophenol, the compound does not undergo molecular association and at the same time cannot form effective hydrogen bonds with water molecules (solvent) also. On the contrary, the *para*-isomers are molecularly associated by formation of intermolecular hydrogen bonding between them; they are capable to form hydrogen bonding also with the water molecules when water is used as solvent. That's why *p*-nitrophenol not only boils at higher temperature than the *o*-isomer, but also has greater water-solubility.

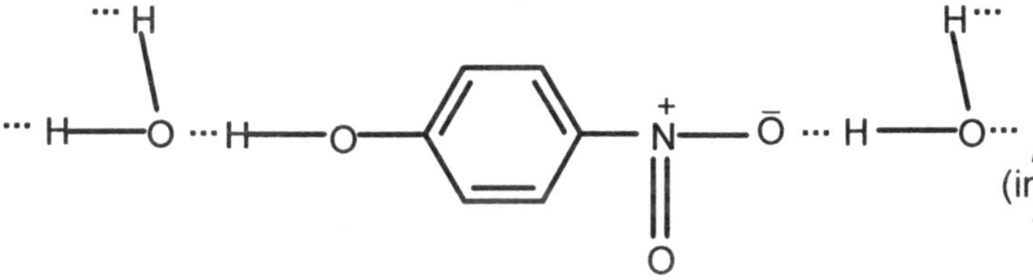

p-nitrophenol in water solution
(intermolecular hydrogen bonding with the solvent molecules)

(c) The solubility of alcohols in water is a consequence of intermolecular hydrogen bonding between alcohol and water molecules. More effective is the hydrogen bonding, greater is the solubility.

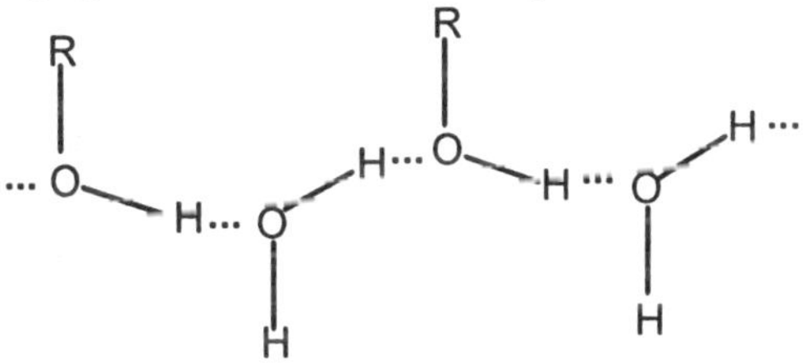

alcohol in water solution
(intermolecular hydrogen bonding with the solvent molecules)

An alcohol molecule bears two parts — non polar (hydrophobic hydrocarbon unit) and polar (hydrophilic hydroxyl unit). Effectiveness of hydrogen bonding (i.e. otherly ease of solubility) depends upon the ratio of the polar part to the nonpolar hydrocarbon moiety. In the lower alcohols the hydroxyl group constitutes a large part of the molecule, while as the molecular weight of an alcohol increases, the hydrocarbon character of the molecule increases, and hence alcohols become progressively less soluble in water. For example, ethanol (CH_3CH_2OH) is highly soluble in water whereas *n*-decanol [CH_3–$(CH_2)_8$–CH_2–OH] is very much less soluble in water since its molecule bears a large hydrophobic hydrocarbon part.

The structure of carbon chain also plays a part; in isomeric alcohols solubility increases with increasing branching of carbon chain. Actually, such branching decreases the relative volume of the hydrophobic hydrocarbon unit and that's why the alcohols become progressively more water-soluble with increase in branching. Thus, *n*-butanol is fairly soluble in water, but *t*-butanol is miscible with water in all proportions.

(d) Because of comparable dipole moments, the strengths of dipole-dipole interactions involved in hydrogen fluoride and ethyl fluoride are almost the same; moreover ethyl fluoride has larger molecular weight. Now it is expected that boiling point of ethyl fluoride would be higher than that of hydrogen fluoride; but the reverse is true. Hydrogen fluoride boils at 19.37^{0}C, whereas ethyl fluoride boils at –37^{0}C. This appreciable difference in boiling points can be explained in the light of hydrogen bonding. Hydrogen fluoride molecules are closely associated together by strong intermolecular hydrogen bonding to form a cluster, $(HF)_n$.

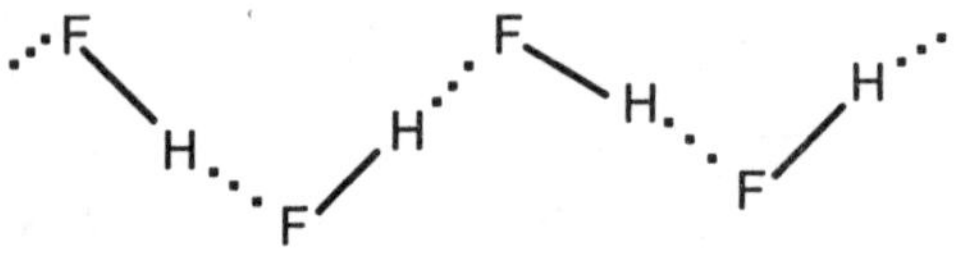

[intermolecular hydrogen bonding in HF molecules (zigzag fashion)]

Such molecular association through hydrogen bonding is not possible in ethyl fluoride because there is no hydrogen atom bonded to an electronegative element. Hence, extra energy is required to break up the hydrogen bonds in hydrogen fluoride as reflected from its higher boiling temperature.

(e) Two monoximes of benzil are known:

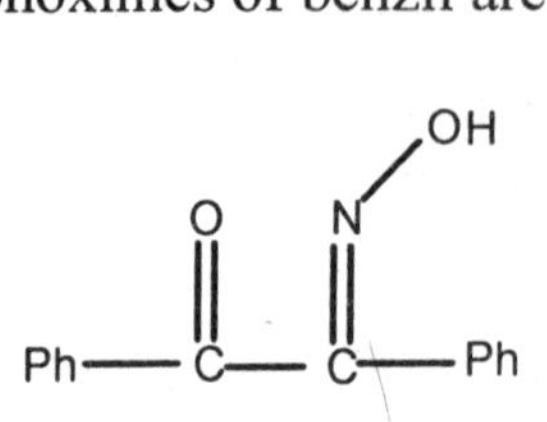

Trans (*anti*)-benzoyl-isomer or *Syn*- phenyl isomer known as benzil α-monoxime (forms intermolecular hydrogen bonding), m.p. 140^{0}C

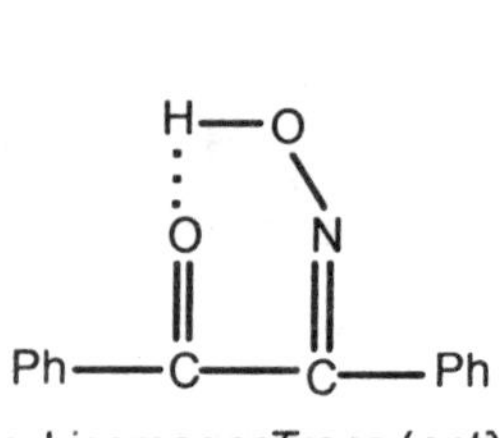

Syn-benzoyl-isomer or *Trans* (*anti*)- phenyl isomer known as benzil β-monoxime (intramolecular hydrogen bonding; chelated form), m.p. 112^{0}C

Syn-(benzoyl)-benzil monoxime (also known as benzil β-monoxime) is intramolecularly chelated through hydrogen bonding, but the *trans*-isomer (also known as benzil α-monoxime) is molecularly associated through intermolecular hydrogen bonding. As a consequence, *syn*-(benzoyl)-benzil monoxime melts at lower temperature in comparison to the other.

(f)

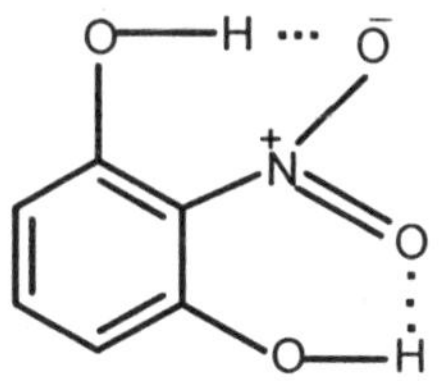

2-nitroresorcinol
(exclusive chelation through intramolecular hydrogen bonding)

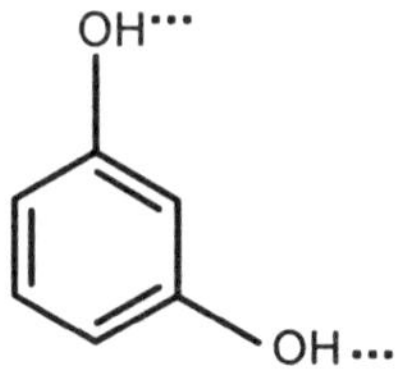

resorcinol
(undergoes molecular association through intermolecular hydrogenbonding)

The two hydroxyls (-OH) placed at the *ortho*-positions of nitro group in 2-nitroresorcinol are intramolecularly hydrogen bonded with the nitro group giving rise to exclusive chelation. Such chelating effect is not observed in the case of simple resorcinol, where the two hydroxyls are in 1,3-relation. Resorcinol forms intermolecular hydrogen bonding among them extending molecular association leading to increase in its boiling point.

Problem 1.15 "Nitration of catechol in the 3-position lowers its boiling point" — justify the statement.

Solution. Nitration of a substance usually brings about an elevation in its boiling point; but if chelation is possible in the nitro-compound, the boiling point is lowered. Like 2-nitroresorcinol, in 3-nitrocatechol both the hydroxyl H-atoms are involved in hydrogen bonding through intramolecular fashion as represented in the following figure.

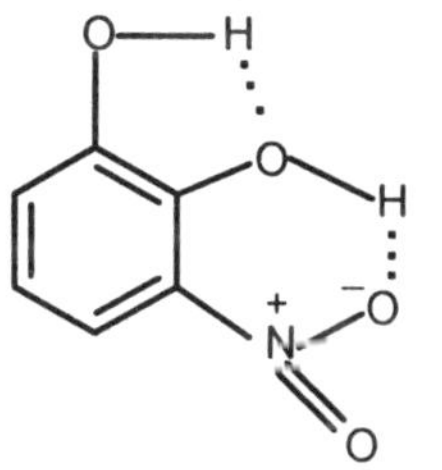

3-nitrocatechol
(exclusive chelation through intramolecular hydrogen bonding)

The phenomenon of this exclusive chelation in 3-nitrocatechol is responsible for lowering of its boiling point. In this connection, it is to be noted that formation of five-membered cyclic rings in 1,2-diols through H-bonding has an obvious boiling point-lowering effect[1].

1. Giles, C. H. *et al.*, *J. Chem. Soc.*, 559(1956).

Problem 1.16 Compare the isomeric distribution of pyridine-2-carboxaldoxime.

Solution. Pyridine-2-carboxaldoxime may exist in two isomeric orientations — *syn* and *anti*-forms.

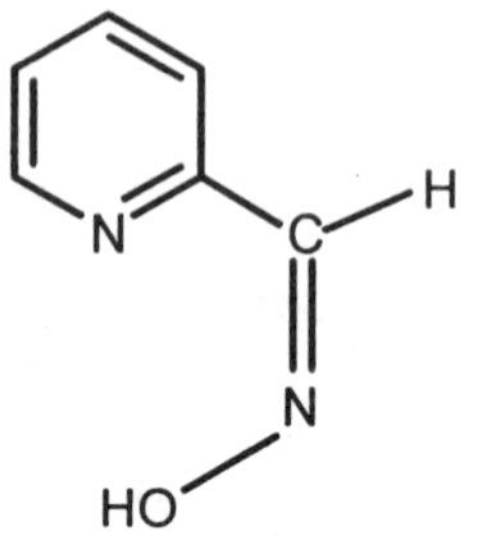

anti -isomer of
pyridine-2-carboxaldoxime

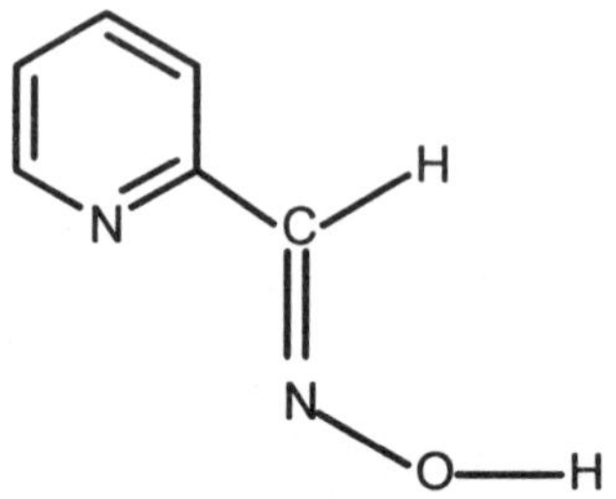

syn-isomer of
pyridine-2-carboxaldoxime

The *anti*-form gets extra stability due to intramolecular hydrogen bonding (chelation) prevailing into the molecule. No such chelating effect is observed in the *syn*-isomer. That's why the *anti*-form is predominant in isomeric distribution of the compound.

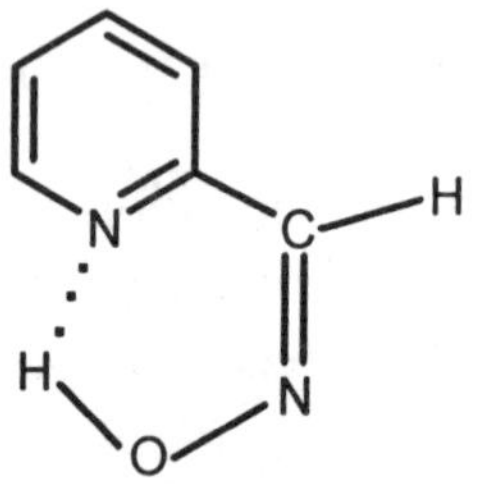

anti -isomer of
pyridine-2-carboxaldoxime
(forms intramolecular hydrogen bonding)

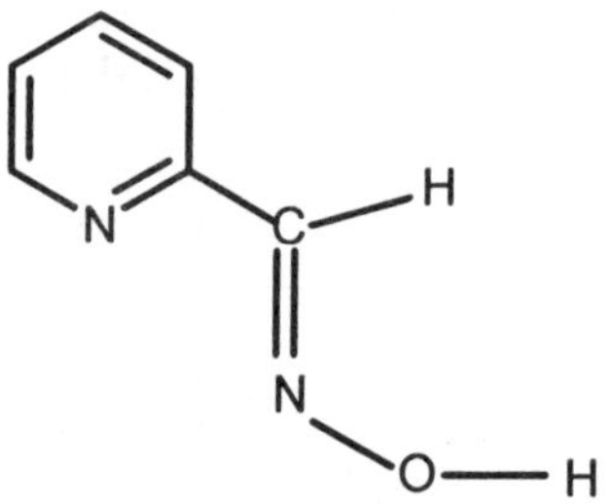

syn-isomer of
pyridine-2-carboxaldoxime

Problem 1.17 Sketch some molecules where formation of H-bonds involve π-electrons.

Solution. In compounds (I – V) intramolecular hydrogen bonds are formed involving an –OH group and the π-electrons of a benzene ring or of an ethylenic unsaturation. An example of intermolecular hydrogen bonding (VI) of such category is also observed between $CDCl_3$ and the benzene ring when $CDCl_3$ is dissolved in benzene.

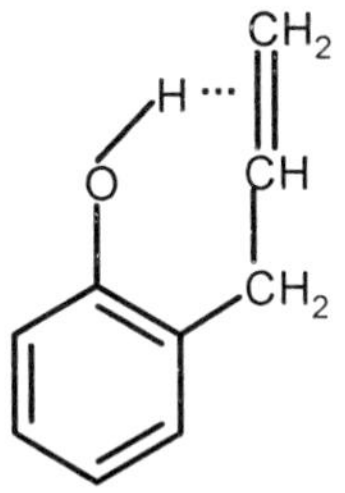

(I)

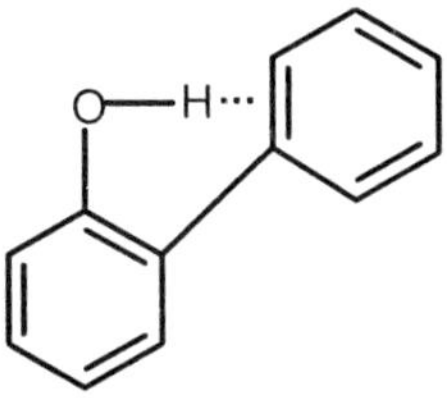

(II)

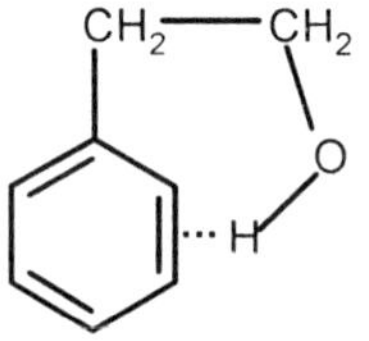

(III)

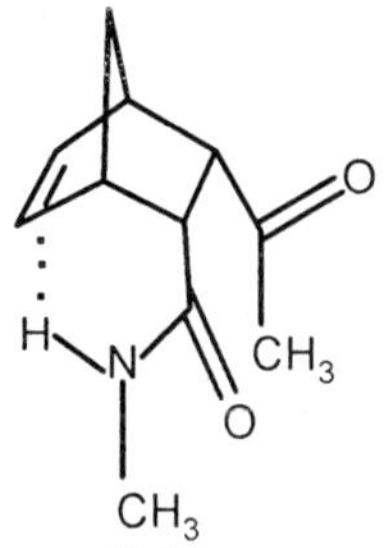

(IV)

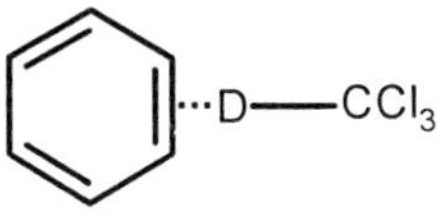

(V)

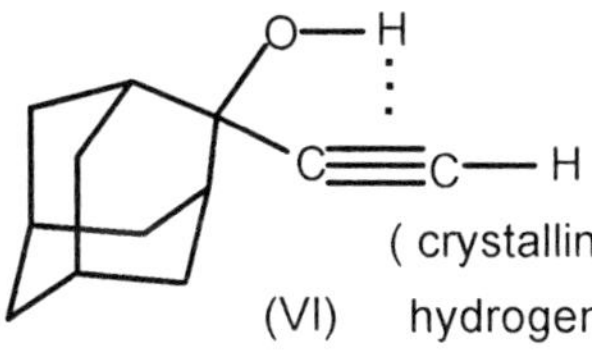

(VI) (crystalline 2-ethynyladamantan-2-ol; unusual O—H ··· π hydrogen bond along with O—H ··· O and C—H ··· O hydrogen bonds)

Electron-releasing substituents in the ethylenic moiety or in the phenyl ring concerned strengthen the hydrogen bond, while electron-withdrawing groups weaken the bond.

Problem 1.18 2-Methylpyrrolidine has higher boiling point than *N*-methylpyrrolidine, although both of them are of identical molecular weight. Why?

Solution.

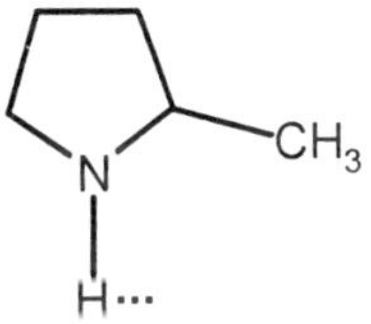

2-methylpyrrolidine (2° amine)
(forms intermolecular H-bonding)
b.p. 100°C

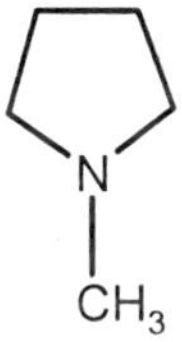

N-methylpyrrolidine (3° amine)
(cannot form H-bonding)
b.p. 79°C

Tertiary amines have usually lower boiling points than isomeric primary or secondary amines, because the latter associate through intermolecular N – H ... N bonds. Thus 2-methylpyrrolidine can form intermolecular hydrogen bonding leading to molecular association, but the isomeric amine, *N*-methylpyrrolidine, can't. Hence, 2-methylpyrrolidine possesses higher boiling point than the other one.

Problem 1.19 Which member in each pair would have the larger wet melting-point depression and why?

(a) *p*-nitroacetanilide; *o*-nitroacetanilide, (b) 2,4-dinitroacetanilide; 2,3-dinitroacetanilide

Solutions. From determination of wet melting-point depression one may get a clear idea about the nature of hydrogen bonding (intra- or inter-) present within a compound concerned. A compound, which is internally chelated through intramolecular hydrogen bonding, exhibits slight depression in its melting-point whereas, if capable of forming intermolecular hydrogen bonding, shows a marked depression.

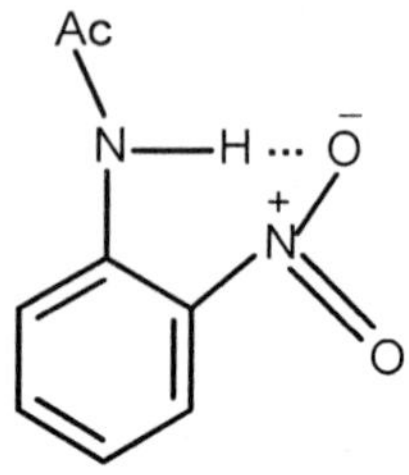

o-nitroacetanilide
(chelated)

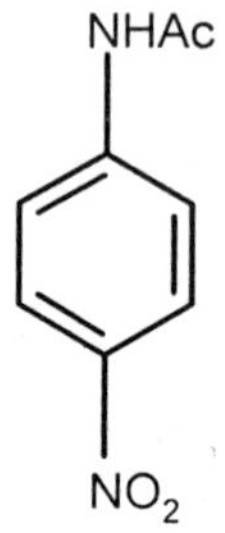

p-nitroacetanilide

In the first case, intramolecular hydrogen bonding between –NHAc and *ortho*-oriented –NO_2 in *o*-nitroacetanilide takes place, but not in the *p*-isomer (–NHAc and –NO_2 groups are far apart). That's why *p*-nitroacetanilide can associate more effectively with water molecules through intermolecular hydrogen bonding than that as the *o*-isomer does, where both the functionalities are blocked. So, *p*-nitroacetanilide would show greater wet melting-point depression.

(b)

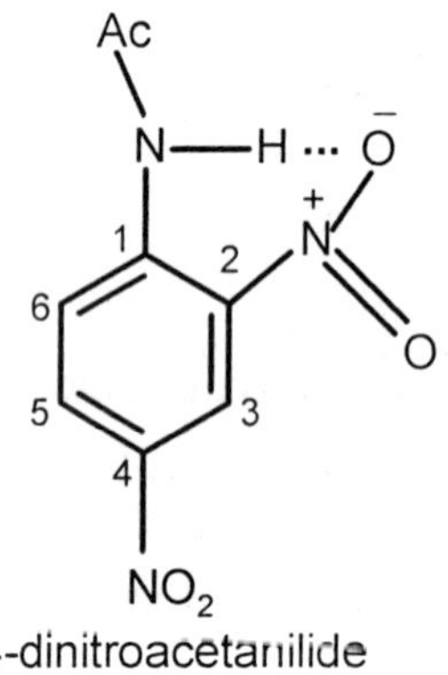

2,4-dinitroacetanilide
(chelated)
wet melting-point depression = 11°C

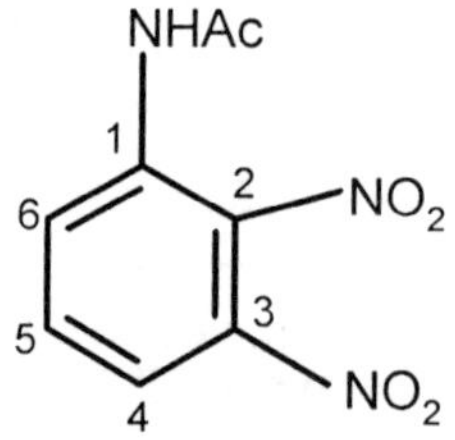

2,3-dinitroacetanilide
wet melting-point depression = 30°C

The bulky nitro group, placed at the *ortho*-position of 2-nitro group in 2,3-dinitroacetanilide, creates a strong steric hindrance as a result of which the 2-nitro group goes out of the plane of the ring so that formation of intramolecular hydrogen bonding is prevented (distance between the bridged electronegative atoms becomes too large for an H-bond). This situation does not arise for 2,4-dinitroacetanilide where only one of the nitro groups (i.e. 2-nitro) is in chelated form. Hence extent of association (intermolecular hydrogen bonding) with water molecules becomes greater in 2,3-dinitroacetanilide (where the two nitro groups take part in the process) as evidenced from its higher wet melting-point depression value.

Problem 1.20 In which one among the following compounds you expect the presence of stronger hydrogen bonding? Give reason in support of your answer.

o-nitrophenol and 6-methyl-2-nitrophenol

Solution. Both the compounds possess intramolecular hydrogen bonding in their molecules.

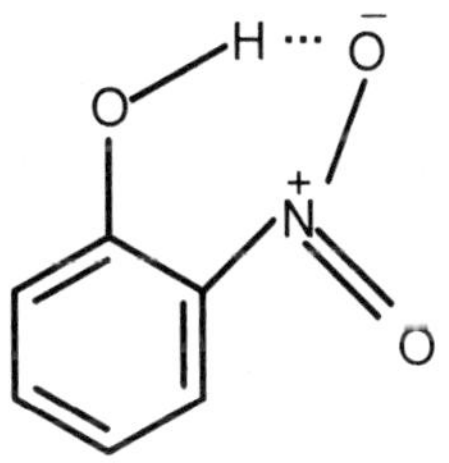

o-nitrophenol

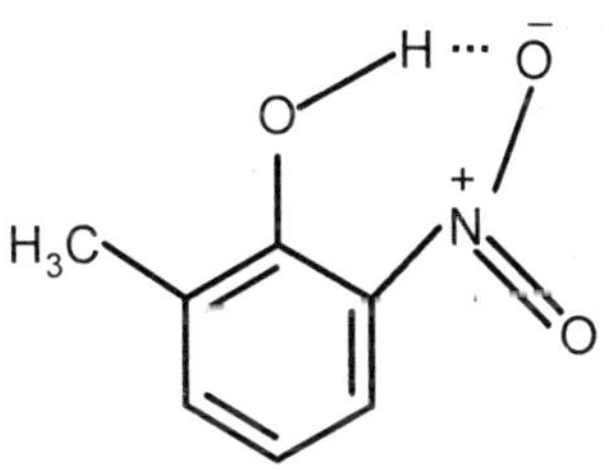

6-methyl-2-nitrophenol

This is an interesting example of steric facilitation rather than steric hindrance to hydrogen bonding. It is observed that introduction of an alkyl group at another *ortho*-position with respect to the –OH function in *o*-nitrophenol, as in the case of 6-methyl-2-nitrophenol, strengthens the intramolecular hydrogen bond. It is supposed that the alkyl group forces the –OH function to form the hydrogen bond with the nitro group, thereby enhancing the strength of that bond.

Problem 1.21 Tropolone does not form a 2,4-dinitrophenylhydrazone. Why?

Solution.

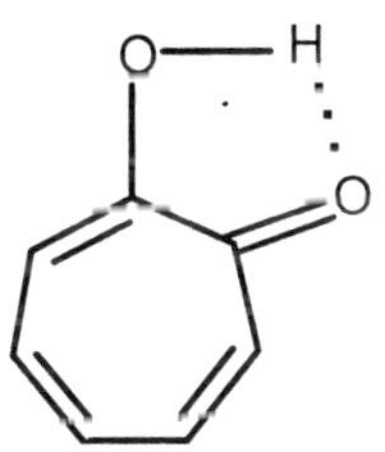

tropolone
(strongly chelated)

The carbonyl oxygen in tropolone forms a strong intramolecular hydrogen bond with the adjacent hydroxyl proton, and so effectively chelated that the carbonyl function can't undergo into reaction with the carbonylic reagent, 2,4–dinitrophenylhydrazine,

— although this particular reaction is regarded as one of the strongest reactions of carbonyl groups. Thus hydrogen bonding not only affects physical properties of compounds, but affects some chemical properties also.

Similar effect of hydrogen bonding in preventing few carnonylic reactions as well as typical reactions of hydroxyl functions is also observed in diaroylmethane.

Problem 1.22 Define tautomerism and tautomers. Give examples. How is tautomerism classified? How it differs from resonance? Discuss the mechanistic aspects of *prototropic* interconversions.

Solution.

Tautomerism: definition

Tautomerism designates the phenomenon that involves reversible and fairly rapid interconversion of two structural isomers of a compound in solution or in liquid state. Such structural isomers of the compound are termed as *tautomers*. Tautomers remain in equilibrium in the solution or liquid phase.

(shifting of 'H' from A to C)

(tautomerism)

Under ordinary conditions, there is usually a sufficient concentration of acid or base present (as catalyst) to cause rapid interconversion. Tautomers differ from each other in regards to electronic distribution, and as well as in the position of a relatively mobile atom or group. In most cases, it is a proton that shifts from one atom of a molecule to another.

Examples of tautomeric interconversions

(i) Keto – enol tautomerism

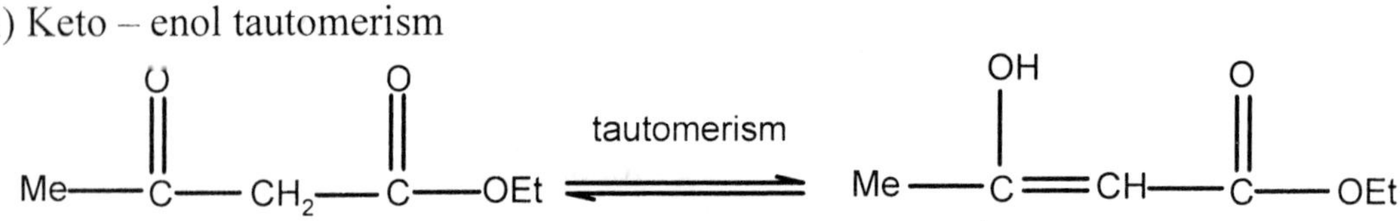

(keto - form) (EAA) (enol - form)

(ii) Nitroso – oxime tautomerism

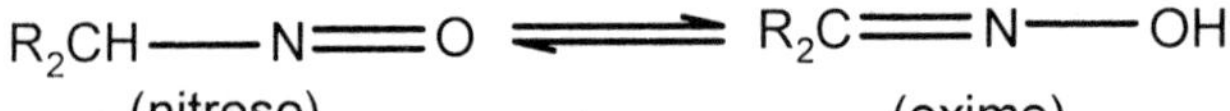

(nitroso) (oxime)

(iii) Nitro – acinitro tautomerism

(nitro-form) (aci-form)

(iv) Imine – enamine tautomerism

(imine) (enamine)

(v) Amide – imidol tautomerism

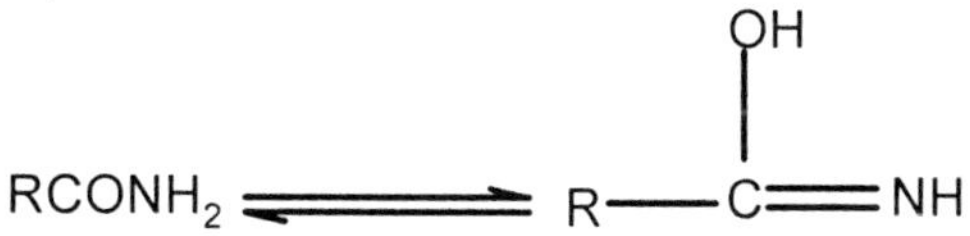

(amide) (imidol)

(vi) Azo – hydrazone tautomerism

(azo-form) (hydrazone)

(vii) Amidine tautomerism

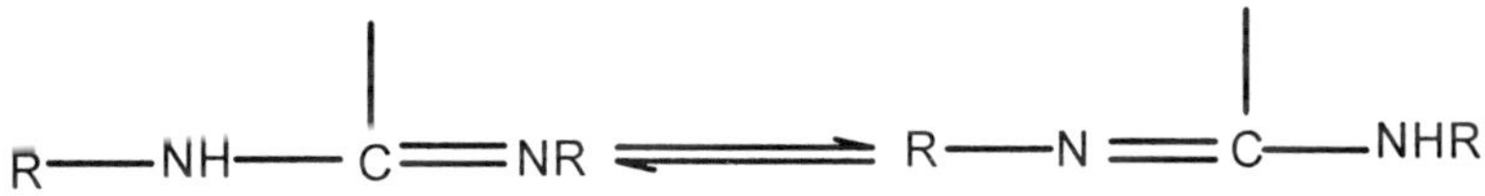

(viii) Lactum – lactim tautomerism

(lactum) (lactim)

Classification

Ingold (1927) classified tautomerism into two categories — *cationotropy* and *anionotropy*. When the shifting species is a cation, the tautomeric process is said to be cationotropy; similarly the process is said to be anionotropy when it involves the shifting of an anion. Lowery (1923) suggested the term, *prototropy* for the tautomeric systems where hydrogen interchanges its position as a cationic species (i.e. H^+) — thus *prototropy* is a special case of cationotropy. Likewise the term, *oxotropy* is in use for anionotropic rearrangements involving the migration of a hydroxyl group.

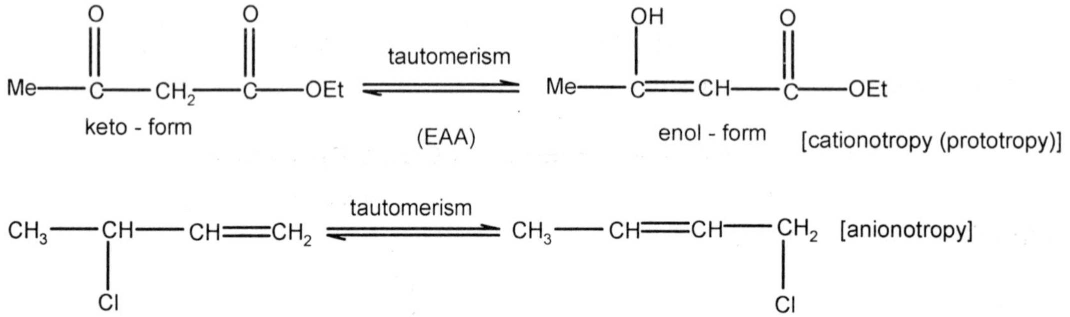

Difference from resonance

There arises very often confusion between *tautomerism* and *resonance* (*mesomerism*), but in practice they are completely different phenomena. In tautomerism the tautomers remaining in rapid equilibrium have real existence and are isolable too in pure states, whereas resonating structures (canonical forms) have no real existence — they are absolutely hypothetical. Let us look into the following sequence of reaction involving base-catalyzed keto – enol tautomerism:

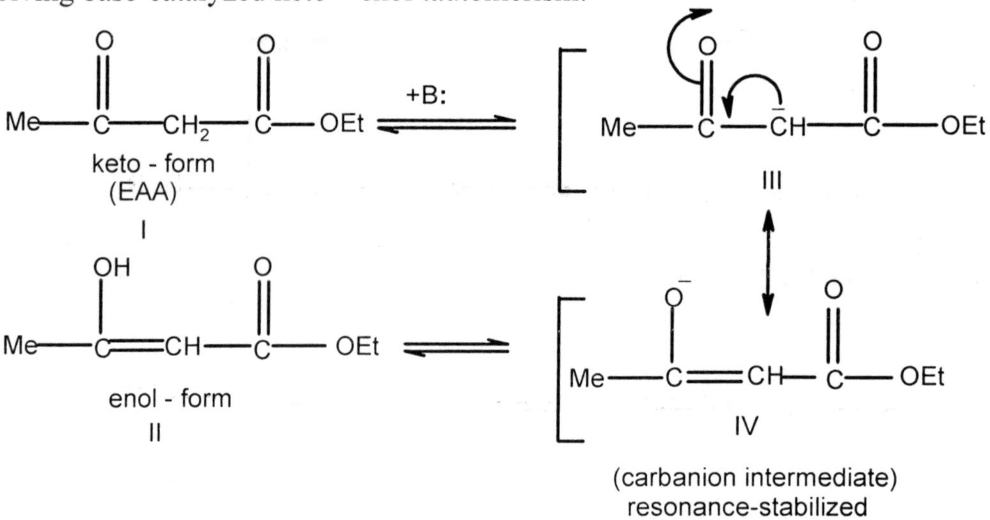

Here, I and II are chemically distinguishable different species which are isolable in pure states. The two structures, III and IV written for the carbanion intermediate are resonating structures (or mesomers), which have no real existence at all — they are merely hypothetical attempts to represent the electronic distribution in the carbanion intermediate that is a single individual species. Thus, the interconvertibility of tautomers is a chemical reaction involving breaking and making of a sigma bond, and this phenomenon is clearly distinguished from resonance that involves the delocalization of π-electrons so that the atomic nuclei remain in identical relative positions in all resonating structures.

Mechanistic aspects of prototropy

Prototropic interconversions may be catalyzed either by base or acid.

a) *Base-catalyzed mechanism*

In most of the cases base-catalyzed prototropic interconversions are found to proceed through intermolecular pathway, but there are evidences in support of intramolecular path also in certain transformations. Both the pathways are presented below:

(i) *Intermolecular Path*

This is a step-wise operation; proton removal and proton acceptance (from solvent) occur in separate steps. A carbanion intermediate is involved into the process:

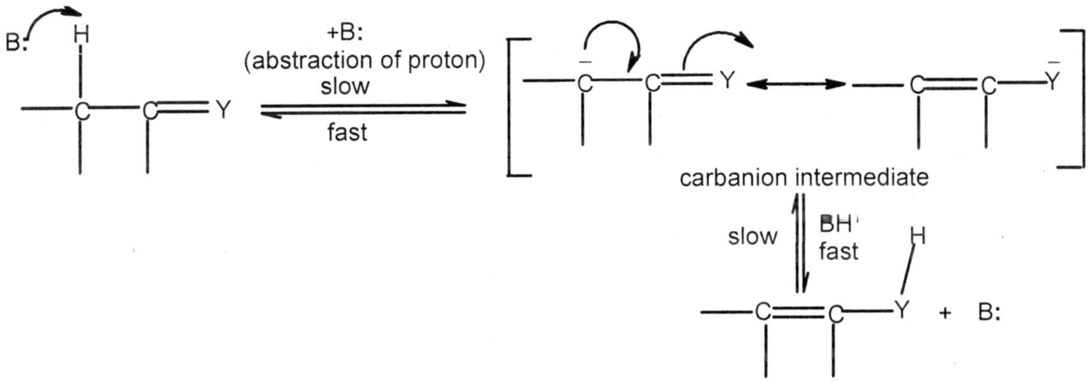

Step-I in the forward reaction is the rate-determining step. Compounds, which are capable to form stable carbanion intermediates, are supposed to follow the intermolecular pathway. Thus, β-ketoesters, 1,3-diketones, aliphatic nitro compounds, etc. undergo ready base-catalyzed prototropic changes.

(ii) *Intramolecular Path*

Shifting of the proton takes place intramolecularly:

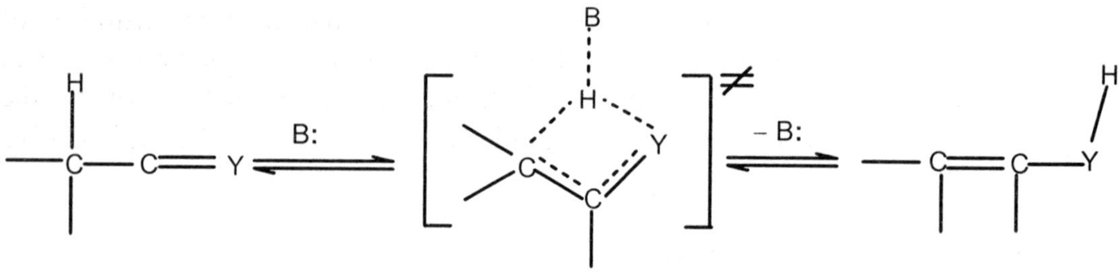

(bridged transition state)

An example obeying the intramolecular path may be cited — the optically active substrate (I) tautomerizes to inactive form (II) under base-catalysis (Et_3N:).

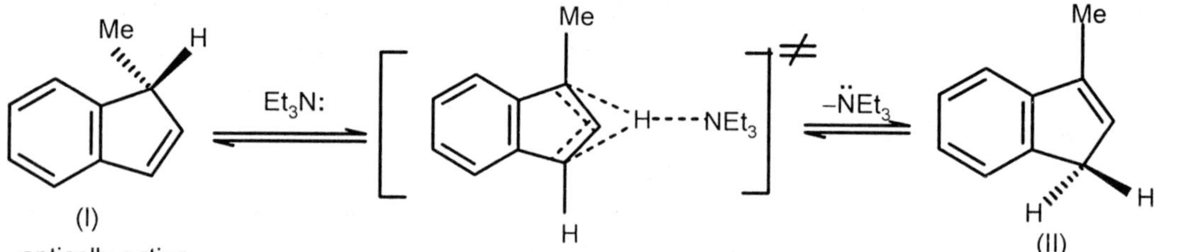

(I)
optically active

bridged transition state

(II)
optically inactive

It is observed that the rate of loss of optical activity and the rate of isomerization are identical, and also observed that if the reaction is carried out in the presence of D_2O, there occurs no deuterium incorporation into the product. Hence, the interconversion does not involve carbanion intermediate, and therefore the reaction follows perfectly the intramolecular path.

However in many substrates, features of both inter- and intramolecular pathways are observed — the extent of relative proportions depends not only on the substrate, but on the nature of base and solvent employed also.

(b) *Acid-catalyzed mechanism*

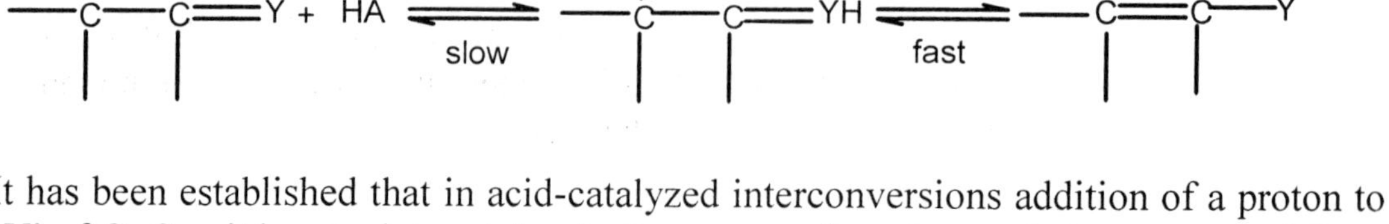

It has been established that in acid-catalyzed interconversions addition of a proton to 'Y' of C=C unit is rate-determining in the *reverse direction*, and proton removal from β-carbon (*w.r.t* Y) is rate-determining in the *forward direction*.

Problem 1.23 Give evidence in favour of the fact that ethylacetoacetate remains as an equilibrium mixture of keto and enol forms.

Solution. Ethylacetoacetate (EAA) tautomerism is a classical example of keto-enol tautomerism. The compound is an equilibrium mixture of two distinct substances — the keto and enol forms.

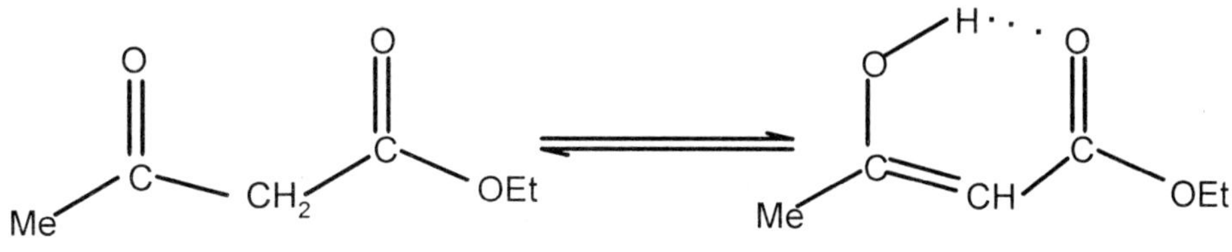

ethylacetoacetate (keto-form 92%) (enol-form 8%)

Chemical tests for both the species are found to be positive. Let us summarize a few of them:

(i) ***Evidences for ketonic species***

Ethyacetoacetate is found to respond the following reactions, which are characteristics of a carbonyl function.

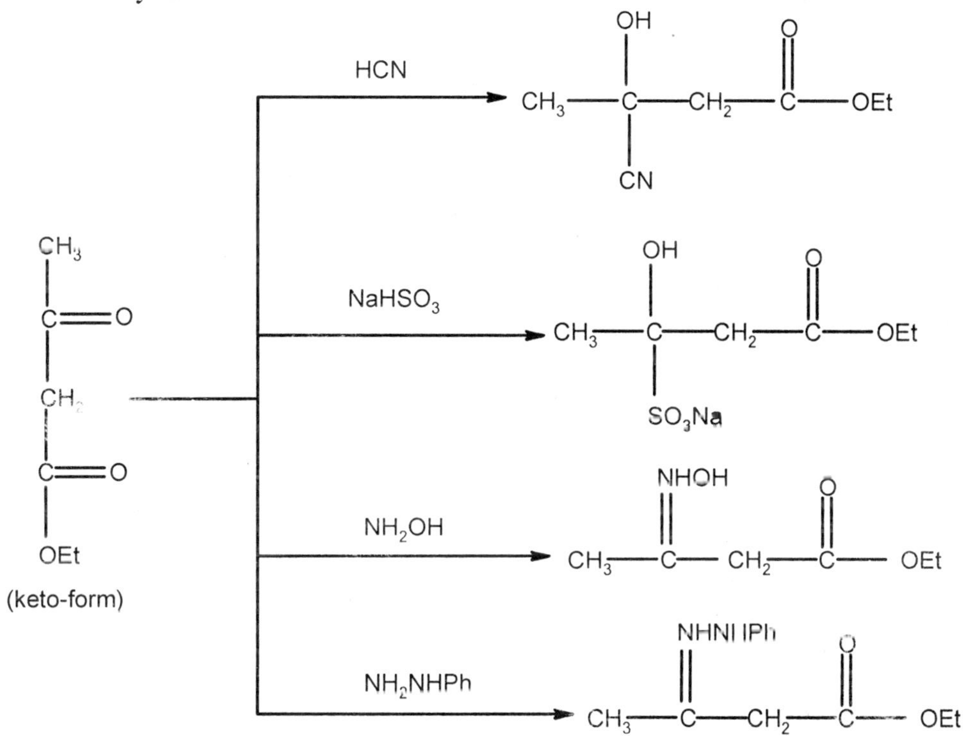

(ii) ***Evidences for enolic species***

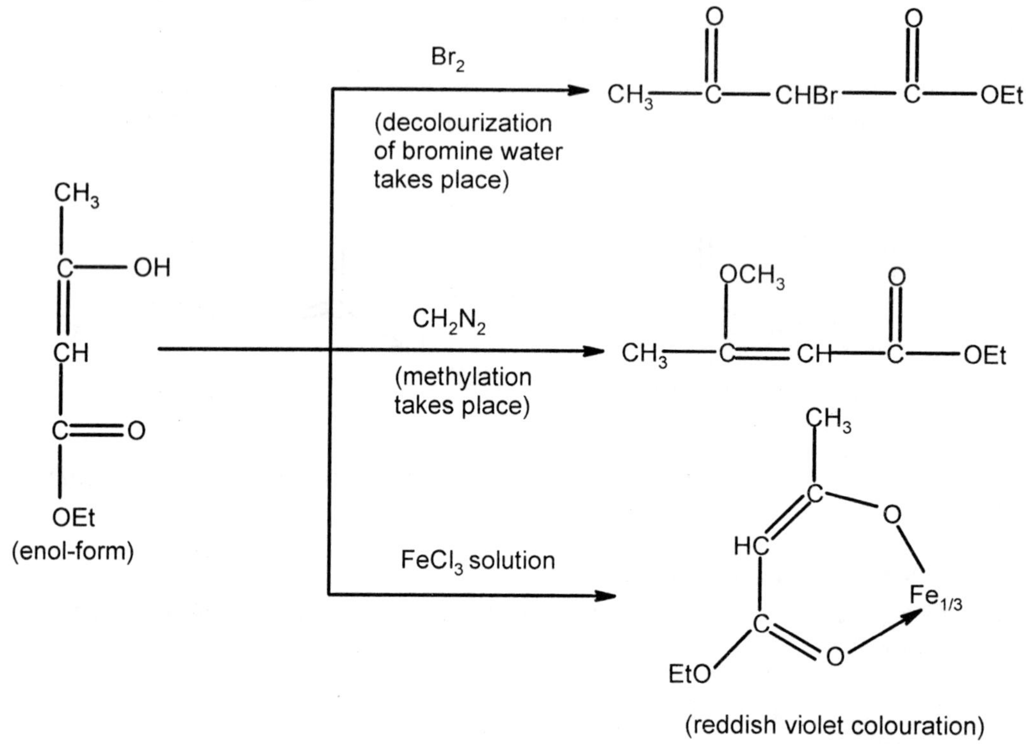

These two components (i.e. tautomers) were first separated by Knorr (1911) in pure states, and latter also by Meyer. The enol content of a keto-enol mixture can also be estimated quantitatively depending upon the fact that enol form absorbs bromine rapidly, whereas the keto form does not.

Problem 1.24 Give a brief account of ring-chain and valence tautomerism.

Solution.

(i) ***Ring-chain tautomerism***

Two constitutional isomers, one is cyclic and another is acyclic, when exist in rapid equilibrium in liquid or in a solution are said to be *ring-chain tautomers*, and the equilibrium reaction is termed as *ring-chain tautomerism*. These ring and chain forms of tautomers are possible when one functional group of a bifunctional acyclic molecule reacts with the other and forms a cyclic system. Ring-chain tautomerism is exhibited by all reducing sugars. Other examples are also known. A few are cited here:

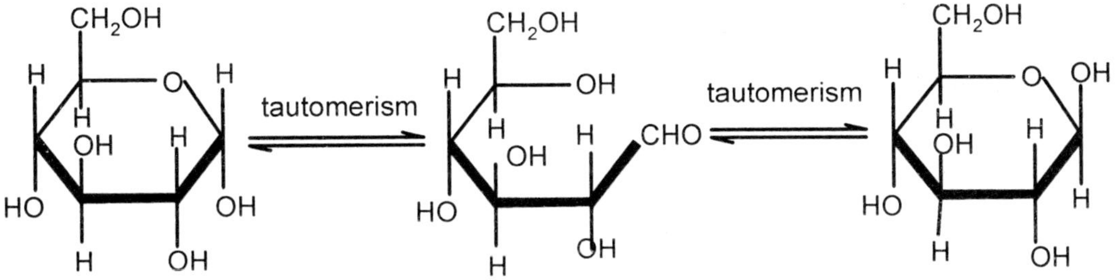

α-D-glucopyranose (ring form) — open chain — β-D-glucopyranose (ring form)

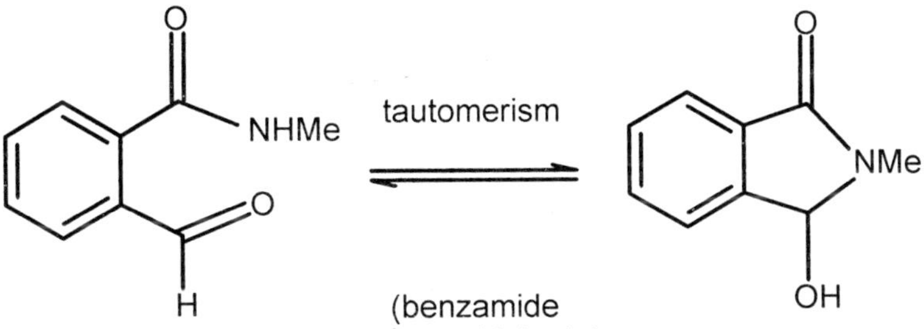

(open chain) (benzamide carboxyaldehyde) (ring form)

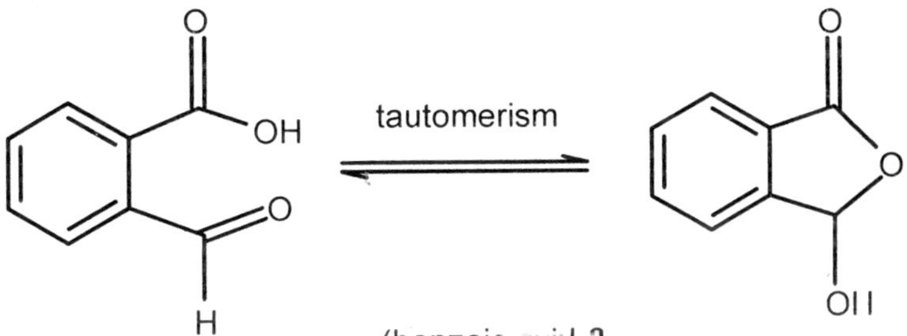

(open chain) (benzoic acid-2-carboxyaldehyde) (ring form)

(ii) ***Valence tautomerism***

Valence tautomerism is quite distinct from resonance, though such isomerization is effected by electron shift; the tautomeric structures are of equivalent constitution and rapidly change back and forth by electron shift. Such valence tautomers are also called *fluxional structures*. They are not resonating structures since atomic nuclei change the positions. The simplest example of this class is:

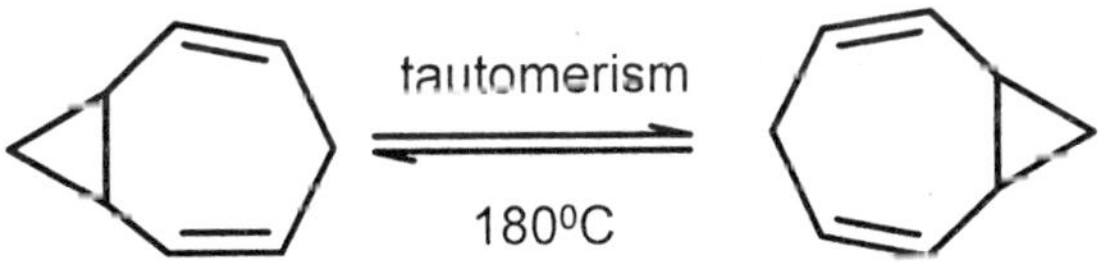

bicyclo[5,1,0]octadiene

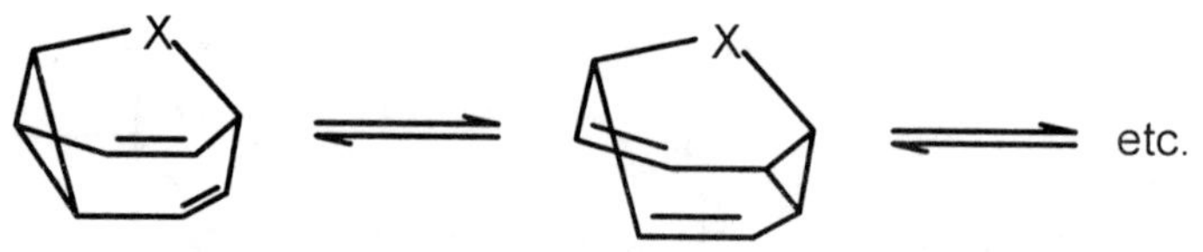

X	Substrate	
CH=CH	Bullvalene	(for bullvalene 12,09,600 possible arrangements are there)
CH_2CH_2	Dihydrobullvalene	
CH_2	Barbaralane	
CO	Barbaralone	
direct bond	Semibullvalene	

1,2,3,4-Tetra-*tert*-butylcyclobutadiene offers also a very interesting act of valence tautomerism. The two valence tautomers are both rectangular, and they exist in a dynamic equilibrium even at -185^0C as evidenced from ^{13}C NMR studies.

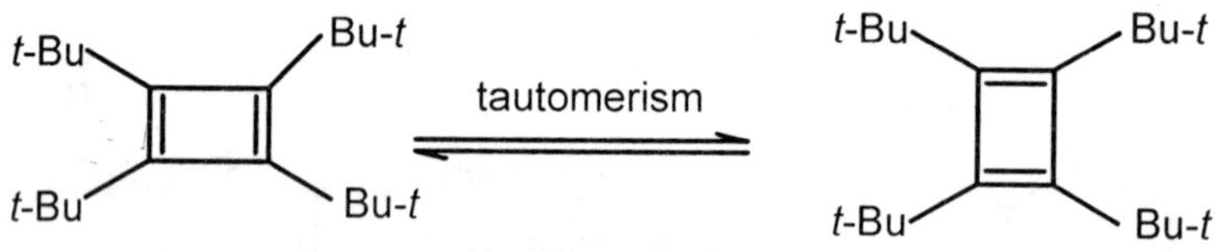

(1,2,3,4-tetra-*tert*-butylcyclobutadiene)

Problem 1.25 Offer proper explanation for each of the following facts:

(i) Enol content of acetone is very small.

(ii) There is no evidence in support of the existence of keto-form for phenol, but phloroglucinol has.

(iii) Enol content of $CH_3COCH_2COCH_3$ is about 76%, whereas the value for CH_3COCH_2COOEt is only about 8%.

(iv) At equilibrium enol content of acetylacetone ($CH_3COCH_2COCH_3$) is greater than that of α–methylacetylacetone.

(v) Butan-2,3-dione exists exclusively in keto-form, but reverse is the situation in case of cyclopentan-1,2-dione.

(vi) Enol content of 2,6-bicyclo[2,2,2]octandione at equilibrium is very small.

(vii) Replacement of methyl(s) in $CH_3COCH_2COCH_3$ by phenyl(s) increases the enol content of the resulting substrate(s) progressively.

(viii) As the polarity of solvent increases, enol content of a compound tends to decrease.

Solutions. (i) It is very much difficult to detect the enol-form in simple aldehydes and ketones where the enol content ranges from 10^{-4} to 10^{-1}. Enol content of acetone is so small that the tautomeric equilibrium mixture consists of almost entirely the keto-form.

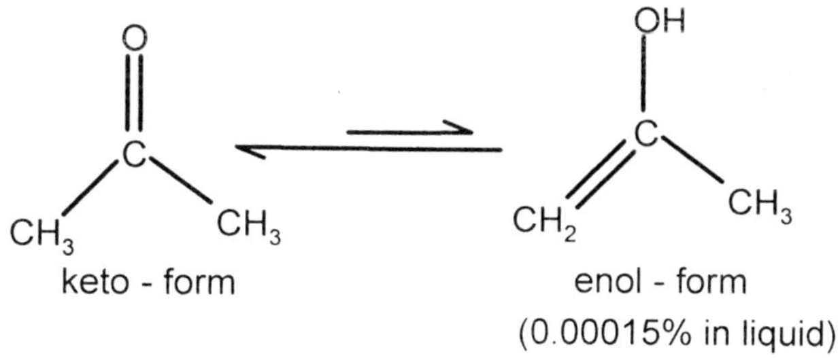

From thermodynamic consideration it is estimated that the enol-form of acetone is 9.8 kcal /mole higher in energy than the keto, and the difference in the gas phase is 13.9 +/- 2 kcal/mole. Thus the keto-form is thermodynamically more stable than the enol-form, and because of that acetone exists exclusively in the keto-form.

(ii) The resonance stabilization of the aromatic ring in phenol is very large, and hence phenol exists entirely in the enol-form; the alternative keto-form can't exist because it would sacrifice the stability associated with the six π-electrons aromatic system.

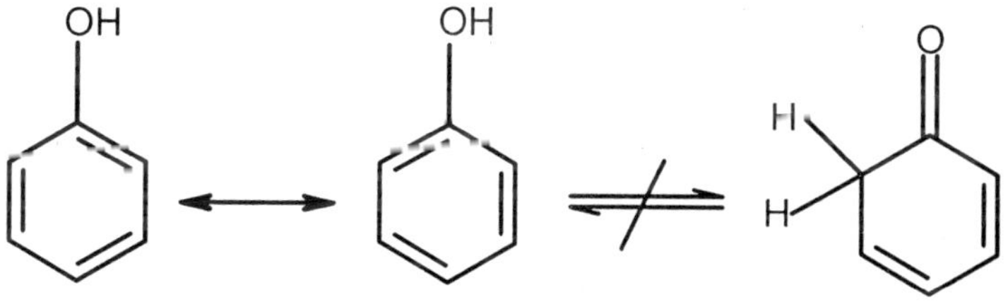

This Kekule′ resonance makes the enol-form more stable than the keto by about 15 kcal/mole.

However, with increase in the number of phenolic –OH groups on aromatic nucleus the difference in stability between the enol and the keto-form progressively decreases; this is because of the fact that the bond energies of several keto groups tends to counterbalance the Kekule′ resonance energy. Thus phloroglucinol (1,3,5-trihydroxybenzene) contains also the ketonic form in its tautomeric equilibrium mixture at an amount that can be detectable — the substrate gives a dark colour with aqueous ferric chloride solution (test for enols) as well as forms a trioxime (test for carbonyls).

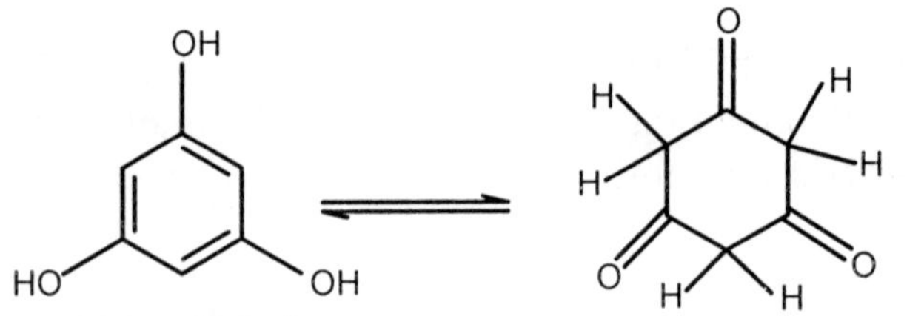

(phloroglucinol) enol-form ⇌ (1,3,5-cyclohexatrione) keto-form

(iii) Presence of strong intramolecular hydrogen-bonding stabilizes enol form of a compound with respect to the keto; it was estimated that intramolecular hydrogen-bonding reduces the difference in potential energy of the keto and enol tautomers by about 7 kcal/mole, and shifts the equilibrium toward the enol side.

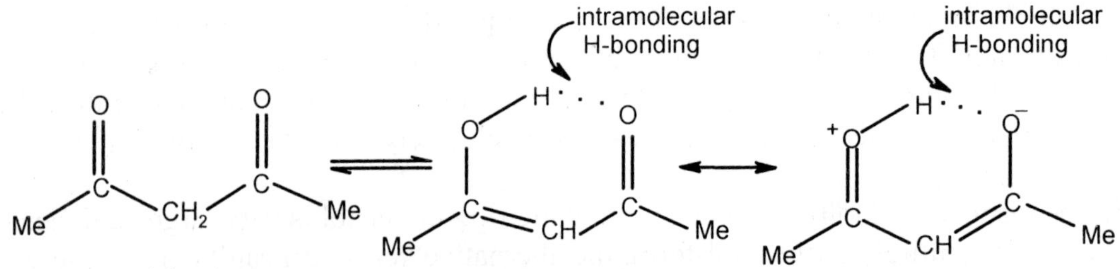

acetylacetone 24% ⇌ 4-hydroxypent-3-en-2-one 76%

The enol-form is stabilized by means of resonance as well as intramolecular hydrogen-bonding. In this connection it should be noted that intramolecular hydrogen-bonding will also lead to the decrease in the polar character of the enol with respect to the keto, and hence the enol shows lower boiling point despite it bears a hydroxyl group.

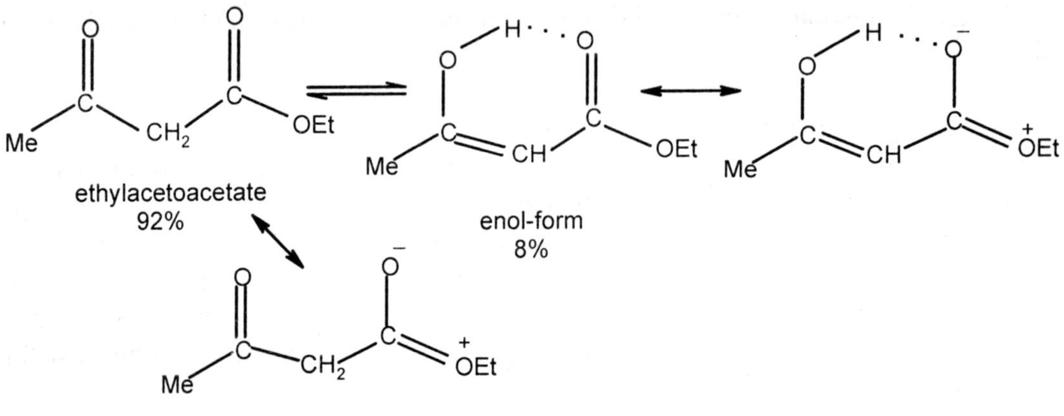

ethylacetoacetate 92% ⇌ enol-form 8%

Though there is a possibility of formation of intramolecular hydrogen-bonding in the enolic tautomer of ethylacetoacetate, it is not so effective due to a cross-conjugation between the ester group and the enol structure. The resonance of the ester group occurs at the expense of the resonance of the enol ring. Thus the enol-form becomes destabilized relatively to the keto tautomer, and equilibrium is shifted towards the ketone.

(iv) Steric factor also plays a significant role in shifting the direction of a tautomeric equilibrium. In case of α–methylacetylacetone the equilibrium is shifted toward the keto-form because the enol is destabilized due to the steric strain imposed by the bulky α–methyl group into the molecule — the α–methyl enters into strong steric interaction with the adjacent methyls. Hence the α–methyl enol possesses greater internal strain than the enol of simple acetylacetone, and so the former becomes less favourable energetically than the latter. Thus, enol content of acetylacetone is much significant than that of α–methylacetylacetone.

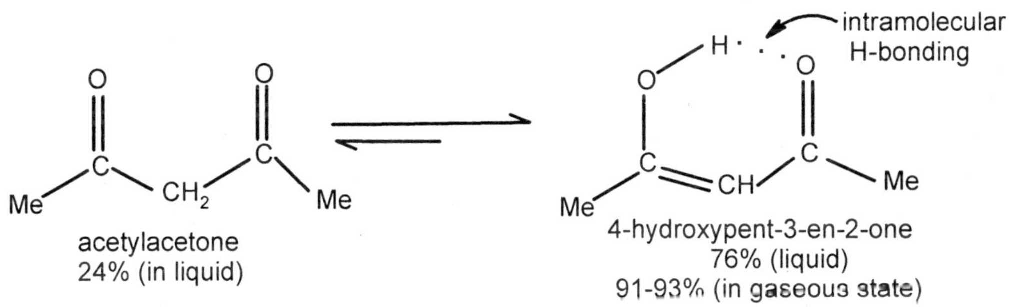

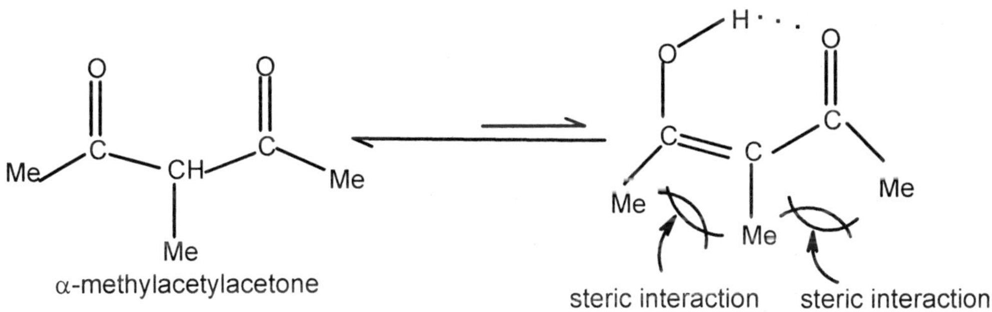

As the size of the alkyl substituents increases, the amount of enol content of the corresponding substrate decreases. In an extreme case, for example, $(Bu^tCO)_2CHBu^t$ is found to exist exclusively in keto form — the corresponding enol form can't exist because it would then be highly energetic due to severe steric repulsive strain developed into the molecule.

(v)

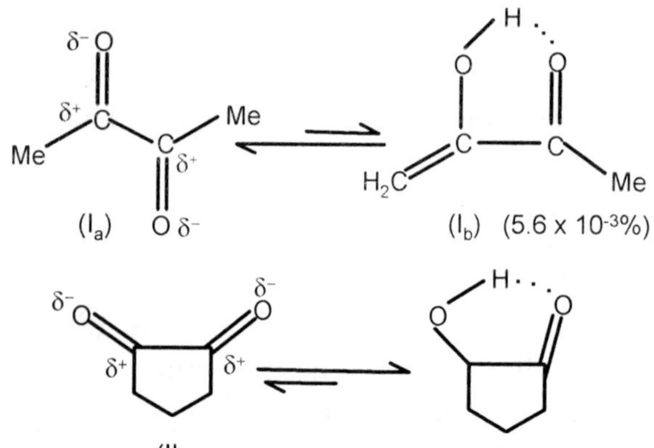

(Ia) (Ib) (5.6 x 10-3%)

(IIa (IIb)

In butan-2,3-dione the keto-form (I_a) takes up the predominant *anti*-conformation such that the two electronegative oxygen atoms are as far as possible, and the two carbonyl dipoles are in opposition. Hence the keto-form becomes so energetically favourable that it exists exclusively over the enolic tautomer (I_b), although the enol can form intramolecular hydrogen-bonding.

On the contrary, in cyclopentan-1,2-dione (II_a) the C=O dipoles can't rotate — they are 'locked' in *syn*-conformation. This conformation offers a repulsive interaction between the carbonyl dipoles, thereby increasing the internal energy of the tautomer. However in the enolic tautomer (II_b), though the –OH and C=O groups within the molecule are in *syn*-conformation, a scope of formation of intramolecular hydrogen-bonding is open. Formation of this intramolecular hydrogen-bonding is probably the driving force for the substrate to remain exclusively in the enolic form.

(v) Two different structures for enolic tautomerization of 2,6-bicyclio[2,2,2]-octandione can be written:

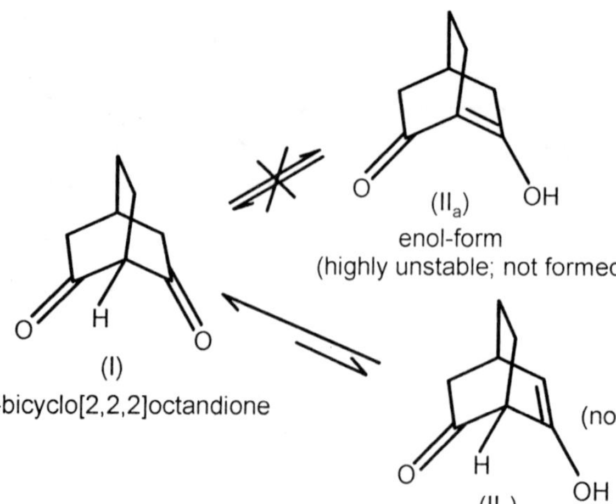

(IIa) enol-form (highly unstable; not formed)

(I) 2,6-bicyclo[2,2,2]octandione

enol-form (not so significant)

(IIb)

The compound (I) shows no tendency to enolize into II_a, because the resulting structure bears a double bond at the bridgehead position, and this kind of structural arrangement in bicyclic system is of high internal energy (*vide* Bredt's rule) making it highly unstable. So this is not formed. Hence only the enolic form (II_b) can be formed, but would not be stabilized through conjugative effect. That's why the enol content of (I) at equilibrium is very small indeed.

(vii) The enol contents of the following compound are observed as:
$CH_3COCH_2COCH_3$ (76%), $PhCOCH_2COCH_3$ (89%), $PhCOCH_2COPh$ (100%).

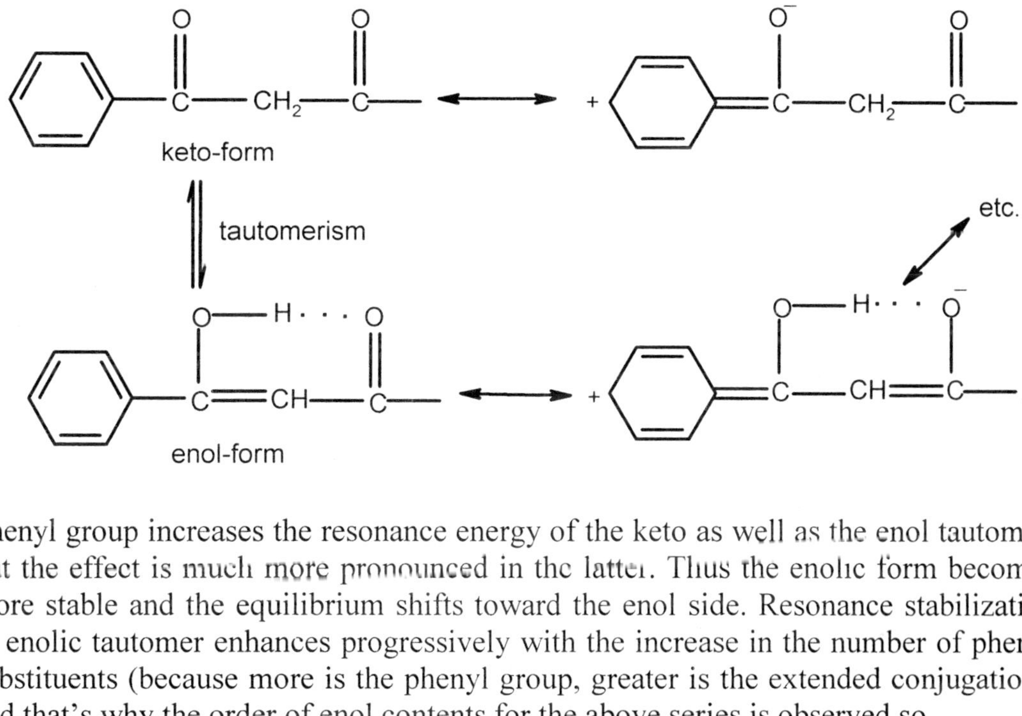

Phenyl group increases the resonance energy of the keto as well as the enol tautomer, but the effect is much more pronounced in the latter. Thus the enolic form becomes more stable and the equilibrium shifts toward the enol side. Resonance stabilization of enolic tautomer enhances progressively with the increase in the number of phenyl substituents (because more is the phenyl group, greater is the extended conjugation), and that's why the order of enol contents for the above series is observed so.

(viii) Nature of solvent has a considerable effect on keto-enol tautomerism. The form, which becomes more solvated, remains in greater proportion at the equilibrium. During solvation there may be considerable energy effects connected with the nature of solvent. A solvent that can form hydrogen bonds with the carbonyl group of the keto form will result a decrease in entropy. On the other hand, the enol is reluctant to form hydrogen bonds with the solvent because intramolecular hydrogen bonding is already present into the molecule, and consequently there will occur a smaller decrease in entropy. Hence, the enol content of a compound should be higher in non-polar solvents since polar solvents relatively tend to stabilize the keto-form more predominantly. Also, hydroxylic solvents would form hydrogen bonds with the solute oxygen atoms to compete with the intramolecular hydrogen bonding in the enol — this would further decrease the enol content of 1,3-diketones. Thus solvents such as water, methanol, acetic acid, etc., tend to reduce enol content while in solvents such as hexane, benzene, etc. the enol content will be larger. These considerations are substantiated by the following data:

Compound	Enol content	Solvent
Ethylacetoacetate	46.4	Hexane
	16.2	Benzene
	10.5	Ethanol
	7.7	Pure liquid
	6.9	Methanol
	5.7	Acetic acid
	0.4	Water
Acetylacetone	92	Hexane
	15	Water

Problem 1.26 Arrange the followings in order of their increasing acidity: *o*-nitrophenol, *m*-nitrophenol and *p*-nitrophenol.

Solution. The increasing order of acidity of the nitrophenols is: *m*-nitrophenol (pKa 8.40) < *o*-nitrophenol (pKa 7.23) < *p*-nitrophenol (pKa 7.15).

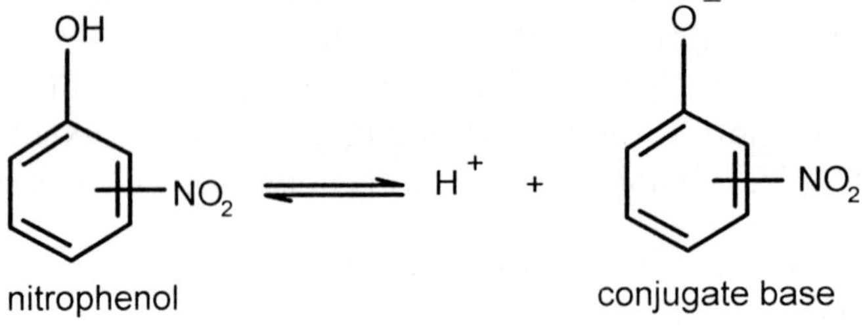

Acidity of a compound depends upon the stability of its conjugate base. More is the stability of conjugate base greater is the acidity. The nitrophenoxide ions get stability due to resonance interaction as well as –I effect offered by the nitro groups attached to the aromatic ring. The negative charge on anionic oxygen atom becomes delocalized into the ring. Now, we have to concentrate on the extent of conjugation in each of them. The corresponding resonating structures are shown below:

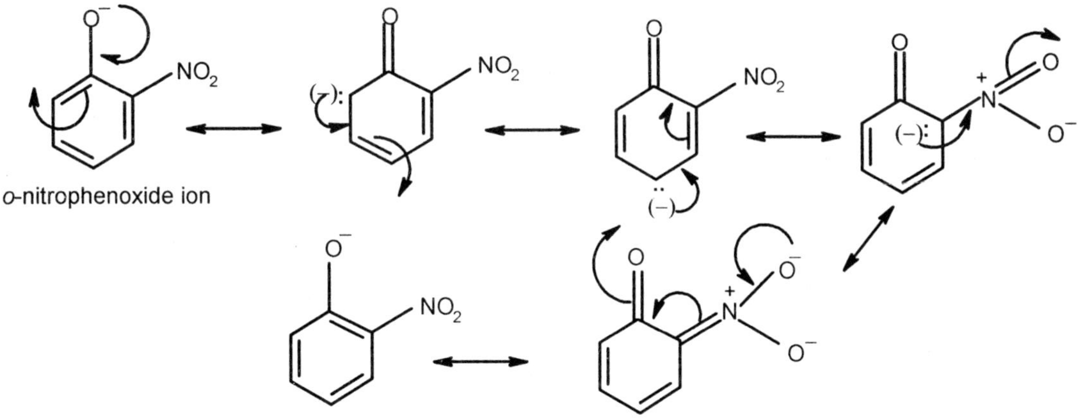

o-nitrophenoxide ion

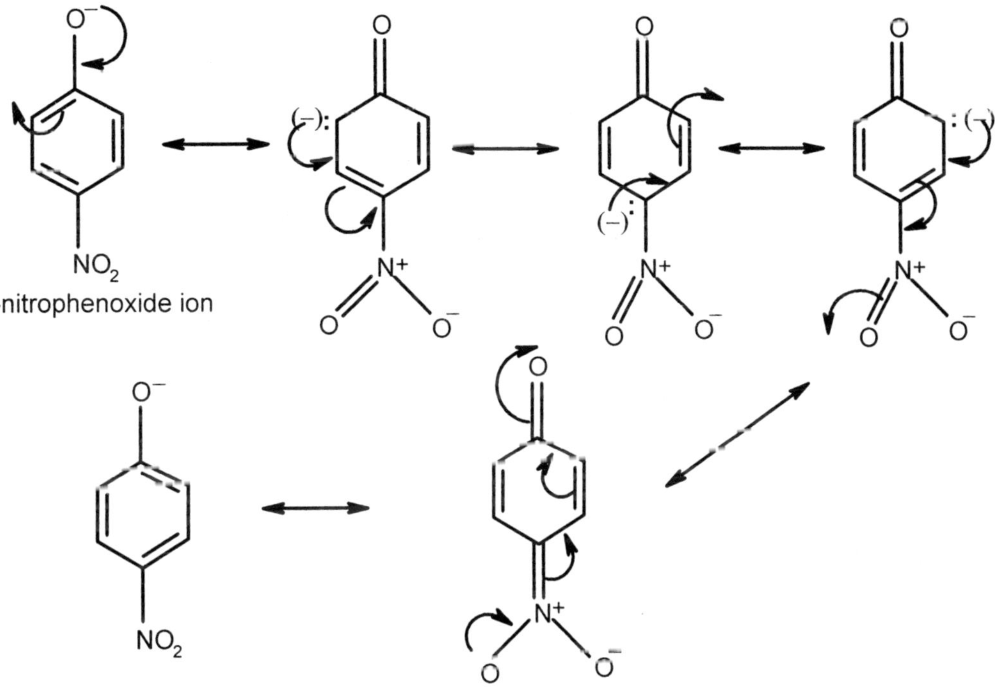

p-nitrophenoxide ion

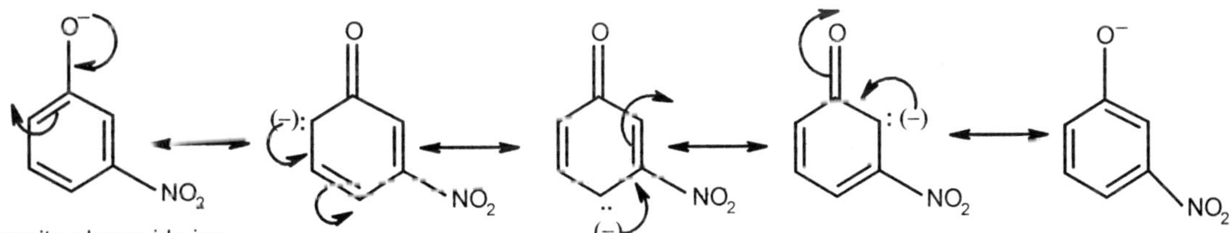

m-nitrophenoxide ion

Besides through its –I effect, the nitro group in both the *o*- and *p*- isomers stabilizes the respective conjugate bases by its –R effect also by taking direct part in extending the deloacalization of negative charge. But the *meta*-oriented nitro group in the *m*-isomer can't take part in resonance — it can offer only –I effect in stabilizing the *m*-conjugate base. That's why conjugate bases of *ortho*- and *para*-isomers of nitrophenol are comparatively more stable than that of the *meta*-isomer. Hence, *o*- and *p*-nitrophenols are more acidic than *m*-nitrophenol.

Although the –I effect of the nitro group is more pronounced in the *o*-isomer than in the *p*-isomer, the former is somewhat less acidic than the latter. This is because of the fact that the hydroxyl group into the *o*-isomer is intramolecularly hydrogen bonded (chelated) with the adjacent $-NO_2$ group, which reduces its proton-releasing tendency to some extent.

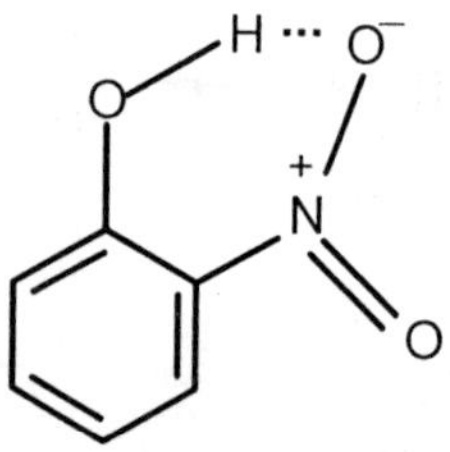

o-nitrophenol
(chelated form)

Problem 1.27 2,4,6-Trinitrophenol possesses appreciable acidic property. How would you account for the fact?

Solution.
2,4,6-Trinitrophenol attains appreciable acidic property (pKa 0.71) due to the presence of strong electron-withdrawing nitro groups at the *para*- and both the *ortho*-positions of –OH group. All the three nitro groups take part in the resonance process (the delocalization process is shown below); extensive delocalization of the anionic charge thereby becomes possible leading to the stabilization of the conjugate base of the compound relatively much higher than that of the parent compound, phenol. That's why 2,4,6-trinitrophenol becomes a very strong acid such that it is well known as picric acid.

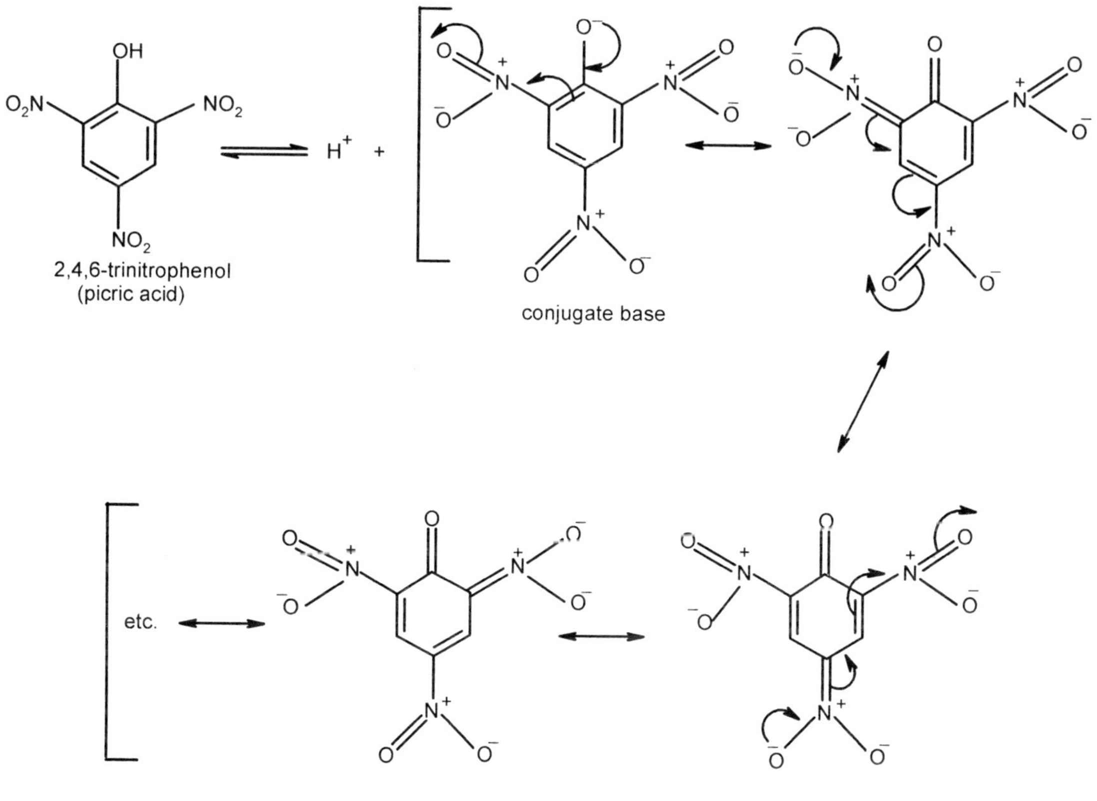

(highly effective delocalization of negative charge)

Problem 1.28 The pKa values for the *ortho*-, *meta*- and *para*-isomers of nitrobenzoic acid are 2.17, 3.49 and 3.43 respectively. Explain the order of acidity.

Solution. Although nitro group takes part in resonance process with the π-electron system of aromatic nucleus of nitrobenzoic acids, it has no direct resonance-interaction with the caboxylate $(-COO^-)$ function in the corresponding conjugate bases. Inductive effect of nitro group attached with the aromatic nucleus determines the acidic order of the compounds concerned. Since extent of inductive effect decreases with distance, an expected acidity order should have to be *o*- > *m*- > *p*-isomer. But in practice it is *o*- > *p*- > *m*-isomer. To understand the reverse order between the *meta*- and *para*-isomer let us draw the resonating structures of the corresponding conjugate bases.

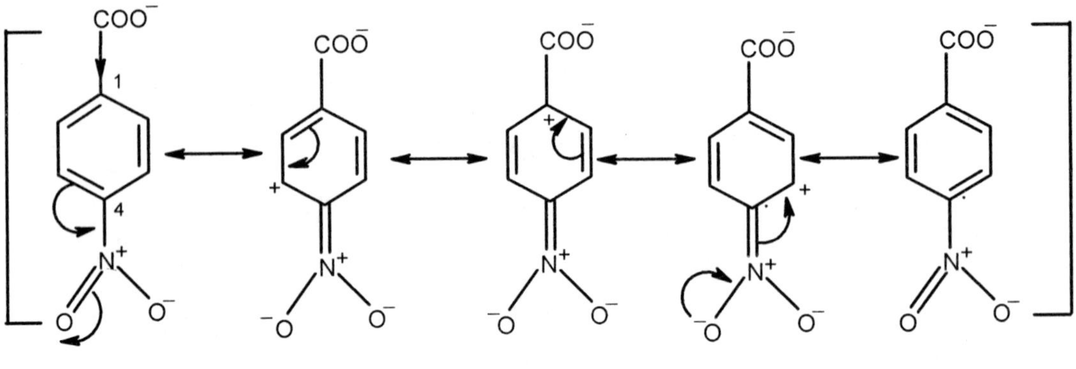

resonating forms of *p*-nitrobenzoate ion

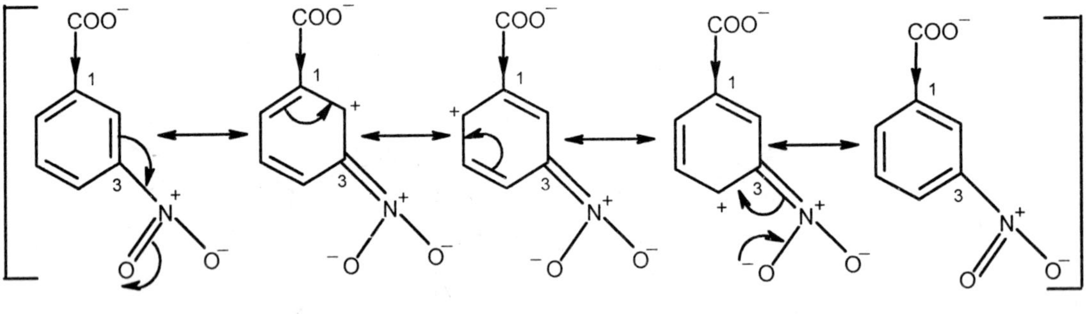

resonating forms of*m*-nitrobenzoate ion

From the resonating structures it is found that the $-NO_2$ group at *para*-position exerts its –R effect to reduce the electron density at the carbon bearing the carboxylate group. A *meta*-nitro also exerts –R effect, but any of its resonating structure does not bear positive charge at the carbon (C-1) bearing the carboxylate function. Thus *para*-carboxylate ion becomes relatively more stabilized over that of the *meta*-isomer giving rise to greater acidity of the *p*-nitrobenzoic acid; but this effect is not so much stronger, and so difference in pKa values of *p*-nitrobenzoic acid and *m*-nitrobenzoic acid is small.

Comparatively much higher acidity of *o*-nitrobenzoic acid is attributed to the powerful –I effect exerted by the nitro group placed at a very short distance with respect to the carboxylic acid function in the compound. In the conjugate base the nitro group withdraws electrons from the $-COO^-$ group by making the carbon (C-1), bearing the functionality, to some extent electron-deficient. Besides, there occurs a more direct electrostatic interaction between the carboxylate and the nitro group — which contributes a lot to the stabilization of the anion.

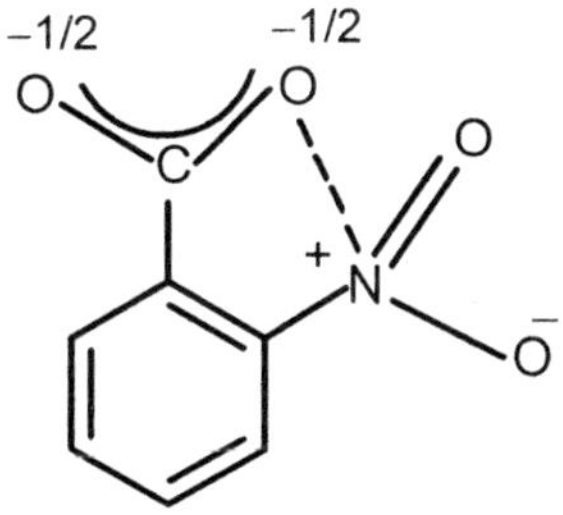

o-nitrobenzoate anion

Problem 1.29 2,6-Di-*t*-butylphenol is less acidic than phenol. Explain.
Solution.

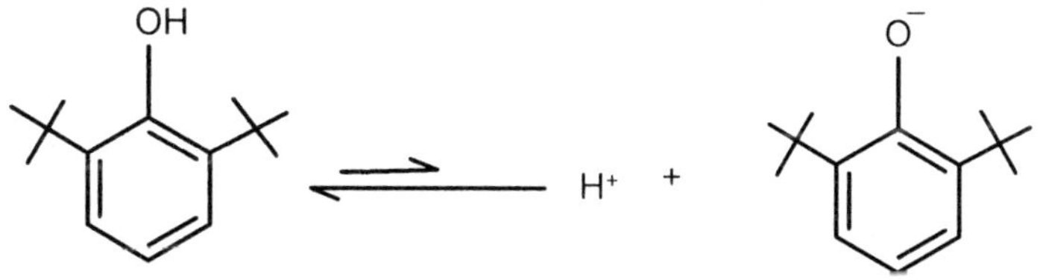

2,6-di-*t*-butylphenol

conjugate base
(not stabilized through conjungation)

Presence of large bulky *t*-butyl groups at the *ortho*-positions produces considerable steric crowding around the C–OH bond in 2,6-di-*t*-butylphenol as a result of which the –OH group goes out of the plane of the ring. This steric hindrance persists in its conjugate base also. Since the $O^{(-)}$ unit remains out of the plane of the ring, lone pair orbitals of oxygen are no longer parallel to the π-orbitals of benzene ring. So, resonance stabilization is not possible in this case, whereas phenoxide ion (formed on deprotonation of phenol) becomes relatively more stabilized through resonance interaction. That's why 2,6-di-*t*-butylphenol becomes less acidic than phenol.

Problem 1.30 Explain why **A** is much stronger than **B**?

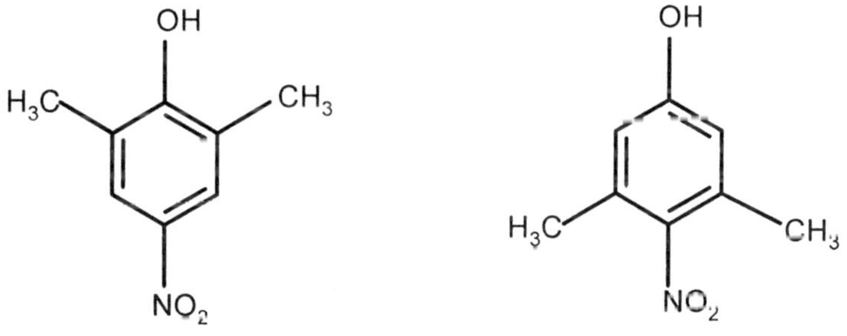

A **B**

Solution.

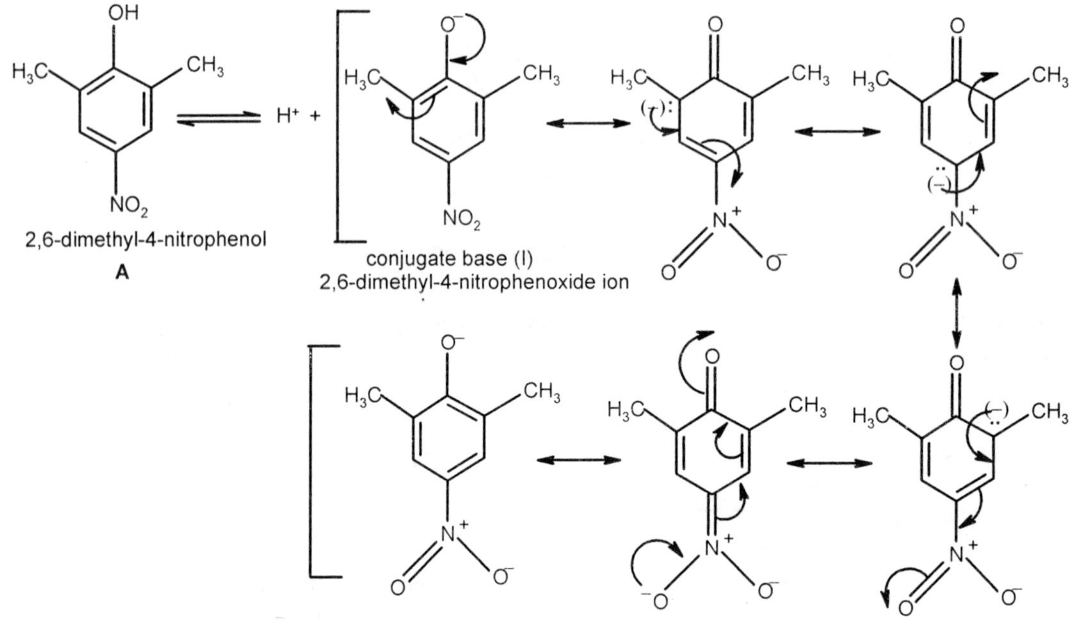

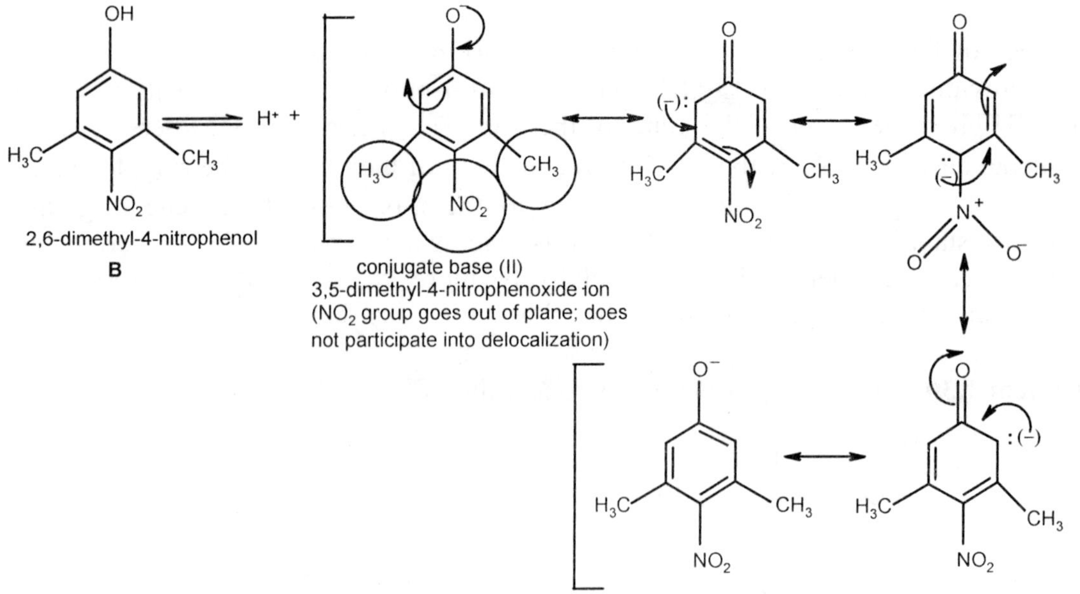

Since $O^{(-)}$ is small, presence of two methyls at the *ortho*-positions in (I) cannot produce so much steric hindrance that the delocalization of lone pair electrons of $O^{(-)}$ with the π-electron system of the aromatic ring may be prevented. The *p*-nitro group also participates into the delocalization process.

On the contrary, nitro group is large enough to be suffered from steric crowding as produced by the two *ortho*-oriented methyls in (II) — the conjugate base of (**B**). Hence the nitro group goes out of the plane of the ring, and can't participate in the delocalization process.

Because of greater delocalization of the anionic charge on oxygen atom of the conjugate base (I) of **A**, it becomes more stabilized than (II). Hence, **A** is more acidic than **B**.

Problem 1.31 Formic acid is stronger than benzoic acid. Why?

Solution.

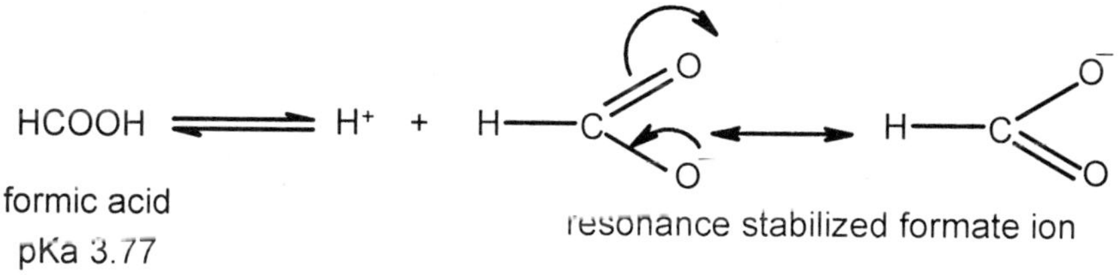

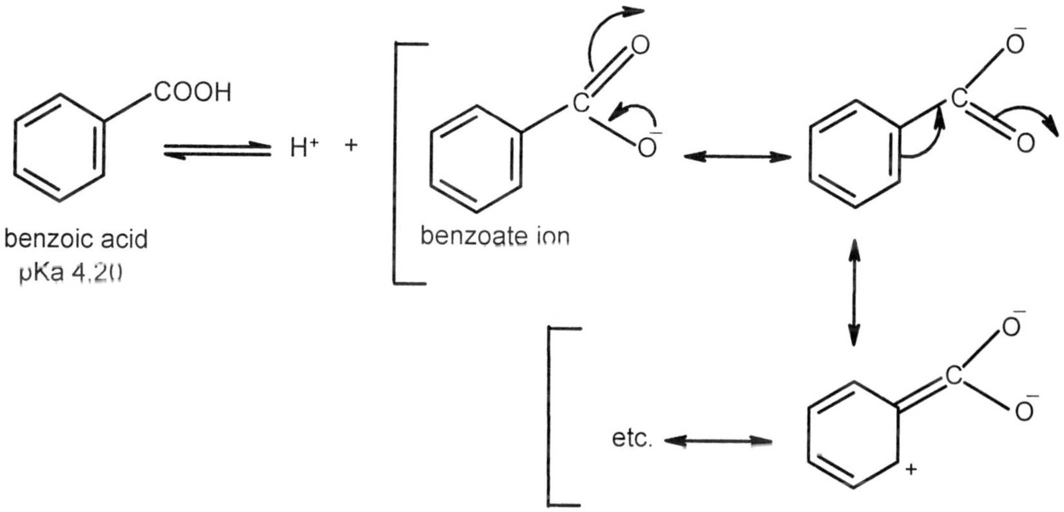

In benzoate ion C=O unit of the $-COO^-$ function is conjugated with the lone pair electrons of anionic oxygen atom as well as with the π-electron system of the benzene ring (cross-conjugation). Because of the cross conjugation, effective resonance stability towards the negative charge on anionic oxygen (shown by the structures as above) becomes less than that in case of formate ion. So formate ion is more stable than benzoate ion, and that's why formic acid is more acidic.

Problem 1.32 In aqueous solution acidity of methanol is greater than that of *t*-butanol. Justify the observation.

Solution.

CH_3OH
methanol

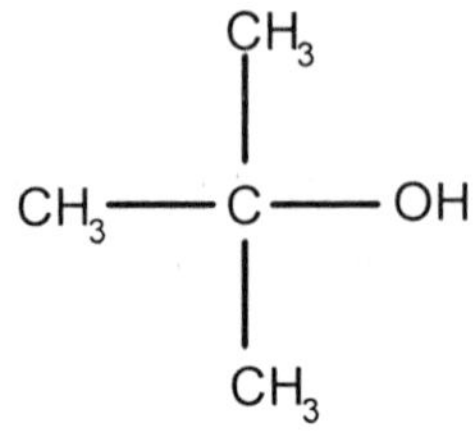

t-butanol

Substitution by successively more bulky groups causes a steady lowering of alcohol-acidity in aqueous solution. As more substituent hydrocarbon groups are present in *t*-butanol in comparison to methanol, the latter would show greater acidity in the aqueous medium; in other words in solution the *t*-butoxide ion (Me_3CO^-) will be stronger base than methoxide or ethoxide. Extent of solvation of conjugate bases directs their acidity — more is the solvation greater is the acidity. Due to presence of a large bulk of hydrocarbon moiety (which is hydrophobic) *t*-butoxide ion becomes less solvated (hydrated) than methoxide ion, and is less stable. Thus methanol (pKa 16) is a stronger acid than *t*-butanol (pKa 19) in aqueous solution. But in gas-phase, where there occurs no solvent interaction, the acidity order is interestingly observed as reversed.

Problem 1.33 The K_1/K_2 ratio for fumaric acid is 23.17, while the value for maleic acid becomes 20,168 —— how would you explain the observation?

Solution.

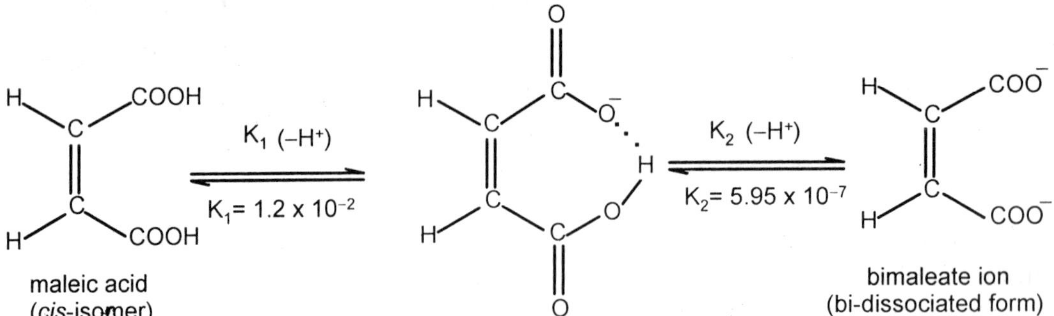

maleic acid
(*cis*-isomer)

mono-maleate ion
(mono-dissociated form)
chelated
(**A**)

bimaleate ion
(bi-dissociated form)

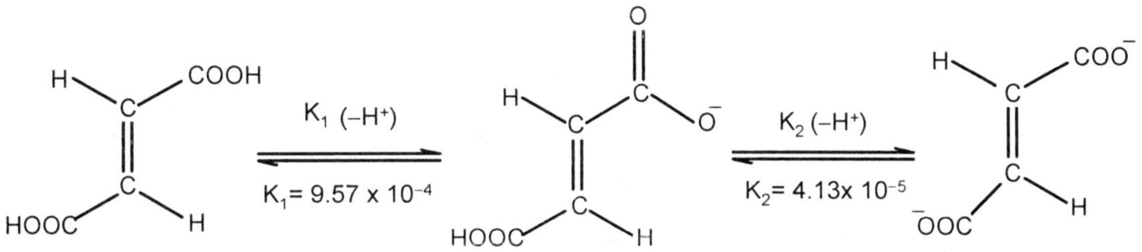

fumaric acid (*trans*-isomer)

momo-fumarate ion (**B**) (mono-dissociated form)

bifumarate ion (bi-dissociated form)

Both the acids are dibasic. From the experimental data it is observed that first ionization of maleic acid is much greater than that of fumaric acid, but during the second ionization reverse is the fact. Maleate ion (**A**, mono-dissociated form) becomes much more stabilized through intramolecular hydrogen bonding (chelation) than the corresponding fumarate ion (**B**) where such chelating effect does not persist. This strong hydrogen bonding between the second COOH group and COO^- moiety in maleate ion (A) retards the rate of second ionization much effectively.

Carboxylic groups approaching to each other exert electrostatic effects on each other. The second COOH group in fumarate ion (**B**) is far apart from the COO^- moiety, and so it suffers from a mild retarding electrostatic effect to its dissociation, thereby decreasing K_2 with respect to K_1. On the contrary, the COO^- group in mono-maleate ion (**A**) has a much larger attractive electrostatic effect on the proton of the second COOH function due to its stereochemical geometry, and the ion is also stabilized by strong hydrogen bonding. Both the effects retard the dissociation of the second proton to give the larger K_1/K_2 ratio for maleic acid.

Problem 1.34 Isocrotonic acid is more acidic than crotonic acid. Why?
Solution.

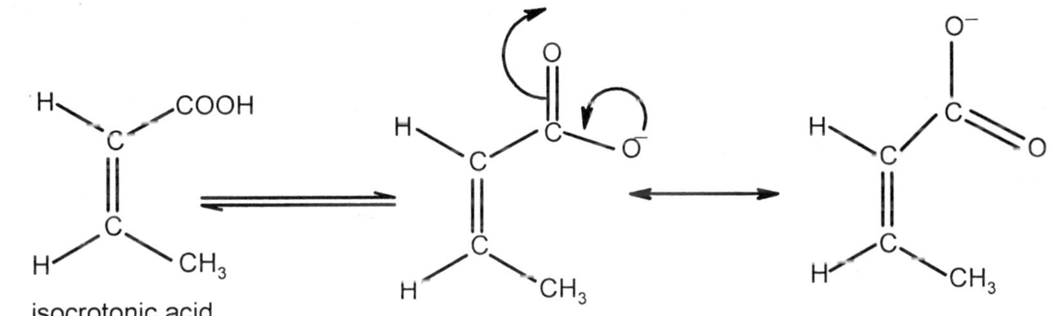

isocrotonic acid (*cis*- but-2-enoic acid) pKa 4.44

conjugate base (I)

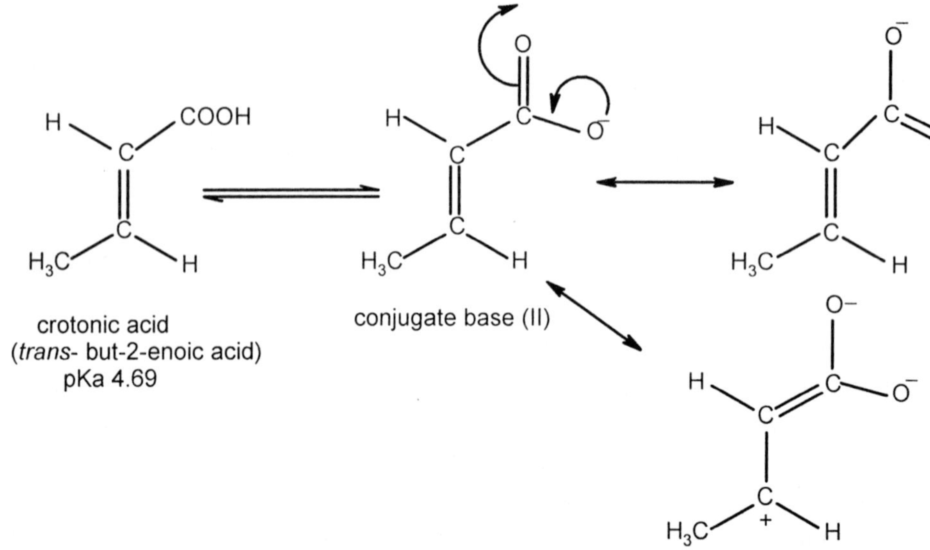

The carbonyl unit of COO⁻ group of the conjugate bases is conjugated with both the anionic oxygen (O⁻) and C=C double bond. Delocalization of the carbonyl unit with anionic oxygen (O⁻) increases the stability of conjugate base, and thus leads to the greater acidity of the corresponding compound. But delocalization with C=C double bond lowers the stability of conjugate base, and also the acidity of parent compound.

In the conjugate base (I) of isocrotonic acid the bulky methyl group and carboxylate moiety (COO⁻) are in close proximity creating a steric crowding that inhibits the effective delocalization between the anionic oxygen (O⁻) and C=C double bond. On the contrary, the respective groups are far apart in the conjugate base (II) of crotonic acid so that more effective delocalization between the anionic oxygen (O⁻) and C=C double bond occurs here leading to decrease in the stability of (II). That's why isocrotonic acid is practically more acidic than its geometrical isomer, crotonic acid.

Problem 1.35 Though phenylacetic acid is stronger than acetic acid, benzoic acid is weaker than formic acid. Explain.

Solution.

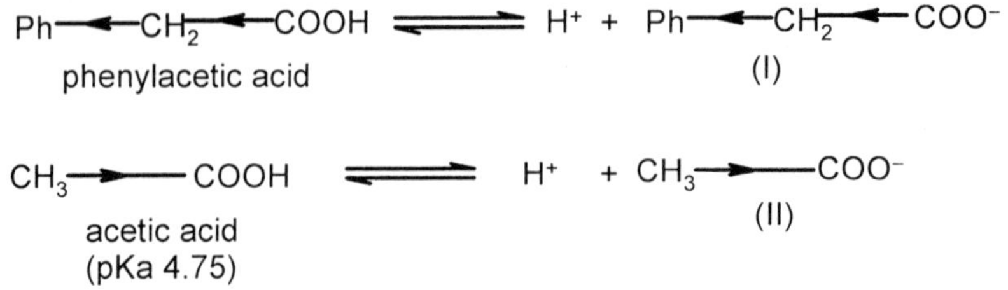

While methyl group offers +I effect, benzyl group has –I effect. Hence benzyl carboxylate (I) is more stable than acetate (II) ion. So, phenylacetic acid becomes stronger than acetic acid.

On the contrary, benzoate ion (III) (conjugate base of benzoic acid) becomes less stable than the formate (IV) because of the delocalization effect of ring π-system with the carbonyl unit of caroxylate (COO^-)[as shown by canonical forms]. Hence benzoic acid is weaker than formic acid.

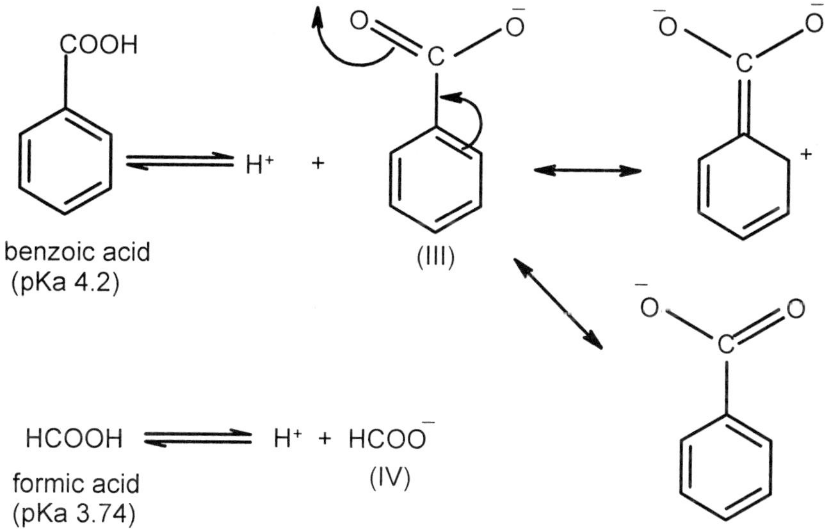

Problem 1.36 How would you account for the greater acidic strength of acrylic acid over that of propionic acid?

Solution.

$$CH_2{=}CH \leftarrow COOH \qquad\qquad CH_3CH_2 \rightarrow COOH$$

acrylic acid (pKa 4.25) | propionic acid (pKa 4.87)

Electronegativity of hybridized carbons is in the increasing order of C(sp^3) < C(sp^2) < C(sp). The α-carbons of acrylic acid and propionic acid are sp^2 and sp^3-hybridized respectively. Thus ethenyl group (or say vinyl group) as a whole shows –I effect, whereas the ethyl group in propionic acid exhibits +I effect. Hence the conjugate base of acrylic acid becomes relatively more stable, and the acid is stronger than the other.

Problem 1.37 Between *p*-chlorophenol and *p*-fluorophenol which one is a stronger acid and why?

Solution.

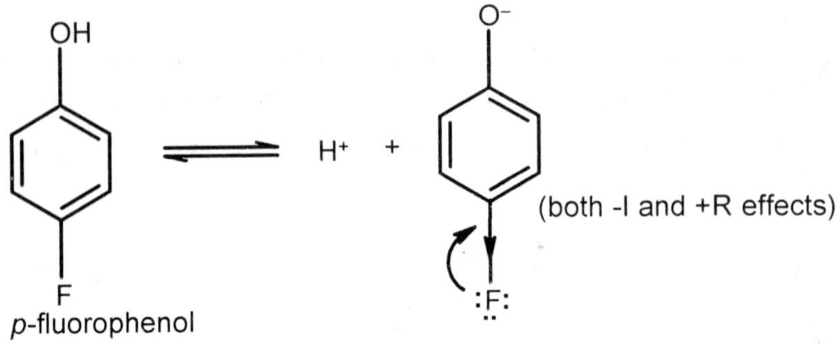

p-fluorophenol

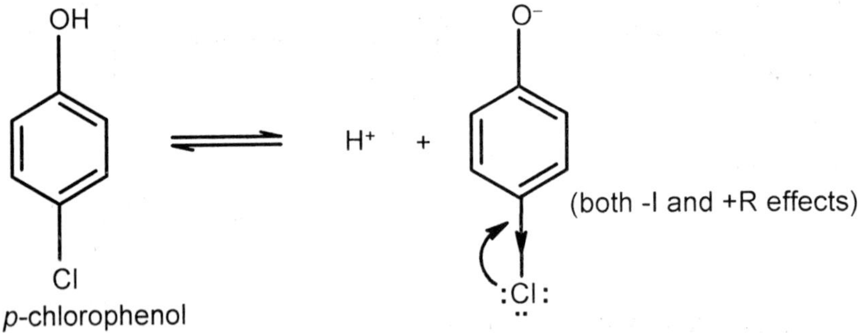

p-chlorophenol

Both the halogen atoms (Cl and F) have –I effect and also +R effect. Inductive effect becomes less pronounced as they are in *para*-relation to –OH function. In this particular case resonance effect becomes the significant directive factor in determining the acidity order. Fluorine being smaller in size in comparison to chlorine undergoes very effective delocalization of its lone pair electrons with the π-electron system of benzene ring, thereby destabilizing *p*-fluorophenoxide ion (conjugate base of *p*-fluorophenol) relatively to a greater extent. Hence, *p*-fluorophenol (pKa 9.81) is less acidic than *p*-chlorophenol (pKa 9.38).

Problem 1.38 Instead of being basic in nature phthalimide and also *N,N*-(2,4,6-trinitrophenyl)amine both exhibit acidic property. Account for the observation.

Solution.

The conjugate base of phthalimide is greatly stabilized due to effective delocalization between the negative charge on 'N' atom and the two electron-withdrawing carbonyl units in the species. Hence it becomes more stable than the neutral species, and that's why the ionization is favoured. Consequently phthalimide shows considerable acidic property and readily dissolves in alkali metal hydroxide solutions to form salts.

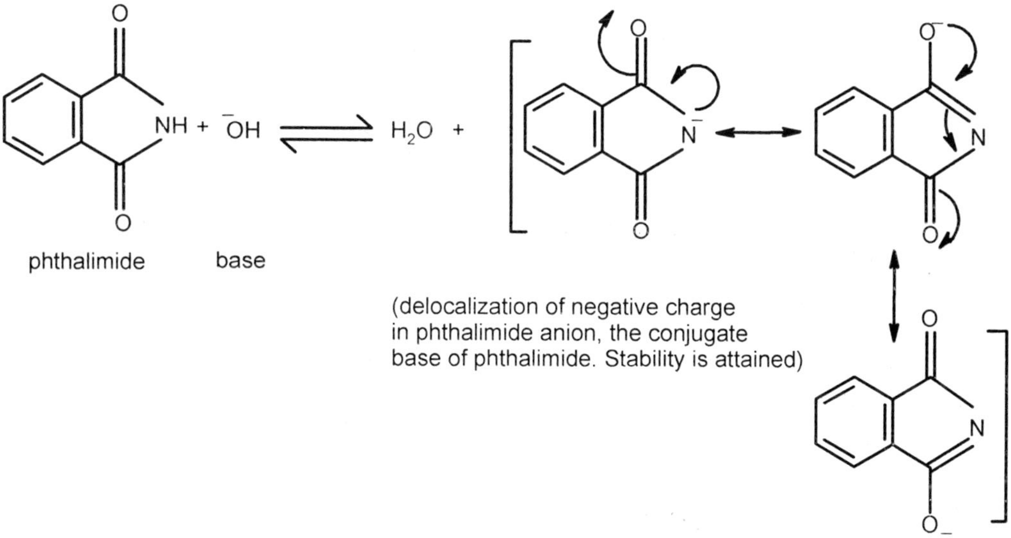

phthalimide base

(delocalization of negative charge in phthalimide anion, the conjugate base of phthalimide. Stability is attained)

Similar situation is also observed in the case of *N,N*-di-(2,4,6-trinitrophenyl)amine; it is acidic about 100 times more than acetic acid. The conjugate base of the parent compound becomes highly stabilized as a result of an effective delocalization of the negative charge on anionic 'N' into the attached phenyl rings each bearing strong electronegative nitro substituents at 2, 4 and 6-positions. The nitro groups take part a major role in the resonance process responsible for the stabilization of the conjugate base concerned.

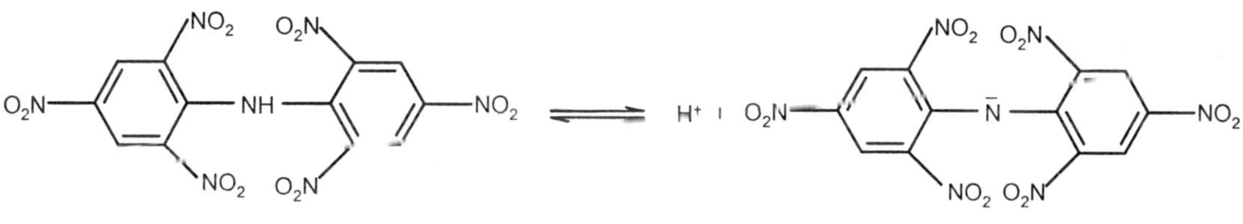

N,N-di-(2,4,6-trinitrophenyl)amine

(conjugate base)
(greatly stabilized resonance stabilized involing the 2,4,6-trinitrophenyl rings)

Problem 1.39 Arrange the following compounds in order of their increasing acidity:

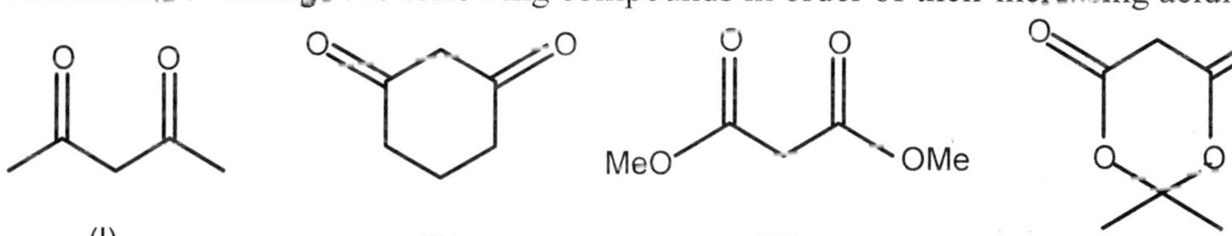

(I) (II) (III) (IV)

Justify your answer.

Solution. Ketones are usually stronger acids than carboxylic esters, and so (I) would be more acidic than (III). But cyclization of (I) to (II), and of (III) to (IV) results noticeable changes in comparative acidity order — conformations of the cyclic derivatives have a profound effect in this regard. Cyclization of (I) to (II) increases the acidity to some extent, but the cyclization effect becomes too much pronounced in (IV). The fact is reflected from their respective pKa values (I to IV) — 13.3, 11.2, 15.9 and 7.3; thus cyclization lowers the pKa value by 8.6 units on going from (III) to (IV). This may be explained by considering the conformational geometry of the cyclic 1,3-diester (IV).

It has been observed that loss of a proton is easier by ~5kcal/mole for the *syn*-isomer of 1,3-diester in comparison to that of the *anti*-conformer.

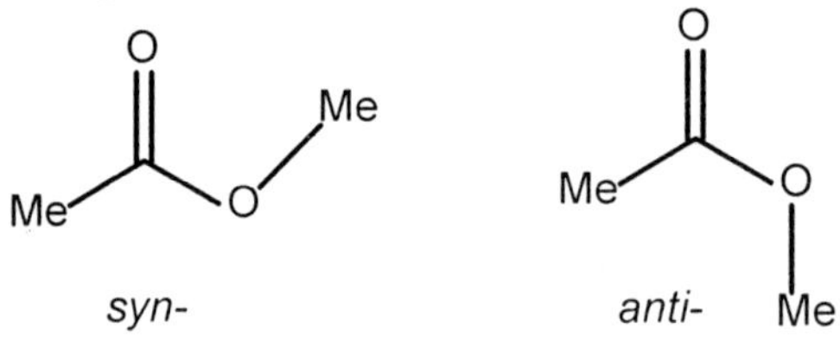

In an acyclic molecule like (III) the preferred conformations are *anti*, but in cyclic ones (IV) the conformation on both the sides is constrained to be *syn*. Hence, molecule (IV) [which is in *syn*-conformation] can remove proton relatively easier leading to its higher acidity. Thus the acidity order may be given as: III (pKa 15.9) < I (pKa 13.3) < II (pKa 11.2) < IV (pKa 7.3).

Problem 1.40 Which one of the following compounds (A or B) would you expect to be the stronger acid? Give reasons.

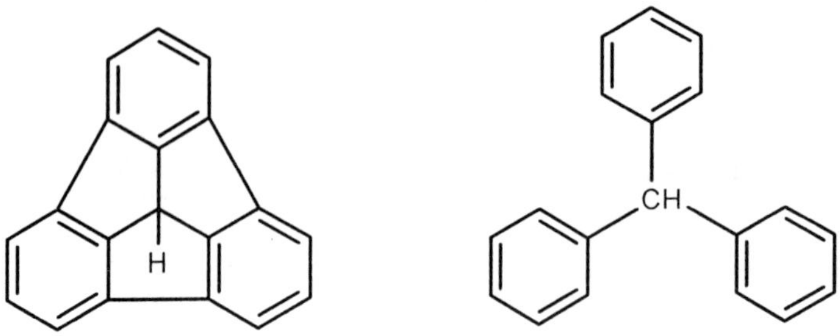

A B

Solution. The conjugate base formed on deprotonation of **A** would be better stabilized because of the fact that the carbanionic charge becomes delocalized into the aromatic rings which are to be coplanar. But in the conjugate base of **B** the phenyl groups would prefer to be in different planes for steric reasons, and hence the carbanion that would be formed from **B** on deprotonation becomes comparatively less stable due to lesser extent of delocalization. Thus compound **A** would show greater acidity than **B**.

Problem 1.41 Offer an explanation for each of the following facts:

(a) Amides are less basic than alkylamines.
(b) Aniline is a weaker base than cyclohexylamine.
(c) Urea acts as a monoacidic base.
(d) Triphenylamine practically shows no basicity.
(e) Guanidine behaves as a strong base.
(f) Pyrrole is less basic than pyridine.
(g) Trimethylamine or triethylamine is more basic than pyridine, which in turn is more basic than acetonitrile.

Solutions.

(a) Lone pair electrons on 'N' atom of $-NH_2$ unit in amide group are delocalized with the carbonyl (C=O) π-electrons, thereby decreasing electron-density on the 'N' atom. Thus electron-donating tendency of an amide molecule is reduced. On the other hand, such reducing effect does not arise in alkylamines; moreover +I effect of an alkyl group enhances the electron-density on 'N' atom of $-NH_2$ group. Thus, amides become less basic than aklylamines. This is reflected from their comparative pKa values — acetamide (CH_3CONH_2, pKa −1.4, very weak base), methylamine (CH_3NH_2, pKa 10.62).

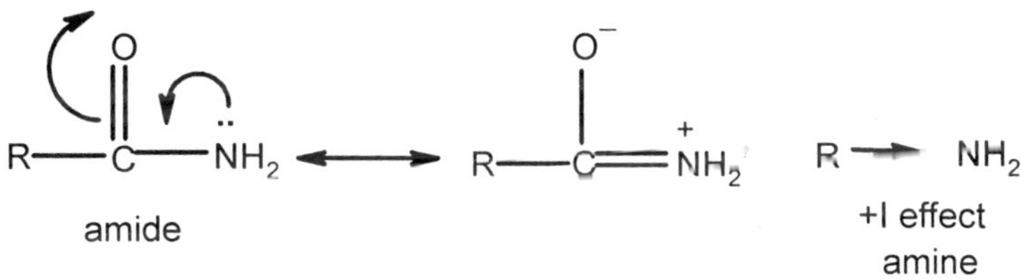

(b) The strength of nitrogenous bases depends on the availability of the unshared electron pair on nitrogen. In arylamines, like aniline, the unshared electron pair on nitrogen is in conjugation with the π-electron system of the aromatic nucleus, and undergoes delocalization into the ring. Thus electron availability on 'N' atom of $-NH_2$ group is decreased considerably making the molecule relatively a weak base. But in cyclohexylamine (or commonly alkylamines) such delocalization does not take place; besides, the electron-releasing inductive effect of alkyl group pushes the bonding electrons towards nitrogen making the unshared electron pair more available for coordination with a proton. Thus generally speaking, aromatic amines are comparatively weaker bases than aliphatic amines.

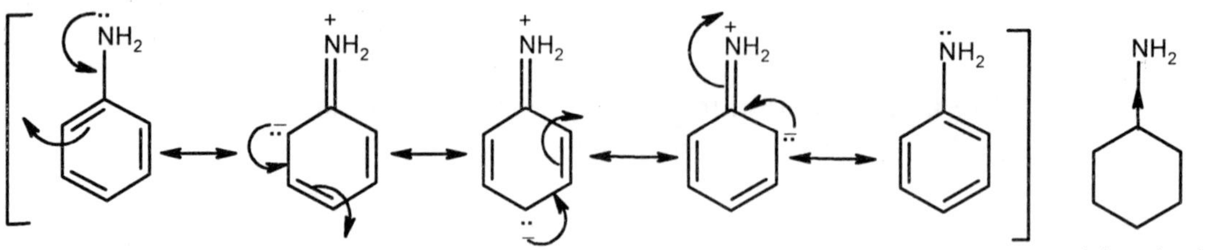

cyclohexylamine
pKa 10.64

aniline as a resonance hybrid
pKa 4.62

Alternatively if aniline is protonated to form anilinium (conjugate acid of aniline) ion, any delocalization of positive charge will not take place as the electron pair on 'N' is no longer available.

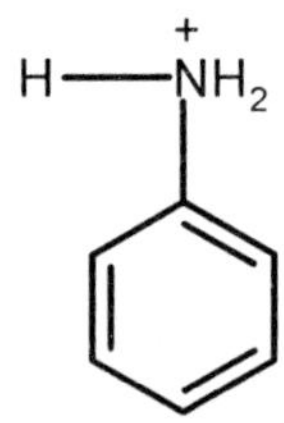

anilinium ion

The aniline molecule is thus more stabilized than its conjugate acid, anilinium ion, and it is therefore *'energetically unprofitable'* for aniline to take up a proton. Consequently, it behaves as a base with utmost reluctance (pKa 4.62) while cyclohexylamine behaves as a strong base (pKa 10.64).

(c) Urea [$(NH_2)_2CO$] is regarded as a resonance hybrid of the following canonical forms:

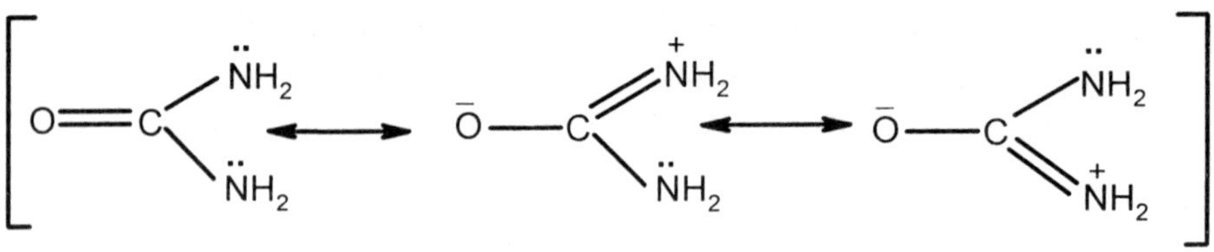

urea as a resonance hybrid

Thus protonation to urea prefers to occur through oxygen atom rather than nitrogen. The corresponding conjugate acid is stabilized to some extent as a result of delocalization of the positive charge over a wide area due to resonance.

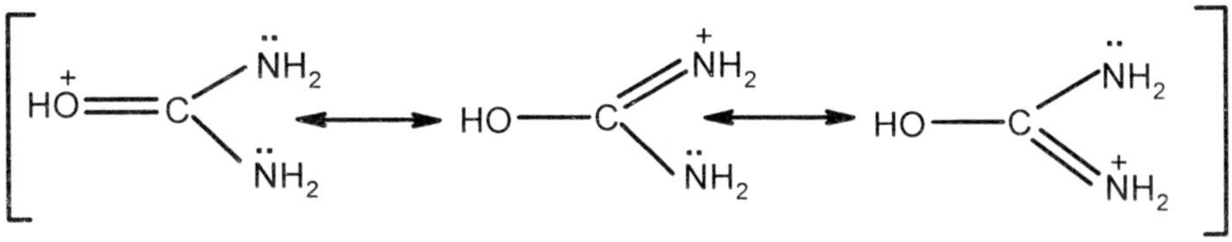

resonance-stabilized conjugate acid of urea

Hence urea shows basic nature (although it is a weak base; pKa 0.18), and obviously it should be 'monoacidic' base. NMR spectra of crystalline salts of urea (and also of thiourea) are consistent with the fact that protonation occurs at the oxygen atom (or sulphur atom in thiourea).

(d) Availability of unshared electron pair on 'N' atom in aromatic amines is reduced effectively due to its delocalization into the aromatic ring as observed in the case of aniline. This base-weakening effect will naturally be more pronounced if further phenyl groups are introduced on the nitrogen atom. Thus diphenylamine (Ph_2NH) is an extremely weak base (pKa 0.8), while triphenylamine (Ph_3N) is by ordinary standard not basic at all.

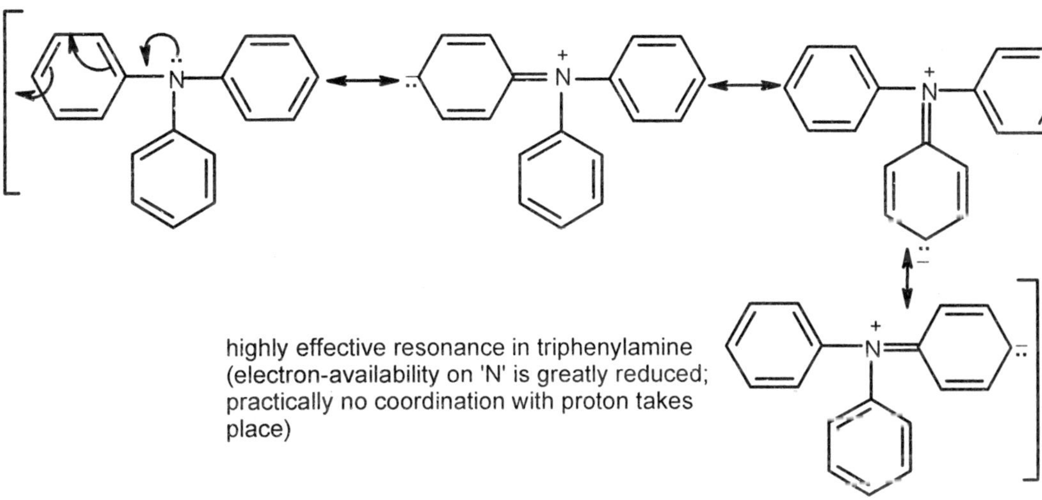

highly effective resonance in triphenylamine (electron-availability on 'N' is greatly reduced; practically no coordination with proton takes place)

(e) Guanidine [$HN{=}C(NH_2)_2$] is an example of the phenomenon where delocalization becomes responsible for increasing the basic strength of a nitrogenous compound (pKa 13.59). Both the neutral molecule and the cation (conjugate acid resulting from protonation) are stabilized by delocalization as shown below:

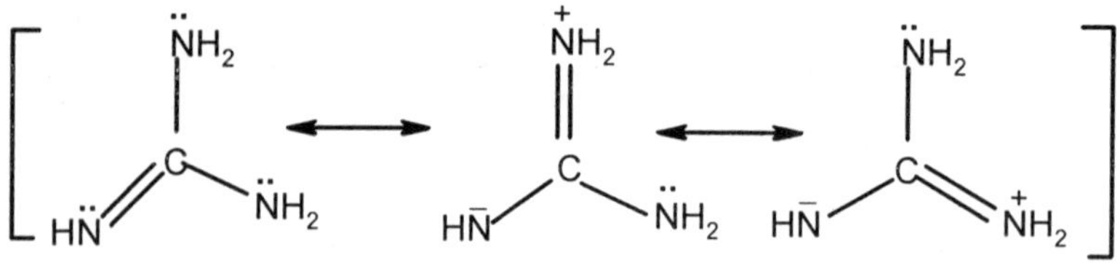

guanidine as a resonance hybrid (neutral molecule)

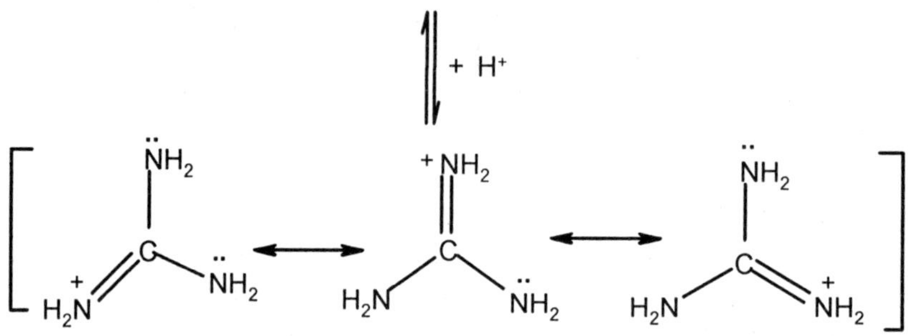

resonance-stabilized conjugate acid of guanidine

On inspection of the above resonating structures it is clear that two of the three canonical forms of the neutral molecule involve charge separation, and hence charge separated structures are of low contribution to the resonance hybrid of guanidine. Delocalization is thus less effective. On the contrary, all the three canonical forms of the conjugate acid of guanidine are equivalent and are of equal energy. So, delocalization of the positive charge becomes very much effective. As a result the cation is greatly stabilized with respect to the neutral molecule, thus making the protonation *'energetically profitable'* and guanidine, thus, behaves as an extremely strong base (pKa 13.59).

(f) In pyrrole all the carbon atoms are sp^2 hybridized and coplanar, while the nitrogen atom is sp^3 hybridized. So one of the sp^3 orbitals of 'N' that contains the lone pair electrons is not primarily parallel to the π-orbitals of carbon. But to attain molecular aromaticity as a whole the lone pair orbital is forcefully rotated through an angle of 90^0 so that effective delocalization may take place between the lone pair electrons and the ring π–electrons. Energy released in the process of delocalization is greater than the energy required to rotate the lone pair orbital of nitrogen through 90^0. Thus electron availability of nitrogen atom of pyrrole is reduced much for coordination to a proton, and hence pyrrole is a very weak base as revealed from its pKa value (pKa = –0.27).

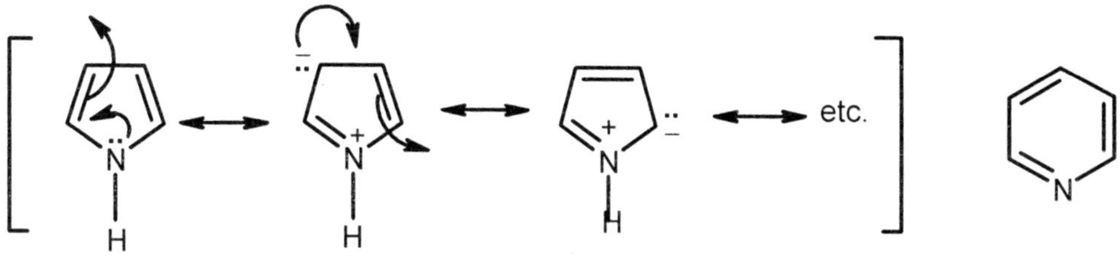

delocalization in pyrrole
(pKa –0.27, very weak base)

pyridine
(pKa 5.17)

Pyridine is aromatic; all the carbons and the nitrogen atom are sp^2 hybridized. They lie in the same plane. But one of the sp^2 orbitals of 'N' that contains the lone pair electrons remains perpendicular to the plane of the π-orbitals, and hence the lone-pair electrons are not delocalized with the ring π-electron system. Thus the electron pair is available on nitrogen atom, and pyridine is found to be more basic (pKa 5.17).

(g) It is the state of hybridization of nitrogen that determines the basicity of nitrogenous compounds. It is observed that basicity of such compounds decreases as the nitrogen atom is bonded in multiple fashion. This is attributed to the fact that as the nitrogen atom becomes progressively more multiply bonded, its lone pair of electrons is accommodated in an orbital that has progressively more s character. Electronegativity of the hybridized orbitals increases in the order of: sp^3 (s character 25%) < sp^2 (s character 33.3%) < sp (s character 75%). Hence nitrogen atom being multiply bonded attracts the lone pair electrons closer to it, and the lone-pair electrons are held progressively more tightly. Availability of these electrons for forming a bond with an incoming proton is, thus, becoming progressively less on going —

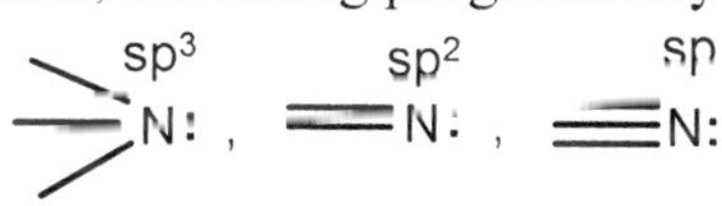

The nitrogen atoms in trimethylamine (Me_3N) or triethylamine (Et_3N), pyridine (C_5H_5N) and acetonitrile (Me–CN) are sp^3, sp^2 and sp hybridized respectively. Consequently, trimethylamine or triethylamine would be more basic than pyridine, which in turn would be more basic than acetonitrile.

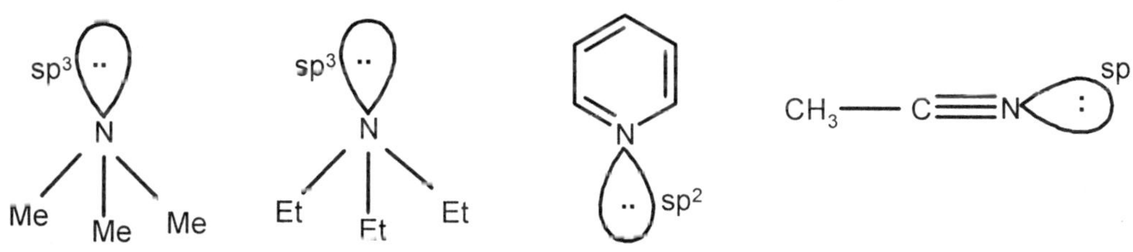

trimethylamine
(pKa 9.79)

triethylamine
(pKa 9.79)

pyridine
(pKa 5.17)

acetonitrile
(pKa –4.3)

Problem 1.42 The following compound is a strong base. Explain why?

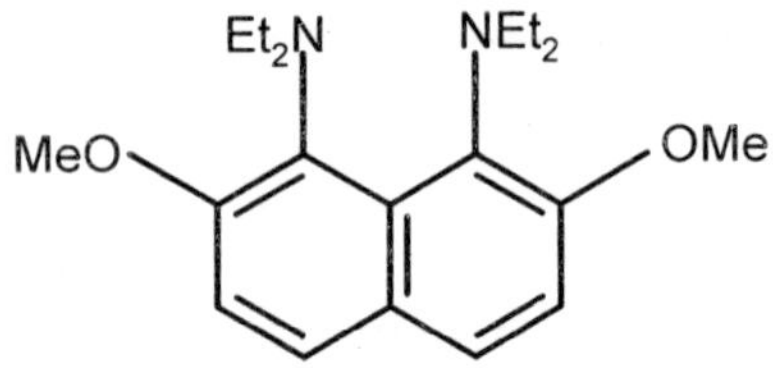

Solution.

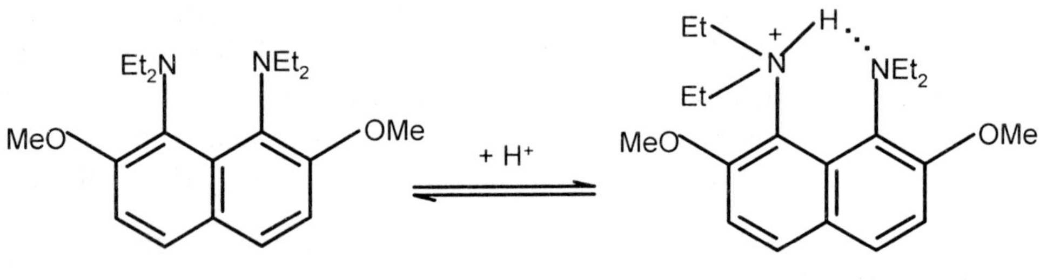

(I) (II)

1,8-Bis-(diethylamino)-2,7-dimethoxynaphthalene (I) is an extremely strong organic base (pKa 16.3). The neutral molecule suffers from severe strain because the two nitrogen lone pairs are forced to be near each other. On protonation the conjugate acid (II, monoprotonated ion) as formed attains stability through intramolecular hydrogen bonding — one lone pair is now connected to hydrogen. As a result the two large groups are now held tightly by H-bridge, and the strain is relieved. Therefore, protonation becomes *'energetically profitable'* making the parent compound strongly basic. This type of compounds is known as *'proton sponges'*.

▲Another examples of *'proton sponges'* are:

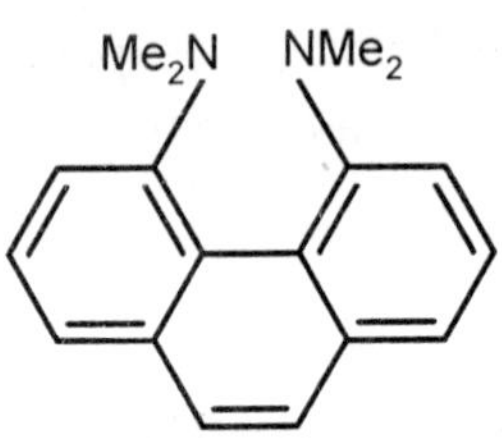

4,5-bis(dimethylamino)phenanthrene

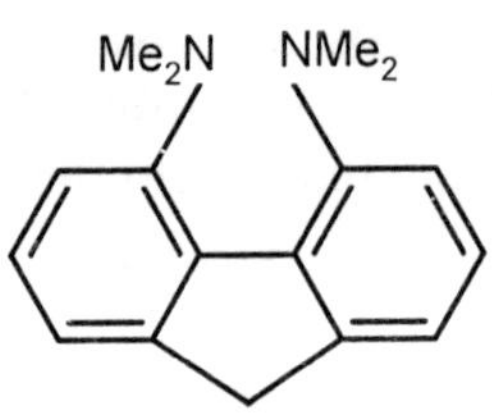

4,5-bis(dimethylamino)fluorene

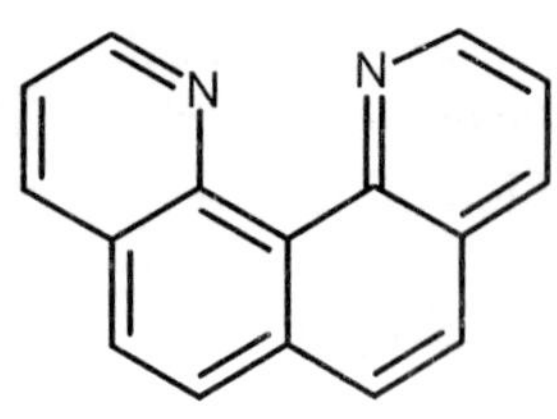

quino[7,8-h]quinoline

Problem 1.43 The observed order of basicity of the primary, secondary and tertiary amines in aqueous medium is $Me_2NH > MeNH_2 > Me_3NH$, whereas in gas phase the order is $Me_3NH > Me_2NH > MeNH$. Justify the observation.

Solution. Basicity of an amine in aqueous medium depends not only upon the electron-availability on amino-nitrogen, but also upon the extent to which the respective conjugate acid (formed on protonation) gets stability by solvation through hydrogen bonding. With increase in the number of electron-releasing groups attached to amino-nitrogen, availability of unshared electron pair increases progressively. At the same time as the number of hydrogen attached to the nitrogen of conjugate acid decreases, the extent of solvation by water molecules (through hydrogen bonding) decreases; thus affecting much to the basic property of the corresponding amines. Though trimethylamine bears three electron-releasing methyl groups attached to nitrogen, its conjugate acid has less capacity to enter into hydrogen bonding with water molecule(s) as it has only one hydrogen atom as compared to two and three hydrogens in the conjugate acids of dimethyl- and monomethylamines, respectively.

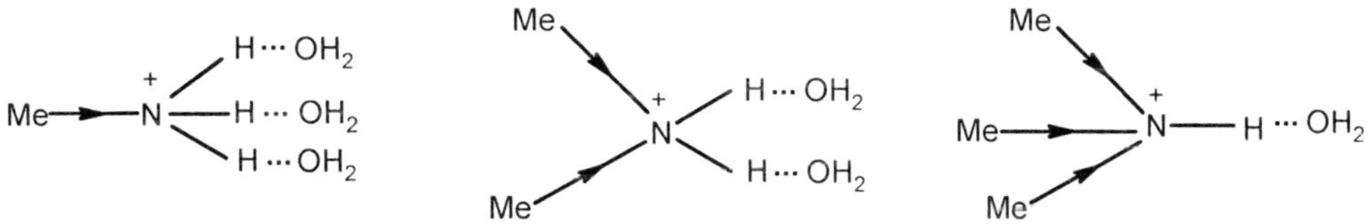

conjugate acid of 1°-amine solvated to a great extent. | conjugate acid of 2°-amine solvated moderately. | conjugate acid of 3°-amine; solvatation less pronounced.

Presence of the third alkyl group in the tertiary amine further reduces the extent of solvation of its conjugate acid due to its hydrophobic nature as well as steric factor. Hence tertiary amine is least basic. The secondary one becomes the stronger among them because of the combined effect (+I effect and moderate solvation), while the primary amine is of intermediate basic strength. Thus in aqueous medium —

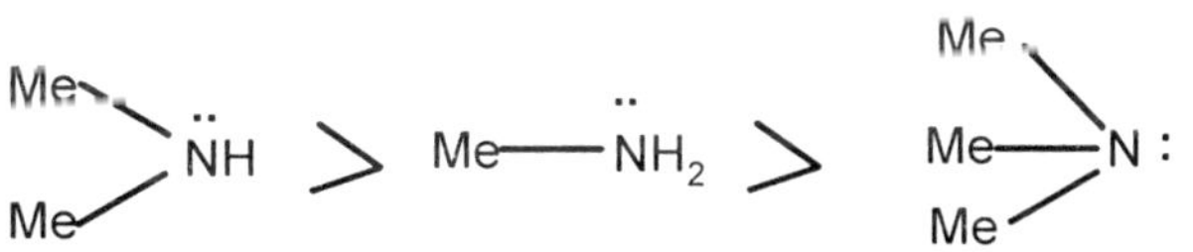

2°-amine pKa 10.87 | 1°-amine pKa 10.62 | 3°-amine pKa 9.79

Since solvation effect does not play in the gaseous state of the amines, the extent of positive inductive effect directs the basic strength. So, in the gaseous phase the basicity order would be —

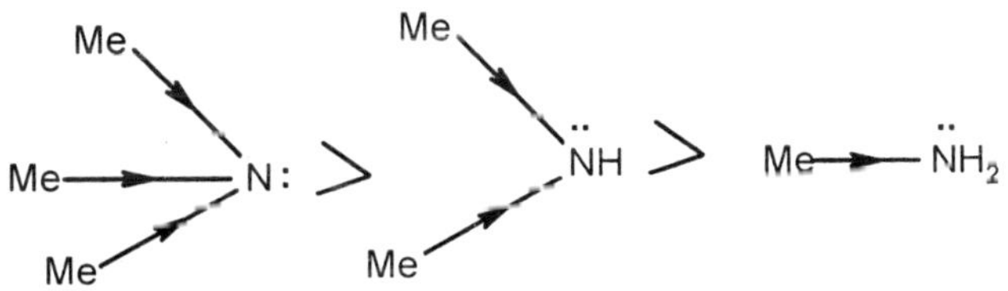

3°-amine | 2°-amine | 1°-amine

Problem 1.44 Which one will show greater basicity — 2,6-dimethyl-4-nitroaniline or 3,5-dimethyl-4-nitroaniline — and why?

Solution.

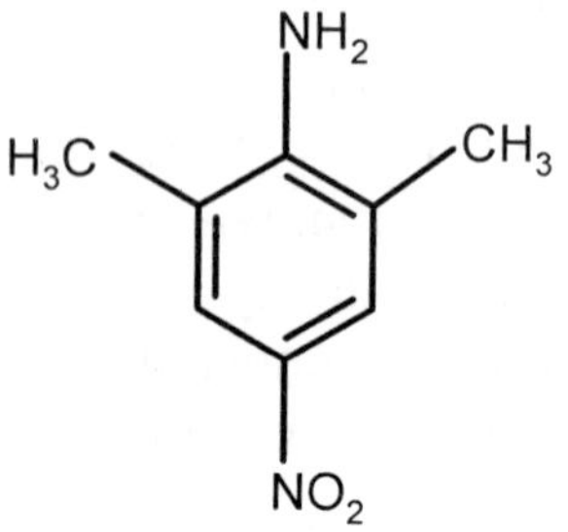

2,6-dimethyl-4-nitroaniline
(I)

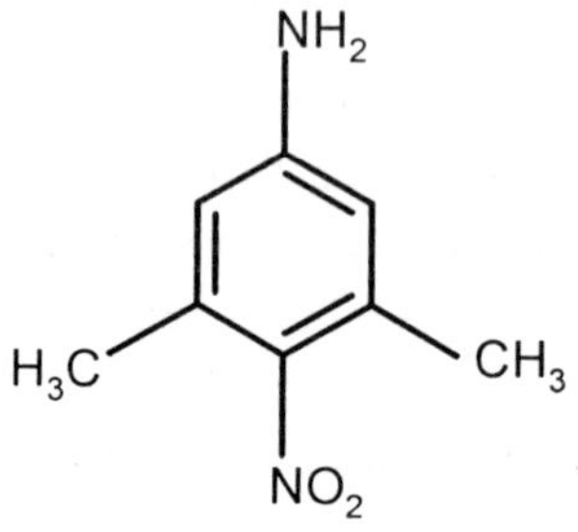

3,5-dimethyl-4-nitroaniline
(II)

Compound (II) will be more basic than (I). The fact may be explained on the basis of steric inhibition of resonance. The two bulky methyls present at the two *othro*-positions of the nitro group (another bulky moiety) in 3,5-dimethyl-4-nitroaniline generate a strong steric repulsion as a result of which the nitro group goes out of the plane of the ring so that it can't enter into resonance with the amino group. On the other hand, in 2,6-dimethyl-4-nitroaniline the nitro group enters into resonance with the amino group because the smaller H-atoms of $-NH_2$ do not interfere sterically with the *ortho*-methyls. Consequently, the unshared electron pair on nitrogen of amino group is more available in 3,5-dimethyl-4-nitroaniline (II) than that in the 2,6-isomer (I). Hence, compound (II) is more basic than (I).

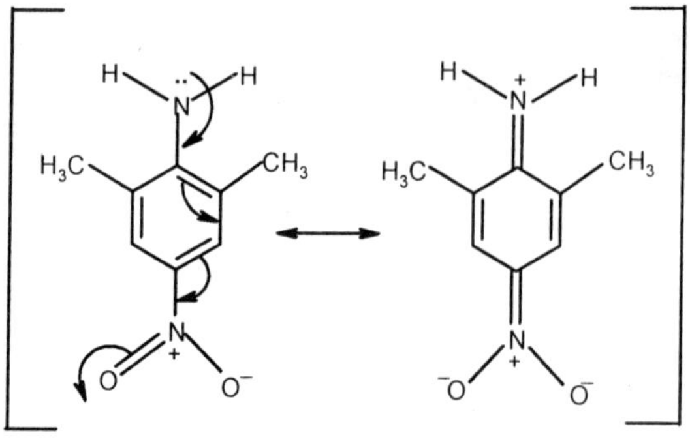

nitro group enters into resonance with the amino group in 2,6-dimethyl-4-nitroaniline (I)

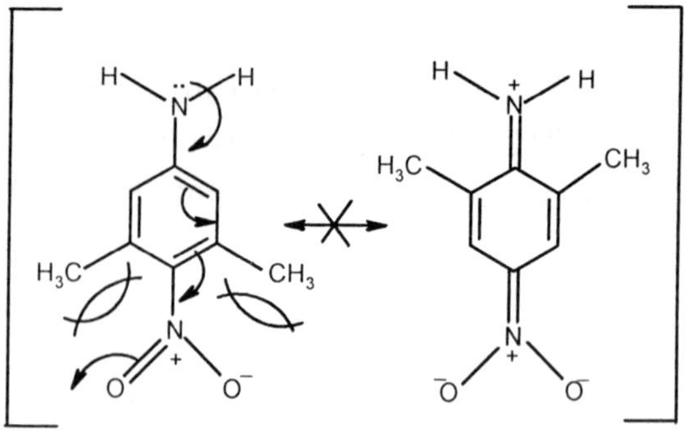

steric inhibition prevents nitro group from entering into resonance with the amino group in 3,5-dimethyl -4-nitroaniline (II)

Problem 1.45 Which one of the compound (I and II) shows greater basicity towards the Lewis acid, trimethylborane.

Et_3N:

(I)

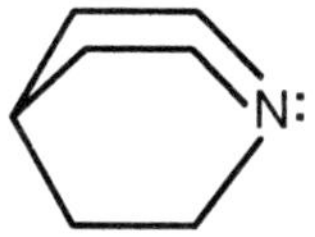

(II)

Solution. Both the compounds (I and II) are of comparable basicity under ordinary conditions, but towards Me3B (Lewis acid) the reactivity differs markedly. In triethylamine, at least two of the three free rotating ethyl groups remain folded towards, and partly covering the unshared pair of electrons so that reactivity of this amino compound towards large Lewis acid is diminished sharply. Adduct (I′) formation is unfavourable as also due to steric repulsion to be impcsed into the salt (of amine) — steric repulsions between the ethyl groups on the nitrogen atom and the methyl groups on the boron decrease the stability of the salt and thereby decrease the apparent basic strength of the amine. This type of steric repulsion at the interface is known as *F-strain* for face or *frontal strain*.

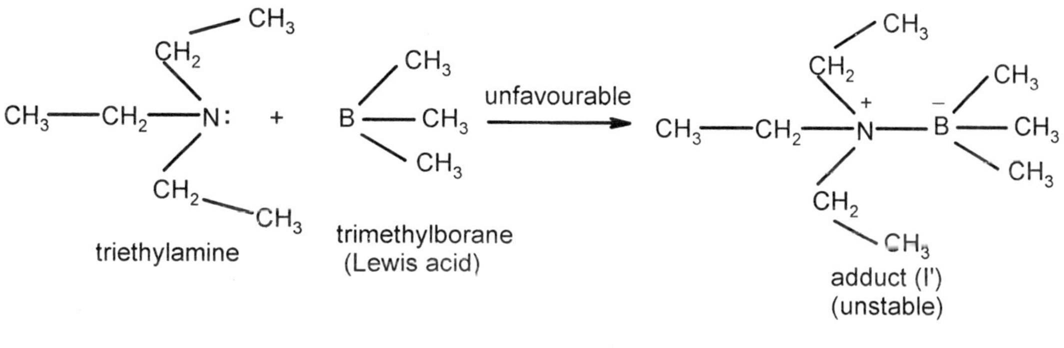

On the contrary, in quinuclidine (II) the substituents on nitrogen are held back by the ring system so that they can't interfere sterically with the methyl groups on the boron. Hence quinuclidine behaves as a strong base, and forms a stable adduct (II′) with the incoming Lewis acid.

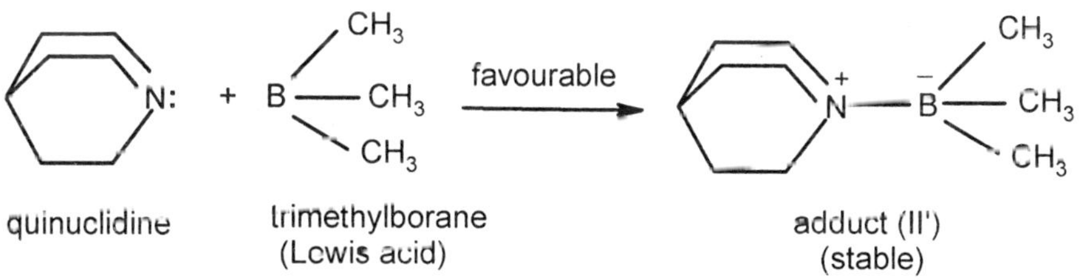

Problem 1.46 Compare the basicity between the compounds (I) and (II).

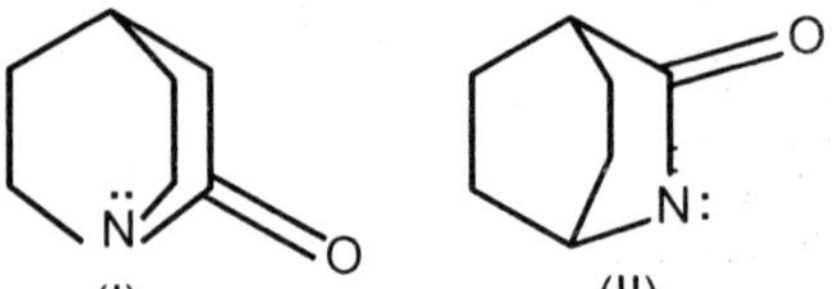

(I) (II)

Solution.

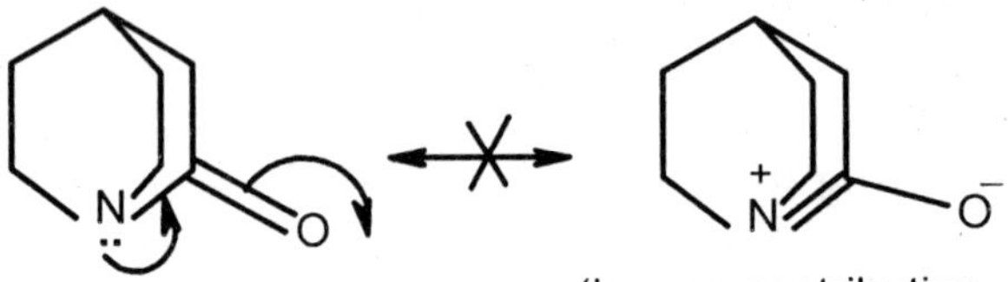

(I) (has no contribution to the resonance hybrid.)

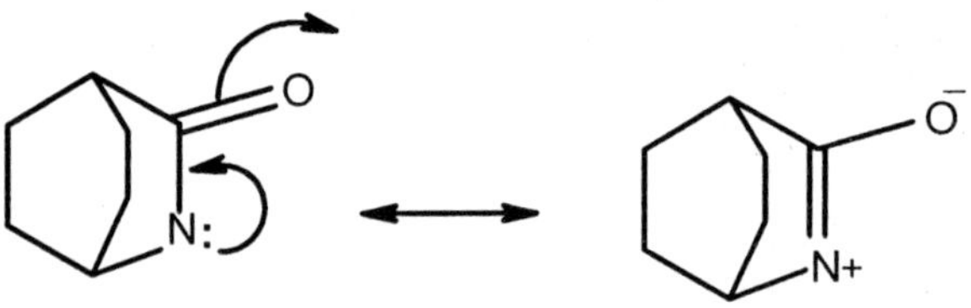

(II)

Compound (I) is more basic than compound (II). Availability of unshared electron pair on 'N' atom of the amide (I) is much more pronounced because it cannot delocalize with the π-electrons of the C=O unit since delocalization forms a double bond at the bridged head that would put much strain to the molecule (Bredt's rule) making it unstable. But this is not the situation for compound (II) — unshared pair of electrons on the 'N' is delocalized with the π-electrons of the C=O unit decreasing the electron-availability on nitrogen for coordinating with incoming proton. Hence, compound (I) is more basic than (II).

Problem 1.47 In picryl iodide the C–N bond distance for the *o*-NO_2 groups is close to the aliphatic C–N length, whereas that for the *p*-NO_2 is of considerable double-bond character. Explain the fact.

Solution. It is revealed from diffraction studies that the *p*-NO_2 group is coplanar with the ring whereas the *o*-NO_2 groups are not. This lack of coplanarity is a consequence of strong steric repulsion between the large iodine atom and the two bulky *o*-NO_2 groups. Hence, only the *p*-NO_2 group enters into the resonance with the lone-pair electrons on iodine atom. Thus C–N bond for *p*-NO_2 group attains considerable double bond character consistent with the reported bond distance of 1.35 ^{0}A (normal C–N is 1.47 ^{0}A; C=N is 1.28 ^{0}A). Since iodine atom prevents the *o*-NO_2 groups from being coplanar with the ring, they are not involved in the resonance process — as a result of which the C–N bond distance (1.45 ^{0}A) for them is close to the aliphatic C–N length.

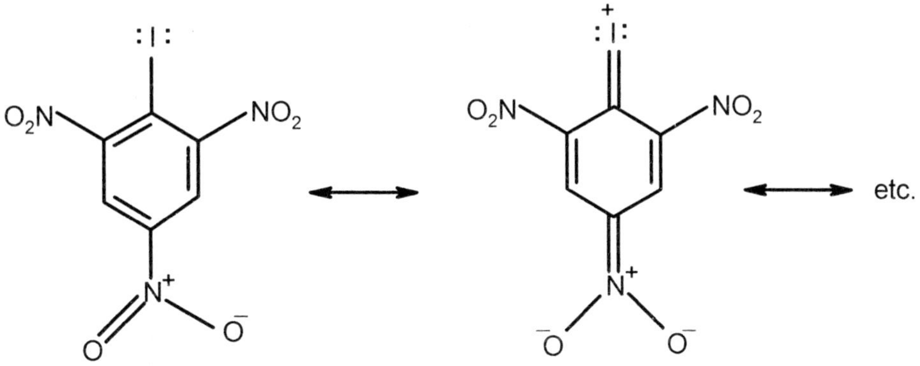

Problem 1.48 Presence of *p*-methoxy group in benzenediazonium ion retards the rate of reactivity of the cation. Why?

Solution.

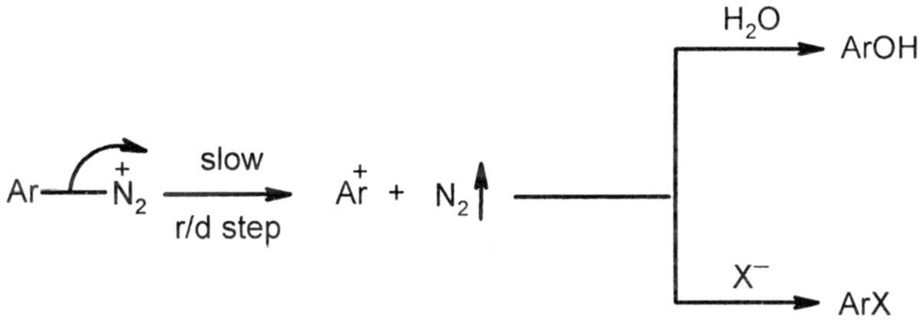

Due to +R effect of *p*-methoxy group, the Ar – N_2^+ bond attains some double bond character, with consequent increase in strength. Since breaking of this bond is involved in the rate-determining step, the activation energy of this step is raised up leading to retardation of the process. Such retardation is observed for groups having electron-releasing +R effect (or by the hyperconjugation effect, e.g. Me). On the other hand, electron-releasing groups in the *m*-position accelerate the reaction — here instead of +R effect only +I effect is operating.

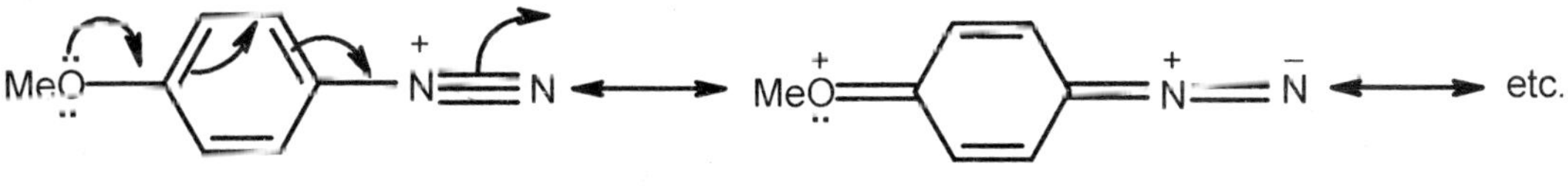

Problem 1.49 Explain the terms critically: aromaticity, antiaromaticity and homoaromaticity.

Solutions.

(a) ***Aromaticity***

Aromaticity i.e. *aromatic character* of a molecule is associated with its unusual stability. A molecule is aromatic if it is thermodynamically more stable than expected for its open-chain analogue. The best-known example of this class of compounds is benzene — in benzene each of the six carbons is sp^2-hybridized, and the remaining six p orbitals overlap to form π orbitals that contain six π electrons. These π electrons are effectively delocalized forming an electronic cloud on either side of the plane of the ring. Thus all occupied bonding molecular orbitals are doubly filled.

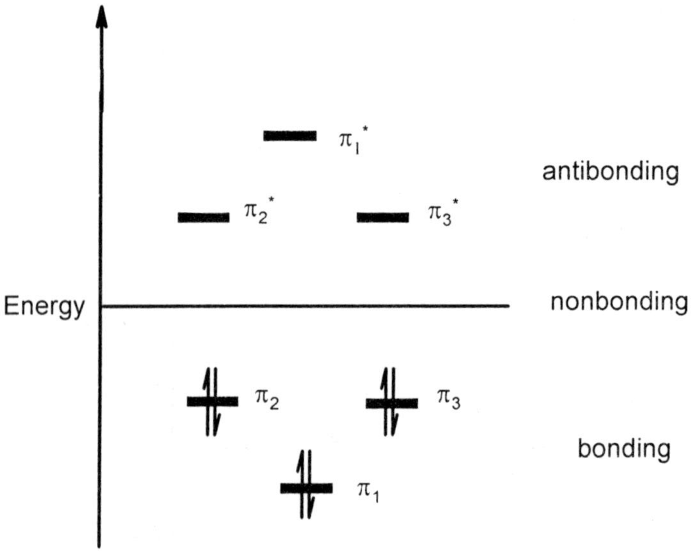

:Six π electrons in π MOs of benzene

This fruitful delocalization of π electrons with a 'closed loop' accounts for the great stability and large resonance energy. Benzene is thermally stable and tends to undergo substitution rather than addition reactions with electrophiles. Thus aromatic substances may be looked upon as planar and cyclic conjugated species that are more stable than related noncyclic analogues.

In 1925, Armit and Robinson first suggested the *"aromatic sextet theory"* that states that *aromatic properties* of compounds are attributed to the presence of a closed loop of six π electrons in a planar ring. This closed loop is not possible in aliphatic and alicyclic compounds, whereas is possible not only in benzene, but in heterocyclics also; in five-membered heterocycles the heteroatom contributes two electrons (nonbonding) to the aromatic sextet.

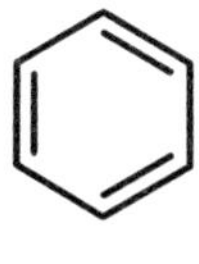

benzene

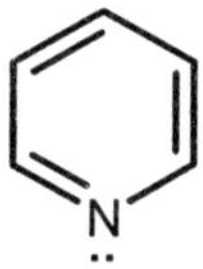

pyridine

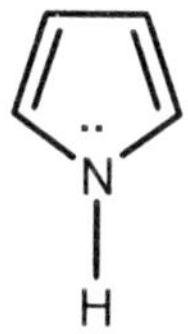

pyrrole
(4π and 2 nonbonding electrons in aromatic sextet)

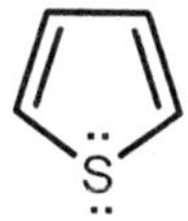

thiophene
(4π and 2 nonbonding electrons in aromatic sextet)

However, a more generalized definition of *aromaticity* was offered by Huckel (1931): any conjugated monocyclic polyene that is planar and has $(4n + 2)\pi$ electrons, where n equals to zero or a whole number, will be *aromatic in character* associated with an extra stability. Though the rule was originally applied to carbon-containing monocyclics in which each carbon is capable of being sp^2-hybridized to provide a p orbital for extended π bonding, latter it has been extended to unsaturated heterocyclics and also to fused-ring polycyclic compounds. Huckel's prediction has given great impetus to the search for new types of aromatic systems and to study of systems that are superficially aromatic, but do not obey the rule.

In modern definition, *aromaticity* is looked upon as the *ability to sustain induced ring current.* An applied magnetic field (external) perpendicular to a conjugated ring that has $(4n + 2)\pi$ electrons (and is planar or nearly planar), will cause the π electrons to behave as though they were circulating around the ring and thereby generating their own magnetic field[1]. This circulation is called *induced ring current.* An aromatic system (the ring current of which leads to a decrease in the magnetic susceptibility) is said to be *diatropic*. Although this definition also has its own flaws, it is widely accepted today[2]. There are several methods of determining whether a compound can sustain a ring current, but most significant one is based on NMR chemical shifts.

1. Pople, J. A. *et al.*, *J. Am. Chem. Soc.*, **88**, 4811(1966).
2. ***Note:*** For more discussion in view of unified theory of *aromaticity* one may consult the references — Haddon, R. C., *J. Am. Chem. Soc.*, **101**, 1722(1979); Haddon, R. C. and Fukunaga, T., *Tetrahedron Lett.*, **21**, 1191(1980); Karritzky, A. R., Karelson, M. and Wells, A. P., *J. Org. Chem.*, **61**, 1619(1996). It should also be noted in this regard that molecular environments as a whole can relatively alter the magnitude of aromaticity.

We can summarize the following features that are attributable to the aromatic character of a molecule:

(i) 'Unusual' thermal stability.

(ii) 'Unexpected' chemical properties, particularly substitution rather than addition reactions with electrophiles (i.e. aromatic unsaturated ring is preserved).

(iii) Resistance to oxidation under ordinary conditions.

(iv) Unique NMR spectral behaviour.

(b) ***Antiaromaticity***

If a compound exhibits neither stabilization nor destabilization with respect to its open-chain analogue, it is said to be *nonaromatic.* An *aromatic* compound shows greater stability in comparison to its open-chain analogue; but in certain cases it is observed that a planar cyclic conjugated species becomes thermodynamically less stable than corresponding acyclic unsaturated species. Such systems are termed as *antiaromatic*. They usually bear a closed loop of $4n\pi$ (n=1,2,3,...) electrons. Both theory and experiment show that they are much less stable. Thus an *antiaromatic* compound may be defined as a compound that is destabilized by a closed loop of electrons. For example, cyclobutadiene is an *antiaromatic* compound, and is found to be extremely reactive. It is able to exist only when immobilized by freezing in an inert matrix at very low temperature, even then is not static[1-4].

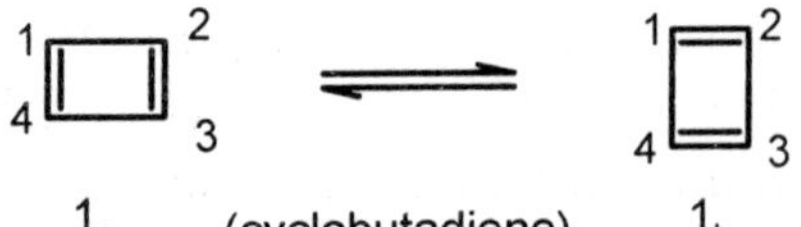

1_a (cyclobutadiene) 1_b

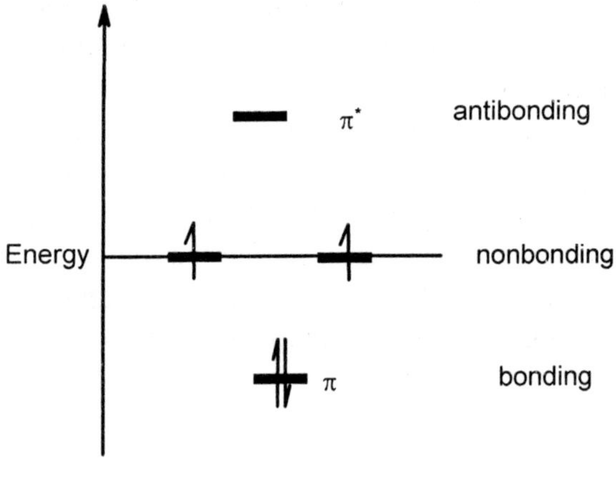

: Four π electrons in π MOs of cyclobutadiene

---

1. Watt, L. *et al., J. Am. Chem. Soc.*, **87**, 3253(1965).
2. Maier, G., *Angew. Chem. Int. Ed. Engl.*, **13**, 425(1974).
3. Malsch, K. D., *et al., J. Am. Chem. Soc.*, **110**, 2648(1988).
4. Arnold, B. R. *et al., J. Am. Chem. Soc.*, **113**, 692(1991).

However, it has been possible to synthesize some cyclobutadiene derivatives that are relatively stable at room temperature for varying period of time. Examples are:

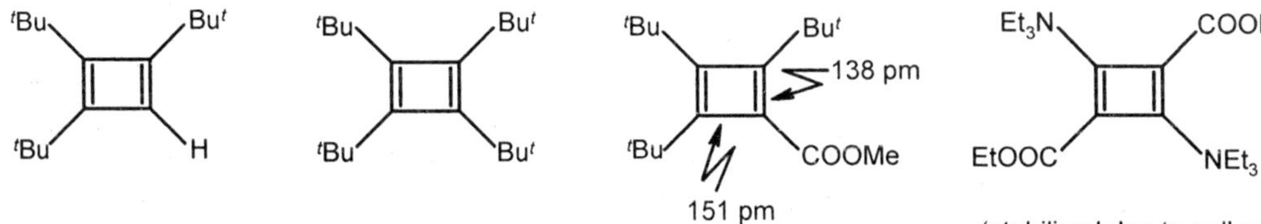

tri-*tert*-butylcyclobutadiene (dimerization is sterically hindered)

tetra-*tert*-butylcyclobutadiene (dimerization is stereically hindered)

methyl 2,3,4-tri-*tert*-butylcyclobutadiene 1-carboxylate

(stabilized due to pull-push or captodative effect between electron-donating and electron-withdrawing groups)

Conjugated rings of 4nπ electrons also could support ring current theoretically, but in the direction opposite to those in (4n + 2)π rings[1]. The ring current in an aromatic system, while leads to a decrease in the magnetic susceptibility (known as diamagnetic shift), the change in an antiaromatic molecule is in the opposite direction and hence termed as *paratropic*. Thus the ring proton in tri-*tert*-butylcyclobutadiene (appearing at δ5.38) is shifted upfield, compared with the signal as expected for a nonaromatic proton (e.g. in cyclopentadiene).

(c) ***Homoaromaticity***

Any system that gains stability by virtue of being cyclic may be classified as aromatic as per theory of *aromaticity*. The *homoaromatics* belong to this category. In a *homoaromatic* compound the π system is interrupted at one or more points by a saturated centre but in which geometry still permits significant overlap of the π orbitals across the insulating gap.

Homotropylium ion (formed on dissolution of cyclooctatetraene in concentrated H_2SO_4) is the earliest and most investigated species under this category[2-4]. For effective

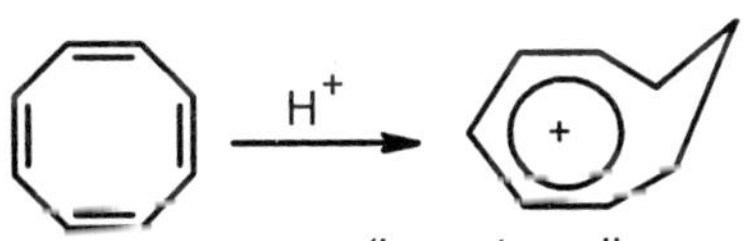

(homotropylium cation)

A proton adds to one of the double bonds of the parent molecule to form the homotropylium ion; an aromatic sextet is spread over seven carbons-- the eighth carbon (sp^3) hybridized can't take part in the aromaticity.

1. Pople, J. A., *et al., J. Am. Chem. Soc.*, **88**, 4811(1966).
2. Rosenberg, J. L. *et al., J. Am. Chem. Soc.*, **84**, 2842(1962).
3. Childs, R. F. *et al., Acc. Chem. Res.*, **17**, 347(1984).
4. Haddon, R, C., *J. Am. Chem. Soc.*, **110**, 1108(1988).

overlapping of the orbitals so as to close a loop, the sp^3 atoms are forced to lie almost vertically above the plane of the aromatic atoms (as evidenced also from NMR studies)[1,2].

Another example may be cited:

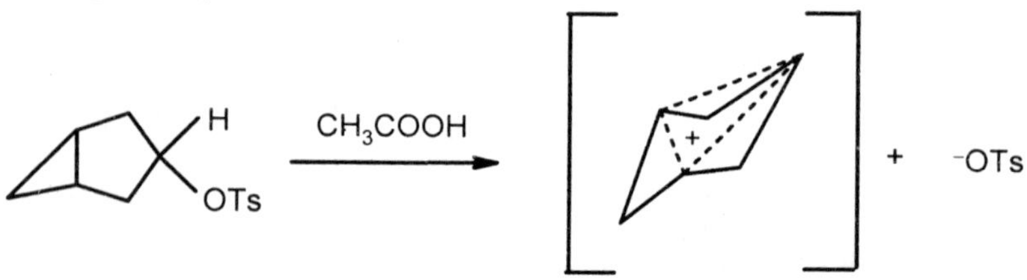

(trishomocyclopropenium ion;
very short-lived reaction intermediate)

Problem 1.50 Designate each of the following species as aromatic, nonaromatic or antiaromatic:

(i) 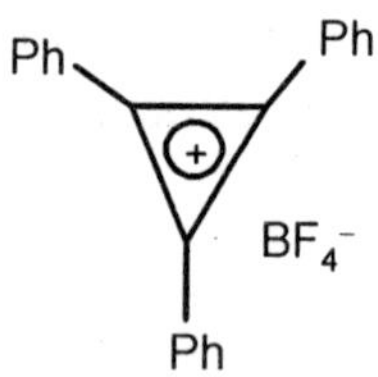

(ii) 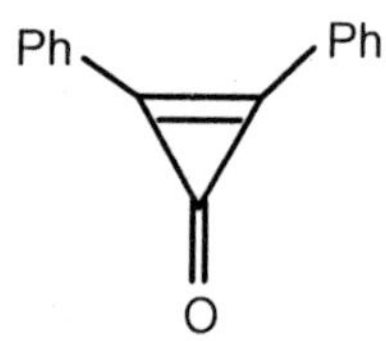

(iii) 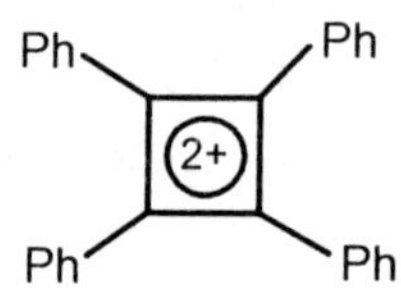

(iv) 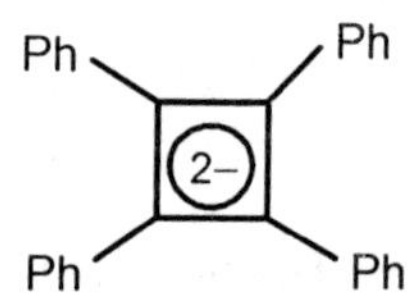

(v) 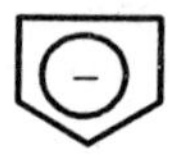

(vi) 

(vii) 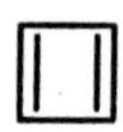

(viii) 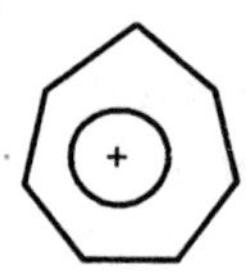

(ix) 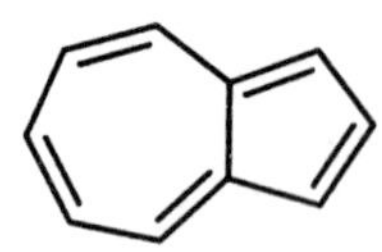

(x) 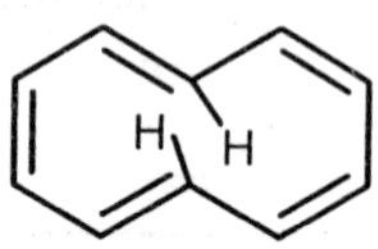

(xi) 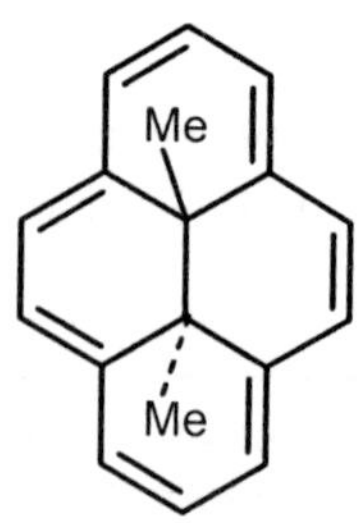

(xii) 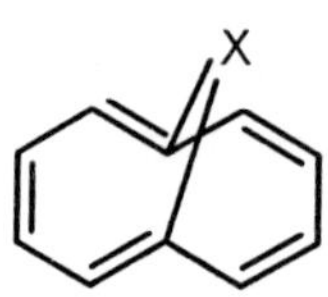

X = CH_2, O, NH

(xiii) 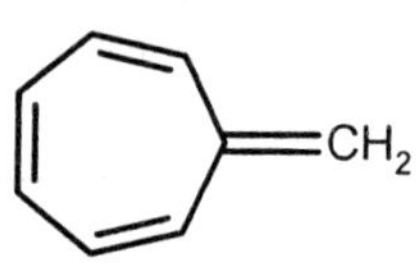

(xiv) Styrene

---

1. Childs, R. F. *et al.*, *J. Am. Chem. Soc.*, **106**, 5974(1984).
2. Childs, R. F. *et al.*, *Acc. Chem. Res.*, **17**, 347(1984).

Solutions. (Hints only)

(i) 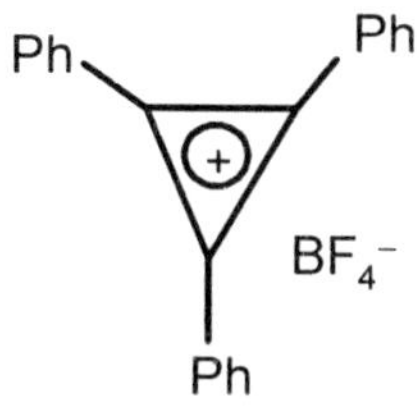

Planar; 2π electrons [$(4n + 2)\pi$ system, n=0]; aromatic

(ii) 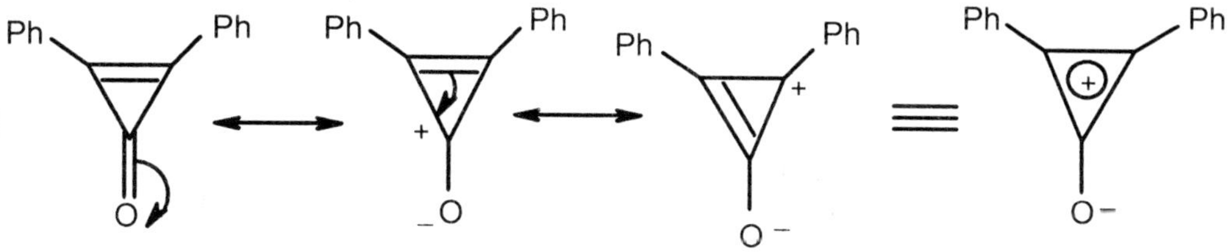

Planar; 2π electrons [$(4n + 2)\pi$ system, n=0]; aromatic

(iii) 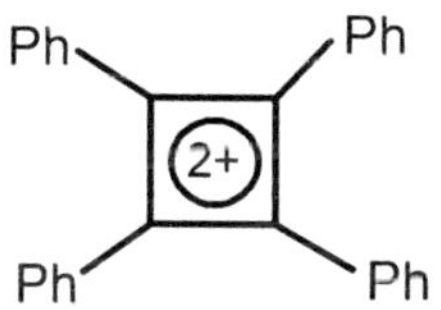

cyclobutadienyl cation

Planar; 2π electrons [$(4n + 2)\pi$ system, n=0]; aromatic

(iv) 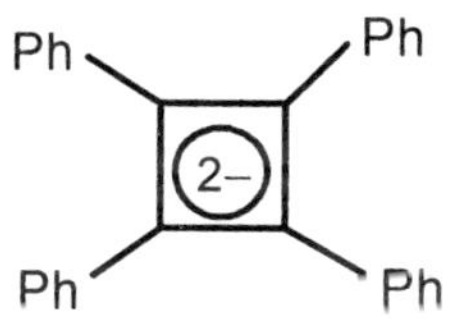

Planar; 6π electrons [$(4n + 2)\pi$ system, n=1]; aromatic

(v) 

Planar; 6π electrons [$(4n + 2)\pi$ system, n=1]; aromatic

(vi) 

Planar; 4π electrons [$4n\pi$ system, n=1]; antiaromatic

(vii) 

cyclobutadiene

4π electrons [$4n\pi$ system, n=1]; antiaromatic

(viii) 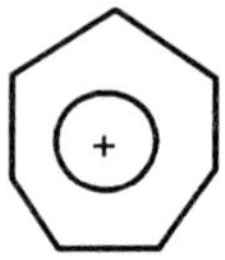

Planar; 6π electrons [(4n + 2)π system, n=1]; aromatic

cycloheptratrienyl cation

(ix) 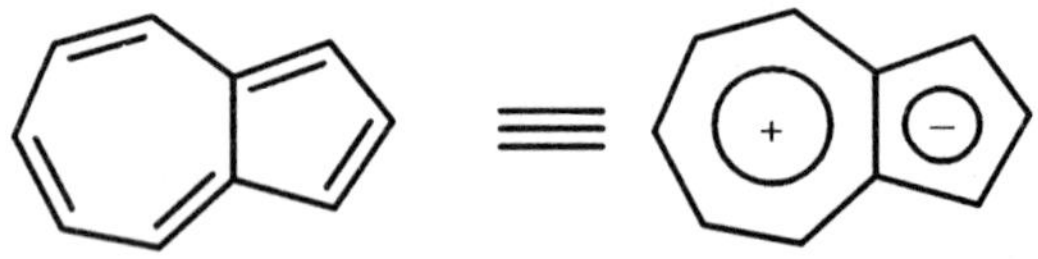

It is aromatic.

azulene

(x) 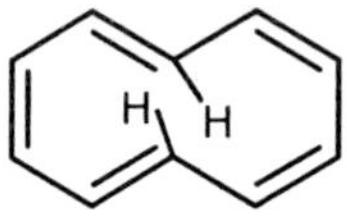

10-annulene

It is a (4n + 2)π system (where n=2), still it is nonaromatic. Because of the non-bonded interaction between the two 'inner' hydrogens within the ring, the molecule buckles in order to supply space to the hydrogen atoms and for doing so the coplanarity of the molecule is lost. Hence, conjugation is disturbed, and it becomes nonaromatic.

(xi) 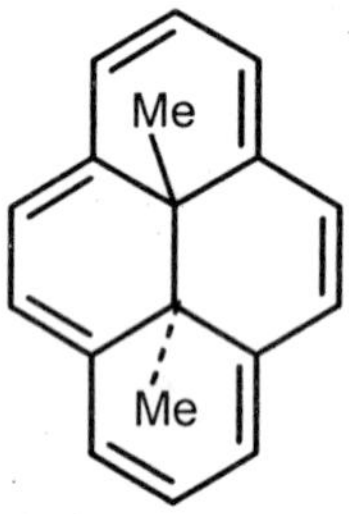

trans-dimethyldihydropyrene

Trans-dimethyldihydropyrene shows aromaticity and is more stable than14-annulene; the non-bonded interactions of the hydrogens inside the planar ring is absent.

(xii) 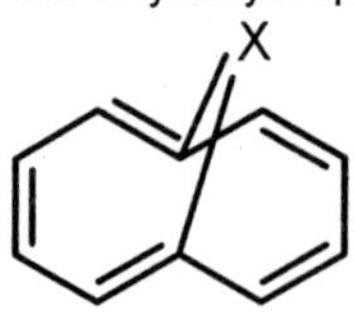

10-annulene

X = CH_2, O, NH Should be aromatric.

(xiii) 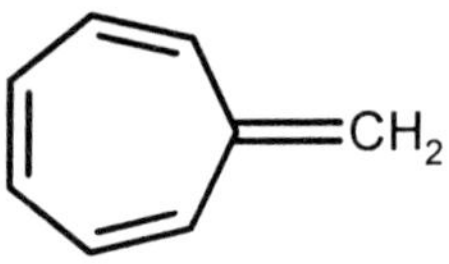

heptafulvene

Though it has a resonance energy of 36 kcal/mole, equal to that of benzene, it is typically olefinic – nonaromatic.

(xiv) Styrene

Styrene (Ph-CH=CH_2) is aromatic. It has much greater delocalization (resonance) energy than that of benzene; this is attributed to the fact that the benzene ring which is conjugated with the C=C double bond, thereby, extending the delocalization of the electron cloud.

Problem 1.51 Offer explanation for each of the following observations:

(i) Cyclooctatetraene possesses 4nπ electrons, but it is not antiaromatic.

(ii) The following cyclic diol (A) is regarded as a strong acid.

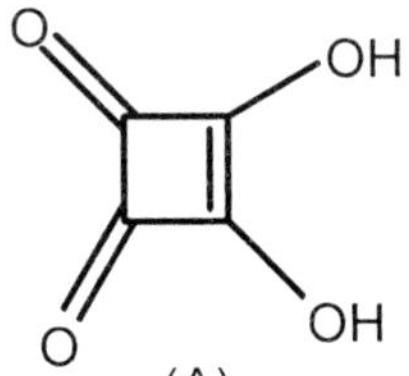

(A)

(iii) The compound (I) undergoes solvolysis when treated with silver perchlorate in propionic acid, but compound (II) does not.

(I) (II)

(iv) Compare the acidity of the following compounds:

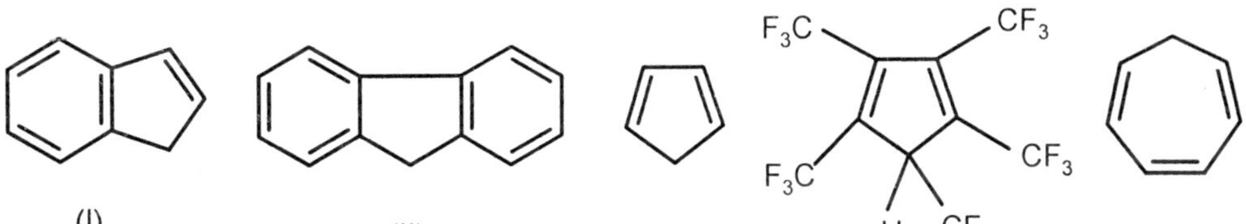

(I) (II) (III) (IV) (V)

(v) The resulting species from the reaction of 2 mole of K with 1 mole of 1,3,5,7-cyclooctatetraene shows a single peak in Proton Magnetic Resonance (^{1}H-NMR) spectrum.

(vi) Tropone possesses high dipole moment.

Solutions.

(i)

(I) (II)
(planar conformation of cyclooctatetraene)

Cyclooctatetraene is a stable compound, although it is a 4nπ electrons (n=2) system. If it attains planar conformation, it should be *antiaromatic* because it would then exhibit cyclic conjugation between the 4nπ electrons (n=2). Besides, this planar form suffers from angle strain also (a regular octagon has angles of 135^0, while sp^2 angles are most stable at 120^0). Thus to avoid the unfavourable delocalization and angle strain, cyclooctatetraene attains a stable nonplanar 'tub-shaped' conformation (III).

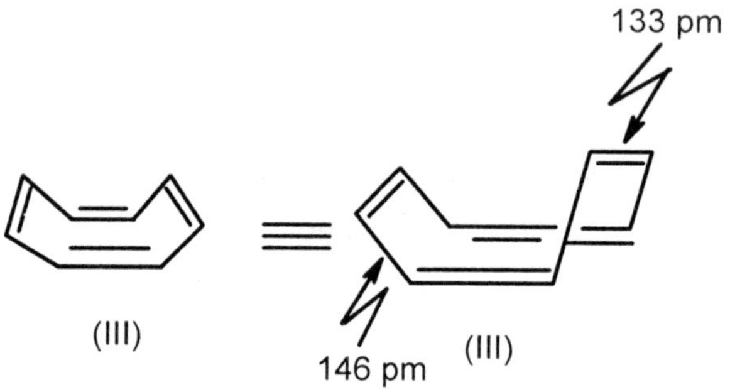

'tub-shaped' conformation

In regards to its planar conformation whatever stabilization we might have expected to gain from the delocalization I → II is evidently not sufficient to inspire the molecule to abandon its 'tub-shaped' nonplanar conformation. Hence, not being antiaromatic, cyclooctatetraene becomes nonaromatic.

(ii) The cyclic diol, squaric acid (A) behaves as a strong acid.

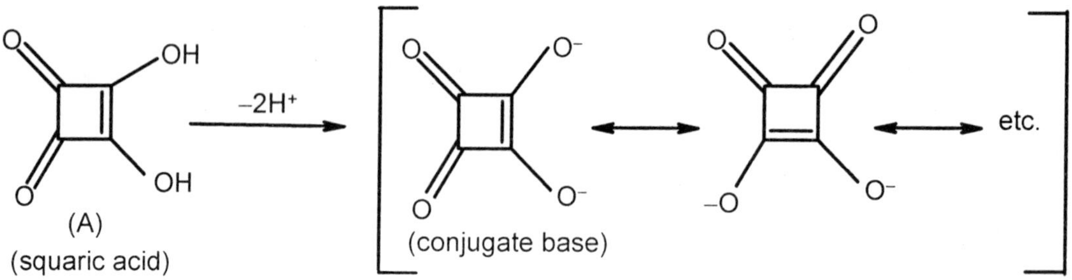

The conjugate base of squaric acid (i.e. the dianion) attains aromaticity [(4n + 2)π system, n=0], and that's why it becomes unusually stable making the parent compound highly acidic. The values for pK_1 and pK_2 of the compound are ~1.5 and ~3.5, respectively; thus even the second proton is given up much more readily than the proton of acetic acid.

(iii) This solvolysis reaction will proceed through S_N1 pathway; thus a carbocation intermediate will be formed.

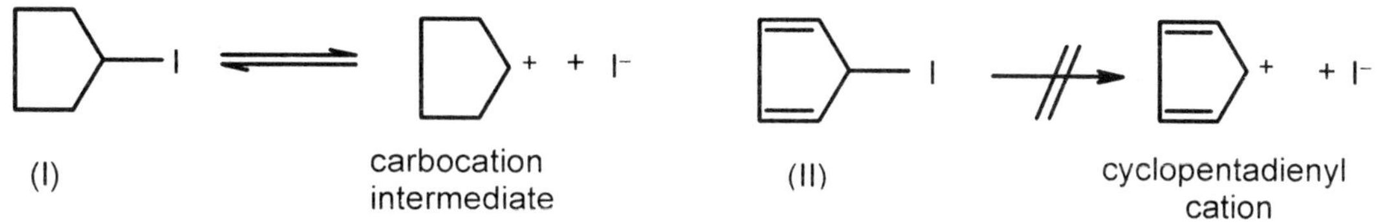

Cyclopentadieneyl cation is very much unstable, because it is an antiaromatic species. Thus ionization of compound (II), leading to the formation of this antiaromatic intermediate, is thermodynamically unfavourable and hence, the parent compound does not undergo solvolysis under the reaction conditions. On the other hand, the cyclopentyl cation is a nonaromatic species, and relatively more stable. That's why compound (I) undergoes solvolysis reaction under the same reaction conditions.

(iv)

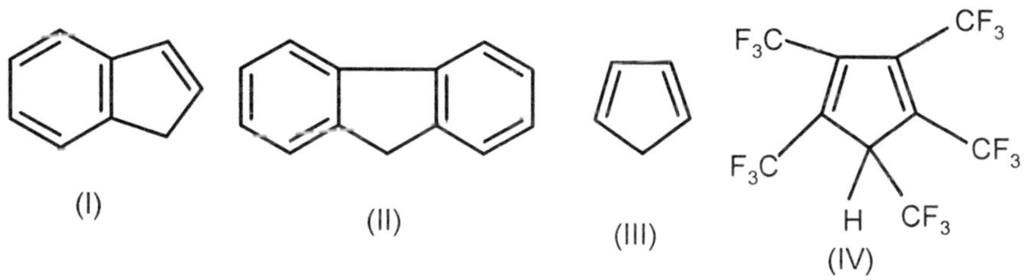

Between cyclopentadiene and cycloheptatriene, the former would show greater acidity than the latter. Because cycloheptatrienyl anion, the conjugate base of cycloheptatriene, is an antiaromatic species and hence becomes less stable while cyclopentadienyl anion (formed on proton removal from cyclopentadiene) is much more stable by virtue of aromaticity.

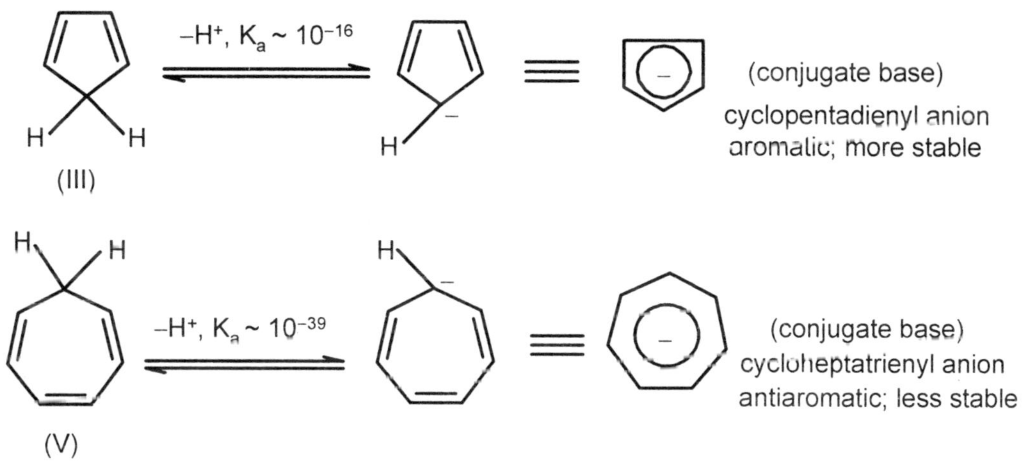

The compounds I and II are acidic ($pK_a \approx 20$ and $pK_a \approx 23$, respectively) but are less than cyclopentadiene (III), since annellation causes the electrons to be less available to the five-membered ring. 1,2,3,4,5-Pentakis (trifluoromethyl) cyclopentadiene (IV) in the above series would show the highest acidity; its acidity order is even greater than that of nitric acid, because of the electron-withdrawing effects of trifluoromethyl (CF_3) groups.

Thus, the acidity order for the above series would be given as: V($pK_a \approx 39$) < II ($pK_a \approx 23$) < I ($pK_a \approx 20$) < III ($pK_a \approx 16$) << IV.

(v) 

cyclooctatetraene (nonaromatic) — salt of cyclooctatetraenyl dianion (stable; aromatic)

In the above reaction no hydrogen gas is evolved out. Potassium being a strong reductant supplies one electron each (i.e. two electrons from two K atoms) to form cyclooctatetraenyl dianion. The anion is a planar regular octagon (as confirmed by the X-ray crystallographic studies) with equal bond lengths. This planar conjugated unsaturated molecule has 10 π-electrons [(4n + 2)π electrons, n=2]. Thus it becomes aromatic. Establishment of the conjugation in this aromatic species is sufficient to overcome the angle strain within the species, which tends to oppose a planar structure. Hence the ring-protons of the aromatic system as developed are in an identical environment, and exhibit a single signal in Proton Magnetic Resonance (^{1}H-NMR) spectrum of the species.

(vi) Tropone would exhibit an aromatic sextet if the two C=O electrons are stayed away from the ring and reside near the electronegative oxygen atom.

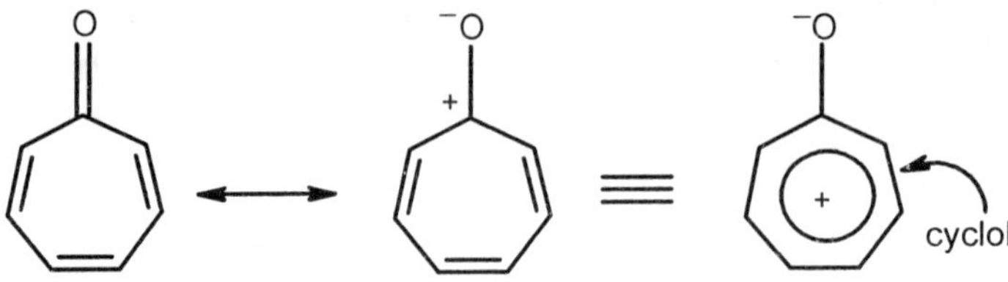

tropone (I) — (I') — cycloheptatrienyl cation moiety

Consequently, the polar structure (I') that bears an aromatic moiety of cycloheptatrienyl cation has significant contribution to the tropone-resonance hybrid. That's why the compound is found to possess unusually high dipole moment ($\mu = 4.17D$), and at the same time the molecule attains much stability.

Supplementary problems

SP-1.1 Draw a picture of the orbitals of the allene, $CH_2=C=CH_2$.

SP-1.2 Would you predict the C–Cl bond in CF_3Cl to be longer or shorter than that in CCl_4? Give reason in support of your prediction.

SP-1.3 The C–N bond length in NH_2CONH_2 is 0.137 nm instead of 0.147 nm. How can you justify the shortening in the bond length?

SP-1.4 Explain why the C–D and N–D bonds are to be cleaved in reactions more slowly than C–H and N–H bonds, respectively?

SP-1.5 Unlike most hydrocarbons, azulene is highly coloured (deep blue) and dissolves in aqueous acid. How would you explain these properties of azulene from its structural feature?

SP-1.6 Chlorine in chlorobenzene cannot be readily replaced but chlorine in *ortho*-nitrochlorobenzene can be. Explain.

SP-1.7 The dipole moment of chlorobenzene is larger than that of fluorobenzene, even though fluorine is more electronegative than chlorine. Explain.

SP-1.8 Predict whether pentalene is stable as a planar hydrocarbon. Offer explanation in support of your prediction.

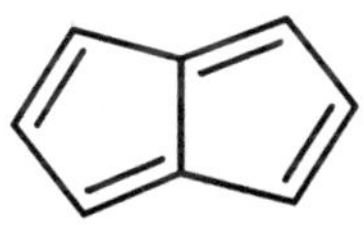

(pentalene)

SP-1.9 From Huckel's rule, would you expect 10-annulene to be aromatic?

SP-1.10 Explain why acetone, although an organic ketone, is completely soluble in water?

SP-1.11 Fatty peracids, RCOOOH, exist as monomeric state in benzene probably due to the formation of intramolecular hydrogen bonding. Has this H-bonding any role on strengthening or weakening the acidity of the compounds? Clarify your answer.

SP-1.12 Predict which compound in each of the following pairs has the larger dipole moment?

(i) 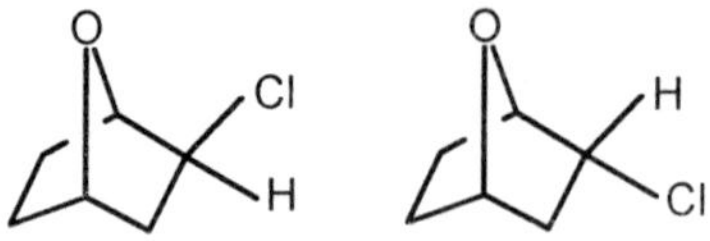 (ii) 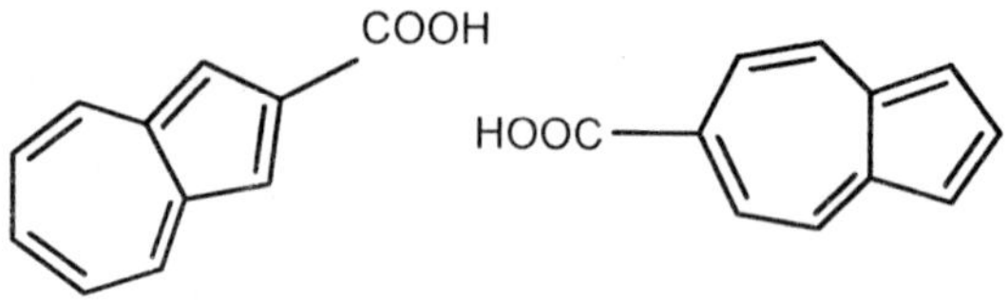

SP-1.13 In the following molecules, which hydrogen would most easily be removed by treatment with one equivalent of base?

(i) 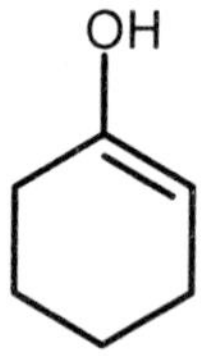 (ii) 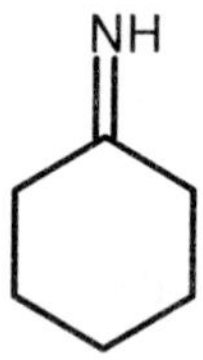 (iii) 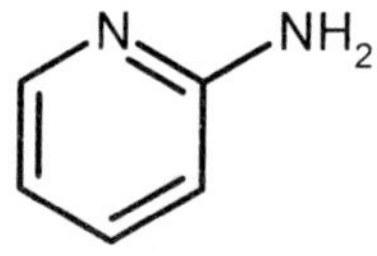 (iv) 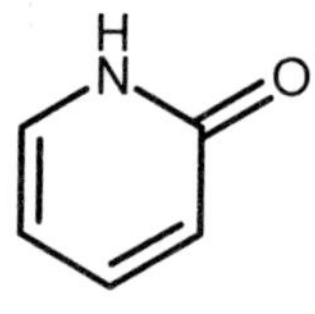

SP-1.14 Arrange the following compounds in order of their increasing acidity with proper justifications:

(i) MeCOOH, Me_3CCOOH, $MeSO_2CH_2COOH$
(ii) C_6H_5OH, *p*-MeC_6H_4OH, *p*-$NO_2C_6H_4OH$, *p*-NCC_6H_4OH
(iii) Conjugate acids of H_2NNH_2, $C_6H_5NH_2$, Me_2NH, *p*-$NO_2C_6H_4NH_2$
(iv) $CH_2{=}CHCH_2COOH$, $CH_2{=}CHCH_2CH_2COOH$, $CH_3CH_2CH_2COOH$
(v) Carboxylic acids of cyclopropane, cyclobutane, and cyclohexane.
(vi) C_6H_5OH, C_6H_5OD
(vii) (*p*-$NO_2C_6H_5)_3CH$, (3,5-dimethyl-4-nitrophenyl)$_3$CH
(viii) $CH{\equiv}CCOOH$, $CH_2{=}CHCOOH$, CH_3CH_2COOH
(ix) 2,6-Dihydroxybenzoic and 2-hydroxybenzoic acid
(x) Benzoic acid, 2,6-dimethylbenzoic acid

SP-1.15 Predict the acidic/basic nature of the following compound:

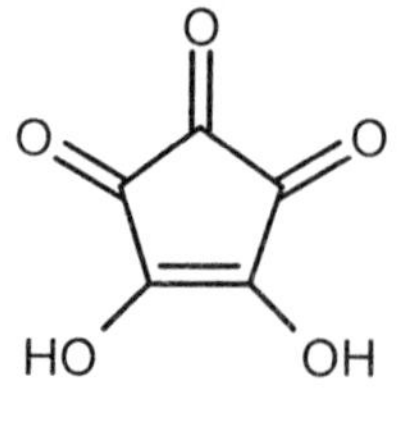

SP-1.16 How can you account for the observation that the ratio K_1/K_2 is much larger for tetramethylsuccinic acid (6135) than for succinic acid (19.2)?

SP-1.17 Tricyanomethane is as strong as mineral acids. Account for the fact.

SP-1.18 Account for the greater acidity of the *aci*-form of nitromethane as compared to nitromethane itself.

SP-1.19 Arrange the following compounds in order of their increasing basicity:

(i) CH_3O^-, NH_2^-, HO^-, HSO_4^-
(ii) NF_3, NH_3, NH_2OH
(iii) NH_3, CH_3CONH_2, $(CH_3CO)_2NH$
(iv) Me_3CO^-, CH_3O^-, PhO^-, $CH_3CO_2^-$
(v) CH_2=CH-NH_2, CH_3NH_2, pyrrole, aniline, pyridine

(vi)

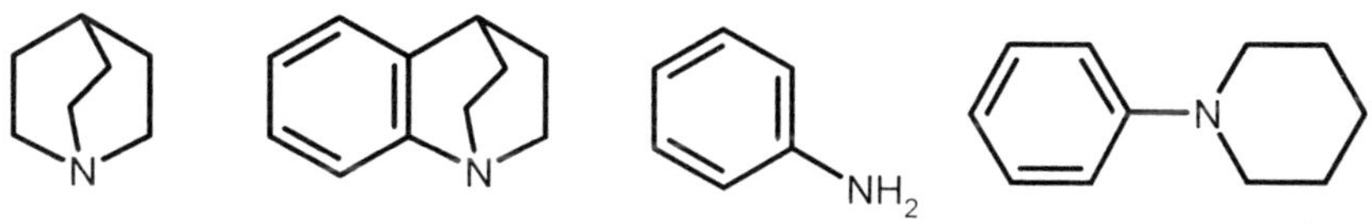

(vii) 1-aminopropane, ethanamide, guanidine, aniline
(viii) aniline, 4-nitroaniline, 4-methoxyaniline, 4-methylaniline
(ix) 2,6-Dimethylnitroaniline, 3,5-dimethylnitroaniline

(x)

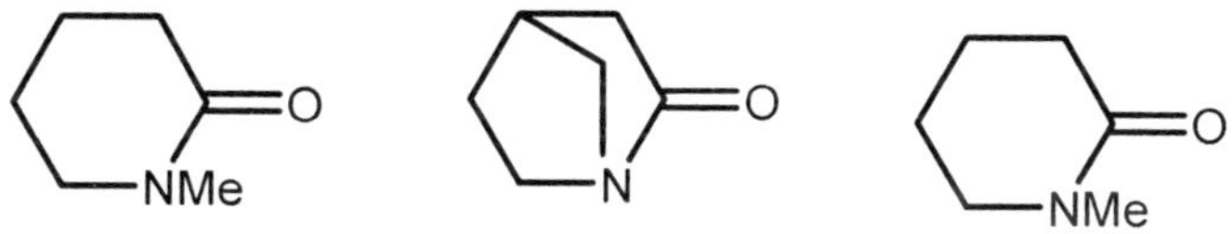

SP-1.20 What is the conjugate acid of CH_3Li? Would you expect it to be a strong base or a weak base?

SP-1.21 Cyclopropanone is less stable than cyclopropenone. Why?

SP-1.22 An aqueous solution of tropylium bromide on treatment with silver nitrate gives precipitate of silver bromide. Account for the observation.

SP-1.23 Compare the stability between I and II:

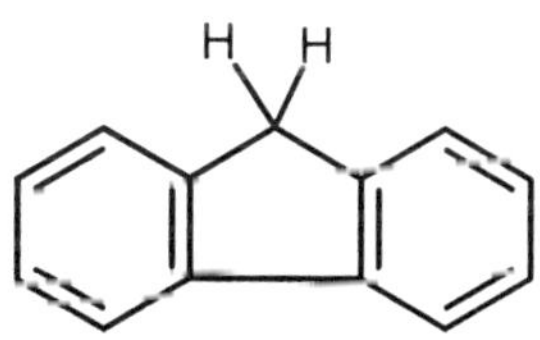

(I)

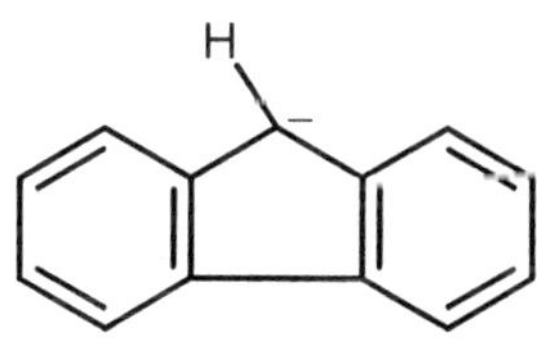

(II)

SP-1.24 The following hydrocarbon has an unusually high dipole moment. Explain how a large dipole moment might arise?

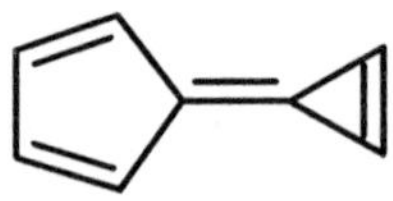

SP-1.25 Designate the following molecule as aromatic, antiaromatic, or nonaromatic. What is the role of *t*-butyl substituents in regards to its molecular stability?

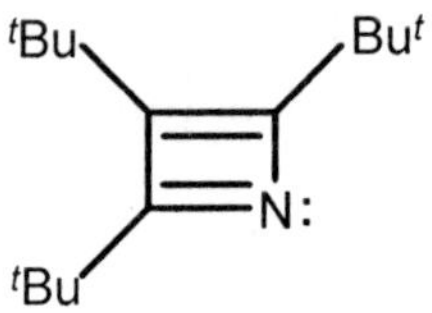

SP-1.26 Compare the rates of proton-losses in hydrogen exchange reaction of the following two compounds. Give reasoning.

(I) (II)

SP-1.27 8-hydroxyquinoline can be steam-distilled for its separation from the 4-hydroxy isomer. Offer explanation.

SP-1.28 How would you justify the cause for the difference in boiling points of propyleneimine ($CH_3CH_2CH{=}NH$, b.p. 67^{0}C) and its isomer, *N*-methylethyleneimine ($CH_3CH{=}NCH_3$, b.p. 27.5^{0}C)?

SP-1.29 2,2-Dimethylpentane is of higher molecular weight than 2-methylhexane. Yet, its boiling point is lower than that of the latter. Why?

SP-1.30 Boiling point of 2-nitroresorcinol is lower than that of resorcinol. Justify.

SP-1.31 Which one of the following pair is much more soluble in water?

$$CH_3CH_2OH,\ CH_3(CH_2)_3CH_2OH$$

SP-1.32 The boiling points of *n*-pentane and its isomer neopentane are 36.2^{0}C and 9.5^{0}C, respectively. Account for this difference.

SP-1.33 The observed boiling points for trimethylamine, ethylmethylamine and propylamine are 3.5^{0}C, 37^{0}C and 49^{0}C, respectively. Account for the fact.

SP-1.34 1-Butanol and diethylether are almost equimiscible with water, but they differ largely in their boiling points. Explain.

SP-1.35 Which one of following pairs of compounds would show greater melting and boiling points?
(i) *p*-Toluidine, *o*-toluidine; (ii) *p*-nitroaniline, *o*-nitroaniline

SP-1.36 Dibenzoylmethane does not form Schiff bases, nor adds to HCN. It also does not give acetylated product with acetic anhydride, nor produces methane on reaction with MeMgI, nor evolves out nitrogen on treatment with diazomethane. Offer explanation for the aforesaid observations.

SP-1.37 Compare the basicity between *N,N*,2,6-tetramethylaniline and *N,N*-dimethylaniline.

SP-1.38 2,6-Di-*t*-butylpyridine is a specific proton scavenger. Account for the fact.

SP-1.39 Which one of the following compounds would show higher basicity?

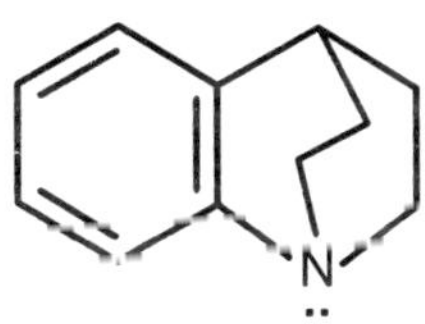
(A)

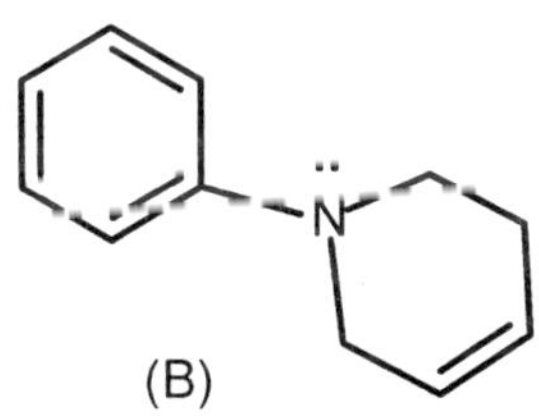
(B)

SP-1.40 *N,N*-Di(2,4,6-trinitrophenyl)amine is 100 times more acidic than acetic acid. Justify the statement.

SP-1.41 Phenylacetone can exist two enolic forms; predict which one will be present in the larger amount at equilibrium.

SP-1.42 Explain why —
(i) An equilibrium between *R*- and *S*-isomers of 2-methylcyclohexanone is established on treatment of either one of the pure isomers with aqueous acid or base.

(ii) When *cis*-2,4-dimethylcylohexanone is dissolved in aqueous ethanol containing a trace amount of base, a mixture of *cis*- and *trans*- isomers results.

(iii) α-Bromination of (*S*)-*sec*-butyl phenyl ketone with Br_2 in acetic acid yields a racemic mixture of products.

(iv) Although the inductive effect of the aceto group ($CH_3CO–$) is considerably larger than that of the methoxyl group, yet the pKa values for acetoacetic acid and methoxyacetic acid are nearly the same.

(v) The enol content of 4,4,4-trifluoro-2-butanone is larger than that of 2-butanone.

SP-1.43 Out of (A) and (B), one forms intermolecular hydrogen bonds whereas the other forms only intramolecular hydrogen bonds. Discuss.

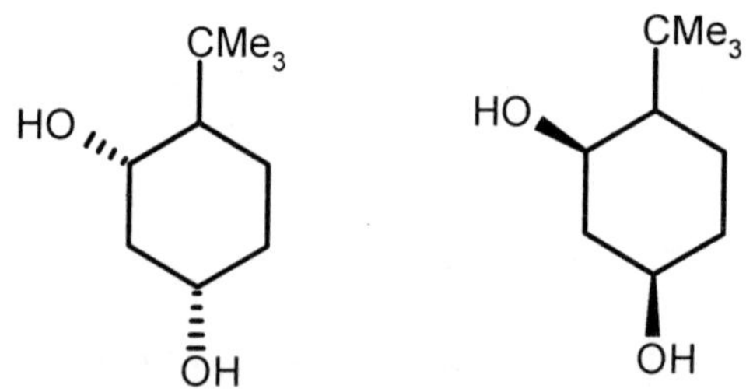

(A) (B)

SP-1.44 Which of the following compounds would remain in the enol form in *n*-hexane solution?

$CH_3COCOCH_3$ or $CH_3COCH_2COCH_3$

SP-1.45 The bromination of acetone is catalyzed by acids and the rate is zero order with respect to bromine. Draw inference from the experimental findings.

SP-1.46 Ethylacetoacetate exists in the enol form much more in hexane than in water. Explain.

CHAPTER2

STEREOCHEMISTRY

Introduction

Stereochemistry is literally meant for chemistry in three dimensions, and describes chemistry as a function of molecular geometry. In the last 30 years, the third dimension has become so much significant in understanding of chemical problems not only in organic, but in all other branches of chemistry along with other subjects relating to chemistry, that all chemists must have to acquire a reasonably detailed knowledge of the subject concerned. Most molecules are three-dimensional in their geometry, and hence stereochemistry, in fact, pervades throughout the subject. All chemical reactions takes place in three dimensions and the spatial arrangement of those atoms can have a profound effect on the outcome of a chemical reaction.

Customarily, stereochemistry is regarded in two perspectives — *static* and *dynamic* aspects. *Static stereochemistry* deals with the counting of stereoisomers, their molecular architecture, energy-content and physical properties, while *dynamic stereochemistry* (may be denoted as stereochemistry of reactions) deals with the stereochemical requirements and the stereochemical outcome of chemical reactions including interconversion of conformers or topomers. *Dynamic stereochemistry* is intimately associated with the study and understanding of reaction mechanisms.

A good understanding of stereochemistry is fundamental to any detailed appreciation and study of organic chemistry. Thus, it is very much essential for the students, who are intending to study chemistry, to conceive a clear idea on this subject of interest and high impact. Under the scope of the present book, we will discuss some basic and advanced problems relating to stereochemistry for better understanding of the subject.

Solved Problems

Problem 2.1 Show interconversions between various projection formulae, taking suitable examples.

Solution.

Example-I

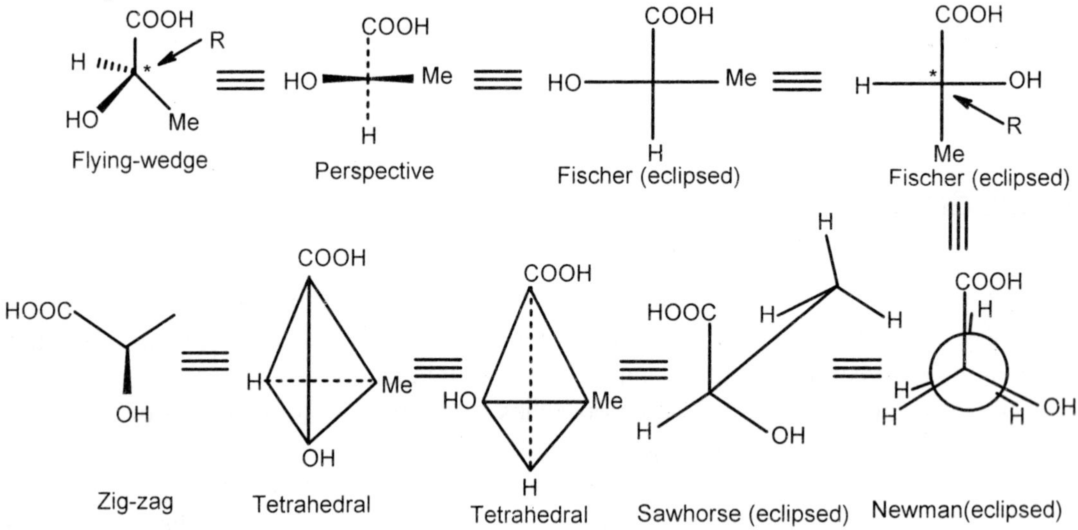

Example-II

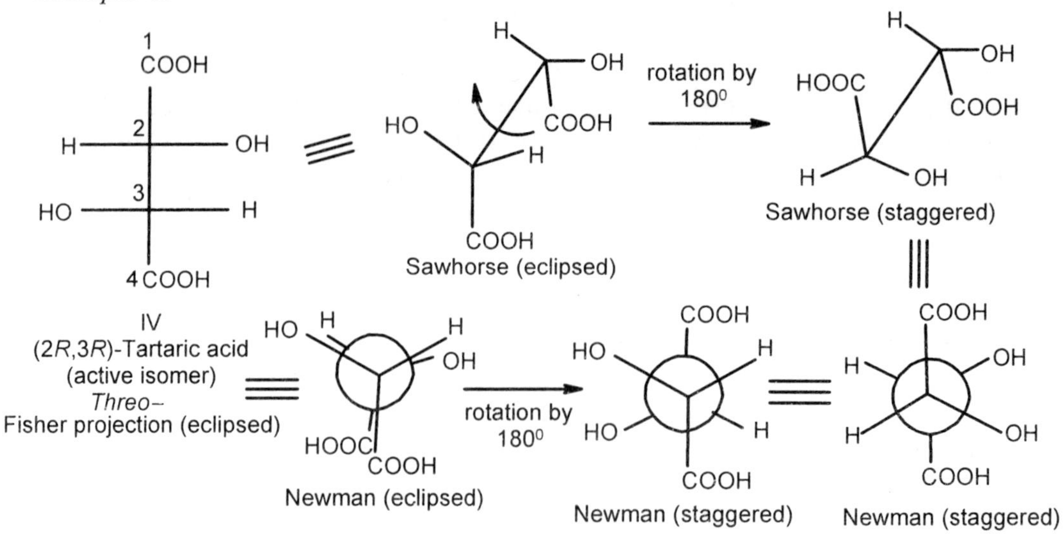

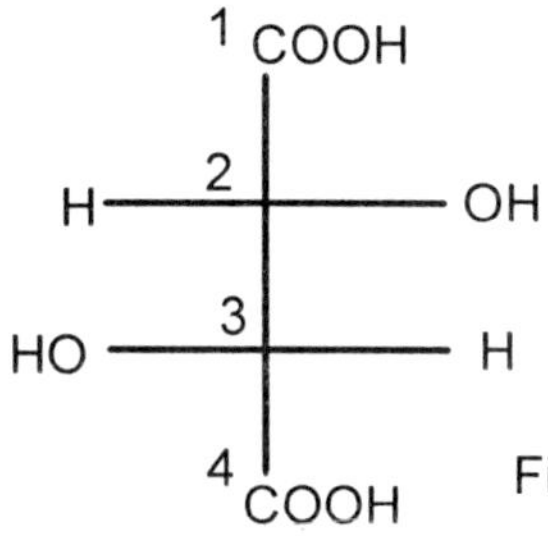

(2*R*,3*R*)-Tartaric acid
(active isomer)
Threo–
Fisher projection (eclipsed)

IV

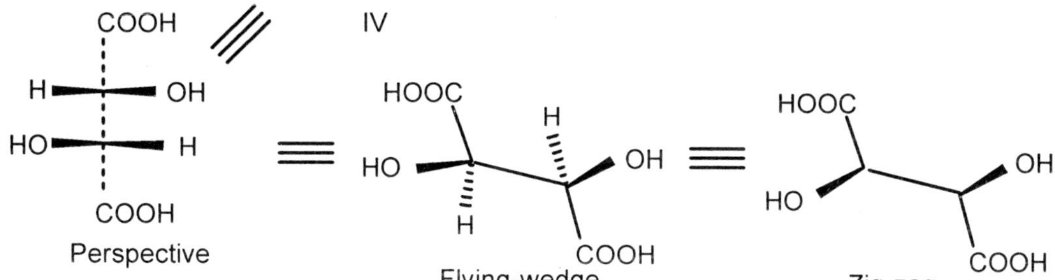

Perspective

Flying-wedge

Zig-zag

Problem 2.2 What do you mean to say 'optical activity'?

Solution. Optical activity is an important property of chiral molecules which, when placed in the path of a beam of plane-polarized light, are capable of rotating the plane of polarization to the right or left. Such *optically active* compounds are dextrorotatory if they rotate the plane of polarization to the right (*i.e.* clockwise direction), and levorotatory if to the left (*i.e.* anti-clockwise direction).

Actually, a plane-polarized light being constituted of two oppositely (right and left) circularly polarized components provides a chiral medium and stereoisomers of a compound may interact differently, thereby, rotating the plane of the polarized light either in a clockwise or in an anti-clockwise direction. A dextrorotatory substance is denoted by (+) or by the prefix, *dextro*- and a levorotatory substance is denoted by (–) or by the prefix, *levo*. Degree of optical rotation is expressed either by '*specific rotation*' ($[\alpha]_\lambda^T$) or by '*molecular rotation*' ($[M]_\lambda^T$).

Problem 2.3 How the terms "asymmetric" and "dissymmetric" differ?

Solution. A molecule lacking all the elements of symmetry (i.e. simple axis, centre, plane and n-fold alternating axis of symmetry) is termed as asymmetric, and molecule of this type usually shows optical activity.

But a molecule may be optically active even though it bears a simple axis of symmetry. Therefore, the terms 'optically active' and 'asymmetric' as applied to molecules should not properly be used in the same context. Hence a new term 'dissymmetric'

mmetric' has been introduced to denote the absence of a centre, a plane and an n-fold alternating axis of symmetry within a molecule, but not necessarily of a simple axis of symmetry.

Term	Alternating axis	Simple axis	Optical activity
Asymmetric	absent	absent	usually active
Dissymmetric	absent	may or may not be present	usually active

Since alternating axis of symmetry is sufficient enough to represent also the plane and centre of symmetry, it would be better to say that a dissymmetric molecule must be devoid of an alternating axis of symmetry. A dissymmetric molecule usually becomes optically active.

Problem 2.4 What is the necessary and sufficient condition for a molecule to be optically active?

Solution. Not the asymmetry, but the dissymmetry is the minimum and necessary condition for a molecule to be optically active. Thus, the necessary and sufficient condition for an optically active molecule is such that its mirror image must be non-superimposable on it; *i.e.* molecular dissymmetry as a whole is required for this purpose.

For examples:

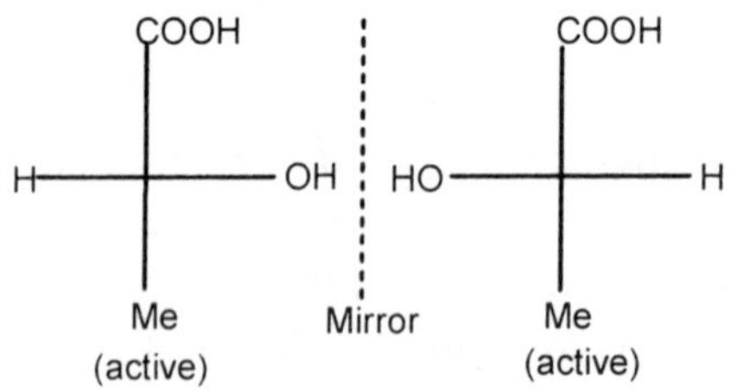

non-superimposable mirror images
(enantiomeric pair)

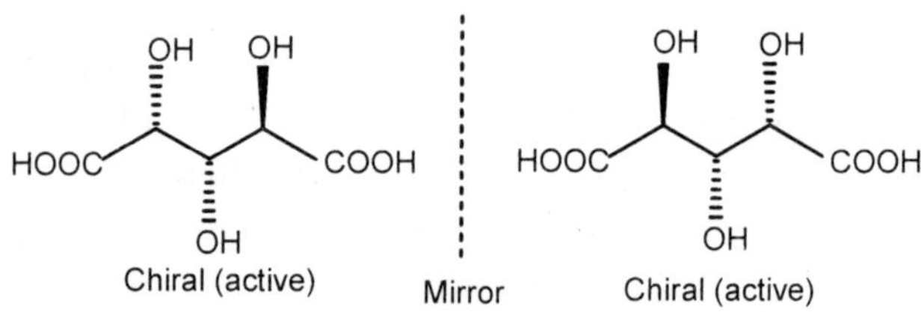

non-superimposable mirror images
(enantiomeric pair)

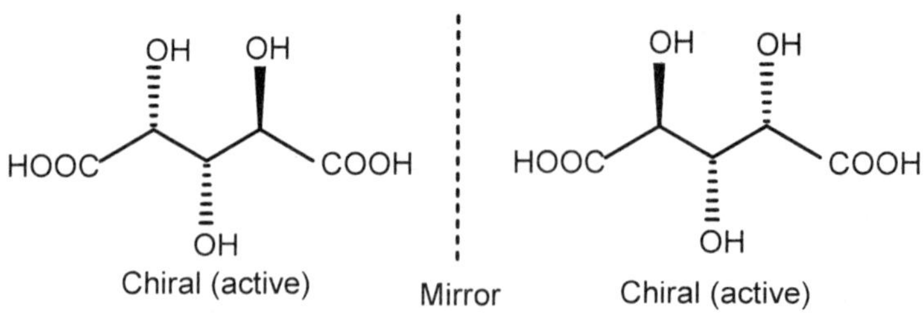

(enantiomeric pair)

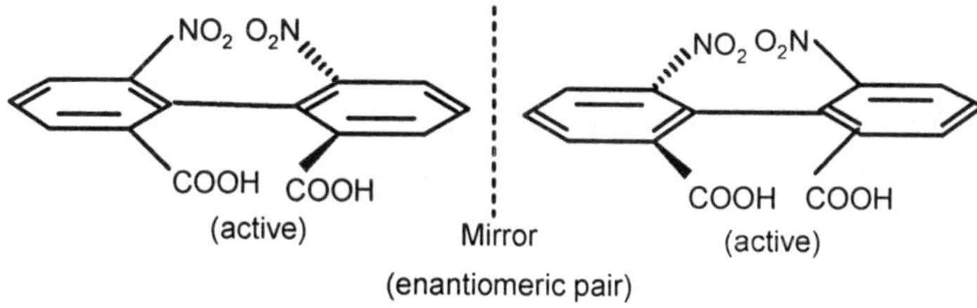

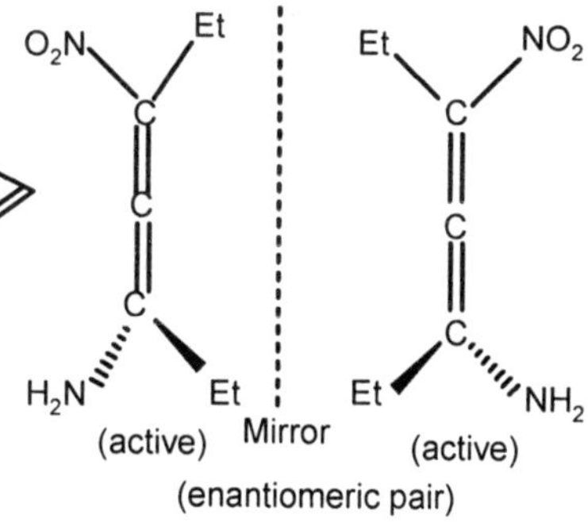

Problem 2.5 " Plane of symmetry and centre of symmetry both may be regarded as the special cases of alternating axis of symmetry" — Explain

Solution. The alternating or rotation-reflection axis of symmetry is sufficient enough to explain the plane of symmetry (σ-plane) as well as the centre of symmetry (i). A plane of symmetry is equivalent to one-fold alternating axis (S_1), while a centre of symmetry is equivalent to a two-fold alternating axis of symmetry (S_2).

An S_1 operation involves two successive steps — *i.e.* a C_1 and a σ-operation. C_1 -operation does not change a molecule, because rotation through an angle of 360^0 about any axis through any molecule gives back an identical structure. Thus S_1 operation becomes only a σ -operation.

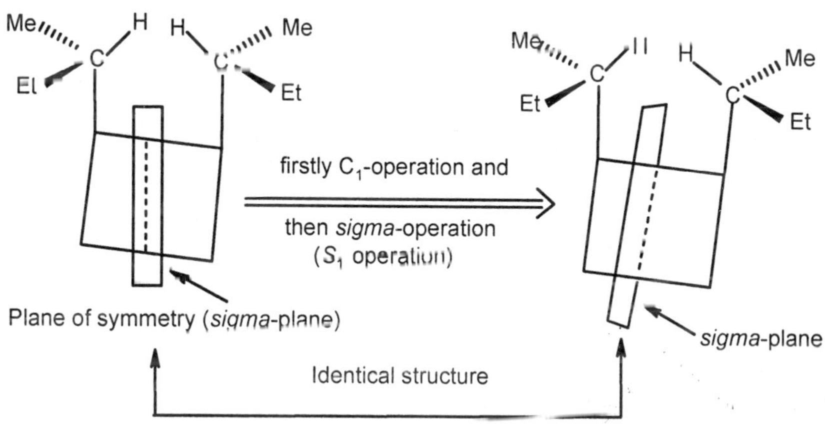

Similarly, S_2 operation involves successively C_2 and σ -operation. A molecule bearing centre of symmetry (i) is found to have two-fold alternating axis of symmetry (S_2) as in the case of α-trucxillic acid.

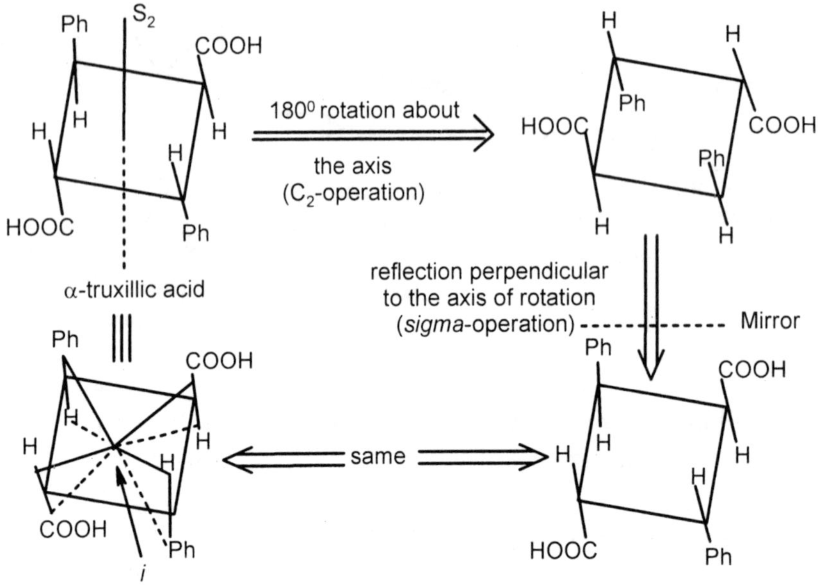

Problem 2.6 Can you cite an example of a dissymmetric molecule, which is optically inactive?

Solution. Dissymmetric molecules are generally optically active, but in 1955 Mislow and Bolstad[1] synthesized a molecule, (*dextro*)-menthyl-(*levo*)-menthyl-2,6,2′,6′-tetra-nitro-4,4′-diphenate (I), which is found to be optically inactive although it is devoid of symmetry elements.

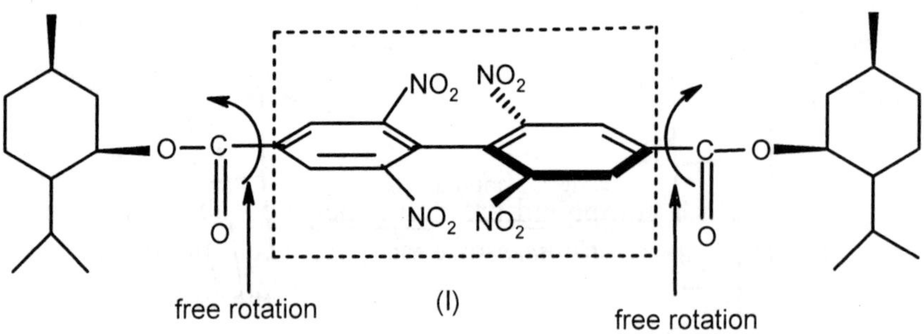

(I)

The inactivity of this molecule (I) is attributed to the fact that the molecule and its mirror image, though not actually superimposable, can readily be made superimposable

1. Mislow, K. and Bolstad, R., *J. Am. Chem. Soc.*, **77**, 6712(1955).

by rotation around the bonds linking the carboxylate groups to the rings[1]. The biphenyl unit as a whole can rotate around two single bonds (relatively unhindered) as mentioned. A rotation of 90^0 converts one enantiomer into the other leading to spontaneous racemization.

Problem 2.7 Sketch a molecule that would show optical activity, although it bears a simple axis of symmetry.

Solution. Molecules of the type as shown below would be optically active, even though they bear simple axis of symmetry.

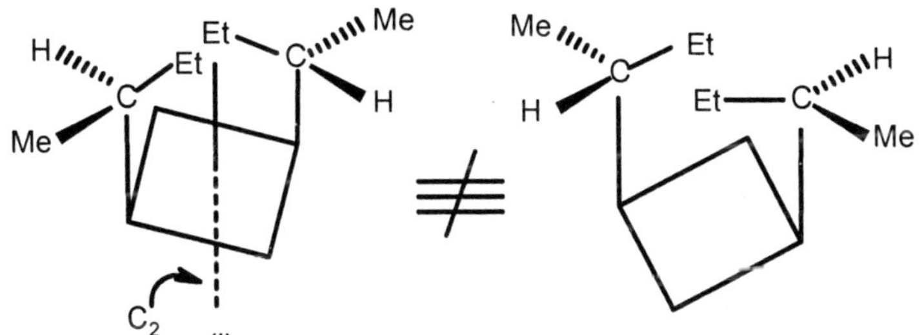

(I) (II)

Structure (I) has a two-fold simple axis of symmetry (C_2), but its mirror image (II) is not superimposable on it. Thus, both I and II would show optical activity.

Problem 2.8 What do you mean by the terms '*enantiomers*' and '*diastereoisomers*'? How do they differ?

Solution.

(a) ***Enantiomers***

If two stereoisomers of a compound are non-superimposable mirror images of each other, they are termed as *enantiomers* of each other (*i.e.* enantiomeric pair), and are said to exhibit enantiomeric relationship. Enantiomers are also called *enantiomorphs*; they are usually optically active and that is why they are termed as *optical antipodes* — because they rotate the plane of a plane polarized light to an equal magnitude, but in opposite directions. Examples of enantiomeric pairs:

---

1. Mislow, K., *Science*, **120**, 232(1954); *Trans. N. Y. Acad. Sci.*, **19**, 298(1957).

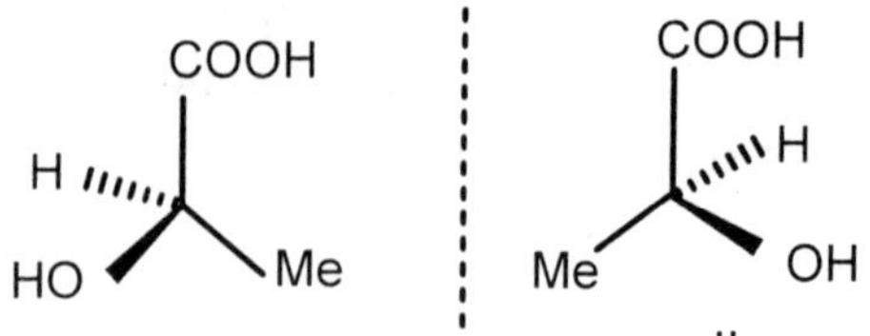

I Mirror II

(2*R*)-Lactic acid (2*S*)-Lactic acid

(I & II are non-super-imposable mirror images)

(two enantiomers of lactic acid)

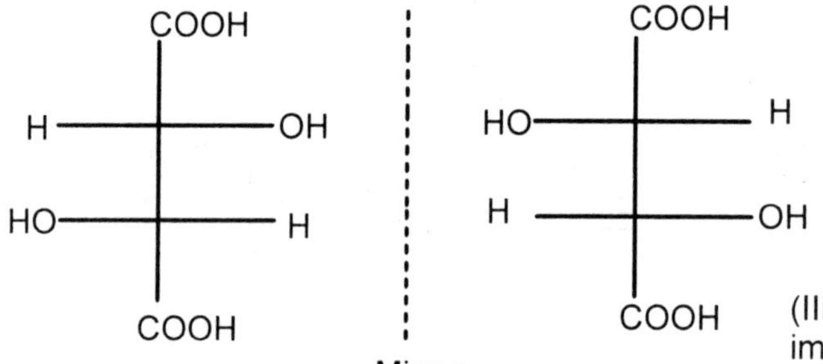

III Mirror IV

(2*R*,3*R*)-Tartaric acid (2*S*,3*S*)-Tartaric acid

(III & IV are non-super-imposable mirror images)

(two enantiomers of tartaric acid)

Enantiomers are of identical connectivity of atoms along with identical relative spatial arrangements of atoms with respect to distance and dihedral angles. That is why they are chemically alike toward achiral molecules, but behaves differently in chiral environments. Enantiomers show different rates of reactions toward chiral reagents. Regarding physical parameters, enantiomers exhibit similar properties like melting points, boiling points, *etc*. But they differ in their behaviour towards plane-polarized light as mentioned earlier. Plane-polarized light consisting of two dissymmetric components (*i.e.* a right handed and a left handed circularly polarized light-beam) offers a chiral environment and that's why enantiomers behave oppositely. This specific property of an enantiomeric pair can enable one to make them differentiate.

(b) ***Diastereoisomers***

If two stereoisomers of a compound do not maintain mirror-image relationship between them, they are known as *diastereoisomers* and the isomeric pair is said to exhibit a *diastereoisomeric relationship*.

For example:

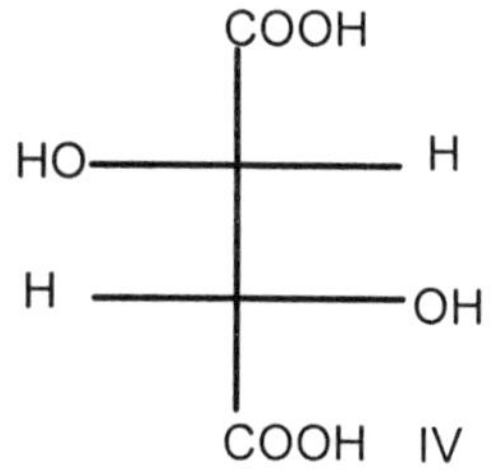

IV

(2*S*,3*S*)-Tartaric acid
active isomer

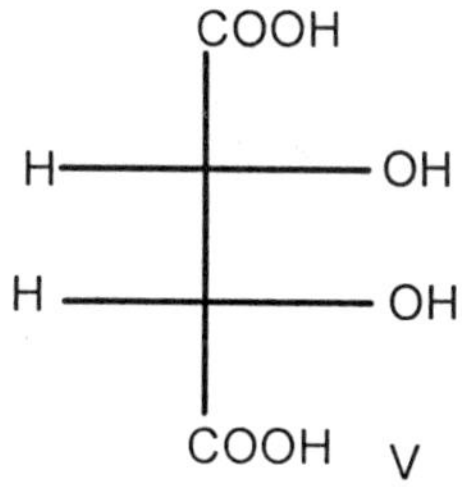

V

(2*R*,3*S*)-Tartaric acid
(*meso*-isomer)

(IV & V are two stereoisomers of tartaric acid; that are not mirror images of each other – hence they are diastereoisomeric pair.)

Similarly, (*Z*)-but-2-ene (VI) and (*E*)-but-2-ene (VII) represent another example of a pair of *diastereoisomers.*

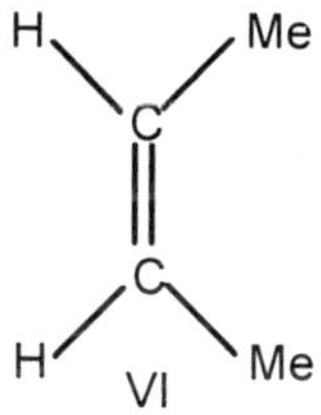

VI

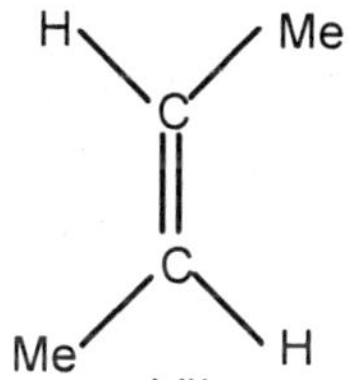

VII

Diastereoisomers bear identical connectivity of atoms in their molecules, but they differ in relative spatial arrangements of atoms with respect to distances and dihedral angles. That is why, diastereoisomers are termed as *anisometric isomers*. For this anisometric property, diastereoisomers differ not only in their chemical properties, but in physical properties also, like behaviours towards plane-polarized light, melting points, boiling points, solubility, dipole moments, heats of formations and standard free energies; so they can in principle be separated by conventional means — for example, fractional distillation or crystallization. If both the isomers of a diastereoisomeric pair show optical activity, they can be expected to have different specific rotations, and their specific rotations will have no relationship. An example, showing the difference in their chemical properties towards the same reagent, may be drawn:

Out of the diastereoisomers of 1,2-cyclopentandiol, only the *cis*-isomer (VIII) forms a ketal with acetone.

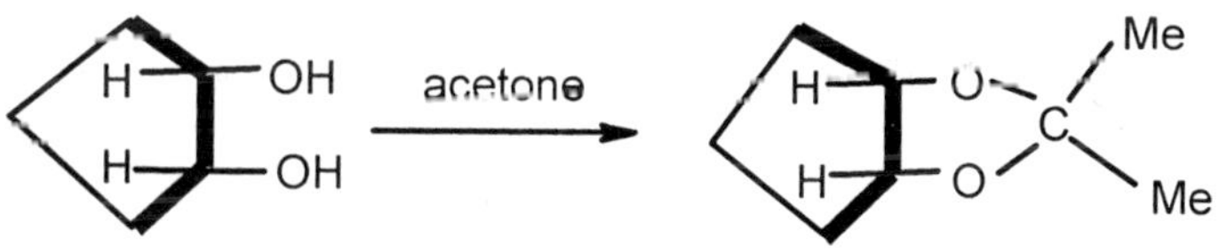

VIII ketal

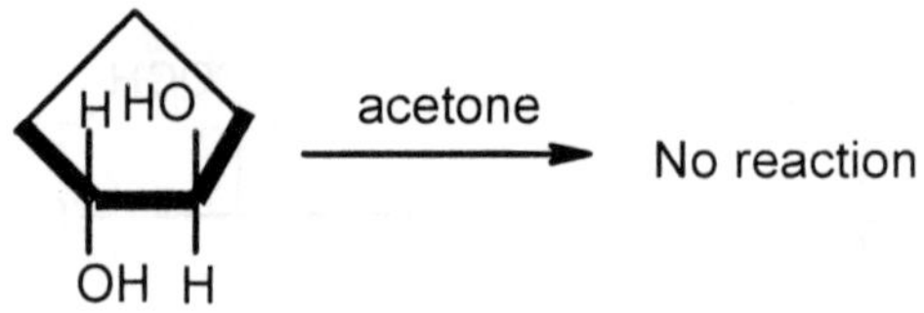

Problem 2.9 Optically active ketone 'A' undergoes racemization in basic solution. Suggest a mechanism for this process. Explain whether ketone 'B' would also racemize or not in basic solution.

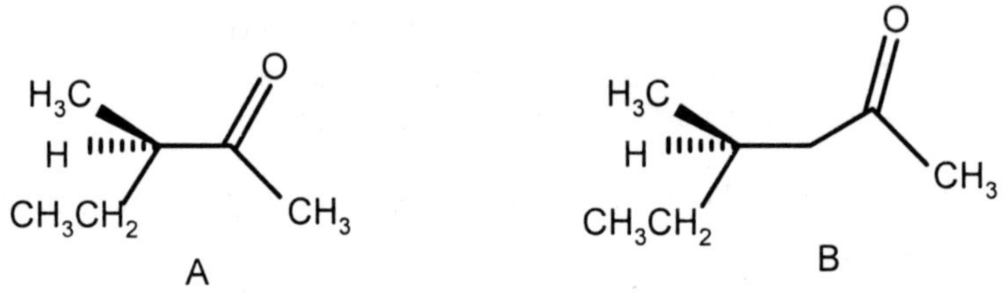

Solution. In basic solution α–proton (acidic in nature) on asymmetric carbon of ketone (A) is abstracted yielding an enolate, that in turn results 1:1 enantiomeric mixture. Hence, resultant is the racemization of the active ketone.

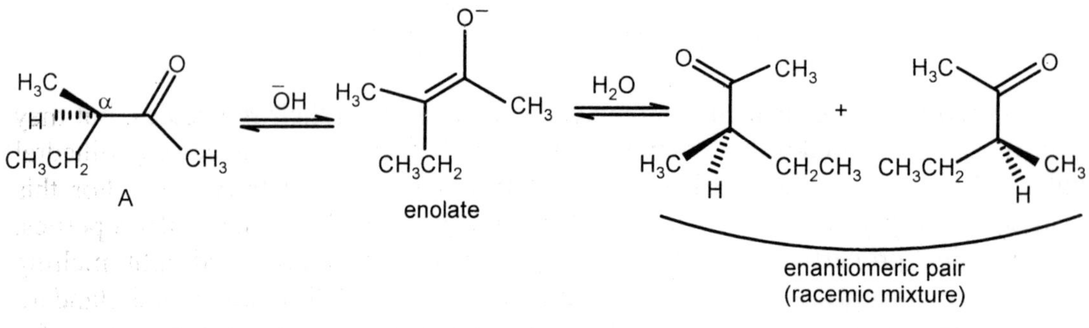

But, ketone (B) would not suffer racemization in basic solution. Because the asymmetric carbon atom is separated from the carbonyl function by a methylene group; this methylenic proton (acidic in nature) would be abstracted by base to form the corresponding enolate. The reaction course has no implication towards the chiral properties of the molecule.

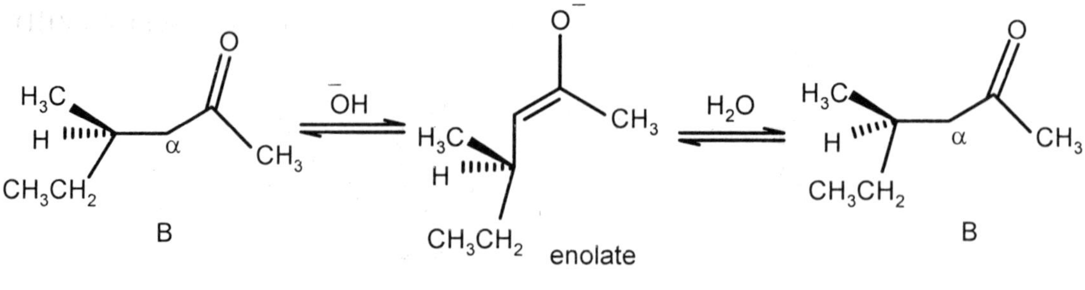

Problem 2.10 Define '*conformation*' and '*configuration*'.

Solution.

(a) ***Conformation***

Spatial orientations of a molecule, which differ only in the dihedral angle and are easily interconvertible through mechanical processes without breaking and making of bonds, are known as *conformations* of that molecule. Mechanical processes include rotation around single bonds or twisting of bonds. Isomers developed due to such orientations are termed as *conformational isomers*. They are also known as *rotational isomers* or *rotamers*.

A molecule may have a number of conformations, but those one, which correspond to energy minima, would be the conformational isomers of that molecule. Each different conformation will have a different energy and the lower-energy conformations will be populated in preference to those of higher energy. Hence a term *conformational analysis* has been introduced, that denotes a study of the physical and chemical properties of a compound in terms of various conformations of the pertinent ground states, transition states and also the excited states.

Conformations of propane are shown here:

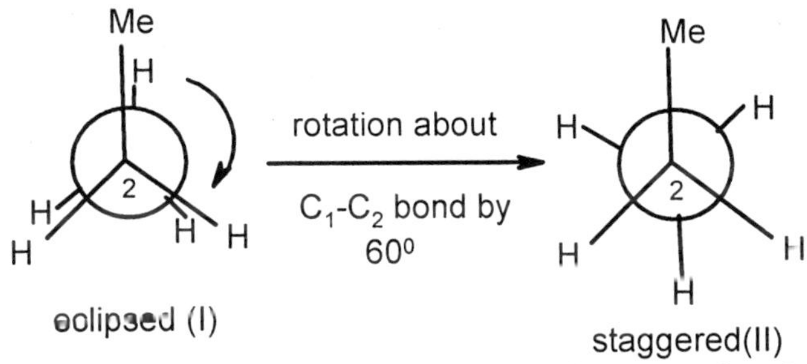

(staggered conformation is more stable and preferred; energy barrier between the two conformations is ~14.0 kJ/mole)

(b) ***Configuration***

The term '*configuration*' denotes a fixed relative arrangement of atoms or groups in space around the dissymmetric or rigid part of a molecule. Stereoisomers thus having the same bonding connectivity but differing in their configurations, are called *configurational isomers*. An example may be cited below:

Lactic acid exists in two stereoisomeric forms — one is (+)-lactic acid and the other is (–)-lactic acid — both are optically active and they are enantiomeric pair. One enantiomer bears a configuration around the central asymmetric carbon atom (C-2) opposite to that of the other.

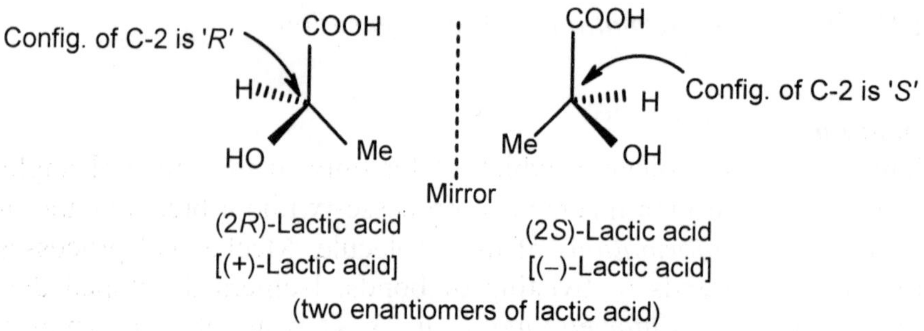

Problem 2.11 Discuss about the conformations of *n*-butane.

Solution. In case of *n*-butane rotations around the C_1-C_2 & C_3-C_4 bonds give rise to homomeric conformations and are not considered; only the rotation around C_2-C_3 bond is considered, that gives rise distinct conformations as shown below. The conformations are arisen out of successive rotations by a dihedral angle of 60^0 each time.

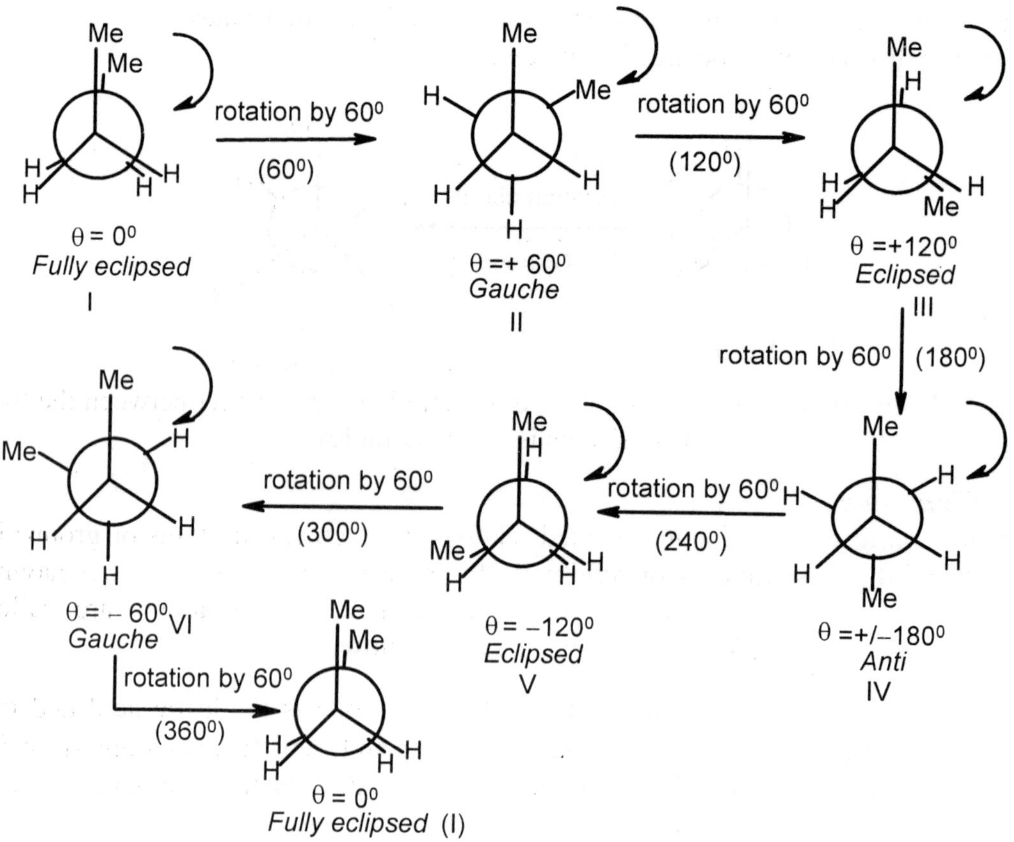

Among the three energy-minima conformations (II), (IV) & (VI), the anti-conformer (IV) with two methyls oppositely oriented ($\theta = 180^0$) is the most stable one. Here torsional strain is absent, and the H–H and Me–H skew interactions are very small. The potential energy of (IV) is arbitrarily taken as zero and those of others are calculated relative to it. The potential energy diagram for *n*-butane may be drawn as (**Fig. 2.1**):

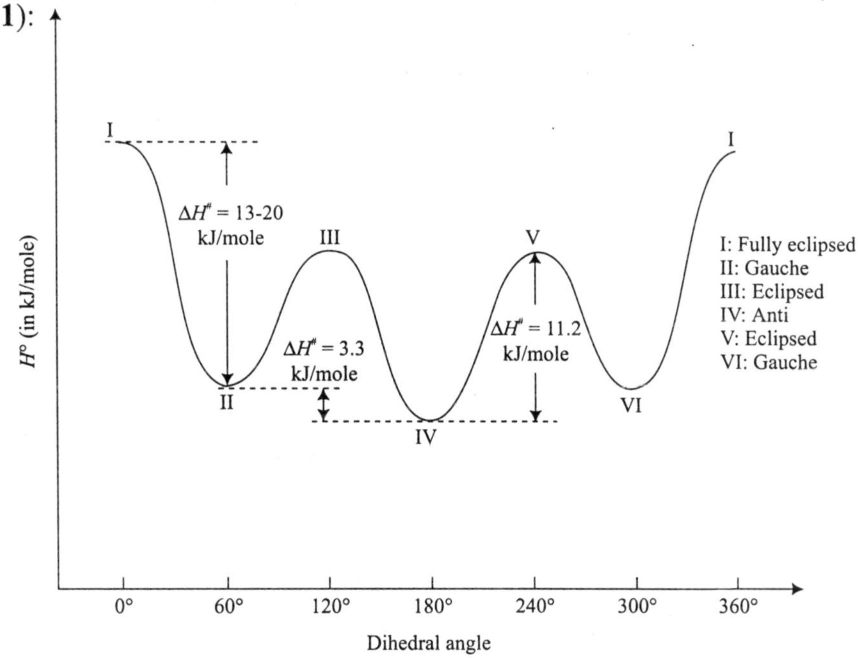

Fig. 2.1: Change in energy due to free rotation about C_2–C_3 bond in *n*-butane

The conformers (II) and (VI) with torsion angle (θ) of $+60^0$ & -60^0, respectively are equi-energetic and have a potential energy barrier of ~3.3 kJ/mole (0.9 kcal/mole) more than that of IV and the barrier for the conformers (III) and (V) is ~11.2 kJ/mole (3.4 kcal/mole).

At room temperature, *n*-butane contains 66% of the anti (IV), 17% of gauche (II) and 17% of another gauche (VI) — the last two forming a racemic mixture since II and VI are non-superimposable mirror images having only a C_2-axis (*enantiomeric pair*). With increase in temperature, population of the less stable conformer increases and eventually becomes equal to that of the stabler conformer. Conversely, at low temperature the stable anti-conformer exists at a much greater extent.

Problem 2.12 What is a *meso*-compound? Meso-tartaric acid is optically inactive due to the presence of a plane of symmetry within the molecule. How far this statement is correct?

Solution. *Meso-* isomer denotes an achiral member of a set of diastereoisomers of a compound that also bears at least one chiral centre.

Meso-isomer of tartaric acid bears two asymmetric carbon atoms of opposite configuration in its molecule:

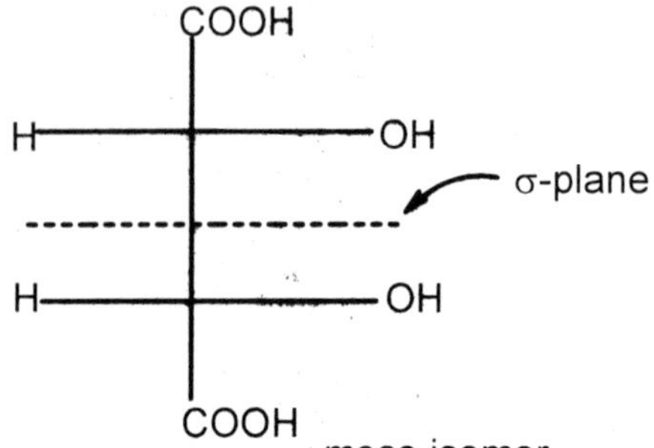

A *meso*-isomer (Fischer projection)

It is observed that this form bears a plane of symmetry, and hence optically inactive. This Fisher projection corresponds to eclipsed conformation of the molecule — which is not a prevalent state of it. Rather, *meso*-tartaric acid exists predominantly in the staggered forms as shown in (I) – (III):

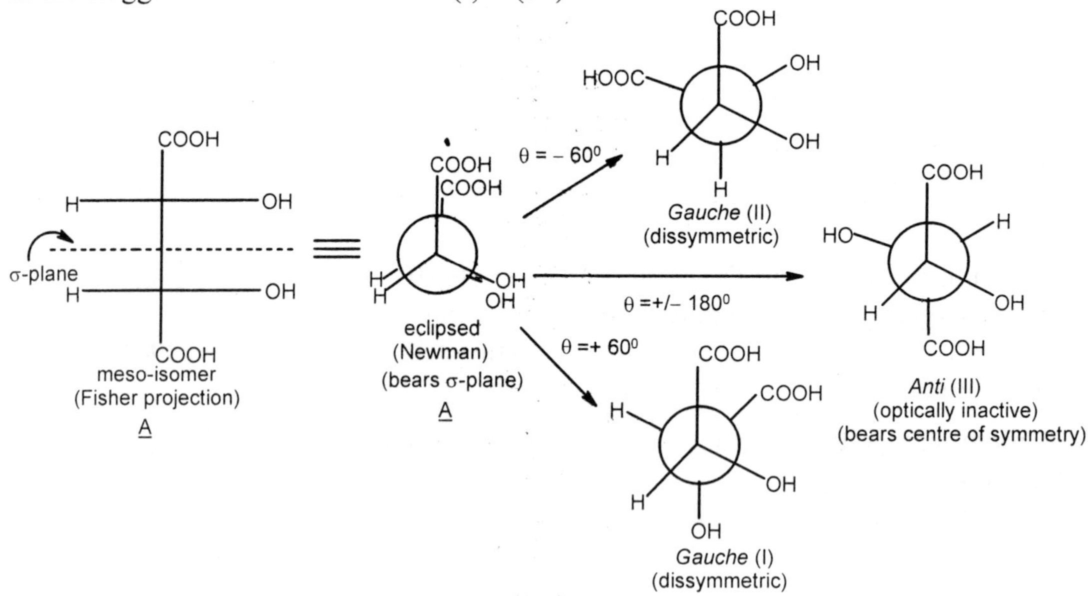

Forms I & II, though dissymmetric, are mirror images of each other and distributed in equal amounts. Thus, the optical inactivity of the stereoisomer as a whole may be ascribed due to the fact that (I) and (II) (enantimeric pair) are present in equal amount.

They are readily interconvertible by internal rotation, and during this rotation each form passes through the form 'A', which does have a plane of symmetry.

Problem 2.13 Explain the term "Optical Purity".
Pure (–)-enantiomer of an optically active compound **A** has a specific rotation $[\alpha]_{20}^{0}$= (–)-51.3^0. What would be the optical purity of a sample **A**, which shows a specific rotation of (–)-35.9^0? How much of each enantiomer is present in the enantiomeric mixture?

Solution. The term 'Optical Purity' (OP) denotes the excess of one enantiomer in a partially resolved material expressed as a percentage of the total. In case of a racemic mixture (d*l*-pair) where none of enantiomers is in excess over the other, the optical purity of the mixture becomes zero; conversely for a completely resolved material (where the excess enantiomer is equal in weight to the total material) optical purity is 100%. It is expressed as:

$$\text{Optical Purity (OP)} = \frac{\text{observed specific rotation of enantiomeric mixture } ([\alpha]_{obs})}{\text{specific rotation of one pure enantiomer } ([\alpha]_{obs})} \times 100$$

If optical purity of an enantiomeric mixture is 60% with respect to d-form, it is meant that the mixture contains (60+20)=80% d-isomer and 20% *l*-isomer. Thus, optical purity corresponds to *enantiomeric excess* (*ee*).

In our present problem–

$$OP = \frac{[\alpha]_{obs}}{[\alpha]_{max}} \times 100\%$$

$$= \frac{(-)35.9^0}{(-)51.3^0} \times 100\%$$

$$= 69.98\% \sim 70\% \text{ with respect to } (-)\text{-enantiomer}$$

Thus, (–)-enantiomer = 70+15 = 85%; (+)-enantiomer = 15%

Problem 2.14 Why in molecules like ethylene glycol and ethylene chlorohydrin *gauche* conformation is more favoured over *anti*-conformation?

Solution. In case of molecules of the type $CH_2OH–CH_2X$, where X= OH, OR, Cl, Br, F, NH_2, NHR, *etc*., *gauche* conformations become more stable than the corresponding *anti*-forms; the extra stability is attributed to the formation of intramolecular hydrogen bonding which is geometrically possible only in *gauche* conformation.

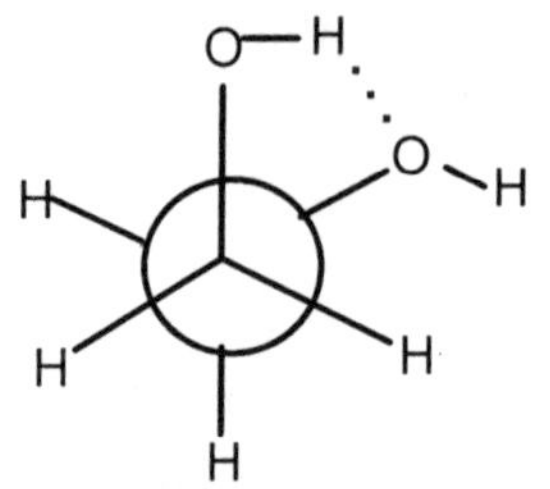

Gauche(ethylene glycol)

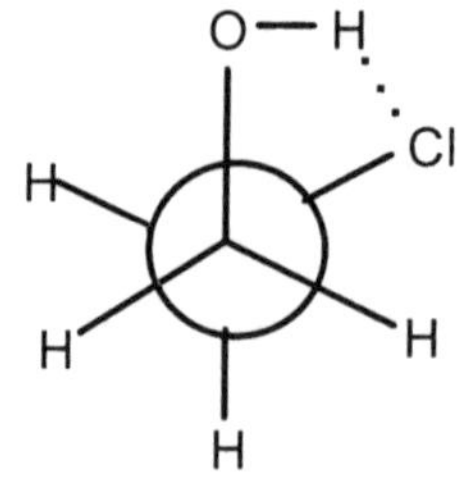

Gauche(ethylene chlorohydrin)

Problem 2.15 What do mean to say 'atropisomerism'?

Solution. 'Atropisomerism' denotes any kind of stereoisomerism developed due to restricted rotation about single bonds, and the isomers can be separated. This type of stereoisomerism is also known as *conformational chirality*; separable atropisomers (*i.e.* enantiomers having energy barrier >80–100 kJ/mole) show chiral behaviour, though devoid of any chiral centre, actually due to attaining molecular dissymmetry as a whole. Properly substituted biphenyls and cumulenes having even number of double bonds are the examples of two significant classes of compounds exhibiting atropisomerism.

Problem 2.16 Comment on biphenyl atropisomerism.

Solution. Properly substituted biphenyl molecules can show atropisomerism when steric interactions between *ortho*-substitutents are sufficient enough to inhibit free rotation of the phenyl rings relative to each other about the sigma bond (may also be called as pivotal bond) joining them. The essential criteria for such biphenyl compounds to exhibit optical activity may be pointed out as:

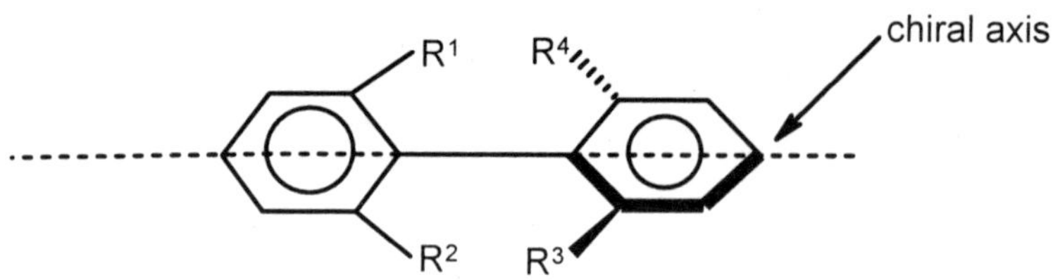

(i) Neither of the rings should possess plane of symmetry (σ_h), and for this purposes not only the R^1 & R^2, but R^3 & R^4 will also be non-identical.

(ii) The *ortho*-substitutents should be large in size so that the two rings must not be coplanar; they should be orthogonal to each other.

A properly substituted biphenyl molecule exits as enatiomeric pair, *i.e.* two separable atropisomers —

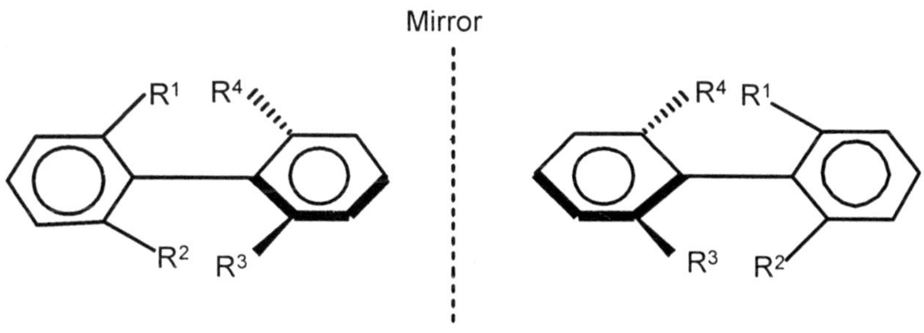

(enantiomeric pair)

Resolvability (*i.e.* ease of resolution) increases with the increase in steric repulsion introduced into the biphenyl molecule by the appropriate bulky substituents. The bulkier the *ortho*-substituents are, the higher is the energy barrier separating the enantiomers and when the amount exceeds 80-100 kJ/mole, the stereoisomers may be separable at room temperature. Thus atropisomerism is developed due to the restricted rotation of the two-phenyl rings about the pivotal bond and the restriction arises out of the steric hindrance caused by the bulky *ortho*-substituents. The order of steric hindrance produced by various groups appears to be Br > Me > Cl > NO_2 > COOH > OMe > F, which roughly corresponds to the order of the respective vanderWaals radii of atoms and groups.

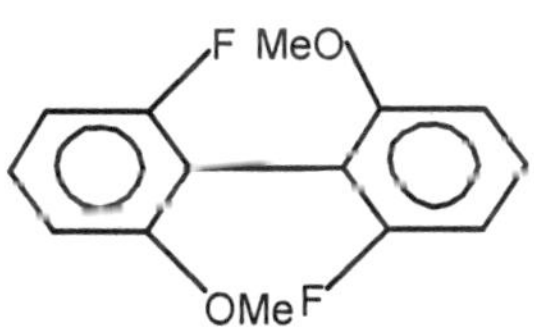

Non-resolvable tetra-*o*-substituted biphenyl derivative

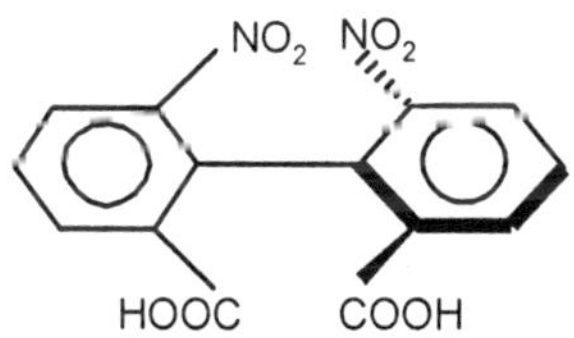

Resolvable (active)

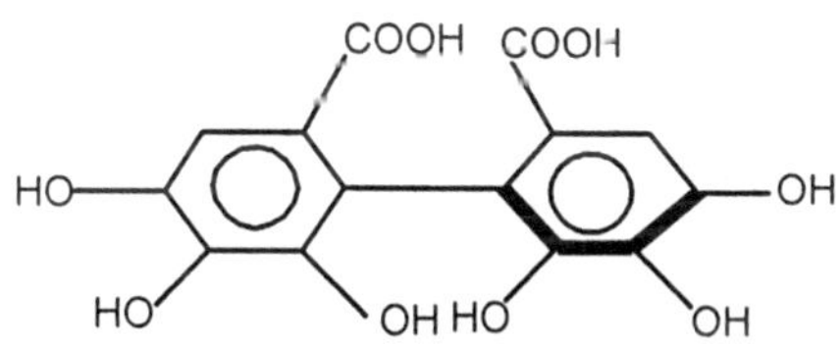

4,4',5,5',6,6'-hexahydroxydiphenic acid; a naturally occurring biphenyl isolated in optically active form

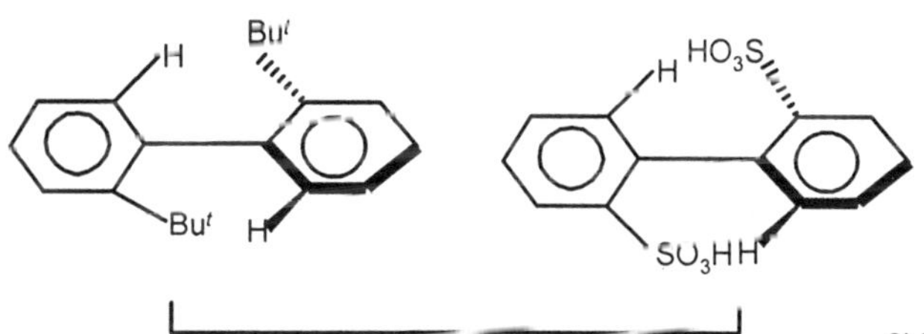

Resolvable di-*o*-substituted biphenyls (active)

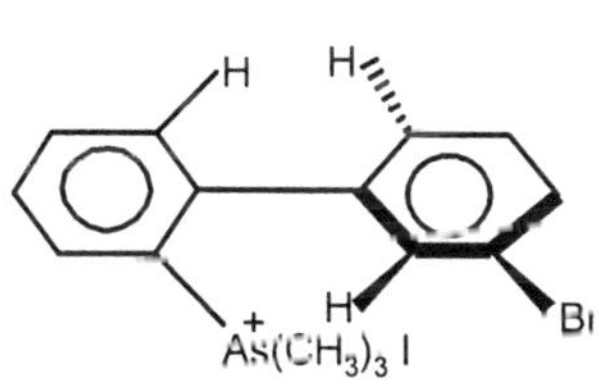

3'-bromobiphenyl-2-trimethylarsoniumiodide
Resolvable into two atropisomers (mono-*O* -substituted)(active)

Configurational stability of atropisomers also depends upon the nature and position of some other substituents other than *ortho*-substituents — known as 'buttressing effect'. A bulky group adjacent to an *ortho*-substituent retards racemization; such buttressing effect (*viz.* order as NO_2> Br > Cl > Me, not in corresponding to the order of steric hindrance of these groups) is observed as in the case of biphenyls (I) & (II):

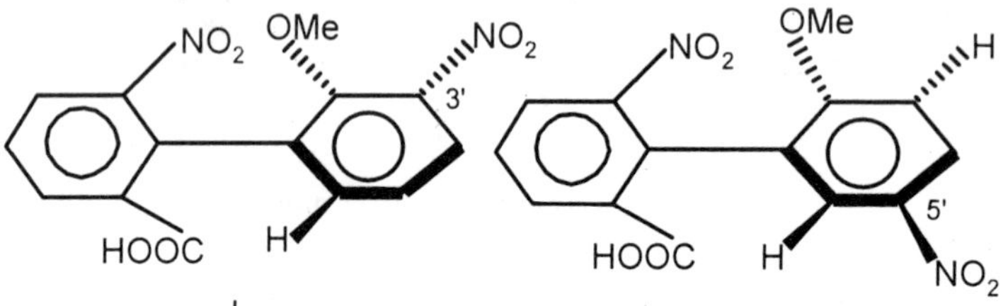

I II

Buttressing effect of NO_2 group adjacent to OMe group in 3′-nitro derivative (I) is more pronounced than that in the 5′-nitro derivative (II), and that is why the rate of racemization of the former is much slower than that of the other. A group at 4- or 4′-position also retards the racemization. This effect is attributed to a change in the entropy of activation. Presence of electron-donating and electron-withdrawing functionalities at the 4-or 4′-positions of the biphenyls also helps in retarding the rate of racemization — this is because of the fact that these groups facilitate interannular resonance so that the pivotal bond assumes more double bond character and the transition state energy is lowered.

Problem 2.17 Compare the rate of racemization between the biphenyl molecules (I) and (II).

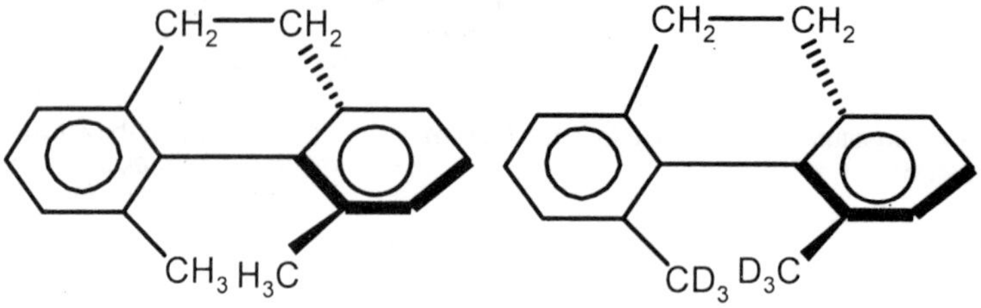

I II

Solution. This is an example of secondary isotope effect. That the rate of racemization of the deuterated biphenyl derivative (II) is about 1.13 times as fast as that of undeuterated one (I). Deuterium is at the level of lower zero-point vibrational frequency and hence, energy-barrier for the racemization of deuterated biphenyl derivative becomes relatively more in amount.

Problem 2.18 Designate *R*/*S* configuration of the following biphenyl derivatives.

Solution.

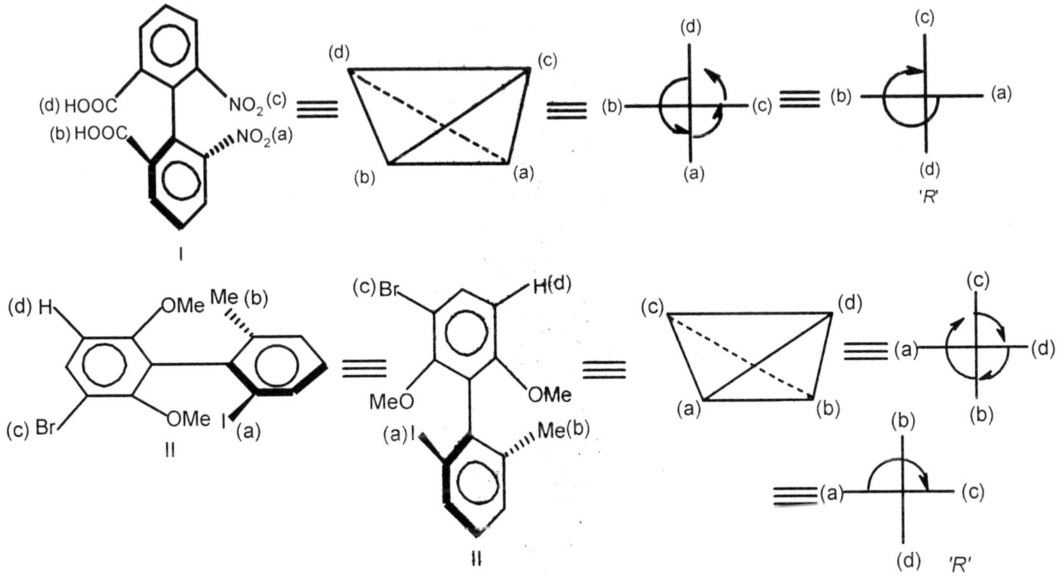

Problem 2.19 Offer evidence for the 'Obstacle Theory' relating to biphenyl isomerism.

Solution. One of the major criteria for biphenyl isomerism is that the two phenyl rings must not be coplanar. Their coplanarity is lost as a result of steric interaction between the *ortho*-substituents in a biphenyl molecule. The bulky groups, at the same time, restrict the free rotation of the rings about the pivotal bond between them. In a word, the properly substituted bulky groups create some 'obstacle' into the molecule, which is partly responsible for its optical activity. These groups are called 'obstacle groups'.

Meisenheimer *et al.* (1927) offered a direct chemical proof for the non-coplanar configuration. They started with (±)-2,2′-diamino-6,6′-dimethylbiphenyl, resolved it and then carried out the following reactions on one of the enantiomers.

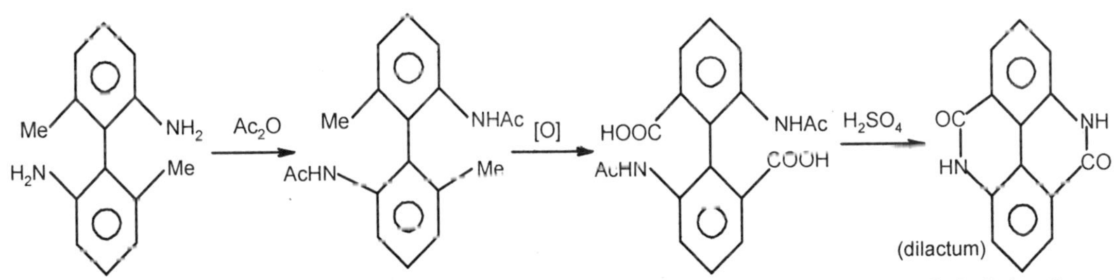

Optically active form (I) Optically active form (II) Optically active form (III) Optically inactive (IV)

Phenyl rings in molecule (I) and its derivatives II & III are orthogonal to each other, and they are optically active. Similarly if the dilactum (IV) becomes non-planar, it would show optical activity — since in that case it would be molecularly dissymmetric as a whole. Conversely if becomes planar, it would bear a centre of symmetry and consequently can't show optical activity. The compound (IV), in fact, was found as optically inactive and so must be planar. This planarity can easily be explained in terms of resonance — all the bonds have double bond character.

Thus, non-coplanarity of the phenyl rings caused by the so called '*obstacle groups*' is one of the essential conditions for optical activity of biphenyl derivatives.

Problem 2.20 Comment on the optical nature of the following biphenyl compounds I and II.

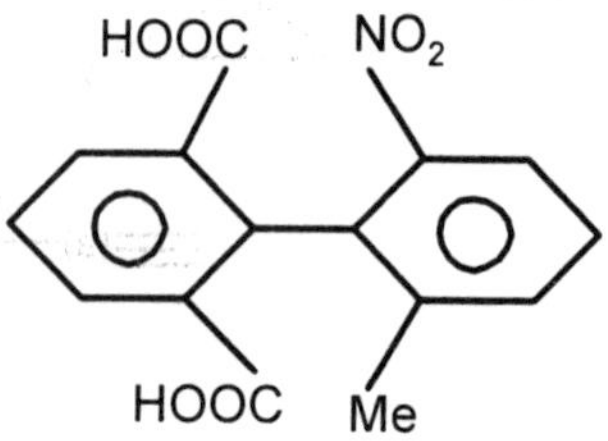

I

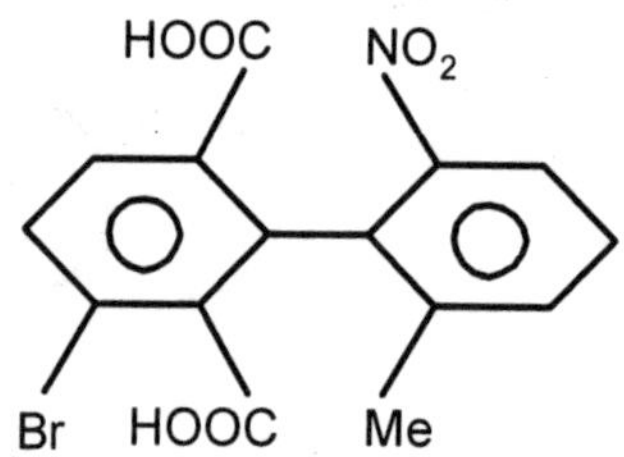

II

Solution.

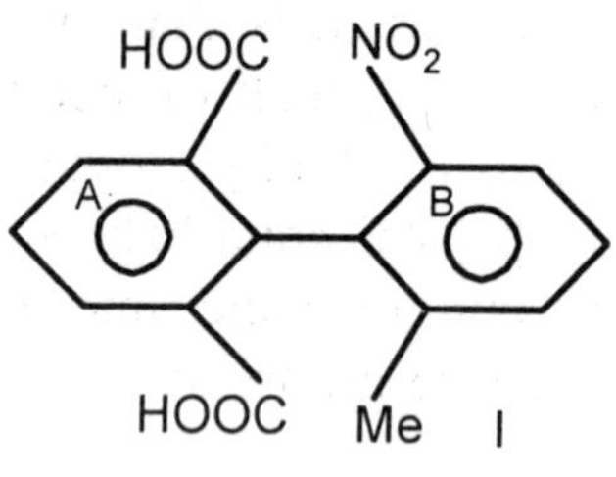

I

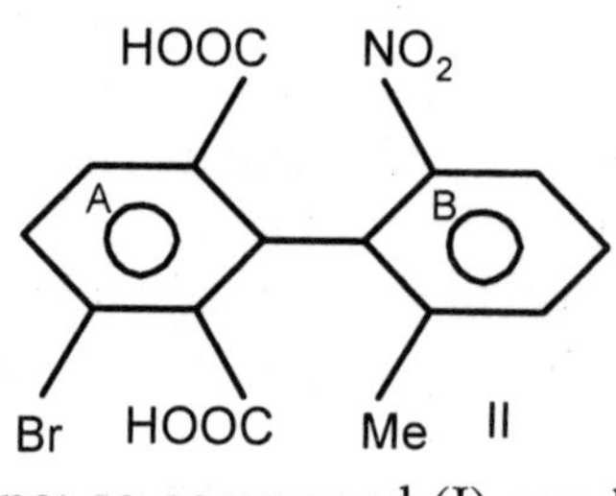

II

Phenyl ring A in compound (I) bears a σ_h-plane; so compound (I) would not show optical activity. But in compound (II) both the rings (A&B) are devoid of σ_h-plane and due to the steric interactions between the bulky *ortho*-substituents, they are also non-coplanar. Consequently, compound (II) would show optical activity.

Problem 2.21 Comment on the stereochemistry of allenes.

Solution. An unsymmetrically substituted allene shows optical isomerism and exits as two enantiomeric forms. A simple case may be given as:

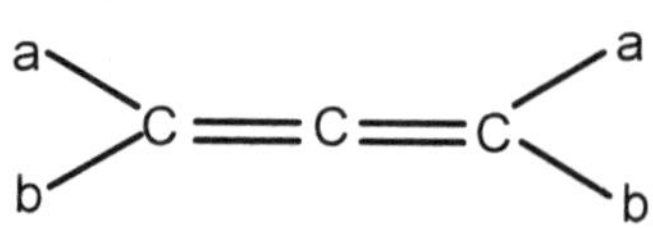

(allene)

The optical activity of such type of allene compounds arises out of *molecular dissymmetry*. The necessary and sufficient condition for such dissymmetry is that the substituted groups (a & b) at the two end carbon atoms should not be identical (*i.e.* 'a' is not equal to 'b'). The reason for the dissymmetry is that the groups 'a' and 'b' at one end of the system lie in a plane at right angles to those at the other end. This can easily be understood from its orbital picture —

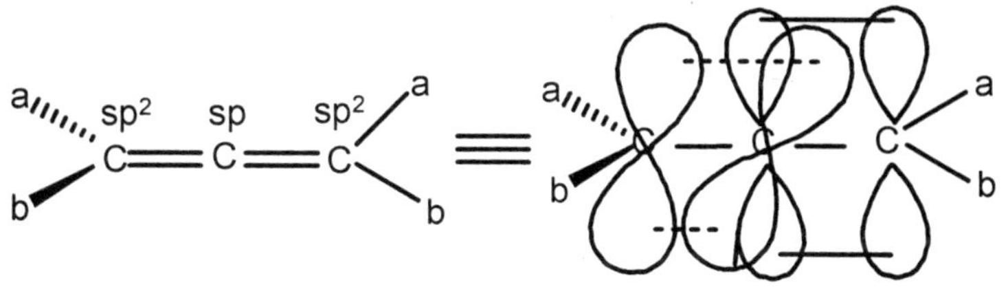

The separate overlapping between two p-orbitals in two separate sets of p-orbitals (shown as dashed and solid lines in the picture) forms π bonds making the two end groups non-planar.

The general formula abC=C=Cab possesses a C_2-axis, but no other symmetry elements and so belongs to C_2-point group. If three or four of the substituents are different as in abC=C=Cad or abC=C=Cde, the C_2-axis also disappears and the molecules become totally asymmetric (C_1 point group).

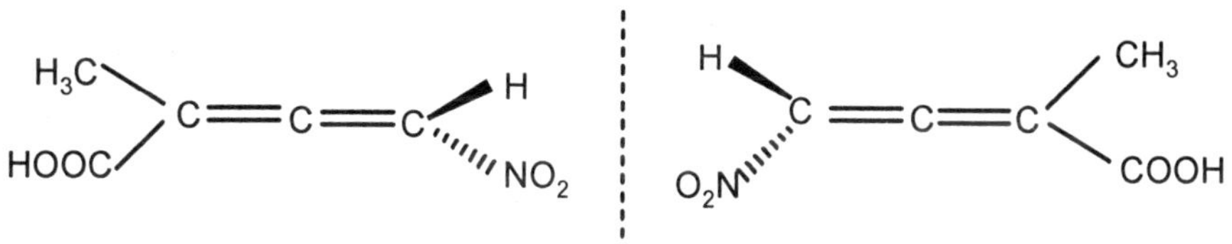

(enantiomeric pair)

Configurational Nomenclature

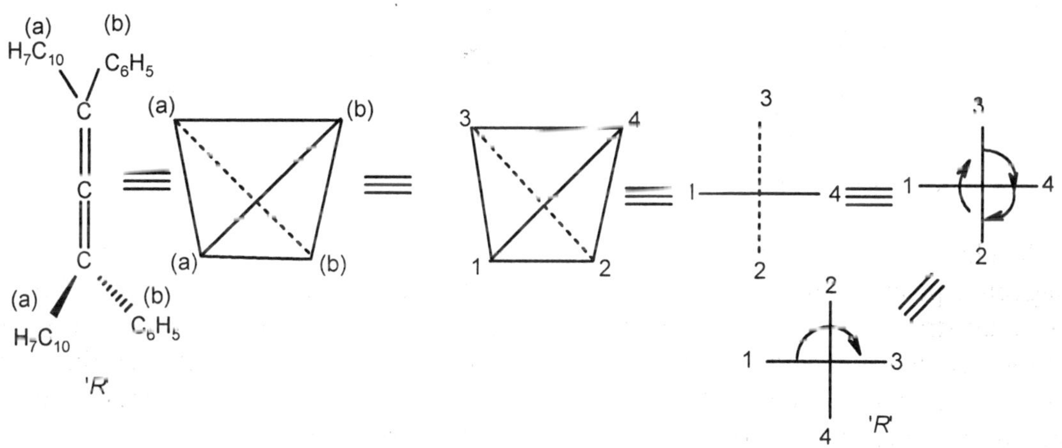

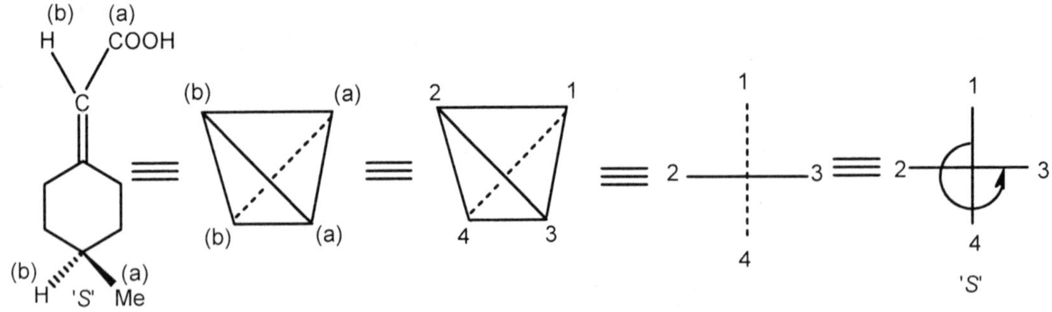

Problem 2.22 Compare the relative stabilities of chair, boat and twist-boat conformations of cyclohexane.

Solution. Cyclohexane may exist in three conformational forms — chair, boat and twist-boat.

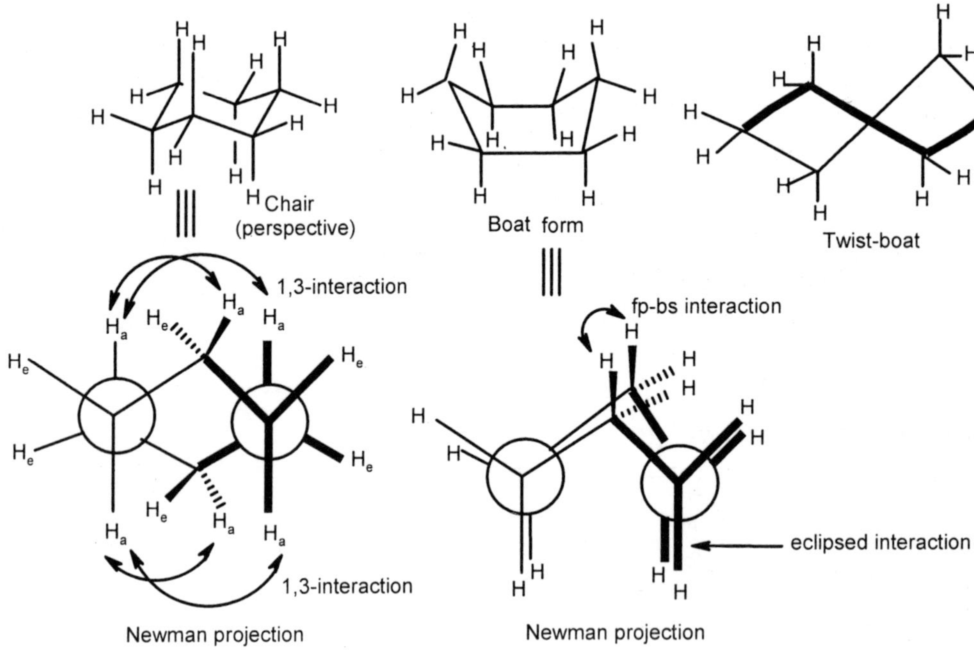

In chair conformation all the C–H bonds on the adjacent carbon atoms are in staggered position; the conformation is not only free from angle strain, but free of torsional strain as well. It suffers only from six butane-gauche interactions and so its entropy content is about 19.8 kJ/mole. Potential energy difference between the chair and boat forms has been found to be 29.4 kJ/mole (minimum) to 45.4 kJ/mole (maximum); hence chair form is more stable than the boat form.

They differ in energy-content because of differences in steric and bond opposition strain in them. The boat form consists of four gauche butane and two eclipsed butane units (corresponding to energy 18–26kJ/mole at per). At the same time, there will be some bond opposition strain for these two pairs of eclipsed bonds along with a non-bonded interaction between the two perpendicular hydrogen atoms at C-1 & C-4, known as *flagpole – bowsprit* interaction.

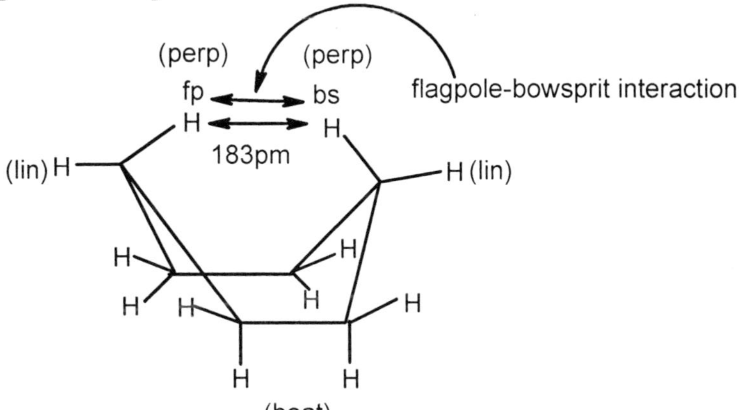

(boat)

In twist-boat conformation, the flagpole and bowsprit hydrogen atoms are little apart minimizing thereby the *flagpole–bowsprit interaction*; the conformation also has less eclipsing of bonds. As a whole twist-boat conformation is relatively more stable than the classical boat by an amount of about 10.3 kJ/mole (or 1.5 Kcal/mole) of energy.

Problem 2.23 "Inversion of cyclohexane chair may be regarded as a special case of diastereotopomerism" — Comment.

Solution. At room temperature cyclohexane chair undergoes very rapid and spontaneous interconversion between two equivalent forms; this conformational change is known as ring inversion, ring reversal or flipping of ring, and is accompanied with the interchange of 1-3-5 and 2-4-6 carbon planes leading to an inverted chair.

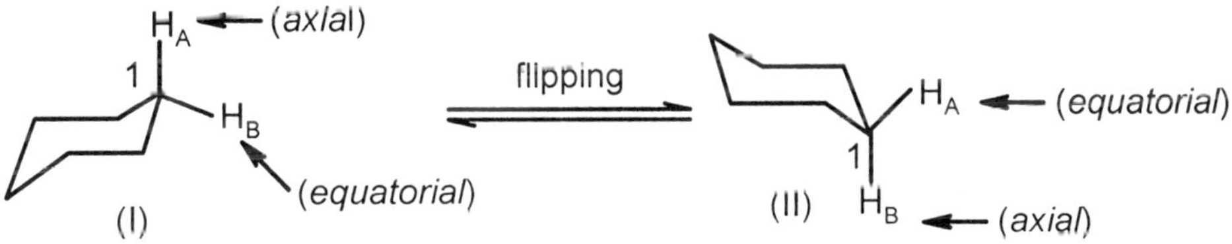

The six axial and six equatorial protons constitute two sets of stereochemically and also chemically non-equivalent protons; thus axial and equatorial protons are diastereotopic in stereochemical relation. During inversion of one chair form into the other equivalent form, axial hydrogens become equatorial of the other and *vice versa*. Such homomeric topomerization involving ring inversion of cyclohexane chair, which accompanies interchange between diastereotopic protons may thus be specifically called as a process of diastereotopomerization.

Problem 2.24 What do you mean by ananchomeric system?

Solution. An 'ananchomeric' or 'conformationally biased' system is one where an equilibrium is maintained between two interconvertible conformers in such a way that the stabler isomer constitutes over 98% of the total at room temperature (Antenius *et al.*,1966). A suitable example of such system may be the case of *tert*-butylcyclohexane.

We know that cyclohexane chair undergoes very rapid ring inversion at room temperature. But the interconversion can easily be slowed down effectively by introducing some bulky substituents like *t*-butyl moiety (Winstein and Holness, 1955). The bulky group can't take up the axial position without effecting considerable deformation of the ring — the phenomenon that is energetically unfavourable. Thus, because of high preference for equatorial disposition of this group, the equilibrium is shifted almost completely to the direction where *t*-butyl would attain equatorial position; such system would be conformationally homogeneous (population of the axial isomer is approximately 1 in 10,000) at ambient temperature.

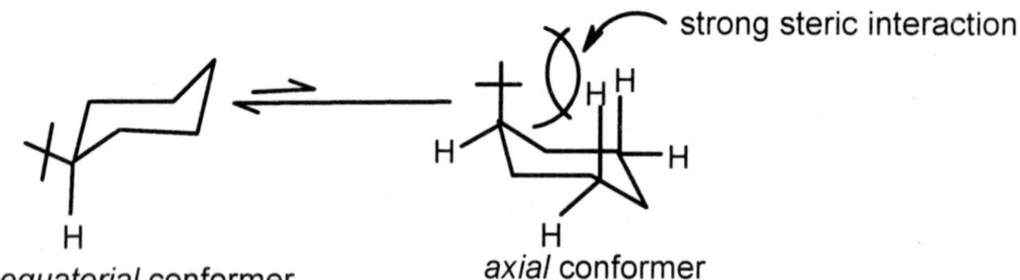

equatorial conformer

(conformationally biased or ananchomeric system)

axial conformer

(energetically unfavourable)

Problem 2.25 Illustrate some situations where molecules are forced to attain exclusively one of the possible cyclohexane conformations.

Solution. Although the chair form of cyclohexane is usually the most stable conformer, but under certain conditions the boat or the twist-boat can also be stabilized in preference to the chair. Various factors like configurational requirement, formation of intramolecular hydrogen bond, electronic effect or in many cases structural features of some molecules force them to exist exclusively in only one or the other of the cyclohexane conformers. Some examples may be cited:

Chair conformations for the molecules I and II are not possible due to configurational requirements.

Camphor (I)

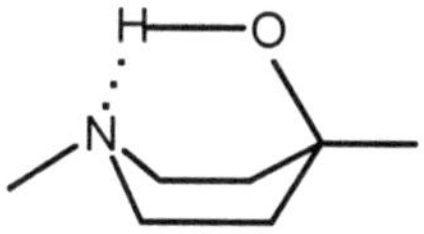

(II)

Similarly, adamantine (III) shows a fixation of chair form, while fixation of boat form in pentaasterane (IV), and fixation of twist-boat in twistane (V).

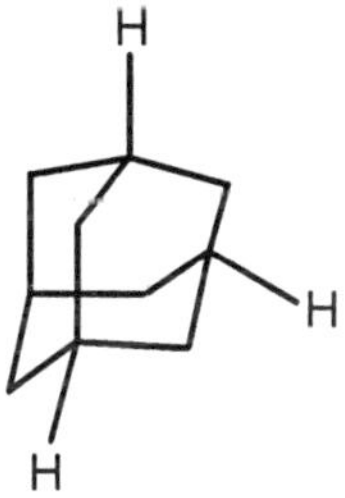

(III)

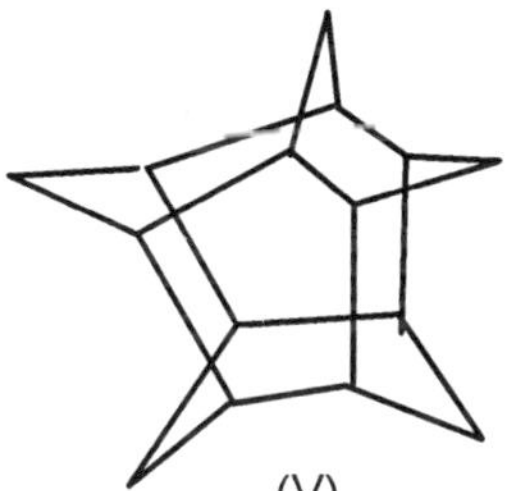

(V)

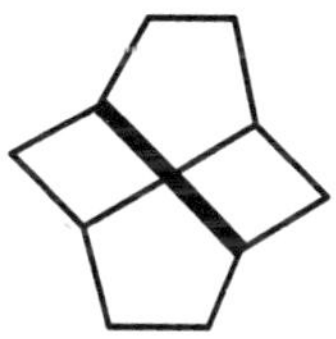

(IV)

Problem 2.26 How would you account for the fact that ester from benzoylformic acid and (–)-2-octanol on reaction with CH_3MgI forms (–)-atrolactic acid as the major product?

Solution.

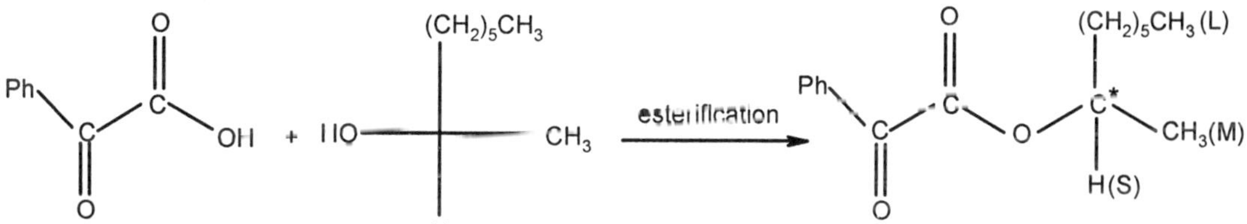

Benzoylformic acid (Phenylglyoxylic acid)

(–)-2-Octanol

Chiral ester (*R*)-configuration

According to the Prelog's rule, the three energetically favourable conformations for the chiral ester having small (H), medium (Me) and large [$(CH_2)_5CH_3$] groups at the generalized form Ph–CO–CO–O–C*SML, would be a, b & c, and the final products after hydrolysis obtained from each isomer can be shown as:

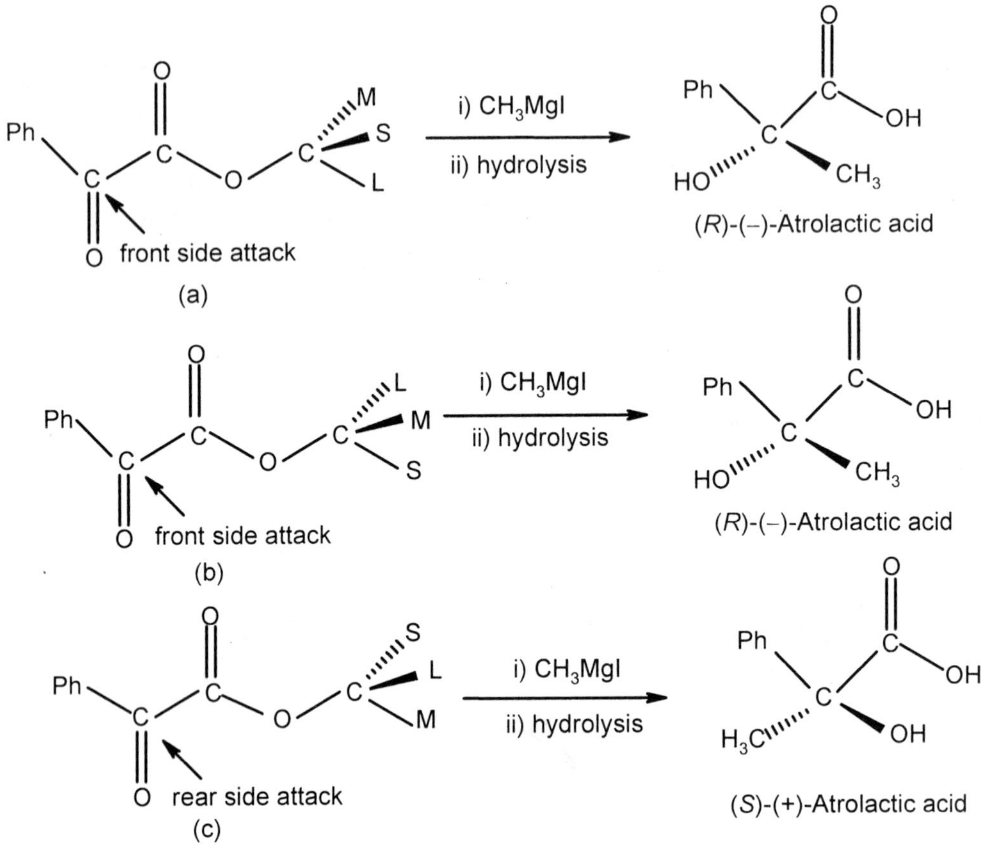

Thus, even when conformations a, b & c are present in equal populations, the major reaction product would be the (*R*)-(–)-atrolactic acid.

Problem 2.27 The relative rates of acetolysis of the following compounds are given as:

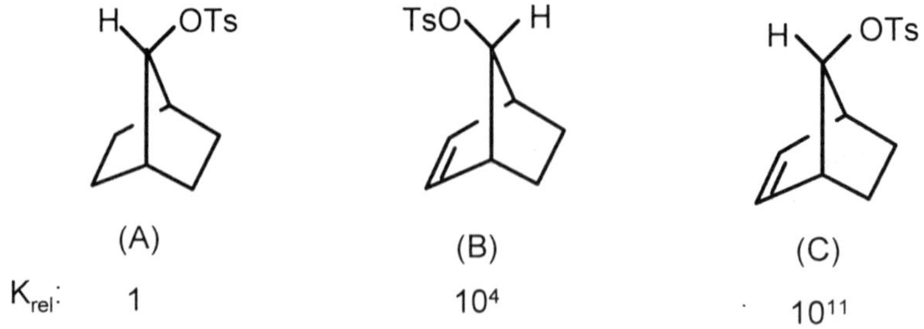

Justify the above observation with acetolysis product(s) in each case.

Solution.

The highest acetolysis rate for compound (C) in the above series is attributed to anchimeric assistance of the double bond present in the nor-bornene skeleton. In this *anti*-tosylate derivative (C) (the double bond and the – OTs group are in *anti*-relation) the removal of the tosylate group (rate determining step) is facilitated from strong anchimeric assistance exerted by the double bond and the resulting non-classical carbonium ion can react only from the right hand side giving an acetate with retained configuration.

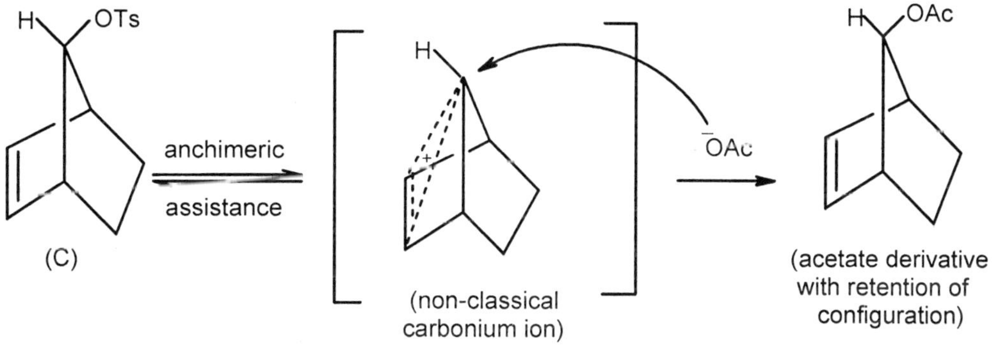

The *syn*-tosylate (B), on the other hand, dissociates (without anchimeric assistance) into a classical homoallylic carbocation, which in turn rearranges to an allylic one; the allylic carbocation then leads to a bicyclic acetate derivative. The relatively higher rate (10^4 times) in comparison to that of the compound (A) may be rationalized from the participation of σ-electrons of $C_1 - C_6$ bond.

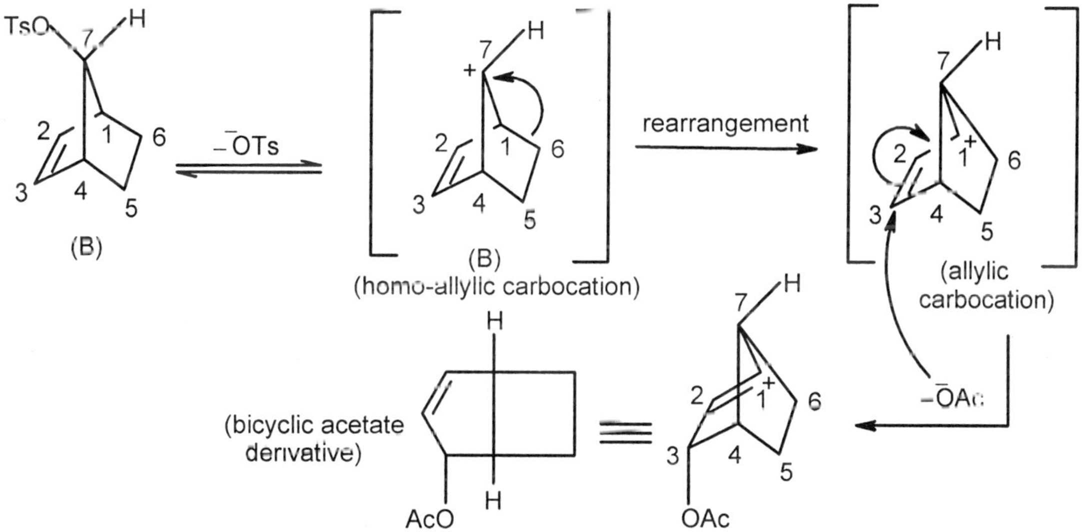

Problem 2.28 Describe the axial chirality feature with the help of a suitable example.

Solution. The chirality (or dissymmetry) of biphenyls, allenes, spirans and alkylidene-cycloalkanes, which is not related to the presence of asymmetric atoms, is sometimes called "axial chirality" (or "axial dissymmetry"). This is because one may look at such molecules in terms of an imaginary axis (running through the C=C=C bond in allenes, through the Ar–Ar bond in biphenyls, etc.), which is asymmetrically surrounded by various substituents. Thus only pairs of vertices around the two ends of the axis need to be distinguished. The structure (I), an elongated tetrahedron representing the allene, abC=C=Cab would be three-dimensionally chiral and hence is eanantiomorphous with its mirror image (I′). The axis along which the tetrahedron (that becomes regular in shape during representation of a three dimensional chiral simplex, *Cabcd) becomes elongated (shown by doted lines) is termed as the *chiral axis* or the *stereoaxis*.

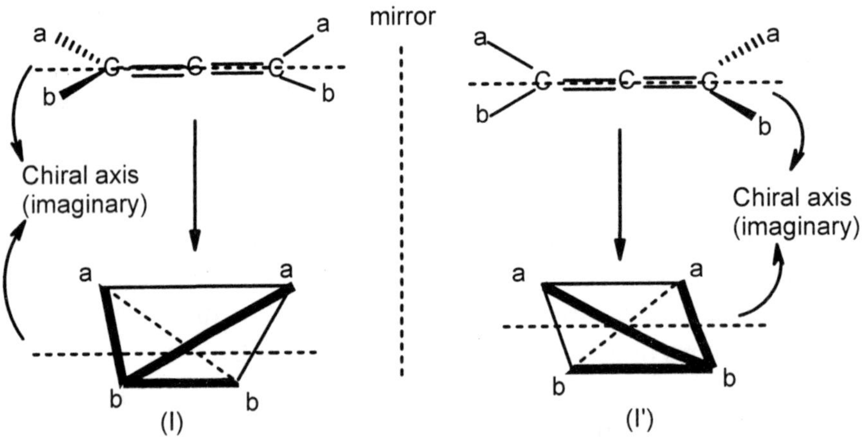

Problem 2.29 Discuss the stereochemical features of the acetolysis reaction for both *exo-* & *endo*-norbornyl brosylates.

or

On acetolysis an optically pure *exo*-norbornyl brosylate gives 100% racemic product, while an optically pure *endo*-brosylate gives 93% racemic exo-acetate. Explain.

Solution. Both *exo-* and *endo*-norbornyl brosylates (Bs= *p*-bromobenzenesulphonyl) undergo acetolysis reaction to give acetate derivatives. During such acetolysis reaction some observations are noted:

i) The acetolysis-rate for *exo*-isomer is about 350 times faster than that of *endo*-isomer.
ii) Both the isomers furnish exclusively the *exo*-acetate.

iii) An optically pure *exo*-brosylate forms 100% racemic mixture of acetates, while an optically pure *endo*-brosylate forms only a 93% racemic *exo*-acetate.

The above observations can easily be rationalized from the following arguments:

In case of *exo*-brosylate (I) the C_1–C_6 σ-bond offers an anchimeric assistance to remove –OBs group by backside attack leading to direct formation of a non-classical carbonium ion (III), while such anchimeric assistance is absent in case of *endo*-isomer (II) and it first dissociates in relatively slower rate into a classical carbocation (IV); the classical carbocation then delocalises to the non-classical one (III) which becomes the common intermediate and achiral in optical nature. Hence the nucleophile can attack at C-1 & C-2 with equal feasibility only from the *exo*-side due to cage structure. Among other arguments, Brown pointed out that the exclusive formation of *exo*-products is due to steric approach control — the concave *endo*-side of the classical norbornyl cation inhibits the *endo*-approach. He also argued that during ionization, the *endo*-substituent experiences steric hindrance to become ionized by three *endo*-hydrogens.

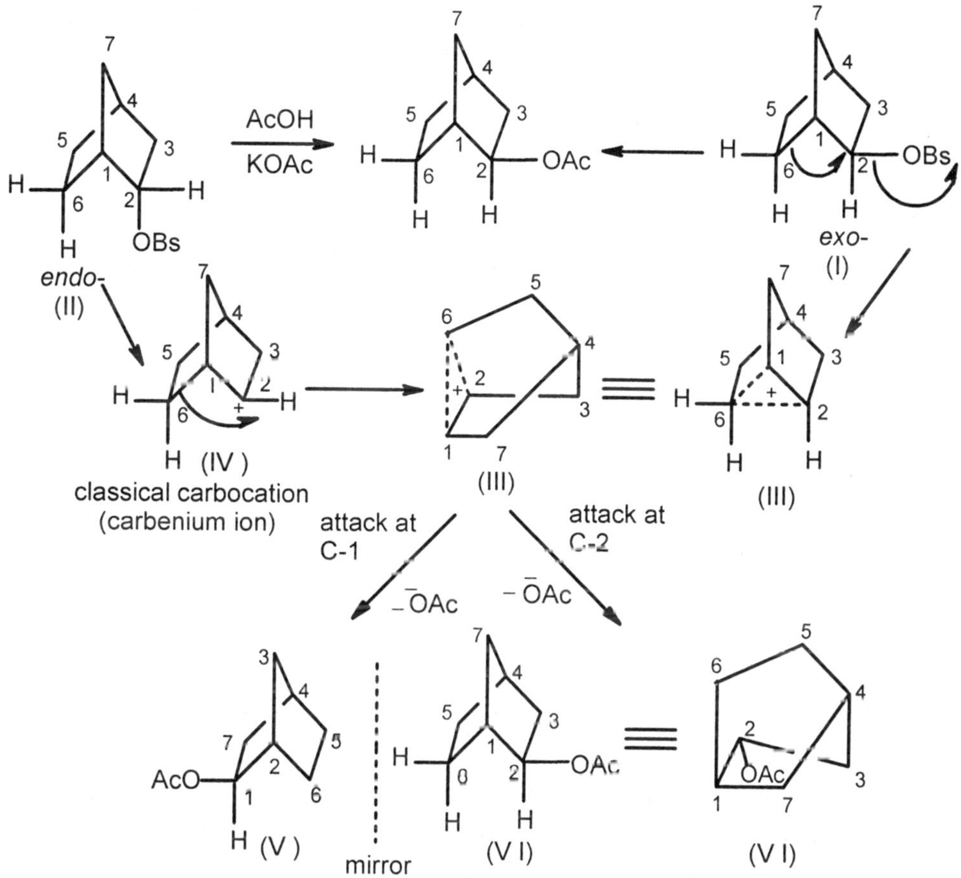

Exo-acetate (V) obtained from the attack at C-1 by acetate ion is just enantiomeric to that one (VI) obtained during attack at C-2. This explains the exclusive formation of the *exo*-isomer in completely racemic form. But in case of *endo*-isomer, S_N2 mechanism also operates to some extent that accounts for incomplete racemization (93%).

Problem 2.30 Predict the stereochemical change of acetolysis of 3-methoxy-2-bromobutane in the presence of silver acetate in acetic acid.

Solution. Acetolysis of 3-methoxy-2-bromobutane in the presence of silver acetate in acetic acid involves neighbouring group participation of lone pair of electrons as well as molecular rearrangement. The 'Br' is pushed back by –OMe group forming oxonium intermediate. The reaction scheme for the substrate is shown considering both the *threo*- and *erythro*-isomers.

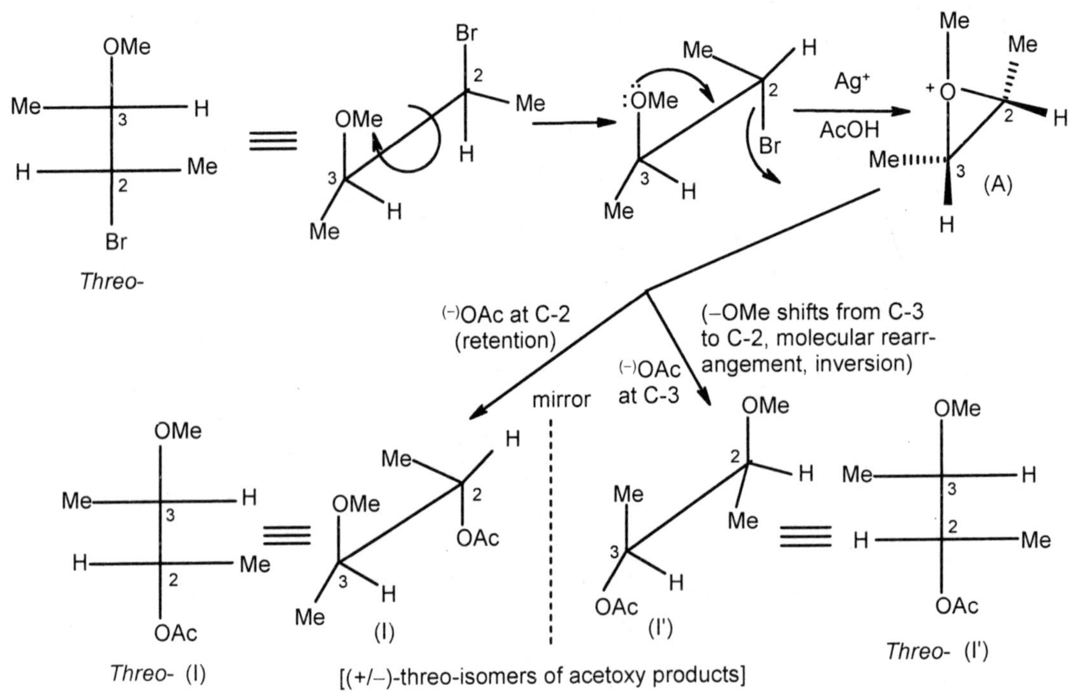

The oxonium ion (A) from the *threo*-isomer is achiral with C-2 and C-3 which are enantiotopically related, so that subsequent nucleophilic attack by $OAc^{(-)}$ results a racemic mixture (I & I' are enantiomeric pair) of *threo*-3-methoxy-2-butylacetate.

Conversely, the intermediate (B) from the *erythro*-isomer is chiral and gives a single *erythro*-product (II=II'). The stereochemistry of *erythro*-product is dependent on the nature of the starting material — if it is optically pure, the product would be optically pure and if it is racemic, the product will also be racemic.

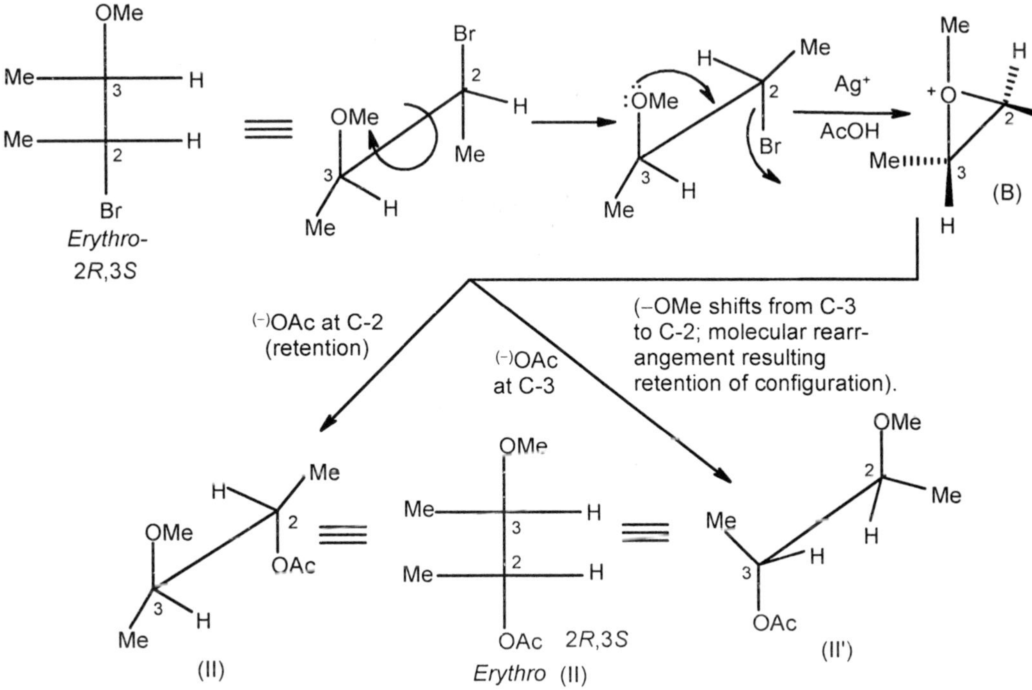

Problem 2.31 Explain the observation that 5-methylcyclohex-2-enone on treatment with lithium dimethylcuprate gives the *trans*-isomer to the extent of 98%.

Solution. 1,4-Nucleophilic addition to an enone system is governed by both the stereoelectronic factor and the steric factor. The incoming nucleophile prefers to approach the β-carbon perpendicular to the enone plane to maintain a continuous deocalization with the extended π-system. The substrate (cyclohexanone) is generally represented by half-chair conformation (A); the incoming nucleophile (reagent) can approach the reaction site in two ways — the parallel and anti-parallel approaches that may be shown in the following scheme:

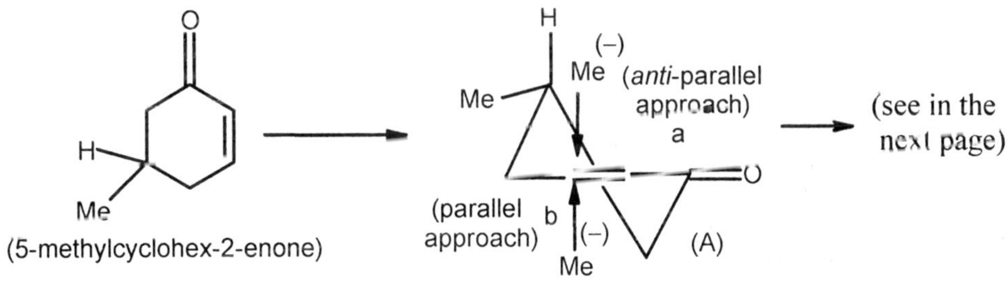

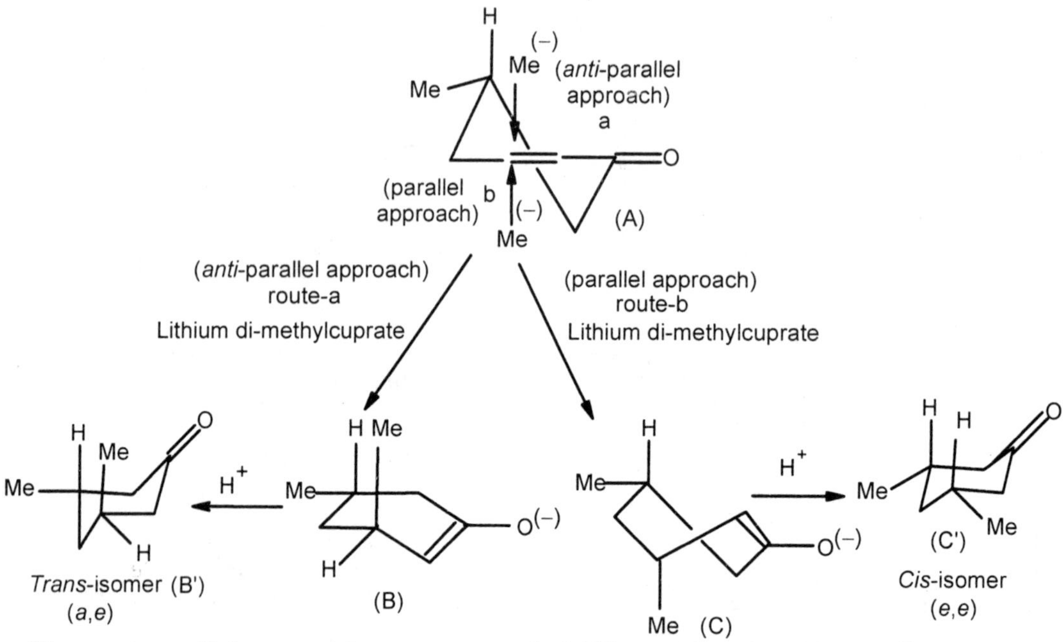

The *anti*-parallel approach generates a chair-like enolate ion (B), while the parallel approach forms a twist-boat enolate ion (C). In absence of any compelling steric effect, the former provides a more facile pathway and the product with axial Me (B′) is formed in a kinetically controlled reaction in preference to that with equatorial Me (C′). Hence, it has been observed that the *trans*-isomer (B′) formed during the reaction between lithium dimethylcuprate and 5-methylcyclohex-2-enone predominates to the extent of 98% in the product.

Problem 2.32 What would be the preferred conformation of *trans*-1,2-dibromocyclohexane? If the conformation is considered in solution, is there any change in conformation with the change of solvent polarity?

Solution. Dipole-dipole interaction has a pronounced effect on the conformation of *trans*-isomers of 1,2-dibromocyclohexanes. In these compounds, the diaxial conformer is populated sometimes in more proportion than the diequatorial one. The percentage of diaxial conformers increases across the series: Cl < Br < I.

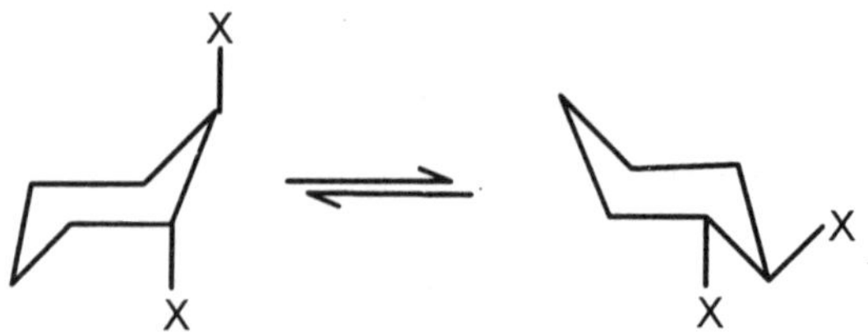

Diaxial conformer (I) Diequatorial conformer (II)

In the diaxial form (I) the two dipoles are oppositely placed ($\theta=180^0$) affecting the dipole moment nearly zero and as a result there is no electrostatic interaction between them. But in the diequatorial conformation (II) the two dipoles are disposed at a dihedral angle of 60^0 as a result of which a considerable electrostatic repulsion exists destabilizing thereby the diequatorial conformer.

For *trans*-1,2-dibromocyclohexane it has been found that % population ratio of *a,a*-conformer to *e,e*-conformer in gaseous state is 95:5 and in liquid state is 65:35. The relative population of the two conformers of *trans*-1,2-dibromocyclohexane depends on both the solvent nature and solvent polarity. In solution, the diequatorial form becomes more stabilized by increasing solvent polarity. Thus the population ratio of *a,a*- and *e,e*- conformers alters in solution and varies with varying solvents — as in the case of CCl_4 solvent the ratio has been found to be 84:16, while in more polar solvent C_6H_6, it becomes 52:48 (*i.e.* relative population of diequatorial conformer increases).

Problem 2.33 The pKa values of *cis*- and *trans*-4-*t*-butylcyclohexane carboxylic acids in 66% aqueous dimethylformamide are found to be 8.23 and 7.79 respectively. Account for the difference in acid strength.

Solution.

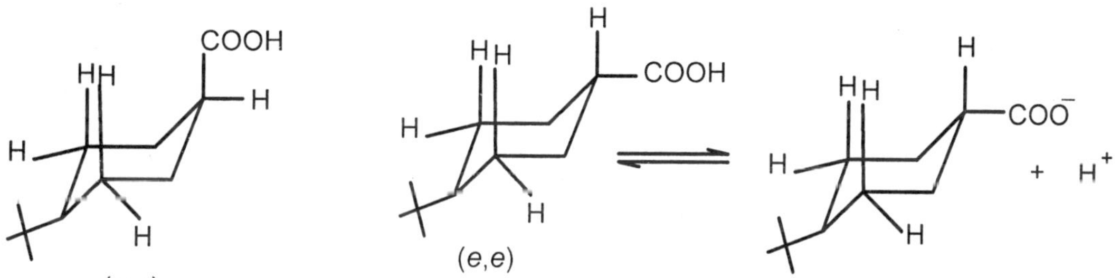

(a,e) *Cis*- pK_a 8.23

(e,e) *Trans*- pK_a 7.79

Conjugate base (A)

The conjugate base (A) of *trans*-isomer would be more stabilized as a consequence of greater solvation of the $-COO^{(-)}$ moiety being disposed at equatorial position at which the solvent molecules can approach with greater ease. On the contrary, the $-COO^{(-)}$ moiety in the respective conjugate base of *cis*-isomer is axially disposed and suffers from 1,3-diaxial interactions; due to such steric repulsion the solvent molecules solvate the anion relatively less effectively than that in the former case. That's why, *trans*-4-*t*-butylcyclohexane carboxylic acid is more acidic in comparison to its *cis*-counterpart.

Problem 2.34 Optically pure (*R*)-enantiomer of 1-phenylethanol is separately treated with (i) *p*-TsCl followed by EtOK and (ii) K followed by EtOTs. Identify the products with their stereochemical relationship.

Solution.

Case-I:

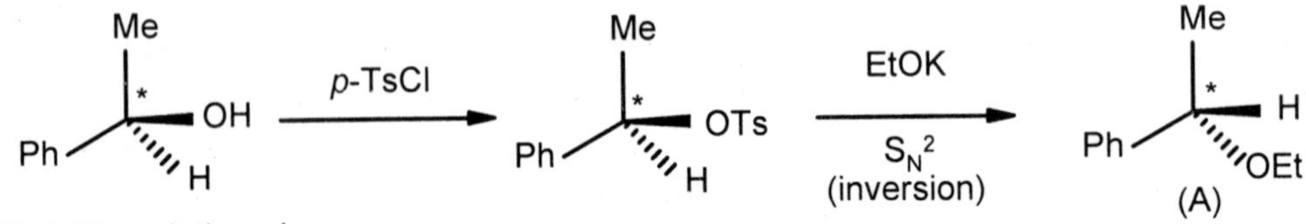

(*R*)-1-Phenylethanol Tosylatederivative (*R*) (*S*)-1-Ethoxy-1-phenylethane

During tosylation there occurs no change in the configuration of the chiral centre, but the change occurs in the second step involving a S_N2 displacement reaction resulting the product (A) with inverted configuration.

Case-II:

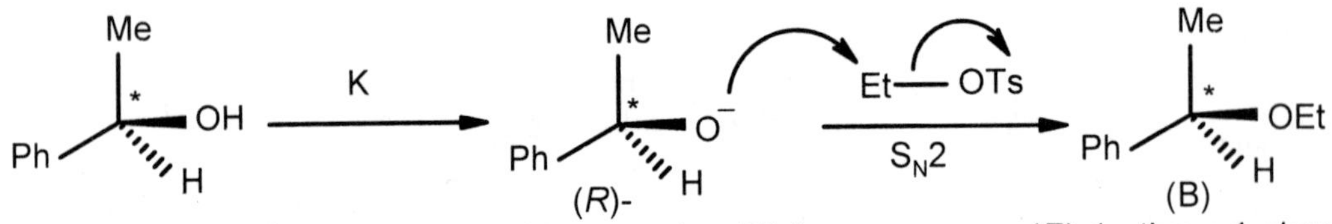

(*R*)-1-phenylethanol (*R*)- (alkoxide as nucleophile) (*R*)-1-ethoxy-1-phenyl-ethane

There is no change in configuration of the chiral centre of the product.

Problem 2.35 1,2-Diphenyl-1-propanol may be prepared by either of the two ways:

i) $LiAlH_4$ reduction of PhCH(Me)COPh .
ii) Reaction of PhCH(Me)CHO with PhMgBr.

Which method would you choose to prepare the *threo*-isomer? Give reason in favour of your answer.

Solution. The second method would be effective in preparing the *threo*-isomer of 1,2-diphenyl-1-propanol as may be rationalized from the following reaction sequence. The nucleophilic attack at the carbonyl carbon obeys Cram's rule and accordingly the products will be *erytho-* as major one from both the (*R*)- & (*S*)-isomers of PhCH(Me)COPh and *threo*-from both the (*R*)- & (*S*)-isomers of PhCH(Me)CHO.

10

Reagent: ***Pyridinium chlorochromate (PCC)***

Structure:

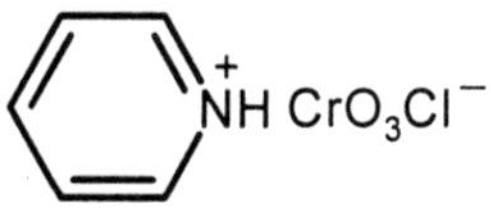

Main objective(s): Useful oxidizing agent for oxidizing alcohols.

Representative illustrations:

(i) $RCH_2CH_2OH \xrightarrow[CH_2Cl_2]{PCC} RCH_2CHO$

(ii)

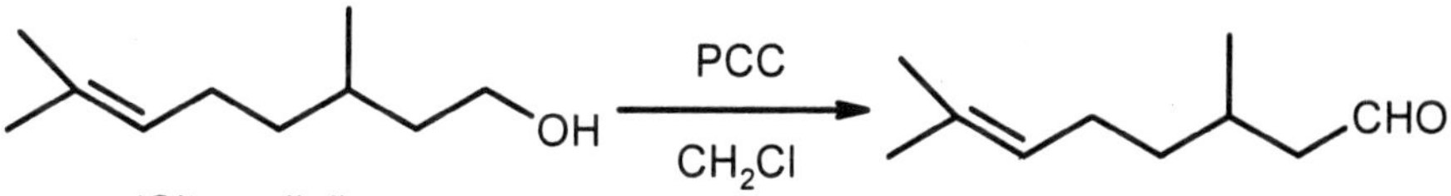

(Citronellol) (Citronellal)

(iii)

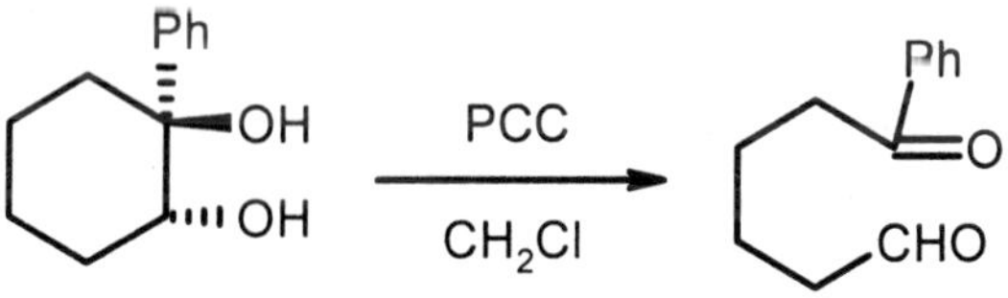

[*Syn. Commun.*, 833(1982)]

(iv)

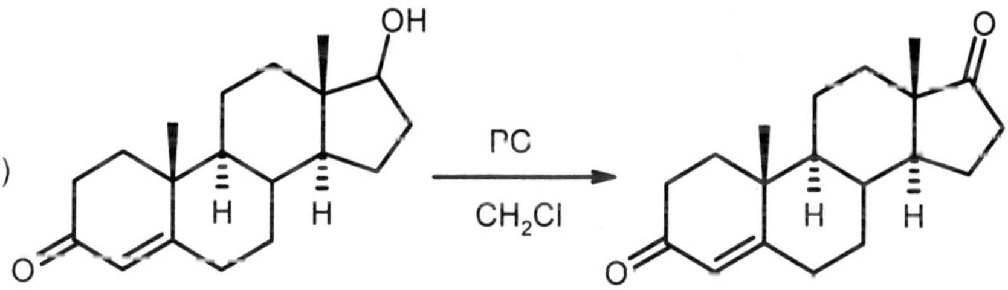

Appendix 2

pKa Values of few organic acids and bases shown in Table 1

Table 1

Compounds (acids)	pKa	Compounds (acids)	pKa
Formic acid	3.74	Benzoic acid	4.17
Acetic acid	4.75	*o*-Fluorobenzoic acid	3.27
Propionic acid	4.87	*m*- Fluorobenzoic acid	3.83
Oxalic acid	pK_1 1.27 pK_2 4.27	*p*- Fluorobenzoic acid	4.14
Malonic acid	2.67	*o*-Chlorobenzoic acid	2.94
Succinic acid	pK_1 4.21 pK_2 5.64	*m*-Chlorobenzoic acid	3.83
Glutaric acid	pK_1 4.39 pK_2 5.50	*p*-Chlorobenzoic acid	3.99
Adipic acid	pK_1 4.42 pK_2 5.41	*o*-Bromobenzoic acid	2.85
Acrylic acid	4.25	*m*-Bromobenzoic acid	3.81
Crotonic acid	4.69	*p*- Bromobenzoic acid	4.00
Isocrotonic acid	4.44	*o*-Iodobenzoic acid	2.84
Vinylacetic acid	4.35	*m*-Iodobenzoic acid	3.86
Methacrylic acid	4.43	*p*- Iodobenzoic acid	3.98
Maleic acid	pK_1 3.92 pK_2 6.23	*o*-Hydroxy benzoic acid	2.98
Fumaric acid	pK_1 3.02 pK_2 4.38	*m*-Hydroxy benzoic acid	4.08
Fluoroacetic acid	2.59	*p*-Hydroxy benzoic acid	4.58
Difluoroacetic acid	1.24	*o*-Nitrobenzoic acid	2.17
Trifluoroacetic acid	0.23	*m*-Nitrobenzoic acid	3.45
Cyclopropane carboxylic acid	4.83	*p*-Nitrobenzoic acid	3.44
Cyclobutane carboxylic acid	4.79	Phthalic acid (1,2-dicarboxylic acid)	pK_1 2.96 pK_2 5.25
Cyclopentane carboxylic acid	4.99	Isophthalic acid (1,3-dicarboxylic acid)	pK_1 3.54 pK_2 4.60

Table 2 (continued)

Compounds (acids)	pKa	Compounds (acids)	pKa
Cyclohexane carboxylic acid	4.90	Terphthalic acid (1,4-dicarboxylic acid)	pK_1 3.51 pK_2 4.82
Cyclohexane dicarboxylic acid		Methoxyacetic acid	3.53
		Acetoxyacetic acid	2.91
-1,2-dicarboxylic		Phenol	10.02
acid (*cis*)	pK_1 4.34	*o*-Chlorophenol	8.48
	pK_2 6.76	*m*-Chlorophenol	9.02
(*trans*)	pK_1 4.18	*p*-Chlorophenol	9.38
	pK_2 5.93	*o*-Nitrophenol	7.21
-1,3-dicarboxylic		*m*-Nitrophenol	8.40
acid (*cis*)	pK_1 4.10	*p*-Nitrophenol	7.14
	pK_2 5.46	2,4-Dinitrophenol	4.11
(*trans*)	pK_1 4.31	2,5-Dinitrophenol	5.22
	pK_2 5.73	2,6-Dinitrophenol	5.23
-1,4-dicarboxylic		2,4,6-Trinitrophenol	0.96
acid (*cis*)	pK_1 4.44		
	pK_2 5.79		
(*trans*)	pK_1 4.18		
	pK_2 5.42		

Compounds (bases)	pKa	Compounds (bases)	pKa
Ammonia	9.21	*o*-Chloroaniline	2.62
Methyl amine	10.62	*m*-Chloroaniline	3.32
Ethyl amine	10.63	*p*-Chloroaniline	3.81
Propyl amine	10.53	*o*-Nitroaniline	–0.28
Isopropyl amine	10.63	*m*-Nitroaniline	2.45
Dimethyl amine	10.87	*p*-Nitroaniline	0.98
Trimethyl amine	9.79	*m*-Aminobenzene sulphonic acid	3.80
Ammonia	9.21	*p*-Aminobenzene sulphonic acid	3.32
1-Ethylpiperidine	10.41	*o*-Phenylenediamine	4.47

Table 2 (continued)

Compounds (bases)	pKa	Compounds (bases)	pKa
Bis-2-hydroxyethyl-amine	8.80	*p*- Phenylenediamine	6.08
Tris-2-hydroxy ethyl-amine	7.65	Pyridine	5.17
Cis-1-amino-2-hydroxy-cyclohexane	9.72	2-Methylpyridine	5.97
Trans-1-amino-2-hydroxy cyclohexane	9.63	3- Methylpyridine	5.68
Hydrazine	–2.95	4- Methylpyridine	6.02
Methylhydrazine	7.87	4-Amino-pyridine-*N*-oxide	3.69
Dimethylhydrazine	7.21	4-(*N*-Methylamino)-pyridine-*N*-oxide	3.85
Acetamide	–1.40	4-(*N,N*-Dimethylamino)-pyridine-*N*-oxide	3.88
Benzamide	–1.85	Quinoline	4.80
Guanidine	13.59	2-Methylquinoline	5.42
Phenylguanidine	10.77	3-Methylquinoline	5.14
s-Diphenylguanidine	10.12	4-Methylquinoline	5.20
Triphenylguanidine	9.10	Isoquinoline	5.46
Nitroguanidine	–0.93	Thiazole	2.5
Aniline	4.62	Oxazole	0.8
N-methylaniline	4.85	Imidazole	7.1
N, N-dimethylaniline	5.06	Pyrazole	2.53
o-Toluidines	4.42	Pyrimidine	1.30
m-Toluidines	4.73	2-Hydroxypyridine	2.24
p-Toluidines	5.08	4-Hydroxypyridine	1.85
m- Phenylenediamine	4.88		

Note: pK_a for base (B) actually measures the acid strength of its conjugate acid (BH^+). Thus stronger the conjugate acid (BH^+), weaker is the base (B) — i.e. lower value of pK_a (for BH^+) indicates weaker base or vice versa. pK_a for a base (B) actually stands for pK_{BH^+}.

[Ref. *Handbook of Structural Analysis*, Yasuhide Yukawa (Editor), W. A. Benjamin, Inc., 1965].

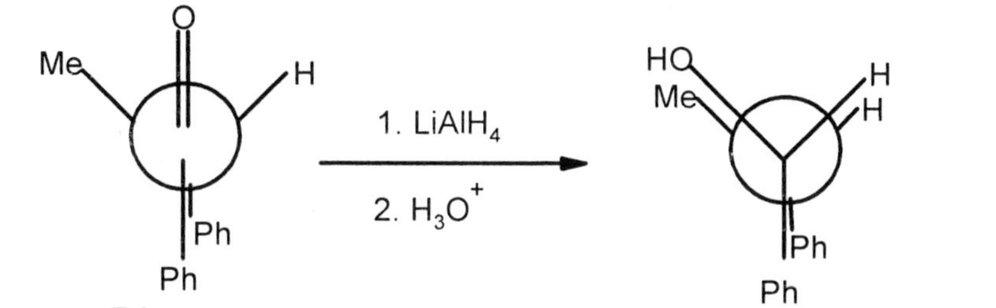

R-isomer
(preferred conformation)

(*R*,*R*)-*erythro*-isomer

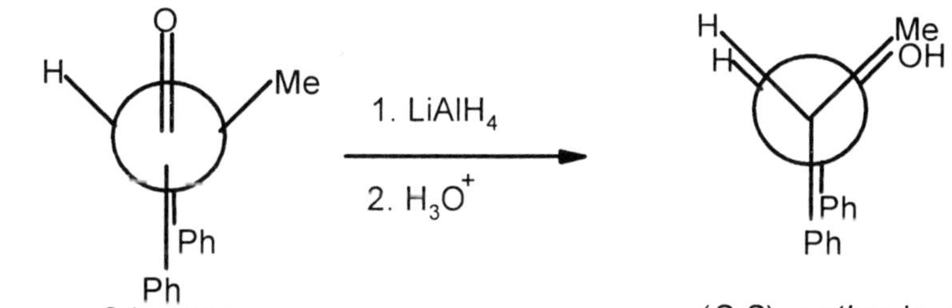

S-isomer
(preferred conformation)

(*S*,*S*)-*erythro*-isomer

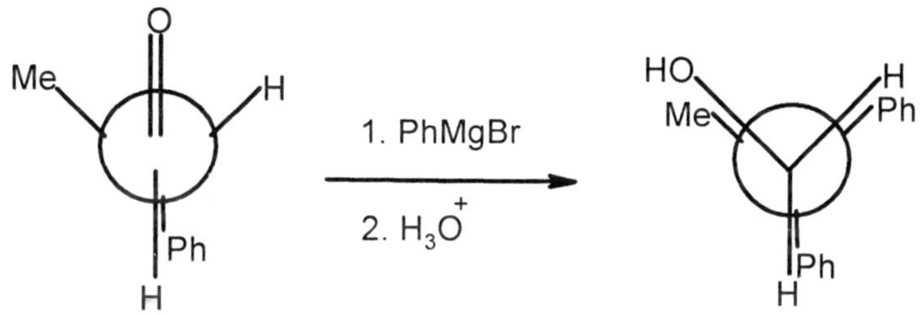

R-isomer
(preferred conformation)

(*S*,*R*)-*threo*-isomer

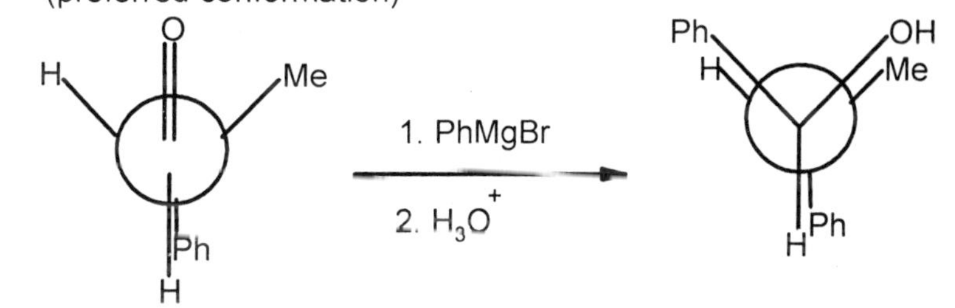

S-isomer
(preferred conformation)

(*R*,*S*)-*threo*-isomer

Problem 2.36 How do you establish the fact that the mechanistic pathway of the S_N2 reaction is governed by stereochemical control?

Solution. A S_N2 reaction is a one step process without intermediate resulting in a Walden inversion and such reaction is governed by stereoelectronic control. Nucleophiles have to approach the reaction centre of the substrate from a 180^0 angle opposite to the leaving group. In the transition state (I) the central carbon is considered to be sp^2 hybridized; one lobe of the remaining *p*-orbital overlaps with the nucleophile and the other with the leaving group.

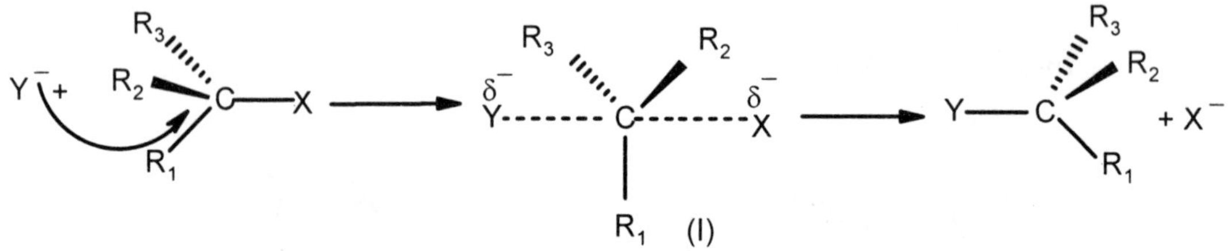

Electronic effects that force the transition state to adopt a specific stereochemistry (linear arrangement of the incoming and leaving groups), therefore, control the mechanism of this reaction. In 1970, Eschenmoser *et al.* offered an experimental proof to establish that in the S_N2 transition state the incoming nucleophile is to be aligned collinearly.

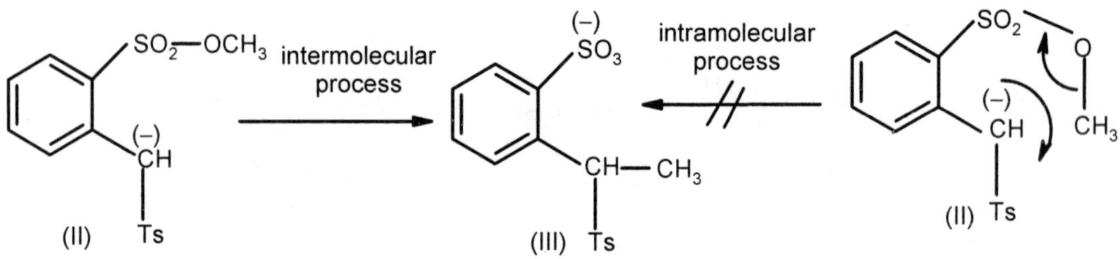

They showed that the anion (II) yielded the expected product (III) *via* an intermolecular process rather than the "formally appealing" intramolecular process; although the intramolecular path may be of thermodynamically favoured, it does not occur because of the fact that during intramolecular fashion the proper alignment of the nucleophile with the substrate cannot be attained in the transition state.

Problem 2.37 The hydroxy-enone (A) fails to undergo cyclization under basic conditions, but on acid-catalyzed cyclisation it gives the corresponding furanone derivative. Account for the observation.

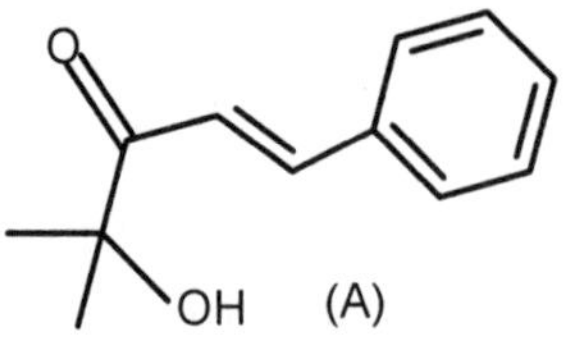
(A)

Solution. According to Baldwin's rule for ring closure in trigonal systems, 5-Endo-Trig process is indeed a geometrically disfavoured reaction and the lack of ring closure is a consequence of a kinetic rather than a thermodynamic barrier. That is why the hydroxy-enone (A) fails to undergo cyclization under basic condition —

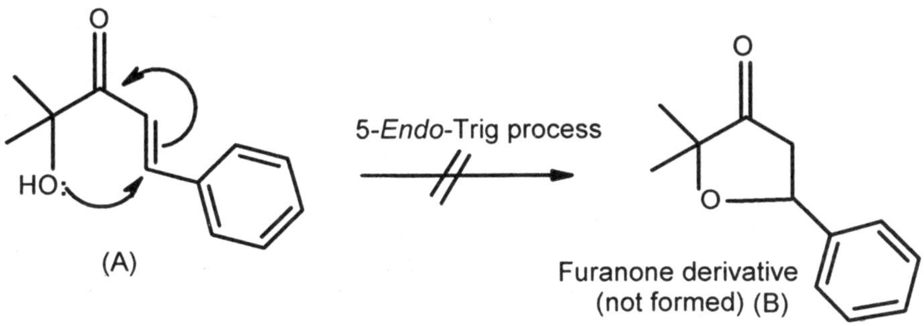

But under acid-catalyzed condition cyclizaion of (A) takes place to give the corresponding furanone derivatives (B). The reaction involves protonation to the enone oxygen in the first step followed by enolization thereby reducing the rotational barrier around the enone double bond, which allows the cyclization. Hence the cyclization process becomes 5-Exo-Trig, which is an allowed path for such ring closure reaction as per so called Baldwin's rule.

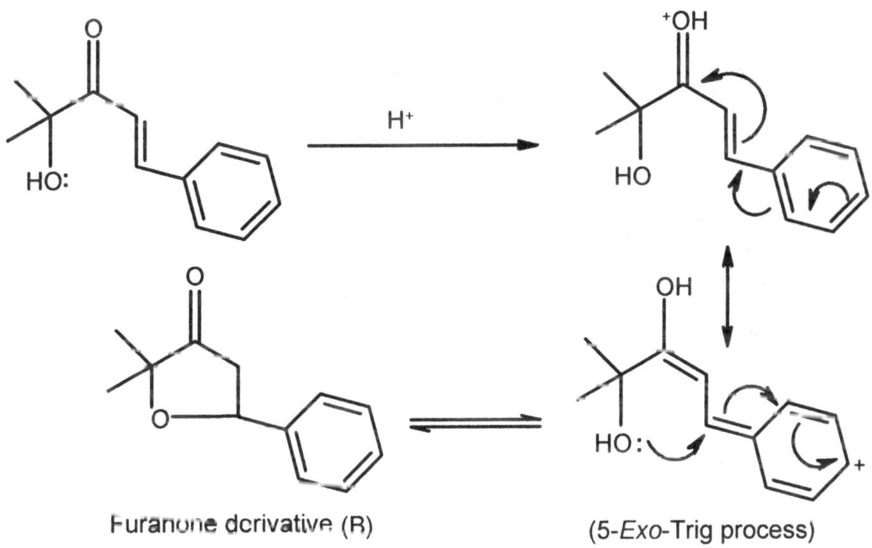

Problem 2.38 The ratio of relative rate of chromic acid oxidation of the compounds (I) and (II) is about 62.7/1.87. Comment on this observation.

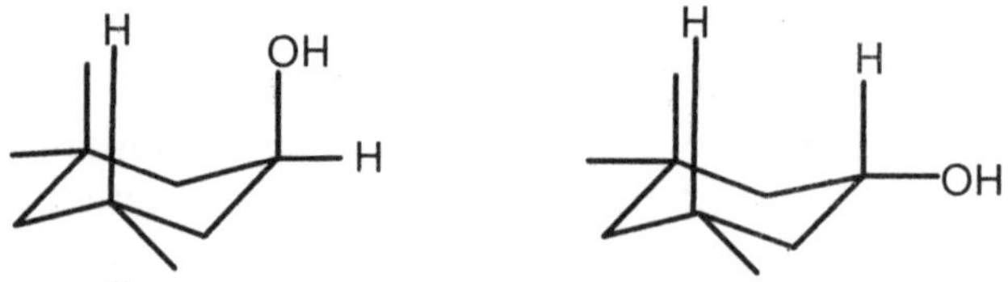

(I) (II)

Solution. The chromic acid oxidation of a secondary alcohol to a ketone is believed to occur in two steps. The first step is the rapid formation of chromic ester and the second step (that is the rate-determining step) involves the decomposition of the rapidly formed chromate ester into the ketone, a chromite ion and a proton. Steric acceleration or assistance is responsible here for the higher rate of oxidation for *axial*-alcohol.

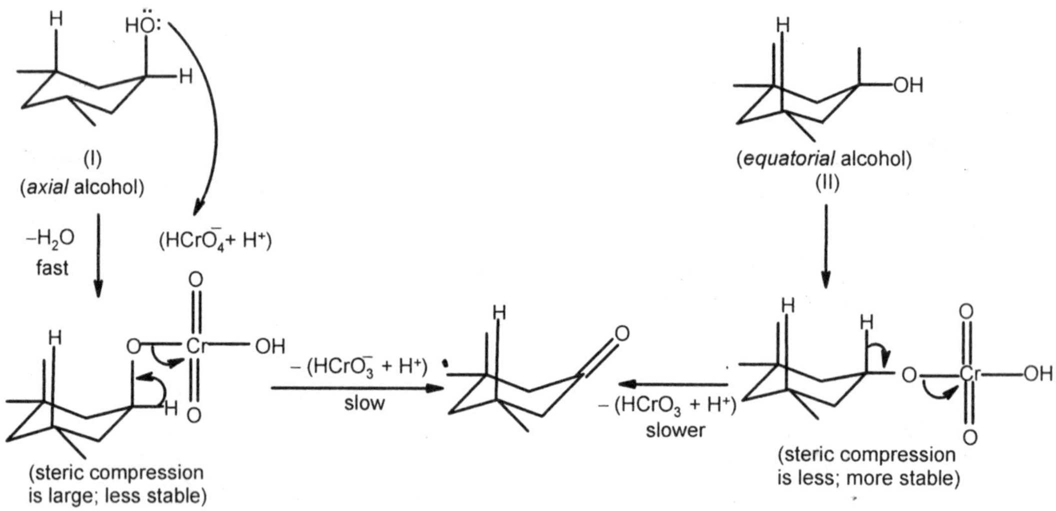

The *axially* disposed –OCrO$_3$H moiety suffers from greater steric repulsion in comparison to the *equatorially* disposed one, such that the difference in free energies between the *axial* and *equatorial* chromate esters in the ground states becomes greater than that in the transition state. Thermodynamically it may be represented as:

$$\Delta G^{\neq}_{\text{axial}} < \Delta G^{\neq}_{\text{equatorial}}$$

That is, the axial isomer with its higher ground state energy would react at a faster rate. Thus it is a typical case of steric assistance — the relief of steric strain in the transition state in going from chromate ester to ketone (product) is more pronounced in the case of *axial* isomer, affecting its greater reactivity.

Problem 2.39 Solvolysis rate of the *cis*-isomer of 4-*tert*-butylcyclohexyl tosylate is greater than that of the *trans*-isomer. Explain.

Solution. *Cis*-isomer of 4-*tert*-butylcyclohexyl tosylate undergoes solvolysis reaction at a faster rate than its *trans*-counterpart ($K_{cis}/K_{trans} \sim 4$); steric interaction plays a vital role in accelerating the reaction rate for the *cis*-isomer.

The reaction is assumed to proceed through S_N1 mechanism. The carbocation intermediate formed for both the isomers would be the same having approximately identical energy-content.

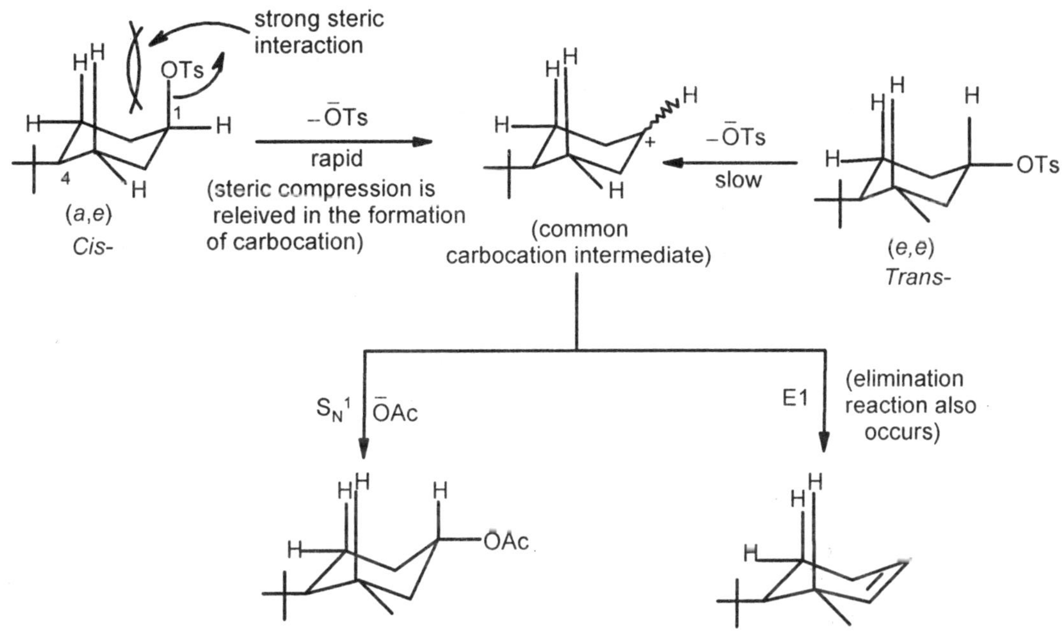

Due to the presence of strong steric repulsion between the *axially* disposed –OTs group and *syn-axial* H atoms at C-3 & C-5, the *cis*-isomer has a higher ground state energy than that of the *trans*-isomer. Accordingly, the energy to be supplied to cross the energy barrier for the *cis*-isomer would be less than that for the *trans*-isomer and the former will react at a faster rate.

Problem 2.40 Explain why camphenonic acid is highly resistant to decarboxylation.

Solution. It is believed that β-ketoacids undergo decarboxylation through a six-membered transition state that eliminates CO_2 to give an enol as the primary product. The enol then tautomerizes into carbonyl compounds.

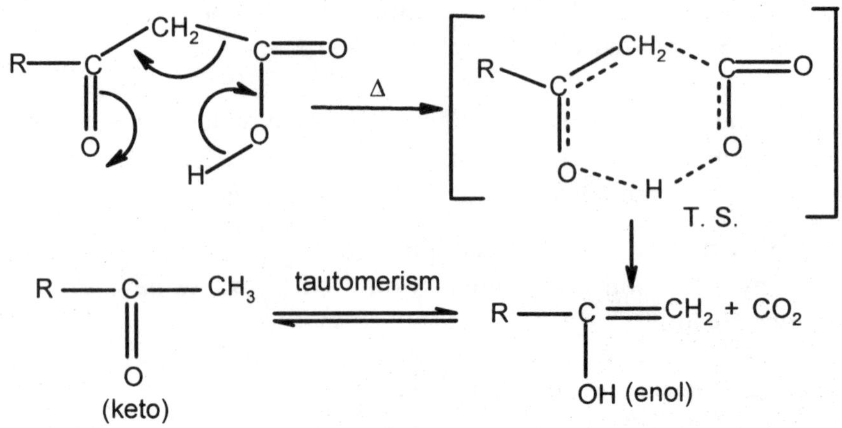

Thus during decarboxylation, an enol is necessary to be formed. Regarding decarboxylation of camphenonic acid (I) or similarly of ketopinic acid (II), such intermediatory enol-formation is restricted —

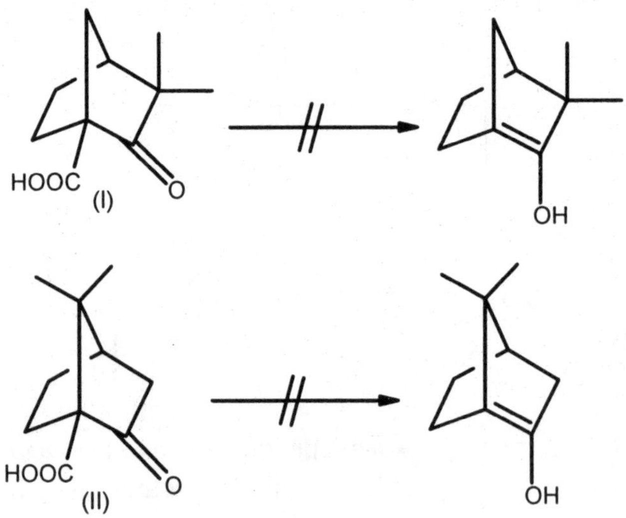

The fact that enols, which presumably would result are not stable, can be understood on the basis of Bredt's rule. The rule suggests that a double bond at the bridgehead of a bicyclic system would put strain on the system. Hence camphenonic acid as well as ketopinic acid is highly resistant to decarboxylation.

Problem 2.41 Explain the terms i) chemoselective and ii) regioselective reactions, with suitable examples.

Solution.

i) ***Chemoselective Reactions***

A reaction is said to be specifically chemoselective only when a reagent would cause a change of only a particular functionality among other functional group(s) present within the substrate molecule. For example, $NaBH_4$ can reduce >CO group in RCOCOOH and not the –COOH function. But LAH can reduce both the functionalities. Hence the former reaction is a *chemoselective reaction.*

$$\text{R-CO-COOH} \xrightarrow{NaBH_4} \text{R-CH(OH)COOH}$$

Some other examples may be cited:

a) $\text{R-CH=CH-CO-R'} \xrightarrow{\text{Reagents}} \text{R-CH=CH-CH(OH)-R'}$

R'= H or alkyl group

Reagents used: AlH_3, $NaBH_4$, $NaBH_3(OAc)$, LAH in presence of lanthanide salts (e.g. $LaCl_3$, $CeBr_3$), di-isobutylaluminiumhydride (DIBALH).

b) $\text{R-CO-CHO} \xrightarrow[\text{Zinc borohydride in THF}]{NaBH_4 \text{ in isopropanol or}} \text{R-CO-CH}_2\text{OH}$

c) $\text{R-CO-CHO} \xrightarrow[-15^0\text{C in presence } CoCl_3]{NaBH_4 \text{ in aq. EtOH at}} \text{R-CH(OH)CHO}$

ii) ***Regioselective reactions***

There are a number of substances which are capable of reacting at more than one centres (polydent molecules), but react at one centre with a higher rate than those at the others. Such reactions are called *regioselective reactions* and hence regioselectivity of the substances.

For examples:

a) $\text{R-CH=CH}_2 \xrightarrow[\text{ii) } H_2O_2/OH^{(-)}]{\text{i) } B_2H_6} \text{R-CH}_2\text{-CH}_2\text{OH}$ (regioselective product) not R-CH(OH)CH_3

Hydroboration oxidation

[-OH group would attach regioselectively on the double bonded carbon atom containing maximum number of H atoms]

b)

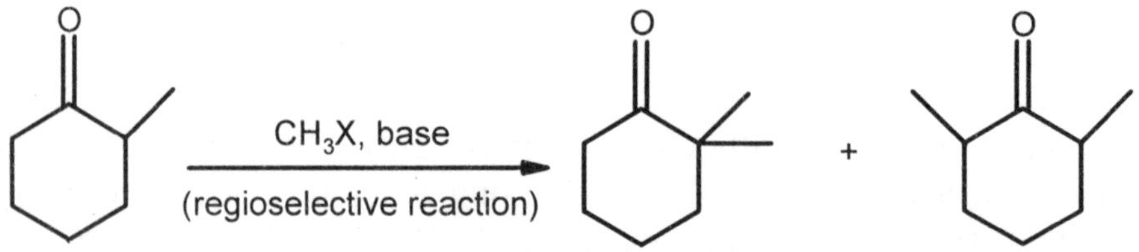

α-Methyl cyclohexanone α,α- Dimethyl cyclohexanone (major) α,α'- Dimethyl cyclohexanone(minor)

c) Addition of HX (halogen acids) to unsymmetrical alkenes (*c.f.* Markownikoff's rule)

Problem 2.42 Account for: *Cis*-1,2-dimethylcyclohexane is optically inactive at room temperature even though the molecule lacks any element of symmetry.

Solution.

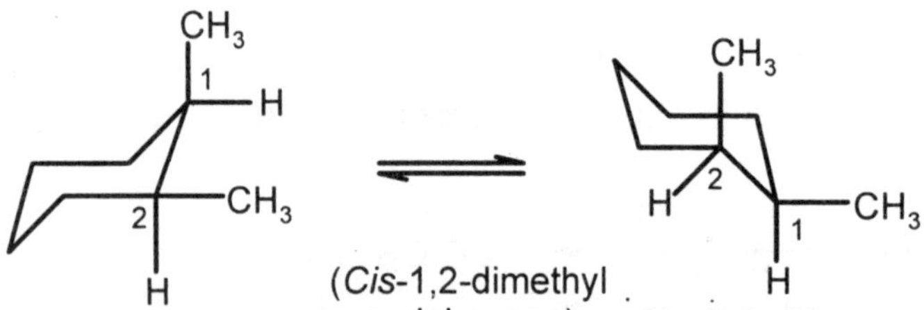

(1*a*,2*e*) (I) (*Cis*-1,2-dimethyl cyclohexane) (1*e*,2*a*) (II)

(I) and (II) are the two conformers of *cis*-1,2-dimethylcyclohexane, which are enantiomers of each other. Since they are identical in energy-content, their populations are equal and consequently this *cis*-isomer becomes optically inactive by external compensation. They can't be resolved as the two enantiomers are readily interconvertible — each enantiomer undergoes very rapid *auto-racemization.*

Problem 2.43 What do you mean by Steric approach control (SAC) and Product development control (PDC)? Taking a particular example show how they affect the stereochemistry of reaction product(s)?

Solution. Steric approach control (SAC) is a purely steric postulate that suggests competitive attack of a reacting species from a favoured (unhindered) and unfavoured (hindered) side, while Product development control (PDC) originates out of thermodynamic consideration involving the relative stability of the products.

For unhindered ketones, the attack of hydride ion on the two sides of >C=O group in a cyclohexanone is about equally favoured, but the more stable equatorial alcohol is favoured predominantly as a result of PDC.

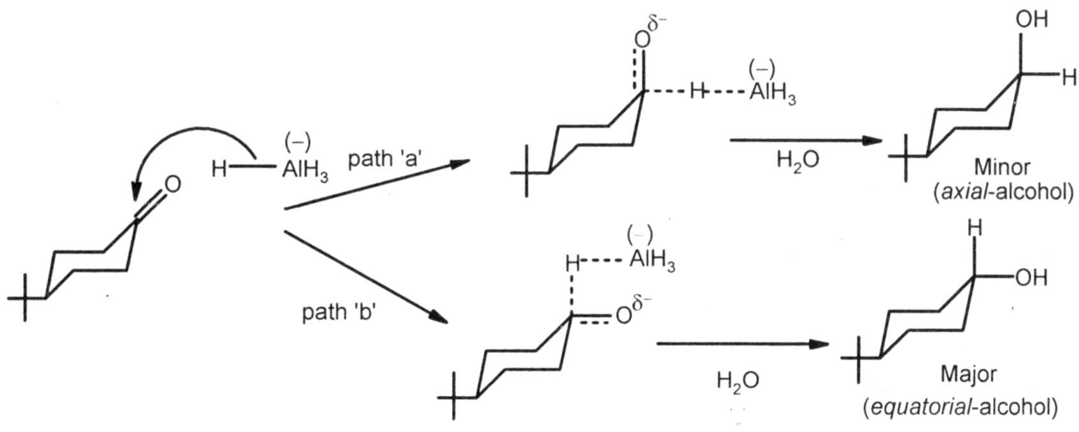

On the other hand, for hindered ketones SAC becomes more important and the axial attack, which is sterically hindered is considerably reduced — forming, thereby, the axial alcohol in large excess.

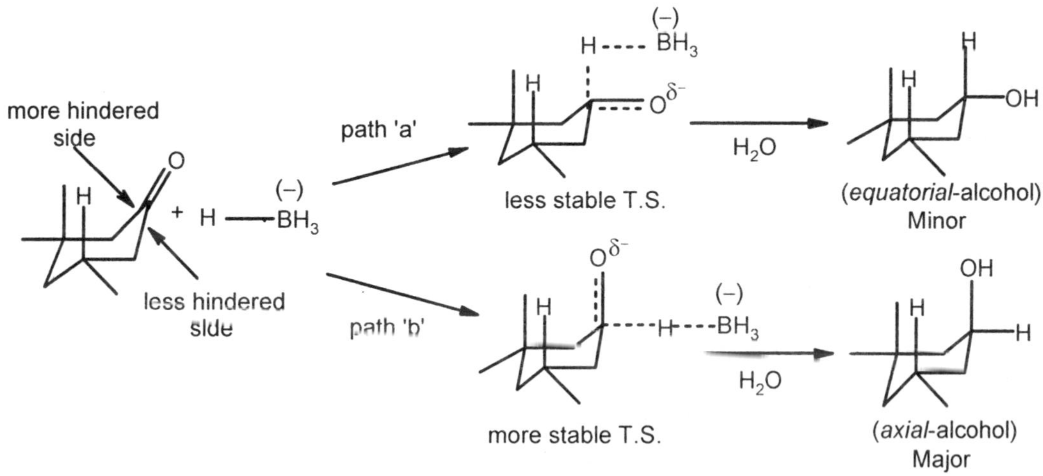

Problem 2.44 Define stereospecific and stereoselective reactions with proper examples.

Solution. A stereoselective reaction is one where one diastereoisomer of the product is formed in excess over the other. But, a stereospecific reaction is one where stereochemically different isomers give stereochemically different products. All stereospecific reactions are necessary to become stereoselective, while the reverse is not true.

Addition of singlet carbene to olefinic double bond is an example of stereospecific reaction —

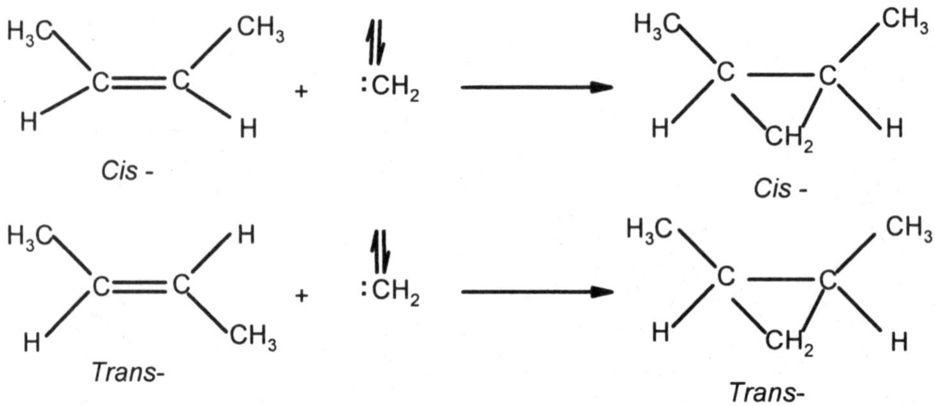

On the other hand, reduction of 4-*tert*-butylcyclohexanone by LAH follows stereoselective path; a mixture of *equatorial* & *axial* alcohols are formed of which the *equatorial* is more preferred and exists as major one.

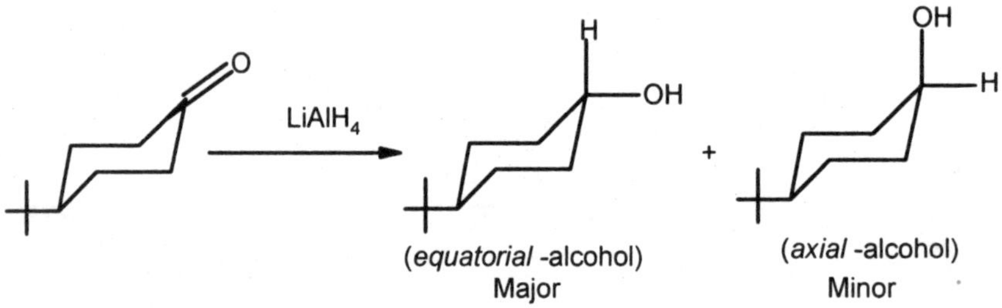

Problem 2.45 Cite examples of two organic molecules whose optical activity arises out of molecular helicity.

Solution. A helix is inherently chiral and non-superimposable. Assignment of chiral designation of helical molecules is represented by the terms *P* and *M*. Helicity (i.e. optical isomerism developed due to helical conformation) often arises at the molecular level from secondary structure and is conformational in origin. But helicity in molecules may also result from molecular overcrowding as in the case of I & II —

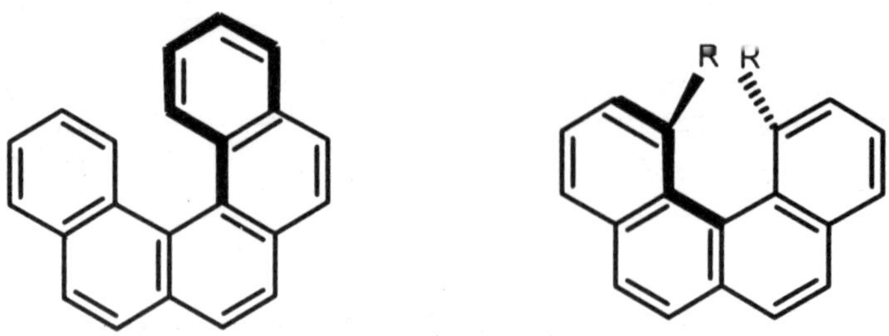

P- (Helicene) I   *M*- (Benzphenanthrene) II

Due to molecular overcrowding the ring structures assume a helical shape — the terminal rings and the substituents are in different planes and both the molecules exist d*l*-pairs (i.e. two helical enantiomers).

Problem 2.46 Write the conformational structures of the two cyclic isomers of glucose, which are responsible for the phenomenon of mutarotation.

Solution. In case of glucose, mutarotation is a consequence of configurational change at C-1 establishing equilibrium between α– and β- forms of glucose (two anomeric forms) through an acyclic aldehydic intermediate. This is basically an epimerization process brought about by an amphoteric solvent like H_2O and is catalyzed by acids and bases both.

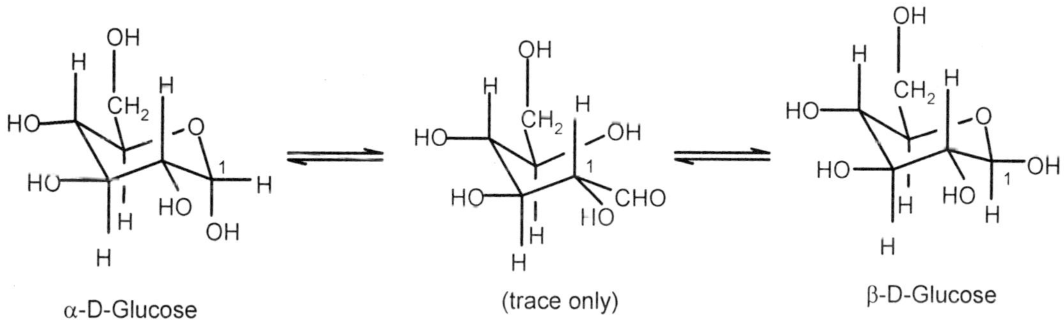

This conformational change of α-D-glucose to the β-isomer or vice-versa in solution phase is an example of *first order asymmetric transformation.*

Problem 2.47 Comment on the conformational preference in case of 1,2-dihaloehanes.

Solution.

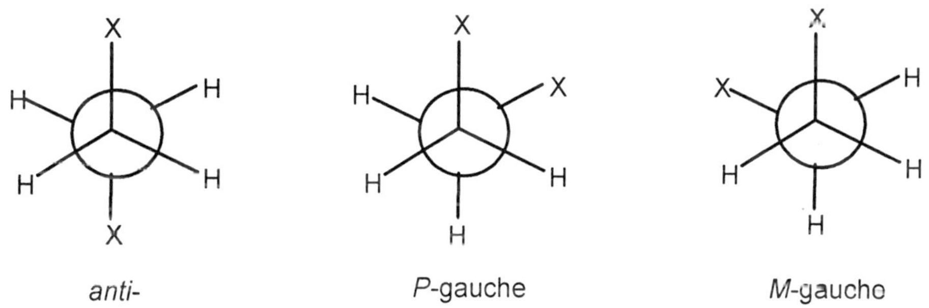

In gaseous state (at 22^{0}C) 1,2-dihaloethanes prefers to attain the *anti*-conformation — the relative stability of *anti*-conformation is a combined effect of a steric factor and an electronic interaction (dipole-dipole interaction). But very interestingly, in the liquid state or in polar solvents the ratio of *gauche* conformers (*P* and *M*) increases because of the fact that high dielectric constant of the medium decreases the electrostatic repulsion between the halogen atoms. Thus in case of 1,2-dichloroethane, the both types of conformers (*anti* and *gauche*) are almost equally populated in the liquid state and/or in polar solvent.

Problem 2.48 What do you mean to say '*anomeric effect*'?

Solution. Let us consider the axial and equatorial isomers of 2-methoxy tetrahydropyran:

From the first sight it is apparent that the equatorial-isomer having anomeric substituent at equatorial position should be more stable than the axial-isomer; but in practice the reverse is the observed phenomenon. This is usually observed when the anomeric substituents are electronegative having non-bonding electron-pair(s). Such preference of electronegative substituents for the axial position at the anomeric carbon is known as ***anomeric effect***. Electronic repulsion between non-bonding electrons on the ring oxygen atom and those on the equatorially oriented methoxyl oxygen is greater in magnitude than that developed in the axial isomer as they are farther apart in the latter geometry.

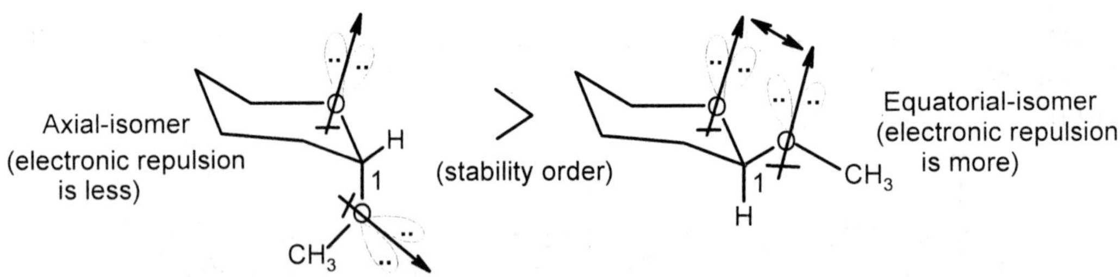

Due to such anomeric effect α-anomer of methyl D-glucopyranoside is more stable than the β-anomer.

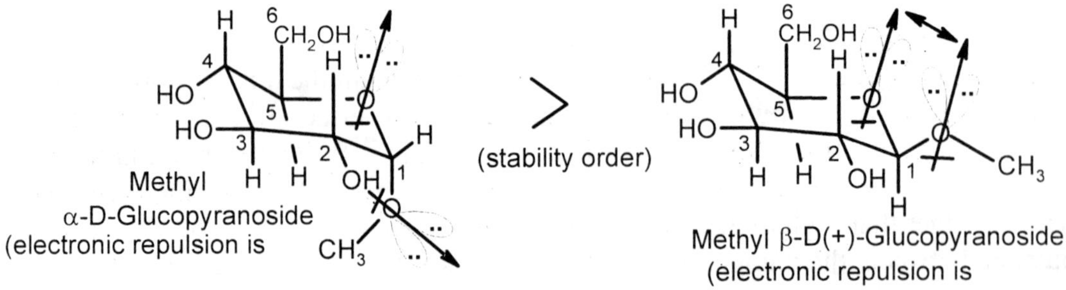

Problem 2.49 Comment on the relative stability between the following pairs:

a) *Exo*-and *endo*-isomers of 2-norborneol

b) *Exo*- and *endo*-isomers of borneol.

Solution.

a)

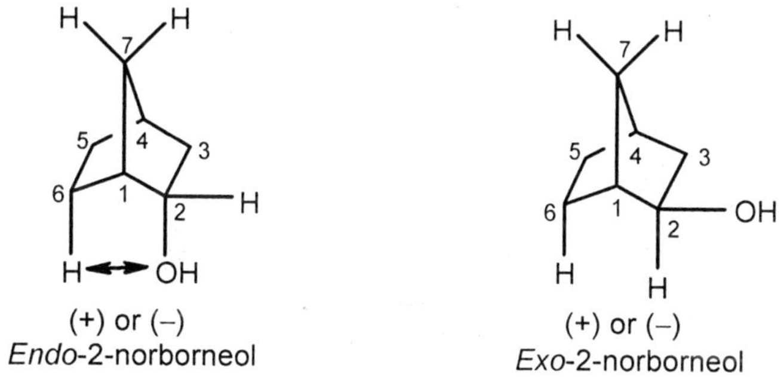

Exo-2-norborneol is more stable than the *endo*-isomer in an equilibrium mixture of two isomers; the former predominates over the latter to the extent of 80%. The destabilization of the *endo*-isomer is a consequence of the phenomenon that the –OH functionality (at C-2) undergoes an unfavourable interaction with the *endo*-hydrogen at C-6.

b) But the situation becomes reverse in the case of 2-borneol in which the C-7 atom holds a *gem*-dimethyl group.

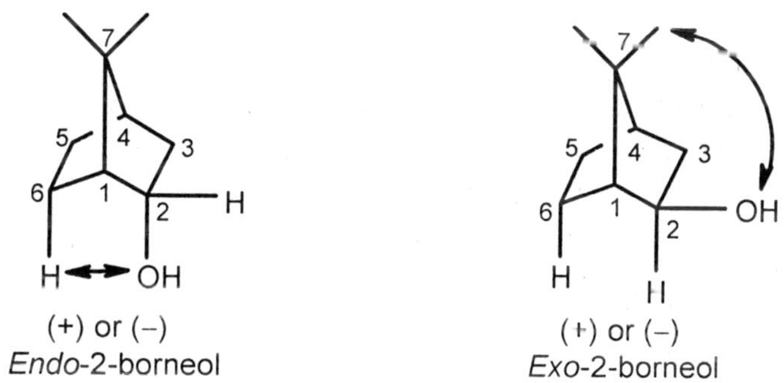

The steric interaction between the overhanging C_7-Me and the *exo*-OH at C-2 in the *exo*-2-borneol (i.e. isoborneol) is greater than that between the *endo*-OH (at C-2) and the *endo*-hydrogen at C-6 in 2-borneol. Hence the *exo*-isomer becomes more energetic and is less stable. In an equilibrium mixture of the two isomers, the *endo*-isomer predominates over the other to the extent of 71%.

Problem 2.50 Explain the term 'memory effect' with a suitable example.

Solution.

Let us consider deamination of *syn*- and *anti*-2-bornenyl-7-carbinyl amines I & II. The *syn*-isomer on deamination followed by concomitant (accompanying) hydrolysis forms a bicyclic allylic alcohol (IV) in greater amount, while the tricyclic saturated alcohol (V) is obtained in greater proportion from the *anti*-isomer.

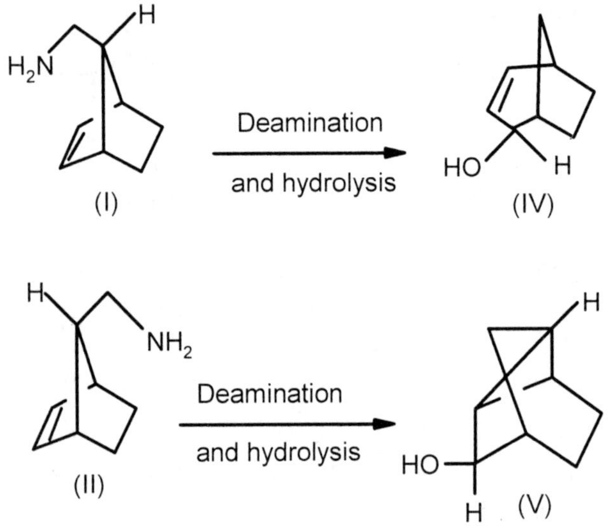

The formation of the respective product may be explained assuming a double rearrangement mechanism. Both the isomers I and II are supposed to furnish the same rearranged carbocation remaining as ion pairs (IIIa and IIIb). The carbocation then can undergo two reactions — i) a second 1,2-shift with concomitant solvolysis to give the bicyclic alcohol (IV) and (ii) participation of the C=C π-electrons to form a σ-bond giving the tricyclic alcohol (V).

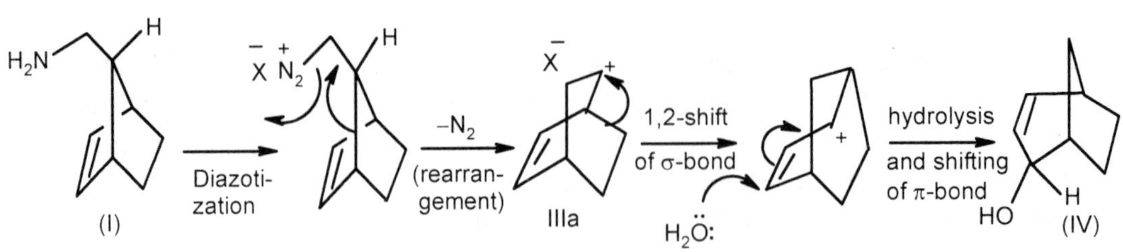

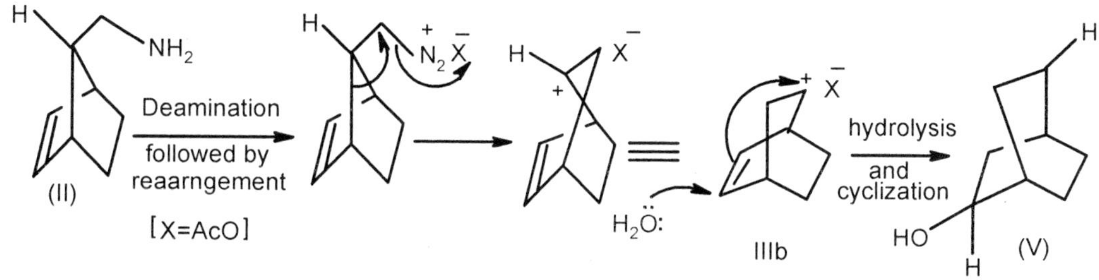

Hence the product ratio depends upon the stereochemistry of the starting amines — the *syn*-isomer (I) and the *anti*-isomer (II) give predominantly (IV) and (V) respectively, a phenomenon known as '*memory effect*'[1].

The plausible explanation for this phenomenon that appears to be more acceptable is Collins' proposition[2]. According to this proposition, the two carbocations (IIIa & IIIb) are not exactly equivalent; they exist as two ion-pairs such that the *syn*- and the *anti*-sides are unequally blocked by the counter anions (X^-) leading to the formation of (IV) and (V) in an unequal amount from each of the ion-pairs.

Problem 2.51 (±)-Stilbene dichloride gives *Z*-chlorostilbene on heating 200^0 with pyridine. But the *meso*-isomer does not react under these conditions. Give stereochemical explanation for these observations.

Solution. *Anti*-elimination is quite facile. (±)-Stilbene on heating at 200^0 in presence of pyridine can easily undergo *anti*-dehydrochlorination to form *trans*-chlorostilbene (as per nomenclature *Z*-chlorostilbene). But the *meso*-isomer does not undergo the reaction. The difference in reactivity is explained on the basis of *'cis-effect'*. This destabilizing effect is absent in the transition state of the former isomer, but severely present in that of the latter. The *'cis-effect'* arises due to eclipsing of two phenyl groups.

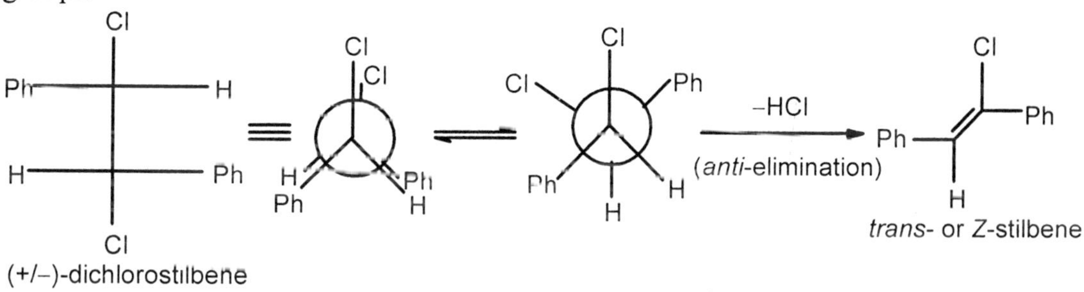

1. Berson, J. A., *Angew. Chem. Int. Edn. Engl.*, **7**, 779 (1968).
2. Collins, C. J., *Quart. Rev. Chem. Soc.*, **4**, 251(1975).

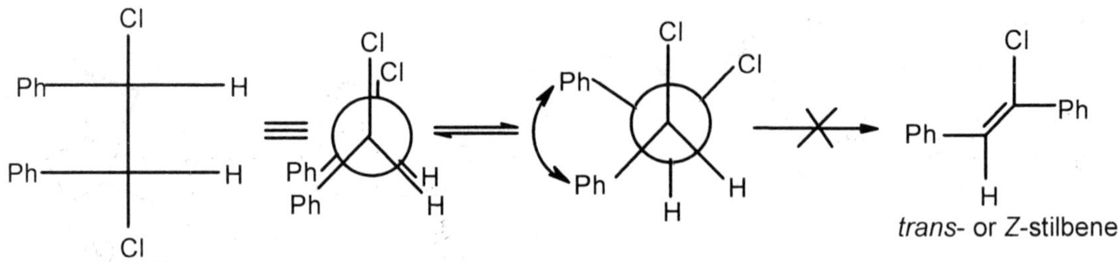

meso-dichlorostilbene

trans- or *Z*-stilbene

Due to the presence of severe '*cis-effect*' in the transition state of the *meso*-isomer, no reaction can take place.

Problem 2.52 The compound **A**, whose absolute configuration is shown below, shows weak positive Cotton effect in methanol but strong negative Cotton effect in octane. Comment on the conformational status.

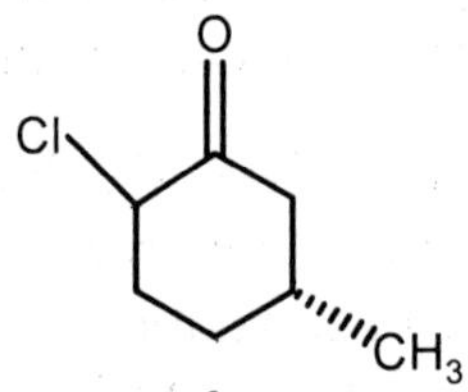

A

Solution. Negative Cotton effect for compound **A** in octane (*i.e.* nonpolar solvent) solution suggests diaxial conformation of the *trans*-configured molecule. In nonpolar solvent (having much lower dielectric constant) the halogen attains the axial disposition to minimize its dipole repulsion with that of the carbonyl group.

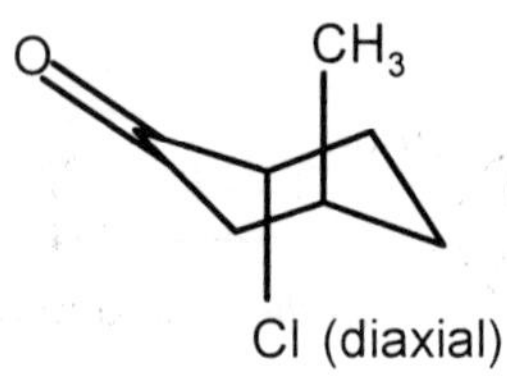

(conformation of *trans*-2-chloro-5-methylcyclohexanone)

But when the rotatory-dispersion curve is measured in methanol, the Cotton effect becomes positive. This indicates that the *trans*-isomer suffers a conformational change from diaxial to diequatorial one, presumably because of the fact that in methanol (polar, having much higher dielectric constant) the dipole repulsion between the carbonyl group and the adjacent equatorial halogen is not as serious as in octane.

Supplementary Problems

SP-2.1 What do you mean to say conformation? How it differs from configuration?

SP-2.2 Assign *R*/*S* configuration to the following molecules:

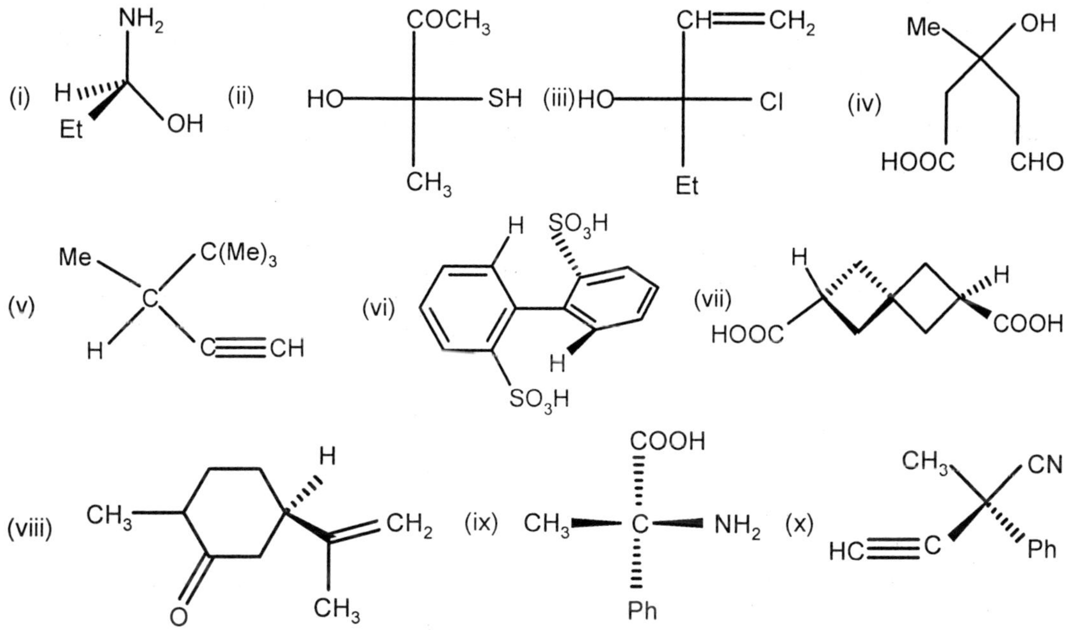

SP-2.3 What are 'skew' conformations? Are all the '*skew*' conformations to be termed as '*gauche*' conformation?

SP-2.4 Draw the potential energy diagram of *n*-butane as a function of dihedral angle and explain the implication of such a curve.

SP-2.5 The following conformer of butane is chiral, then why is butane optically inactive?

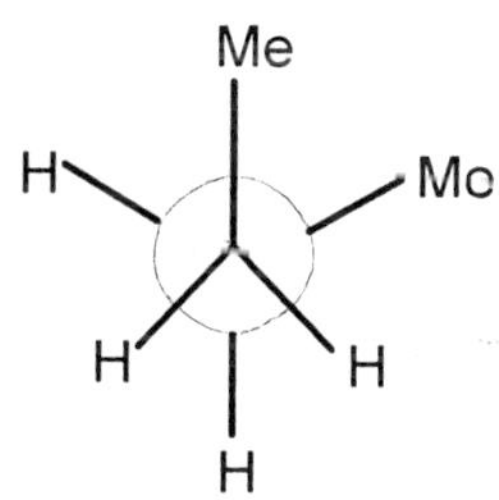

SP-2.6 What is *gauche-butane* interaction? Draw the conformers of 2-methylbutane for rotation about C_2–C_3 bond in Newman projection formula and compare their relative stabilities.

SP-2.7 At room temperature, the *anti* conformer of butane is more stable than the *gauche* conformer by about 1 kcal/mole of energy; this results an equilibrium constant of about 4.6 favouring the *anti* isomer. Will there be any change to the value of this equilibrium constant as the temperature is raised? Give reasons.

SP-2.8 Deduce relationship between the members of the each pairs:

(i)
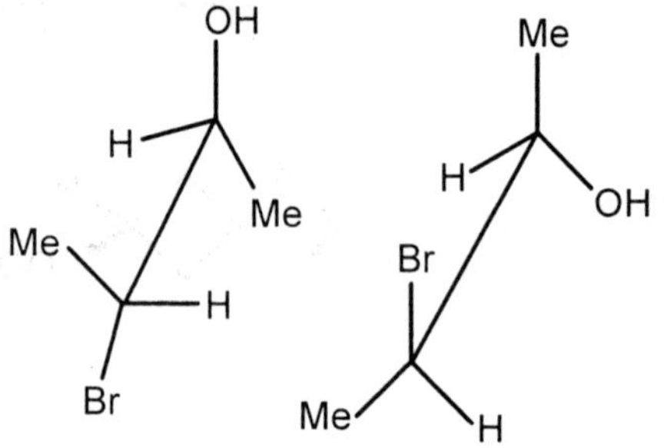

(ii)
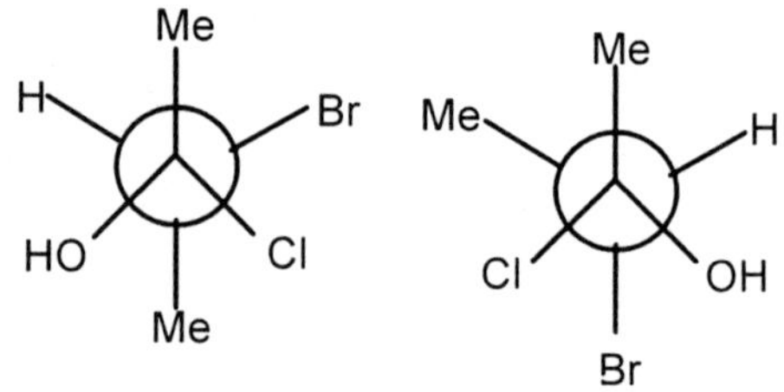

(iii)
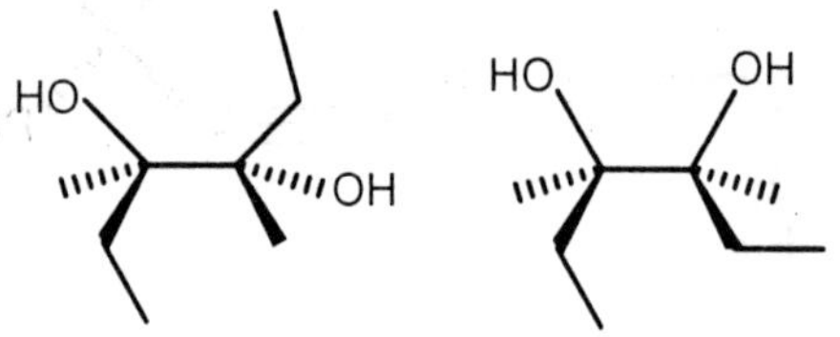

(iv)
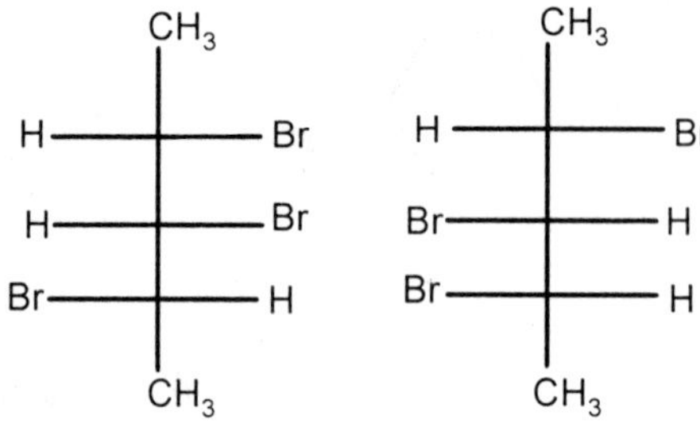

(v)
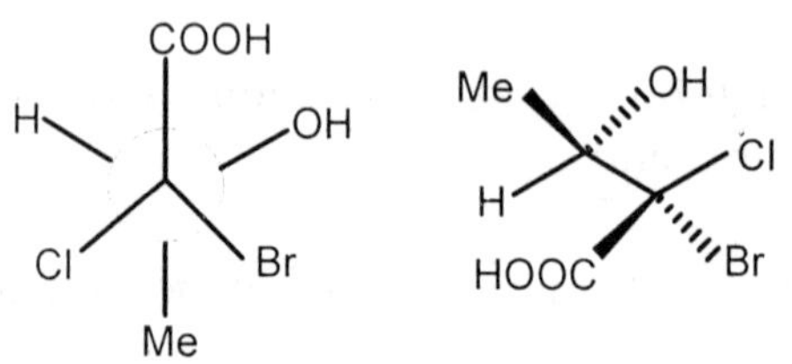

(vi)
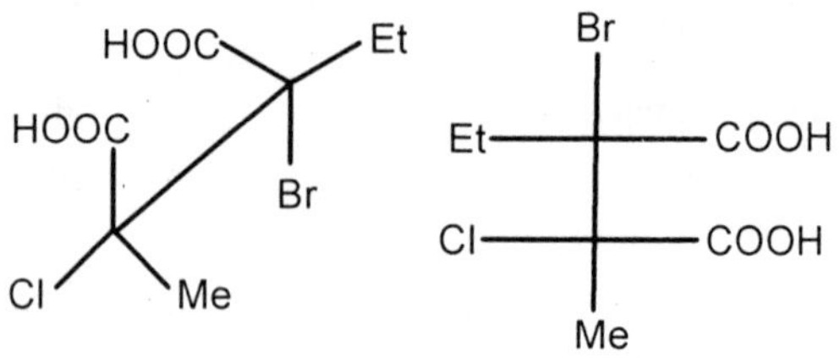

(vii)
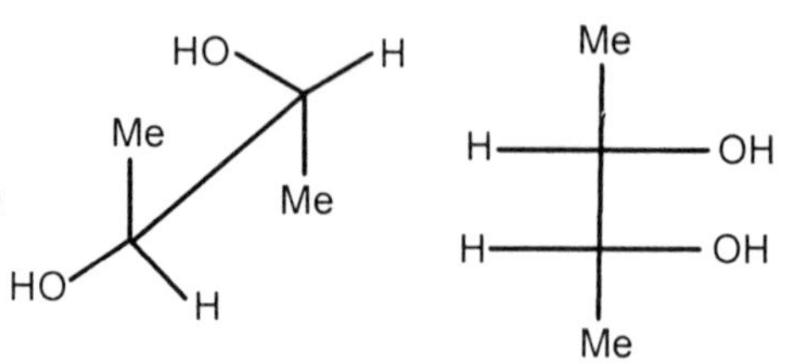

(viii)
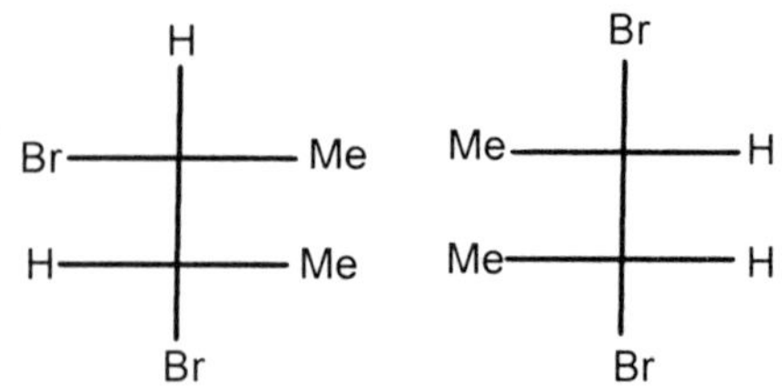

SP-2.9 What are *meso* compounds? Are the *meso* compounds optically active or inactive? Explain your answer through a suitable example.

SP-2.10 Write down the possible isomers of 3-bromo-2-butanol. Draw their configurations using Fischer projection and indicate their relationship.

SP-2.11 Draw all the stereoisomers of the following compound:

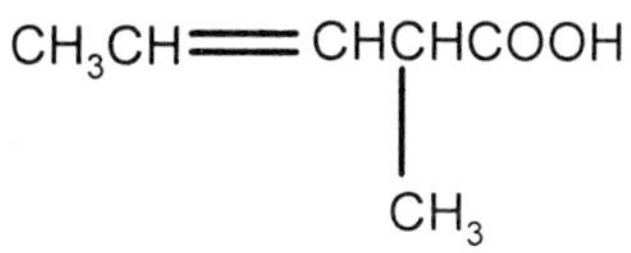

SP-2.12 Comment on the physical and chemical properties of enantiomers and diasteroisomers.

SP-2.13 In monosubtituted cyclohexanes, why does a substituent prefer to occupy the equatorial position?

SP-2.14 Show how solvation would account for the difference in acid strength of *cis* and *trans*-4-*t*-butyl cyclohexane carboxylic acids.

SP-2.15 Designate *E* and *Z* to the following compounds:

(i)

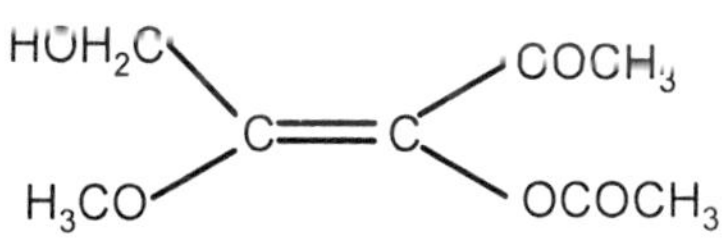

(ii)

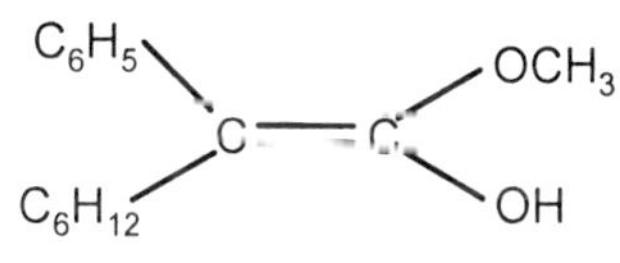

(iii)

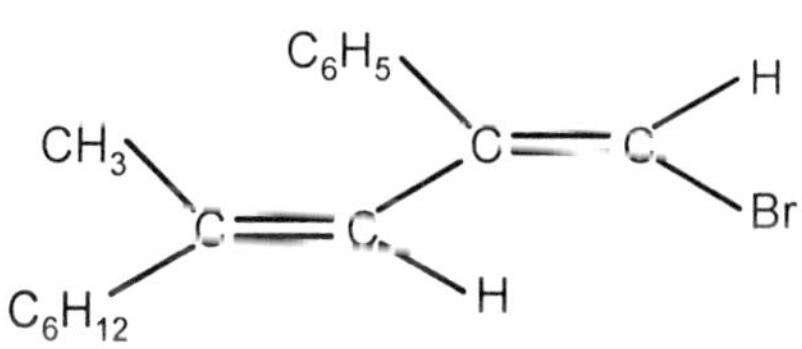

(iv)

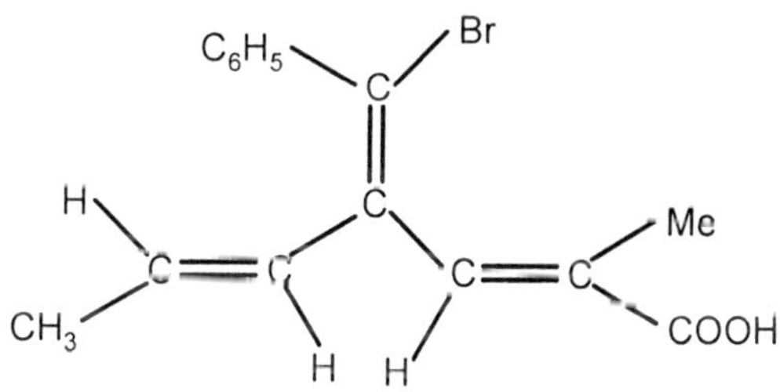

SP-2.16 Isomers of RCH=CHR differ widely in chemical properties, but those of XYC=C=CXY do not. Account for the fact.

SP-2.17 Define asymmetric synthesis and give one example.

SP-2.18 An optically active dicarboxylic acid $C_5H_4O_4$ on catalytic hydrogenation gives an optically inactive tetrahydro derivative $C_5H_8O_4$. How would you explain the reaction and the optical activity in the parent acid?

SP-2.19 Define enantiotopic and diastereotopic atoms with suitable examples.

SP-2.20 An optically active compound 'A' has the molecular formula C_6H_{10}. This compound gives a precipitate when treated with a solution containing $Ag(NH_3)_2OH$. On catalytic hydrogenation 'A' yields 'B' (with molecular formula C_6H_{14}) that is optically inactive and cannot be resolved. Propose structures for 'A' and 'B' and explain the observations.

SP-2.21 Draw all the possible stereoisomers of Et-CH=CH-CHBr-CH=CH-Et and mention whether they are *R* or *S* and optically active or not.

SP-2.22 What is meant by stereogenic centre? Are centers of stereogenicity always centers of chirality? Explain with suitable examples.

SP-2.23 Draw the Fischer projection formulae for all the possible stereoisomers of 2,3,4-trihydroxyglutaric acid. Comment on the stereogenicity of C-3 centre in the active and the *meso* isomers.

SP-2.24 Comment on the stereoisomerism of 1,2-dichlorocyclopropane.

SP-2.25 Comment on the chiroptic nature of *trans*-cyclooctene.

SP-2.26 What are the essential conditions for optical activity in biphenyls? What direct proof you can provide to establish that optically active biphenyls bear non-planar configuration?

SP-2.27 2,3-Pentadiene does not possess any chiral carbon but is resolvable into enantiomers. Draw an orbital picture that accounts for its chirality.

SP-2.28 Write the most stable conformer of ethylene glycol with reasons.

SP-2.29 Draw the preferred conformation of 2-hydroxypyran.

SP-2.30 Give an example where a molecule bears both enantiotopic and diastereotopic H-atoms. Label them with prochirality descriptors.

SP-2.31 Predict the optical nature of Troger's base.

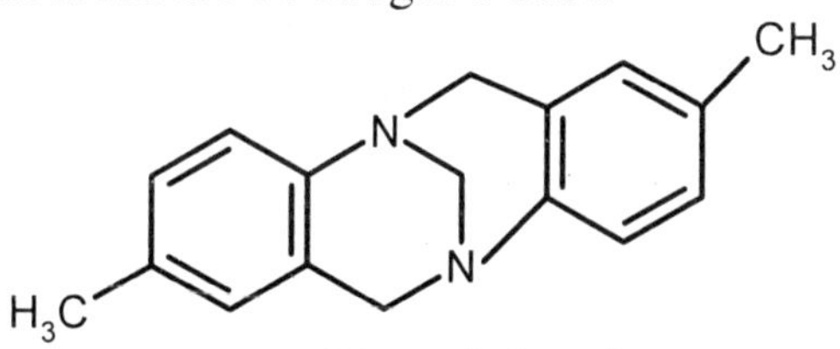

(Troger's base)

SP-2.32 "Addition of bromine to maleic acid gives (±)-2,3-dibromosuccinic acid whereas addition of bromine to fumaric acid gives *meso*-2,3-dibromosuccinic acid" – justify the observation.

SP-2.33 Explain the stereochemistry of the product with suitable mechanism when both the *erythro*- and *threo*-diastereoisomer of PhCHMeCHMeOTs is treated separately with acetic acid in presence of sodium acetate.

SP-2.34 What happens when *threo*-3-bromo-2-butanol is treated with HBr? Explain the stereochemistry involved.

SP-2.35 When dehydrochlorination of the *meso*- and optically active form of 1,2-dichloro-1,2-diphenylethanes is carried out by heating with pyridine, only the active form reacts to give the product of *Z*-configuration — the *meso*-isomer does not react. Account for the fact.

SP-2.36 Write down all the possible stereoisomers of 1-bromo-1,2-diphenylpropane with *R/S* and *threo/erythro* designations. Draw the Fischer projections as well as sawhorse-representation of the stereoisomers.

SP-2.37 Optically active 2-benzoylpropanoic acid undergoes racemization when treated with sodium ethoxide in ethanol. Account for the fact.

SP-2.38 Iodide ion catalyzed debromination of 2,3-dibromobutane is stereospecific. Justify the statement showing the mechanism of debromination.

SP-2.39 Compare the relative acid strengths of *cis*- and *trans*-4-*t*-butylcyclohexane-carboxylic acids in 66% aqueous DMF medium.

SP-2.40 When (–)-$CH_3CH(OH)CH_2CH_3$ is treated with aluminium *tert*-butoxide in acetone, the resulting product was found to be a racemic mixture of the alcohol. Offer explanation.

SP-2.41 The following compound racemizes in base. Offer explanation.

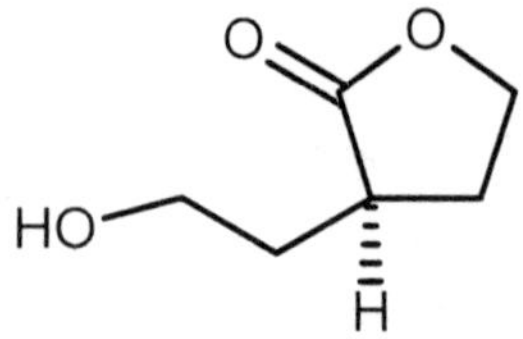

SP-2.42 *Numerical calculations:*

(a) Calculate the optical purity of an enantiomeric mixture in which the specific rotation of one pure component is $+200^0$ and the observed rotation of the mixture is $+50^0$, assuming concentration and cell-length both are unity.

(b) An optically pure sample of *R*-(−)2-butanol shows a specific rotation of -13.6^0. What relative molar proportion of *S*-(+)-2-butanol and *R*-(−)-2-butanol would give a specific rotation of $+6.8^0$?

(c) Ten grams of a 9:1 mixture of enantiomers is dissolved in 100 mL of solvent. Specific rotation of the major enantiomer is $+100^0$. Calculate the observed rotation of the mixture, assuming that a standard 1-dm cell is used.

SP-2.43 *Cis*-4-*t*-butylcyclohexyl tosylate undergoes solvolysis at a faster rate than the corresponding *trans*-isomer. Account for the fact.

SP-2.44 The molecule (I) on treatment with nitrous acid affords a tertiary alcohol (II). Account for the stereochemical outcome of this reaction.

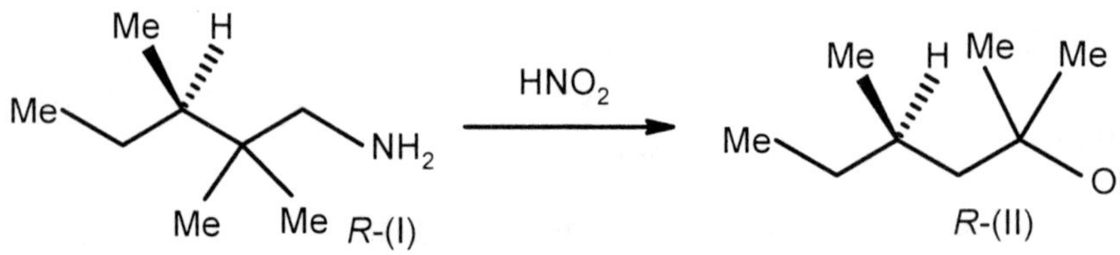

SP-2.45 Account for the stereochemical implication of the following transformation:

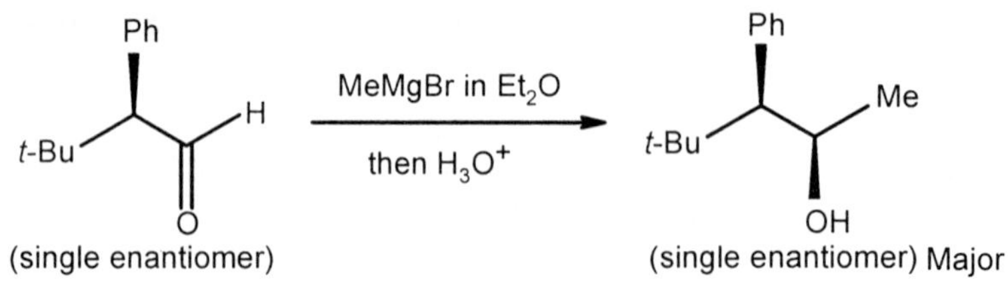

CHAPTER 3

REACTION INTERMEDIATES

Introduction

A chemical reaction that occurs in several steps involves the formation of one or more intermediates — the species which are short-lived, and are never present in high concentrations because they react as quickly as they are formed. These species are termed as *reaction/reactive intermediates*. In a particular reaction, a reactive intermediate must have to overcome some energy-barrier (though small) to proceed onward to product (or backward to the starting material). These energy-barriers offer these intermediates a measurable lifetime and form the walls of the energy well in the reaction profile — hence these species exist at a local energy minimum i.e. at the bottom of a potential-energy well.

Although reactive intermediates are not so stable compounds, they play a significant role in our study of organic chemistry. A great majority of organic reactions are multi-step processes involving reactive intermediates. Thus detailed understanding of the structures, stability, and properties of such reactive intermediates is of profound importance to the understanding of organic reaction mechanisms.

Reactive intermediates include normal covalent compounds, ions (positively/negatively charged) or free radicals — such structural difference arises out of the fact that whether they are formed either by *homolysis* or *heterolysis* of a chemical bond. In our present study, we will consider all the important reactive intermediates, namely carbocations, carbanions, free radicals, carbenes, nitrenes, and benzynes.

Solved Problems

▲On Carbocations

Problem 3.1 Define carbocations. Offer a critical view on its nomenclature. Sketch the geometry of a carbocationic species. Give examples of reactions that form carbocations, and also of a few undergone by them.

Solution.
Definition
A carbocation is a reactive intermediate species that contains a carbon atom bearing a positive charge and is trivalent; thus the species has only six bonded electrons. Hence, a carbocation may be represented as:

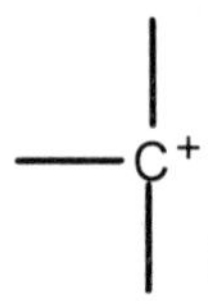

(carbocation)

Examples are:
$\overset{+}{C}H_3$, $CH_3\overset{+}{C}H_2$, $(CH_3)_3\overset{+}{C}$, etc.

Nomenclature
For many years these species were known as *'carbonium ions'*, although the term was realized to be non-systematic and inappropriate. But the most confusing situation was developed when Olah *et al.* reported another type of carbocationic intermediate in which the formal covalency of the positively charged carbon atom is five rather than three. The simplest example of this category is methanonium ion (CH_5^+; three centre-two electron bond).

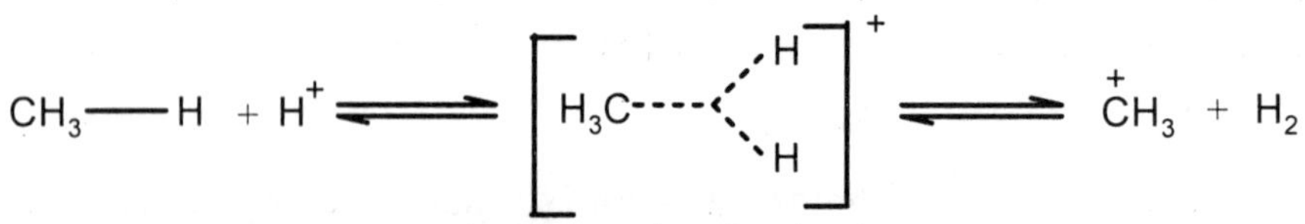

(methanonium ion having a three-centre, two electron bond) [detected spectroscopically]

Olah *et al.* (1971) then proposed the term *'carbonium ion'* to be reserved for such penta-coordinated positive ions, and the term *'carbenium ion'* for the tri-coordinated ions bearing the positive charge on carbon. He also proposed the term *'carbocation'* to encompass both the types — *carbonium* and *carbenium ions*. The IUPAC has accepted these definitions. Since pentavalent species (*carbonium ions*) are much rare than trivalent species (*carbenium ions*), the term *'carbocation'* can be used with very little or no confusion. Still some authors use the term *'carbonium ion'* for the trivalent species and some ones use the term *'carbenium'*, but the general tendency is to refer them as *'carbocations'*.

Geometry

Carbocations are *electron-deficient* species. The positively charged carbon atom has only six electrons in its valence shell, and is bonded to three other atoms. The central carbon is sp^2 hybridized, with a planar geometry and bond angles are of about 120^0. The unhybridized p orbital is vacant, and lies perpendicular to the plane.

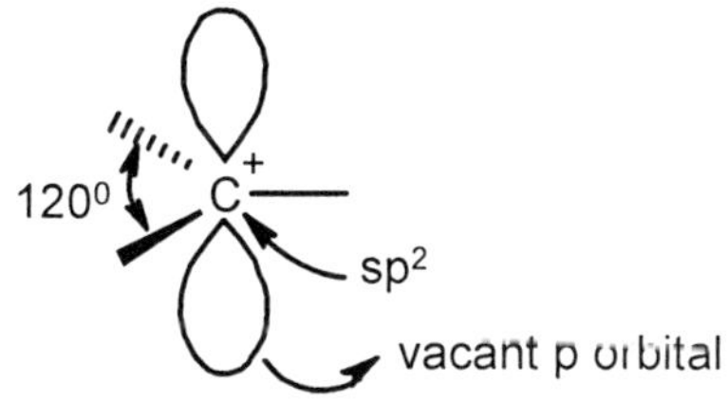

planar
(side view)

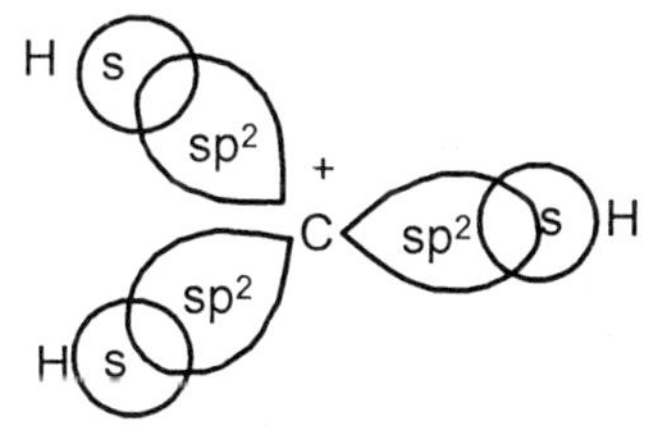

(methyl cation)
(top view)

Generation

Carbocations, stable or unstable, may be formed by the following reactions:

(i) Direct ionization

$$R{-}X \xrightleftharpoons{\text{solvent/Ag}^+} \overset{+}{R} + X^-$$

(ii) Addition of a proton or an electrophile to an unsaturated system

(iii) Decomposition

$$R{-}\overset{+}{N}{\equiv}N \xrightarrow{\Delta} \overset{+}{R} + N_2$$

(iv) Protonation to alcohols

$$ROH + H^+ \rightleftharpoons \overset{+}{R} + H_2O$$

(V) Reaction of alkyl fluorides with superacids (FSO_3H/SbF_5)

$$R{-}F + SbF_5 \longrightarrow \overset{+}{R}\ \overset{-}{S}bF_6$$

Reactions undergone by carbocations

(i) Carbocations behave as powerful electrophiles (Lewis acids), and react with any nucleophile they encounter.

$$\overset{+}{R} + \overset{-}{Y} \longrightarrow R-Y \quad \text{[Lewis acid - base recation]}$$

(ii) Loss of a proton

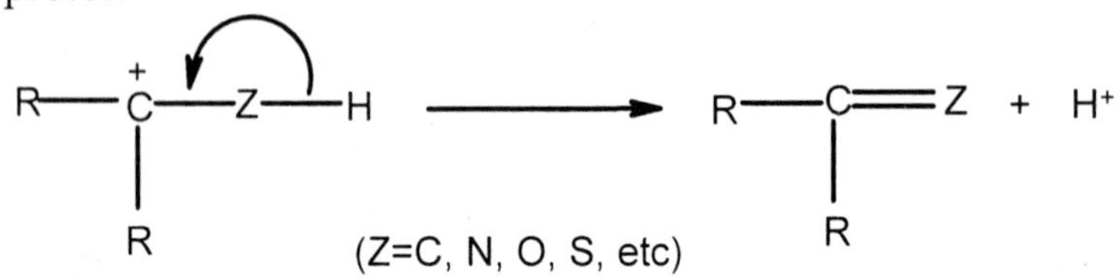

(iii) Rearrangement reaction

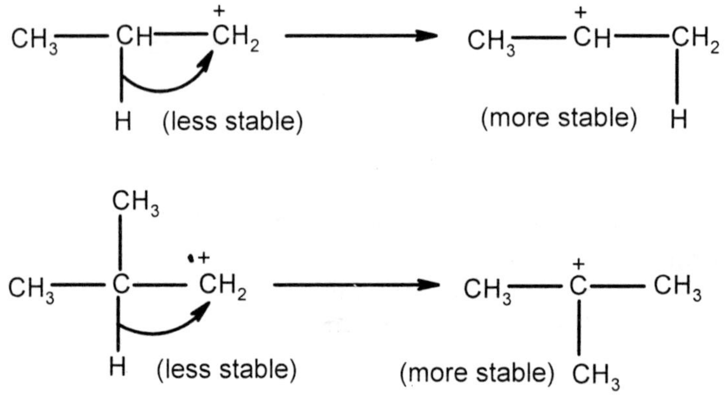

(iv) Addition reaction

Problem 3.2 Have it been possible to isolate carbocations? If so, give examples.

Solution. Carbocations are highly reactive species and mostly are very unstable. But it has been possible to keep many of such species indefinitely as stable solutions in the mixtures of fluorosulphuric acid (FSO_3H) and antimony pentafluoride (SbF_5). The original experiment, carried out by Olah *et al.* (1964), involved the addition of alkyl fluorides with antimony pentafluoride (as a Lewis acid) in either liquid SO_2 or excess SbF_5 as solvent.

$$R{-}F + SbF_5 \rightleftharpoons \overset{+}{R}\ \overset{-}{S}bF_6$$

[Formation of simple alkyl cations in these conditions allows their detailed study by NMR spectroscopy and other means.]

Mixtures of fluorosulphuric acid (FSO_3H) and antimony pentafluoride (SbF_5), usually dissolved in SO_2 (liq.) or SO_2ClF, are known as *superacids* (also known as '*magic acids*'). The use of *superacid* allows the generation of carbocations not only from alcohols (in *superacid*-SO_2 at -60^0C) and from alkenes at low temperatures, but even from alkanes also –

$$Me_3CH \xrightarrow[\text{(superacid)}]{FSO_3H\text{-}SbF_5} Me_3\overset{+}{C}\ SbF_5F\overset{-}{S}O_3 + H_2$$

(*tert*-butyl cation in stable solution)

In SO_2 solution, the existence of the following equilibrium is known:

$$Ph_3CCl \rightleftharpoons Ph_3\overset{+}{C} + \overset{-}{Cl}$$

Both triphenylmethyl and diphenylmethyl cations have been isolated as solid salts; $Ph_3C^+BF_4^-$ and related salts are available commercially. Methoxymethyl cation ($MeOCH_2^+\ SF_6^-$) is also obtained as a stable solid. *p*-Bromocyclohepta-2,4-6-triene (is known as tropylium bromide) exists as a crystalline solid (m.p. 208^0C), which is highly soluble in water yielding bromide ions in solutions,; i.e. the compound exists as an ion-pair.

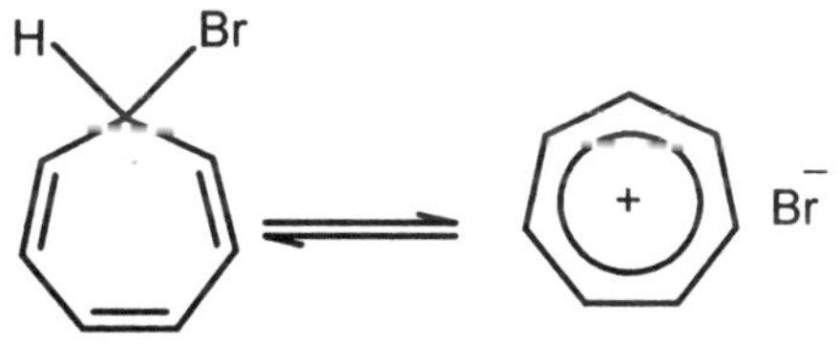

(tropylium ion attains high stability due to aromatization)

tropylium bromide ion-pair

Tris-cyclopropylmethyl cation has also been prepared from its corresponding alcohol in 96% H_2SO_4.

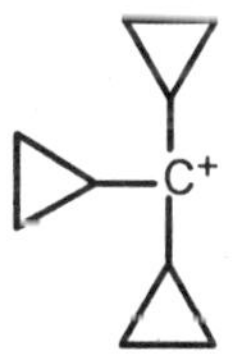

(*tris*-cyclopropylmethyl cation; even more stable than benzyl cations)

It is to be noted that till to date, no primary cation has survived so long enough for its detection.

Problem 3.3 Discuss the factors affecting the carbocation stability.

Solution. The factors responsible for carbocation stability are:

(i) Inductive effect
(ii) Hyperconjugative effect
(iii) Resonance effect
(iv) Steric effect
(v) Solvation effect

(i) Inductive effect: The electron-releasing inductive effect (+I) exerted by an alkyl group attached to the positively charged carbon of a carbocation reduces the charge density on the atom to some extent, and thereby stabilizes the carbocation.

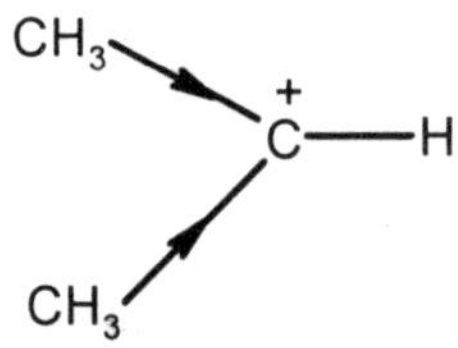

(+ I-effect of methyl groups)

Thus the more the +I effect, the more stable is the carbocation.

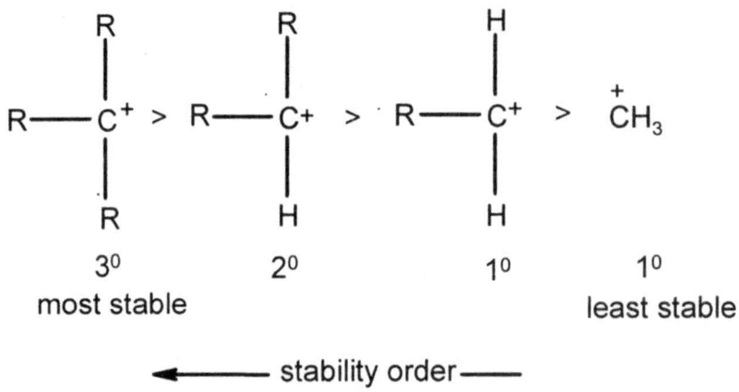

(ii) Hyperconjugative effect: An alkyl substituent may also reduce the positive charge density on carbocationic carbon by means of *hyperconjugative effect*. Alkyl groups have filled sp^3 orbitals that can overlap with the empty p orbital on the positively charged carbon atom, thereby stabilizing the species. Even though the attached alkyl group rotates, one of its sigma bonds is always aligned with the empty p orbital on the carbocation. The pair of electrons in this sigma bond spreads out into the empty p orbital, stabilizing the electron-deficient carbon atom.

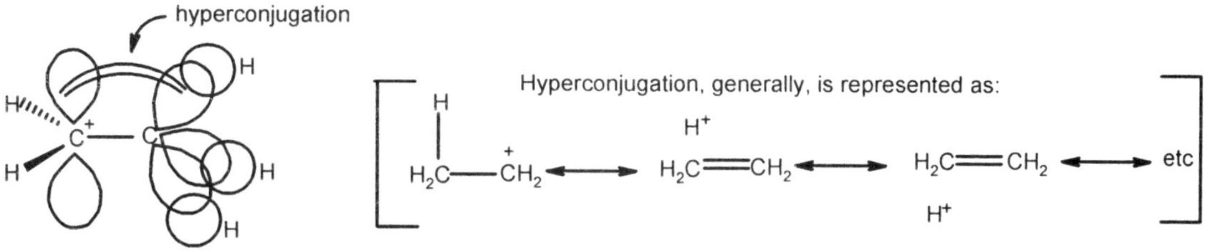

carbocation (ethyl cation) (overlap between a sigma bond and a p orbital; i.e. hyperconjugation)

With increase in number of α-hydrogens (that increase the number of hyperconjuga-tive forms) the stability of a carbocation also increases. Hence, stability order of the following series would be given as:

$$^{+}CMe_3 > {}^{+}CHMe_2 > {}^{+}CH_2Me > {}^{+}CH_3$$

(iii) Resonance effect: Resonance stabilization is a major influencing factor. If a π-bond is adjacent (α) to the positively charged carbon of a carbocation, the filled p orbitals of the π-bond will overlap with the empty p orbital of the central carbon atom. The result is a delocalized ion having positive charge shared by two atoms. This is said to be resonance stabilization, and is of prime importance in stabilizing carbocataions.

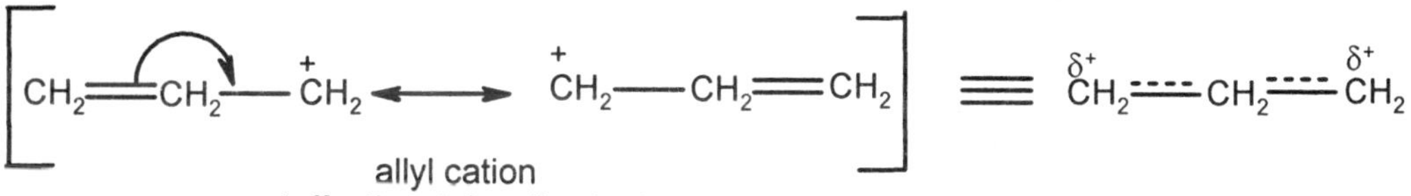

allyl cation
(effective delocalization)

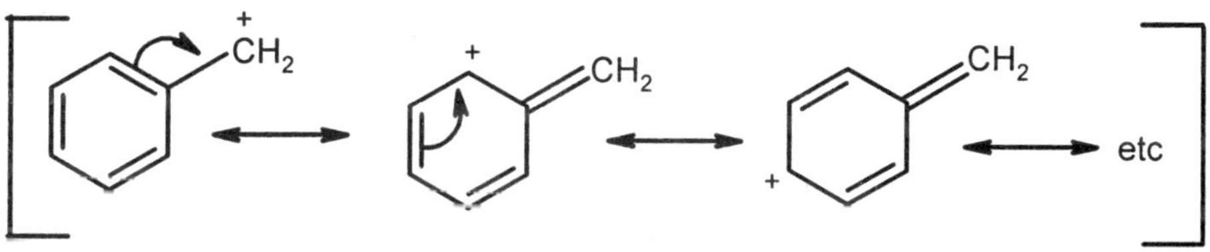

(benzyl cation; resonance stabilized)

(iv) Steric effect: Steric effect results an increase in stability of tertiary carbocations having bulky alkyl groups. As for example, isopropyl groups in triisopropyl cation (having a planar geometry of bond angle ~120^0) are far apart from each other, thus, creating no appreciable steric repulsion between them. But if they have to attain a tetrahedral geometry (when hybridization state of central carbon changes from $sp^2 \rightarrow sp^3$; bond angle ~109^0) on addition with a nucleophile, there occurs a strong steric repulsion between the bulky groups and as a result of this steric strain (B-strain) the carbocation is much reluctant to undergo this reaction. Hence, steric reason also plays an important role in stabilizing carbocations.

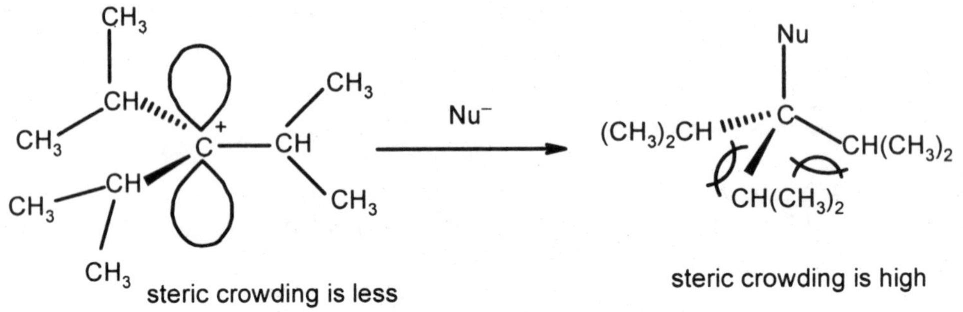

(v) Solvation effect: Solvation of a carbocation becomes effective in polar solvents, provided there is no chemical reaction between them. Thus the stability of carbocations is enhanced as the polarity of solvents increases.

Problem 3.4 Justify the following facts:

(i) Triphenylmethyl cation is so stable that some of its salts can be stored for months.

(ii) A solution of triphenylacetic acid in concentrated H_2SO_4 affords $MeOCPh_3$ when poured into methanol.

(iii) Cyclopropylmethyl cations are relatively stable.

(iv) The k_H/k_D value for the following solvolysis reaction is found to be 1.34.

$$(CH_3)_2CHBr + H_2O \xrightarrow{k_H} (CH_3)_2CHOH$$

$$(CD_3)_2CHBr + H_2O \xrightarrow{k_D} (CD_3)_2CHOH \qquad k_H/k_D = 1.34$$

Solution. (i) Greater stability of the triphenylmethyl cations is attributed to effective resonance. The central positively charged carbon is attached with three phenyl rings,

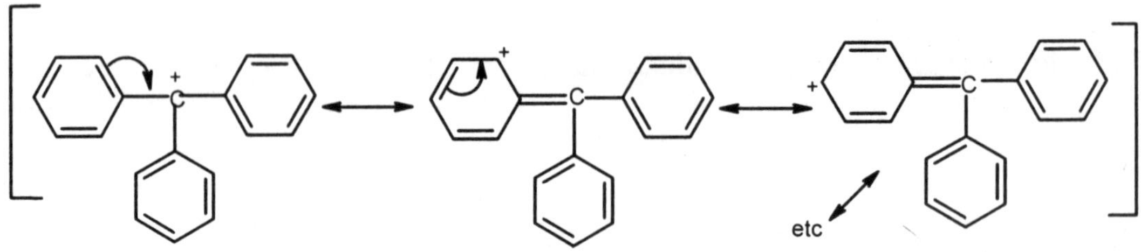

(triphenylmethyl cation)

and the positive charge on this carbocationic carbon is effectively delocalized with the π-systems of the aromatic rings. Hence the intermediate becomes very much stabilized.

In fact, $Ph_3C^+BF_4^-$ and related salts are available commercially.

(ii) Triphenylacetic acid reacting with concentrated H_2SO_4 forms a stable solution of triphenylmethyl cation (this cation gets high stability due to the consequence of effective delocalization of the positive charge with the aromatic systems). When this solution is poured into methanol, the carbocation intermediate undergoes reaction with methanol leading to the formation of methyl triphenylmethyl ether ($MeOCPh_3$).

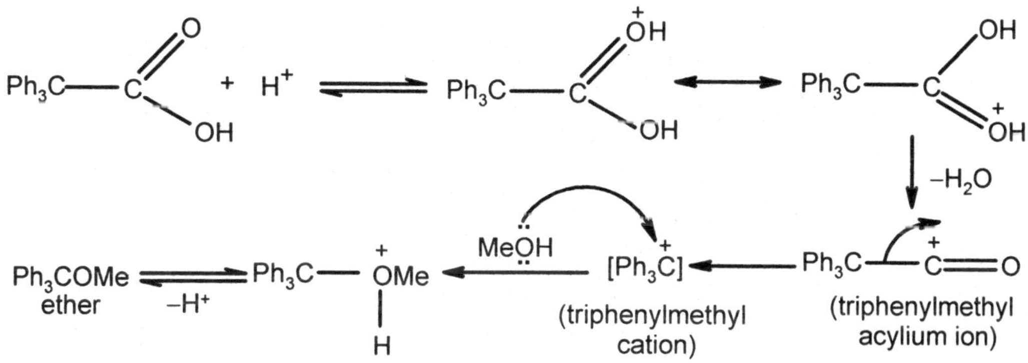

(iii) Cyclopropylmethyl cations get stability as a result of conjugation between the bent orbitals of the cyclopropyl ring and the vacant p orbital of the cationic carbon — the vacant p orbital lies parallel to the $C_2–C_3$ bond of the ring.

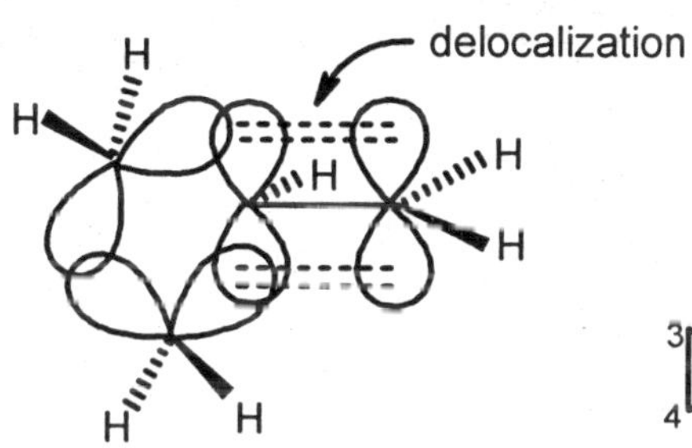

(cyclopropylmethyl cation)

(cyclopropylmethyl cation)

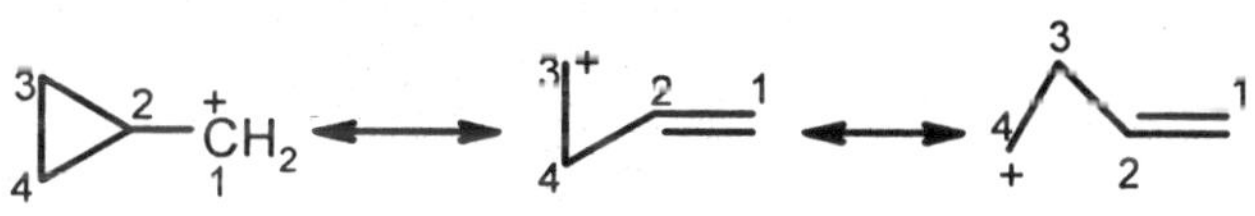

Both the $C_{2,3}$ and $C_{2,4}$ σ-bonds help to stabilize the positive charge. In this connection it should be noted that a cyclopropyl group stabilizes an adjacent positive charge even better than a phenyl group. This special stability increases with each addtional cyclopropyl group. Thus,

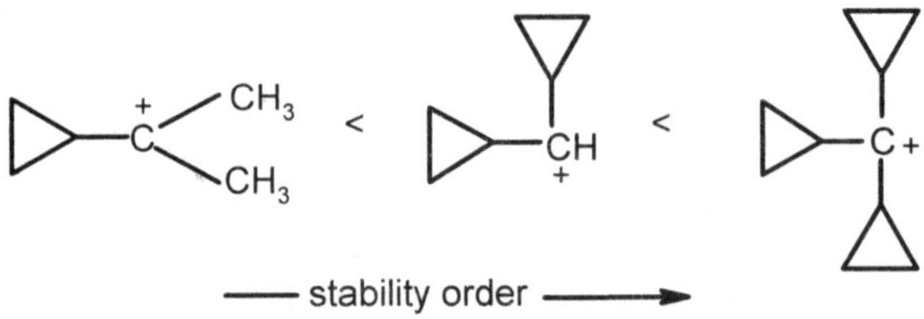

(iv)

$$(CH_3)_2CHBr + H_2O \xrightarrow{k_H} (CH_3)_2CHOH$$

$$(CD_3)_2CHBr + H_2O \xrightarrow{k_D} (CD_3)_2CHOH \qquad k_H/k_D = 1.34$$

The difference in solvolysis rate for the two substrates is attributed to β secondary isotope effect. In such effect, inclusion of deuterium in place of hydrogen β to the position of bond, which undergoes cleavage, slows down the reaction rate. The solvolysis reaction for both the above substrates is supposed to proceed through carbocation intermediate following the S_N1 path. Thus stability of carbocation intermediate has a direct impact on the reaction rate.

The cause of β secondary isotope effect is most likely due to hyperconjugation effects in the transition state having considerable carbocation character. Although the C–H bond in question is not broken in the transition state, the carbocation is stabilized by hyperconjugation involving this bond. Deuterium is at a level of lower zero-point vibrational energy and hence C–D bond is relatively more stable than the C–H bond. As a result, hyperconjugation in $(CD_3)_2CH^+$ is less than that in $(CH_3)_2CH^+$. Consequently, k_H is greater than k_D.

Similar situation is noticed in the following case:

$$\underset{\text{I}}{(CD_3)_3\overset{+}{C}} + (CH_3)_3CH \rightleftharpoons \underset{\text{II}}{(CH_3)_3\overset{+}{C}} + (CD_3)_3CH$$

The equilibrium constant (k at 25^{0}C) is observed as about 1.98. Here also due to β secondary isotope effect carbocation (II) becomes more stable than (I), because extent of hyperconjugative effect is greater in (II) than that in (I).

Problem 3.5 Arrange the carbocations in each of the following series in order of their stability:

(i) Me_3C^+, Me_2CH^+, $MeCH_2^+$, CH_3^+

(ii) 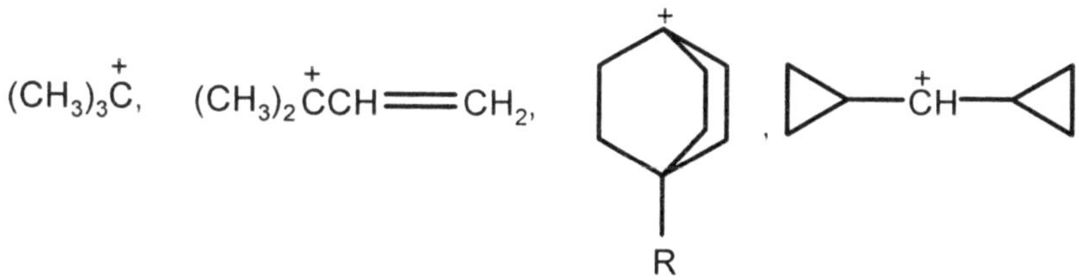

Solution. (i) The decreasing order of stability of the carbocations in the given series would be:

$$Me_3C^+ > Me_2CH^+ > MeCH_2^+ > CH_3^+$$

Two factors are to be considered to explain the give order:

(a) *Hyperconjugative effect*

Greater is the number of contributing structures of comparable energy in a resonance hybrid, greater is its stability. In *tert*-butyl cation (Me_3C^+), there are nine C–H σ-bonds and hence its resonance hybrid consists of ten hyperconjugative forms nine of which do not bear positive charge on carbon atom. There are six similar structural forms for isopropyl cation (Me_2CH^+), three such forms for ethyl cation ($MeCH_2^+$) and none for methyl cation (CH_3^+). Consequently, the order of stability appears as given above.

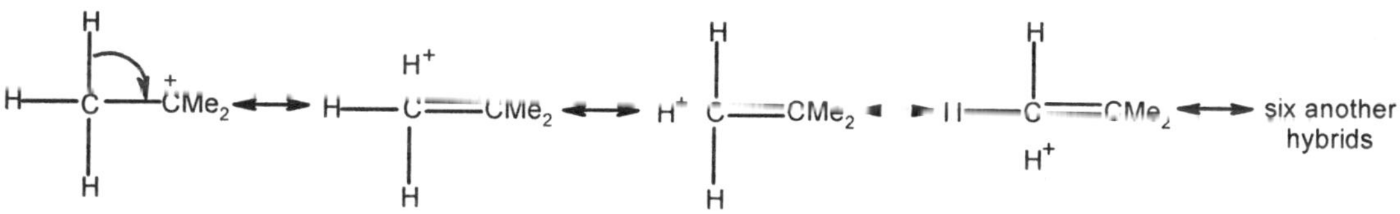

Similarly, six such hyperconjugative structures for Me_2CH^+ and three for $MeCH_2^+$.

(b) *Inductive effect*

The given order can also be explained by positive inductive effect (+I effect) of the methyl groups. Thus, due to the +I effect of the three methyl groups attached to the positively charged carbon atom in *tert*-butyl cation, its positive charge-density is reduced to a greater extent than that in the isopropyl cation which possesses only two methyl groups for such an effect. This charge-reducing effect is still less in the case of ethyl cation and it is least in methyl cation because it does not have any such electron releasing group attached to the positively charged carbon atom.

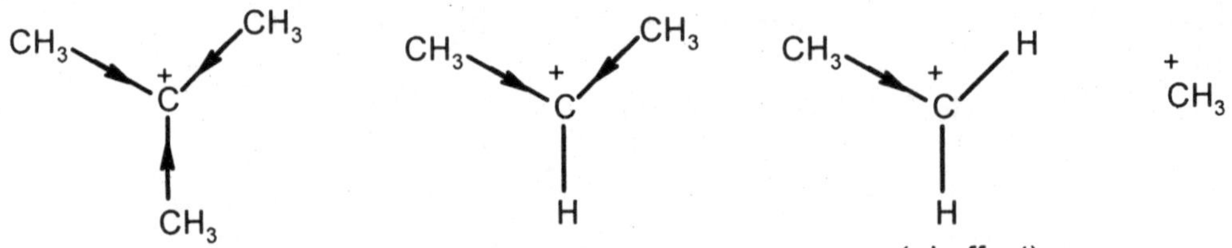

(+I effect) (+I effect) (+I effect)

(ii) $(CH_3)_3\overset{+}{C}$, $(CH_3)_2\overset{+}{C}CH{=}CH_2$,

I II

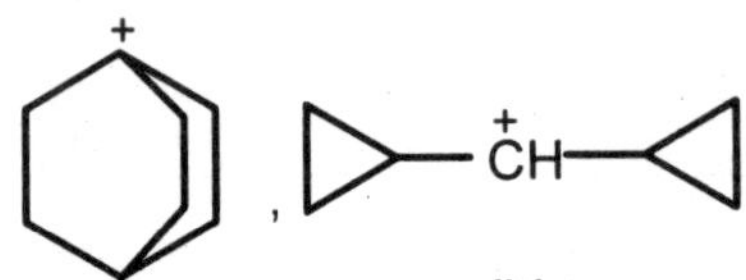

III IV

The carbocation (I) is stabilized by the +I as well as hyperconjugative effects exerted by the three methyl groups attached to the positively charged carbon atom. The carbocation (III) is similarly stabilized by +I effect of three ring bonds, but the hyperconjugative stabilization would be less effective though it contains three α-hydrogens because the bridged-head carbocation cannot collapse to the stable planar state. Besides, the carbocation III suffers from angle strain since the angle between bonds is somewhat less than the sp^2 bond angle. Thus, III would be less stable than I.

The carbocation (II) is stabilized by more influencing resonance factor as well as the +I and hyperconjugative effects of the two attached methyls.

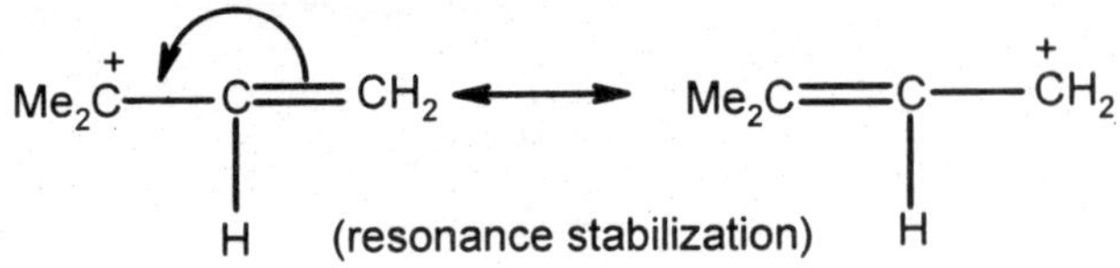

In the series, the carbocation (IV) is the most stable one due to greater conjugation offered by cyclopropyl rings.

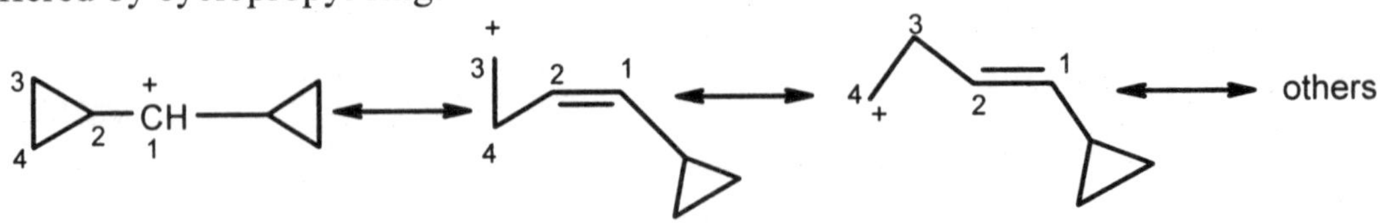

Hence, the stability order would be given as:

IV > II > I >III

Problem 3.6 What do you mean by "*nonclassical carbocations*"? In what respect they differ from "*classical carbocations*"? Give few examples of "*nonclassical carbocations*".

Solution. By older usage "*nonclassical carbocations*" or "*bridged carbocations*" are known as "*nonclassical carbonium ions*". A *classical* carbocation (i.e. carbenium ion) is represented by a single Lewis structure involving only two-electron, two-centre bonds, and the positive charge is localized on one carbon atom or delocalized by resonance involving an unshared pair of electrons or a double or triple bond in the allylic position. On the contrary, a *nonclassical* carbocation cannot be represented by a single Lewis structure — it bears one or more carbon or hydrogen bridges joining two electron-deficient centres. Coordination numbers of the bridging atoms are higher than the usual. The positive charge is not localized on a particular carbon atom — the charge is delocalized by a double or triple bond that is not in the allylic position or by a single bond.

norbornyl cation
(nonclassical carbocation/carbonium ion)

$CH_3-\overset{+}{C}H-CH_3$

isopropyl cation
(classical carbocatin)

Few examples of *nonclassical carbocations*:

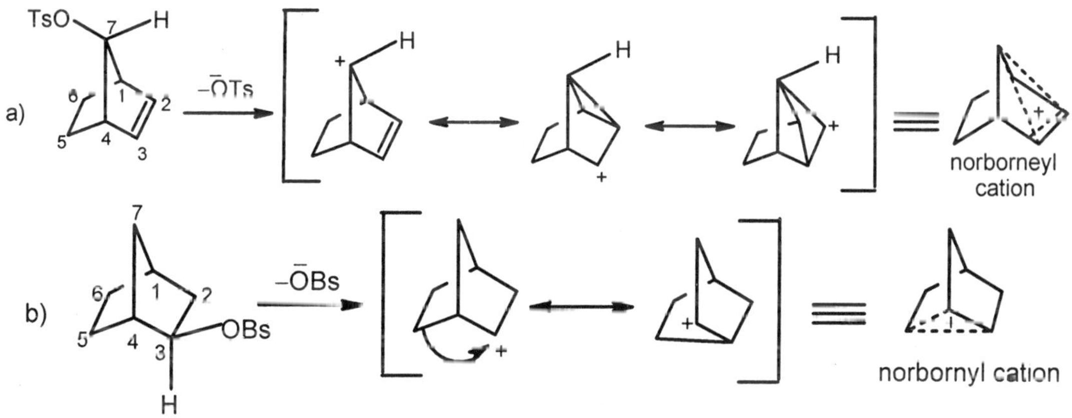

▲On Carbanions

Problem 3.7 Define carbanions. How they can be generated? Give examples of reactions undergone by them.

Solution.

Definition

A carbanion is a species that bears a negatively charged trivalent carbon atom and an unshared pair of electrons. Thus there are eight electrons around the central carbon making it electron rich; so it behaves as a strong nucleophile (Lewis base). It may be represented as:

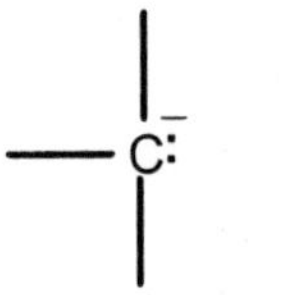

(carbanion)

Examples are:

H_3C^{-}: (methyl anion)

$PhCH_2^{-}$: (benzyl anion)

$Cl_3C—\bar{C}H_2$ (trichloroethyl anion)

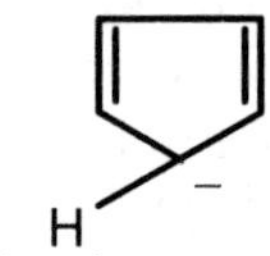

(cyclopentadienyl anion)

Generation

Carbanions are generated by the following reactions:

(a) Hydrogen-abstraction by base:

$$R—H \underset{}{\overset{\text{base (B)}}{\rightleftharpoons}} R^{-} + \overset{+}{B}H$$

e.g.

$$\underset{\text{pKa 8.6}}{CH_3CH_2NO_2} \xrightarrow{\text{base}} CH_3\bar{C}HNO_2$$

$$\underset{\text{pKa 12.7}}{CH_2(COOEt)_2} \xrightarrow{\text{base}} \bar{C}H(COOEt)_2$$

$$Ph_3C—H \xrightarrow{\text{base}} Ph_3\bar{C}$$

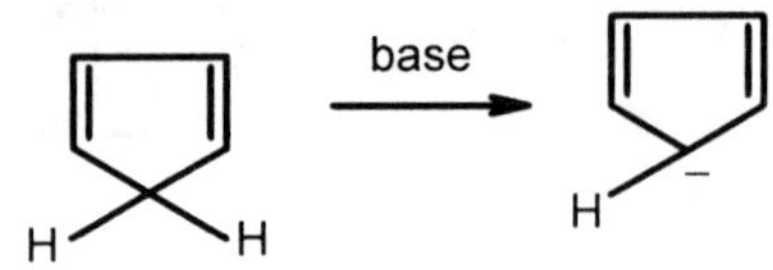

pKa 15

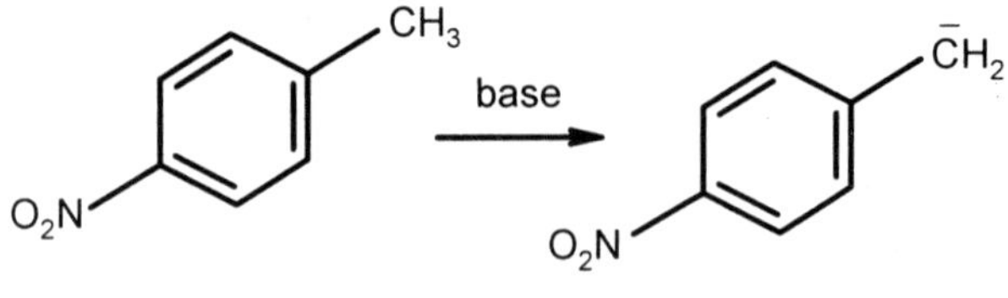

Commonly used bases are — NaH/ether; NaOEt/EtOH; K^+*t*-BuO$^-$/*t*-BuOH; $NaNH_2$/ liq. NH_3, etc.

(b) Metal halogen interchange:

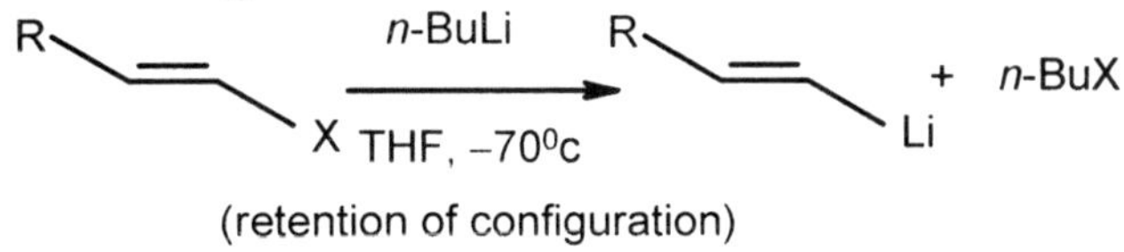

(c) Nucleophilic addition to an unsaturated C – C bond:

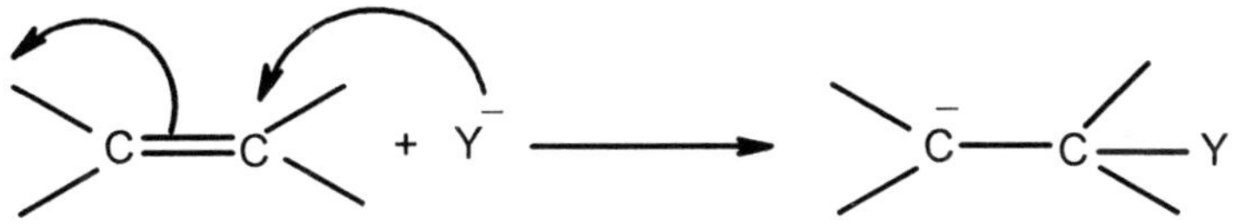

(d) Decomposition reaction:

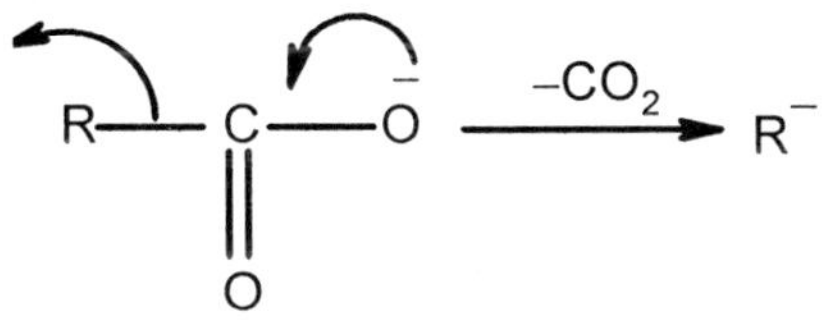

Reactions

(a) Combination with a positive species (Lewis acid – base reaction):

$$\bar{R} + \overset{+}{Z} \longrightarrow R—Z$$

(b) Displacement reaction:

$$(COOEt)_2\bar{C}H + R—X \longrightarrow (COOEt)_2CH—R + \bar{X}$$

(c) Addition to a multiple bond:

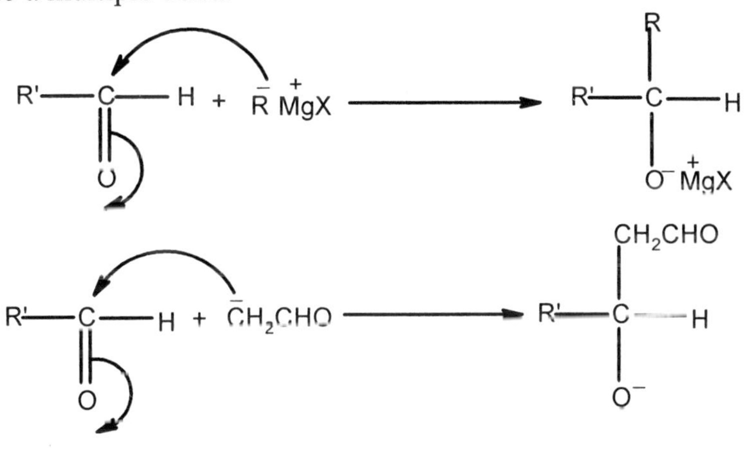

(d) Elimination reaction:

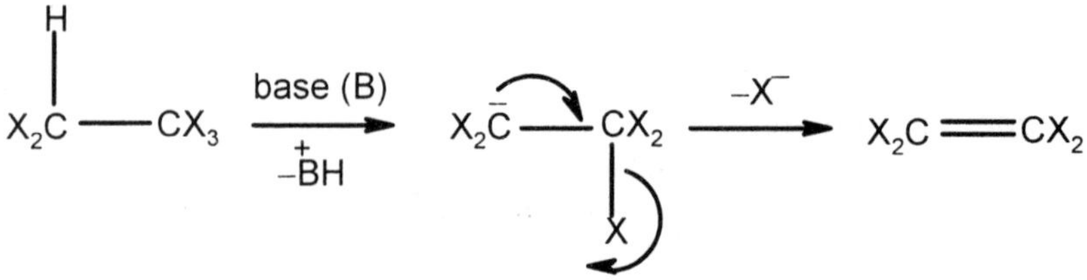

(e) Rearrangement reaction:

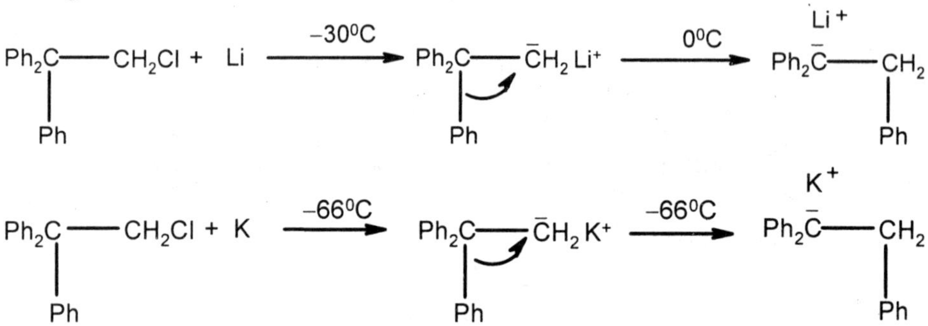

When there is a choice between a phenyl and a tolyl group, the phenyl group migrates in preference.

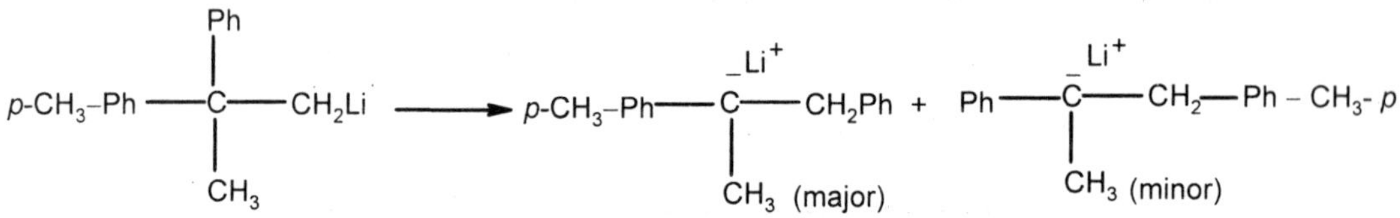

Problem 3.8 Comment critically on carbanion configuration.

Solution. It is evidenced that the central carbon of simple carbanions is sp^3 hybridized, with the unshared pair of electrons occupying one apex of the tetrahedron. Carbanions would thus attain pyramidal shape similar to those of isoelectronic amines. This configuration is preferred from energy consideration, because the unshared electron pair would then be accommodated in a hybridized sp^3 orbital rather than in the higher- energy unhybridized p orbital of the planar (sp^2) configuration.

(Pyramidal carbanion) (I)

Methyl anion (CH_3^-) in gas phase has been reported to have pyramidal configuration. Besides, easy formation of stable bridgehead carbanions supports sp^3 hybridized states of the bridgehead carbons; such carbons ordinarily can't attain sp^2 hybridized state and hence are extremely reluctant to undergo reaction in which they must be converted to carbocations.

This structure (I) undergoes rapid *pyramidal inversion* and remains in equilibrium.

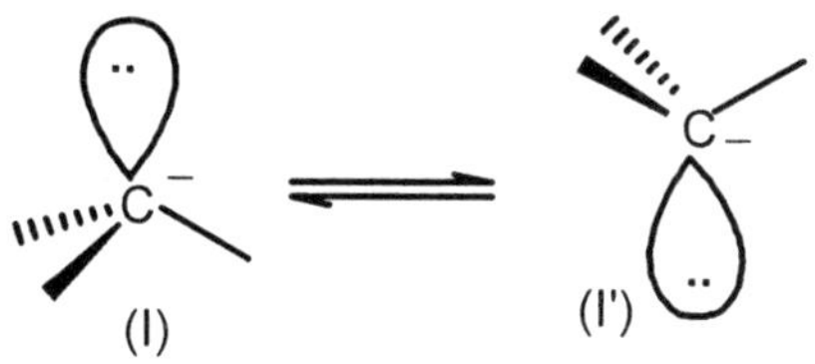

(rapid pyramidal inversion)

Planar carbanions, where the central carbon is sp^2 hybridized, may arise when they bear the substituents capable of stabilizing the negative charge through conjugative delocalization of the unshared electron pair. Such resonance-stabilized carbanions must be planar or nearly planar to allow the maximum orbital overlap of the p orbital with those of the substituents. Few examples are:

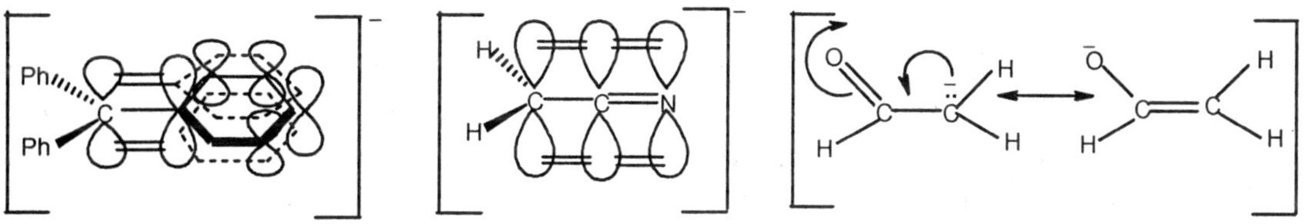

($Ph_3\bar{C}$) ($\bar{C}H_2CN$) (resonance stabilized)

Problem 3.9 Discuss the factors that serve to stabilize carbanions.

Solution. The following structural features tend to stabilize a carbanion:

(i) increase in s character of the carbanionic carbon,
(ii) inductive effects exerted by electron-withdrawing substituents,
(iii) conjugation of the carbanion unshared electron pair with a polarized multiple bond, and
(iv) constitution of an aromatic system.

(i) s-Character of orbitals: The s orbitals can hold negative charge more tightly than the corresponding p orbitals since the former are closer to the nucleus; this is reflected also into the hybrid orbitals resulting from their deployment. Thus the electronegativity order (capacity of holding negative charge) of such hybrid orbitals is: sp (50% s-character) > sp^2 (33% s-character) > sp^3 (25% s-character). This trend is reflected in the stability order of the following carbanions:

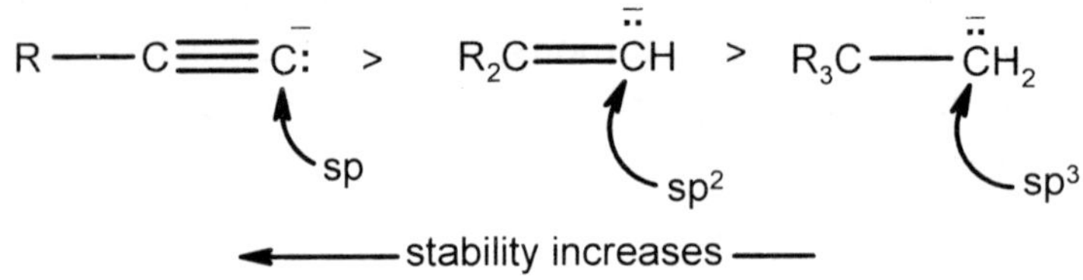

(ii) Electron-withdrawing substituents: Electron-withdrawing substituents attached to a carbanionic carbon stabilize the anion through dispersing the negative charge by –I effects. Examples of carbanions experiencing such stabilization are:

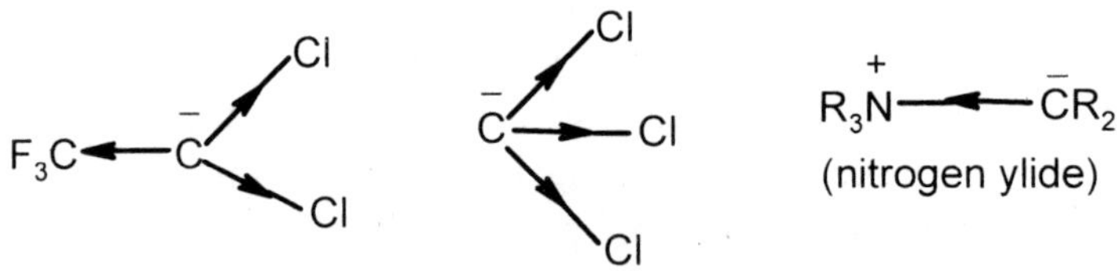

(carbanions stabilized by –I effects of electron-withdrawing groups)

(iii) Conjugative effect: Carbanions become greatly stabilized due to effective conjugation of the negative charge (i.e. unshared electron pair) with a polarized multiple bond located at α to the carbanionic carbon. Thus, allylic and benzylic carbanions and the carbanions having the substituents like $-NO_2$, –COR, –COOR, –CN, $-SO_2$, etc. at the α-position with respect to carbanionic carbon are greatly stabilized.

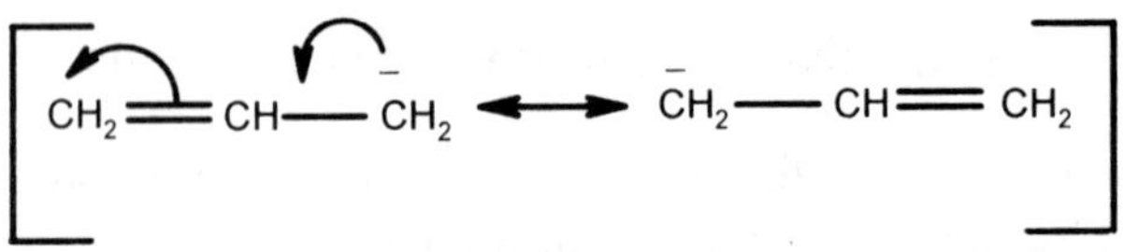

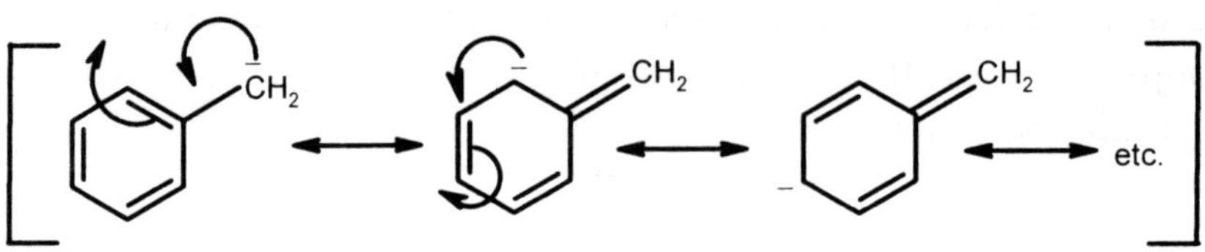

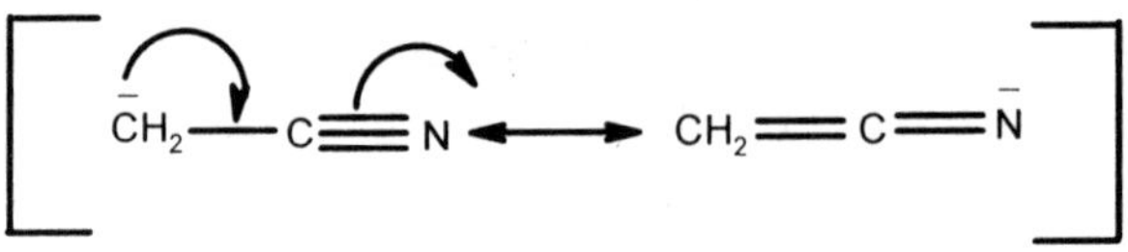

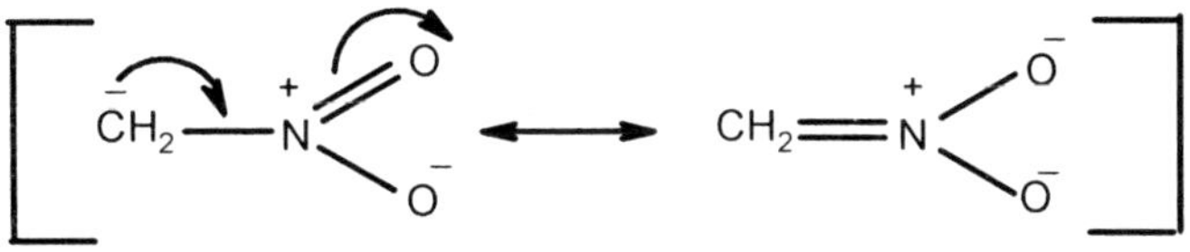

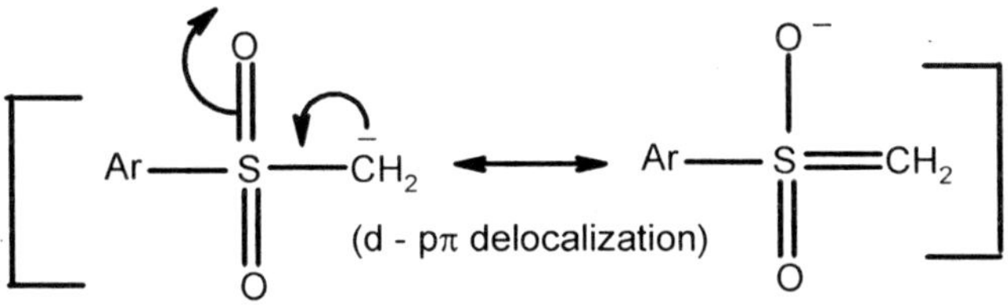

(iv) Development of aromaticity: If the unshared electron pair of a carbanion is involved in constituting an aromatic system [i.e. planar $(4n + 2)\pi$ electron system], the carbanion becomes greatly stabilized. As for example, cyclopentadienyl anion becomes stabilized due to its aromatic properties.

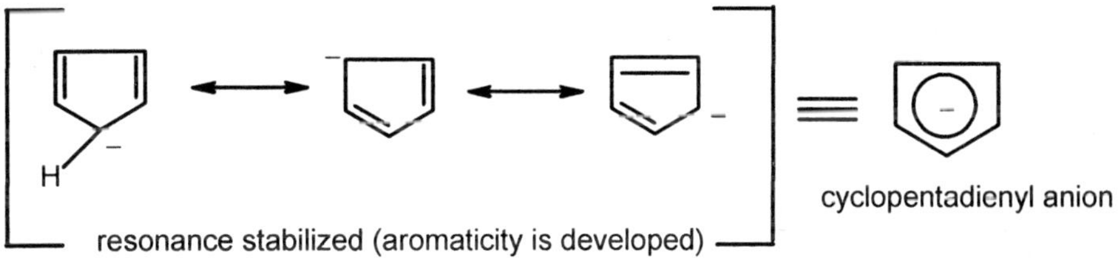

Problem 3.10 Discuss the stereochemical features of chiral carbanions and the stereochemical outcome of the products resulting from them.

Solution. A carbanion in which the negatively charged central carbon atom is attached with three different atoms/groups should be chiral, and reactions involving such chiral carbanions should maintain retention of configuration. But it is very much difficult to isolate, even to demonstrate, chiral carbanions. This is because of rapid *pyramidal inversion.* There occurs a very rapid interconversion between two enantiomeric pyramidal forms so that they can't be separated into two enantiomers, and hence exists as a racemic mixture exhibiting zero optical rotation. Theoretical calculations predict a barrier of only about 1– 2 kcal/mole for free methyl anion.

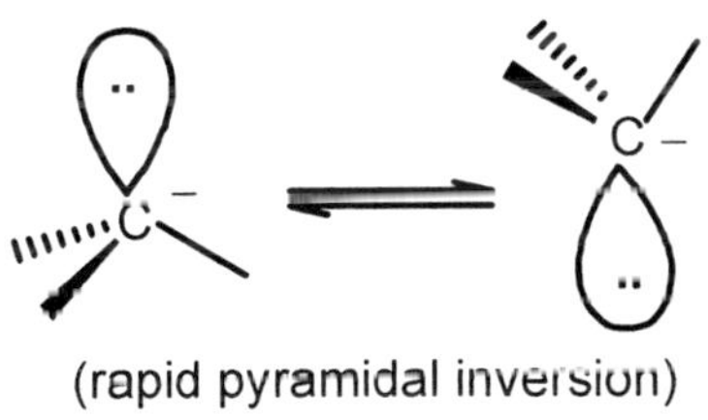

[The unshared pair and the central carbon rapidly oscillate from one side of the plane to the other]

However, strong electronegative substituents and π–electron acceptor groups raise the inversion barrier. The experiment carried out by Hargreaves *et al.*[1] demonstrated at best only very short-term configurational stability of a chiral carbanion:

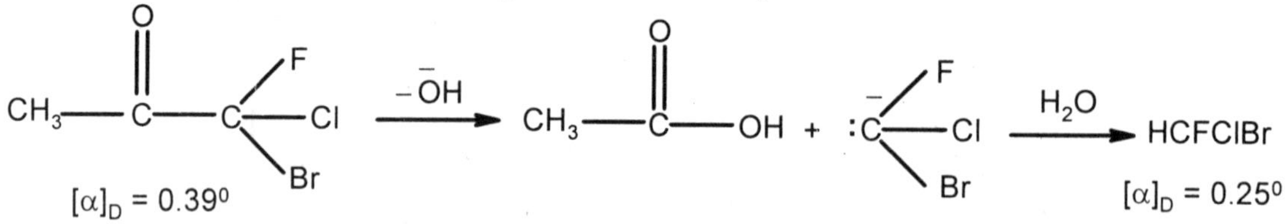

The scheme shows that the chiral carbanion is apparently trapped (by water) before it has time to racemize.

The energy barrier for the inversion can be enhanced by incorporating the carbanion in a small ring (usually cyclopropyl ring) or coordinating with lithium — in the first case, a substantial increase in ring strain associated with the planar transition state for inversion (called I-strain) raises the barrier while in the second case, there is a bond between carbon and lithium, although is of ionic character. As a result inversion in these situations does not take place, which is substantiated by the fact that carbonation of *cis*-2-methylcyclopropyl lithium gives rise to stereospecifically *cis*-2-methylcyclo-propyl carboxylic acid, none of the *trans*-isomer[2].

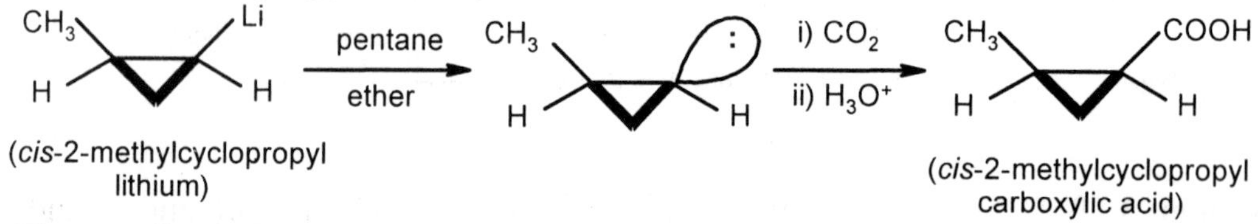

The sodio-derivatives of the cyclopropyl system also show significant stereochemical stability to give optically active products. Besides, vinyl anions show appreciable configurational stability and undergo reaction with retention. The carbanions stabilized by adjacent sulphur or phosphorus seem to be planar and retain the configuration (s) intact, and their inherent chirality is caused by lack of rotation about the C–S or C–P bond. Stereochemical outcome of formation and protonation of carbanions stabilized by π-acceptor groups (like phenyl, carbonyl, nitro, cayano, etc.) is found to be strongly influenced by the nature of solvent and also of cations used. Thus, the species (**A**) is planar with K^+ or Cs^+ while pyramidal with Li^+.

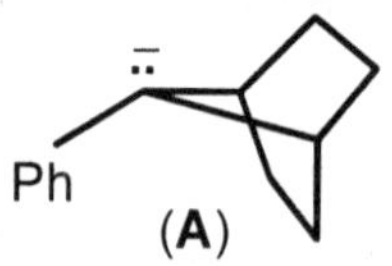

(**A**)

1. Hargreaves *et al.*, *J. Chem. Soc. C*, 1013 (1971).
2. Applequist *et al.*, *J. Am. Chem. Soc.*, **83**, 862 (1969).

Problem 3.11 Compound (I) is readily soluble in aqueous NaOH, and gives a red colour with $FeCl_3$ solution while compound (II) does neither. Justify the observation.

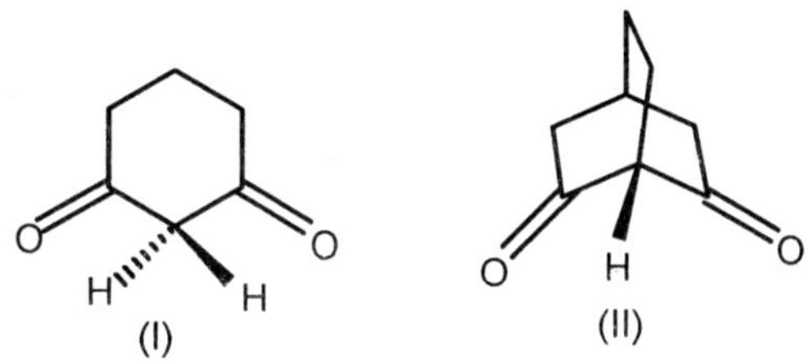

Solution.

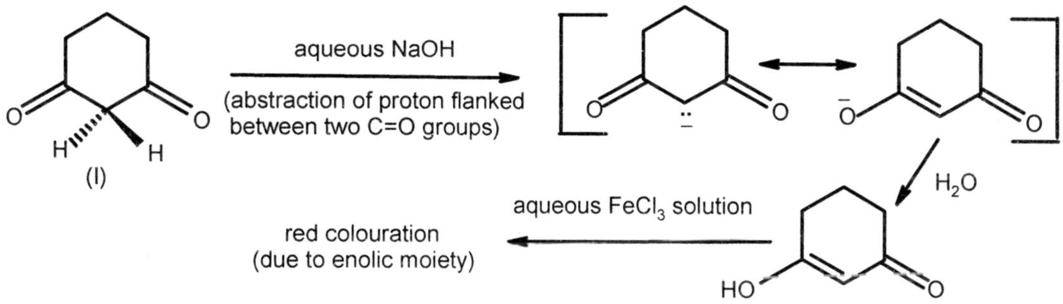

1,3-cyclohexadione (I) generates the carbanion, negative charge of which is capable to undergo delocalization with the adjacent C=O giving rise to enol. This enolic form responds positively toward ferric chloride solution.

On the other hand, the hydrogen atom flanked by the two C=O groups in compound (II) is extremely reluctant to release itself (as proton) forming a negative charge on the bridgehead carbon, because in that case the carbanion's lone pair would be in sp^3 orbital more or less at right angles to the p orbitals on each of the adjacent carbon atoms — hence there occurs no sp^3-p overlap. Consequently, there is no scope at all for the negative charge to become stabilized through delocalization, and so the carbanion is not formed. That's why compound (II) remains insoluble in aqueous base and does not respond toward $FeCl_3$ solution.

Problem 3.12 Arrange the following carbanions in order of increasing stability:

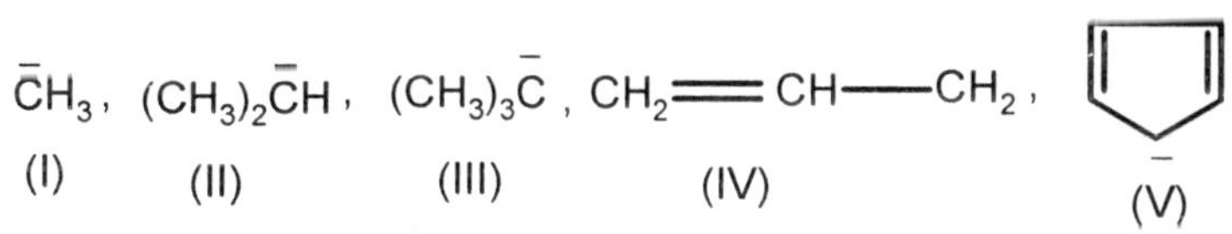

Solution. +I effects of electron-releasing methyl groups destabilize a carbanion. That's why (III) would be the least stable in the series. Due to same reason (I) is more stable than (II). In (IV) the negative charge on carbanionic carbon is delocalized with the allylic double bond raising the stability of the carbanion. But the carbanion (V) [cyclopentadienyl anion] is greatly stabilized due to cause of aromatization.

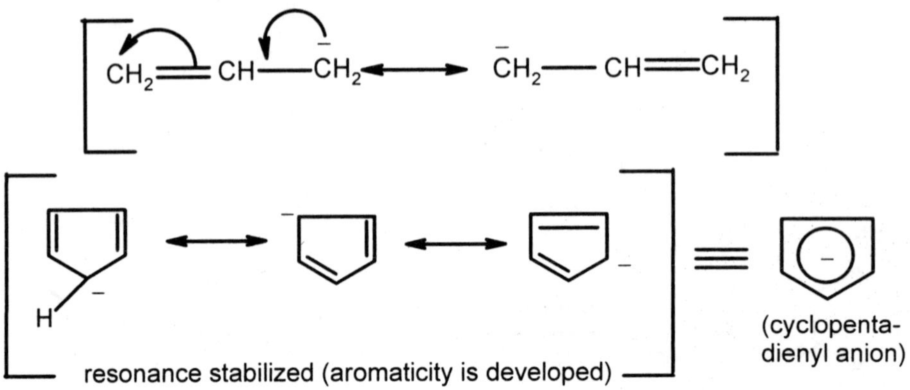

Therefore, the stability order may be given as: (III) < (II) < (I) < (IV) < (V).

▲On Free Radicals

Problem 3.13 Define carbon free radicals. How they can be detected? Comment on their geometry.

Solution.

Definition

Free radicals are species that contain one or more unpaired electrons. Carbon free radicals have three bonds on the carbon atom that contains one unpaired electron. A carbon radical may be represented as:

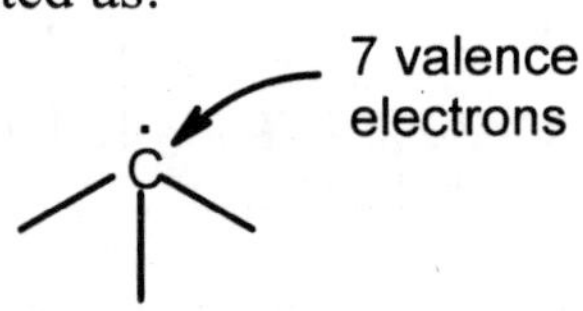

(carbon radical)

It is a neutral intermediate and is electron-deficient because it lacks an octet around the central carbon atom (has only seven valence electrons).

Detection

A carbon radical bears an unpaired electron; so there is a net magnetic moment and the species becomes paramagnetic. Hence it can be detected by electron spin resonance (ESR) spectroscopy.

Geometry

For simple carbon radicals two possible structures are of consideration — one is planar where the central carbon is sp^2 hybridized bearing the odd electron is in an unhybridized p orbital, and the other is pyramidal where the odd electron occupies one of the sp^3 hybrid orbitals.

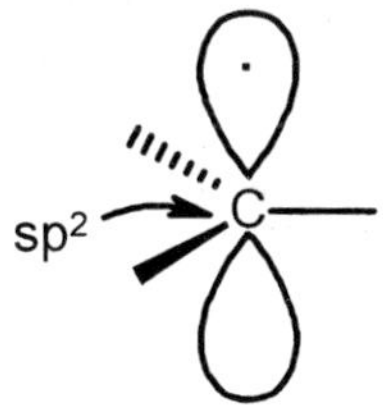

(planar)

(pyramidal)

Observed evidences support the planar structure for methyl radical ($^{\bullet}CH_3$) and other alkyl radicals. However, free radicals in which carbon is attached with strong electronegative atoms/groups (e.g $^{\bullet}CF_3$) prefer to attain a pyramidal shape — increase in the electronegativity increases the deviation from planarity. Cyclopropyl radicals are found to bc pyramidal. Actually, the energy difference between the two forms is not too large. Brigehead radicals are preferably pyramidal.

Problem 3.14 How carbon radicals could be generated? Mention some reactions of carbon free radicals.

Solution.

Generation

Carbon free radicals are formed on homolysis of covalent bonds. Following methods are in use to generate carbon free radicals:

(a) *Thermal and Photochemical Cleavage*

The energy necessary to break a covalent bond in homolytic fission may be supplied either by thermal or by photochemical means.

(i) *Thermolysis*

Decomposition of metal alkyls in the vapour phase yields alkyl radicals. For example:

$$(CH_3)_4Pb \xrightarrow{600^{\circ}C} Pb + 4\dot{C}H_3$$

(tetramethyl lead) (methyl radical)

Dissociation of –C–H, –O–O– and –N–N– bonds may also occur.

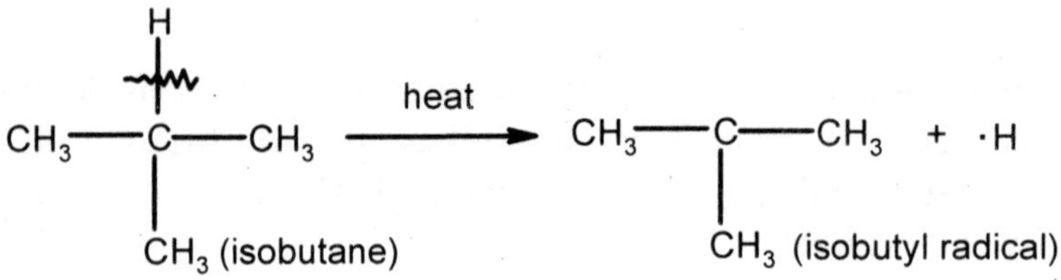

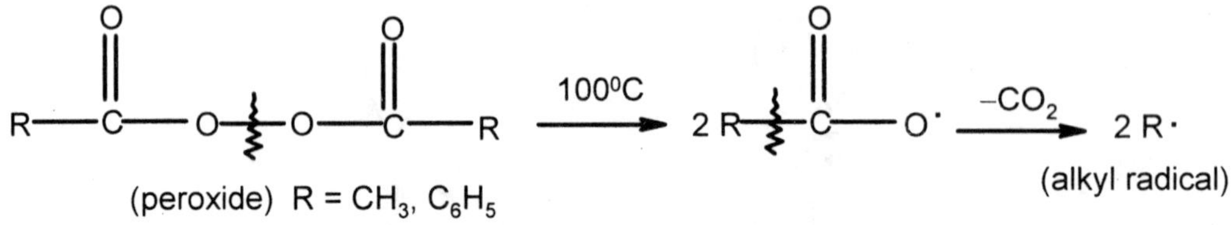

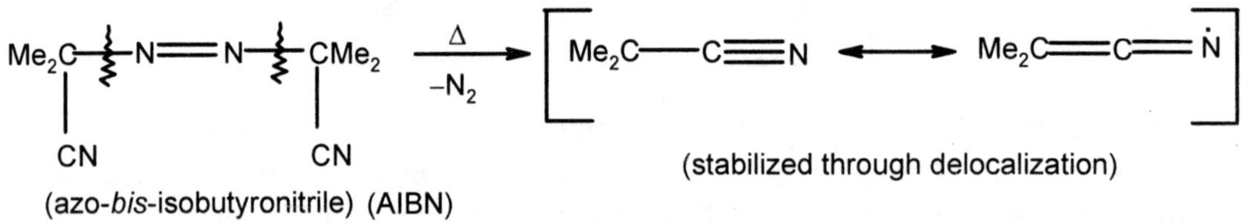

(ii) *Photolysis*

Photo-irradiation of wavelength in the range of 300–600 nm (amounts to 48–96 kcal/mole) can break covalent bonds to give rise free radicals. Acetone in the vapour phase is decomposed on absorption of light having wavelength of 320 nm.

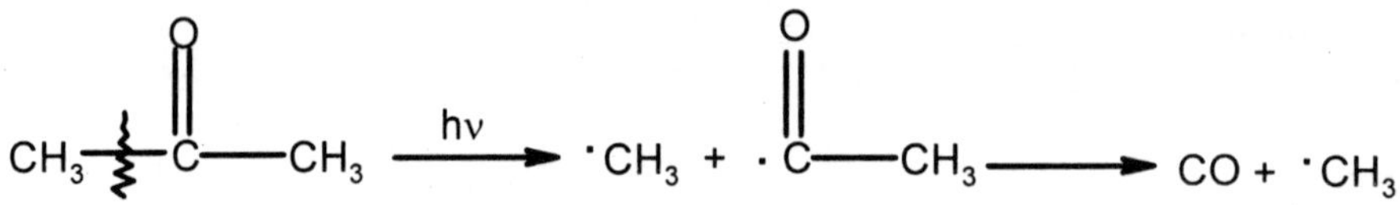

Another very useful photolytic homolysis is that of halogens:

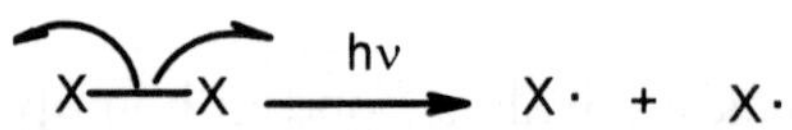

(X = Cl, Br, I)

Photolysis of azoalkanes and diacyl peroxides are also known:

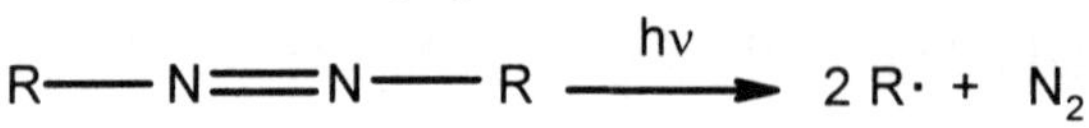

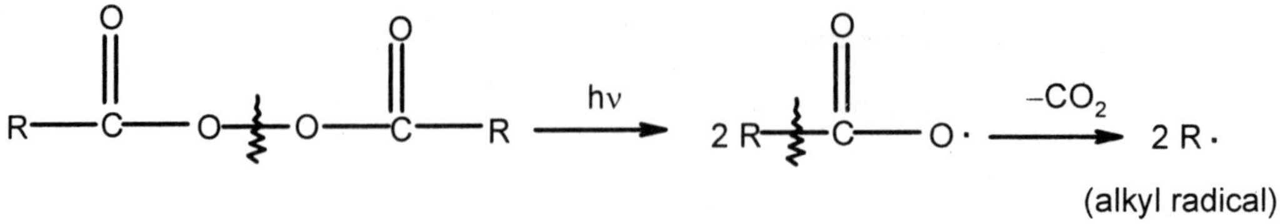

(b) *Redox Reactions*

A useful method of generating free radicals is to carry out redox reactions of appropriate substrates. Generation of benzoyl radical may be done from the auto-oxidation of benzaldehyde catalyzed by Fe^{3+}.

Cathodic reduction of ketones (on electrolysis) forms radical anions –

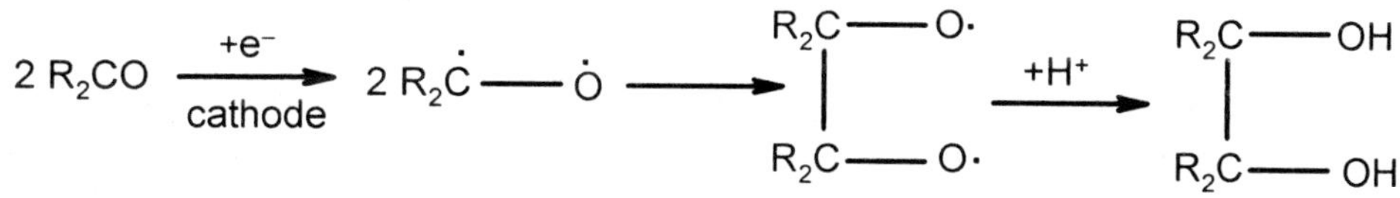

Similarly, in Kolbe's electrolysis anodic oxidation of carboxylate ions on subsequent decomposition yields alkyl radicals, which dimerize to hydrocarbon.

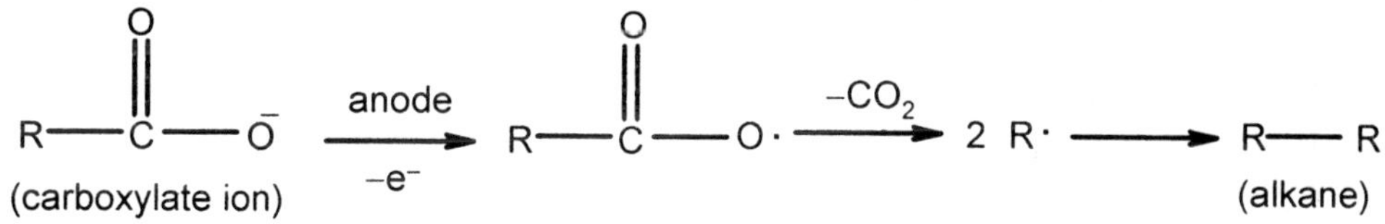

Reactions undergone by free radicals

Carbon free radicals undergo four major types of reactions:

(a) Abstraction of hydrogen

(b) Addition Reaction

(addition between two moles)

(addition with multiple bonds)

(d) Rearrangement

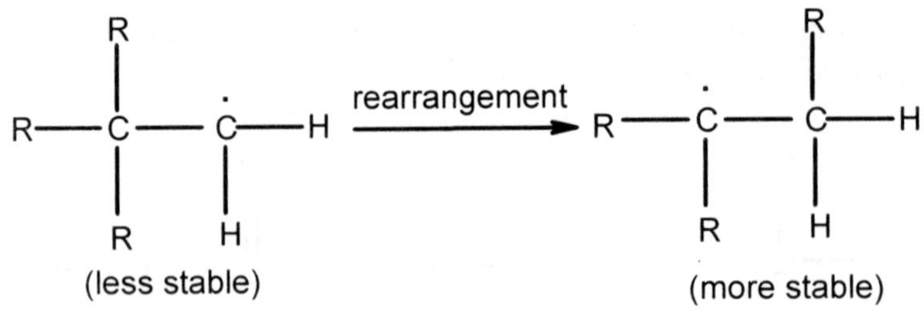

Problem 3.15 Discuss the factors responsible for stabilizing a carbon free radical.

Solution. The lifetime of a radical depends not only on its inherent stability, but also on the conditions under which it is generated. A *persistent* free radical is one that has relatively long lifetime under the conditions at which it is generated, although it may not be very stable; on the other hand a *stable* free radical is meant for its inherent stability. Stability of free radicals is influenced by the following factors:

(i) Hyperconjugative effect
(ii) Resonance effect
(iii) Captodative effect, and
(iv) Steric effect

(i) Hyperconjugative effect: Simple alkyl radicals are stabilized by hyperconjugation, and hence their relative stability is found to follow the sequence: $R_3C^{\cdot}$ (tertiary) > $R_2CH^{\cdot}$ (seconadry) > $RCH_2^{\cdot}$ (primary) > $H_3C^{\cdot}$ (methyl radical).

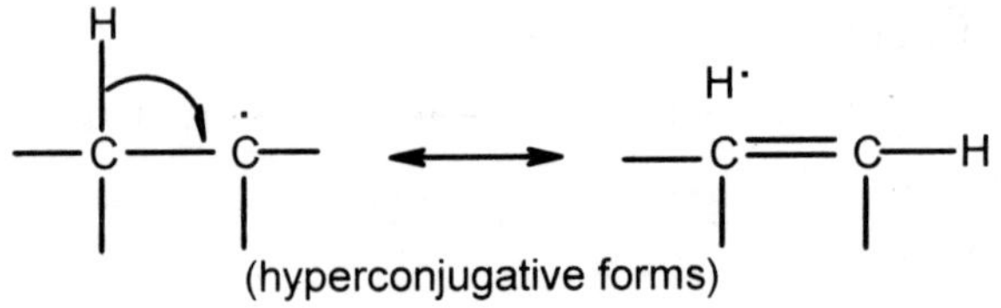

Similarly, the decreasing stability order in the series will be:

$$Me_3C^{\cdot} > Me_2CH^{\cdot} > MeCH_2^{\cdot} > H_3C^{\cdot}$$

(ii) Resonance effect: Resonance plays a major role in stabilizing free radicals. Delocalization of the unpaired electron (on the radical carbon atom) over a π orbital system contributes a much to the stability of a free radical. For examples, allylic and benzylic radicals become stabilized due to resonance effect :

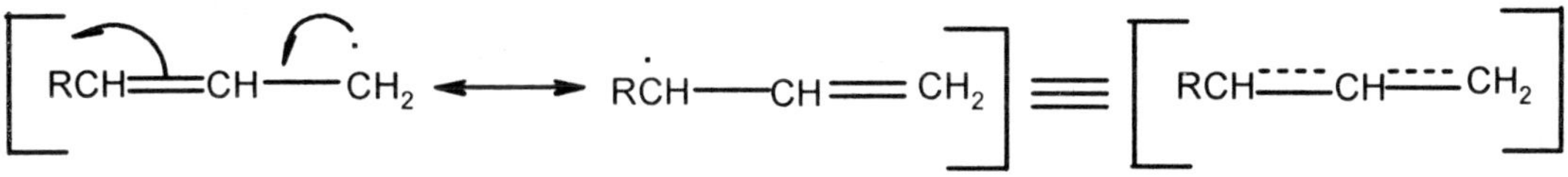

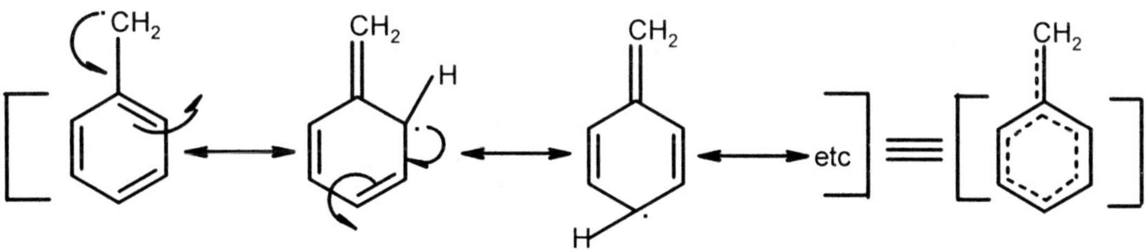

The stability of such radicals increases as the extent of effective delocalization increases, and for maximal overlapping the radical should be planar, i.e. essentially be sp^2 hybridized.

(iii) Captodative effect: If the central radical carbon is attached with both an electron-donating and electron-withdrawing groups, its stability will increase due to *captodative effect* (or *push-pull effect*) resulting from enhanced resonance.

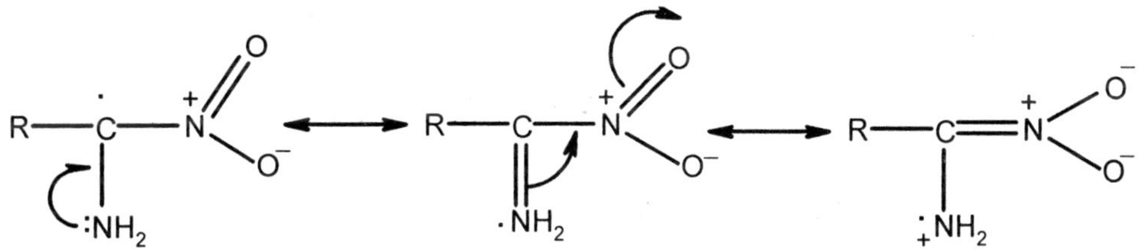

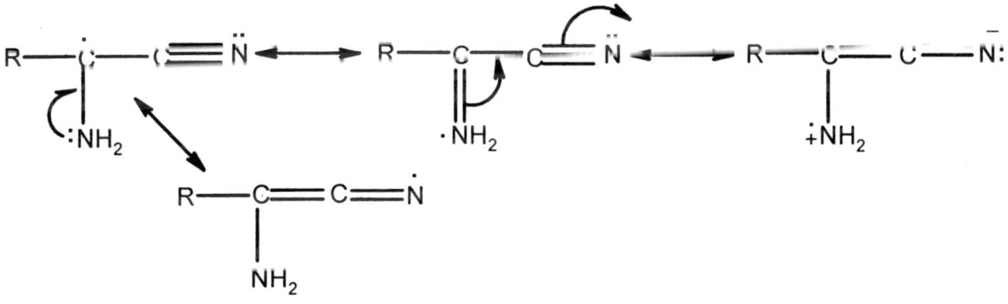

However, this effect is of much controversy.

(iv) Steric effect: Steric repulsive factor sometimes becomes very much significant in stabilizing a carbon free radical. For instance, triphenylmethyl radical ($Ph_3C^{\bullet}$) is stable enough to exist in solution at room temperature, though in equilibrium with a dimeric form.

$$2\ Ph_3C^{\bullet} \rightleftharpoons (Ph_3C)_2$$

(monomer) (dimer)

Available evidences support the following structural pattern for the dimer:

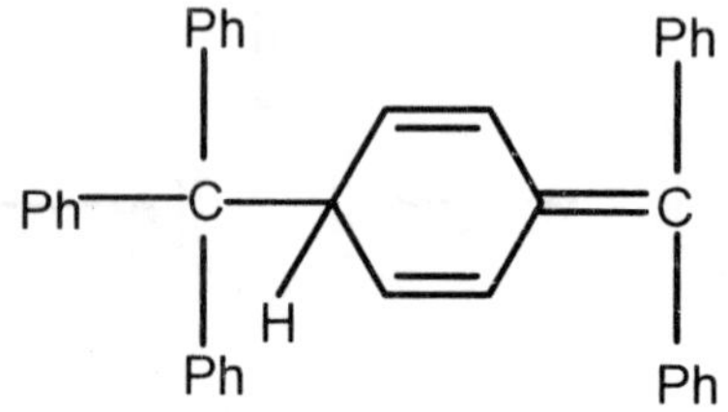

(hexaphenylethane)

Though resonance has an effect on its stabilization, it is the steric factor, which appears to be the more significant contributor towards the stability concerned. All the three phenyl rings in $Ph_3C^{\cdot}$ are not perfectly coplanar due to steric interaction between the *ortho*-hydrogens of adjacent bulky phenyl rings; in fact the triphenylmethyl radical attains a propeller shape — the phenyl rings being aligned at about 30^0 out of the common plane.

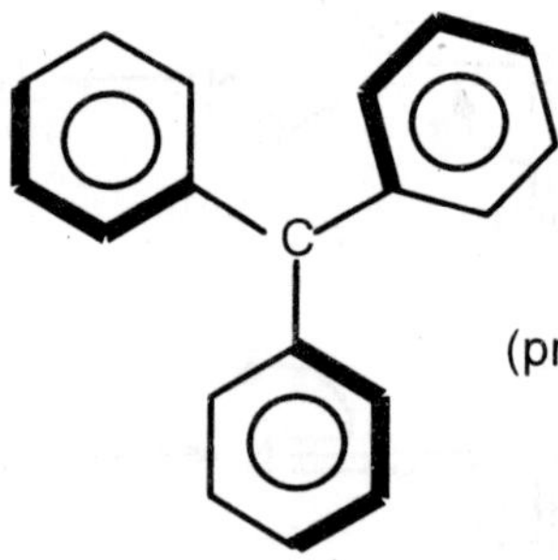

(propeller shape)

Hence, it is clear that delocalization between the unpaired electron and the overall π orbital system is too some extent disturbed. Therefore, the major reason for the greater stability of $Ph_3C^{\cdot}$ is its greater reluctance towards dimerization — because dimerization will introduce a considerable amount of steric strain (developed due to steric repulsion between the bulky phenyl groups) into the resulting molecule. Thus not the resonance, but steric hindrance to dimerization is the major cause for the stability of triarylmethyl radicals.

This conclusion is also reflected in the following phenomena — rate of dimerization of p-$PhCOC_6H_4(Ph)_2C^{\cdot}$ is less than that of $Ph_3C^{\cdot}$, and $(p\text{-}XC_6H_4)_3C^{\cdot}$ [where X= F, Cl, NO_2, CN, etc] are kinetically stable and do not dimerize.

Problem 3.16 Between (I) and (II) which one would show higher rate of dimerizaton?

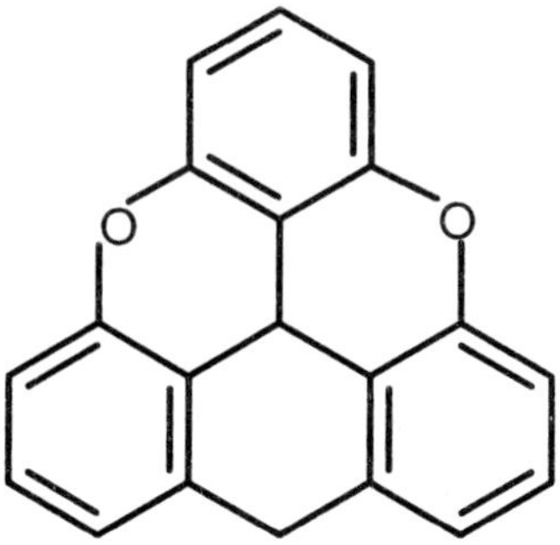

(I)

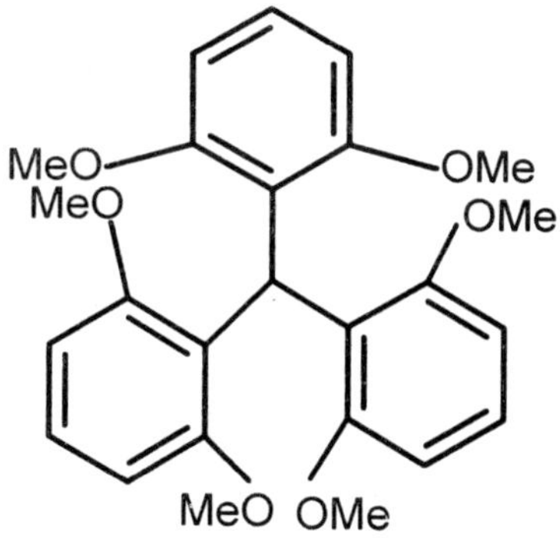

(II)

Solution. The radical (I) has much less steric hindrance and will essentially be planar; so extent of delocalization into the radical is greater than that in the radical (II), which suffers from large steric repulsive strain. Thus, apparently it may be expected that (I) would be more stable and would show relatively lower rate of dimerization. But in practice, reverse is the true. The radical (I) exists primarily in dimeric form — dissociates only to a small extent in solution. On the contrary, (II) does not undergo dimerization even in the solid state. Hence not the resonance, but steric hindrance is the major cause behind the greater relative stability of (II) and its reluctance to dimerize. Actually dimerization will lead to incorporate much steric repulsive strain (developed due to steric interactions between the *ortho*-substituents of the three adjacent bulky phenyl rings) into the dimmer — the phenomenon that is energetically unfavourable.

Problem 3.17 Name a vitamin that plays an important role in stabilizing a carbon free radical.

Solution. It is vitamin B_{12} that assists certain enzymes (*viz.* Homocysteine methyltransferase) in the production and stabilization of methyl radicals. These methyl radicals then take part in the synthesis of important cellular components.

▲ On Carbenes

Problem 3.18 Define carbenes. What are the *singlet and triplet* carbenes? Discuss their geometry.

Solution. Carbenes are uncharged reactive intermediates containing a divalent carbon atom; the central carbon is covalently bonded to two atoms/groups and bears two non-bonded electrons, either paired or unpaired. Thus a carbene carbon has only six valence electrons and is electron-deficient. The simplest carbene is methylene (:CH_2). Some common carbenes are dichlorocarbene (:CCl_2), dibromocarbene (:CBr_2), diphenylcarbene (:CPh_2), etc. Carbenes are too reactive to be isolated and stored, but have been trapped in frozen argon for spectroscopic study at very low temperature.

If the two non-bonded electrons of a carbene species are paired, it is called *singlet*, while the species is called *triplet* when the non-bonded electrons remain unpaired. A *singlet* carbene is diamagnetic, while the *triplet* is paramagnetic.

(singlet; diamagnetic) (triplet; paramagnetic)

Triplet carbenes bear unpaired electrons, and thus are observable by ESR spectroscopy; but *singlet* carbenes are not observable by ESR spectroscopy since the non-bonded electrons in *singlet* carbenes are paired up.

Geometry

In *singlet* carbene the central carbon atom is presumably sp^2 hybridized with trigonal geometry. An unshared pair of electrons occupies one of the sp^2 hybrid orbitals and the other two are utilized in forming two covalent bonds. There is an empty p-orbital extending above and below the plane.

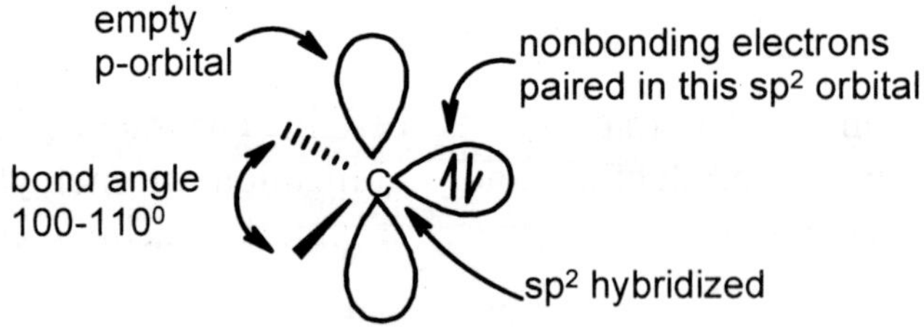

(singlet carbene trigonal shape bond angle less than120⁰)

On the other hand, the central carbon of a *triplet* carbene is sp hybridized. The two hybrid orbitals form covalent bonds with two atoms/groups, and the two unhybridized p-orbitals contain one electron each.

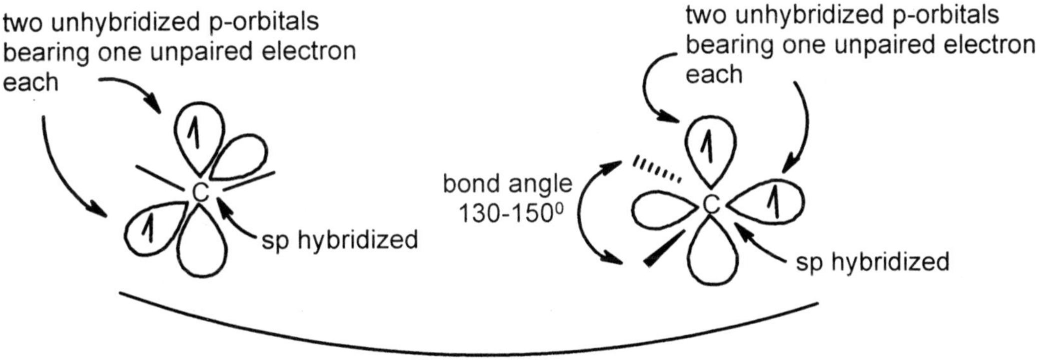

(triplet carbene)

The sp-hybridized carbenes are supposed to be linear or near-linear, but in most of the cases these are nonlinear, because bending lowers the energy of an occupied orbital. Structural information is available for a number of carbenes (e.g halocarbenes, arylcarbenes) from experimental observations (e.g. rotation-vibration spectra, absorption spectra). The H–C–H bond angle of singlet methylene (**:**CH_2) was determined as 102.4^0, whereas for triplet state the value is 133.8^0. Spectroscopic studies of a variety of carbenes showed that halomethylenes are ground-state singlets with bond angles in the range $100\text{-}110^0$, whereas methylene, arylmethylenes, and alkylmethylenes are ground-state triplets with bond angles $130\text{-}150^0$ and have excited singlets with angles of $100\text{-}110^0$.

Problem 3.19 Comment on the relative stability between *singlet* and *triplet* carbenes.

Solution. Although one might expect that a *single*t carbene would be less energetic (i.e. is of lower energy) since the unshared electron pair is in the hybridized orbital, but the ground state remains in *triplet* state. This is because of the fact that there is an energy cost to overcome the electronic repulsive force developed in pairing up the two electrons in a single orbital; unless the energy gap between hybridized orbital and unhybridized p-orbital is larger than this cost, the electrons will remain in the separate orbitals with spins parallel. It is, therefore, supposed to believe that the *triplet* state of a carbene is more stable than the *singlet* state, and is expected to be the ground state. However, depending on the mode of generation, a carbene may initially be formed in either state, irrespective of its stability.

It has been established that aryl, alkyl and dicyano carbenes remain in *triplet* ground state; on the contrary carbenes substituted by π-donor atoms (N, O, halogens) usually have *singlet* ground state. Biarylmethylene (I) has been reported to have *triplet* ground state, while (II) and vinylidine carbenes (III) attain *singlet* ground state.

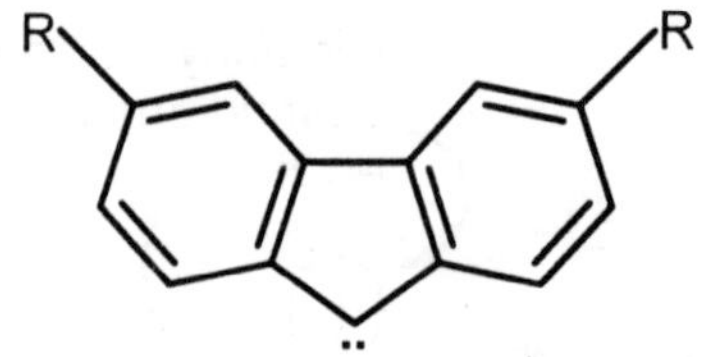

I: R= H; II: R= OCH_3

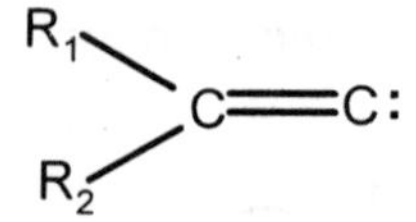

(III)

Problem 3.20 Mention few reactions by which carbenes can be generated.
Solution. Carbenes can be generated as reaction intermediates by the following reactions:

(a) Action of a base on alkyl halides:
Haloforms on treatment of base forms dihalocarbenes; e.g.

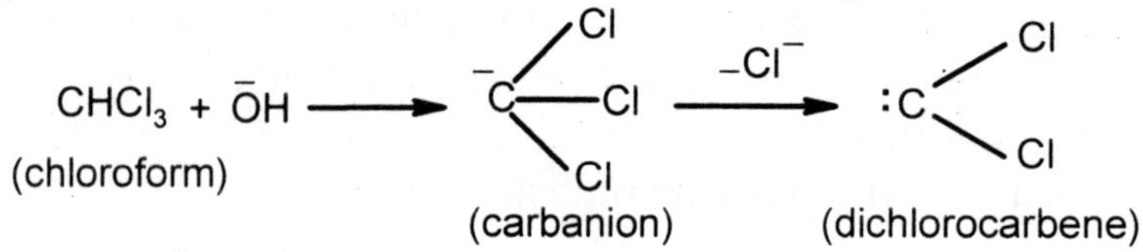

(b) Decomposition of diazoalkanes and ketenes:
Diazoalkanes as well as ketenes on photochemical or thermal decomposition yield carbenes.

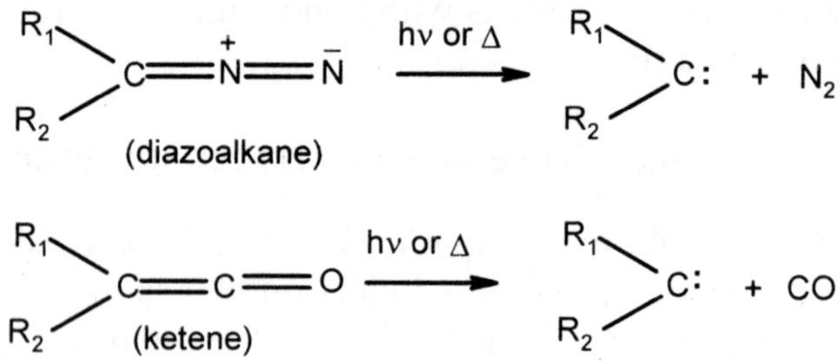

(c) From cycloeliminations:
Thermal or photochemical cleavage of various cyclic compounds offers a method for generating carbenes.

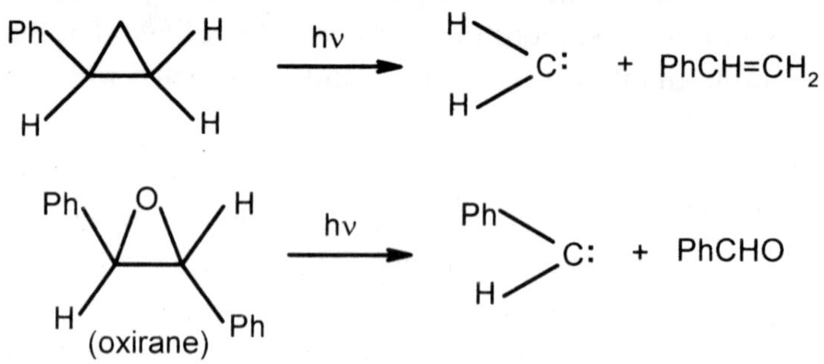

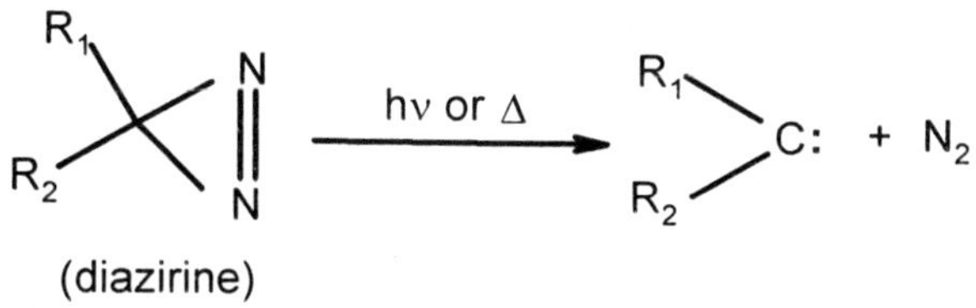

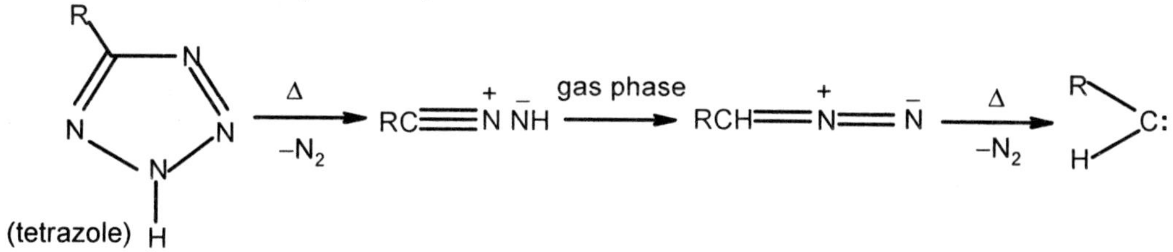

Problem 3.21 Mention few reactions undergone by carbenes.
Solution. Carbenes undergo following reactions having high synthetic utilities.
(a) Cycloaddition reactions:
Carbenes add to alkenes giving rise to cyclopropane derivatives; the stereochemistry of these cycloadditions is so specific that it may be used as a diagnostic test for identification of *singlet* and *triplet* carbenes. Singlet carbenes add to double bonds in a *stereospecific manner* so that stereochemistry of the alkene in its product remains intact.

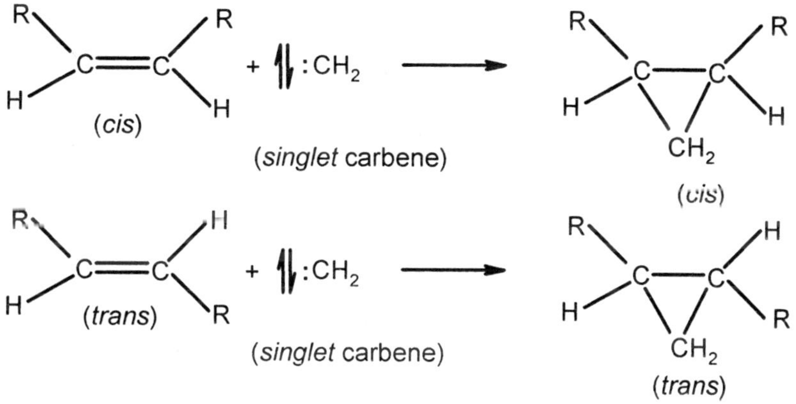

On the contrary, *triplet* carbenes add non-stereospecifically giving rise to a mixture of products.

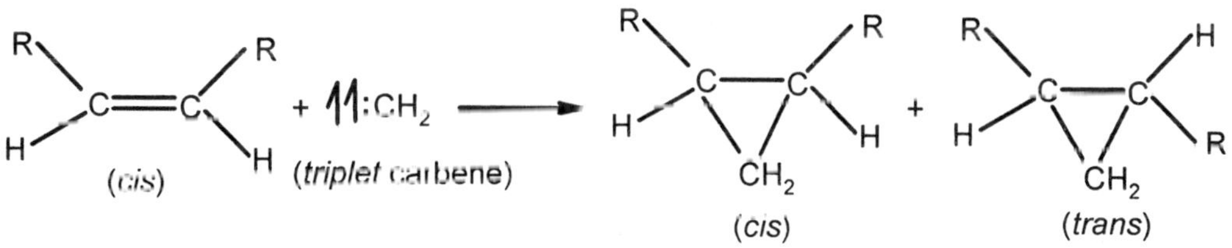

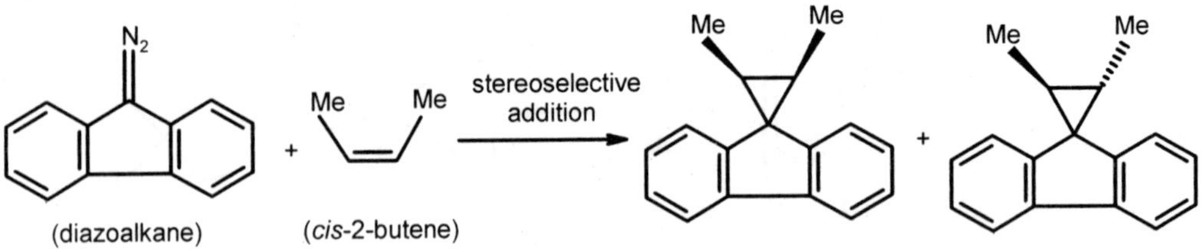

The observed difference in behaviour between these two electronic states can be explained in the following way. The addition of a *singlet* carbene to an alkene takes place in a one step *concerted process* as there is no spin restriction on the simultaneous formation of two new σ-bonds of the cyclopropane, and hence it is a stereospecific process.

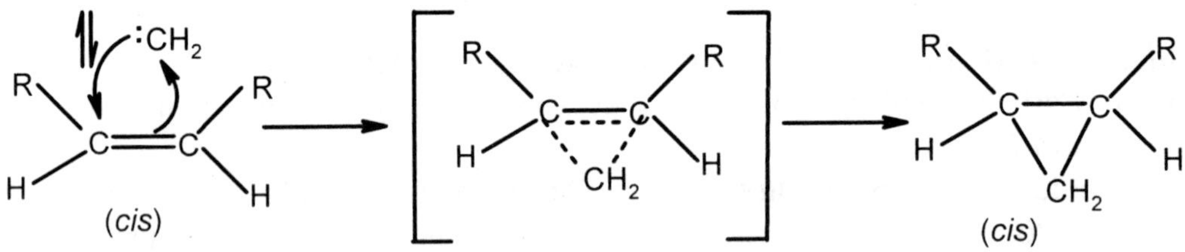

But the cycloaddition of a *triplet* carbene is a step-wise process. In the first step one of the unpaired electrons forms a bond with the electron from the double bond that has the opposite spin, leaving two unpaired electrons with parallel spins (i.e. gives rise to a *triplet* diradical intermediate); the diradical intermediate now must have to change the spin state of one of the unpaired electrons (i.e. to undergo spin inversion), by some collision process, before it collapses to a cyclopropane. If this process takes long enough for rotation to occur about bonds in the intermediate, a mixture of products should result. These steps can be visualized as follows:

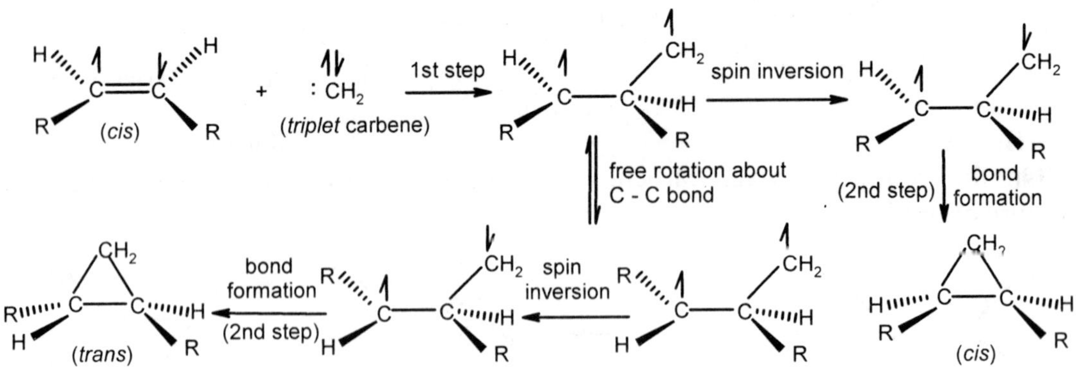

Dibromocarbene undergoes addition reaction with conjugated dienes to give predominantly 1,2-addition product.

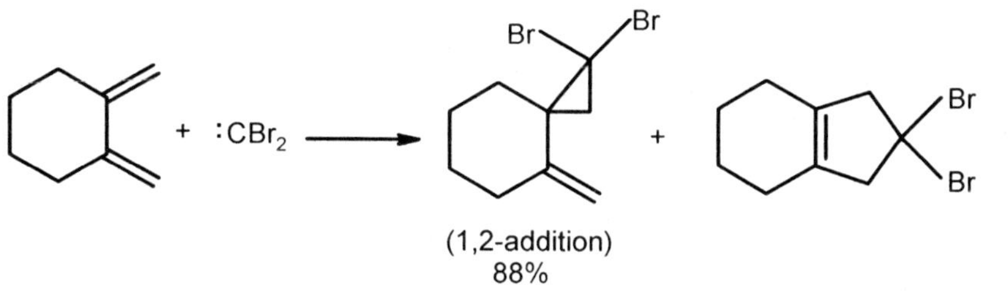

Another interesting cycloaddition reaction undergone by dichloro or dibromomethylene is cited below:

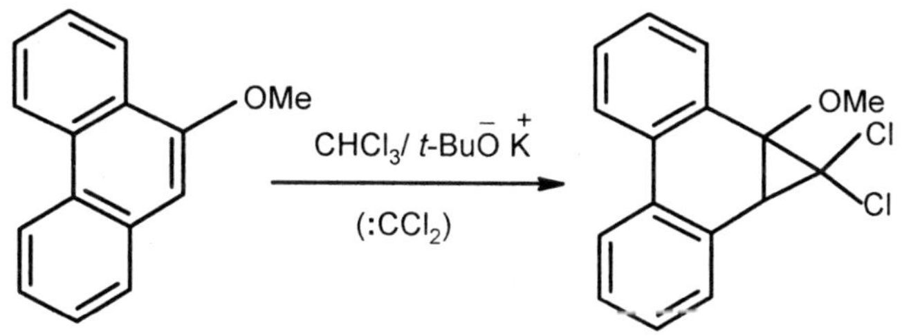

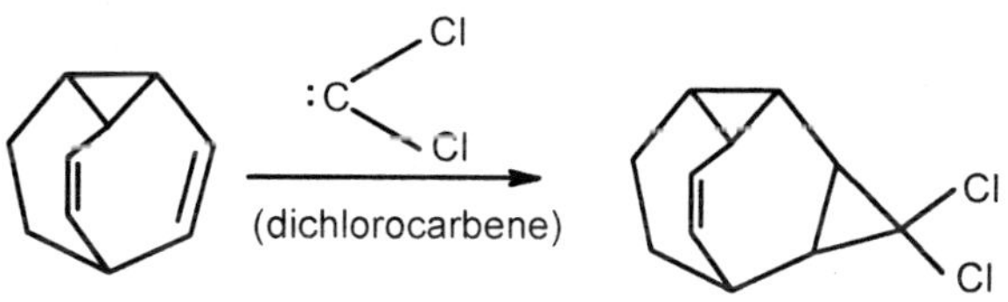

(b) Insertion and hydrogen abstraction:

A *singlet* carbene inserts through *concerted* process into C- H bond with retention of configuration.

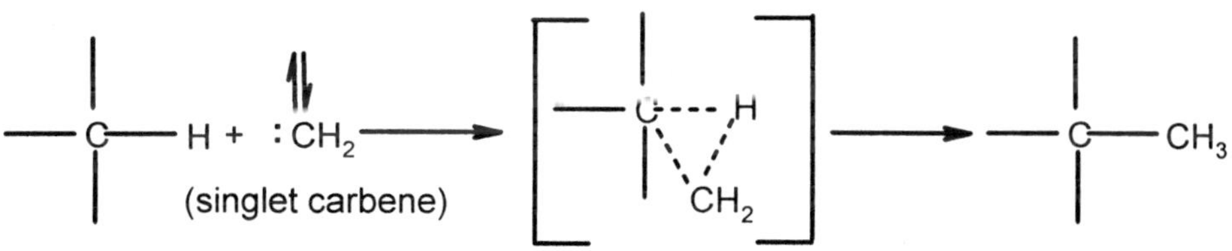

But the insertion mechanism for a triplet carbene is believed to abstract hydrogen first to form a radical pair, which subsequently recombines to make a new C–C bond.

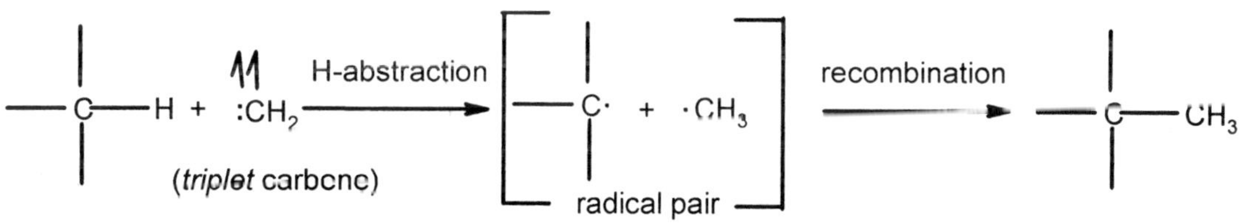

Thus the above insertion reactions are intermolecular one; similarly intramolecular insertion reactions are also known. Such insertion reactions lead to rearranged products.

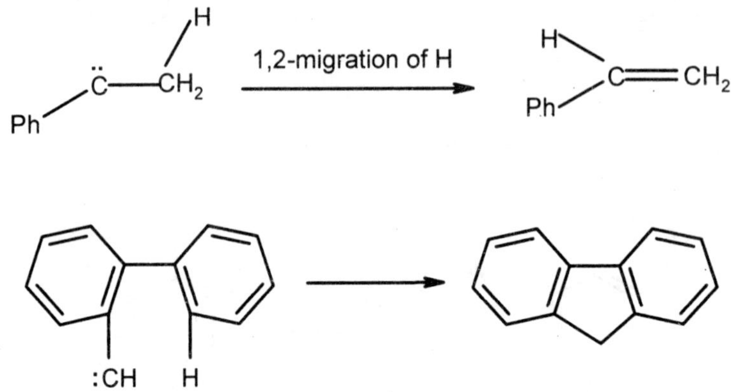

If the carbene carbon becomes β with respect to the C–H bond, a cyclopropane ring is formed along with other products.

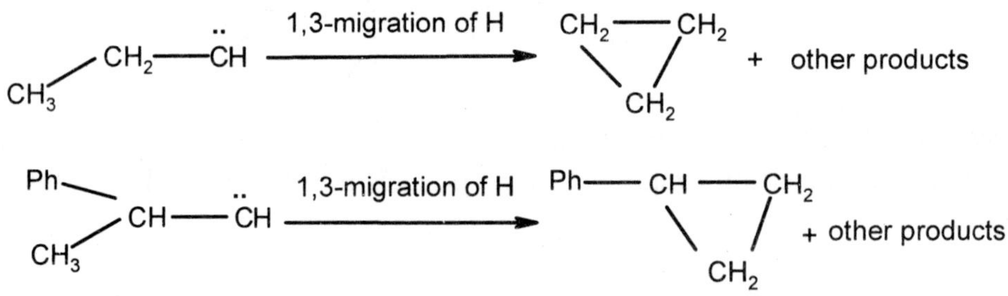

Insertion reactions are very useful for preparing strained molecules. For example:

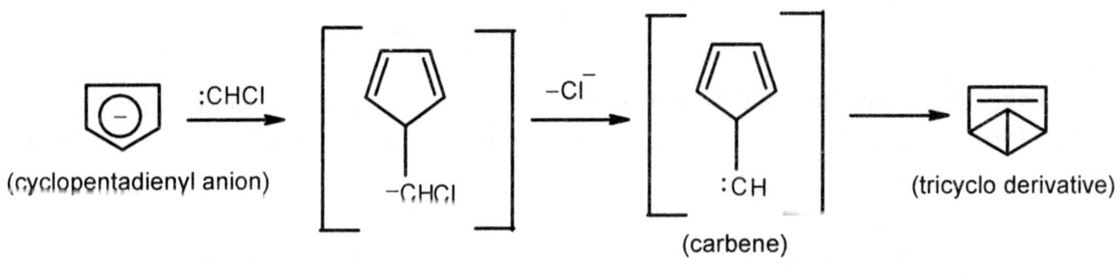

(c) Ring expansion and contraction reactions:

By using carbene intermediates ring expansion as well as ring contraction reactions, which are of highly synthetic significance, may be carried out. Examples are:

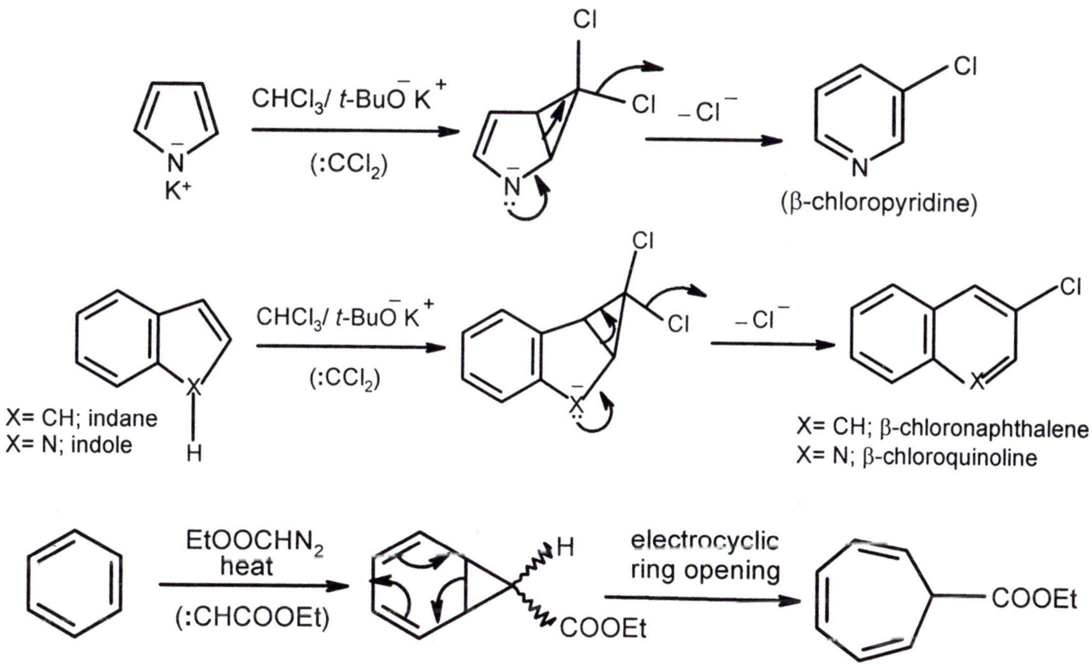

When a cyclic diazoketone is decomposed, a rearrangement results giving rise to ring contraction product.

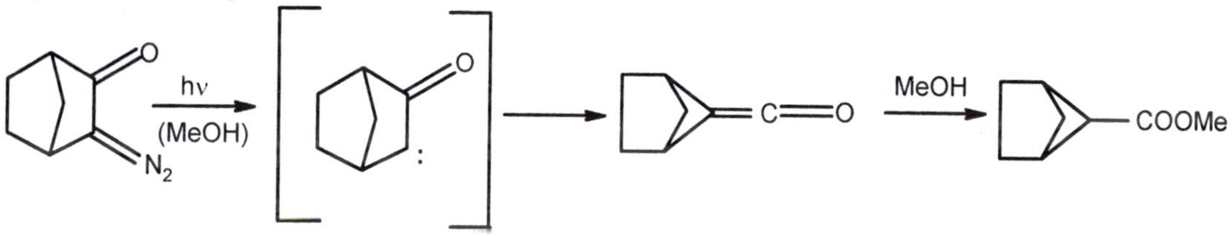

Problem 3.22 Stereospecificity of the addition reaction of a carbene to a double bond is regarded as a diagnostic test for identification of its multiplicity. Is there any limitation to this method?

Solution. Skell and Woodworth developed an ingenious method to distinguish between *singlet* and *triplet* carbenes, based on the common but separate addition of these two types of carbenes to double bonds to form cyclopropane derivatives. *Singlet* carbene undergoes stereospecific *cis*-addition, while the other adds non-stereospecifically. So identification of reaction products reveals their multiplicity.

However the fact that addition is stereospecific does not mean that the carbene must be a singlet. In some cases, bond rotation may be quite slow, and spin-inversion rapid, leading to stereospecific addition. This method suffers from another limitation.

Actually carbenes are generated in *situ*; when such species are generated by photolytic means, they are often in a highly excited *singlet* state. Under this condition they add to double bond in a stereospecific manner to form a cyclopropane derivative of expected configuration; but the cyclopropane derivative as formed also carries excess energy and remains in the excited state — hence there is a chance for the cyclopropane derivative to undergo *cis-trans* isomerization. It has been observed that under certain conditions (low pressures in the gas phase) excited cyclopropane undergoes such type of reaction. In that situation *triplet* carbene may seem to be involved although in reality the *singlet* carbene was formed. Thus, during application of this method one should consider the matter carefully.

Problem 3.23 Do you cite examples of carbenes that exist as stable compounds?

Solution. Very few examples of stable carbenes are known; carbenes **(I–IV)** are among them.

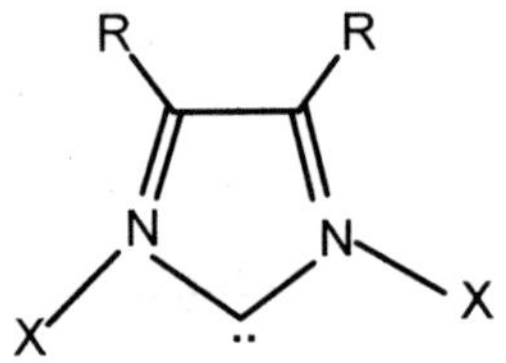

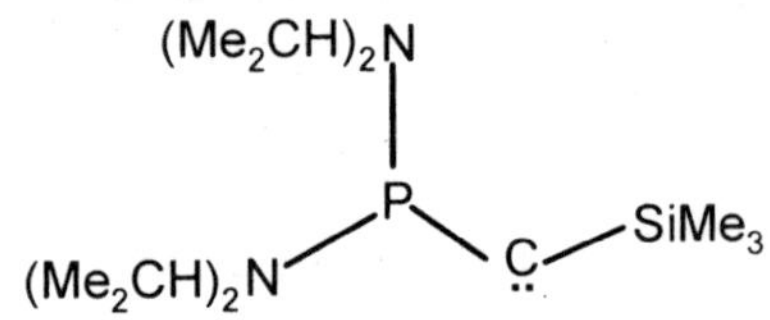

(IV)

(a red liquid)

I: R = H; X = 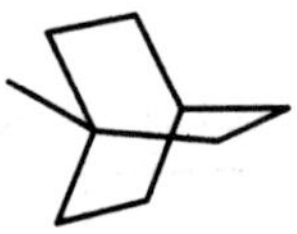 ; II: R = H; X = 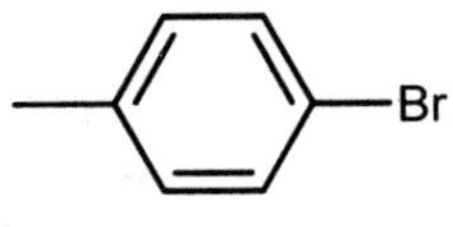

(colourless crystals; singlet state)

III: R = X = Me

It should be noted that these carbenes are the exceptions — in general carbenes are very short-lived and highly reactive species.

Problem 3.24 Carbenes are necessary electrophilic species. Compare the electrophilicity between (I) and (II).

:CHCOOEt (I) :C(R)OMe (II)

Solution. Carbenes are electron-deficient species, and behave as electrophiles. But carbenes that have strong elctron-donating substituents become less electrophilic. In (II) the electron-deficient carbene carbon is attached with methoxyl function; non-bonded electrons on methoxyl oxygen are delocalized — localized electrons on oxygen move into the orbitals shared between carbon and oxygen.

(II)

On the contrary, in (I) the carbene carbon is attached with a strong electron-withdrawing ester (COOEt) substituent that enhances the electrophilicity of the carbene concerned through its powerful (–I) effect.

Hence, (I) would show greater electrophilicity than (II).

Problem 3.25 *Cis*-2-butene on photoirradiation in presence of diazomethane yields *cis*-1,2-dimethylcyclopropane stereospecifically. If the reaction is carried out in presence of large excess of perfluoropropane, then what will happen?

Solution. In the first situation *singlet* carbene ($:CH_2$), formed on photodecomposition of diazomethane, undergoes *cis*-addition with the double bond stereospecifically. If the reaction is carried out in presence large excess of perfluoropropane (C_3F_8), the *singlet* carbene suffers from more collisions before it reacts and thereby the chance of spin-inversion from singlet state to triplet state is increased. Addition to double bonds is then less stereospecific.

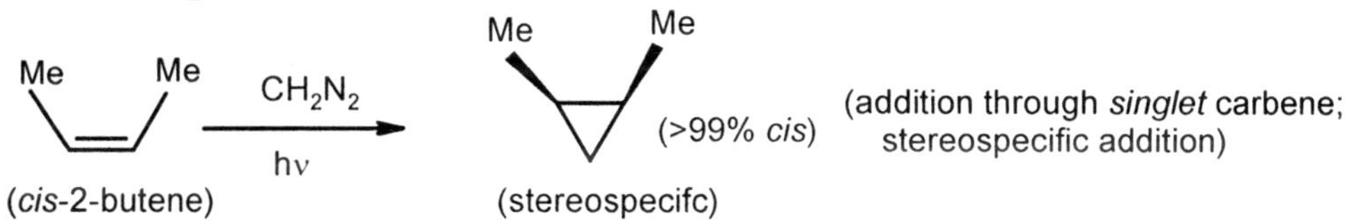

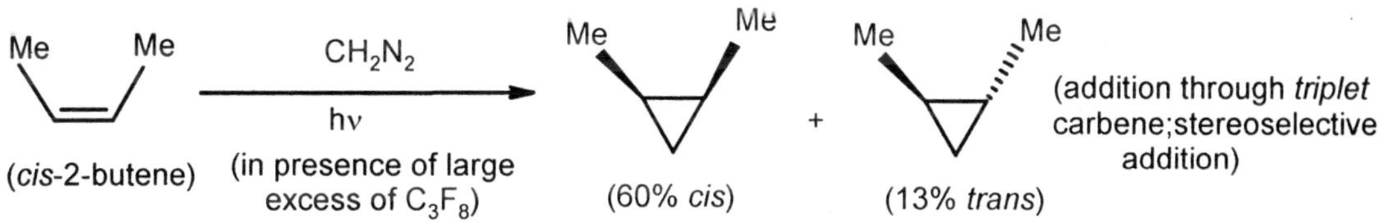

▲On Nitrenes

Problem 3.26 Write a short account on *nitrenes*.

Solution. A nitrene is a nitrogen-containing neutral but electron-deficient species in which the nitrogen has only a sextet of valence electrons. Nitrenes may be represented as:

R—N̈: (nitrene)

(R= H; halogens; $COCH_3$; CH_3; C_6H_5, etc)

The species are *iso*-electronic with carbenes, and are also of two types — *singlet* nitrenes and *triplet* nitrenes.

Like carbenes, *singlet* nitrenes add to double bonds in a stereospecific manner while triplet nitrenes react non-stereospecifically. Nitrenes are high-energy species, and very much difficult to isolate them under ordinary conditions; however alkyl nitrenes have been isolated by trapping in matrices at -269^0C (4^0K). Aryl nitrenes are relatively less reactive, and have been possible to trap at -196^0C (77^0K).

Nitrenes have been postulated as intermediates in several reactions. One such example is shown:

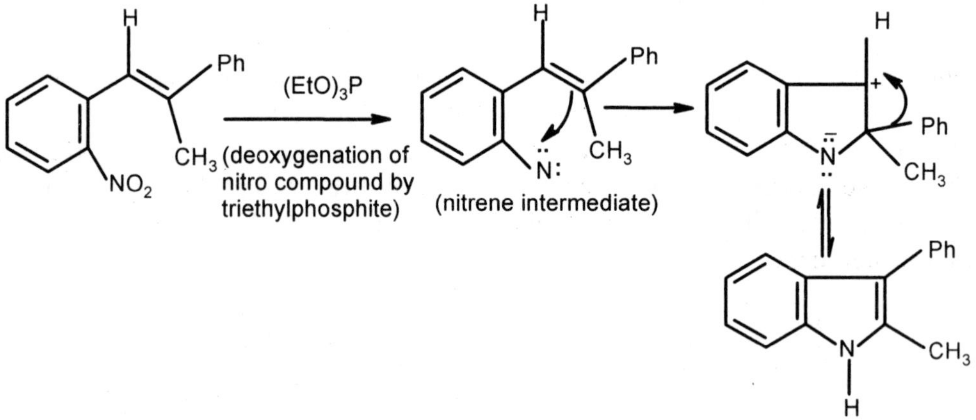

Generation

Few methods of generating nitrenes, and some of their reactions are summarized below:

(a) Decomposition of azides:

Photolytic or thermal decomposition of azides leads to the generation of nitrenes.

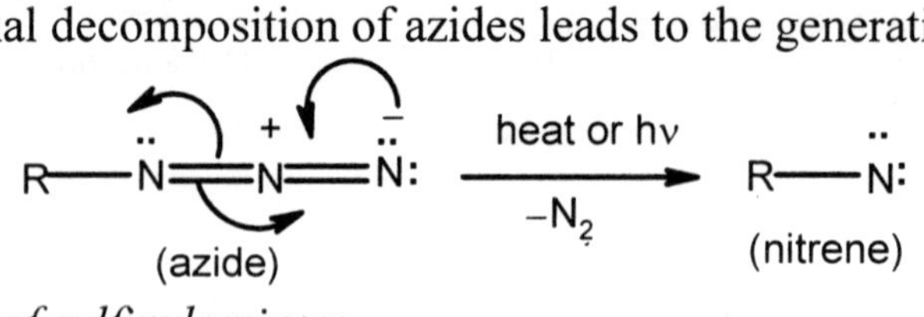

(b) Decomposition of sulfinylamines:

Thermal decomposition of sulfinylaniline yields an arylnitrene.

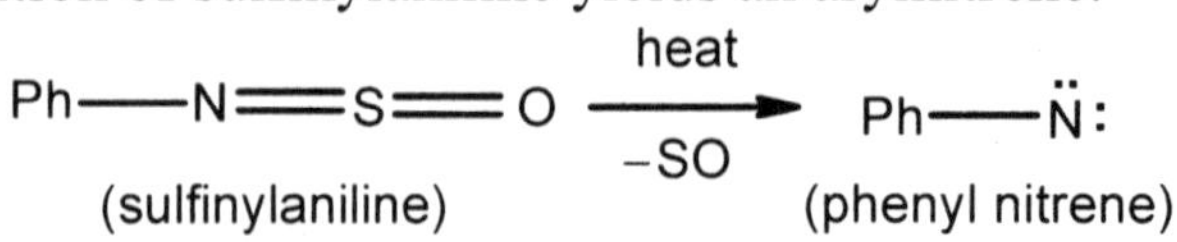

(c) Deoxygenation:

Nitrenes may also be generated by deoxygenation of nitro and nitroso compounds carried out in presence of triethyl phosphite.

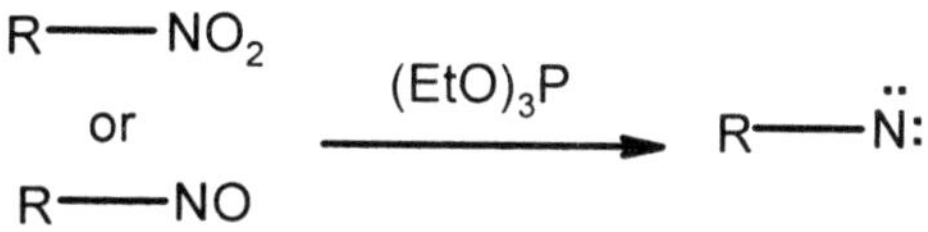

(d) α-Elimination:

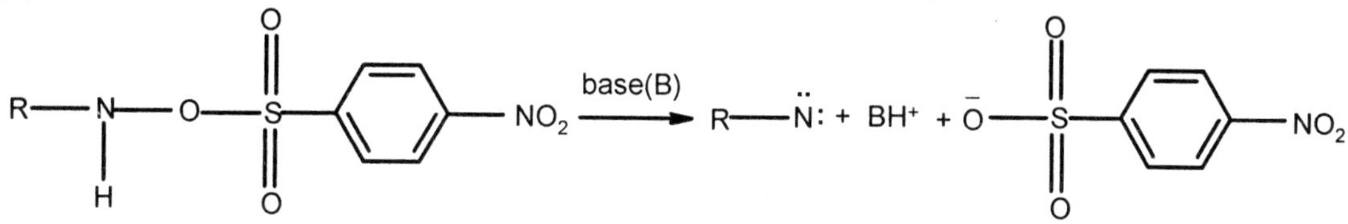

Reactions undergone by nitrenes

Nitrenes undergo chemical reactions similar to those of carbenes. Significant reactions are:

(a) Addition to C=C double bond:

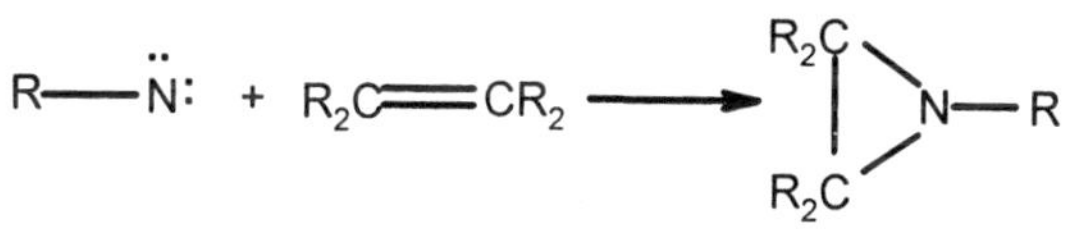

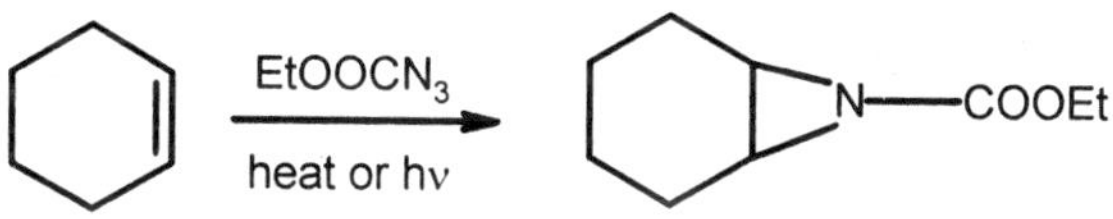

(b) Insertion reaction:

Nitrenes, particularly acyl nitrenes and sulphonyl nitrenes, can insert into C – H bond:

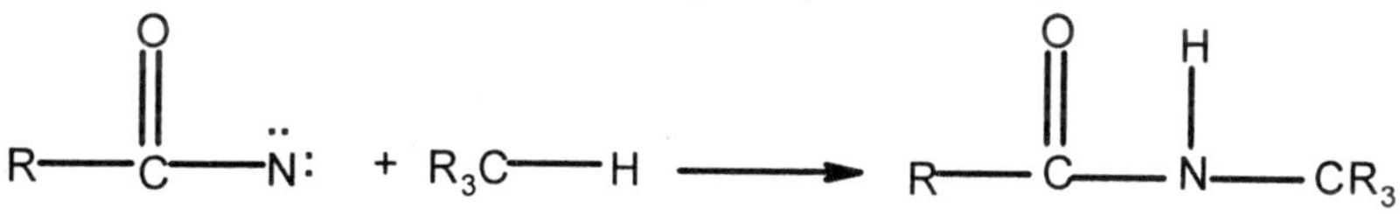

(c) Rearrangement Reaction:

Nitrenes undergo rearrangement reaction so rapidly that it is usually difficult to come in conclusion whether it was present in a free state or the migration was taken place at the time of its generation.

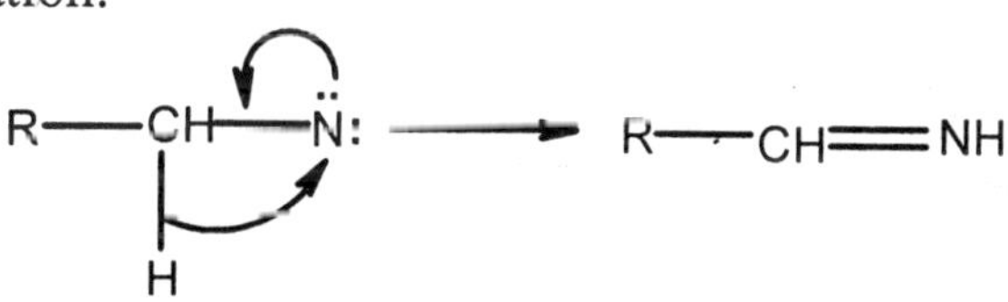

▲ On Benzyne

Problem 3.27 Offer evidence in support of the existence of benzyne intermediate.

Solution. Benzyne has not yet been isolated under ordinary conditions, but it has been made possible to isolate the intermediate in an argon matrix at 8^0 K. Evidence for the existence of benzyne has been received both from '*trapping*' experiments as well as spectroscopic technique.

'Trapping' experiment

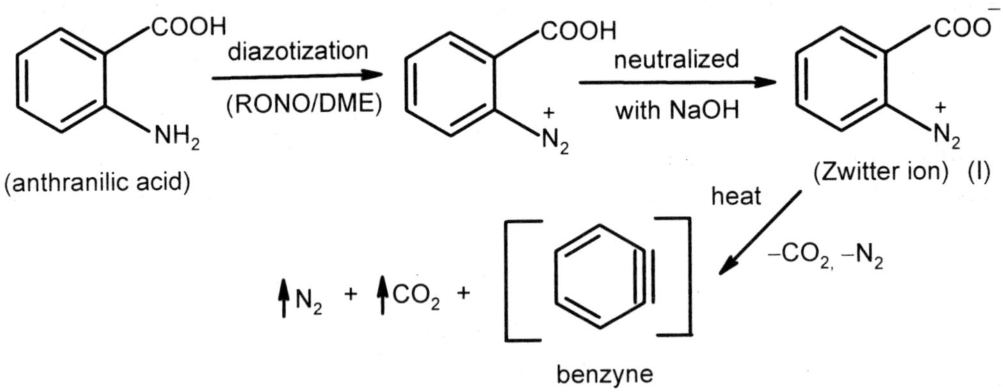

Benzyne can readily undergo Diels-Alder addition; thus generation of benzyne in the presence of furan or anthracene gives the respective adduct:

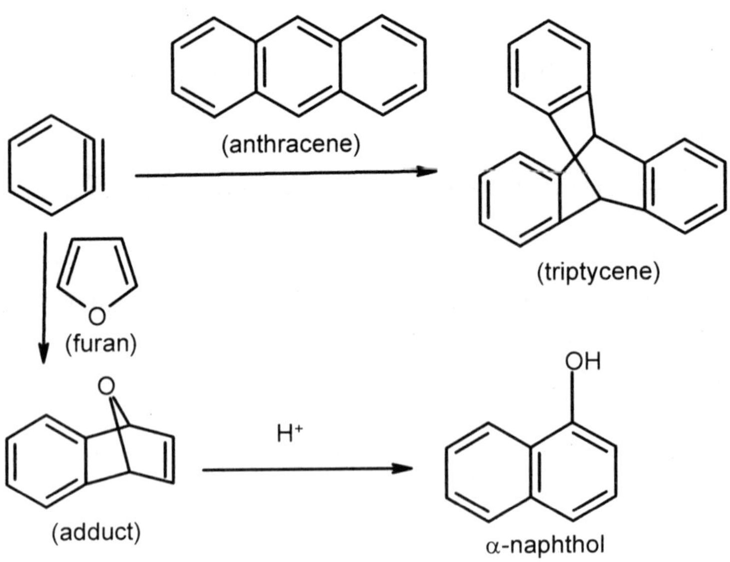

Spectroscopic evidence

A more convincing demonstration of the existence of benzyne was received from mass spectral analysis of the data obtained on introduction of the zwitter ion (I) into the heated inlet of a mass spectrometer. The resulting mass spectrum was found to show significant mass peaks at *m/z* 28, 44, 76 and 152. The peak at *m/z* 152 appears for the biphenylene dimer (II) developed on dimerization of benzyne intermediate that exhibits a strong peak at *m/z* 76 in the mass spectrum. The mass peaks at *m/z* at 28 and 44 correspond to evolving out of nitrogen and carbon dioxide gas.

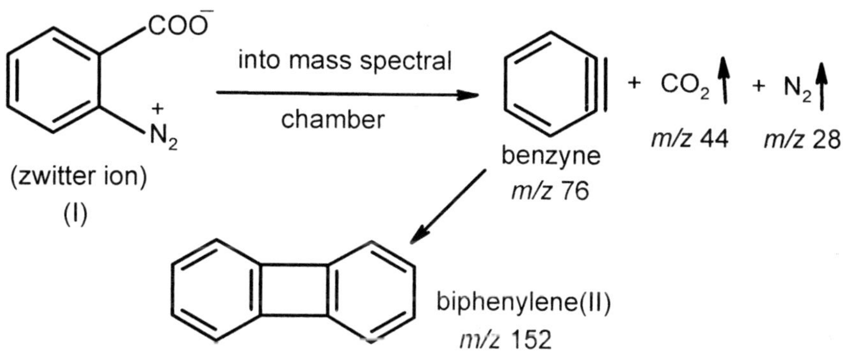

The lifetime of a particle in the spectrometer is about twenty nanoseconds, so benzyne can exist for at least that long in the gas phase.

Problem 3.28 How benzyne triple bond differs from alkynic triple bond?

Solution. Nature of benzyne triple bond is totally different from alkynic triple bond that is of linear geometry; benzyne triple bond can't be linear, as this would require tremendous deformation of benzene ring in order to accommodate the 180^0 bond angle required by the sp-hybridized carbons. The extra pair of electrons does not affect the aromaticity. In fact one π-bond is normal and just a part of the aromatic nucleus, whereas the other one is abnormal and is formed by overlapping of two sp^2 orbitals outside the ring.

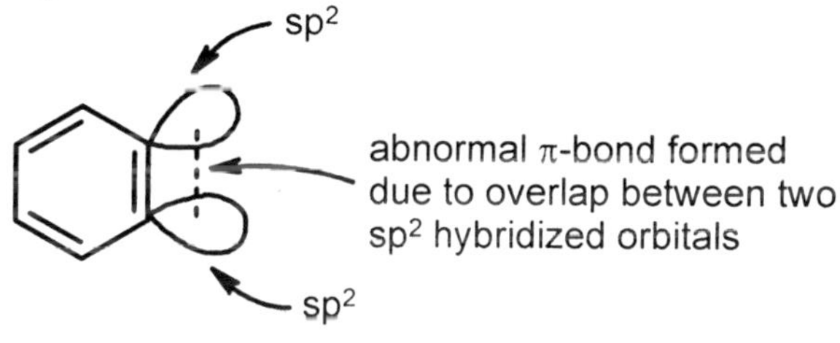

Unlike alkyne, benzynes do not have formal triple bond because the following two canonical forms (I & II) contribute to the resonance hybrid of benzyne.

(I) (II)

The fact has also been substantiated by the appearance of IR stretching band for the benzyne triple bond at about 1846 cm^{-1}, which is intermediate between those of double bond (ethylene, ~ 1660 cm^{-1}) and triple bond (ethyne, ~2150 – 1980 cm^{-1}).

Problem 3.29 Suggest few methods of generating benzyne intermediates.

Solution. Benzynes can be *generated* by the following methods:
(a) Oxidation of 1-aminobenzotriazole with lead tetraacetate:

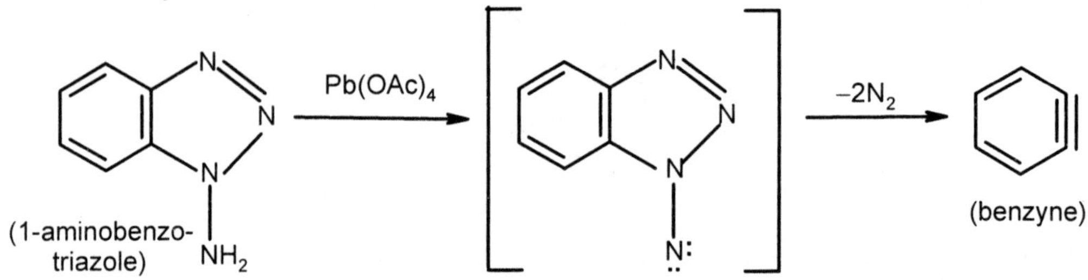

It is a very useful method for generating benzynes.

(b) Thermal decomposition of phthalic anhydride and also of diazo-derivative of anthranilic acid:

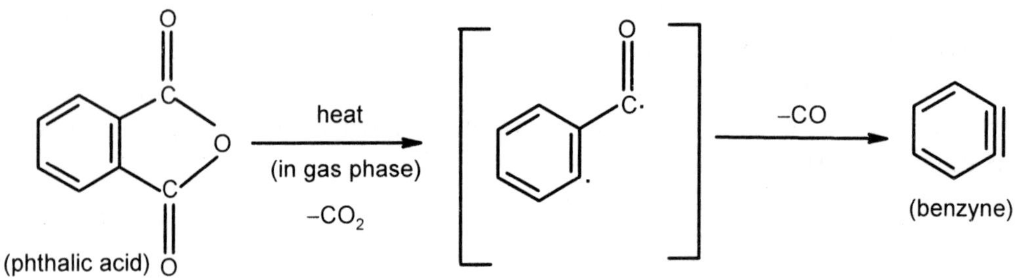

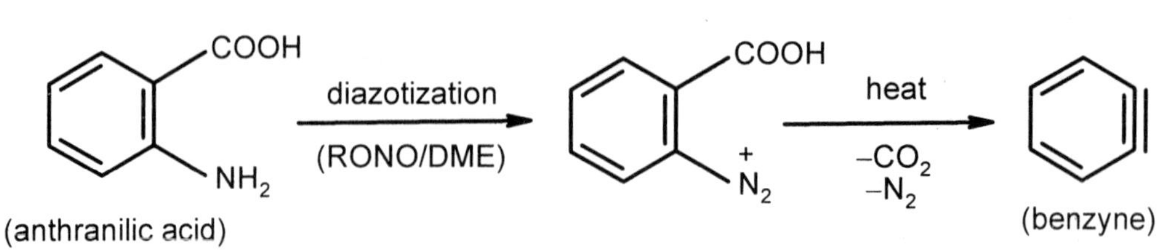

(c) Photolytic decomposition method:

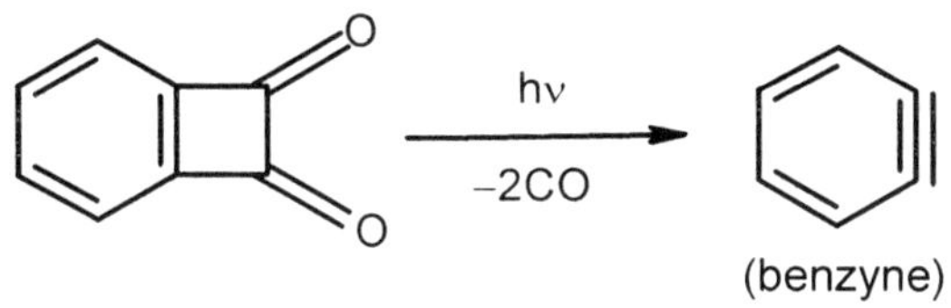

(d) From halobenzenes:

Halobenzenes on treatment with strong bases (e.g. alkoxides, amide anions, etc.) generate benzynes as intermediates.

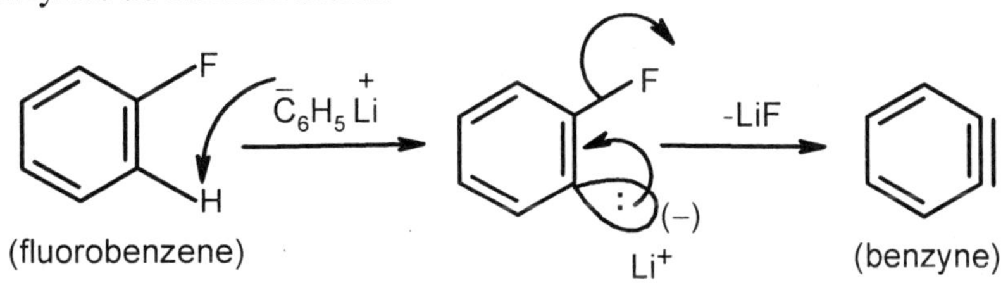

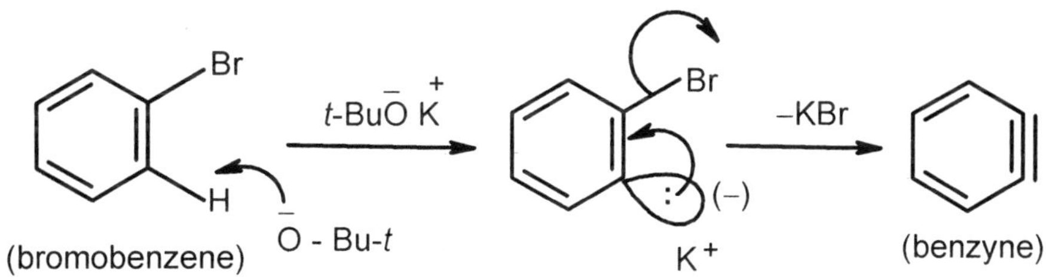

Problem 3.30 The following reactions proceed through benzyne mechanism. Predict the product(s) in each case.

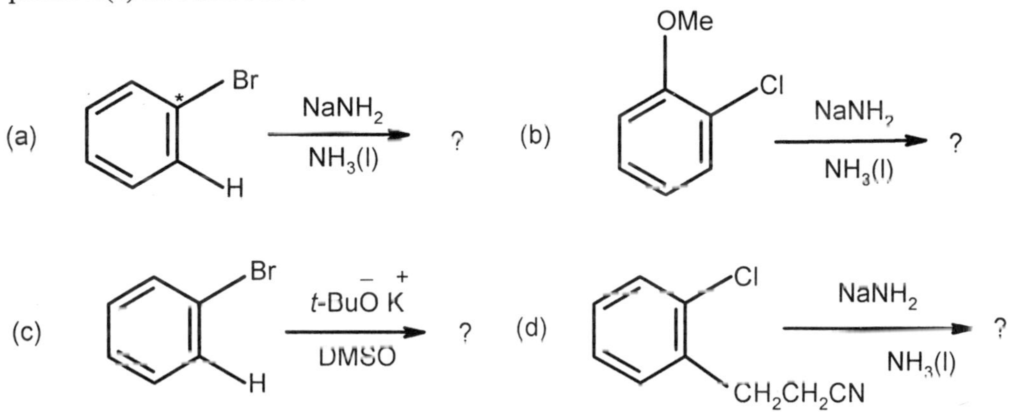

(e) 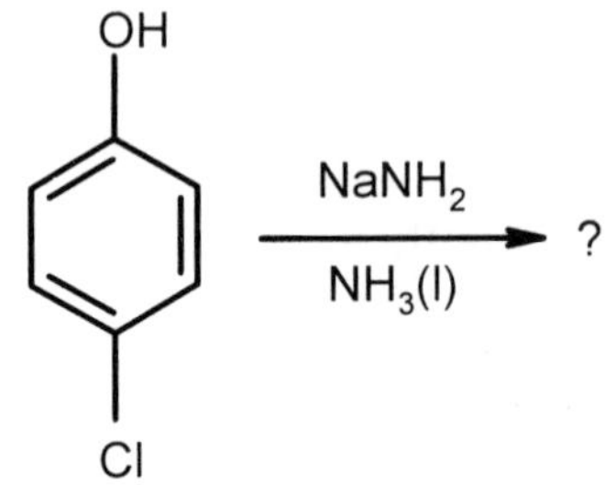

Solution.

(a) 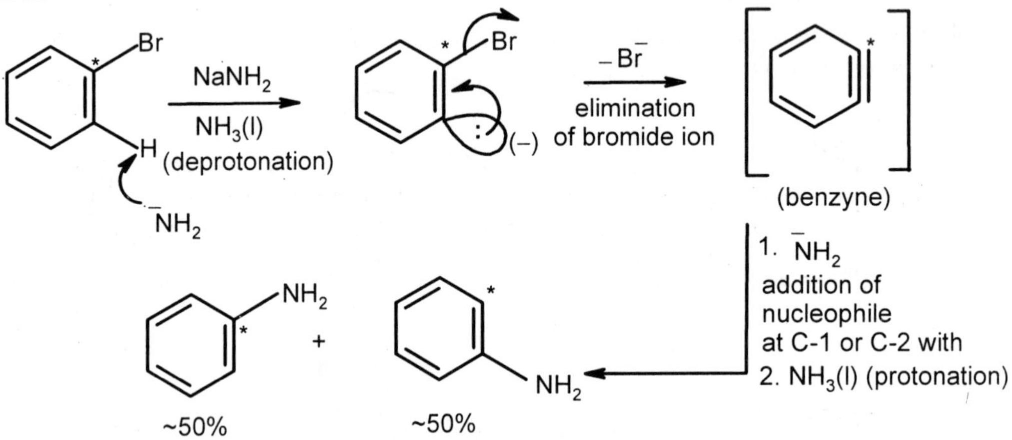

b) 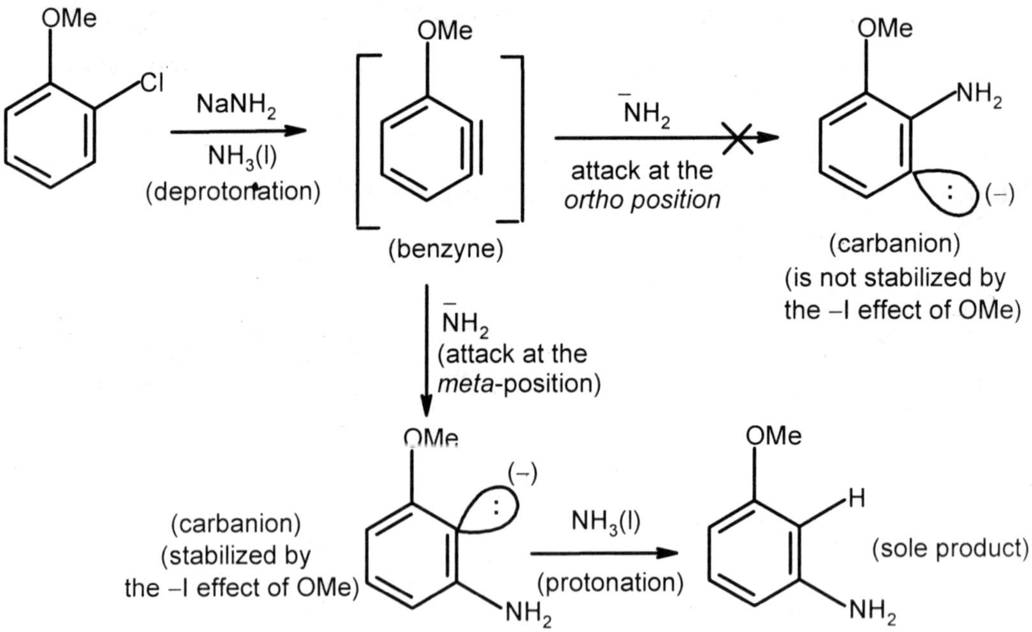

(c)

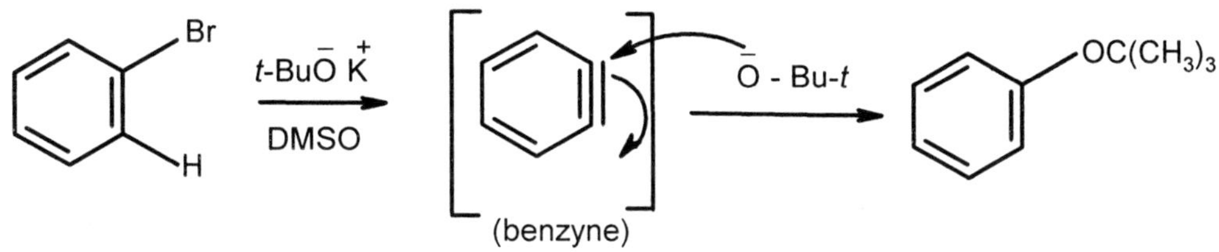

DMSO solvates K^+ but not the alkoxide (RO^-); so the RO^- is left as a "*naked anion*" with higher reactivity. That's why DMSO, a dipolar aprotic solvent is used to maximize reactivity.

d)

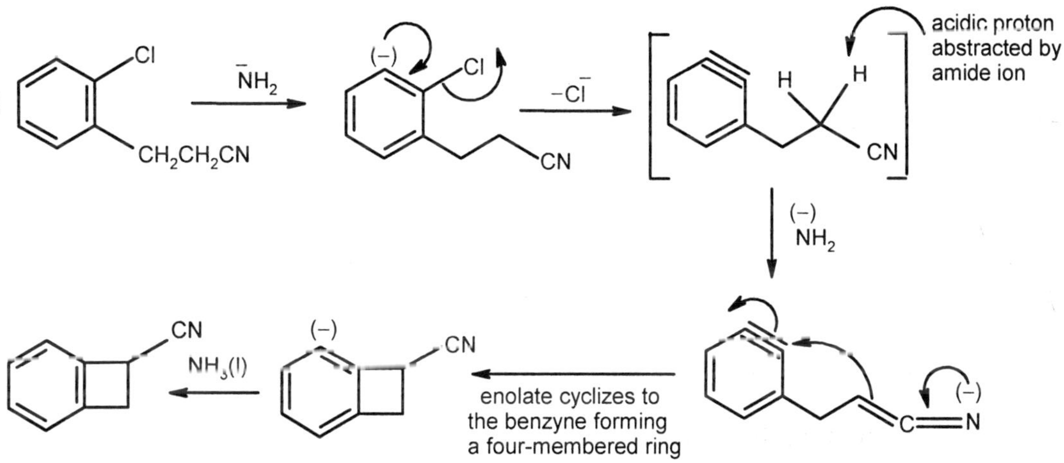

e)

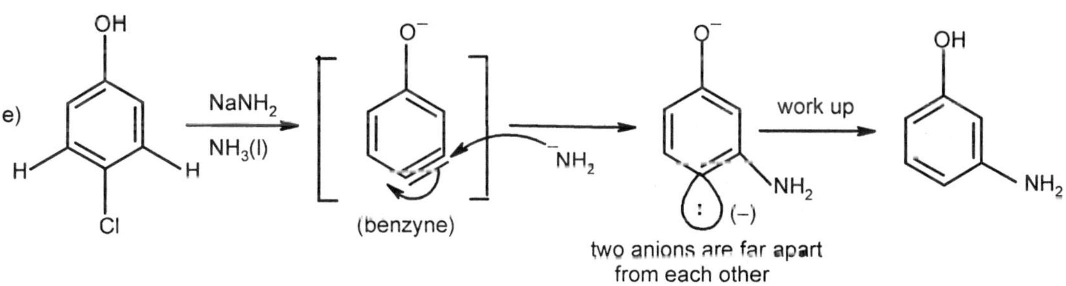

Supplementary Problems

SP-3.1 Arrange the following species in order of their increasing stability in each case:

(i) $(CH_3)_2CHCH_2\overset{+}{C}H_2$, $(CH_3)_2CH\overset{+}{C}HCH_3$, $(CH_3)_2\overset{+}{C}CH_2CH_3$

(ii) $Ph_3\overset{+}{C}$, $(CF_3)_3\overset{+}{C}$, $(CH_3)_3\overset{+}{C}$, $(CCl_3)_3\overset{+}{C}$, $Ph(CH_3)_2\overset{+}{C}$

(iii) 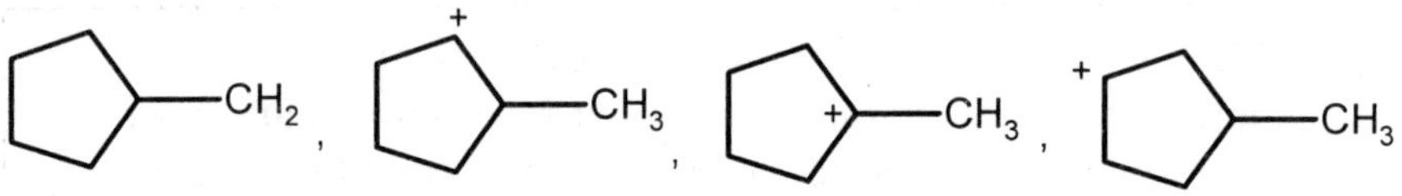

(iv) $Ph\overset{-}{C}H_2$, *p*-$MeOC_6H_4\overset{-}{C}H_2$, *p*-$NO_2C_6H_4\overset{-}{C}H_2$, *p*-$MeC_6H_4\overset{-}{C}H_2$

(v) $Ph\dot{C}H_2$, $CH_2{=}CH\dot{C}H_2$, $Me_3\dot{C}$, $Me_2\dot{C}H$, $CH_3\dot{C}H_2$, $\dot{C}H_3$

(vi) $:CH_2$, $:CH_2CH_3$, $C_6H_5CH:$, $Ph_2C:$

SP-3.2 Acetylacetone reacts with sodium hydroxide to give water and the sodium salt of a carbanion. How this carbanionic species is formed and attains stability?

SP-3.3 Discuss the planar and pyramidal structure of carbanions. Illustrate with suitable examples how these intermediates are detected?

SP-3.4 Give a short account on the 'mercury method' of making alkyl radicals.

SP-3.5 Comment on the structure of free radicals. How are they detected?

SP-3.6 How the carbenes react with an alkene in their singlet or triplet state?

SP-3.7 The compound, $Ph_2CHC(Me)_2OH$, on treatment with SbF_5/SO_2 produces a carbocation intermediate having the ^1H-NMR spectral properties as: δ1.48 (d, 6H), 4.45 (m, 1H) & 8.0 (m, 10H). Assign the structure of the carbocation, and show schematically how it is formed.

SP-3.8 Account for the following facts:

(i) Benzylic hydrogen undergoes bromination reaction at a faster rate than that of a hydrogen atom in ethane.

(ii) Allylic chloride adds halogens about thousand times faster than the corresponding vinylic chloride.

(iii) α-Naphthoxyl radical is more stable than the β-isomer.

(iv) The benzhydryl carbocation is more stable than the benzyl carbocation.

(v) During the decomposition of benzoylperoxide in trichlorobromomethane, some *para*-trichlorobenzoic acid is formed.

(vi) 2,4,6-Tri-*t*-butylphenoxy radical is highly reluctant to undergo dimerization.

(vii) Benzoylperoxide decomposes at a normal rate in methyl iodide solvent, but the decomposition-rate of *ortho*-nitrobenzoylperoxide is much faster in the same solvent.

(viii) Butadiene reacts almost twenty times faster than 1-octene towards trichloromethyl radical.

(ix) The radical anion of benzophenone is more stable than that of cyclohexanone.

(x) When diazomethane is decomposed in the gas phase by light, ethylene and cyclopropanone are formed.

(xi) Decomposition rate of the compound (I) is 20 times greater than that of (II).

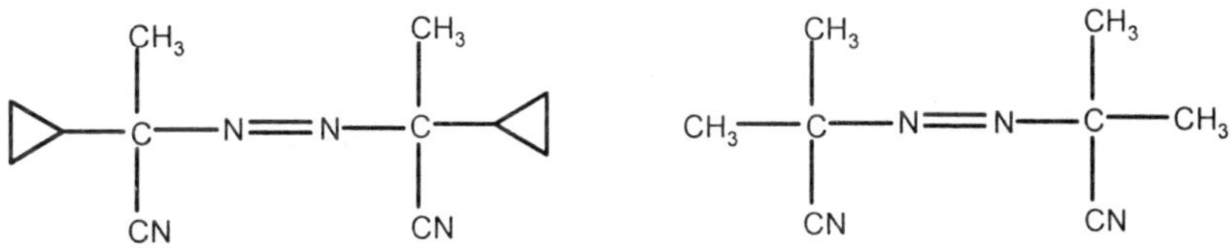

(I) (II)

SP-3.9 Ph_3COH reacts with HBF_4 in acetic anhydride solvent to give a bright orange-yellow salt that is ionic in character. When water or alcohol is added to the salt, the colour is discharged. Explain for all these observations mentioning the reactions involved.

SP-3.10 Chlorobenzene on treatment with sodamide in liquid ammonia in the presence of furan affords a 1,4-addition product. Draw the structure of the adduct with the detailed reaction scheme.

SP-3.11 Could you cite an example of a highly stable and at the same time almost inert free radical?

(**Hints:** Few examples of such species are known. one of these is the perchlorotriphenylmethyl radical. This radical is highly stable and inert; does not react with oxygen at temperatures up to 300^{0}C. The unreactivity of the species is attributed to the fact that the large chlorine atoms shield the electron-deficient central carbon atom, thereby, preventing the incoming reagents to attack the reaction site.)

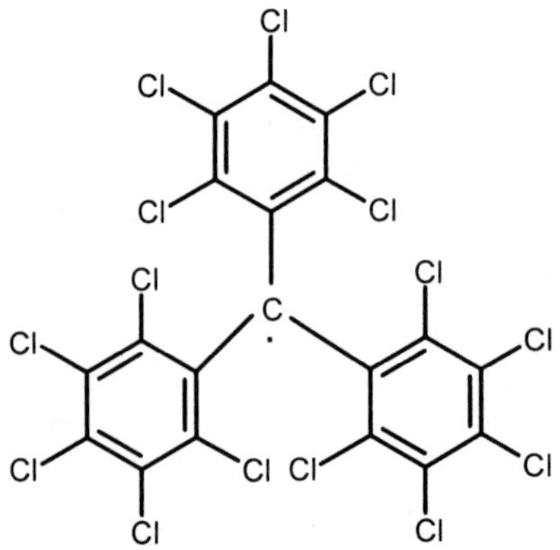

(Perchlotriphenylmethyl radical; an inert free radical)

SP-3.12 Predict the product(s) indicating the formation of reactive intermediates in each of the following cases:

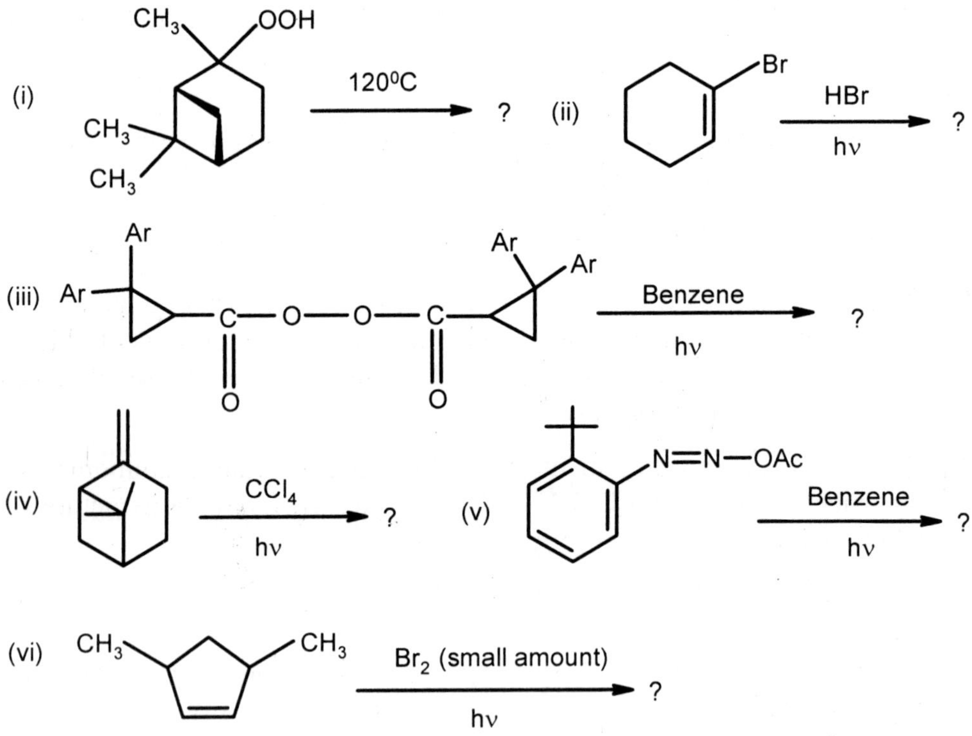

C H A P T E R 4

ORGANIC REACTION MECHANISM

Introduction

In organic chemistry, exciting new discoveries are being made at an ever-increasing pace, and the concept of the subject has faced a tremendous change in the last few years. Organic reaction mechanism is an essential part of chemistry courses in almost all universities. A throughout knowledge of reaction mechanism is very much essential for understanding chemical processes, *i.e.* to appreciate as to how and why reactants go to products; theories about reaction mechanism can explain diverse chemical phenomena in terms of a relatively small number of general principles, and hence it is an important organizing device for the subject of organic chemistry as a whole.

Reaction mechanism is usually defined as the elaboration of step-by-step paths of conversion of reactants into products. It might not be forgotten that the study of detailed stepwise changes occurring in each and individual substrate is not only an up-hill task, but also practically impossible. Fortunately, a careful survey on the vast number of organic reactions reveals certain similarities among themselves and accordingly organic reactions may be categorized into several groups — each group has been found to have some common criteria. Generally, organic reactions are classified on the basis of two criteria — the first one is the operational criterion of the reaction (*i.e.* whether they are substitution, elimination, addition, or rearrangement reactions) while the second criterion is related to the electronic character of the reagent, *i.e.* its nucleophilicity or electrophilicity. Each of the operational classification may be combined with electronic classification to describe the overall reaction path.

In this present chapter a sincere effort has been made to collate significant aspects of organic reaction mechanism for better understanding of the subject.

Solved Problems

Problem 4.1 What are the nucleophiles and electrophiles?

Solution.

Nucleophiles

Nucleophiles are electron pair donating electron-rich species, and thus are "nucleus-loving" or "nucleus-seeking" ones. Nucleophiles may either be negatively charged (anions) or neutral molecules. They must either contain an unshared electron pair or a bonding electron pair that can easily be donated from the bond involved and thus be made available to the electron-deficient reactive sites. Nucleophiles are generally represented as:

$$\overset{(-)}{\text{Nu}}\text{:} \quad \text{(regardless of charge)}$$

Few examples of nucleophiles are:

$\overset{(-)}{\text{HO}}$, $\overset{(-)}{\text{Cl}}$, $\overset{(-)}{\text{Br}}$, $\overset{(-)}{\text{I}}$, $\overset{(-)}{\text{RS}}$, $\overset{(-)}{\text{R}}\ \overset{(+)}{\text{Mg}}\text{X}$, $\ddot{\text{N}}H_3$, $H_2\ddot{\text{O}}$:, aromatic π-electron systems (benzene, naphthalene), etc.

Electrophiles

Electrophiles are electron-deficient species that can form a covalent bond by accepting an electron pair from an electron-rich site. Thus electrophils are "electron-loving" or "electron-seeking" species. Electrophiles may either be positively charged (cations) or neutral molecules. They are usually represented as E^+, regardless of charge. Few examples of electrophiles are:

$$\overset{(+)}{\text{N}}O_2,\ \overset{(+)}{\text{H}},\ \overset{(+)}{\text{Cl}},\ \overset{(+)}{\text{Br}},\ \overset{(+)}{\text{C}}R_3,\ BF_3,\ AlCl_3, \text{ etc.}$$

Thus, in brief nucleophiles are electron-pair donors and electrophiles are electron-pair acceptors.

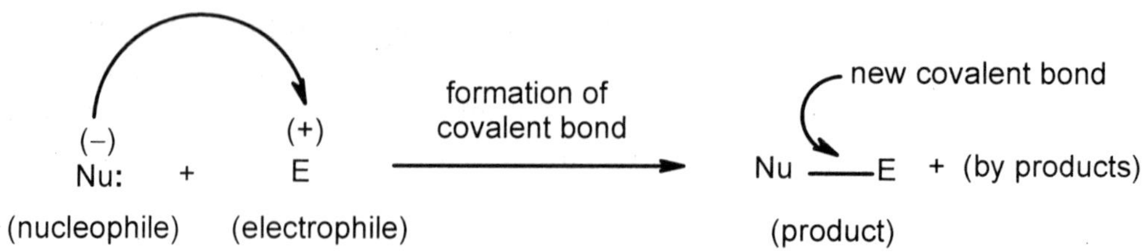

Problem 4.2 Give a short account on "*ambident nucleophiles*".

Solution. *Ambident* or *ambifunctional nucleophiles* are those one that have more than one suitable sites through which they can attack the substrate (electrophilic centre). Examples of such nucleophiles are nitrite $[NO_2]^-$, cyanide $[CN]^-$, enolate and aza-enolates, etc. — each of these nucleophiles can react with electrophiles either at the heteroatom or at the carbanionic carbon atom. *Ambidoselectivity* occurs if one of these alternative modes of reaction dominates.

It is a well-known fact that alkyl bromides form alkyl cyanides (R-CN) when reacts with NaCN, but predominantly yields isocyanides (R-NC) when treated with AgCN. Cyanide ion can act as an *ambident nucleophile*.

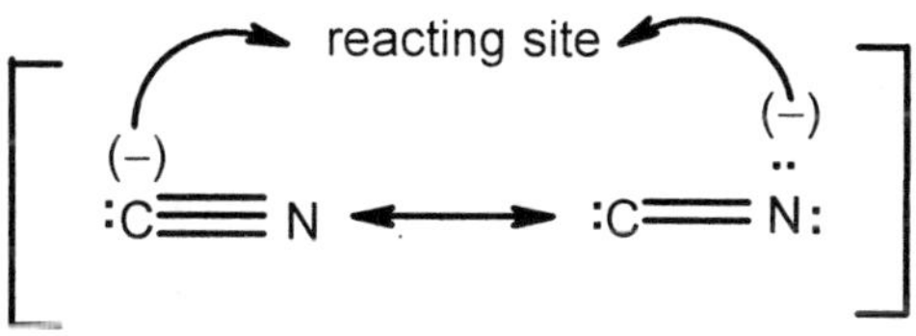

If a particular substitution reaction attains an increasing degree of 'S_N1 character', the incoming nucleophile tends to attack with its most electronegative atom; but if the reaction is of greater 'S_N2 character', the incoming group attacks with a less electronegative centre. In effect, the most nucleophilic site in an ambident nucleophile is the least electronegative atom having an unshared electron pair. When the reagent is silver cyanide, $Ag^+[CN]^-$, the Ag^+ ion promotes the formation of carbocation by removing the halide from the substrate (R–Br) — the carbocation (R^+) formed is preferentially attacked by the nitrogen centre of $[CN]^-$ ion forming the isocyanide.

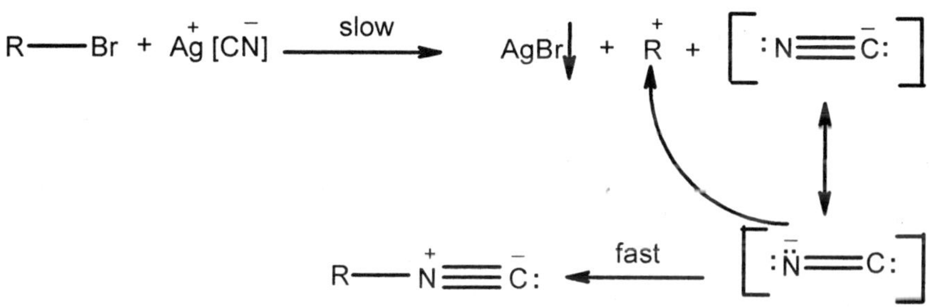

In the absence of such promotion by Ag^+, the resulting S_N2 reaction is found to proceed with preferential attack by the least electronegative atom of the nucleophile —thus R – Br on reaction with NaCN forms predominantly R – CN.

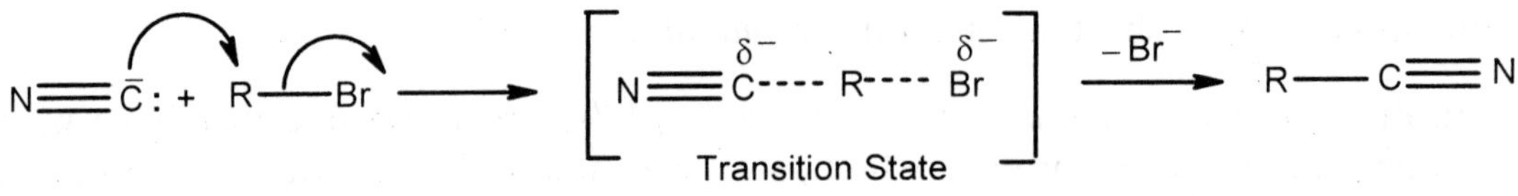

The phenomenon can also be explained on the basis of the principle of 'hard and soft acids and bases', which states that hard acids prefer hard bases, and soft acids prefer soft bases. The more electronegative atom of an ambident nucleophile is a harder base than the less electronegative atom. In S_N1 path, the intermediate carbocation (a hard acid) prefers to add with hard bases. Since Ag^+ ion promotes S_N1 pathway, the R^+ intermediate as formed is preferably attacked by the more electronegative nitrogen atom (a hard base) of $CN^{(-)}$ forming alkyl isocyanide. On the other hand, in the absence of Ag^+ ion the substitution reaction tends to attain S_N2-like path – hence the partially positively charged alkyl part (soft acid) of the substrate is preferably attacked by the less electronegative carbon atom (a soft base) of $CN^{(-)}$ yielding alkyl cyanide in major amount.

Similar effect is also found in:

$$R{-}Br + AgNO_2 \longrightarrow \underset{\text{nitroalkane (less)}}{R{-}NO_2} + \underset{\text{alkylnitrite (more)}}{R{-}ONO}$$

$$R{-}Br + NaNO_2 \longrightarrow \underset{\text{nitroalkane (more)}}{R{-}NO_2} + \underset{\text{alkylnitrite (less)}}{R{-}ONO}$$

Some other examples may also be cited:

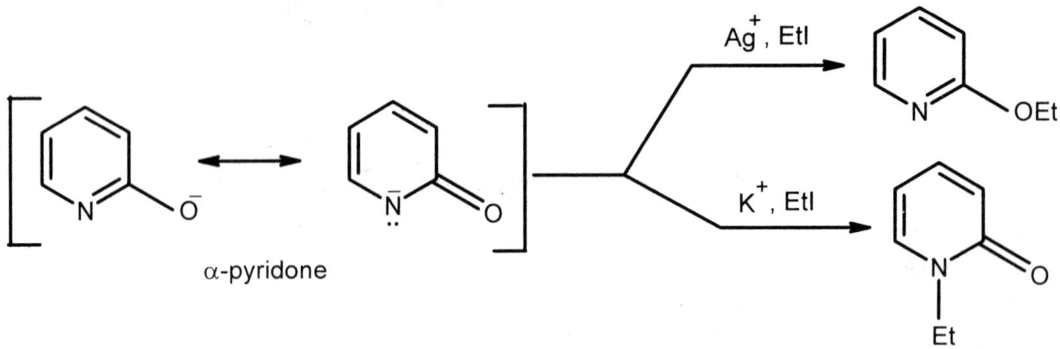

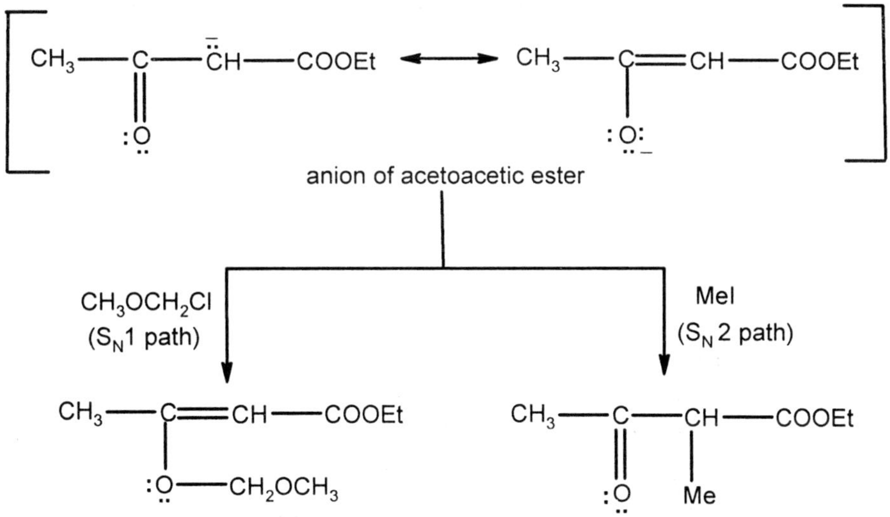

Problem 4.3 Comment on '*nucleophilicity*'. Compare the term with basicity.

Solution. *Nucleophilicity* is a measure of how fast a nucleophile displaces a leaving group within a substrate; it is indeed the relative strength of the nucleophile, usually estimated by measuring reaction rates. Nucleophilicity trends may be summarized as:

- In case of the same central atom, higher electron density increases the nucleophilicity; thus HO^- is more powerful than H_2O, NH_2^- is more powerful than NH_3, and so on. This trend is true for basicity also.

- Across a row in the periodic table, nucleophilicity is approximately in order of basicity; both the properties decrease along the row since increasing electronegativity decreases the lone pair availability. Thus, $R_3C^- > R_2N^- > RO^- > F^-$.

- On going down along a group in the periodic table, nucleophilicity increases while the corresponding basicity decreases. Thus the usual nucleophilic order of the halide is $I^- > Br^- > Cl^- > F^-$ (solvent dependent). Similarly, any sulphur nucleophile is more powerful than its oxygen analogs, and the same is true for phosphorus versus nitrogen. This is because of polarizability factor that increases down the group.

Relative nucleophilicity of some common Nucleophiles

Reactivity	Nucleophiles
Very good	I^-, HS^-, RS^-
Good	Br^-, HO^-, RO^-, CN^-, N_3^-
Fair	NH_3, Cl^-, F^-, RCO_2^-
Weak	H_2O, ROH
Very weak (poor)	RCOOH

Difference with basicity

Although both nucleophilicity and basicity involve availability of an electron pair to form a covalent bond, there are some basic differences between the two properties. Basicity is thermodynamically controlled, while nucleophilicity is kinetically controlled. Nucleophilicity of a reagent is its tendency to attack an electrophilic carbon (usually) atom with the available electron pair resulting in the formation of a new C– Nu bond in a kinetically controlled fashion.

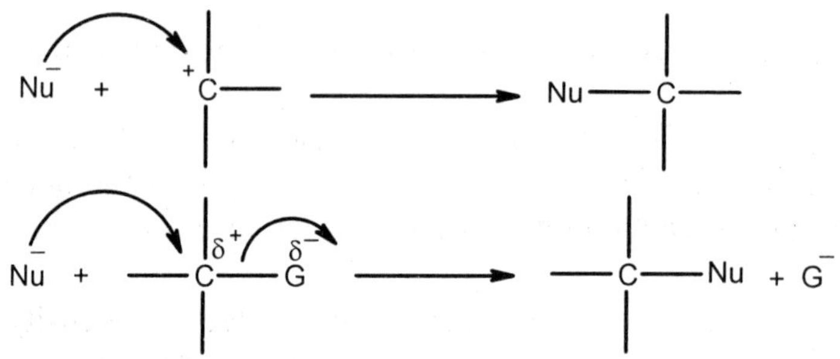

Basicity of a reagent (electron pair donor) usually is the tendency to undergo an equilibrium (thermodynamic) reaction with a proton, forming a new H – X bond.

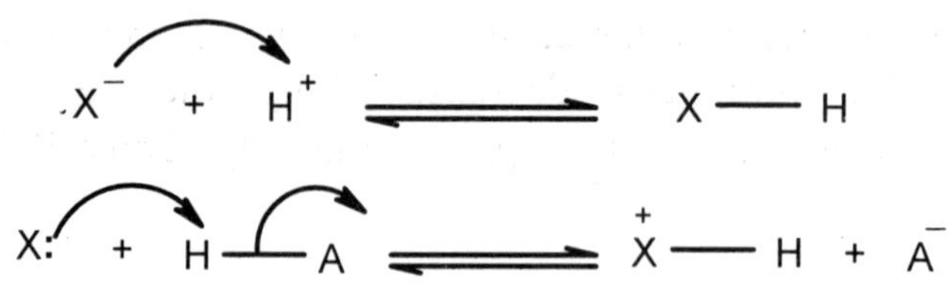

Problem 4.4 Compare the nucleophilicity of CH_3O^- with that of F^-

Solution. When the nucleophilic atoms of two reagents are in the same row of the periodic table, nucleophilicity order becomes approximately in the order of basicity of the two reagents. Along a row (from left to right) basicity decreases due to the increase in electronegativities of the corresponding atoms, thereby lowering the lone pair availability. Since 'O' and 'F' are in the same row of the periodic table, basicity of the methoxide ion would be much more than that of the fluoride ion. This can be best illustrated by considering the pKa values of the corresponding conjugate acids — the values are 4.77 and ~16, respectively for HF and CH_3OH. Lower pKa values indicate stronger acids corresponding to the weak conjugate bases, and *vice versa.* Hence, basicity of F^- is lower than that of CH_3O^-. Consequently, the methoxide ion would show higher nucleophilicity than the fluoride. *In general, the higher the pKaH (i.e. pKa of the corresponding conjugate acid), the better is the nucleophile.*

Problem 4.5 Discuss the solvation effect on nucleophilicity order of the halide ions.

Solution. In polar protic solvents the usual order of halide ion nucleophilicity is $I^- > Br^- > Cl^- > F^-$ — although the basicity order is reverse. The cause of this distinction between basicity and nucleophilicity is originated from the fact that a polar protic solvent reduces the reactivity of X^- through solvation by making hydrogen bonds — a relatively smaller negatively charged nucleophile is more tightly surrounded by a shell of solvent molecules that creates hindrance into the reaction. Hydrogen-bonding would be weakest with the largest ion. That's why in protic solvents, the nucleophilicity order is observed as $I^- > Br^- > Cl^- > F^-$. This is the order also in nonpolar or very weakly polar aprotic solvents — where the salts of nucleophiles are present as ion-clusters (ion-pairs) in which the nearby cations reduces the reactivity of the counter anions. With a given cation, ion-pairing becomes strongest with the smallest ion (F^-), and weakest with the largest one (I^-).

On the contrary, the situation is totally different in case of highly polar aprotic solvents. There is no hydrogen-bonding at all. Furthermore, they solvate only the cations leaving the anions free. Polar aprotic solvents, thus, enhance the reactivity (nucleophilicity) of all anions — but the effect is more pronounced with smaller anions. Hence, the halide ion nucleophilicity order is noticed as the reverse — $F^- > Cl^- > Br^- > I^-$ — which is also the charge density order.

[In gas phase, in absence of solvent, an approximate order of nucleophilicity was found to be $HO^- > F^- \sim MeO^- > MeS^- >> Cl^- > CN^- > Br^-$, giving an evidence that solvation is responsible for the effect in solution[1,2].]

1. Olmstead, W. N. and Brauman, J. I., *J. Am. Chem. Soc.*, **99**, 4219(1977).
2. Tanaka *et al.*, *Canadian J. Chem.*, **54**, 1643(1976).

Problem 4.6 Suggest an explanation for the fact that the order of reactivity of the halides towards *n*-butyl brosylate in acetone is $Cl^- > Br^- > I^-$ when $(C_4H_9)_4N^+$ is the cation of the halide salt, but $I^- > Br^- > Cl^-$ when Li^+ is the cation.

Solution. The difference in reactivity of the halide ions arises due to the difference in size between tetrabutylammonium $[(C_4H_9)_4N^+]$ and Li^+ cations. The former is larger in size, as a result of which the halide ion in $(C_4H_9)_4N^+\ X^-$ is much less associated than in LiX in acetone. Thus halide ions are much freer in $(C_4H_9)_4N^+\ X^-$ providing the reactivity order as $Cl^- > Br^- > I^-$, where normal order is observed in case of lithium halides.

Problem 4.7 What do you mean to say S_N1 reactions? Discuss briefly.

Solution. A nucleophilic substitution reaction at a saturated carbon involving only one species in the transition state of the rate-determining step is said to follow S_N1 mechanistic pathway (*substitution nucleophilic unimolecular*). Such a reaction takes place in two steps:

Step-I: In the first step slow ionization of substrate occurs with the formation of carbenium ion intermediate. This is the rate-determining step.

[heterolysis of the bond between the C-atom and the leaving group 'G']

Step-II: The nucleophile rapidly attacks the carbenium ion intermediate in the second step of the reaction yielding the product.

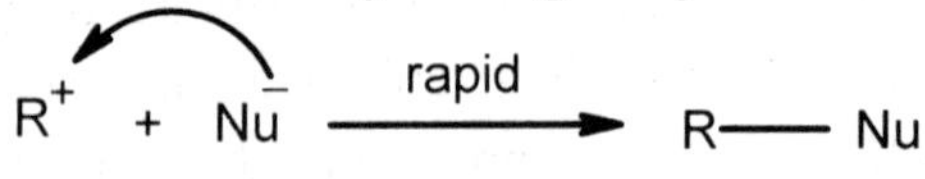

[formation of new bond between the C-atom and the nucleophile]

The overall reaction may be represented as:

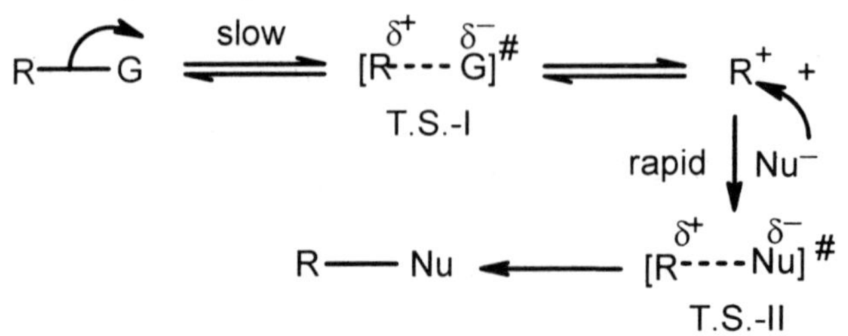

The first step is endothermic, while the second step is exothermic. The energy profile diagram for the process may be drawn as:

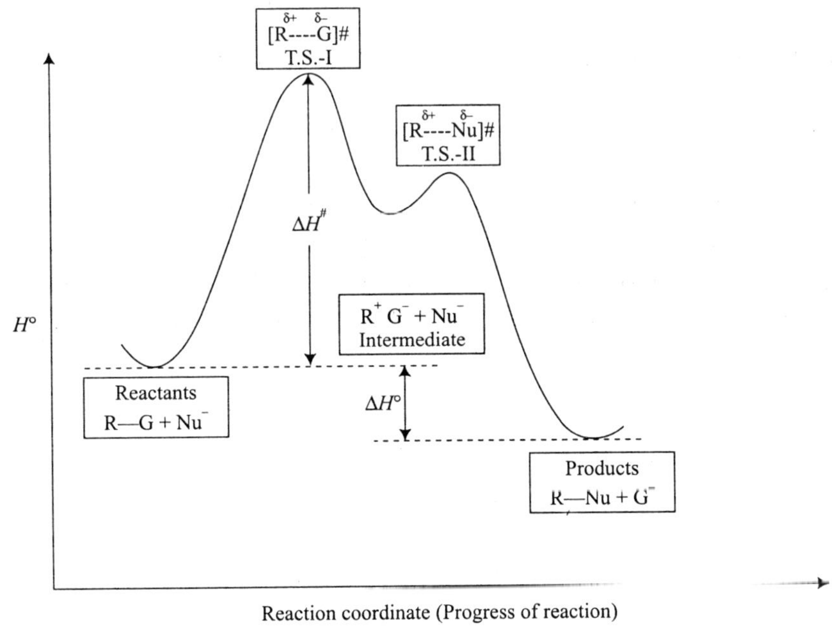

Fig. 4.1: Energy profile diagram of an S_N1 reaction

The rate-determining step of an S_N1 reaction is the slow ionization of substrate molecule, and that's why nature and concentration of nucleophiles have no effect on the overall rate of the reaction. It depends only on the concentration of substrate:

$$\text{Rate} = k[R\text{–}G]$$

Normally, an S_N1 reaction follows first order kinetics. A true example is:

$$t\text{-BuCl} + \bar{O}H \xrightarrow[S_N1]{\text{alkaline hydrolysis}} t\text{-BuOH} + \bar{Cl}$$

Nature of solvent has a profound influence on S_N1 reaction rate. Polar protic solvents greatly enhance the reaction rate because of the fact that the amount of energy required to apply externally to carry out an S_N1 reaction is minimized due to evolution of solvation energy caused by the solvation of intermediate carbenium ion and the nucleofuse, thereby facilitating ionization of the substrate. For example, the ionization of *t*-butylchloride to *t*-Bu^+ and Cl^- in the gas phase requires 630 kJ/mole, while in water this ionization requires only 84 kJ/mole. The difference is the solvation energy.

Problem 4.8 Comment critically on the rate law of S_N1 reactions.

Solution. The simplified rate law for S_N1 reactions is usually given as:

$$\text{Rate} = k[\text{R–G}] \quad \text{............Eqn. 1.} \quad [\text{RG} = \text{substrate}]$$

Many cases are known where pure first-order kinetics is followed, but in many cases more complicated kinetics are observed. This can be explained by taking into account the reversibility of the first step. The G^- formed in this step competes with the incoming nucleophile (Nu^-) for the cation, and hence the rate law should be modified as:

$$\text{R—G} \underset{k_{off}}{\overset{k_{on}}{\rightleftharpoons}} R^+ + G^-$$

$$R^+ + Nu^- \xrightarrow{k'} \text{R—Nu}$$

$$\text{Rate} = \frac{d[\text{RNu}]}{dt} = \frac{k_{on}[\text{RG}] \times k'[Nu^-]}{k_{off}[G^-] + k'[Nu^-]} \quad \text{........................ Eqn. 2}$$

Now, the following situations may be considered:

(a) At the beginning of the reaction, the concentration of G^- is very small compared to that of nucleophile, and hence $k'[Nu^-] >> k_{off}[G^-]$. The eqn.2 is so reduced to eqn.1 — in practice S_N1 reactions usually follow first-order kinetics in their initial stages.

(b) When Nu^- is present in high concentration, as in the cases of S_N1 solvolysis reactions where the solvent is the nucleophile, $k_{off}[G^-]$ term becomes negligible compared to $k'[Nu^-]$, and again first-order kinetics is observed.

$$k'[Nu^-] >> k_{off}[G^-]$$

(c) If the reaction is carried out in presence of a large excess of the leaving group, G^-, then $k_{off}[G^-]$ becomes significant and competitive with $k'[Nu^-]$. As a result, the overall rate of the reaction would decrease; this retardation effect by addition G^- is called *common ion effect* or *mass law effect*.

Problem 4.9 Discuss the *ion pair* mechanistic approach of S_N1 reactions.

Solution. An S_N1 reaction involves the formation of a carbenium ion in the slow and rate-determining first step. If there is a free carbenium ion, it would be planar and achiral; the incoming nucleophile (Nu^-) subsequently should attack the species with equal ease from either side of the plane leading to 50:50 mixtures of two enantiomers (*i.e.* a racemic substitution product from an enantiometrically pure substrate *via* achiral carbenium ion).

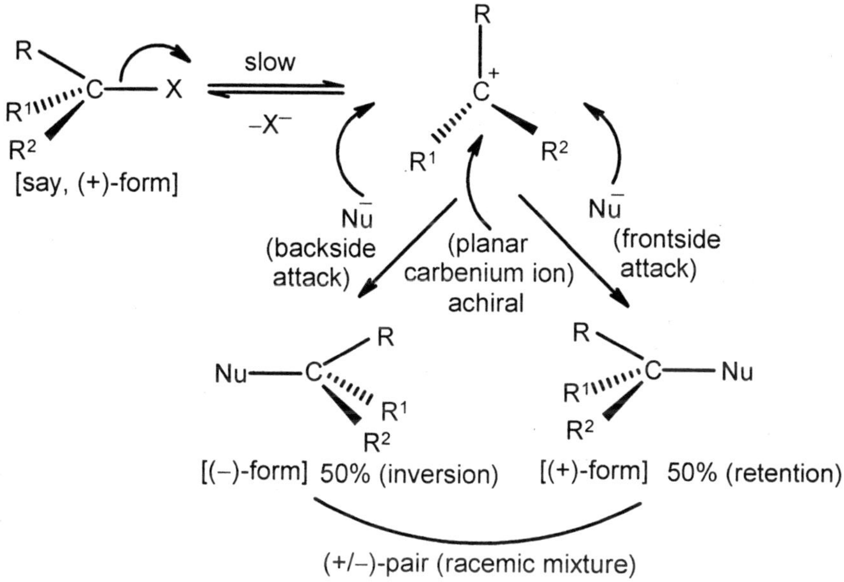

In practice, although many first order substitution reactions yield completely racemized product, many other are known not to do so — in most of the cases there is always the inverted product in excess (5-20%) over the other isomer. However, the relative proportions of two enantiomers are found to depend on:

(i) relative stability of the intermediate carbenium ion — the more stable the carbenium ion, the greater is the proportion of racemization.

(ii) the solvent, particularly on its ability to act as a nucleophile — the more nucleophilic the solvent, the greater is the proportion of inversion.

All these observations led to the conclusion that in many S_N1 reactions at least some of the products are not formed from free carbenium ions but rather from *ion pairs*.

According to this concept proposed by Winstein *et al.*[1], S_N1 reactions proceed in this manner:

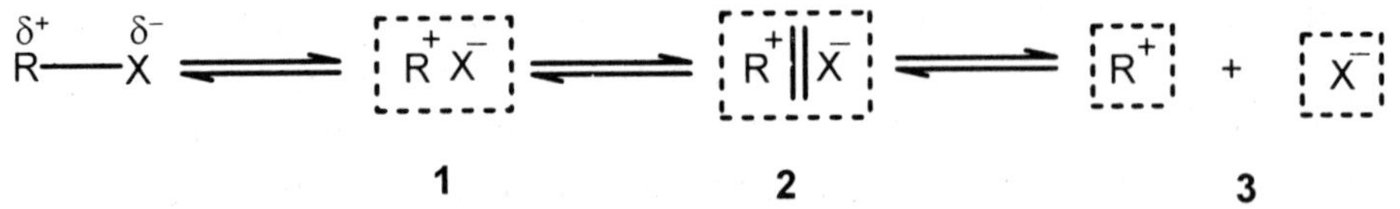

Here (**1**) is an *intimate, contact* or *tight* ion pair, (**2**) is a *loose* or *solvent-separated* ion pair, and (**3**) is the completely *dissociated, free* and *separately solvated* ions. These ion pairs may be attacked at their any stage by the incoming nucleophile to yield the substitution products.

In *intimate* ion pair (**1**) the carbenium ion is unlike the free one — it remains chiral (if originates from an enantiometrically pure substrate) and the leaving group (X^-) adjacent to carbenium ion centre partially protects one side of this ion from the attack of the nucleophile; consequently, the nucleophile attacks from the side that lies opposite to the departing group leading to *inverted* product. Attack at the *solvent-separated* ion pair (**2**) is more likely to lead to attack from either side to produce racemic mixture, while attack on (**3**) can clearly occur with equal ease from either side. Thus, longer the life of carbenium ion (R^+), greater is the proportion of racemization that is expected to take place. A schematic representation of the possible solvolysis reactions in a solvent, 'SH' is shown below:

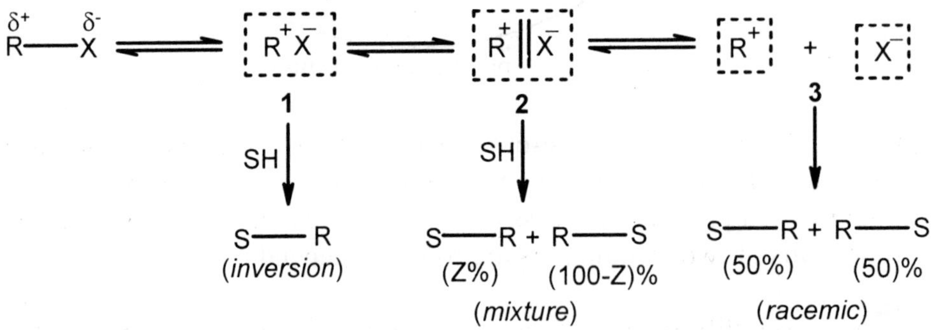

For example

It has been observed that solvolysis of *R*-2-octylchloride, where there is no comparable stabilization can occur, yields only 34% racemized product — i.e. 83% of the solvolysis product is formed with inversion of configuration and 17% with retention.

---

1. Winstein, S., Clippinger, E., Fainberg, A. H., Heck, R. and Robinson, G. C., *J. Am. Chem. Soc.*, **78**, 328(1956).

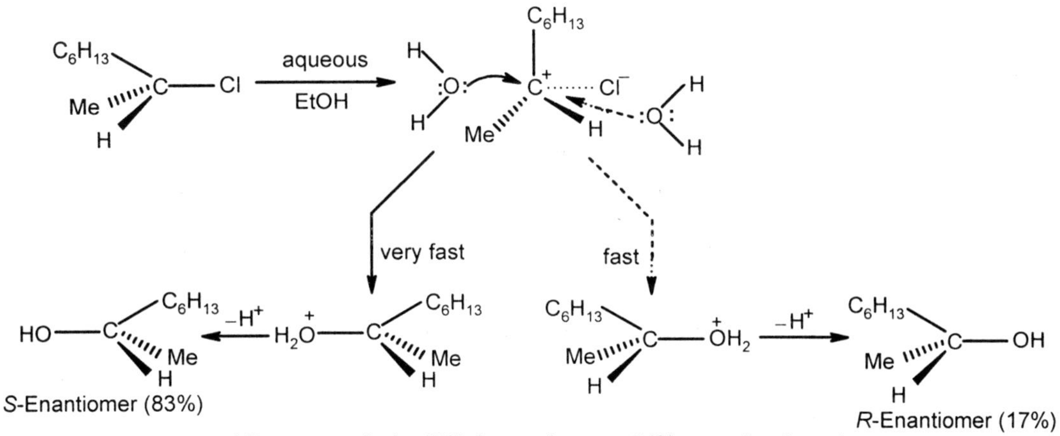

(Procuct analysis: 66% *S*-enantiomer + 34% racemic mixture)

On the other hand, solvolysis of *R*-phenylethylchloride, where a more stabilized α-methylbenzyl action is to be formed, offers 98% racemization in acetone/water (4:1) solvent system — the proportion of racemized product has been found to decrease to 80% when more nucleophilic water alone is used.

Problem 4.10 "Addition of a low concentration of $LiClO_4$ causes an initial steep rate acceleration of the acetolysis of *threo*-3-*p*-anisyl-butyl-*p*-bromobenzenesulphonate" – how do you clarify the observation?

Solution. Addition of a low concentration of lithium perchlorate ($LiClO_4$) or lithium bromide (LiBr) in the acetolysis of certain tosylates/brosylates causes an initial steep rate acceleration that ultimately levels off to the normal linear behaviour of ordinary salt effect. Such initial sharp hike in the rate of reactions is attributed to *special salt effect,* the term as introduced by Winstein *et al.*[1-3]. Prof. Winstein suggested the brosylate substrate (I) solvolyzes *via* the ion pair mechanism:

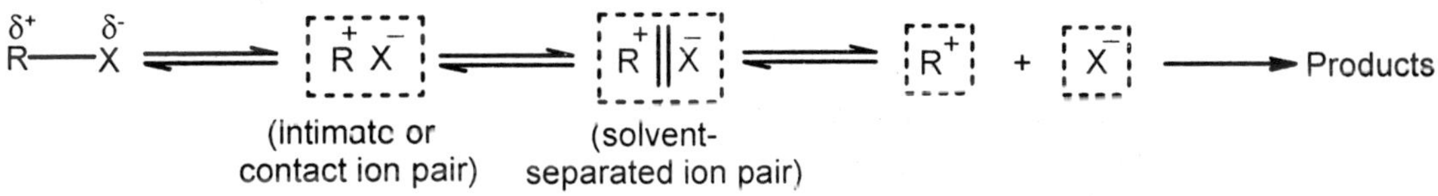

1. Winstein, S., Clippinger, E., Fainberg, A. H., Heck, R. and Robinson, G. C., *J. Am. Chem. Soc.*, **78**, 328(1956).
2. Winstein, S. and Robison, G. C., *J. Am. Chem. Soc.*, **80**, 169(1958).
3. Winstein, S., Klinedinst, Jr., P. E. and Clippinger, E., *J. Am. Chem. Soc.*, **83**, 4986(1961).

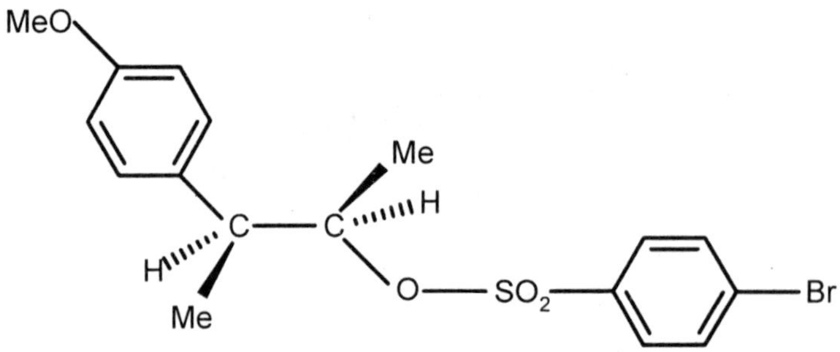

(*threo*-3-*p*-anisyl-butyl-*p*-bromobenzenesulphonate) (I)

The ordinary salt effect enhances the rate of ionization and rate of separation of the cation and the leaving group from each other. The perchlorate ion (ClO_4^-), being highly nonnucleophilic, interacts specifically with the solvent-separated ion pair, and exchanges with the leaving group to give raise a new solvent-separated ion pair:

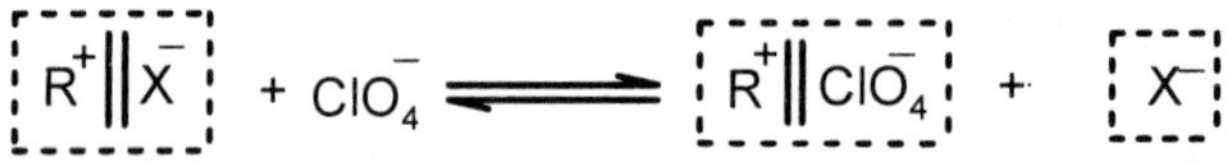

The newly formed solvent-separated ion pair is very much unstable under the reaction conditions, and goes to the product. Thus, the amount of solvent-separated ion pair that would have returned back to the starting compound is thereby reduced, leading to the increase in the rate of overall all reaction.

Problem 4.11 Discuss in brief the factors influencing the rate of S_N1 reactions.

Solution. The major factors influencing the rate of S_N1 reactions are mentioned below:

(a) Nature of substrate: For an S_N1 reaction the ionizing step is rate determining — the rate increases as the stability of carbenium intermediate increases. So the substrates bearing electron-donating groups/atoms tend to undergo S_N1 reaction. That's why the reactivity order of alkyl halides (for example) is found as: R_3CX (3^0) >> R_2CHX (2^0) > RCH_2X (1^0) > CH_3X (1^0).

(b) Nature of solvent: It is generally observed that the rate of an S_N1 reaction is enhanced considerably with increase in the ionizing power of the solvent. In an S_N1 reaction a highly polarized transition state is involved originated from a neutral molecule — an effectively polar solvent stabilizes both the substrate and the transition state, but such stabilization is far greater for the more polarized transition state than that of the less polar substrate. Thus, activation energy for ionization of the substrate is reduced and the rate is increased. Between protic and aprotic polar solvents, the protic polar ones are more preferred for enhancing the reaction rate because they can solvate both cations and anions, respectively through donation of the unshared electron pairs and also through hydrogen bonding.

(c) Nature of nucleophile: Since nucleophile is not involved in the rate-determining first step of S_N1 reactions, the reaction-rate is not influenced by the nature of nucleophiles. However, if better nucleophile is added into an S_N1 reaction mixture, the rate remains unchanged but the product contains the added nucleophile.

(d) Nature of leaving group: The better the leaving group, the faster the rate of both S_N1 and S_N2 reactions — since the leaving group is becoming more negatively charged in the transition state of a nucleophilic substitution reaction, a good leaving group must readily take up the negative charge. For example, sulphonates would be much better leaving groups relative to halogens.

Problem 4.12 Give an outline of the nucleophilic substitution bimolecular (S_N2) mechanistic approach.

Solution. The concerted (one step) mechanism with no intermediate for the displacement of a leaving group by a nucleophile in aliphatic and alicyclic compounds involving both nucleophile and substrate in the transition state is termed as substition nucleophilic bimolecular, and symbolically designated as S_N2. The nucleophile approaches the substrate from an 180^0 angle opposite to the leaving group and by donating its electron pair the nucleophile begins to form a new bond to carbon while at the same time the carbon–leaving group bond begins to break. The energy necessary to break the C–G bond (G = leaving group) is supplied by simultaneous formation of the C–Nu bond (Nu = nucleophile).

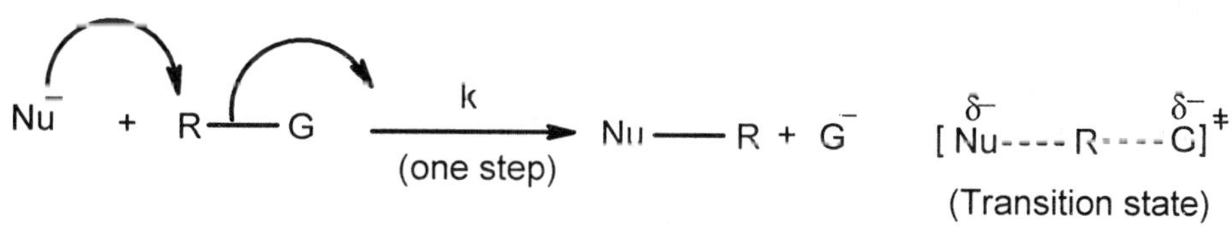

The rate law for such single-step reaction involving only one transition state is given by the equation:

$$\text{Rate of the reaction} = \frac{d[Nu-R]}{dt} = k[\bar{Nu}]\,[R-G]$$

$$\text{where } k = A\exp(-E_a/RT)$$

Thus S_N2 reactions are bimolecular and follow second order kinetics in general. However, it is to be remembered that molecularity of a reaction may or may not the same with its order; in this present case, for example, if there is a large excess of nucleophile (when solvent can act as nucleophile), the mechanism would still remain as bimolecular, but experimentally determined kinetics will be of first order (R= k[RG]) — specifically such kinetics are known as *pseudo-first order*.

The energy profile diagram for an S_N2 reaction is shown as:

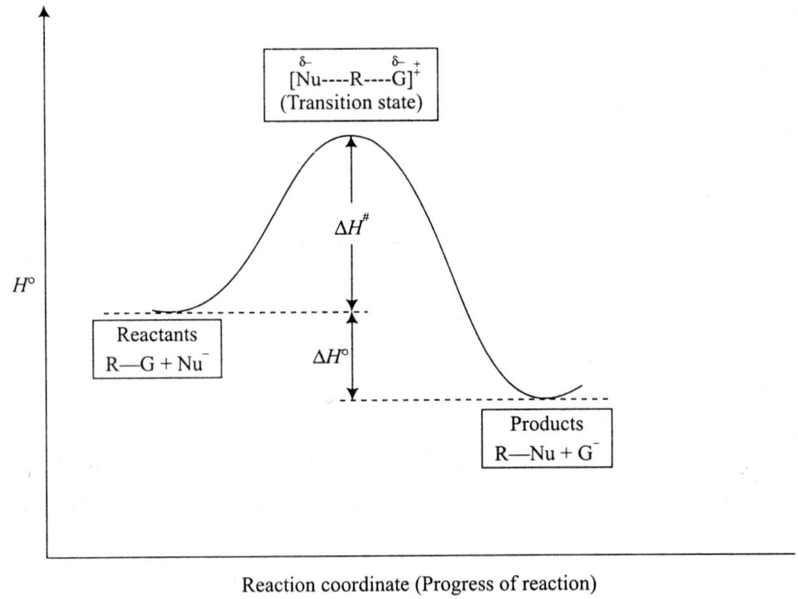

Fig. 4.2: Energy profile diagram of an S_N2 reaction

Stereochemistry of S_N2 reaction

S_N2 reactions take place with stereoelectronic controls involving a backside attack by the nucleophile on the bond between the C-atom and the leaving group in a concerted process without intermediate — the spatial arrangement of the three residual groups attached to the carbon atom attacked are turned inside out as the reaction goes along, just like an umbrella in a high wind. Thus, the mechanism leads to inversion of configuration at the chiral center under attack.

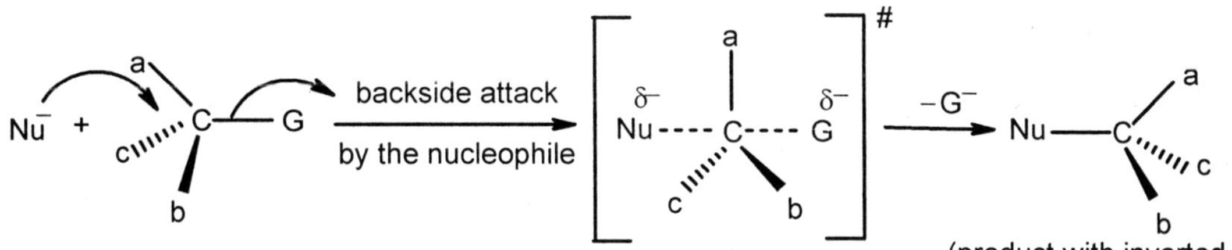

(Transition state)

This inversion of configuration is called the Walden inversion, as observed by P. Walden long before (1893) the S_N2 mechanism was formulated by Hughes *et al.* (1935).

In the transition state (I) the central carbon atom is partially bonded to both nucleophile (Nu) and leaving group (G) — the hybridization state becomes sp^2 (from the initial sp^3) with an approximately perpendicular p-orbital, one lobe of which overlaps with the nucleophile and the other with the leaving group.

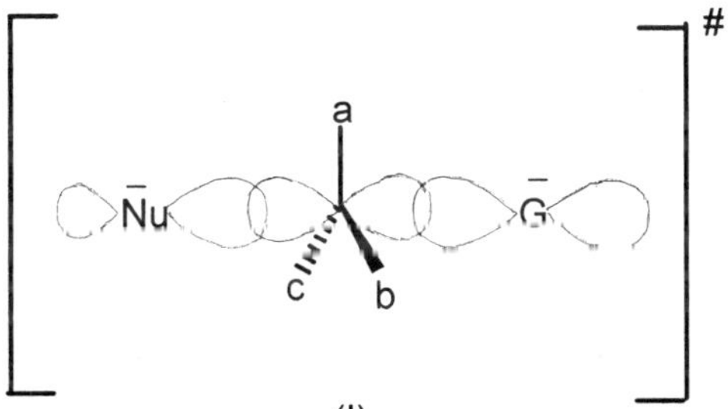

(I)

Problem 4.13 Offer experimental evidence in favour of inversion of configuration at the reaction site in an S_N2 reaction.

Solution. That an S_N2 reaction occurs with inversion of configuration at the attacking centre was first demonstrated by Hughes and his coworkers[1]. They studied the exchange reaction of optically active 2-iodooctane with radioactive iodide ($^{128}I^-$) in acetone.

$$*I^- + C_6H_{13}CH(I)CH_3 \xrightarrow{\text{acetone}} C_6H_{13}CH(*I)CH_3 + I^-$$

$$(*I = {}^{128}I)$$

If inversion of configuration at the central chiral atom takes place, as required by S_N2 mechanism, the optical activity of the resulting solution will reduce gradually to zero, i.e. racemization will occur — because one molecule of pure (*R*)-isomer (say) on exchange reaction is inverted to one molecule of pure (*S*)-isomer that pairs off with a second molecule of (*R*)-enantiomer to form a (±)-racemate.

The investigators observed that — (a) the reaction follows second order kinetics overall (first order with respect to 2-iodoocatane and to $^{128}I^-$) and is thus bimolecular, and (b) the rate of racemization (measured polarimetrically) is twice the rate of inversion. Thus, it may be expected that pure dextrorotatory starting material would be 100% racemized when only 50% of it has been converted to the levorotatory isomer; and hence expected rate of racemization must be twice the rate of inversion.

The striking point of the investigation was to correlate the rate of inversion with that of uptake of radioactive $^{128}I^-$. As per their experimental results the rates were noted as:

Rate of excahnge: $3.00 \pm 0.25 \times 10^{-5}$ (at 30⁰C)

Rate of inversion : $2.88 \pm 0.03 \times 10^{-5}$ (at 30⁰C)

The rate of displacement and of inversion is thus identical within the limits of experimental error — but rate of racemization is twice the rate of inversion. These results unequivocally reveal that each act of S_N2 displacement is an act of inversion.

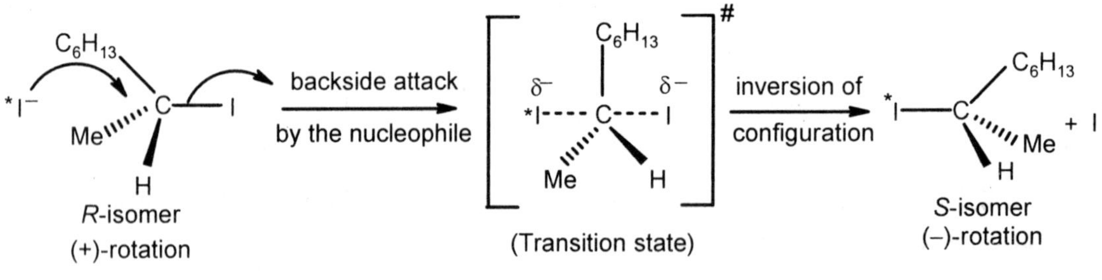

Another classic example supporting the S_N2 mechanistic approach has been obtained from bicyclic compounds bearing good leaving groups at bridgehead carbons; here is also noted that a site incapable of undergoing inversion is at the same

1. Hughes *et. al.*, *J. Chem. Soc.*,1225(1935);1173(1936); 209(1938).

time incapable to undergo S_N2 reaction. The following bicyclic compounds remain completely inert when treated with nucleophile under S_N2 conditions. The reason is that the attacking nucleophile cannot approach from the rear — thus the bridgehead carbon of rigid bicyclic system can't be inverted without fragmentation of the molecule.

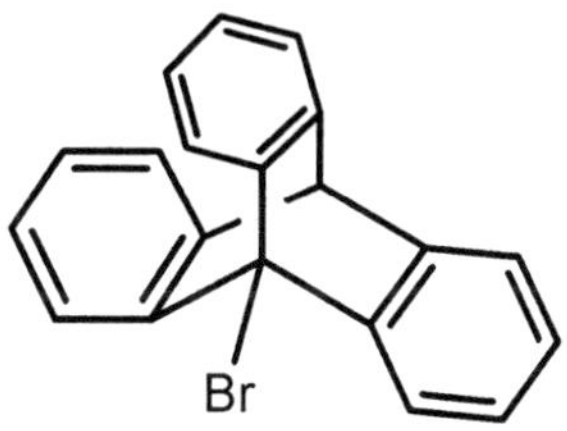

[Ref. 1]

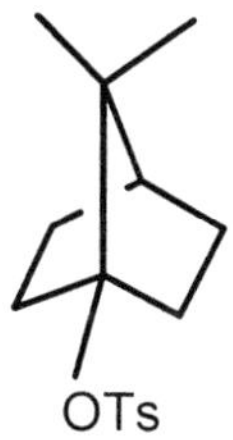

[Ref. 1]

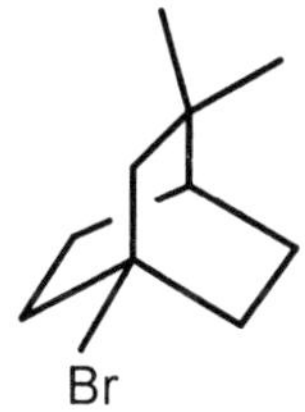

[Ref. 2]

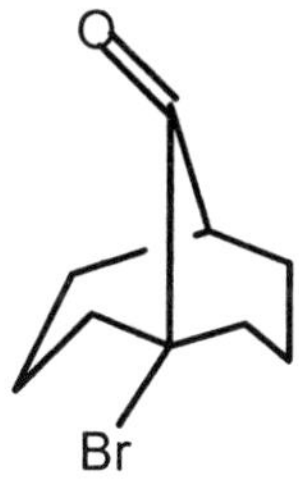

[Ref. 3]

Problem 4.14 Why the backside attack in an S_N2 reaction is feasible?

Solution. In an S_N2 reaction the attacking nucleophile chooses a backside displacement (corresponding to transition state-I) rather than coming from the front (corresponding transition state-II).

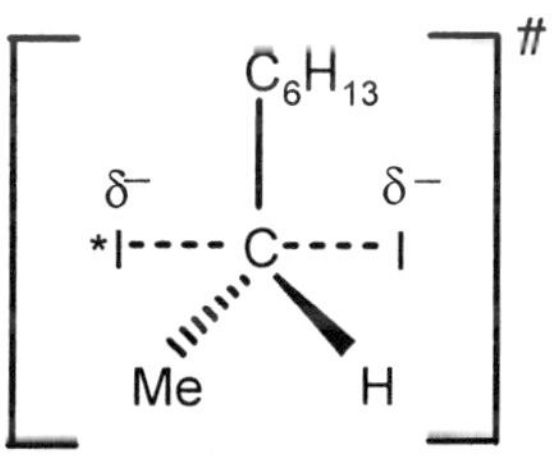

(Transition state-I)
backside attack

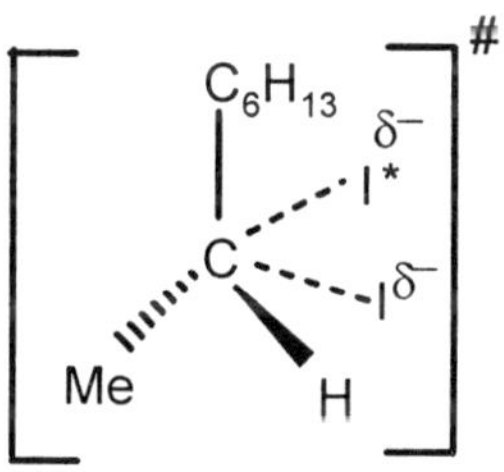

(Transition state-II)
front attack

1. Bartlett *et al*., *J. Am. Chem. Soc*., **61**, 3184 (1939); *ibid*, **72**, 1005 (1950).
2. Doering *et al*., *J. Am. Chem. Soc*., **75**, 1088 (1953).
3. Cope and Synerholm, *J. Am. Chem. Soc*., **72**, 5228 (1952).

The reason might be expected like that the partial negative charges on the incoming and outgoing nucleophiles in transition state-I are as far away from each other as possible, thus minimizing the electrostatic repulsion. However, this assumption has been discarded on the basis of the following experimental observation (**Scheme 4.14**)[1,2], where entering and leaving groups bear opposite charges, but inversion is the result confirming the backside attack. Hence, actual reasoning is somewhat different.

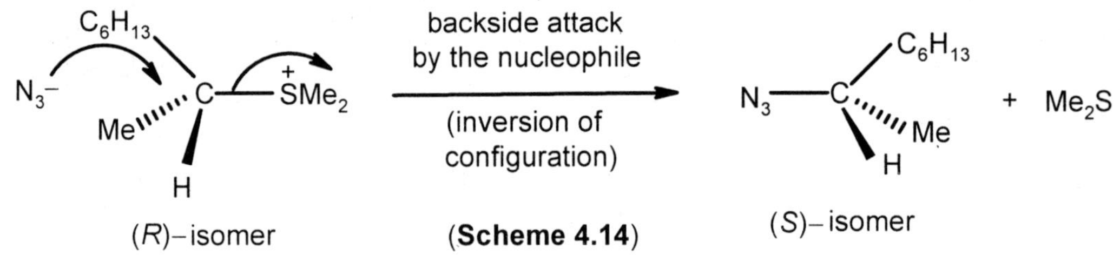

(**Scheme 4.14**)

Preference for backside displacement can be best explained from *Frontier Orbital Theory* (FMO Approach). The electron-filled HOMO of the incoming nucleophile overlap with the empty antibonding σ* orbital (LUMO) between the substitution centre and the leaving group. The backside mechanism involves the maximum amount of overlap throughout the course of the reaction. In course of front side attack, the nucleophilic HOMO will be brought onto the node of the antibonding LUMO, thus resulting the interaction as minimal. That's why the former path becomes the favoured route.

Problem 4.15 Offer an experimental proof in support of linear transition state in an S_N2 reaction.

Solution. Eschenmoser and his coworkers provided strong evidence in support of the fact that, in the S_N2 transition state, the incoming nucleophile is aligned collinearly with the attacked central carbon atom and the leaving group. They found that the substrate, methyl α-tosyl-*o*-toluenesulphonate (I) on treatment with base gives *o*-(1-tosylethyl) benzenesulphonate (III).

1. Harvey *et al.*, *J. Chem. Soc.*, 800(1960).
2. Hoffman and Hughes, *J. Chem. Soc.*, 1252, 1299,1964(1964).

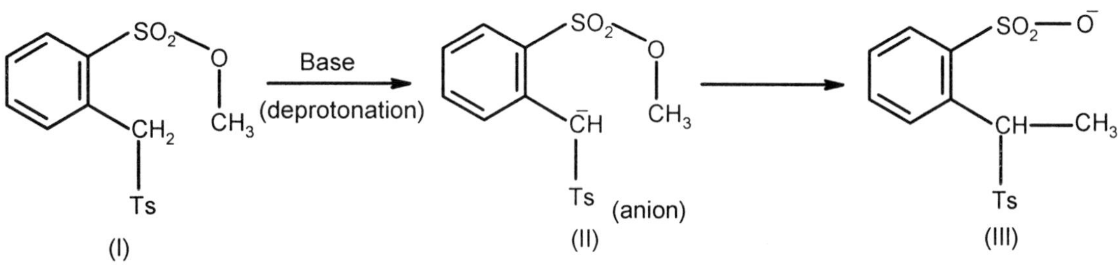

It might be assumed that the negatively charged carbon of the anion (II) would attack the methyl group *via* intramolecular S_N2 process. But this is not the situation. The crossover experiments[1] have demonstrated that the conversion is completely intermolecular — the negatively charged carbon attacks the methyl group of another molecule — not the intramolecular despite of being entropically more favoured.

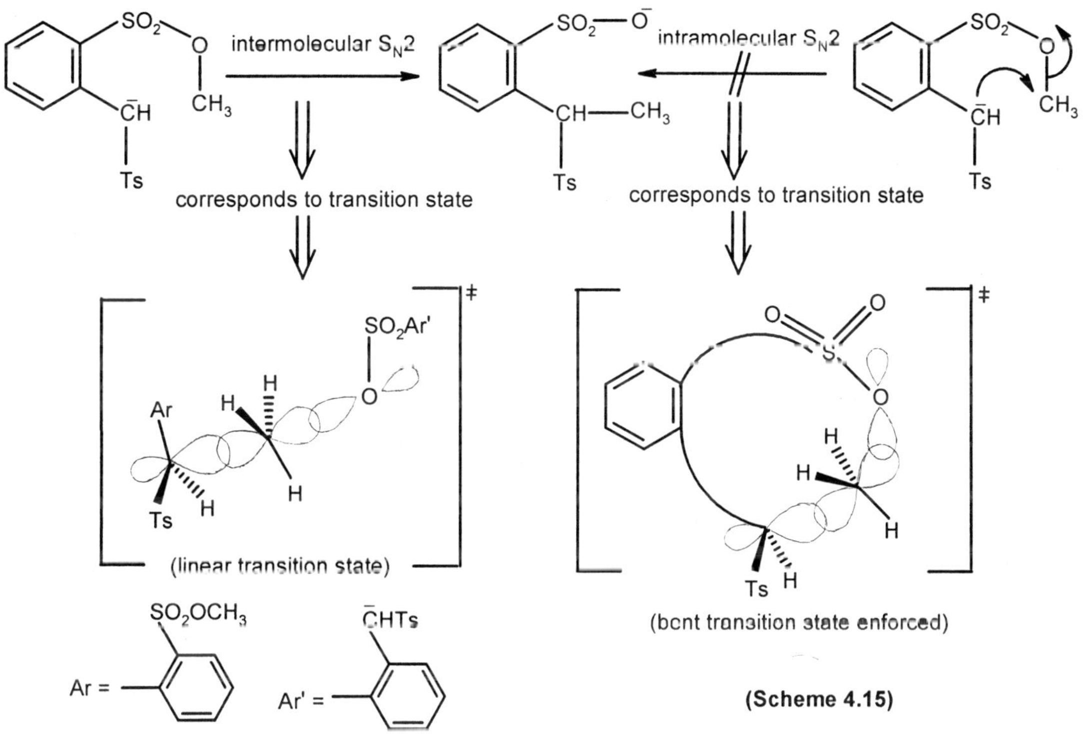

(Scheme 4.15)

1. Eschenmoser *et al.*, *Helv. Chim. Acta*, **53**, 2059 (1970).

This is because of the fact that proper alignment of the nucleophile with the substrate cannot be attained in the transition state undergoing intramolecular process where a bent transition state is to be formed (**Scheme 4.15**). Bent bonds are weaker than linear bonds because of smaller orbital overlap. Thus, one may conclude that in an S_N2 reaction the nucleophile must approach the attacking centre in a fashion collinear with the bond between the attacked carbon site and the leaving group.

Problem 4.16 The S_N2 reaction, $Nu^- + RX \rightarrow NuR + X^-$, when carried out in gas phase shows two potential minima which do not exist for the reaction in solution. Justify.

Solution. This is a very embarrassing feature of gas phase substitution reactions in which a negatively charged nucleophile attacks a neutral molecule. To explain this paradox Olmstead and Brauman[1] proposed double potential well mechanism according to which potential energy of an ion and a neutral molecule in the gas phase decreases as ion and molecule draw near to each other, and becomes minimum (first transition state) when a *loose ion-molecule complex* is formed (at this stage stabilizing ion-dipole forces become important). For the substitution, when they come more close to each other, the potential energy again increases until a second transition state, corresponding to another energy minima, is appeared due to formation of a *new ion-molecule complex* between leaving group and product.

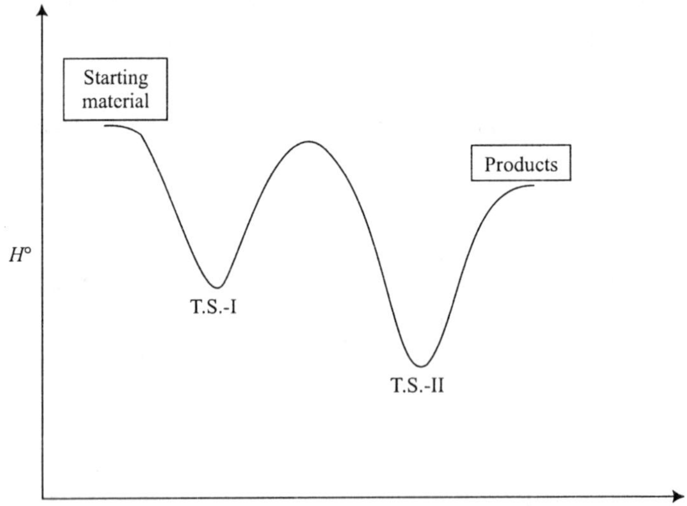

Fig. 4.3: Energy profile diagram for an exothermic S_N2 reaction in the gas phase (interestingly the transition state is at a lower energy than starting material)

1. Olmstead and Brauman , *J. Am. Chem. Soc.*, **99**, 4219 (1977)

This is the energy profile diagram for the reaction (**Fig. 4.3**) where the transition state is at a lower energy than starting material. The rate of the reaction depends both on the difference in entropy between starting material and transition state, and the energy barrier.

That in solution phase these two energy minima do not exist can be explained in the this way: there must be a decrease in potential energy when the nucleophile and substrate molecule come near to each other as a result of ion-dipole interactions, but the decrease is compensated by the increase in the potential energy accompanying the desolvation of the nucleophile as it approaches to the substrate molecule[1].

In contrast to solution phase, such a gas phase reaction not only bears two energy minima but corresponding transition state generally becomes much lower in free energy than that for the same reaction in solution. Caldwell and his coworkers[2] showed that the transition state energy for the reaction between chloride ion and methyl bromide is 2.5 kcal/mole *below* than that of the reactants in the gas phase but 17.7 kcal/mole *above* in DMF (*N,N*-dimethylformamide) solvent. The large difference in potential energy between the ground state and transition state in solution is due the fact that in the starting material the nucleophile is effectively solvated and stabilized, but in the transition state it is considerably desolavated accompanying the increase in potential energy.

Problem 4.17 Discuss the factors that influence the S_N2 reaction rate.

Solution. The rate of S_N2 reactions largely depends upon the following major factors:
(a) *Nature of substrate:* The electrical nature (electron withdrawal or electron release) of substituents attached to the reaction site almost does not affect the rate of S_N2 reactions, but the steric crowding around the central carbon atom undergoing displacement drastically influences the reaction rate. Bulky substituents prevent the nucleophile from approaching the reaction centre; thus as hydrogen atoms are replaced by the larger alkyl groups, nucleophile finds it increasingly difficult to approach the reaction centre. Besides, bulky groups as well as branching around the reacting carbon make an S_N2 transition state more crowded, thereby destabilizing it much, non-bonding interactions between various substituents would raise the transition state energy and so, the reaction rate decreases. Therefore, S_N2 reactions prefer to take place at relatively unhindered sites. Thus the S_N2 reaction rates of alkyl halides follow the order: $CH_3X > RCH_2X\ (1^0) > R_2CHX\ (2^0) >> R_3CX\ (3^0)$.

1. Chandrasekhar *et al., J. Am. Chem. Soc.*, **106**, 3049(1984); *ibid*, **107**, 154(1985).
2. Caldwell *et al., J. Am. Chem. Soc.*, **106**, 959(1984).

(b) *Nature of solvent:* The rate of S_N2 reactions suffers from a slight decrease on increasing polarity of solvents (*i.e.* increasing in dielectric constants); because here a polar solvent stabilizes the reactant (usually negatively charged nucleophile) more effectively than the transition state where the charge is more spread out. However, the reaction rate depends markedly upon the extent of solvation, and hence use of protic or aprotic polar solvents plays a vital role in determining the proper mechanistic pathway. In an S_N2 reaction, a protic polar solvent (*e.g.* H_2O, aq. MeOH, etc.) is likely to stabilize the attacking nucleophile (bearing a concentrated negative charge) by means of effective solvation through hydrogen bond formation, thereby lowering the reactivity of the attacking nucleophile. In contrast, nucleophilic strength is increased in polar aprotic solvents (*e.g.* DMF, DMSO, etc., which preferentially solvate the cations — the counterparts of the nucleophiles) because of the fact that these are not effective solvating agents for nucleophiles due to incapability of forming hydrogen bond.

Thus, an S_N1 reaction can be shifted to S_N2 on changing from a polar protic solvent (*e.g.* H_2O, aq. MeOH, etc.) to a polar aprotic one (*e.g.* DMF, DMSO, etc) and *vice versa.*

(c) *Nucleophilicity:* The rate of an S_N2 reaction increases on increasing the nucleophilic strength. Nucleophiles with greater nucleophilicity always facilitate the reaction rate.

(d) *Nature of leaving group:* The better the leaving group, the faster is the rate of an S_N2 reaction and also of S_N1 reactions as C–G bond cleavage is involved in both the rate determining steps.

Problem 4.18 Alkaline hydrolysis of RCH_2Cl to RCH_2OH is slow, but the reaction becomes rapid if catalytic amounts of KI are added to the reaction mixture.

Solution. The iodide ion being a good nucleophile and also a good leaving group is usually used as a catalyst to speed up the rate of a slow S_N2 reaction; this is known as *nucleophilic catalysis*. The chloride anion is not a so good leaving group.

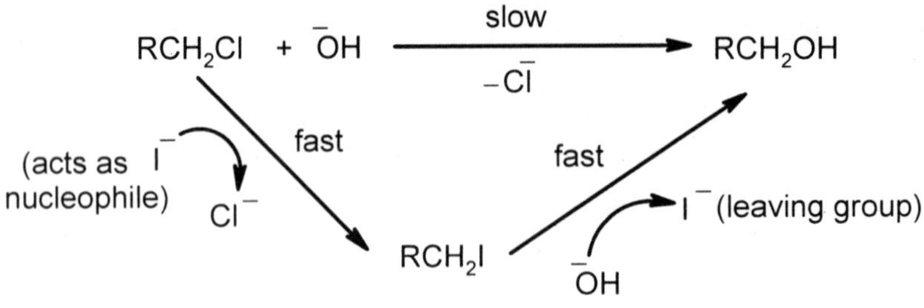

(Two consecutive displacement reactions take place; the iodide ion is regenerated and hence behaves as a catalyst)

Problem 4.19 The reaction rate of MeI with N_3^- at 0^0C is increased 4.5×10^4 fold on transfer from methanol to dimethyl formamide as solvent. Account for the observation.

Solution. Methanol is a protic polar solvent, while dimethyl formamide is an aprotic polar solvent. Polar protic solvents are capable to solvate negatively charged nucleophiles through formation of hydrogen bonds, thereby lowering the reactivity of the attacking species much effectively as the case with azide ion in methanol. On the contrary, the situation is different in aprotic solvents (incapable to form hydrogen bond) where not the nucleophile but its counterion (cationic part) is solvated by an aprotic polar solvent making the nucleophile much freer to react resulting in the enhancement of reaction rate as observed in the mentioned case.

Problem 4.20 Explain why (+)-2-iodooctane reacts with iodide ion in acetone to give a racemic mixture of (±)-2-iodooctane instead of forming only (–)-2-iodooctane. Draw an energy diagram for this racemization process.

Solution. The exchange reaction between 2-iodooctane and iodide ion in acetone proceeds through S_N2 mechanistic path. Each act of S_N2 displacement is associated with inversion of configuration of the starting compound. Thus, (+)-2-iodooctane on reaction with I^- ion would give (–)-2-iodooctane (**Scheme 4.20**) showing opposite optical rotation. As the concentration of (–)-2-iodooctane increases, it also tends to react with the iodide ion forming the (+)-isomer. These are the enantiomers of each other. Thus a situation comes when the reaction mixture contains equimolar amount of (+)- and (–)-2-iodooctane, and this composition does not change since both of the enantiomers react with the iodide ion at equal rate. That is why either (+)- or (–)-2-iodooctane when treated with iodide ion in acetone, the resulting mixture becomes always a racemic one.

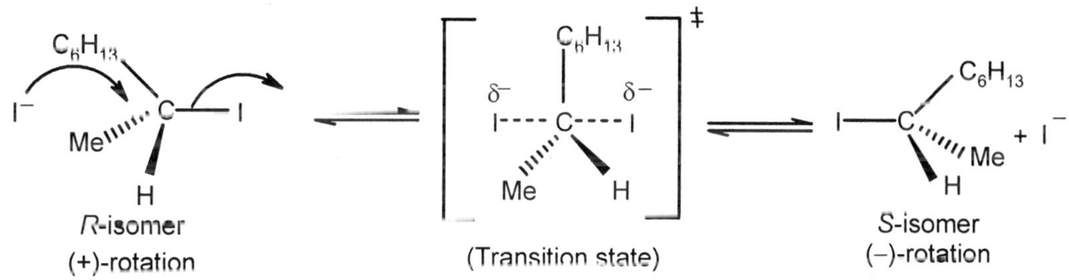

(**Scheme 4.20**)

The energy profile diagram for the reversible reaction may be drawn as follows:

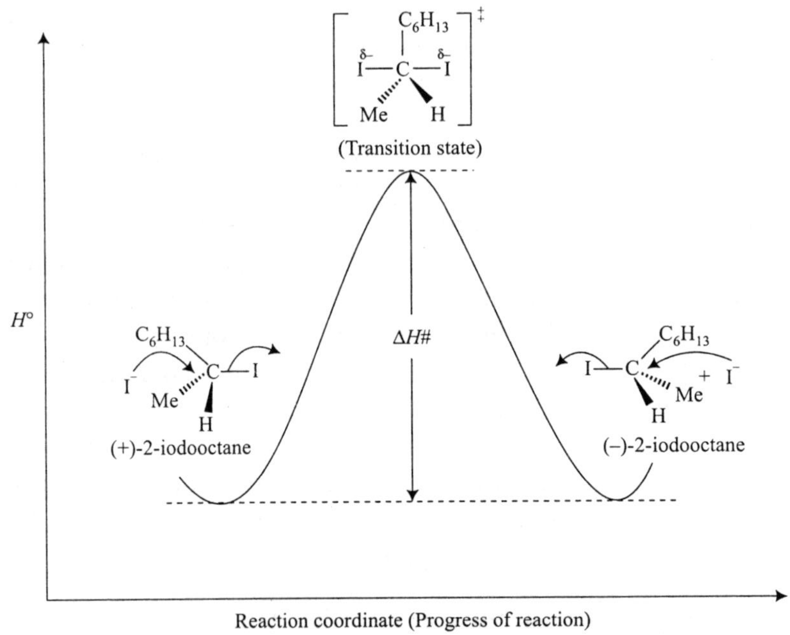

Fig. 4.4: Energy profile diagram for self-exchange reaction, *e.g.* (+/–)-2-iodooctane and iodide ion

Problem 4.21 Why halocyclopropanes are reluctant to undergo S_N2 reaction?

Solution. At the transition state of an S_N2 reaction the reacting carbon atom assumes sp^2 hybidization with bond angle of 120^0. In a cyclopropyl system, where the bond angle is 60^0, the ring carbon atom concerned cannot adopt such configuration at the transition state being distorted a much to avoid the huge bond angle strain that would increase the energy of the transition state at a level where it cannot sustain. Thus halocyclopropanes are reluctant to undergo S_N2 reaction.

Problem 4.22 Cite an example of substitution reaction where both the S_N1 and S_N2 mechanisms operate simultaneously.

Solution. Hydrolysis of *p*-methoxybenzyl chloride in 70% aqueous acetone proceeds through S_N1 mechanistic pathway.

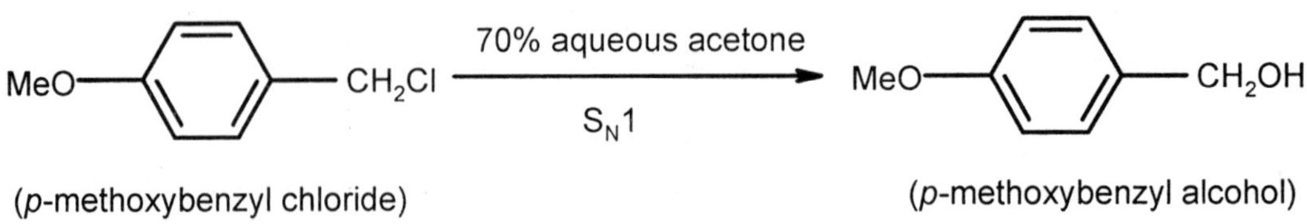

But on addition of azide ions, *p*-methoxybenzyl azide is also formed along with the alcoholic product (*p*-methoxybenzyl alcohol). It has been found that addition of azide ions enhances the rate of ionization as expected due to *salt effect*, but at the same time decreases the rate of hydrolysis. Greater ionization means greater number of intermediate carbenium ions, but the decrease in the rate of hydrolysis suggests that some of the carbenium ions combine to the azide ions through S_N1 path. However, it has been observed that the rate of ionization is always *less* than the rate of the overall reaction – the phenomenon logically indicates that some azide derivatives must also be formed by S_N2 path. Thus, it may be concluded that in this case both the S_N1 and S_N2 mechanisms may operate simultaneously.

Problem 4.23 Give a brief account on $S_{RN}1$ mechanistic pathway.

Solution. For the aliphatic nucleophilic substitution reactions,

$$RX + \bar{Nu} \longrightarrow RNu + \bar{X}$$

a third mechanistic pathway beyond simple S_N1 and S_N2 mechanisms was proposed[1] in 1966 — that involves radical intermediate as evidenced from the observation that free radical scavengers like molecular oxygen or cupric salts completely inhibit the reaction. The mechanistic path is termed as $S_{RN}1$ (*substitution radical nucleophilic unimolecular*). Substrates, which are very unreactive to the classical S_N1 or S_N2 reactions, undergo nucleophilic substitution following this mechanism. The popular example of this type:

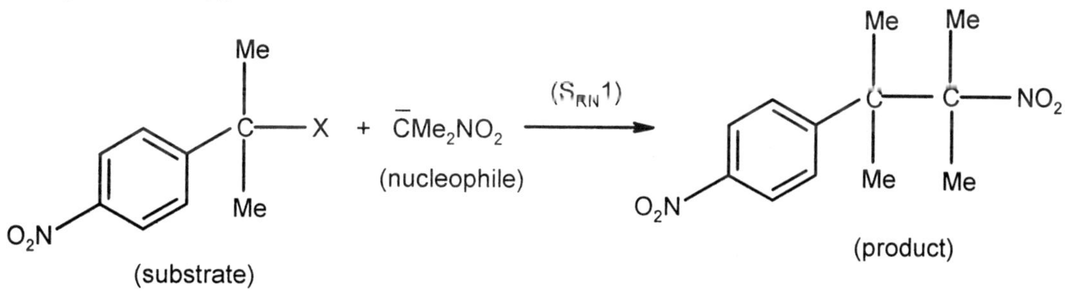

The proposed mechanistic scheme for an $S_{RN}1$ reaction can be shown as:

Initiation: Chain initiation is usually induced by electron-pluse radiolysis (or photo-stimulation) leading to the formation of a radical-anion of the neutral substrate.

1. Kornblum, N., Michel, R. E. and Kerber, R. C., *J. Am. Chem. Soc.*, **88**, 5662(1966); Russell, G. A. and Danem, W. C., *J. Am. Chem. Soc.*, **88**, 5663(1966).

$$R\text{—}X + e^{-} \longrightarrow R\dot{\text{—}}X^{-}$$
(radical-anion)

Propagation: The radical-anion then dissociates into an organic radical and the anionic leaving group. Now, the incoming nucleophile attacks the organic radical to form a new anion-radical species, which in the next step gives up its extra electron to another molecule of neutral substrate to form the product and to develop a new chain to propagate[1].

$$R\dot{\text{—}}X^{-} \longrightarrow R\cdot + X^{-}$$

$$R\cdot + Nu^{-} \longrightarrow R\dot{\text{—}}Nu^{-}$$

$$R\dot{\text{—}}Nu^{-} + R\text{—}X \longrightarrow R\text{—}Nu + R\dot{\text{—}}X^{-}$$

(product) (new chain)

followed by termination steps.

Besides aliphatic $S_{RN}1$ reactions, examples of aromatic ones are also known. For instance:

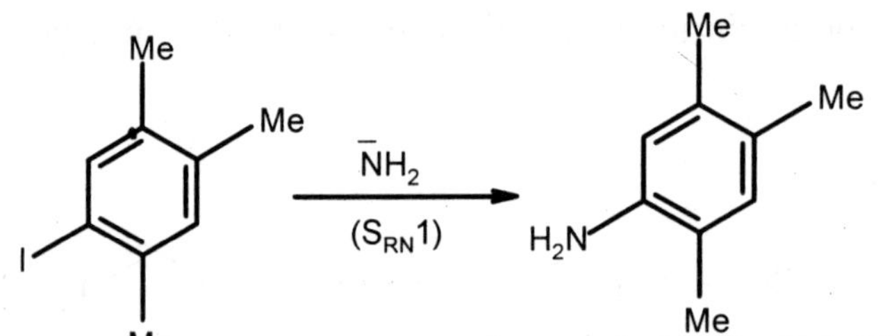

(one product in the mixture)

Aliphatic reactants (I & II) bearing $-NO_2$ group attached either to the α-carbon or to a phenyl moiety that is attached to the α-carbon are found to proceed this reaction smoothly — the nitro group provides a low-energy LUMO orbital to take up the extra electron into the substrate[2] and once this electron is transferred to the carbon–leaving group bond (C – X), it undergoes fragmentation.

1. Mishra, S. P. and Symons, M. C. R., *J. Chem. Res. (S)*, 147(1977); Symons, M. C. R., *J. Chem. Res. (S)*, 360(1977).
2. Bigot, B., Roux, D. and Salem, L., *J. Am. Chem. Soc.*, **103**, 5271(1981).

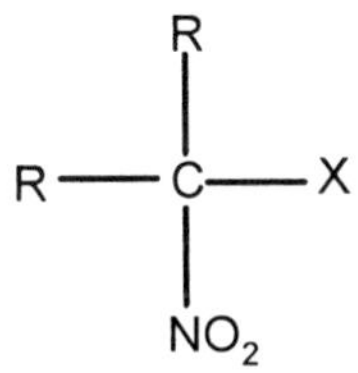

(I)

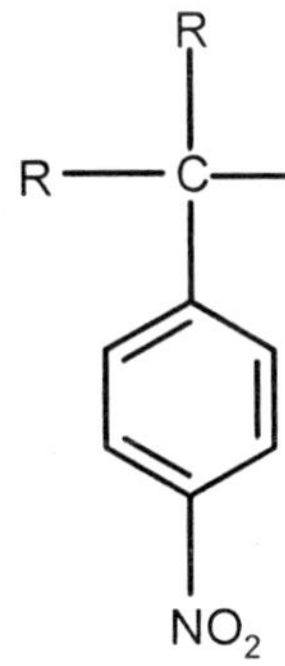

(II)

[X= halogens, and many others including N_3, NO_2, $PhSO_2$, Me, Ph, etc.]

It is interesting to note that an α-CH_3 group enhances the rate $C\overset{\cdot}{-}X^{-}$ bond dissociation while an α-*t*-butyl group reduces the corresponding rate; Norris *et al.*[1] suggested that *t*-butyl group must destabilize the three-electron bond once it is formed, but actual cause of the retardation arises out of the fact that this tertiary alkyl moiety intends to keep the third electron from being transferred to the C – X bond in the first phase.

It has also been observed that the substrates like (III) and (IV) that remain unreactive to S_N1 or S_N2 substitutions undergo nucleophilic substitution through $S_{RN}1$ mechanism[2].

(III)

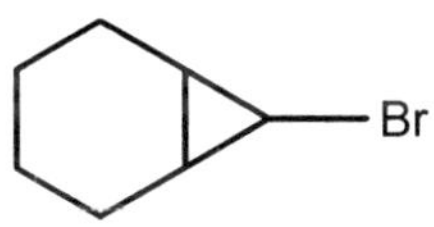

(IV)

Problem 4.24 Discuss S_Ni mechanism along with evidences in favour of the proposed mechanistic pathway.

Solution. The S_Ni mechanism (*substitution nucleophilic internal*) involves the nucleophilic attack of a part of departing group at the reaction site of a substrate molecule from the same side as the leaving group departs; the *internal* nucleophile is then detached from the rest of the leaving group forming the product with *retention of configuration*. This mechanism is believed to play in the reaction of alcohols with thionyl chloride ($SOCl_2$) in the absence of a nitrogen base.

1. Norris, R. K., Barker, S. D. and Neta, P., *J. Am. Chem. Soc.*, **106**, 3140(1984).
2. Rossi, R. A. and Palacios, S. M., *Tetrahedron*, **49**, 4485 (1993).

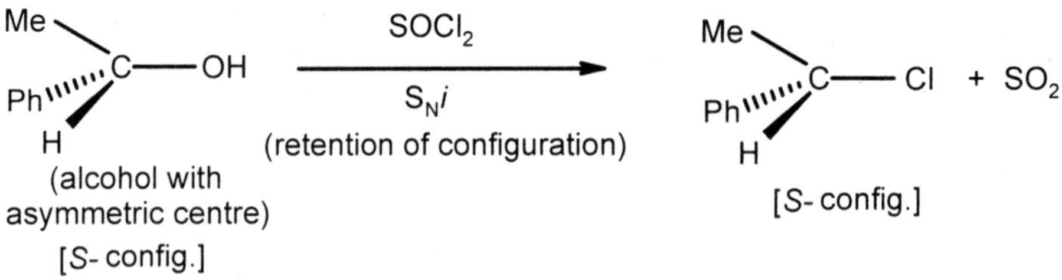

The reaction has been shown to follow a second order kinetics, rate of the reaction=k [alcohol] [$SOCl_2$], but clearly does not proceed through S_N2 path that would lead to inversion of configuration in the product, which is not observed. Hence a different mechanistic path, called S_Ni, has been proposed involving the following steps:

Step-I: Formation of alkylchlorosulphite as a true intermediate that is isolable.

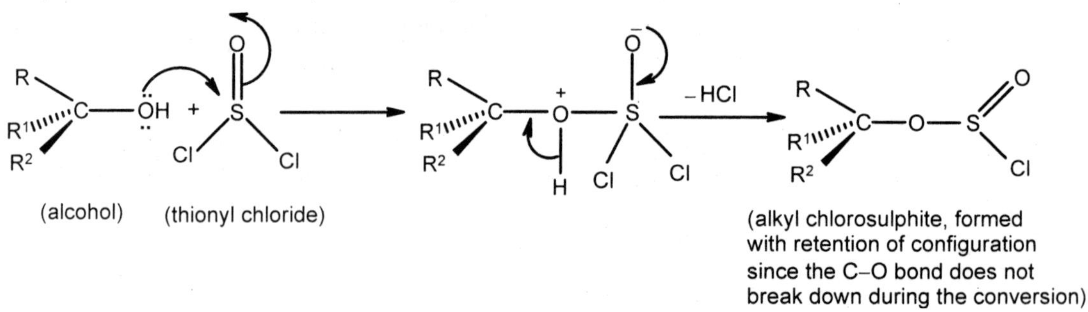

Step-II: The chlorosulphite intermediate dissociates slowly into a tight (or intimate) ion pair: a carbenium ion and $^{(-)}$OSOCl anion — a step alike to S_N1 mechanism; the tight ion pair resides within a solvent cage.

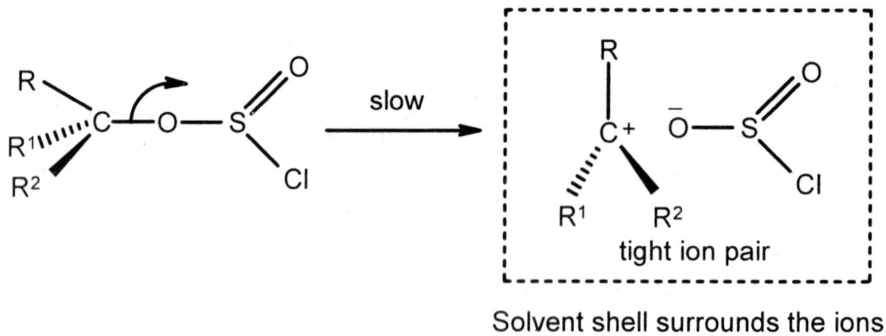

Step-III: In this step the $^{(-)}$OSOCl anion then collapses to give SO_2 and Cl^-, and this *internal* nucleophile (Cl^-), a part of the leaving group, attacks the positively charged carbocationic carbon on the same side as $^{(-)}$OSOCl departs since it is unable to get to the rear, resulting in the *retention of configuration* in the product.

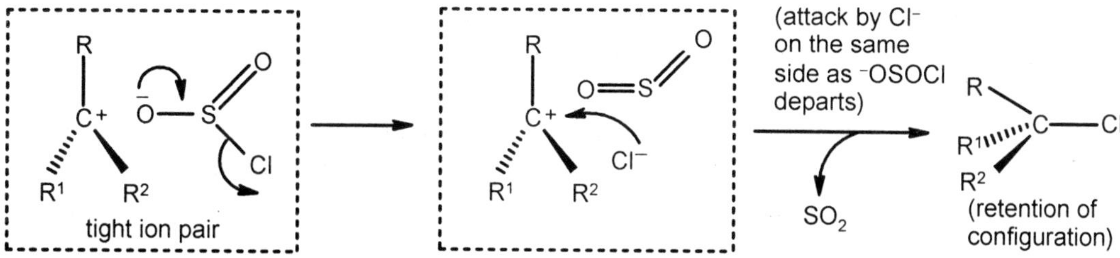

Evidences in support of the proposed mechanism:

(a) It has been possible to isolate alkylchlorosulphite intermediate when the reaction is carried out under milder conditions.

(b) The formation of intimate ion pair has been proved by the fact that 3-methyl-2-butanol (**1**) on reaction with thionyl chloride gives 2-chloro-2-methyl butane (**5**), a rearranged product that is possible only when a carbenium intermediate (within the tight ion pair **3**) is formed and undergoes subsequent rearrangement to the more stable one (in **4**).

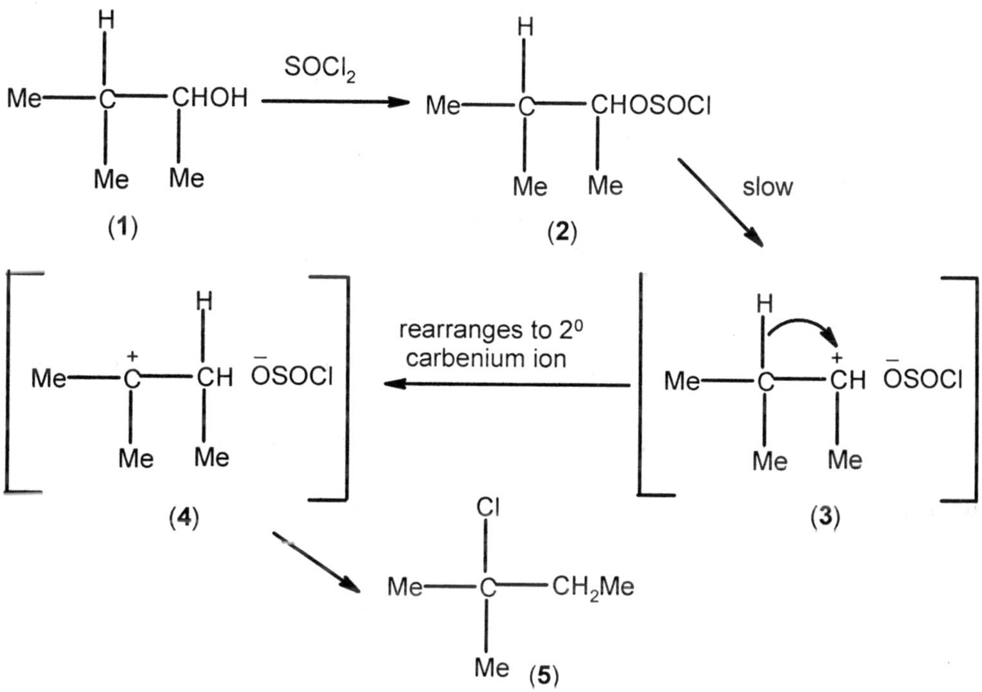

(c) That the rate of the reaction increases on increasing ionization power of solvent is suggestive of intimate ion pair formation.

(d) The most interesting observation is that the addition of pyridine (a nitrogen base) to the mixture of alcohol and thionyl chloride results in the formation of alkyl chloride with *inverted configuration*. This occurs because of the fact that the liberated HCl in the first step (*i.e.* formation of alkyl chlorosulphite), and also the chlorosulphite, ROSOCl react with pyridine producing free chloride ions in the reaction mixture — this Cl^- freed in this way can then act as a nucleophile with greater efficacy and attacks the reaction site only from the rear side in a simple S_N2 mode.

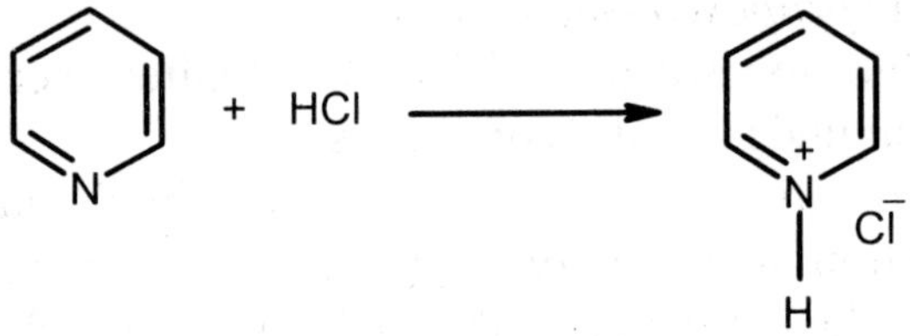

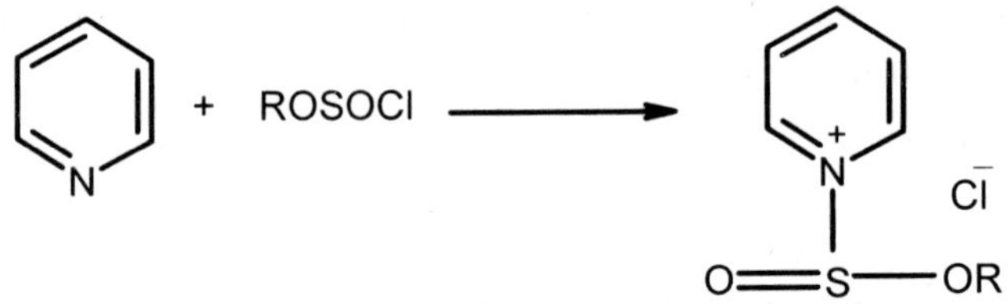

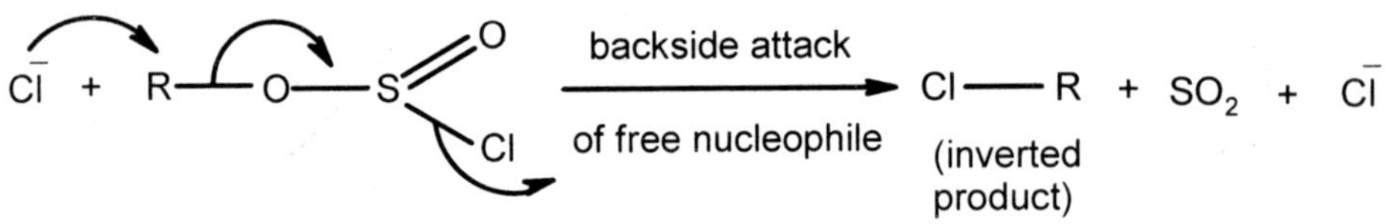

The S_Ni mechanism is relatively rare. Another example of this mechanism to play is the decomposition of alkyl chloroformates (ROCOCl) into alkyl chloride and CO_2.

Problem 4.25 Give a short account of S_N1', S_N2' and S_Ni' mechanistic pathways.

Solution. Allylic substrates undergo nucleophilic substitution reactions giving both the normal substituted product (developed due to attack at α-carbon) and the rearranged substituted product (developed due to attack at γ-carbon). This rearrangement is known as *allylic rearrangement*.

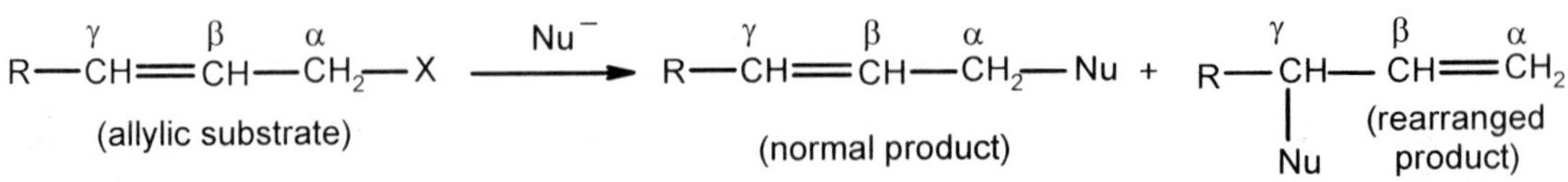

When nucleophilic substitutions on allylic substrates accompanying such rearrange-ment proceed through S_N1 conditions, the mechanistic pathway is termed as S_N1'; similarly S_N2' and S_Ni' mechanistic pathway resemble respectively S_N2 and S_Ni except the change in the position of the double bond in the products. The schematic representations for each of them are shown below:

S_N1' *Mechanism*

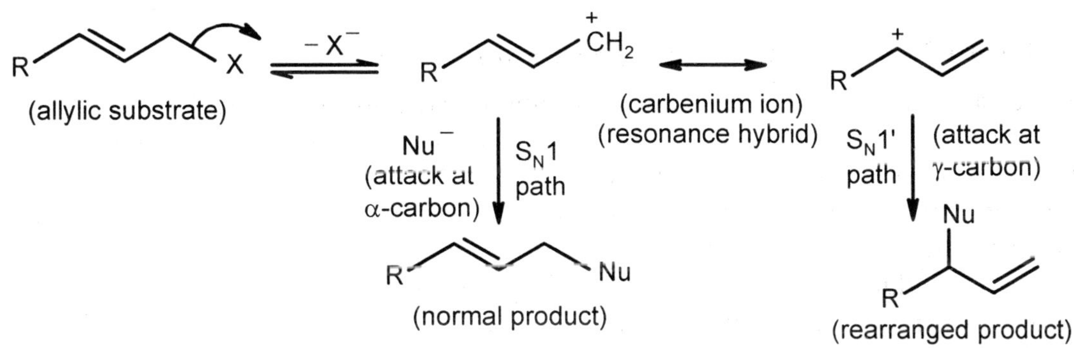

S_N2' *Mechanism*

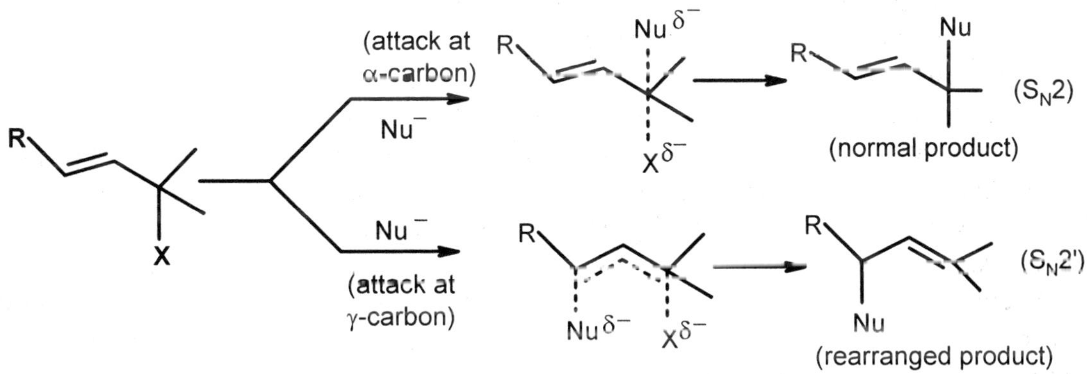

On investigation of the stereochemistry of S_N2' reactions, it has been found that both *syn* and *anti* approaches can take place, depending upon the nature of leaving group and incoming nucleophile, although the *syn*-pathway is frequent in most of the cases.

S_Ni' Mechanism

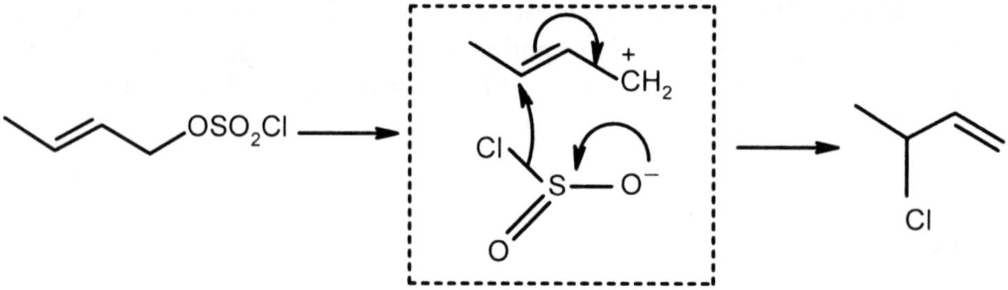

(S_Ni')

Problem 4.26 What do mean to say *phase-transfer catalysis*?

Solution. Rates of nucleophilic substitution reactions occurring between water-soluble salts (containing anionic nucleophiles) and the substrates soluble only in nonpolar organic solvents (insoluble in water) become very low — because the reactants cannot attain homogeneous mixing, and remain in two separate phases called as the organic-water system. *Phase-transfer catalysis* can effectively solve this problem — addition of small amounts of a *phase-transfer catalyst* to the two-phase organic-water system helps in transferring the water-soluble ionic reactant (nucleophile) across the interface to the organic phase where a homogeneous reaction can take place, thus, leading to the enhancement of reaction rate.

Quaternary ammonium salts, Q^+X^- [*viz.* $(n\text{-Bu})_4N^+X^-$], for example, act as good *phase-transfer catalysts*. Q^+ bears a water-soluble ionic part (hydrophilic portion) as well as nonpolar alkyl groups (hydrophobic portion) that tend to be soluble in the nonpolar medium — hence Q^+ shuttles between the two immiscible phases. As a result, it can transport Nu^-, as Q^+Nu^- (ion-pairing is loose since +vely charged nitrogen atom is surrounded by alkyl groups), into the nonpolar solvent where the nucleophile detaches itself and reacts quickly with the organic substrate. The Q^+ then returns back to the water medium effecting transport of the leaving group G^- as Q^+G^-. In this way, a *phase-transfer catalyst* plays its role in enhancing the reaction-rate of a two-phase nucleophilic substitution reaction.

A representative example is cited here. The reaction scheme as given below shows how tetra butyl ammonium chloride, $(n\text{-Bu})_4N^+Cl^-$, facilitates the reaction between 1-heptyl chloride and cyanide ion:

$n\text{-BuN}^+\text{Cl}^- + \text{Na}^+\text{CN}^- \longrightarrow n\text{-BuN}^+\text{CN}^- + \text{Na}^+\text{Cl}^-$ (in water phase)

$n\text{-C}_7\text{H}_{15}\text{Cl} + n\text{-BuN}^+\text{CN}^- \longrightarrow n\text{-C}_7\text{H}_{15}\text{CN} + n\text{-BuN}^+\text{Cl}^-$ (in non-polar phase)

Similarly, quaternary phosphonium salts can also be used.

Problem 4.27 "Crown ethers can also act as phase-transfer catalysts in carrying out certain nucleophilic substitution reactions in organic solvents in the absence of water". Comment critically.

Solution. Crown ethers are effective phase-transfer catalysts in carrying out certain nucleophilic substitution reactions in organic solvents; in these cases, there is no need of an aqueous phase at all — reaction takes place on simple addition of the salt (e.g. KCN, KF, KBr, KI, etc., where the counter anions are the attacking nucleophiles) to the organic phase; the crown ether gets the anion(s) into the organic medium. 18-Crown-6 (I) and dicyclohexano-18-crown-6 (II) are the two examples of such species:

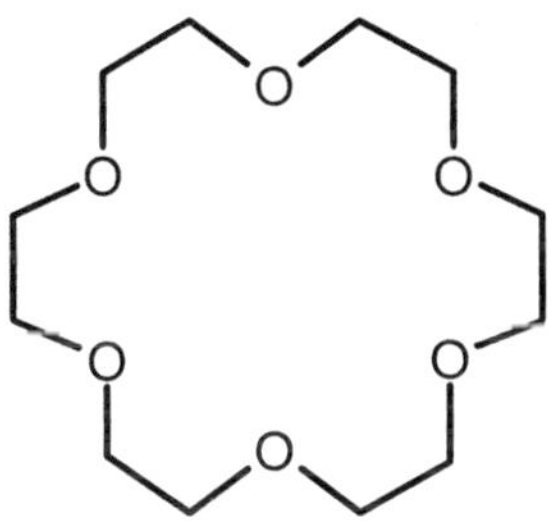

(I)

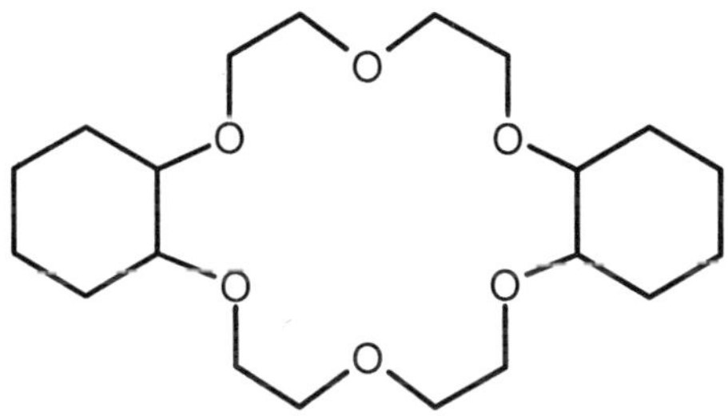

(II)

The crown ether (say, I) coordinates effectively with potassium ions as the cavity size is well fitted and the six oxygens are ideally situated to donate their electron pairs to the central ion. Thus, on coordination with the metal cation the crown ether converts the metal ion into a much larger species with the positive charge spread over a large volume and hence much less concentrated. This larger cation with a hydrocarbon like exterior becomes much more attracted to organic solvents along with the counter anion. That's why, KCN, although is generally insoluble in organic solvents, its cryptate salt is soluble in many of them.

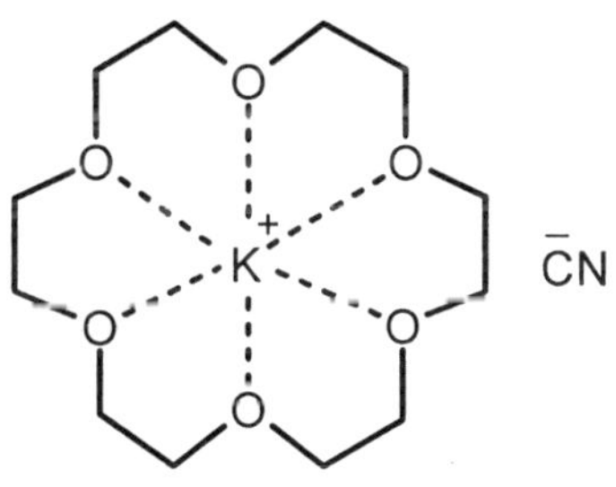

(A crwon ether complex with potassium ion soluble in organic solvents)

Thus, salts like KCN, KF, AcOK can be transferred into organic solvents by using catalytic amounts of 18-crown-6; in the organic phase the relatively unsolvated anions of these salts bring about nucleophilic substitution reaction on an organic substrate.

Problem 4.28 "Mustard gas ($ClCH_2CH_2SCH_2CH_2Cl$) undergoes rapid hydrolysis" — account for the fact.

Solution. The reason for the higher rate of hydrolysis of mustard gas arises out of neighbouring group effect — participation of a free electron pair in a nonbonding orbital on the neighbouring sulphur atom in the reaction facilitates the removal of the chloride ion forming a cyclic sulphonium ion, which being highly strained, undergoes rapid hydrolysis.

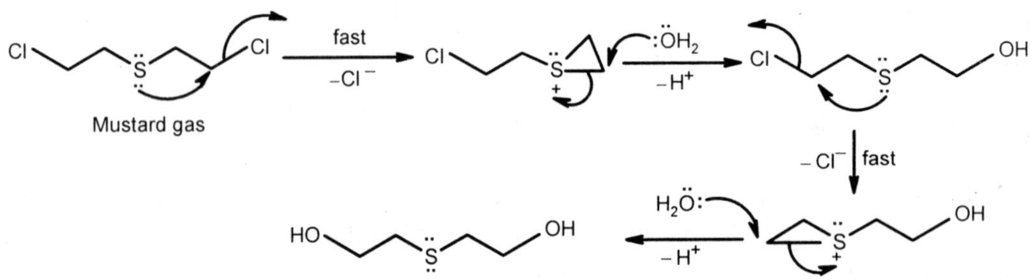

Problem 4.29 The ratio of the rates of solvolysis of ethyl tosylate and phenylethyl tosylate in CF_3COOH is about 1: 3040. Account for the fact.

Solution. Much greater solvolysis rate of phenylethyl tosylate in trifluoroacetic acid is attributed to the neighbouring group participation that is absent in the case of ethyl tosylate. The neighbouring phenyl ring in the former can make a π-electron pair available to form a phenonium ion intermediate, thereby, facilitating the removal of tosylate group. The reaction schemes are shown below:

$$CH_3CH_2OTs \xrightarrow[-TsOH]{CF_3COOH} CH_3CH_2OCOCF_3 \qquad k_{rel} \simeq 1$$

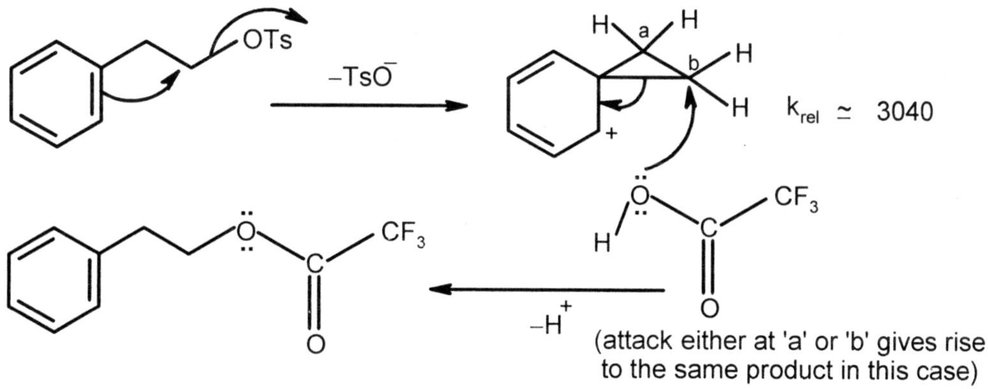

[acceleration of trifluoroacetolysis of phenyl tosylate by neighbouring group participation]

Problem 4.30 What are elimination reactions?

Solution. In elimination reaction two groups are removed from a molecule, neither being replaced by another group. Most commonly a proton is lost from one carbon whereas a nucleophile is lost from the adjacent carbon, and this type of elimination is called β- or 1,2-elimination.

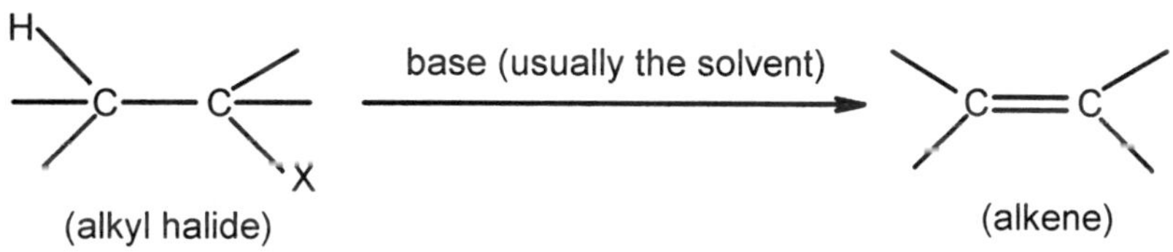

Besides, α- (or 1,1-) and γ- (or 1,3-) eliminations are also known. β-Elimination is the common elimination reactions to take place, and can occur *via* three distinct mechanisms — E1 (*elimination, unimolecular*), E2 (*elimination, bimolecular*), and E1cB (*elimination, unimolecular, conjugate base*) pathways.

Problem 4.31 Discuss E1 mechanism.

Solution. The E1 (elimination, unimolecular) mechanism is a two-step process in which the first step is the rate-determining step involving heterolysis of the substrate to form a carbenium ion intermediate that rapidly loses a β-proton in the second step.

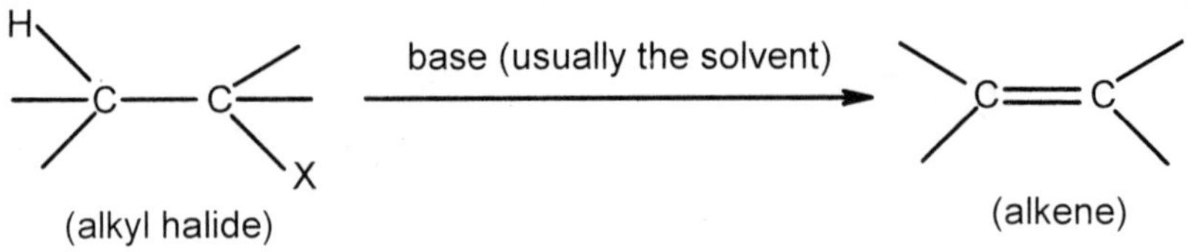

Mechanistic path for the E1 process can be represented as:

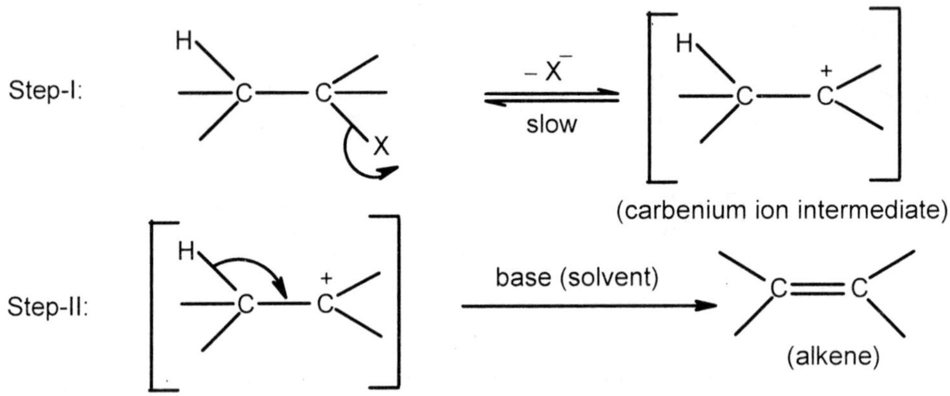

The reaction is of first order kinetics and unimolecular, as the rate depends only on the concentration of the substrate (and not the base/solvent).

$$\text{Rate} = k[\text{substrate}]$$

Tertiary halides react more rapidly than primary halides, because tertiary carbenium ion intermediates become relatively more stable, and therefore are formed more readily.

Reactivity order of alkyl halides undergoing E1 reactions: tertiary> secondary> primary

Thus, the E1 mechanism is analogous to and competes with S_N1 mechanism. In fact, the first step for both the mechanisms is exactly the same; the difference is in the second step — the base, usually the solvent, takes up a proton from the β-carbon of the intermediate carbenium ion leading to E1 product rather than attacking the intermediate at the positively charged carbon, as occurs in S_N1 process. Usually good nucleophiles favour the nucleophilic substitution, and poor nucleophiles and high reaction temperature favour E1 at the expense of S_N1 reaction.

The energy profile diagram for an E1 elimination reaction may be represented as:

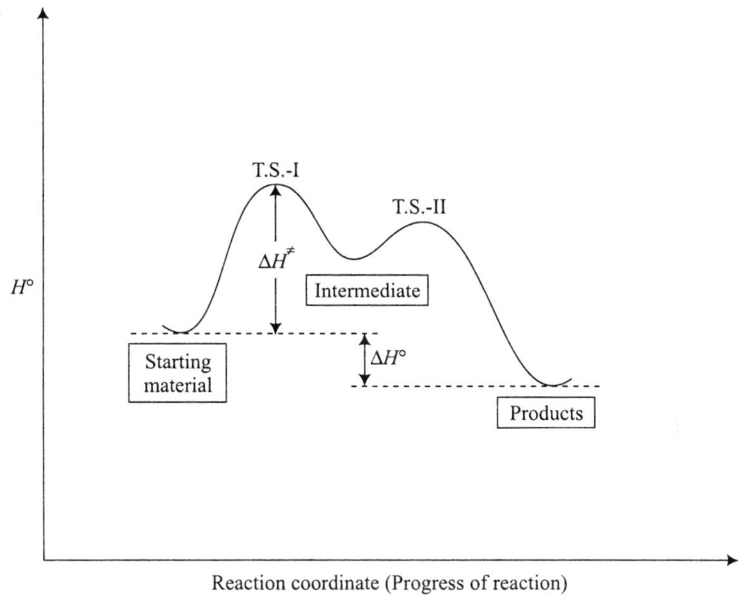

Fig. 4.5: Energy profile diagram of an E1 reaction

E1 reactions maintain *regio-* and *stereoselectivity*:

(i) ***Regioselectivity***

E1 eliminations afford predominantly the thermodynamically more stable alkenes (i.e. more substituted alkenes; Sayzeff elimination). The transition state of the second step (i.e. loss of a β-proton from intermediate carbenium ion) leading to a more substituted alkene will be lower in energy.

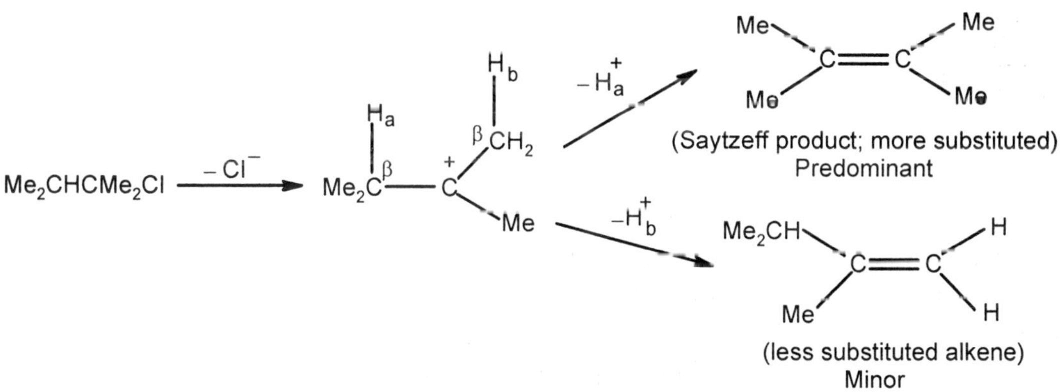

(ii) ***Stereoselectivity***

E-alkenes become more stable than the *Z*-alkene because of steric reasons — bulky substituents are further apart in the former isomer. So, the transition state leading to *E*-alkenes will be lower in energy, and hence E1 eliminations give predominantly the *E*-alkene rather than the *Z*-isomer.

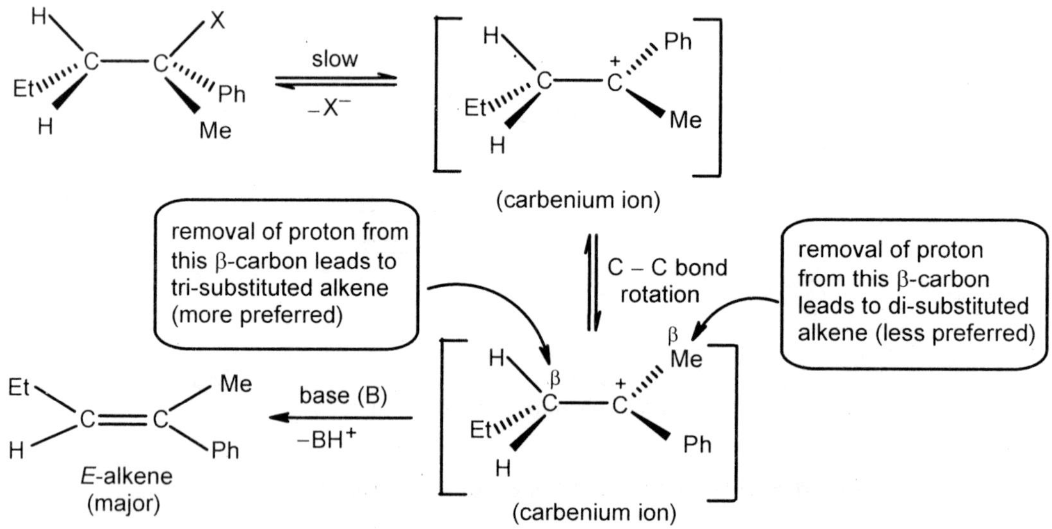

Problem 4.32 Discuss E2 mechanism.

Solution. The E2 (*elimination, bimolecular*) mechanism is a *concerted* (one-step) process involving the simultaneous removal of two groups, one of which is usually a proton that is pulled off by an added base. Thus in case of an alkyl halide undergoing E2 reaction, a new C=C (π) bond begins to form at the same time as the C–H and C–X (two σ) bonds begin to break.

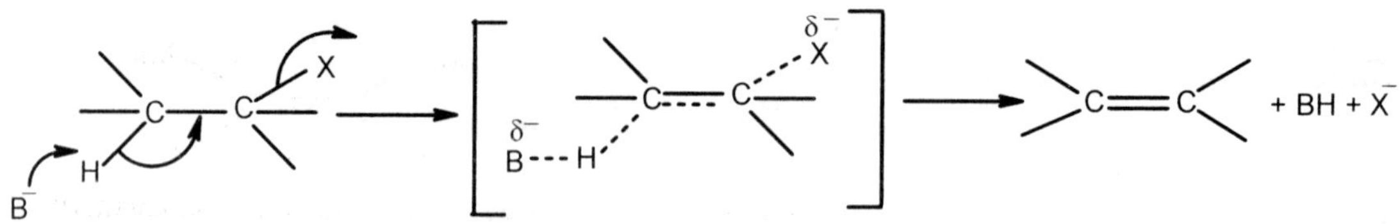

Since the transition state of the single *concerted* rate-determining step of an E2 reaction involves both the substrate and the base, it is bimolecular and kinetically second order — first order with respect to the substrate and first order with respect to the base.

$$\text{Rate} = k[\text{substrate}][\text{base}]$$

The energy profile for the E2 reaction is given below:

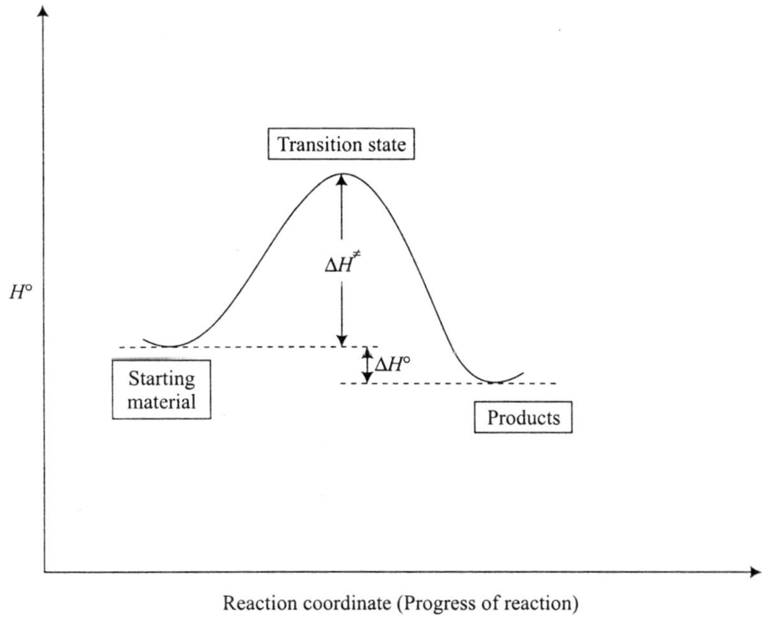

Fig. 4.6: Energy profile diagram of an E2 reaction

E2 mechanism is analogous to the S_N2 path and often competes with it. The difference in the two mechanistic pathways is related to the mode of action of the attacking species — whether it attacks the carbon centre (and thus acts as a nucleophile) or abstracts the proton (and thus acts as a base).

Stereochemistry of E2 path

A stereochemical requirement is maintained in undergoing E2 mechanistic path; the five atoms involved (including the base) in the transition state must lie in the same plane, and this situation can be satisfied in two ways — when the substrate molecule, say alkyl halide (RX), adopts *anti-periplanar* conformation in which the departing groups/atoms (H & X) are on the opposite sides (*trans* relationship) with a dihedral angle of 180^0, or when the substrate molecule adopts *syn-periplanar* conformation in which which the departing groups/atoms (H & X) lie on the same side (*cis* relationship) with a dihedral angle of 0^0.

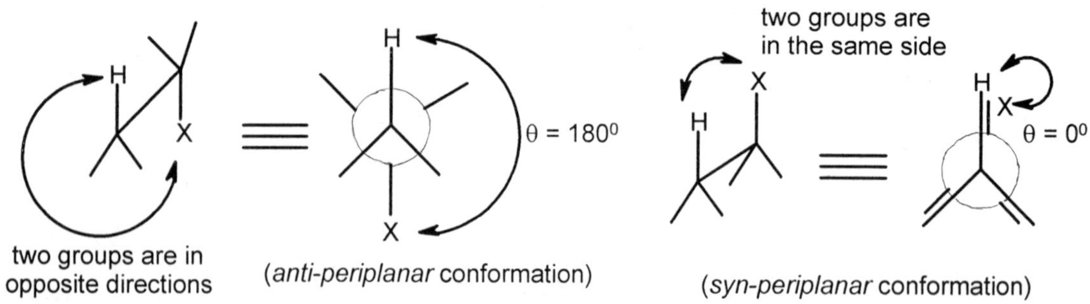

The *anti-periplanar* arrangement is lower in energy than *syn-periplanar* arrangement having an eclipsed conformation. In most cases, *anti*-elimination usually becomes more favoured over *syn*-elimination because of the fact that the substrate molecules concerned require less energy to reach the staggered transition state in comparison to the high energy eclipsed transition state involved in the latter case. However, *syn*-eliminations have also been observed to occur particularly when special structural features of the substrate molecules retard *anti*-elimination.

Anti-elimination:

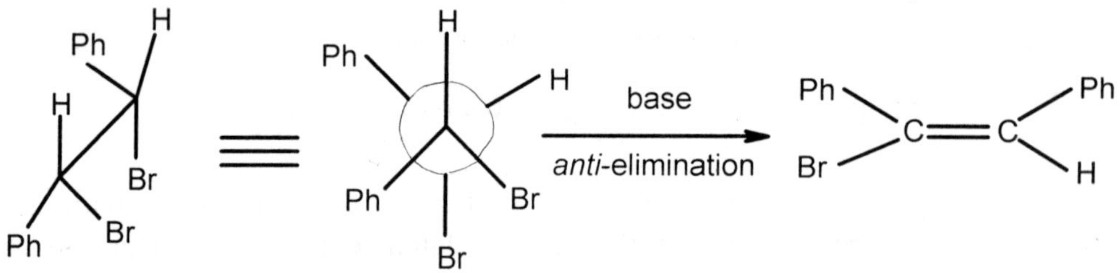

Syn-elimination:

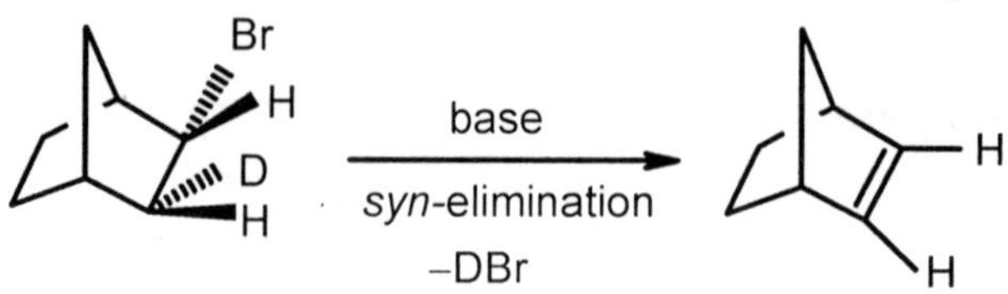

E2 elimination reactions are *stereospecific* — different stereoisomers of a substrate give different stereoisomers of the alkene. For example, the two diastereoisomers of 2,3-dibromobutane on reaction with base afford two different diasteroisomeric alkenes — the *meso*-isomer gives *cis*-2-bromo-2-butene, while the (+) or (–) isomer gives the *trans*-alkene.

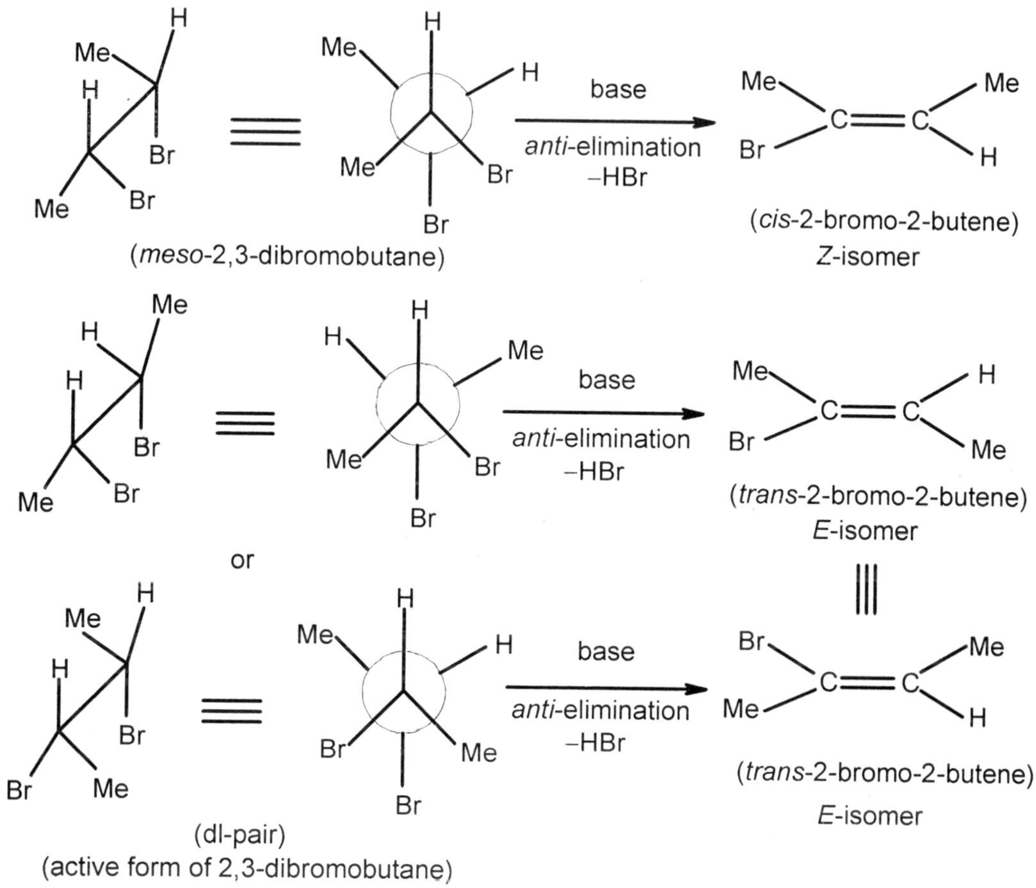

E2 mechanism is also evidenced from the facts that (i) it displays second order kinetics, and (ii) when the departing hydrogen atom is replaced by deuterium, an isotope effect is observed consistent with breaking of the bond concerned in rate-determining step.

Regioselectivity: Saytzeff vs. Hofmann

The regioselectivity, *i.e.* the position of the double bond in alkenylic product, of E2 elimination reactions depends on the nature of the leaving group (X). In substrates having alternative β-hydrogens, there is a possibility of obtaining two different alkenes — one is *more* substituted alkene, called the *Saytzeff* product and the other is *least* substituted, called the *Hofmann* product. The predominance of the product largely depends on the nature of the leaving group (X). It has been observed that when X is neutral (e.g. X = OTs, Br, etc.), elimination leads to more substituted

Saytzeff product (i.e. the thermodynamically more stable alkene) predominantly, and when X is positively charged bulky species (e.g $X = {}^{+}SMe_2$, ${}^{+}NMe_3$, etc.), elimination leads to the formation of least substituted (i.e. thermodynamically less stable alkene) Hofmann product as the major one. Schematic mechanistic representations for both of the eliminations are given below:

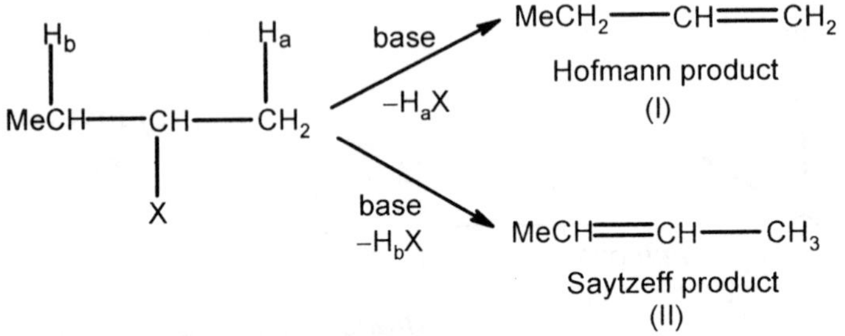

Leaving group X	% Yield I	II
Br	19	81
$\overset{+}{S}Me_2$	74	26
$\overset{+}{N}Me_3$	95	05

Path-a: Hofmann elimination

The elimination proceeds through a '*carbanion-like transition state*', not through '*alkene-like transition state*'. Strong electron-withdrawing capability of the leaving group, ${}^{+}NMe_3$ makes both the β-H atoms sufficiently acidic — thus the H–C bond breaks before the $C–{}^{+}NMe_3$ bond. The powerful electron-withdrawal by ${}^{+}NMe_3$ moiety stabilizes the incipient carbanion forming by abstraction of an acidic β-proton, more preferably the H_a by the added base — the δ^- is thus situated on a carbon bearing the least number of electron-donating alkyl groups. The Hofmann elimination is also normally expected when using bulky bases; this is because H_a (less hindered) is more accessible than H_b.

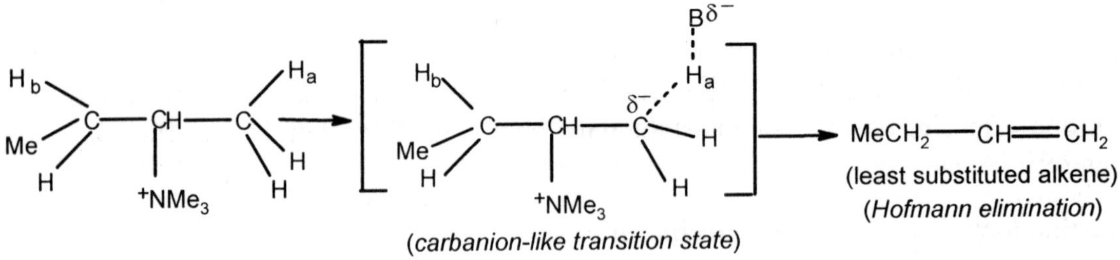

Path-b: Saytzeff elimination

In Saytzeff elimination the H–C bond breaks at the same time as the C–Br bond cleaves (in one step E2 path), thus, forming an '*alkene-like transition state*' — the alkyl substituents thereby being able to begin exerting their stabilizing effect quite early leading to the most highly substituted alkene.

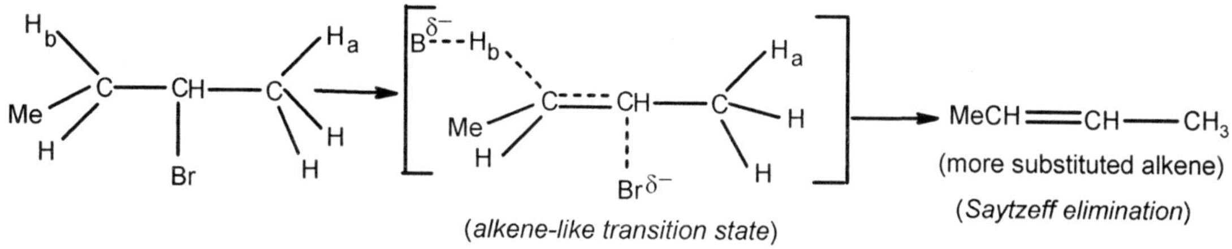

Problem 4.31 Discuss E1cB mechanism.

Solution. The E1cB mechanism (*elimination, unimolecular, conjugate base*) involves two steps like the E1 mechanism, but in this case the C(β)–H bond initially ruptures to form a carbanion intermediate (i.e. *conjugate base* of the substrate) followed by the subsequent loss of the leaving group; the first step is fast and reversible, while the second step is slow and rate-determining. Such a mechanism would be expected to operate if the intermediate carbanion becomes stabilized.

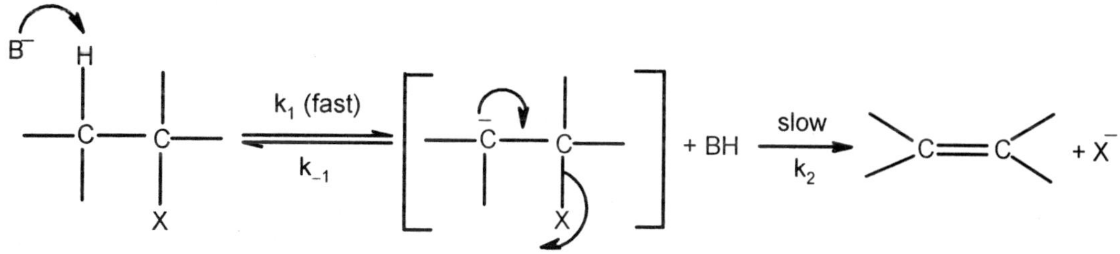

(carbanion intermediate)

An E1cB reaction follows the rate law:

$$\text{Rate} = \frac{k_1k_2[RX][B]}{k_{-1}[BH] + k_2} \simeq \frac{k_1k_2[RX][B]}{k_{-1}[BH]} \quad (\text{since } k_{-1} >> k_2)$$

Therefore, it is much essential to determine the exact course of pathway — whether perfectly E1cB or E2 — through which such type of a reaction proceeds. The distinction between the two mechanistic paths (E2 & E1cB) can be made very easily by means of labeling experiments, depending on the fact that the reversal step takes place at a rate greater than the second step, i.e. $k_{-1} >> k_2$.

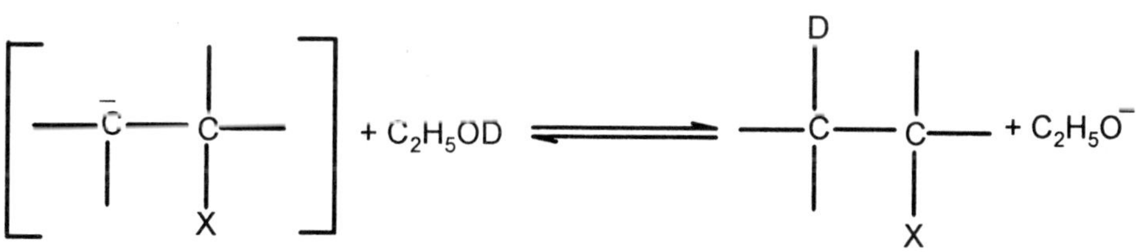

If the reaction proceeds through E1cB path, some carbanions would revert to the starting material, which should accompany the incorporation of deuterium if the reaction is carried in a deuterated solvent. On the contrary, there should be no incorporation of deuterium if E2 mechanism is operative.

Skell and Hauser (1943) carried out such experiment with β-phenylethyl bromide, a compound likely to undergo elimination by a reversible carbanion formation due to the presence of resonance-capable adjacent benzene ring) in NaOEt/EtOD; the products were analyzed after a partial transformation to styrene for their deuterium content.

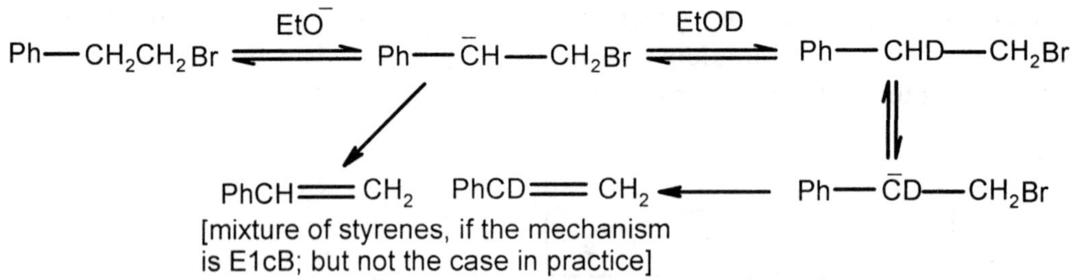

It was observed that there was no uptake of deuterium by either substance — so, there was no reversible formation of carbanion and this particular reaction must proceed through E2, not through the E1cB mechanistic path.

In practice, reactions following this carbanion pathway are exceedingly rare; actually the energy of activation for E2 is generally more favourable than that for E1cB. But the existence of an E1cB mechanism cannot be so easily ruled out — it takes place in certain specific systems where the β-hydrogen atom is highly activated by strong-electron withdrawing atoms/groups and the leaving group is poor, or the systems are under certain steric and geometric requirements of the eliminating substrates. The strong electron-withdrawing species will also stabilize the intermediate carbanion. Examples of reactions undergoing E1cB path are cited:

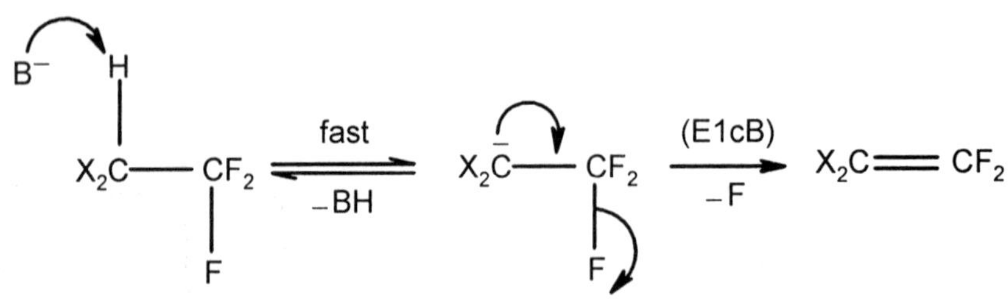

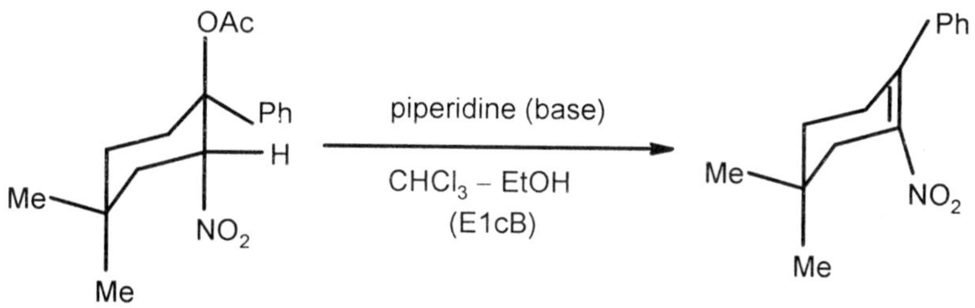

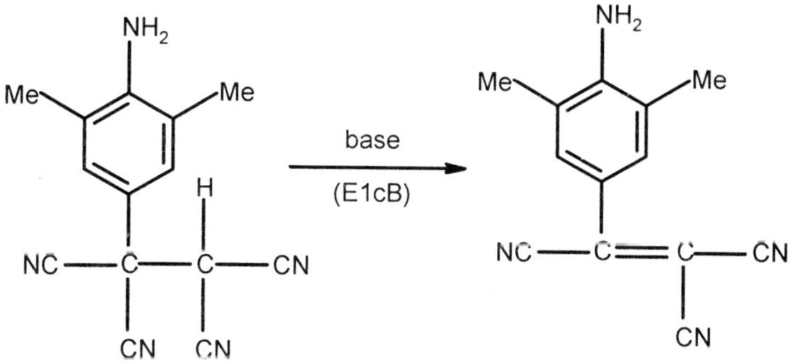

Thus, a substrate must have to satisfy the following conditions to undergo E1cB reaction:

(i) the substrate should bear strong electron-withdrawing atoms/groups on the β-carbon to make the β-H more acidic,

(ii) stabilization of the intermediate carbanion through electron-withdrawal effect of the substituents on carbanionic carbon, and

(iii) the leaving group should be poor.

Some other examples are:

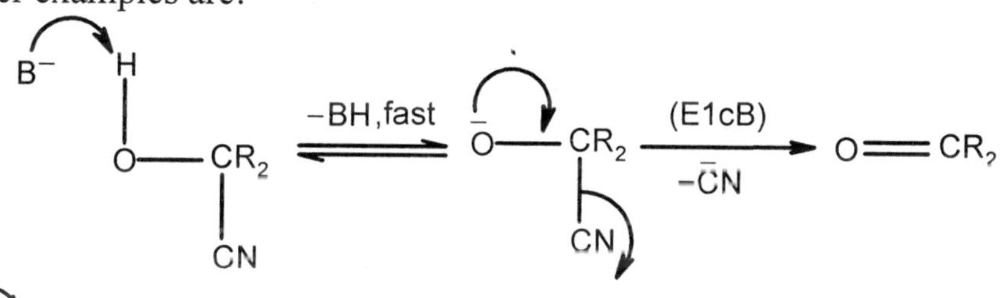

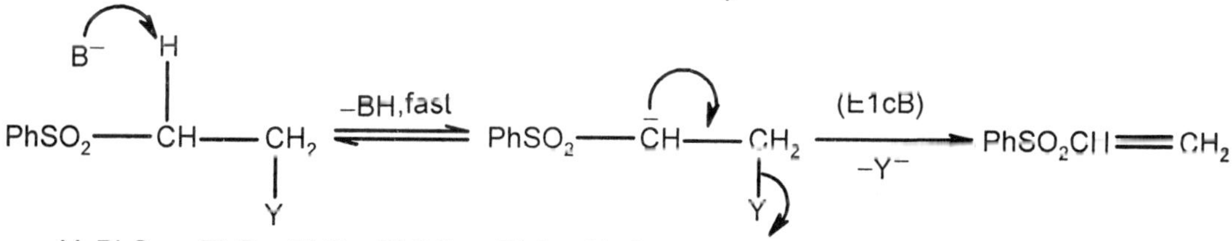

Y: PhSe > PhO > PhS ~ $PhSO_2$ > PhS > MeO >> CN (relative leaving group ability)

Problem 4.34 Discuss in brief the factors influencing E1 and E2 reactions.
Solution. The mechanism of an elimination reaction is greatly influenced by the following major factors (*E1 versus E2*):
(a) *The nature of the substrates*
For both the E1 and E2 pathways the reactivity order of simple alkyl substrate molecules is tertiary > secondary > primary; however the reasons for each pathway are different. For E2 reactions this preference is attributed to the greater stability of highly substituted alkenes formed from substrates of tertiary nature. Again, this preferential order for E1 reactions is ascribed to the increasing stability of the intermediate carbenium ion (formed in the rate-determining step) as per the sequence.

(b) *Nature of leaving group*
The better the leaving group, the faster the rate of both E1 and E2 reactions. Because a better leaving group favours the facile ionization leading to C – G (G = leaving group) bond cleavage.

(c) *Nature of base*
The base is involved in the rate-determining step of an E2 reaction but not in an E1 reaction. That's why increase in the strength and concentration of base will increase the rate of an E2 reaction, while the rate of an E1 reaction will not be influenced appreciably by this change. It has been observed that on increasing the strength and concentration of the base used, an E1 reaction can be made to follow E2 path.

(d) *Nature of solvent*
Increasing polarity of the solvent results in a significant increase in the rate of an E1 reaction, while an E2 reaction is favoured by a decrease in the polarity of the solvent. This is because E1 reactions involve more charge-separated species in their transition states than those in the transition states of E2 reactions.

Problem 4.35 Discuss in brief the factors that govern the proportions of elimination to substitution products.
Solution. Like S_N1 and E1 reactions, S_N2 and E2 reactions also occur simultaneously — the extent of elimination versus substitution depends upon a number of factors as summarized below:
(a) *The nature of the substrates*
E2 versus S_N2: While the rate of an E2 reaction of simple alkyl halides follows the order primary < secondary < tertiary, the rate of corresponding S_N2 reaction follows the reverse order. Branching at the β-carbon and steric crowding at the attacking site facilitate E2 elimination over S_N2 substitution.

E1 versus S_N1: Both the mechanisms involve formation of carbocation intermediate — thus extent of E1 elimination depends upon how fast the intermediate leaves up a proton, whereas the extent of an S_N1 substitution depends upon how fast the incoming nucleophile adds with the cabocationic intermediate.

With increasing branching in the substrate molecules, proportion of elimination tends to increase; this is because of two reasons — (i) more substituted alkenes will then be formed, and (ii) with increase in branching the carbocation intermediates (planar, bond angle 120^0) will show more resistance (to avoid steric strain) to the decrease in bond angle ($120^0 \rightarrow 109^0 28'$) brought about by the substitution whereas elimination does not involve the change in the bond angle ($120^0 \rightarrow 120^0$).

(b) *Nature of base*

E2 versus S_N2: E2 reactions are favoured over S_N2 reactions by increasing strength and concentration of the base involved; however if the attacking reagent becomes weakly basic but strongly nucleophilic, the substitution reaction predominates.

E1 versus S_N1: Unimolecular mechanisms are found to operate in solutions where the concentration of base is low or where the solvent acts as the base; thus substitution is favoured over elimination because of weak basic nature of the solvent.

(c) *Nature of solvent*

Usually bimolecular reactions are favoured by a decrease in the polarity of solvent, and this is more pronounced in case of E2 reactions. Thus less polar solvents not only favour bimolecular reactions, but also favour E2 over S_N2. In contrast, solvents with high polarity facilitate unimolecular mechanisms; but increase in polarity favours the substitution (S_N1) over elimination (E1).

(d) *Effect of temperature*

Elimination reactions are greatly favoured over substitutions on increasing reaction temperature. This is because an elimination reaction has to attain a higher activation energy than the accompanying substitution reaction, as more bonds need to be broken in order to form the alkene product.

Problem 4.36 How do you distinguish E1 from E2 path?

Solution. Differentiation between E1 and E2 mechanistic paths can be possible by studying the rates of elimination in substrates that differ only in that one or more of the protons have been replaced by deuterium atoms, *i.e.* the study of *kinetic isotope* *effects* that involve the changes in rate observed when a hydrogen atom ($^{1}H_{1}$) is replaced by a deuterium atom (D, $^{2}H_{1}$) in the same reaction.

$$\text{Kinetic isotope effects (KIE)} = \frac{k_H}{k_D}$$

k_H = Rate with substrate containing ^{1}H
k_D = Rate with substrate containing ^{2}H

Because the zero point energy of a C–H bond is higher than that for a C–D bond, the C–H bond requires less energy for its cleavage than that required for the latter. So, reactions involving the cleavage of C–H bonds go faster than those involving the cleavage of C–D bonds, provided such bond breaking occurs in the rate-determining step. Between E1 and E2 pathways, only in the E2 elimination cleavage of a C–H (or C–D) bond is involved during the rate-determining step. Thus, kinetic isotope effect (KIE) will be noticed, on changing the C–H bond concerned into C–D bond, only in the case of E2 reactions, not in E1 reactions.

Problems: 4.37 Explain why:

(a) *Neo*-menthyl chloride when treated with NaOEt/EtOH undergoes elimination (dehydrochlorination) reaction at a faster rate than does the menthyl chloride under the same conditions.

(b) 3-Bromopropanal undergoes base-catalyzed dehydrobromination at a faster rate than does the 2-bromopropanal.

(c) The rate of elimination reaction of isopropyl bromide is greater in ^{t}BuOK/DMSO than that in KOH/EtOH.

Solutions.

(a) Menthyl chlolride exists all most exclusively in conformation (I) where all the three substituents are at the equatorial position. For a facile E2 elimination, the departing groups (H & Cl) must have to be oriented *anti*-periplanarly; thus (I) flips to (I') where these two groups attain the required stereochemistry leading to the elimination product, 2-menthene exclusively. But the conformation (I') is of high energy, and less stable – that's why (I) is reluctant to undergo ring inversion, thereby, retarding the rate of dehydrochlorination.

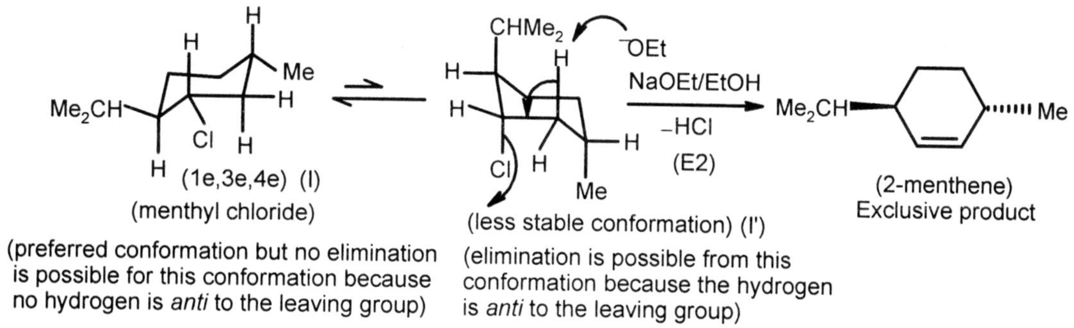

On the contrary, the situation is different in *neo*-menthyl chloride (II), where the two departing groups (H & Cl) are anti-periplanar in its preferred conformation. Thus, it undergoes facile E2 reaction at a much higher rate (under the similar conditions) yielding a mixture of alkenes of which the mostly substituted Saytzeff product predominates.

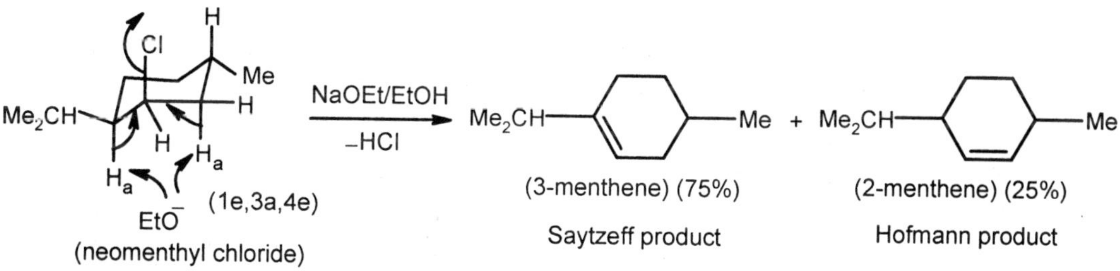

(b)

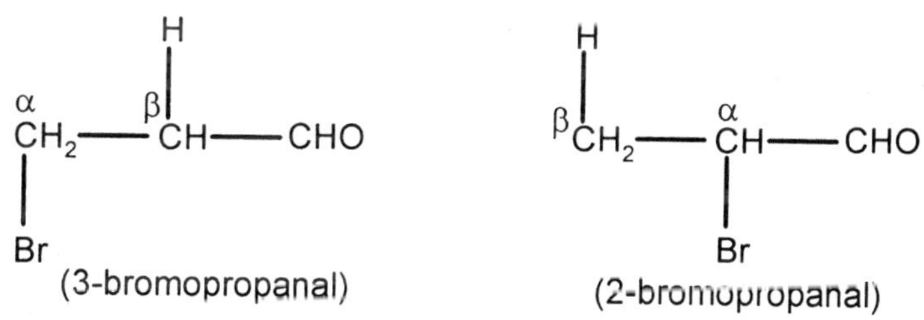

The departing β-H in 3-bromopropanal is more acidic due to strong electron-withdrawal inductive (–I) effect of –CHO group directly attached to that carbon. Hence, it can easily be abstracted by base in comparison to that in 2-bromopropanal. In both the cases, the –CHO group could stabilize the incipient double bond by conjugation. Thus, the transition state for dehydrobromination of 3-bromopropanal is more stable than that for dehydrobromination of 2-bromopropanal, and the former undergoes the elimination reaction more rapidly.

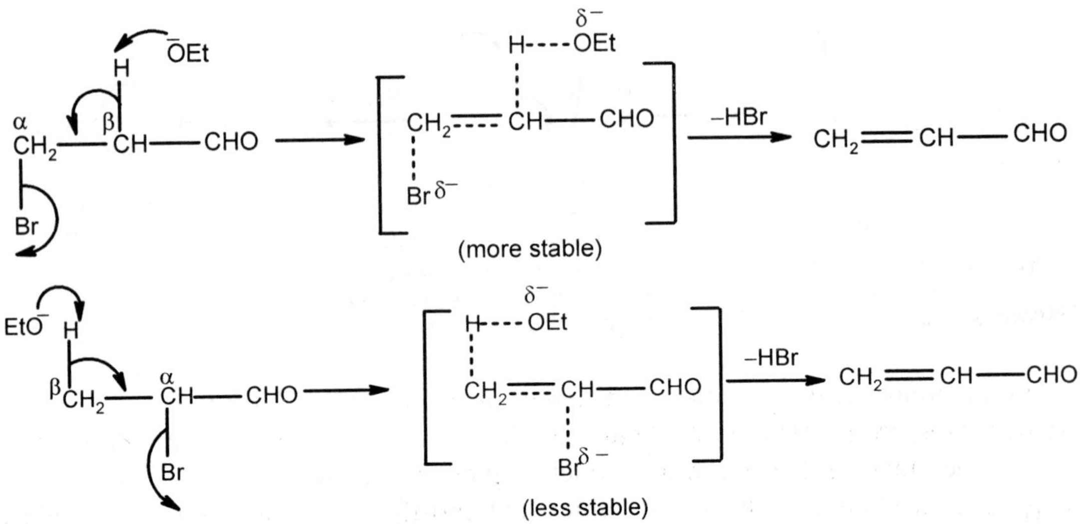

(c) Ethanol (ε=24) is less polar than DMSO (ε=45); since less polar solvent favours E2 over S_N2, thus it may be expected that the rate of dehydrobromination of isopropyl bromide would be greater in KOH/EtOH than that in *t*-BuOK/DMSO. But in practice, the reverse is true. This is because of the fact that strength of the base used is more predominating factor that promotes elimination over substitution. *t*-BuO⁻ is a stronger base than ⁻OH and in the aprotic solvent DMSO, its effectiveness as a base increases considerably because solvation through hydrogen-bonding does not occur. The use of *t*-BuOK/DMSO thus favours elimination much. Elimination is also favoured because the bulkiness of *t*-BuO⁻ prevents it from approaching the carbon centre to bring out the substitution.

Problem 4.38 Stereoselectivity of *anti*-addition of bromine to alkenes is observed to decrease in the following sequence:

MeCH=CHMe > PhCH=CHMe > PhC(Me)=CHMe.

Offer proper explanation in regard to mechanistic pathway.

Solution. Electrophilic addition of bromine to carbon-carbon double bonds may take place either through Ad_E2 or Ad_E3 mechanism. Alkenes that cannot form stable carbenium ions follow Ad_E3 pathway to afford *anti*-addition product, while conjugated alkenes might form stable carbenium ions via Ad_E2 pathway to give nonstereoselective addition. As suggested by Roberts and Kimball (1937)[1] such addition reactions proceed through a positively charged intermediate that may either be bridged 'bromonium ion' (I) or open carbenium ion (II).

1. Roberts, I. and Kimball, G. E., *J. Am. Chem. Soc.*, **59**, 947(1937).

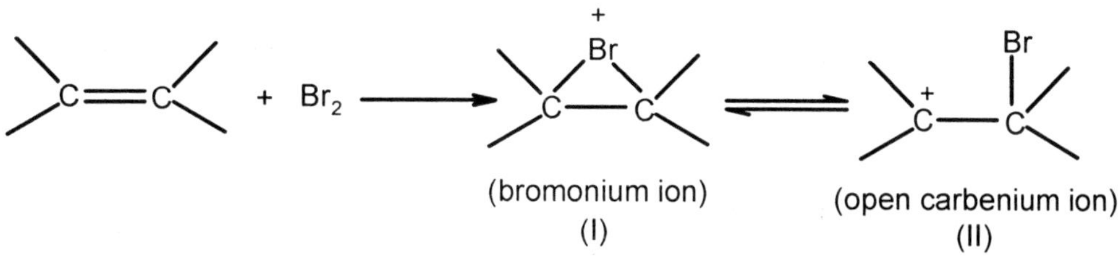

In 'bromonium ion' rotation about the C_α–C_β bond is restricted, and thus Br^- would attack backside from the Br^+ to give *anti*-addition.

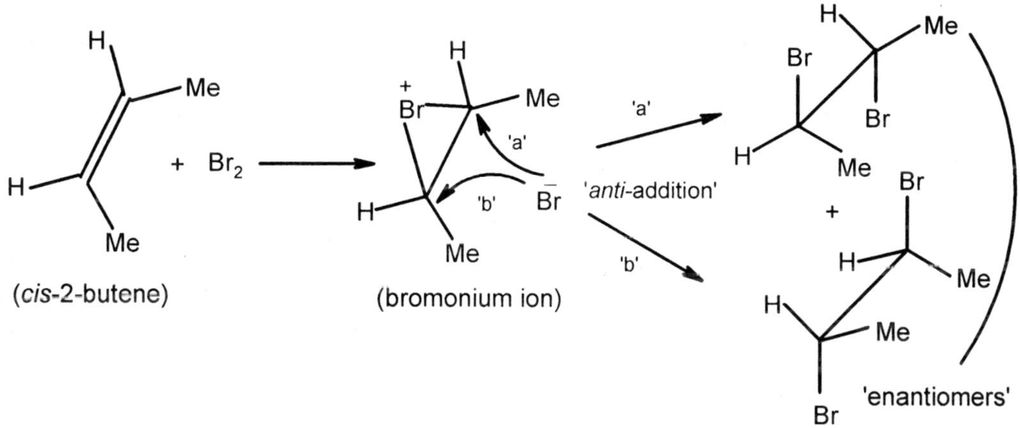

[(+/–) -2,3-dibropmobutane)]

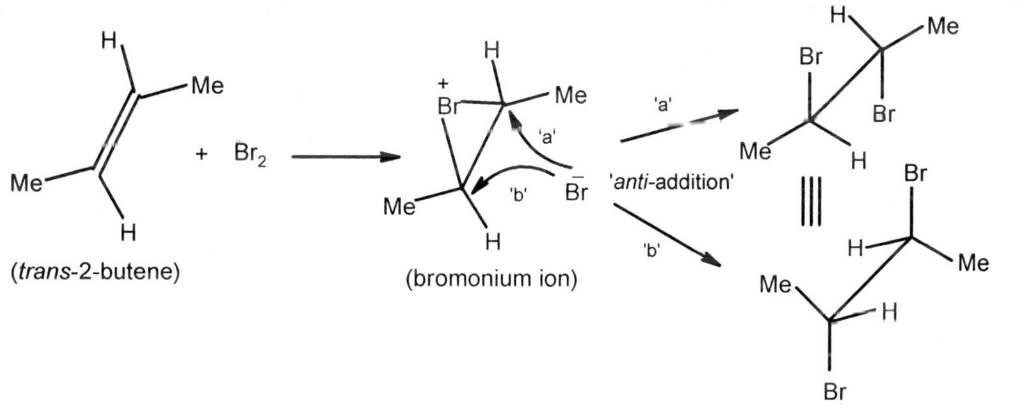

(*meso* -2,3-dibropmobutane)

But in case of conjugated alkenes, like PhCH=CH_2, a competition between an bromonium ion and an carbenium ion occurs — as one of the double-bonded carbons can hold the positive charge more effectively (due to conjugative effects), bridging would be less important leading to the decrease in stereoselectivity; free rotation is possible as well as the bromide ion can now able to attack from both the faces giving a mixture of *syn*- and *anti*-addition products.

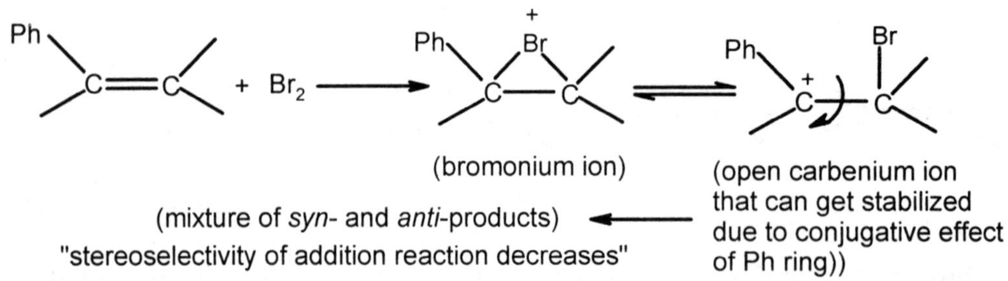

Actual course of reaction pathway depends upon the nature of substrates and reaction conditions[1,2]. Proportion of ‘open carbenium ion intermediate’ is more pronounced in PhC(Me)=CHMe in the given series, and less pronounced in MeCH=CHMe. That’s why stereoselectivity of *anti*-addition of bromine to alkenes is observed to decrease in the following sequence: MeCH=CHMe >> PhCH=CHMe > PhC(Me)=CHMe.

Problem 4.39 Offer strong evidence in favour of the existence of bromonium ion.
Solution. The most striking evidence in support of formation of ‘bromonium ion’ is that it has been possible to isolate bromonium ion when adamentylideneadmentane, a hindered alkene, is treated with bromine, because the corresponding bromonium intermediate (as tribromide salt) (I) is stable and cannot be attacked from the back side due to its cage like structure to give dibromo derivative[3,4].

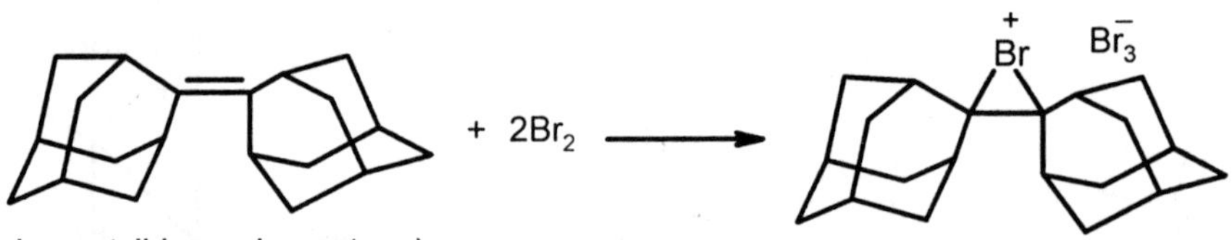

(adamentylideneadamentane) 'stable bromonium ion' (I)

Bromonium ions have also been observed to form in the gas pahse[5]. Detection of such intermediate(s) in SbF_5-SO_2 solution on dissolving of α-bromohalides at low temperatures by means of NMR spectroscopy has also been reported by Olah *et al.*[6-8].

---

1. Fahey, R.C. and Schneider, H.-J., *J. Am. Chem. Soc.*, **90**, 4429(1968).
2. Dubois, J.-E. and Mouvier, G., *Bull. Soc. Chim. Fr.*, 1426(1968).
3. Strating, J., Wieringa, J. H. and Wynberg, H., *J. Chem. Soc. D*, 907(1969).
4. Slebocka-Tilk, H., Bal, R. G. and Brown, R. S., *J. Am. Chem. Soc.*, **107**, 4504(1985).
5. Angelini, G. and Speranza, M., *J. Am. Chem. Soc.*, **103**, 3792(1981).
6. Olah, G.A., Westerman, P.W., Melby, E.G. and Mo, Y.K., *J. Am. Chem. Soc.*, **96**, 356(1974).
7. Olah, G.A., Schilling, P., Westerman, P.W. and Lin, H.C., *J. Am. Chem. Soc.*, **96**, 3581(1974).
8. Olah, G.A., Prakash, G.K.S. and Krishnamurty, V.V., *J. Org. Chem.*, **48**, 5116(1983).

Problem 4.40 What are π- and σ-complexes in aromatic electrophilic substitution? Offer evidences in favour of their formation.

Solution. In aromatic electrophilic substitution (Ar-S_E) reactions, two types of complexes as reaction intermediates are supposed to be involved — one is π-complex and the other is σ-complex. Let us have a brief discussion on each of them:

(i) ***π-Complex***

A π-complex is formed by the loose association of an attacking electrophile with the delocalized π-electron cloud of benzenoid ring; it is a 1:1 complex.

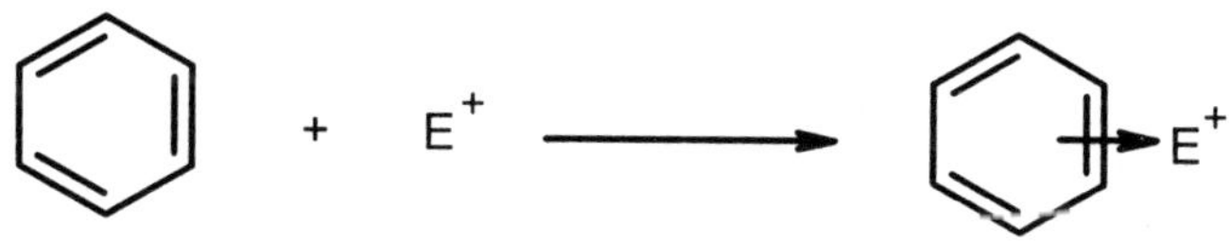

(benzene nucleus) (electrophile) (π-complex)

π-complexation involves a weak interaction between the π-electrons of the aromatic ring and the electrophile — there is no actual bond formation between a ring-carbon and the attacking electrophile. The stability of such complex is developed as a consequence of inductive effect of the substituents on the aromatic ring concerned. That there is no formation of new covalent bond in a π-complex may be understood from the following fact:

It has been observed that toluene forms a 1:1 complex with HCl at -78^0C, and the reaction is readily reversible. If the reaction is carried out with DCl instead of HCl, formation of corresponding π-complex occurs — but its formation (or decomposition since it is a reversible reaction) does not lead to the exchange of deuterium with any of the ring-hydrogen atoms. This establishes that no C–D covalent bond has been formed in the complex. It also does not conduct an electric current as expected.

Evidence

There are so many examples of formation of π-complex[1]. For instance, stable π-complexes are well-known to be formed when aromatic hydrocarbons are treated with Br_2, I_2, picric acid, Ag^+, or HCl. Existence of π-complexes has been evidenced from UV spectral measurements. Thus, benzene absorbs at 292 nm while benzene-iodide complex (1:1) absorbs at 340 nm; similarly toluene absorbs at 302 nm and its iodide complex at around 365 nm.

1. Banthrope, D. V., *Chem. Rev.*, **70**, 295(1970).

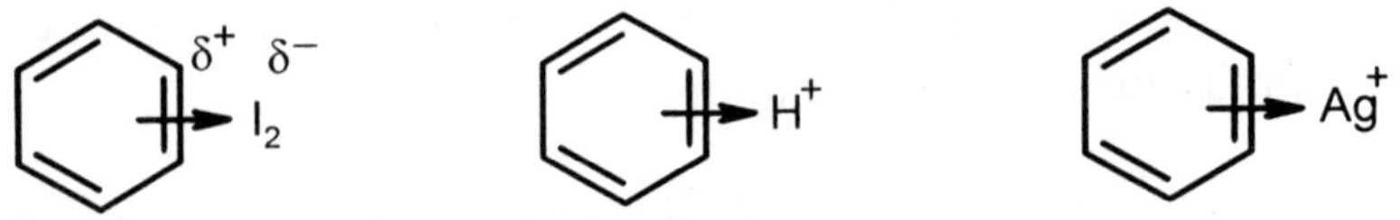

(π-complexes)

Nakane et al.[1] established from detailed spectroscopic study that toluene reacting with ethyl fluoride-boron trifluoride ($C_2H_5F.BF_3$) forms a termolecular π-complex of the type, $CH_3C_6H_4.C_2H_5F.BF_3$, where the un-ionized ethyl fluoride-boron trifluoride is oriented near *ortho* and *para* positions of toluene.

(ii) ***σ-Complex***

A substituted cyclohexadienyl cation is formed when an electrophile attacks an aromatic nucleus removing a pair of electrons from the sextet to form a new covalent bond between a ring carbon and the attacking electrophile — the carbocation exists as a resonance hybrid and is usually represented as (I). The carbocationic intermediates of this type are generally referred as σ-complexes[2] or Wheland complexes[3] or arenium ions.

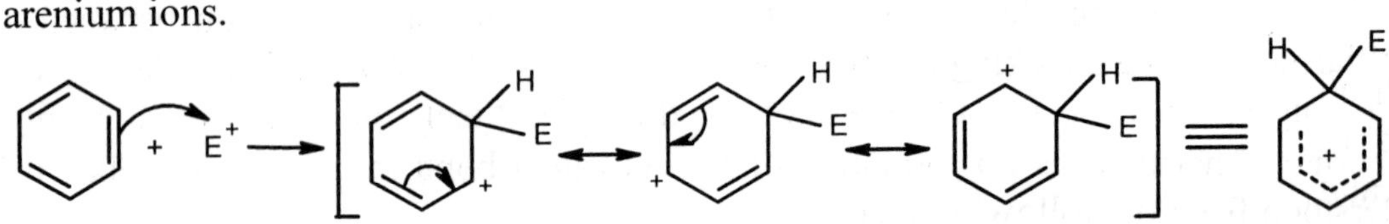

A σ-complex is formed destroying an aromatic sextet (that is associated with the great stability of aromatic compounds) as present in the starting material and also in the product; hence such species becomes high-energy intermediate and highly reactive. However, possible formation of a σ-complex and its stability as an intermediate may be ascribed to two factors:

(i) an appreciable amount of energy is liberated due to complete formation of a new covalent bond between a ring carbon and the attacking electrophile, and
(ii) an effective delocalization of the positive charge with the ring π-electrons contributes a lot to its stability.

---

1. Nakane, R et al., *J. Am. Chem. Soc.*, **87**, 3597(1965); *J. Org. Chem.*, **34**, 449(1969).
2. von Doering, W. et al., *Tetrahedron*, **4**, 138(1958).
3. Wheland, F. W., *J. Am. Chem. Soc.*, **64**, 900(1942)

Evidence

At very low temperature (–80^{0}C) toluene on treatment with HCl and a Lewis acid, e.g. $AlCl_3$ forms a σ-complex intermediate (II) as evidenced from the appearance of green colour of the solution, its ability to conduct an electric current (indicating the presence of positive and negative ions) and also from the fact that deuterium exchange[1] occurs when DCl is used instead of HCl. It has also been possible to prepare simple benzenonium ion in HF-SbF_5-SO_2ClF-SO_2 at the temperature –134^{0}C, where it could be studied spectroscopically[2].

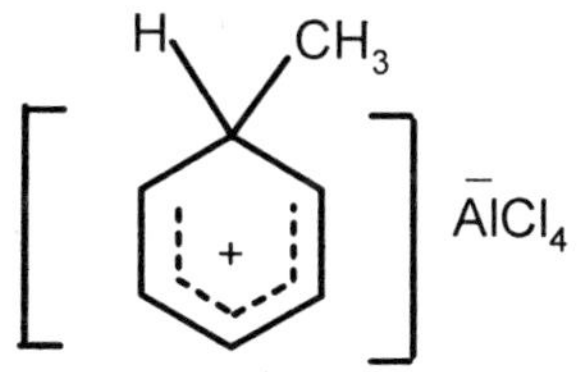

(II)
(σ-complex or Wheland complex)

The most fascinating evidence in favour of the existence of σ-complex intermediates is that in many of the cases arenium ions were isolated[3-5]. For example, Olah and Kuhn[6] prepared and isolated the arenium ion (III as an orange-coloured solid, m.p. –15^{0}C) formed on the treatment of mesitylene with ethyl fluoride and BF_3 (as catalyst) at –80^{0}C.

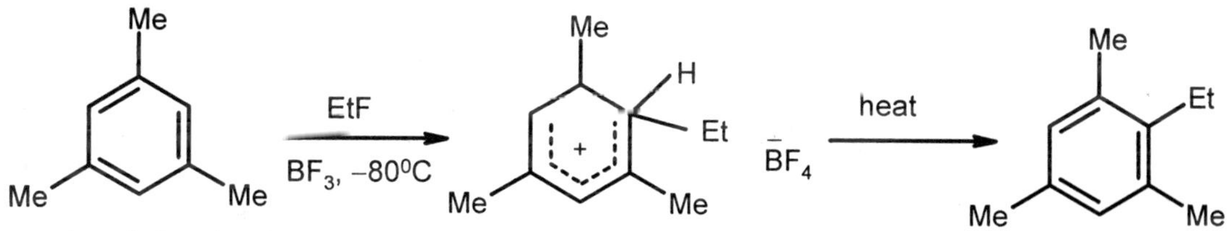

1. Brown, H. C. and Brady, J. D., *J. Am. Chem. Soc.*, **74**, 3570(1952).
2. Olah, G. A. et al., *J. Am. Chem. Soc.*, **94**, 2034(1972)
3. Koptyug, V. A., *Top. Curr. Chem.*, **122**, 1(1984).
4. Knoche, W. et al., *J. Am. Chem. Soc.*, **110**, 7484(1988).
5. Effenberger, F., *Acc. Chem. Res.*, **22**, 27(1989).
6. Olah, G. A. and Kuhn, S. J., *J. Am. Chem. Soc.*, **80**, 6541(1958).

Olah group also reported that the benzotrifluoride-nitryl fluoride-boron trifluoride system forms an isolable σ-complex intermediate (IV, yellow crystal stable up to 15°C).

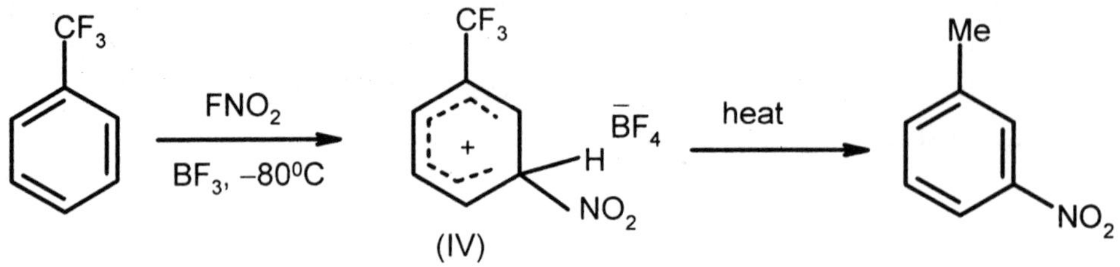

Problem 4.41 Outline the mechanistic path for aromatic electrophilic substitution reactions.

Solution. Aromatic electrophilic substitution reactions follow second order kinetics, and rate of this type of reactions depends on the concentrations of both the substrate and the attacking electrophile. Generally accepted mechanism[1] for such reactions, designated as Ar-S_E2, is suggestive of two discrete steps:

Step-I: In the first step formation of a σ-complex (or Wheland complex) as the reaction intermediate occurs due to the interaction between the aromatic ring and the attacking electrophile (E^+). This process requires high activation energy and leads to the first transition state. It is a slow step, and hence is the rate-determining step.

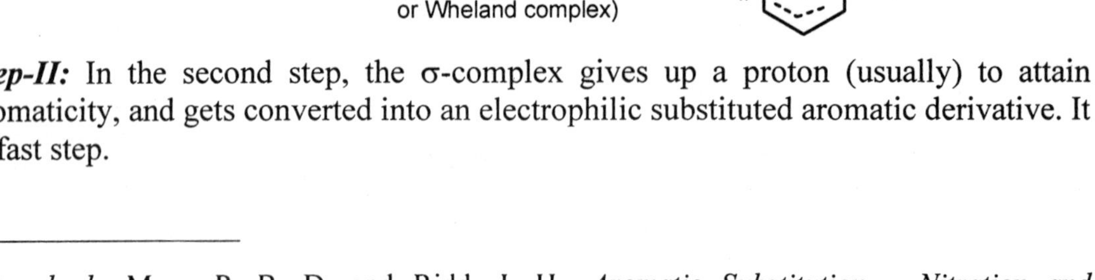

Step-II: In the second step, the σ-complex gives up a proton (usually) to attain aromaticity, and gets converted into an electrophilic substituted aromatic derivative. It is fast step.

1. *de la* Mare, P. B. D. and Ridd, J. H., *Aromatic Substitution – Nitration and halogenation*, Academic Press, New York (1959).

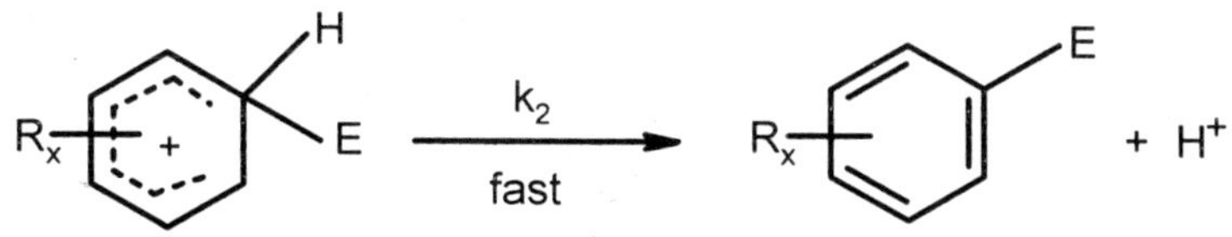

Formation of σ-complexes as reaction intermediates in Ar-S_E2 reactions has received confirmation from their direct isolation in a number of occasions.

The energy profile diagram for this mechanistic pathway may be represented as (**Fig. 4.7**):

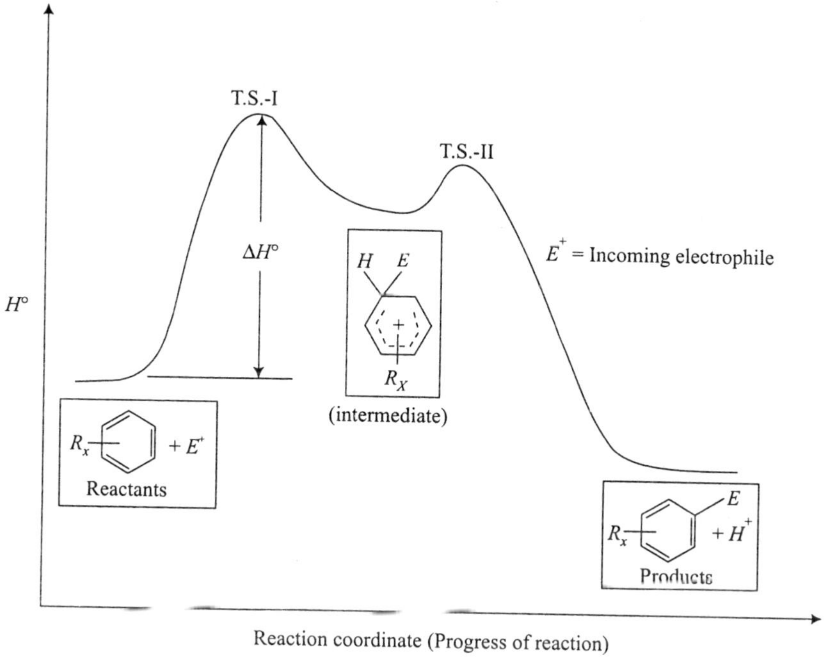

Fig. 4.7: Energy profile diagram for an aromatic electrophilic substitution reaction involving σ-complex as the reaction intermediate

Kinetic isotope effect and rate-determining step

Kinetic isotope effect is very much indicative of the rate-determining step of a multi-step reaction. A reaction in which a C–H bond is broken in the rate-determining step, there occurs a substantial isotope effect (k_H/k_D is of appreciable magnitude), *i.e.* deuterated substrates should undergo substitution more slowly than the corresponding nondeuterated substrates. In contrast, if there is no isotope effect then a C–H bond cleavage is not involved in the rate-determining step. In a reaction undergoing Ar-S_E2 mechanism there occurs no kinetic isotope effect in most cases — suggesting thereby that cleavage and/or weakening of a C–H bond does not take place in the slow step (*i.e.* the first step that generates the σ-complex). The pioneering studies were carried out by Melander[1] where he showed that the rates of nitration of benzene and benzene-d_6 (fully deuterated benzene) are almost equal (*i.e.* no kinetic isotope effect) reflecting the fact that formation of the σ-complex in the first step is the rate-determining step. Similarly, there is no kinetic isotope effect in several other halogenations, as for example, in iodine-catalyzed bromination of toluene[1].

Alternative Pathway

An alternative mechanistic scheme involving both π- and σ-complexes has also been suggested[2,3]. An initially formed π-complex is converted into a second π-complex through the σ-complex. The total scheme is sketched below:

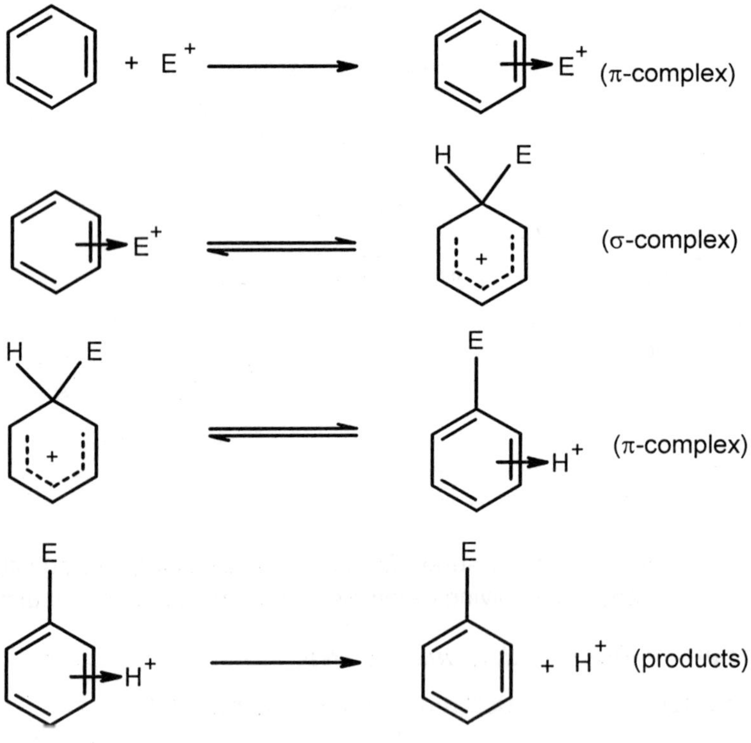

1. Melander, L., *Ark. Kemi*, **2**, 213(1950).
2. Dewar, M.J.S., *The Electronic Theory of Organic Chemistry*, Oxford University Press, London, 1949.
3. Kosower, E.M., *Progress Phys. Org. Chem.*, **3**, 81(1965).

The energy-profile diagram for this mechanism is shown in **Fig. 4.8**.

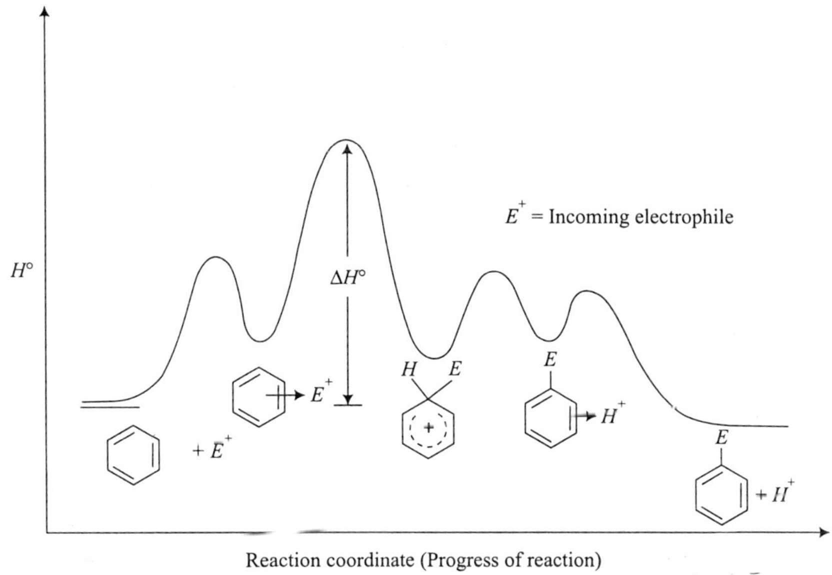

Fig. 4.8: Energy profile diagram for an aromatic electrophilic substitution reaction involving both the π- and σ-complexes

There is a controversy relating to the actual rate-determining step[1-3] — the two types of complexes differ largely in their properties, and to establish which complex is formed in the rate-determining step, it is necessary to study the rate of substitution with the complex stability[4]. From the available experimental data[5-9] a good correlation between the substitution-rate and the σ-complex stability, not with the π-complex stability, is observed — hence the transition state of the rate-determining step closely resembles a σ-complex.

---

1. Dewar, M. J. S. and Marchand, A. P., *Ann. Rev. Phys. Chem.*, **16**, 321(1965).
2. DeHaan, F. P. et al., *J. Am. Chem. Soc.*, **101**, 1336(1979).
3. Santiago, C. et al., *J. Am. Chem. Soc.*, **101**, 1337(1979).
4. Condon, E. F., *J. Am. Chem. Soc.*, **74**, 2528(1952).
5. Howe, F. R., *Chem. Commun.*, 868(1970).
6. Kilpatrick, M. and Luborsky, F. E., *J. Am. Chem. Soc.*, **75**, 577(1953).
7. Brown, H. C. and Brady, J. D., *J. Am. Chem. Soc.*, **74**, 3570(1952).
8. Olah, G. A. et al., *J. Am. Chem. Soc.*, **83**, 4571(1961).
9. Perkampus, H. H., *Adv. Phys. Org. Chem.*, **4**, 195(1966).

Problem 4.42 Correlate partitioning effect and kinetic isotope effect

Solution. In many aromatic electrophilic substitution reactions kinetic isotope effects have been observed. A large value for k_H/k_D (~ 4–7) is suggestive of an S_E1 or a synchronous mechanism, but exhibition of relatively lower value for k_H/k_D (~1–3) may be explained considering *partitioning effect*[1].

From the mechanistic scheme of an Ar-S_E2 reaction it is evident that the first step is slow and reversible, while the second step is fast ($k_2 > k_{-1}$). Since the rate-determining first step does not involve C–H bond cleavage, the rate at which the σ-complex $ArHE^+$ reverts to ArH should be essentially the same as that at which $ArDE^+$ reverts to ArD. However, conversion-rate of the σ-complex $ArHE^+$ to the product (ArE) by the second step would expected to be higher than that of $ArDE^+$ — hence $k_2 >> k_{-1}$ and there is no kinetic isotope effect since the rate is determined by the slow step (i.e. $k_1[ArH][E^+]$), and a large majority of the intermediates is converted to the product.

But if a situation arises where k_2 becomes comparable to k_{-1} (i.e. $k_2 \sim k_{-1}$) or k_2 becomes less than k_{-1} (i.e. $k_2 < k_{-1}$) then reversal of the σ-complex intermediate to the starting materials becomes significant (*i.e.* the σ-complex is *partitioned* between the product formation and reversal to reactants), and a positive kinetic isotope effect is noticed. Hence, the k_2 for the conversion of $ArDE^+ \rightarrow ArE$ would be lower than the k_2 for the conversion of $ArHE^+ \rightarrow ArE$, but k_{-1} is the same, and a larger portion of $ArDE^+$ reverts to starting compounds. Now the *partition factor* (k_2/k_{-1}) for $ArDE^+$ is of lower value in comparison to that for $ArHE^+$ giving rise to positive kinetic isotope effect (this k_H/k_D value is attributed to the second step)[2]. Thus from the value of partition factor of a particular reaction of this kind one may expect whether kinetic isotope effect will be observed for the reaction or not.

Problem 4.43 Discuss the *ipso* substitution.

Solution. Besides the *ortho*, *meta*, and *para* positions in monosubstituted benzenes, attack at one more position by the incoming electrophile may sometimes take place — it is the position which already bears the mono-substituent group that would be substituted by the attacking species; this particular site of attack is called the *ipso* position, and the process is known as *ipso* substitution or *ipso* attack[1,2].

---

1. Hammett, L. P., *Physical Organic Chemistry, 2nd ed.*, p.122, McGraw-Hill, New York (1970).
2. Zollinger, H., *Helv. Chim. Acta.*, **38**, 1597(1955); *Experientia,* **12**, 165(1956).
3. Perrin, C. L. et al., *J. Am. Chem. Soc.*, **93**, 33898(1971)
4. Traynham, J. G., *J. Chem. Educ.*, **60**, 937(1983).

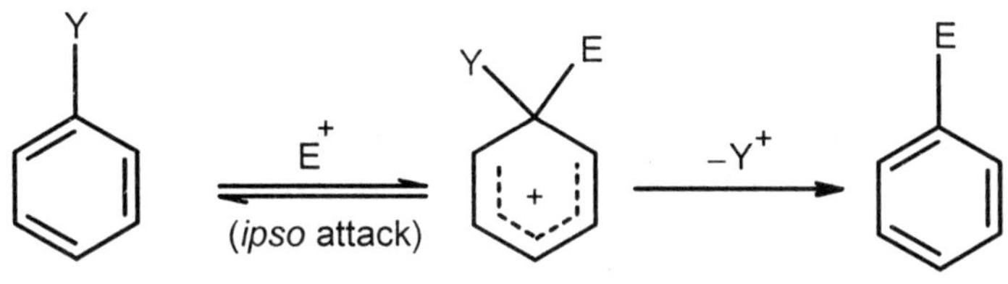

A number of reactions undergoing such substitution pattern are known, e.g.

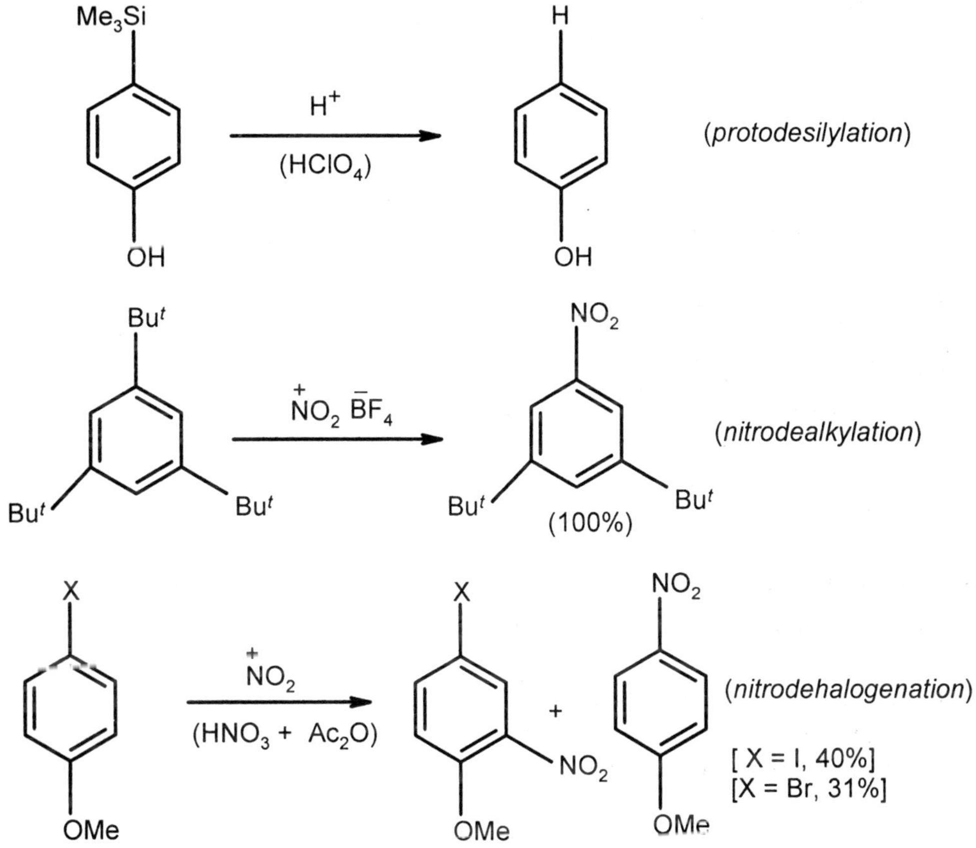

Ipso attack has mostly been studied for nitration[1]. When nitronium ion attacks at the *ipso* position in a monosubstituted benzene, there at least five possible outcomes of the σ-complex intermediate may arise:

---

1. Moodie, R. B. and Schofield, K., *Acc. Chem. Res.*, **9**, 287(1976).

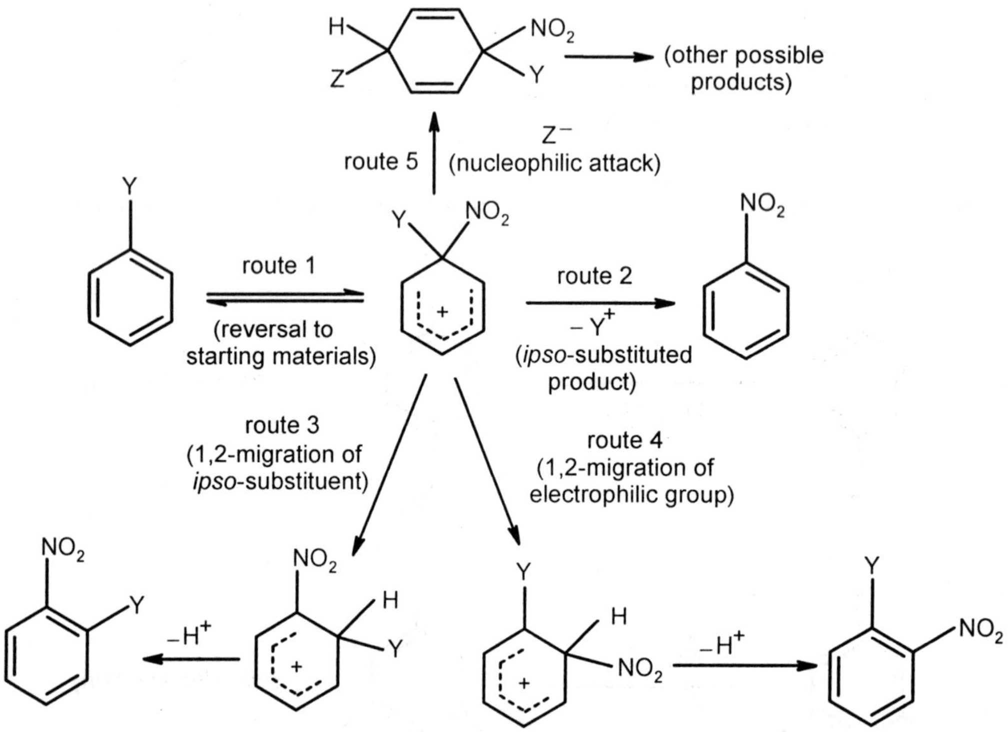

Problem 4.44 Discuss briefly the mechanistic pathway for an aromatic nucleophilic substitution (S_NAr) involving Meisenheimer complex.

Solution. An aromatic nucleophilic substitution (S_NAr or S_N2Ar) occurs smoothly in an activated system involving a bimolecular process — the reaction takes place in two steps and follows a second order kinetics being first order in the concentration of each reactant.

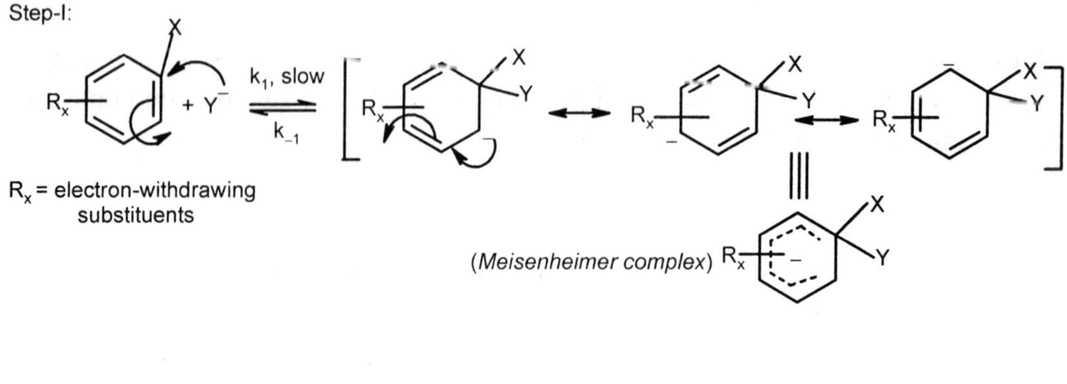

Step-II:

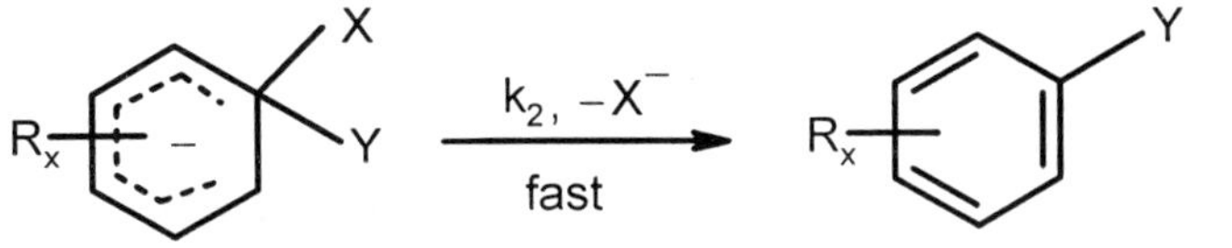

In most of the cases, the first step is the rate-determining step — the negatively charged intermediates formed in the first step of such reactions are known as *Meisenheimer* or *Meisenheimer-Jackson complex*[1].

The leaving group (X^-) may be halide ions, $C_6H_5O^-$, RO^-, NO_3^-, etc., and the attacking nucleophile (Y^-) may ^-OH, $^-OCH_3$, piperidine, aniline, etc. Strong electron-withdrawing groups, like $–NO_2$, activate the aromatic ring in undergoing the reaction; these groups take up important role in stabilizing the negatively charged intermediate (Meisenheimer complex) by strong –I and/or –R effects.

The evidence in favour of this two-step mechanism is that in many cases *Meisenheimer intermediates* have been isolated, and studied (by means of NMR spectroscopy, and X-ray crystallography)[2-5]. For example, the following *Meisenheimer complex* was isolated in the reaction between 2,4,6-trinitrophenetole and sodium methoxide[6].

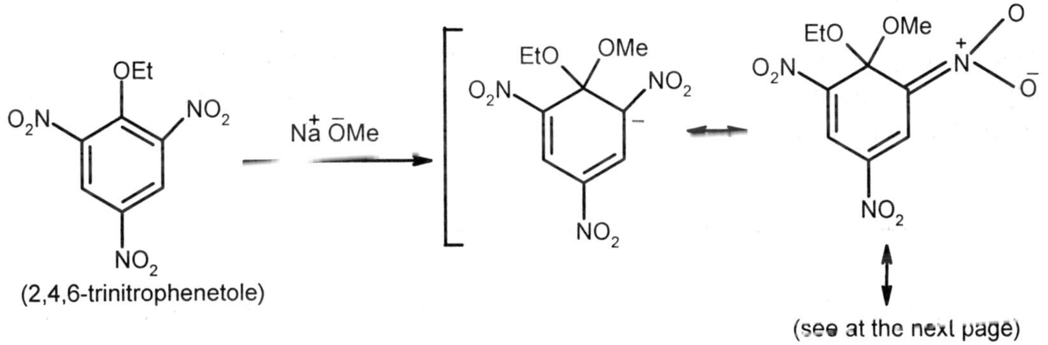

---

1. Buncel *et al.*, *Electron Deficient Aromatic- and Heteroaromatic-Base Interactions*, Elsevier, New York, 1984.
2. Illuminati, G. and Stegel, F., *Adv. Heterocyl. Chem.*, **34**, 305(1983).
3. Terrier, F., Chem. Rev., **82**, 77(1982).
4. Artamkina, G. A. et al., *Chem. Rev.*, **82**, 427(1982).
5. Destro, R. *et al.*, *Acta Crystallogr.*, 24, 1369(1968)].
6. Meisenheimer, J., *Leibigs Ann. Chem.*, 323, 205(1902).

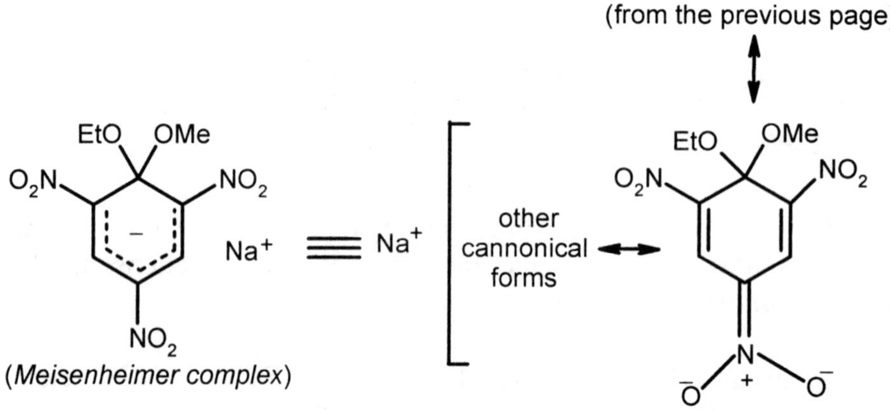

(*Meisenheimer complex*)

The complex is isolable (as salt) because of its attained stability — the strong electron-withdrawing nitro groups located at two *ortho* positions and also the *para* position take part in extending delocalization of the negative charge giving rise to the stability of the carbanion. Another factor may also contribute a part to the complex stability — the ethoxide (EtO^-) being a poor leaving group (because ethoxide ion is a high-energy species) tends to stay in the molecule instead of detaching itself under re-aromatization.

Problem 4.45 Outline the free-radical mechanistic path for halogenation of alkanes.

Solution. Halogenation of alkanes through free-radical intermediates is termed as homolytic substitution reaction, and such type of reactions requires light or heat for initiation. A general scheme may be shown as:

$$\text{R—H} + \text{X—X} \xrightarrow{\text{light or heat}} \text{R—X} + \text{H—X}$$

The mechanism involves a radical chain reaction that consists of the following major steps:

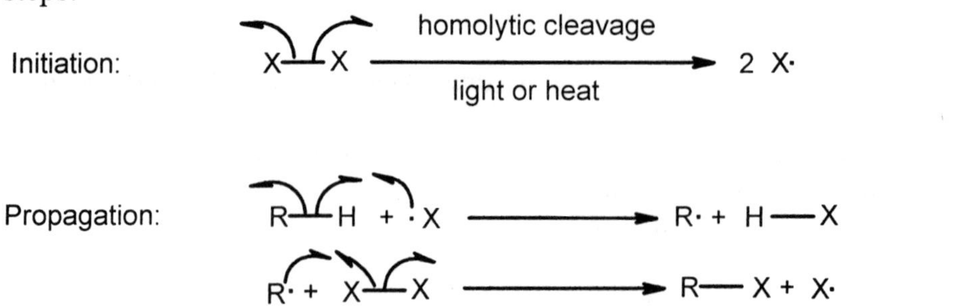

Termination:

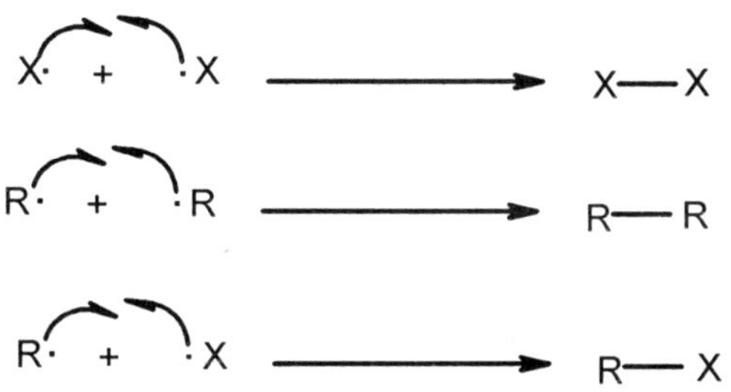

Factors like structural pattern of substrate molecules, nature of solvents, and bond strengths influence the yield of the desired products[1,2]. It has also been observed that less reactive halogens maintain relatively greater selectivity in halogenation. In aromatic solvents, chlorine is believed to form a complex with the solvent molecule, thereby, becoming less reactive and more selective. Presence of bulky groups in the substrates has been found to reduce the rate of halogenation because the steric crowding causes an inhibition of resonance[3].

Problem 4.46 Discuss in detail the chlorination steps of ethane.

Solution. In the free-radical chlorination of ethane, a C–H bond is replaced by a C–Cl bond.

$$CH_3CH_2\text{—}H + Cl\text{—}Cl \xrightarrow{\text{light or heat}} CH_3CH_2\text{—}Cl + H\text{—}Cl$$

Bond energy (kcal/mole): $\Delta H^o_{C-H} = -100$; $\Delta H^o_{Cl-Cl} = -59$; $\Delta H^o_{C-Cl} = -80$; $\Delta H^o_{H-Cl} = -103$

ΔH^o (overall reaction) $= -(80 + 103) + (100 + 59) = -24$ kcal/mole

Thus, this homolytic substitution is an exothermic reaction. The reaction steps are shown below:

1. Tedder, J. M., *Quart. Rev.*, **14**, 336(1960).
2. Russel, G. A., *Tetrahedron*, **8**, 101(1960).
3. Kooyman, E. C. and Strong, A., *Rec. Trav. Chim.*, **72**, 329(1953).

Initiation Step: In this step, the covalent bond of Cl_2 molecule is homolytically cleaved to produce two chlorine radicals — the energy required to break the bond is supplied as either heat or light.

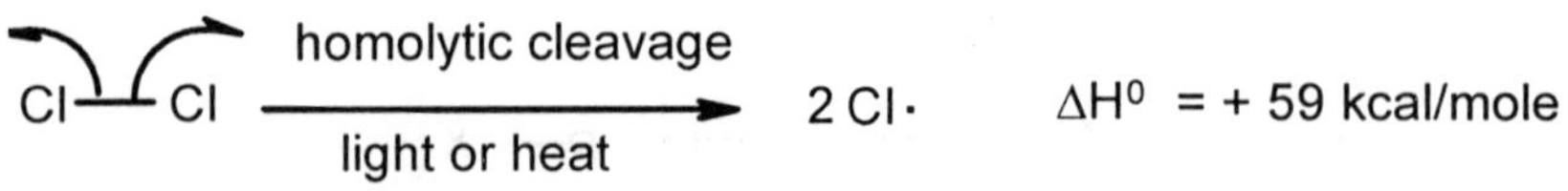

Propagation Steps: A chlorine radical formed in the initiation step interacts with a neutral ethane molecule to produce an ethyl radical that reacts with Cl_2 in the next step yielding the product and another chlorine radical. Both the propagation steps are exothermic and proceed efficiently. Thus the propagation steps in a free-radical chlorination consume one radical while producing another. The reactive intermediate (*i.e.* chlorine radical) that initiates the first propagation step is formed as a product in the second propagation step. This chlorine radical can then serve as a reactant to repeat the first step — hence the propagation steps in this radical chain reaction alternate until the substrate alkane or chlorine or both, are consumed.

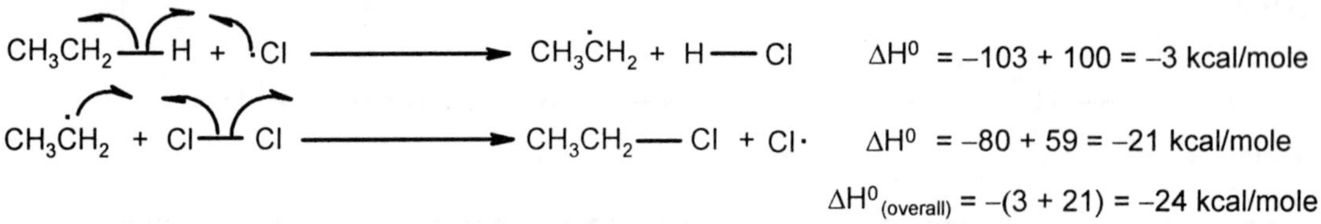

Termination Step: Termination reactions stop the radical chain reaction by consuming reactive intermediates without producing more.

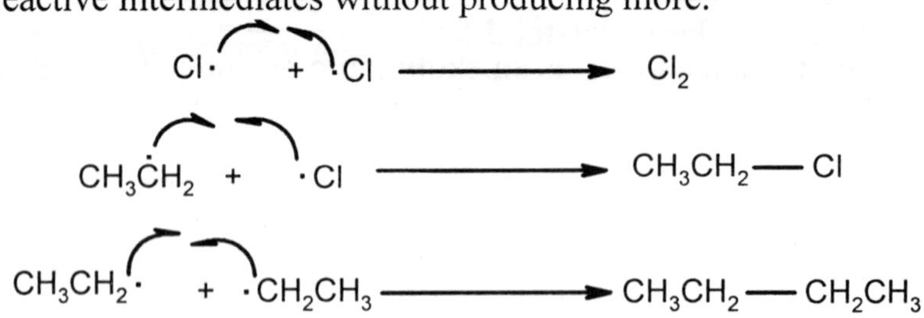

The energy-profile diagram for the chlorination of ethane is shown below (**Fig. 4.9**):

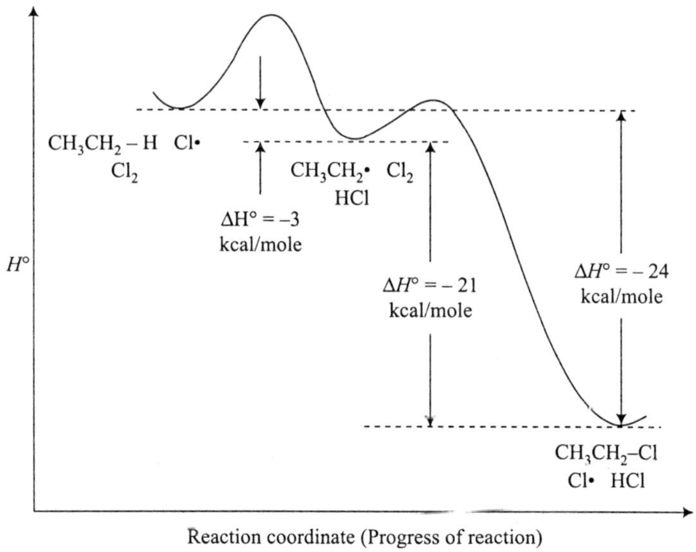

Fig. 4.9: Energy profile diagram for chlorination of ethane (involvement of two propagation steps)

Problem 4.47 Comment on the relative reactivity between chlorination and bromination of ethane under free-radical mechanism with the help of energy-profile diagrams.

Solution. Under free-radical mechanism, abstraction of hydrogen atom from ethane either by chlorine or bromine radical in the first propagation step is the rate-determining step. This may be understood from the energy-comparison between the propagation steps of each halogenation process:

Chlorination:

$CH_3CH_2-H + \cdot Cl \longrightarrow CH_3\dot{C}H_2 + H-Cl$ $\Delta H^0 = -103 + 100 = -3$ kcal/mole

$CH_3\dot{C}H_2 + Cl-Cl \longrightarrow CH_3CH_2-Cl + Cl\cdot$ $\Delta H^0 = -80 + 59 = -21$ kcal/mole

Bromination:

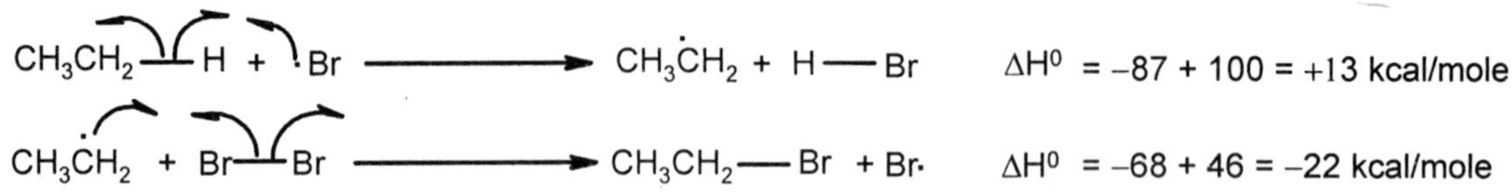

From the energy consideration, it is clear that abstraction of hydrogen from ethane is slightly exothermic ($\Delta H^0 = -3$ kcal/mole) for chlorination, while this step is endothermic ($\Delta H^0 = +13$ kcal/mole) for bromination. The second propagation step is exothermic for both the halogenation processes, and they are of about equal efficiency ($\Delta H^0 = -21/-22$ kcal/mole). Thus, the transition states for both the processes differ in the first propagation step in regard to their nature and energy-content. In chlorination, the first transition state would be reactant-like whereas in bromination this would be product-like. These differences can also be shown by the following energy-profile diagrams (**Fig. 10a** and **Fig. 10b**):

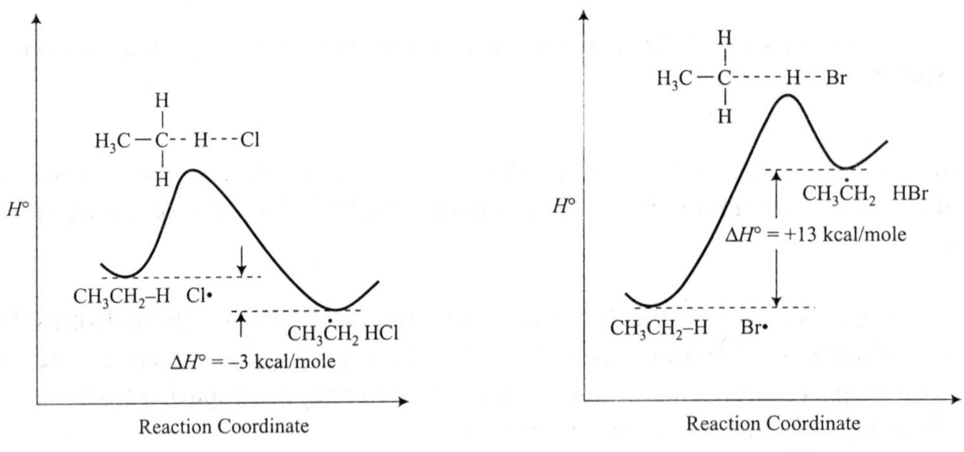

Fig. 4.10(a) **Fig. 4.10(b)**

Fig. 4.10: Energy profile diagram for hydrogen abstraction from ethane by (a) chlorine and (b) bromine

Problem 4.48 Sketch a mechanistic pathway for the $FeBr_3$ catalyzed reaction of benzene with bromine.

Solution. The Lewis acid catalyst ($FeBr_3$) reacts with bromine in the initial step to produce a positive bromine ion (Br^+) that attacks the benzene ring to form an arenium ion (known as σ-complex). Finally, a proton is removed from the arenium ion by $FeBr_4^-$, at the same time the catalyst, $FeBr_3$, is regenerated.

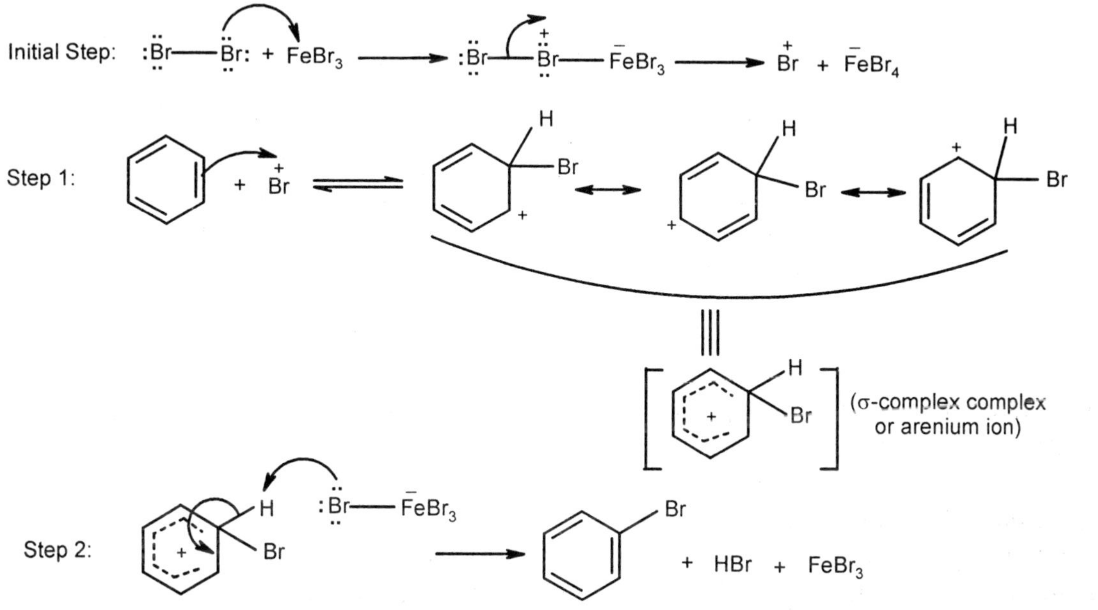

The mechanism of the chlorination of benzene in the presence of $FeCl_3$ is analogous to that of bromination. The catalyst assists in generating and transferring a positive halogen ion.

Problem 4.49 "Vinyl and phenyl halides are usually unreactive towards nucleophilic substitution reaction" – Explain.

Solution. Both the vinyl and phenyl cations are unstable and tend not to be formed readily – hence the corresponding halides are reluctant to undergo S_N1 reaction. The substrates also remain unreactive under S_N2 conditions because the carbon – halogen bond is relatively stronger (attains partial double bond character) due to conjugative effect, and also at the same time the π-electrons of C=C bond or benzene ring repel the approach of a nucleophile from the back.

(vinyl halide)

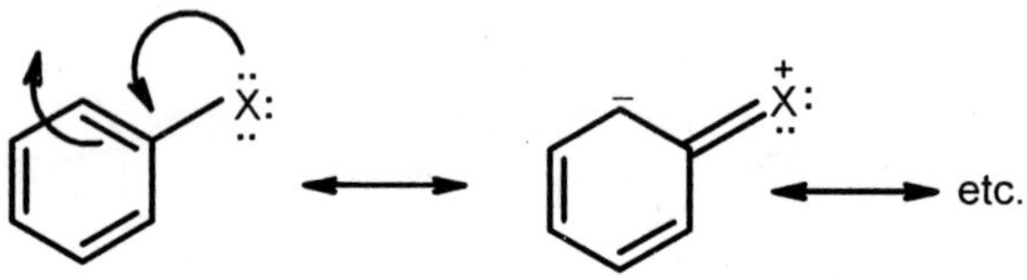

(phenyl halide)

Problem 4.50 "1-Chloroapocamphane or 1-bromotriptycene is inert towards hydrolysis" – Explain.

Solution. 1-Chloroapocamphane cannot be hydrolyzed even upon reflux with alcoholic potassium hydroxide for more than twenty hours. The inversion of configuration at the bridgehead carbon would not be possible. Moreover, the nucleophile cannot approach from the rear due to steric hindrance. That's why S_N2 displacement does not occur. Further, the bridgehead carbons cannot attain coplanarity because of steric requirements of the bridge ring structure — ionization leading to the formation of a carbocation is also thus prevented. Hence the substrate remains inert to undergo S_N1 reactions. Similar arguments are also applicable to 1-bromotriptycene to justify the cause of its failure to undergo hydrolysis reaction.

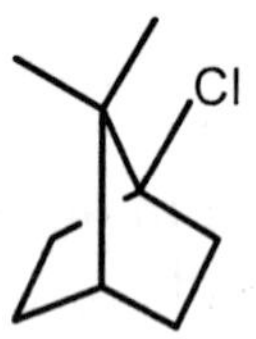

(1-chloroapocamphane)

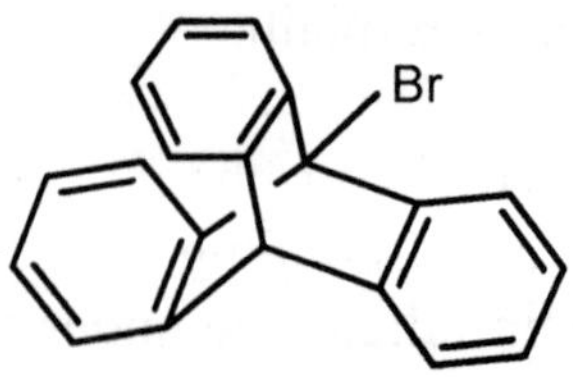

(1-bromotriptycene)

Supplementary problems

SP-4.1 Account for the fact that in dimethylsulphoxide the order of reactivity of halide ions with methyl bromide is $F^- > Cl^- > Br^- > I^-$, which is opposite to that observed in methanol solution.

SP-4.2 "Inversion is the stereochemical requirement for S_N2 reactions, but in comparison racemization is not so much rigid for S_N1" — comment.

SP-4.3 "In many cases S_N1 reactions usually afford the inverted enantiomer in excess by 5-20% over the other isomer" — justify the statement.

SP-4.4 Solvolysis of *R*-2-octylchloride yields only 34% racemized product. Account for the fact.

SP-4.5 Solvolysis of *R*-phenylethylchloride offers 98% and 80% racemization respectively in acetone/water (4:1) and water alone. Account for the fact.

SP-4.6 Acetolysis of *t*-butyl tosylate in the presence of lithium perchlorate or lithium bromide shows an initial steep acceleration of the reaction rate. Account for the fact.

SP-4.7 How would you prove experimentally that inversion of configuration occurs at a saturated carbon atom during S_N2 reaction.

SP-4.8 What is the stereoelectronic requirement needed for a substrate molecule to undergo S_N2 mechanism? Can you justify why *neo*-pentylbromide cannot undergo an S_N2 displacement? Discuss the role of leaving group in an S_N2 mechanism.

SP-4.9 An S_N1 reaction can be shifted to S_N2 on changing from a polar protic solvent to a polar aprotic one and *vice versa*. Offer explanation.

SP-4.10 An α-methyl substituent increases the rate of $S_{RN}1$ reaction of a substrate while an α-*t*-butyl group reduces the corresponding rate. Offer explanation.

SP-4.11 1-Chlorooctane dissolved in methylene chloride does not react with an aqueous solution of NaCN. But if a catalytic amount of tetrabutylammonium chloride is added to the mixture, the reaction proceeds smoothly. Account for the fact.

SP-4.12 Optically active phenyl *sec*-butyl ketone racemizes on treatment with a base in ethanol. Explain.

SP-4.13 Account for the gradual loss of O^{18} content in the unreacted ester during acid catalyzed hydrolysis of $RCH_2CO^{18}–OCH_3$.

SP-4.14 When the compounds I and II were treated with $KHCO_3$ in benzene-methanol-water at 20^0C for 65 hours, hydrolysis occurred to the extent of 18% with I and 70% with II. Explain the difference in reactivity.

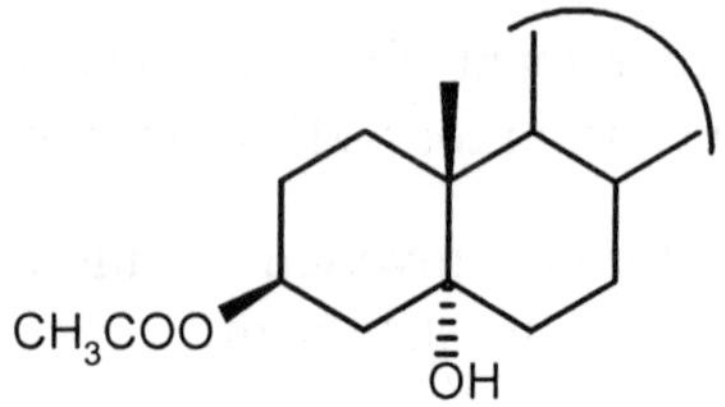

(I)

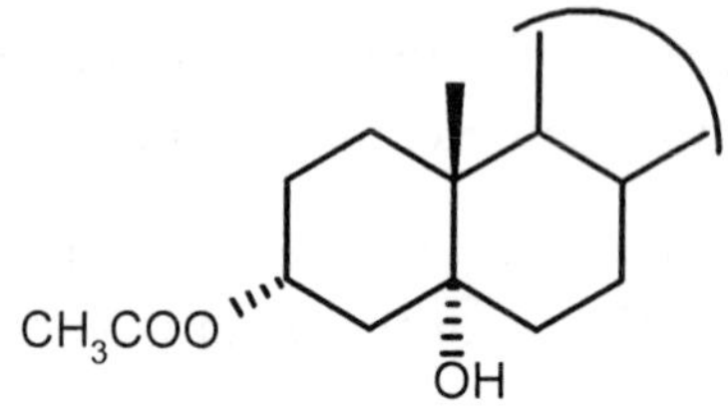

(II)

[Hint. The 5-OH is hydrogen-bonded only to the axially oriented ester oxygen in II that is feasible due to the stereochemical geometry of the molecule. Formation of this hydrogen bond reduces the donation of unshared oxygen-electrons to the attached carbonyl group as a result of which stabilization of the carbonyl group in the ground state is decreased leading to the increasing ease of attack by base. Electron-deficiency on carbonyl carbon (of ester moiety) in II would be greater than that in I.]

SP-4.15 Acid-catalyzed dehydration of *neo*-pentyl alcohol produces two isomeric alkenes. Predict their structures and discuss a suitable mechanistic pathway for the reaction. Which alkene do you think shall be the major products? Give reason.

SP-4.16 The reaction of an optically active alcohol with $SOCl_2$ proceeds with retention of configuration but the same reaction in presence of pyridine proceeds with inversion of configuration.

SP-4.17 Explain why base-catalyzed bromination of acetone leads to the formation of a tribromo derivative with all the bromines at the same carbon. What product(s) do you expect if the reaction is carried out in acidic medium, and why?

SP-4.18 Addition of HCl to 1-butene or 2-butene involves the same intermediate but the reaction of 1-butene is faster than that of 2-butene. Explain the observation with an energy diagram.

SP-4.19 Show the mechanism of addition of bromine to *cis*- and *trans*-2-butene and write down the products indicating their stereochemistry and optical activity.

SP-4.20 When treated with bromine and water, allyl bromide gives chiefly (80%) the primary alcohol $CH_2BrCH_2BrCH_2OH$, in contrast to propene which gives the secondary alcohol, $CH_3CHOHCH_2Br$. Suggest an explanation for the difference.

SP-4.21 What happens when (*R*)-1-phenylethyl acetate is treated with acid? Discuss the mechanism of the reaction indicating stereochemical implications.

SP-4.22 Discuss critically, with reaction mechanism, the easy formation of the oxime of a ketone with hydroxylamine hydrochloride in the presence of sodium acetate.

SP-4.23 What happens when *threo*-3-bromo-2-butanol is treated with HBr? Explain the stereochemistry involved.

SP-4.24 In the presence of a base, *trans*-2-chlorocyclohexanol reacts readily to give *cis*-cyclohexene oxide. On the other hand, *cis*-2-chlorocyclohexanol yields cyclohexanone. Explain the phenomena delineating probable mechanistic pathway in each case.

SP-4.25 (*S*)-2-Bromo-*n*-propylmethyl ether on hydrolysis gives (*S*)-2-hydroxy-*n*-propylmethyl ether, and the solvolysis of the substrate is much faster than that of 2-bromopentane. Account for the observations. Also predict the expected stereochemical course if sodium cyanide in acetone is used as the nucleophile instead of water.

SP-4.26 Treatment of both α- and β-bromoketones with base results in loss of HBr to form α,β-unsaturated ketones. However, the former react much more slowly and require much stronger bases. Account for this differences in reactivity by providing a complete analysis of both the reaction mechanism.

SP-4.27 What product would you expect from the treatment of 1,3-butadiene with HBr in peroxide-containing ether? Offer mechanism for the conversion.

SP-4.28 Chlorine adds to *trans*-2-butene in methylene chloride to give *meso*-2,3-dichlorobutane. However, if the reaction is conducted in water, a chlorohydrin is obtained. Suggest a mechanism for this reaction indicating also the stereochemical aspects.

SP-4.29 Explain why 2-butyne is less reactive than *trans*-2-butene toward the electrophiles, and why it is nonetheless possible to stop after a single equivalent of bromine has been added to the alkyne.

SP-4.30 Discuss the radical mechanism of halogenation of alkanes with particular reference to the energies involved in the initiation, propagation and termination steps. What is the role of light (or heat) in the halogenation of alkanes?

SP-4.31 What are the characteristic differences between E2 and E1cB reactions? How would you establish that the following reaction follows E2 and not E1cB mechanism?

$$PhCH_2CH_2Br \xrightarrow{\overline{O}Et} PhCH{=}CH_2$$

SP-4.32 Discuss various factors, which influence the extent of elimination and substitution.

SP-4.33 What are π and σ-complexes in aromatic electrophilic substitution?

SP-4.34 Account for the stability of Meisenheimer complexes.

SP-4.35 Draw a detailed mechanism for the $FeBr_3$ catalyzed reaction of ethylbenzene with bromine, and show why the sigma complex is lower in energy for substitution at the *ortho* and *para* positions than it is for substitution at the *meta* position?

SP-4.36 How do you explain the inertness of chlorine in chlorobenzene and vinyl chloride in substitution reaction?

SP-4.37 It has been shown that chlorination of activated aromatic substrates can be affected with hypochlorous acid in the presence of a proton catalyst. Suggest a suitable mechanistic scheme showing the role of proton in this reaction.

SP-4.38 What is meant by primary kinetic isotope effect? Explain with an example how this effect can be used in the investigation of reaction mechanism.

SP-4.39 Explain why nitration of aniline is so sluggish and why it gives mostly *meta* product?

SP-4.40 The rates of bromination (using $FeBr_3$ as catalyst) of C_6H_6 and C_6D_6 are the same. Discuss.

SP-4.41 When treated with bromine-water phenol forms 2,4,6-tribromo derivative but gives monobromo derivative (*ortho* and *para*) when treated with bromine in a nonpolar solvent like CS_2 or CCl_4. Explain.

SP-4.42 Draw energy profiles for nitration and sulphonation of benzene in terms of the comparative rate constants as given below:

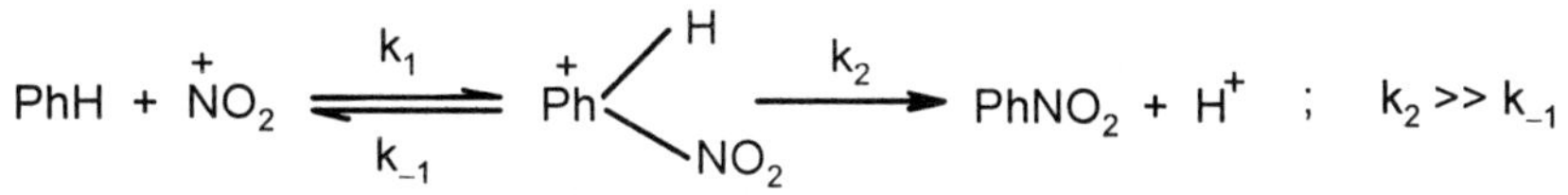

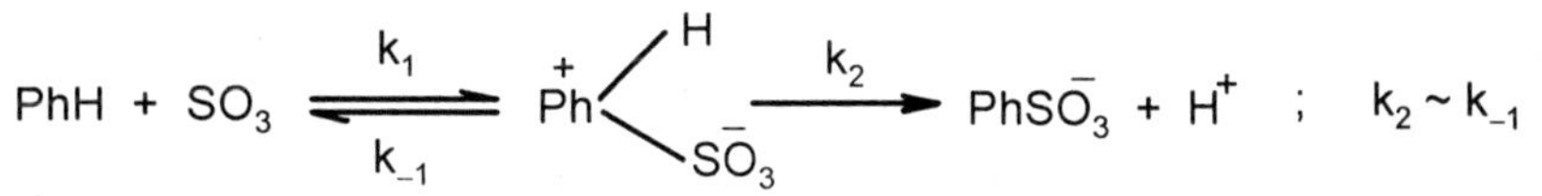

Also account for the fact that a small kinetic isotope effect is observed in sulphonation reaction but not in the nitration.

SP-4.43 Account for the following observations:

a. Towards S_N2 reaction the reactivity order of alkyl halides follows the order: EtBr > PrBr > ^{t}BuBr, while towards E2 reaction the order is exactly the opposite.

b. Addition of NaBr to the aqueous solution of *tert*-butyl bromide has been found to have no effect on the hydrolysis rate, but the addition retards the hydrolysis rate of 1,1-diphenylmethyl bromide.

c. *n*-Butyl chloride reacts with 0.01M NaCN in ethanol to give primarily *n*-butyl cyanide, whereas under the same conditions *tert*-butyl chloride reacts to give primarily Me_3COEt.

d. Acetylene is less reactive than ethylene towards bromine addition.

e. During benzylation of phenols with benzyl chloride in acetone in the presence of K_2CO_3, addition of a small amount of NaI accelerates the reaction.

f. Potassium *tert*-butoxide is often used to promote E2 reaction at the expense of S_N2 reactions.

g. Dehydrohalogenation of (1*S*,2*R*)-1-bromo-1,2-diphenylpropane yields *(E)-1,2*-diphenylpropene exclusively.

h. Treatment of 3-iodo-2,2-dimethylbutane with silver nitrate in ethanol gives 2,3-dimethyl-2-butene as the major product.

i. Acylation of alcohols with acetyl chloride is catalyzed by pyridine.

j. Hydrolysis of ethyl vinyl ether in dilute aqueous acid takes 10^3 times faster than the hydrolysis of diethyl ether.

k. Chloride ion in $Bu_4N^+Cl^-$ in acetone is a better nucleophile than that in LiCl in the same solvent.

l. α,β-Unsaturated ketone does not undergo epoxidation reaction with hydrogen peroxide in acidic medium.

m. The S_N reaction of $EtSCH_2CH_2Cl$ with ethanol proceeds at a rate many fold faster than the similar reaction of $EtOCH_2CH_2Cl$.

n. Allene reacts with HBr to afford two isomeric bromopropenes of which one is obtained as the major product.

o. The rate of bromination of acetone is independent of the concentration of bromine under acid catalyzed as well as base catalyzed conditions.

p. The rate of the reaction of 1-bromobutane with azide ion is increased 5 x 10^3 fold on changing the solvent from methanol to acetonitrile.

q. The solvolysis rate of triphenylmethyl chloride increases by addition of lithium perchlorate.

r. Optically active 1-phenylethyl chloride racemizes when dissolved in sulphur dioxide.

s. Phenacyl chloride reacts more than 12,000 times as fast as 2-phenylethyl chloride with KI in acetone.

t. Relative rate of ethanolysis of the following compounds I and II are in the order of:

Me_3CCl
(I)

(II)

k_{rel} (I/II): 1 : 1.96 x 10^{-12}

u. Solvolysis of 2-octyl-*para*-toluenesulphonate in 75% aqueous acetone at 65^{0}C furnished 2-octanol of 77% optical purity. However, when solvolysis was conducted in the presence of 0.307M sodium azide, the resulting alcohol was found to be 100% optically pure.

v. A benzylic hydrogen is 64 times more reactive in bromination than a hydrogen atom in ethane.

w. Benzoylperoxide decomposes at a normal rate in methyl iodide as a solvent but the rate of decomposition of *ortho*-nitrobenzoyl peroxide is much faster in the same solvent.

x. Electrtophilic attack in pyrrole occurs at the 2-position while in indole it takes place at the 3-position.

y. *Trans*-2-methylcyclohexanol on acid-catalyzed dehydration forms 1-methyl-cyclohexene as the major product, whereas *trans*-1-bromo-2-methylcyclohexane gives 3-methylcyclohexene as the major product on dehydrohalogenation.

z. On bromination *E* or *Z*-1-phenylethane yields both *syn* and *anti* addition products.

CHAPTER 5

ORGANIC SYNTHESES

Introduction

In this chapter we will consider diversified problems along with their solutions, dealing with a variety of organic synthetic strategies. Students are advised to follow very carefully each of the problems as designed for this purpose so that they can conceive proper idea regarding various synthetic methodologies as well as the most plausible mechanistic explanation for them.

Solved Problems

Problem 5.1 Predict the product (s) in each of the following reactions with plausible mechanism in each case:

Problem 5.1.1

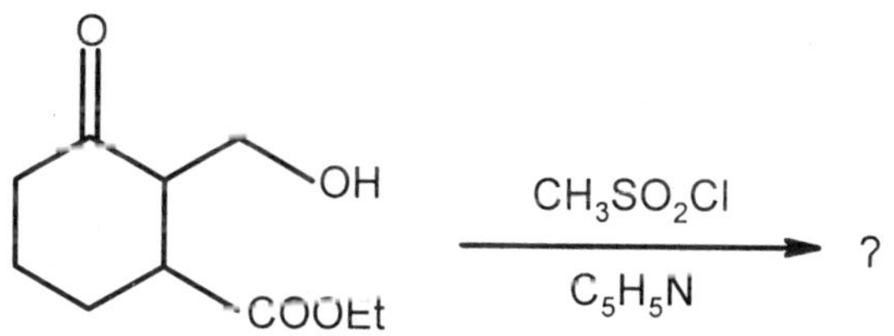

Solution. The reaction follows E1cB mechanistic pathway as shown below:

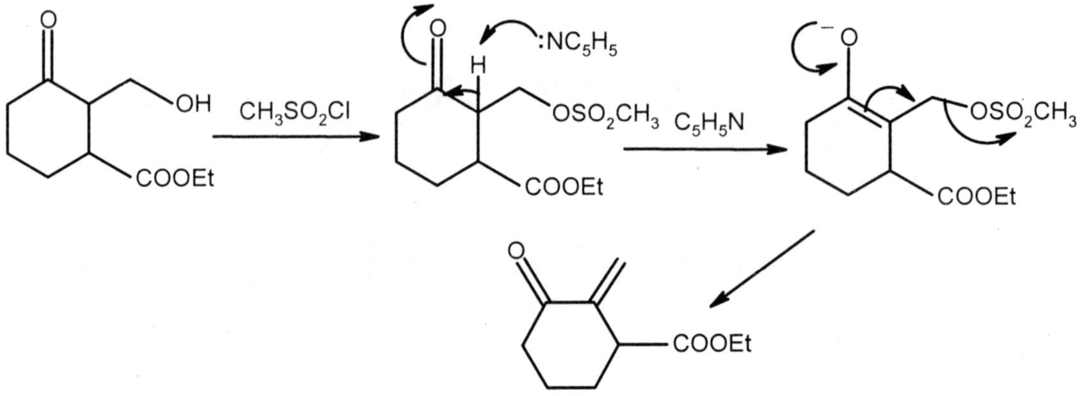

Problem 5.1.2

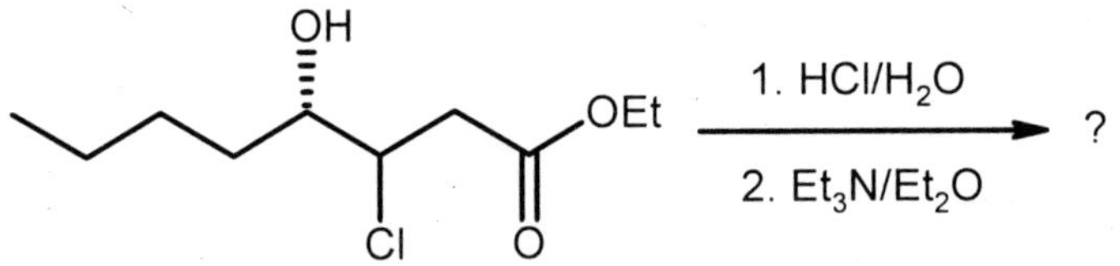

Solution.

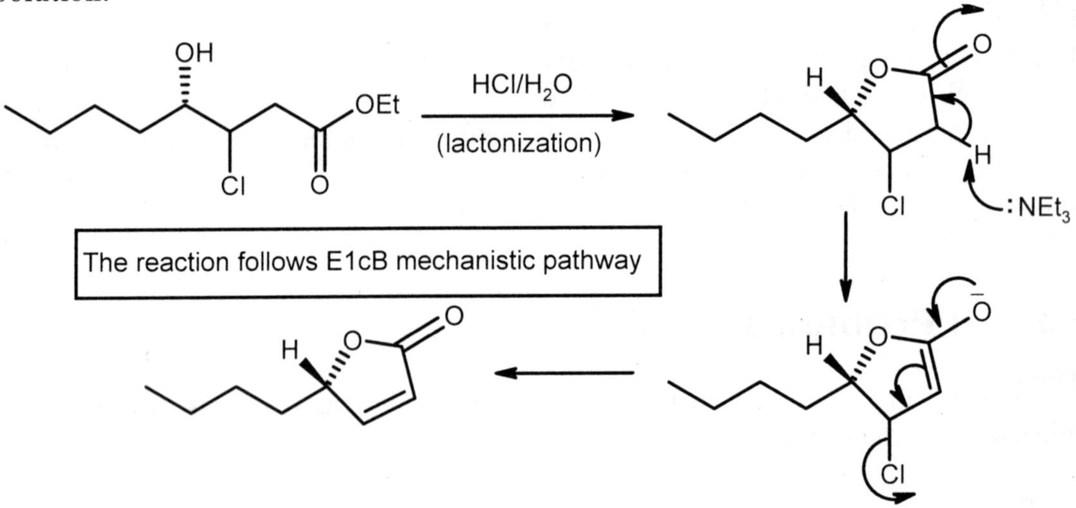

Problem 5.1.3

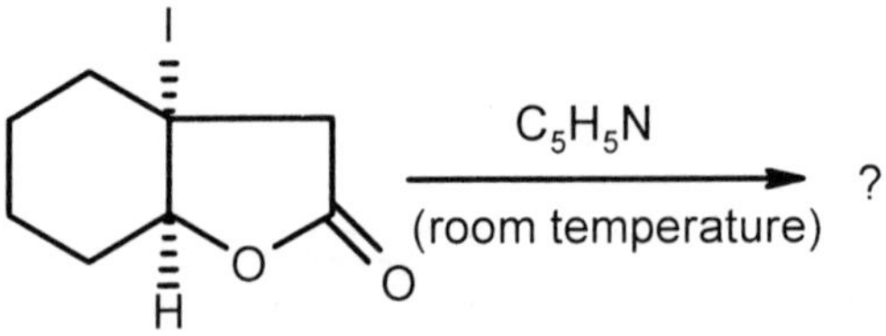

Solution. Facile *trans*-elimination of HI (following E2 pathway) is attributed to the fact that the proton to be removed by the base is α to the carbonyl group, and is thus much acidic and can easily be abstracted.

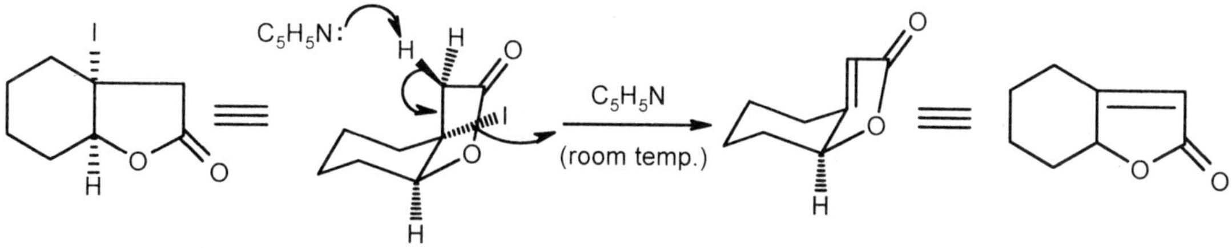

Problem 5.1.4

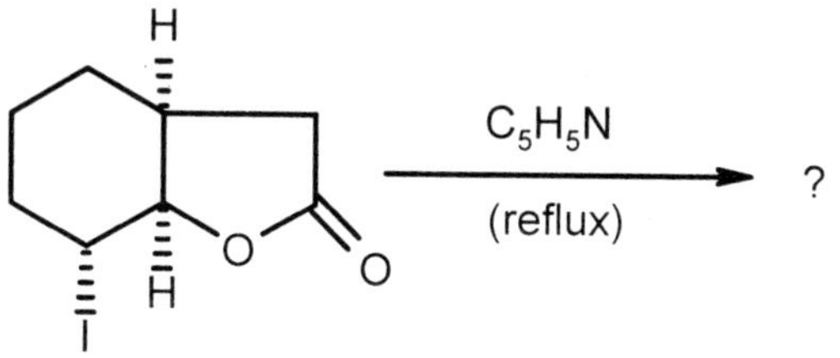

Solution.

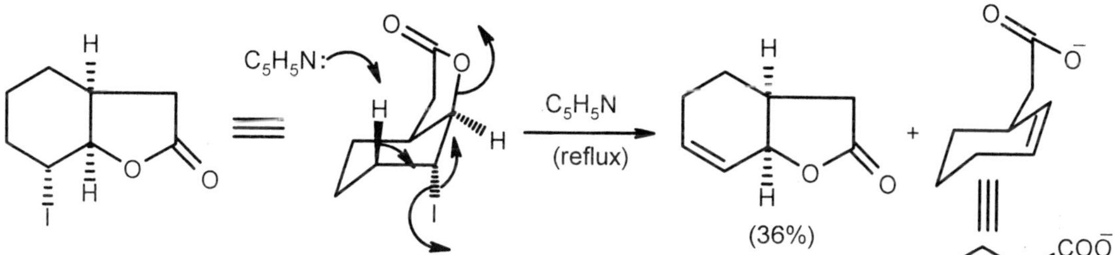

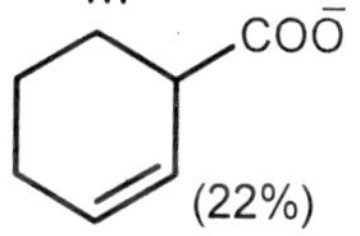

Due to *trans*-eliminations (following E2 pathway) a mixture of products is obtained.

Problem 5.1.5 $ArCOCMe_3 \xrightarrow{\bar{N}H_2}$?

Solution.

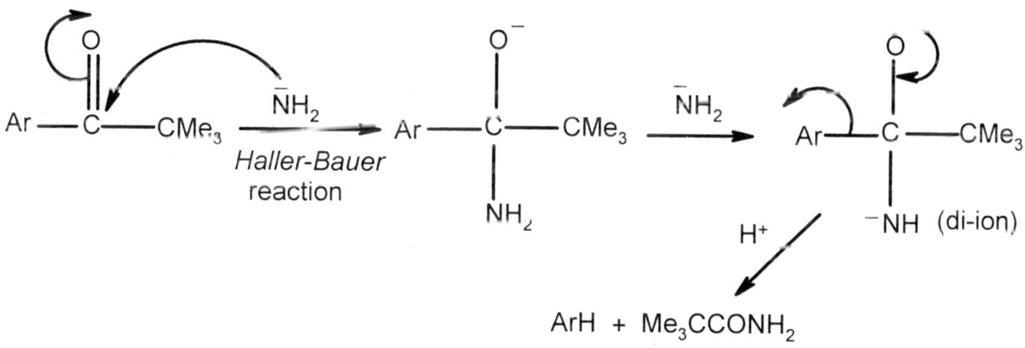

Problem 5.1.6

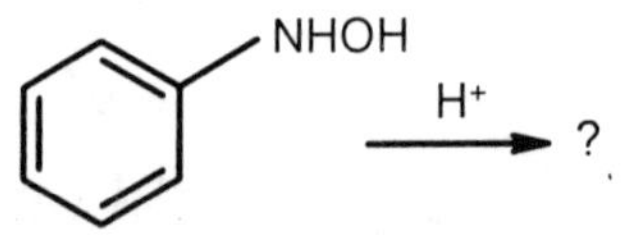

Solution.

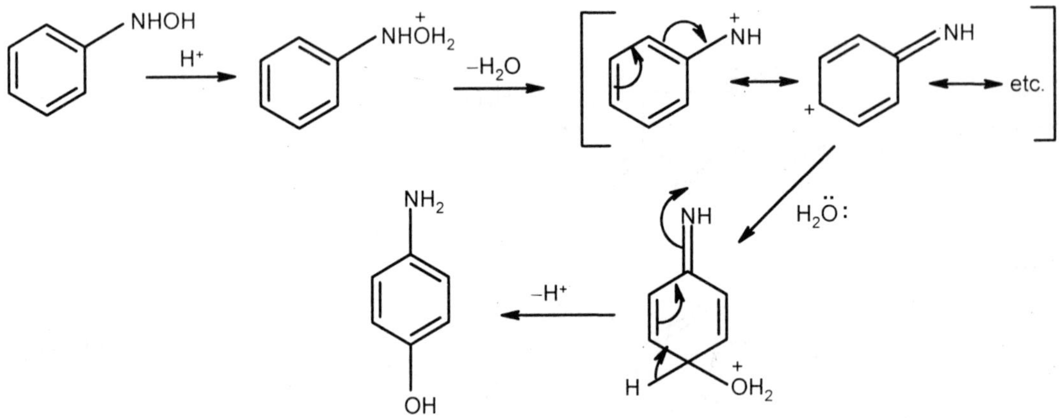

Problem 5.1.7

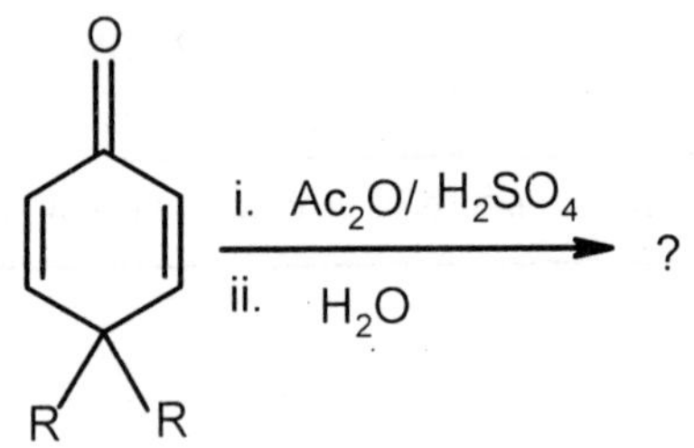

Solution. This is *Dienone-phenol rearrangement.*

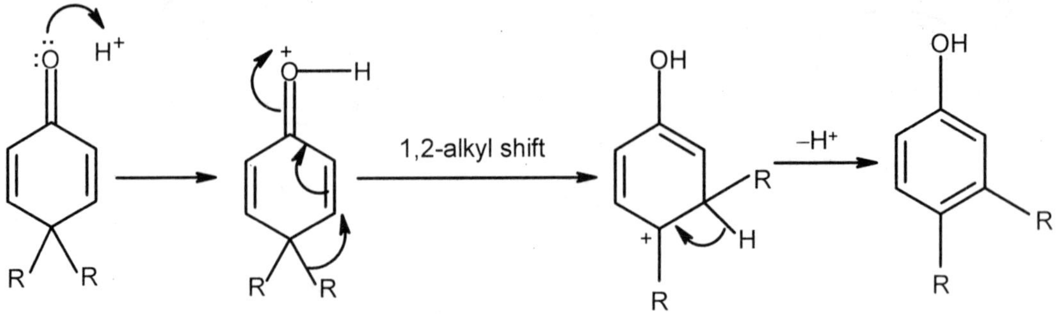

Problem 5.1.8

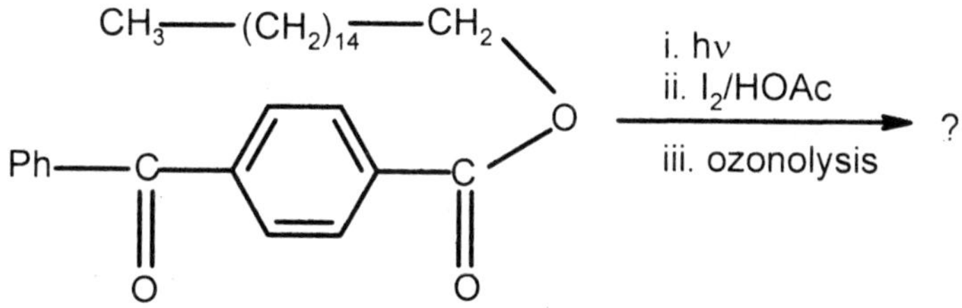

Solution.

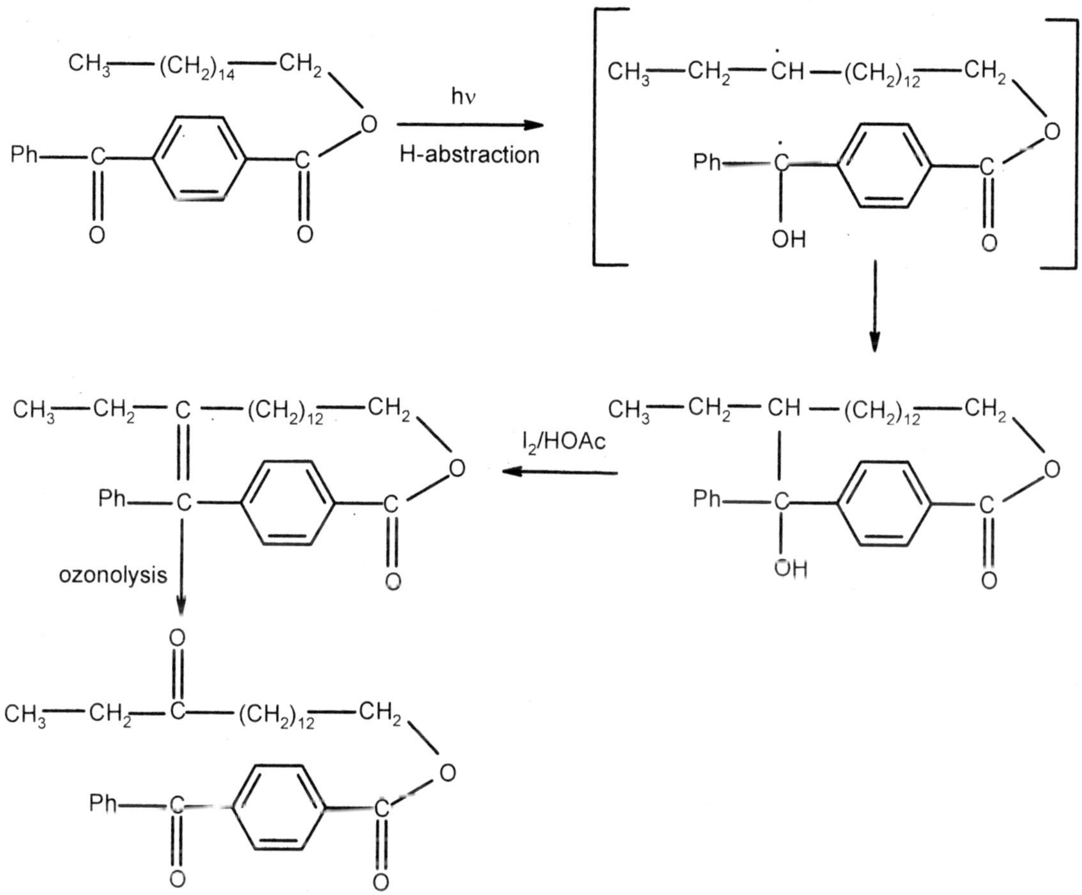

Problem 5.1.9

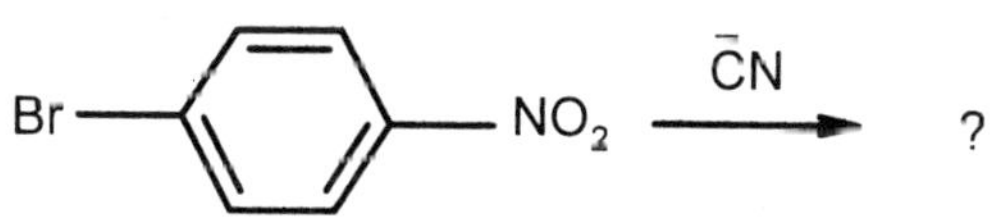

Solution.

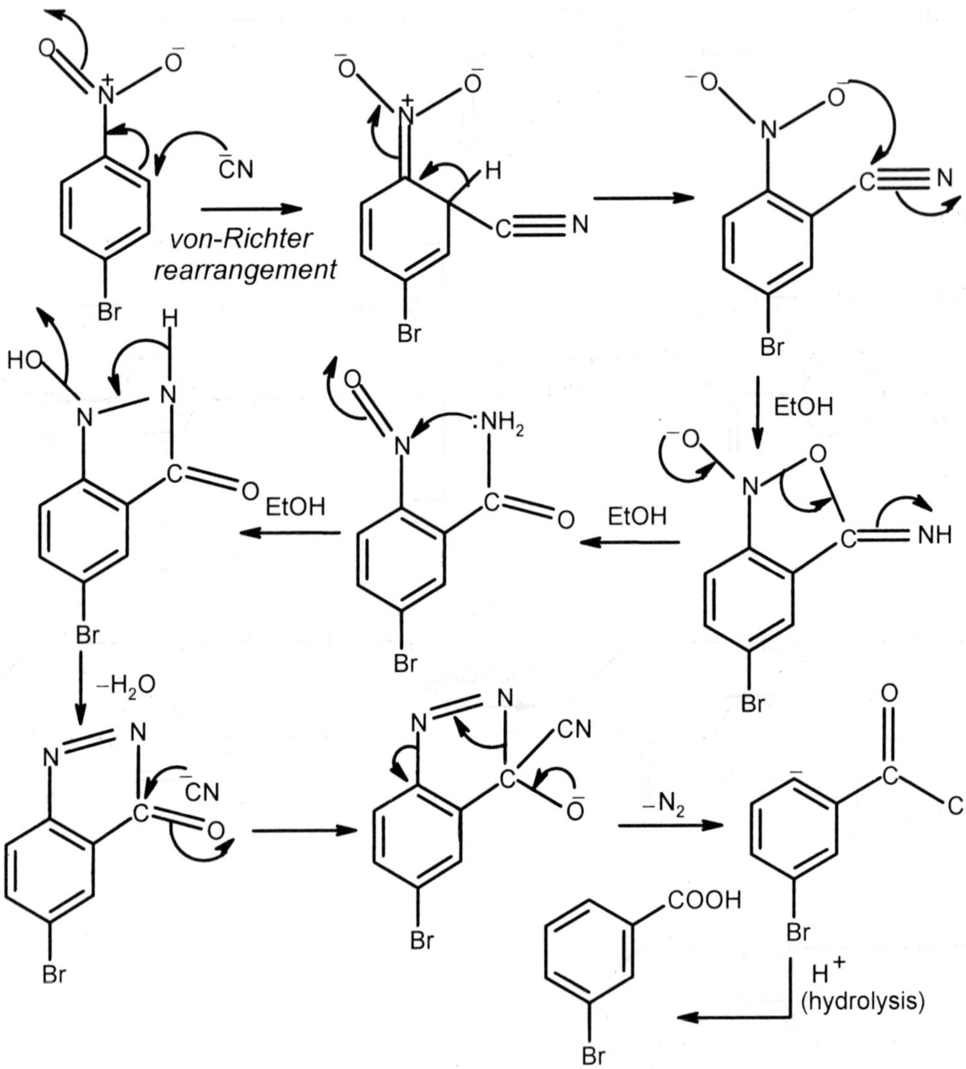

Problem 5.1.10

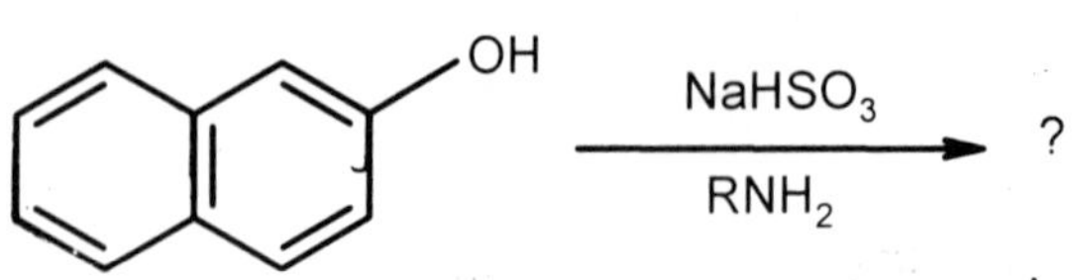

Solution. It is an example of *Bucherer reaction.*

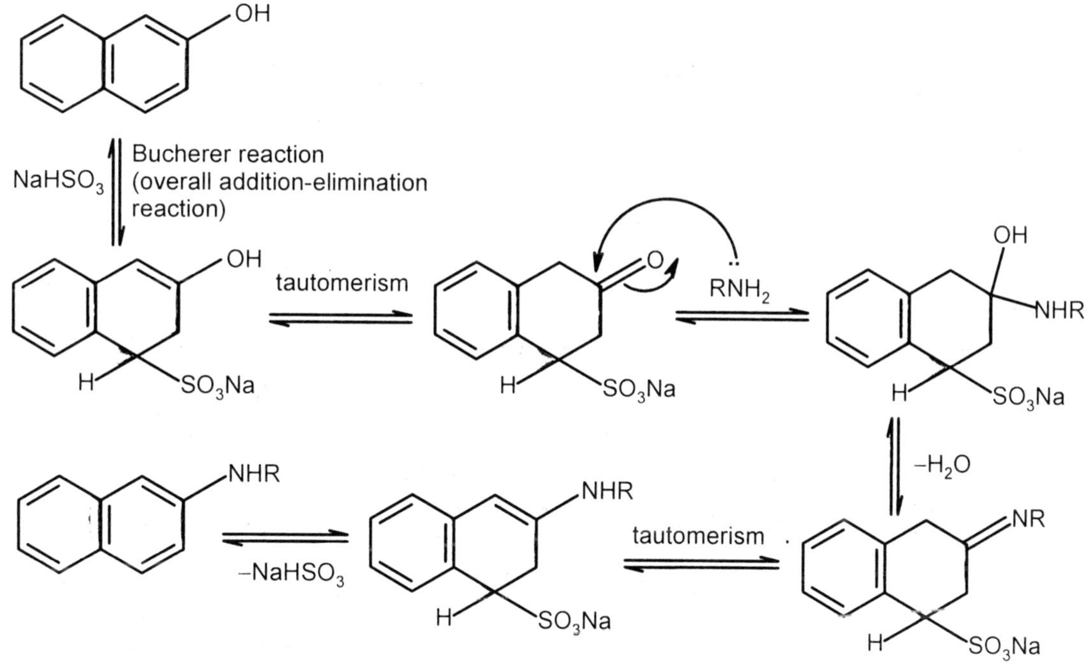

Problem 5.1.11

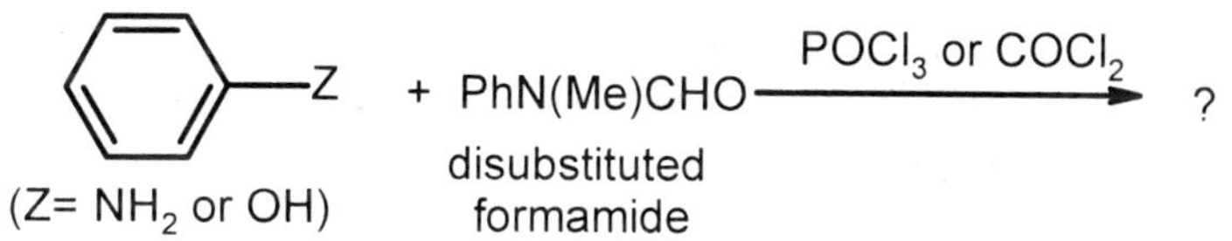

Solution. This is *Vilsmeier-Haack* formylation reaction.

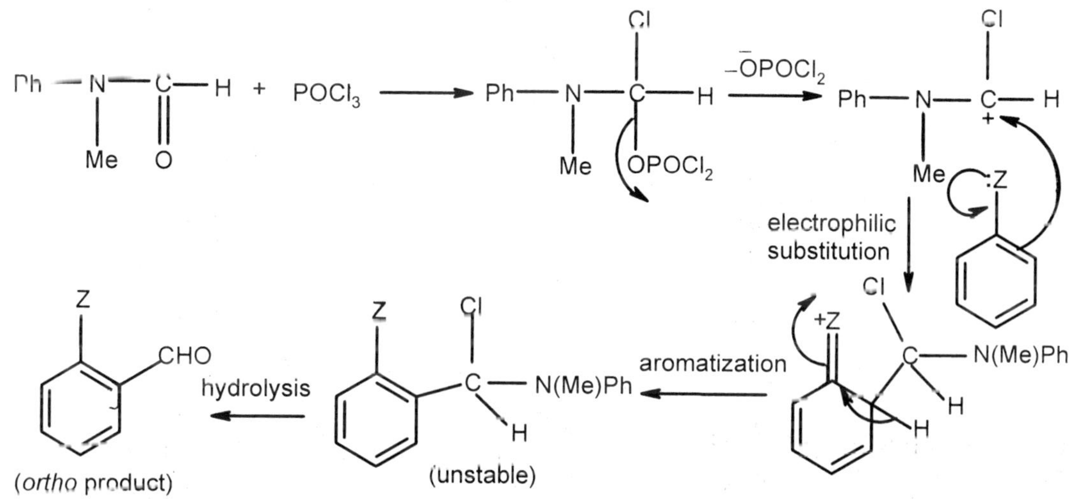

Similarly, *para*-product is also formed. In the mixture, the latter predominates.

Problem 5.1.12 $HOMe_2CC{\equiv}CCMe_2OH \xrightarrow[H_2SO_4]{Hg^+,\ H_2O}$?

Solution.

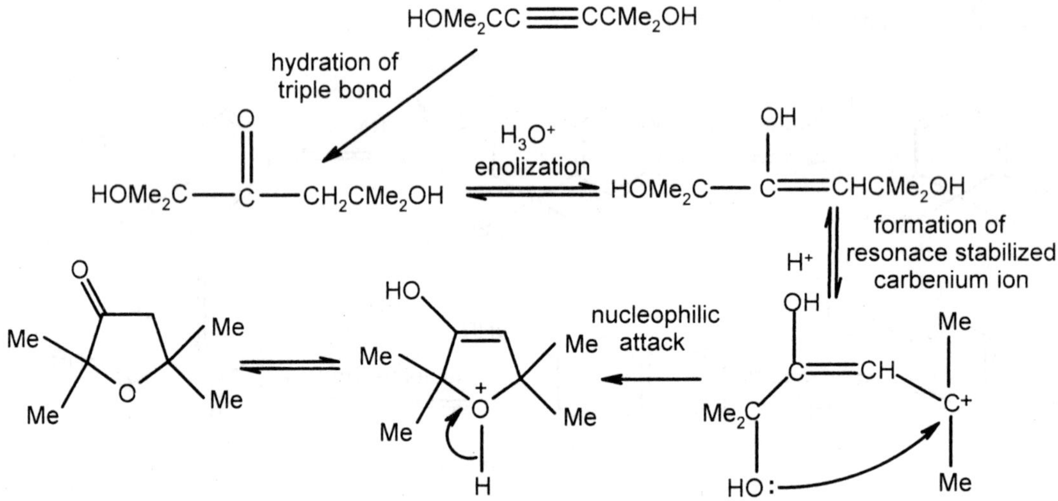

Problem 5.1.13

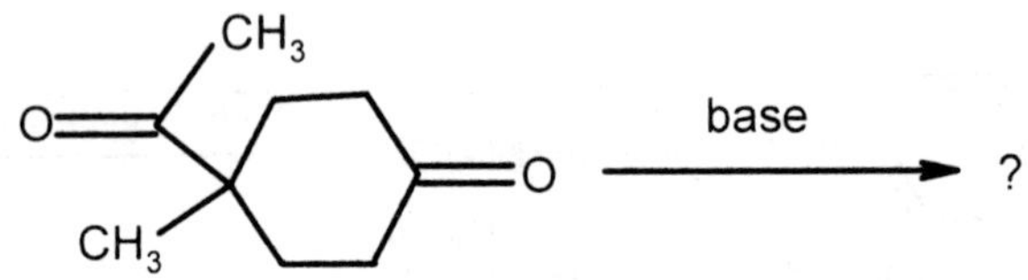

Solution.

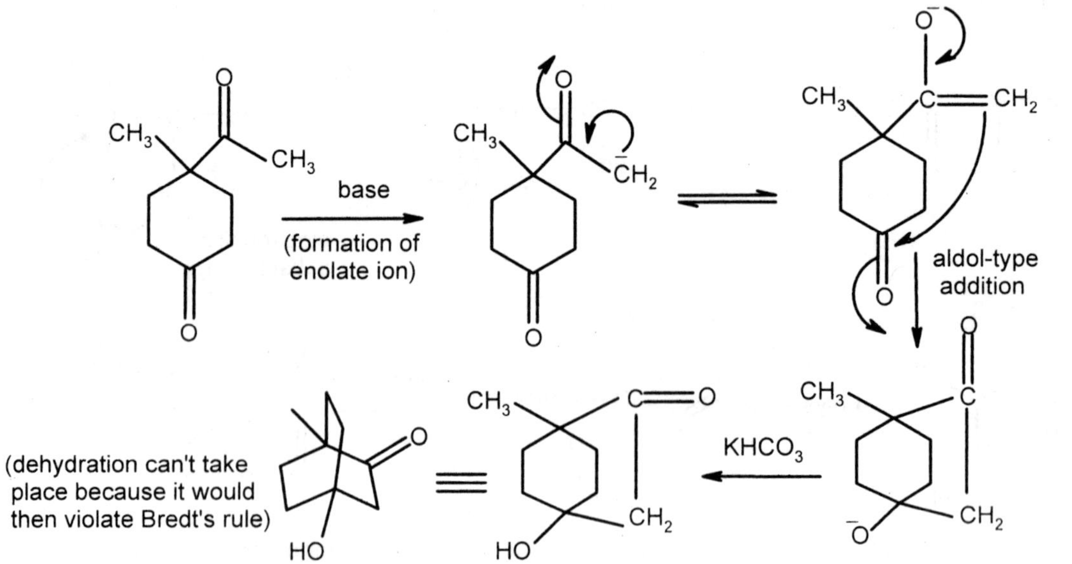

Problem 5.1.14 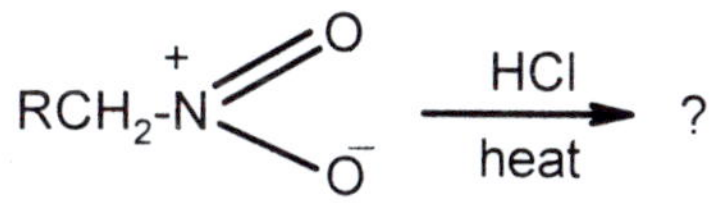

Solution.

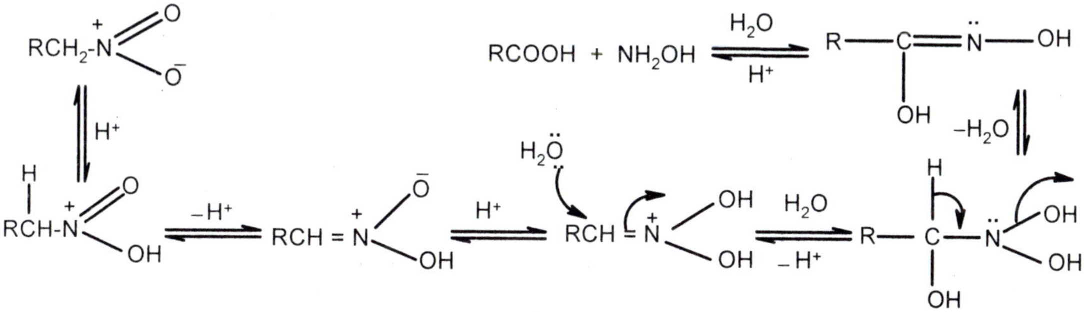

Problem 5.1.15 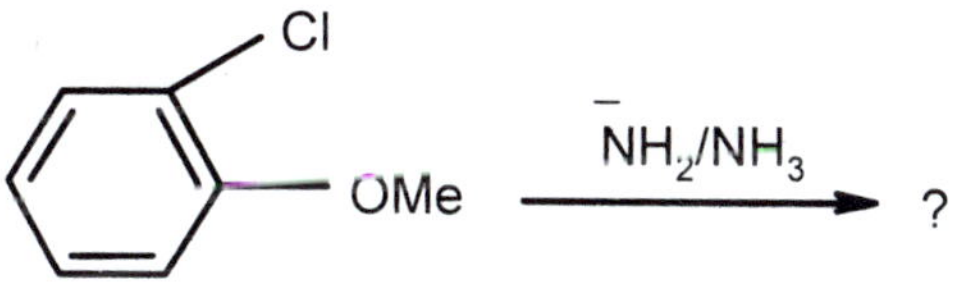

Solution.

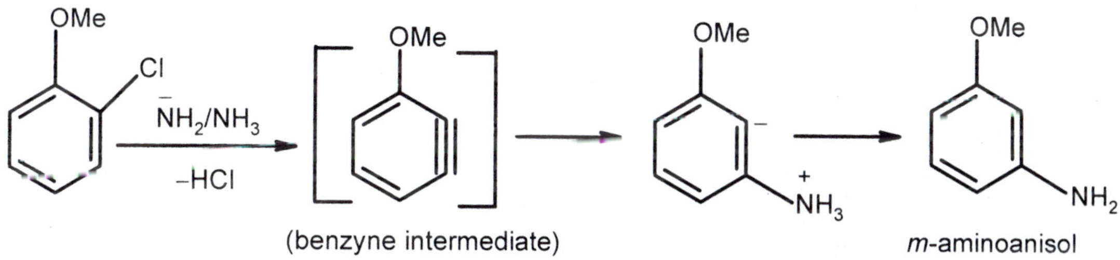

Problem 5.1.16 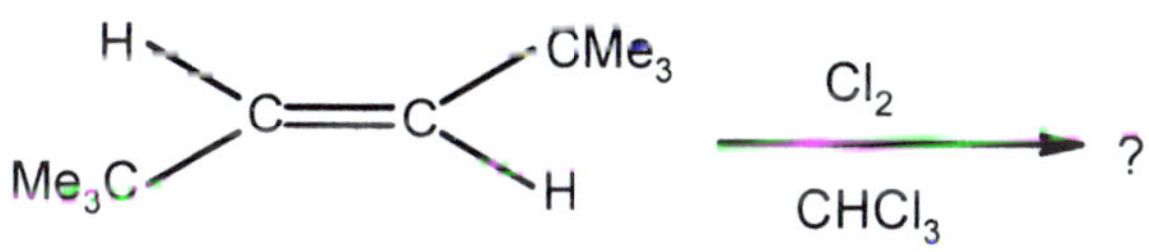

Solution. High degree of steric crowding around the positively charged carbon atom in (I) prevents its solvation, and thereby it becomes destabilized. Rapid rearrangement to (II) relieves steric strain to a good extent as well as leads to more stable intermediate.

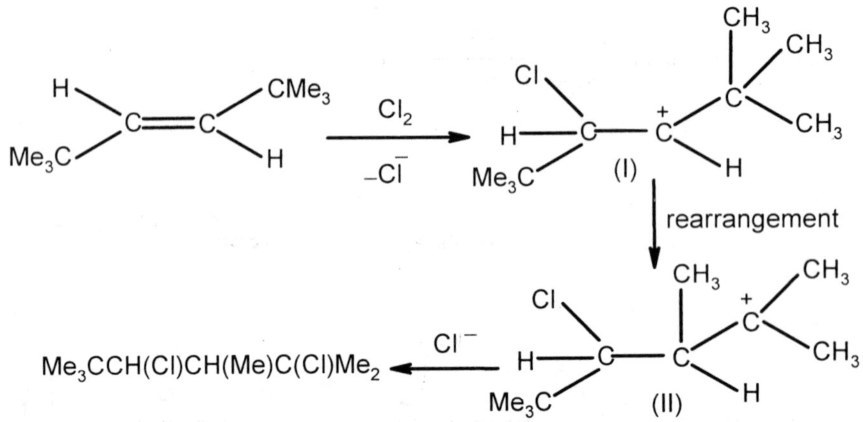

Problem 5.1.17

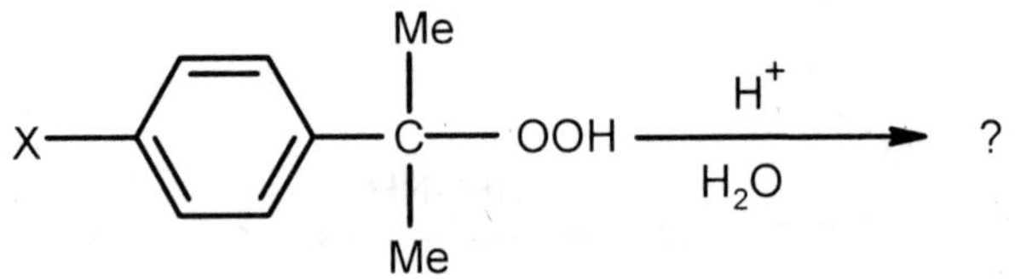

(X=electron-donating substituents)

Solution.

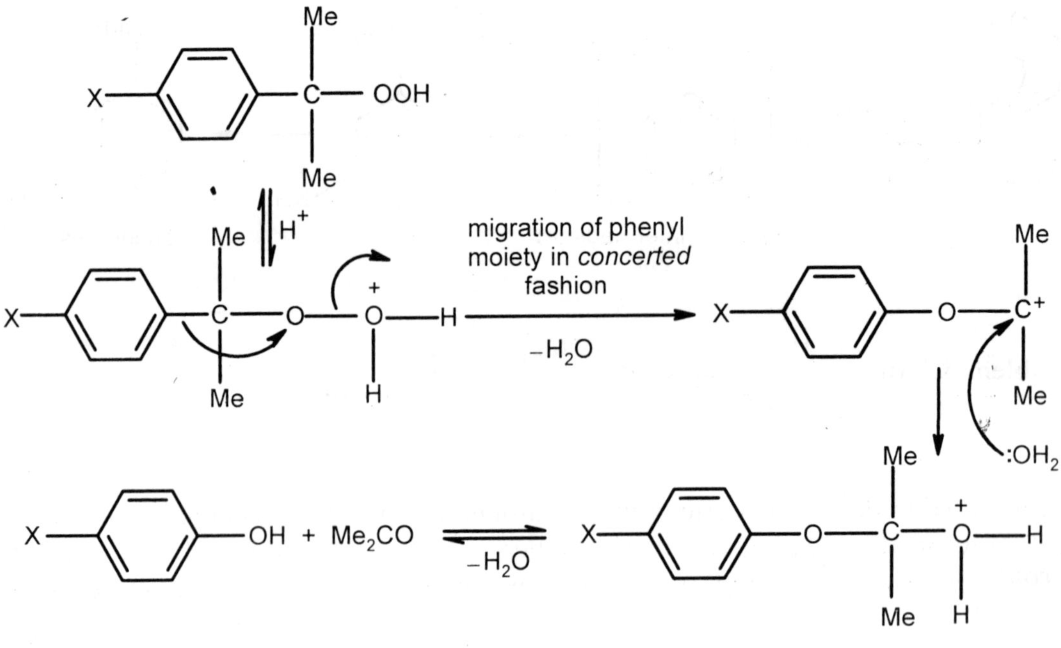

Similarly,

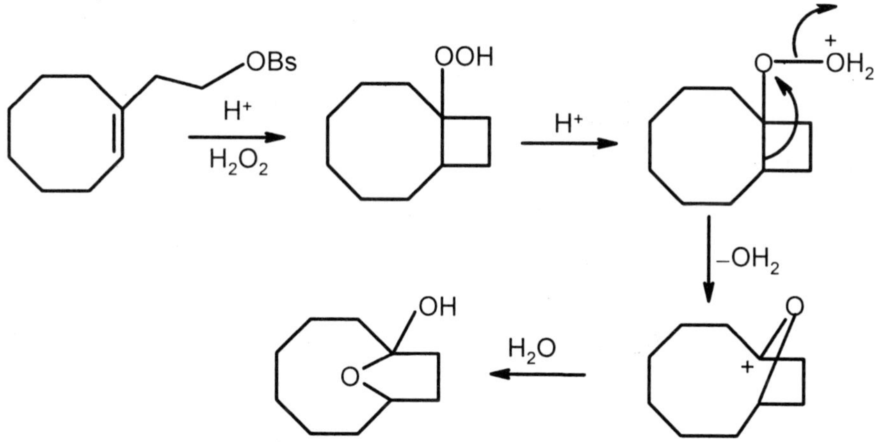

Problem 5.1.18

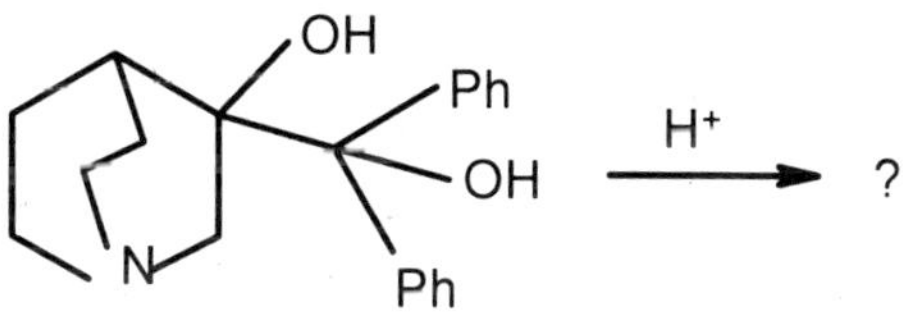

Solution.

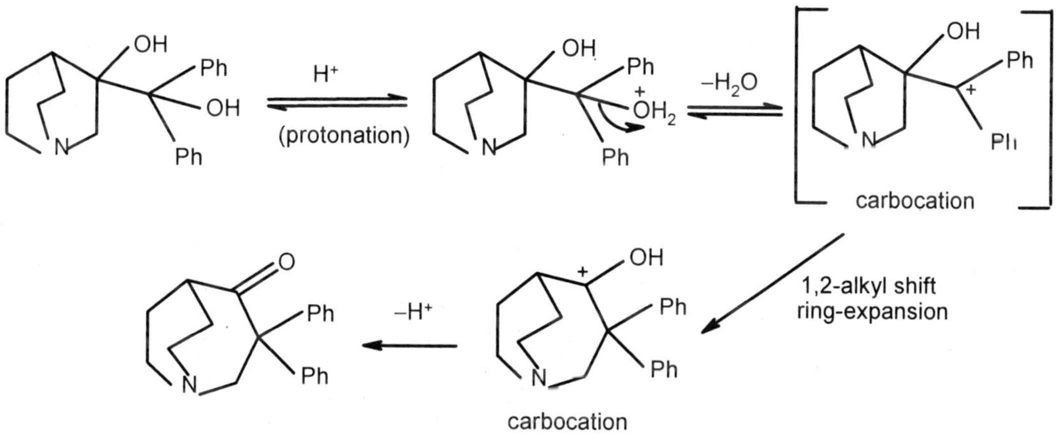

Problem 5.1.19

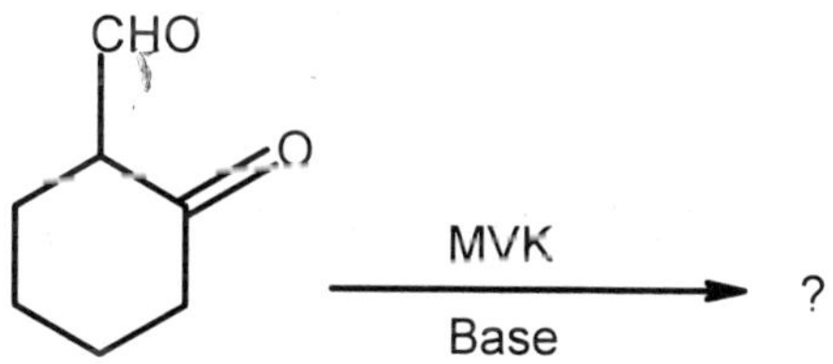

Solution.

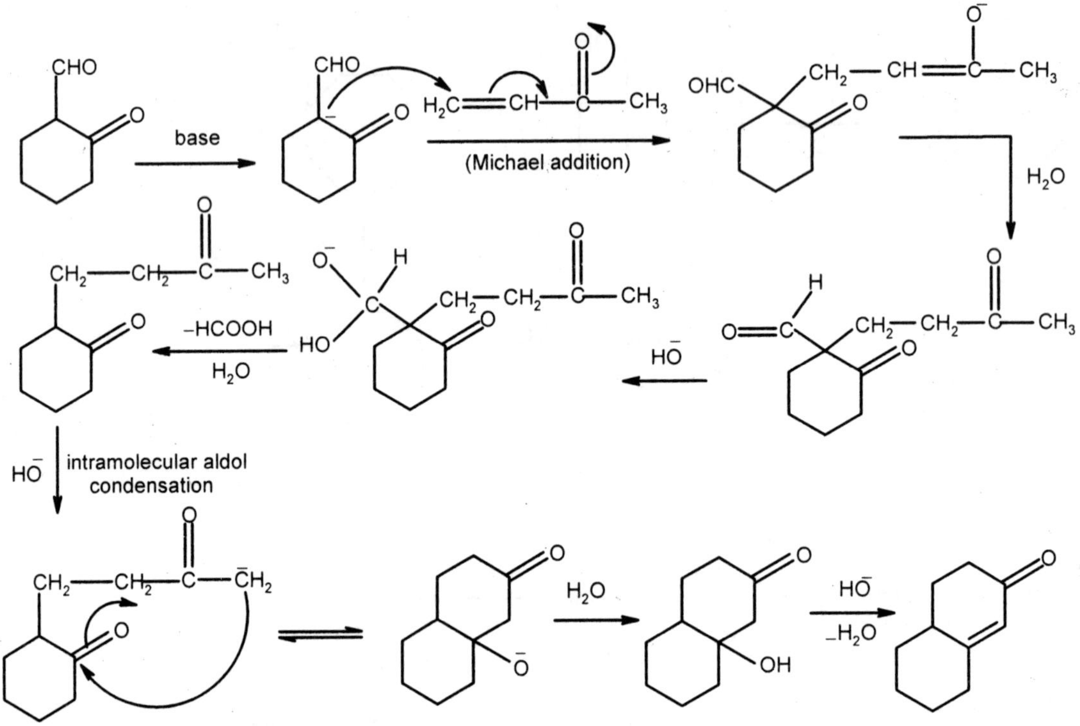

Problem 5.1.20

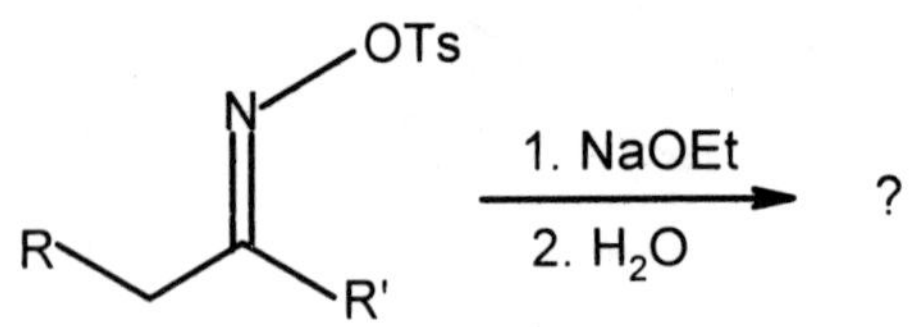

Solution.

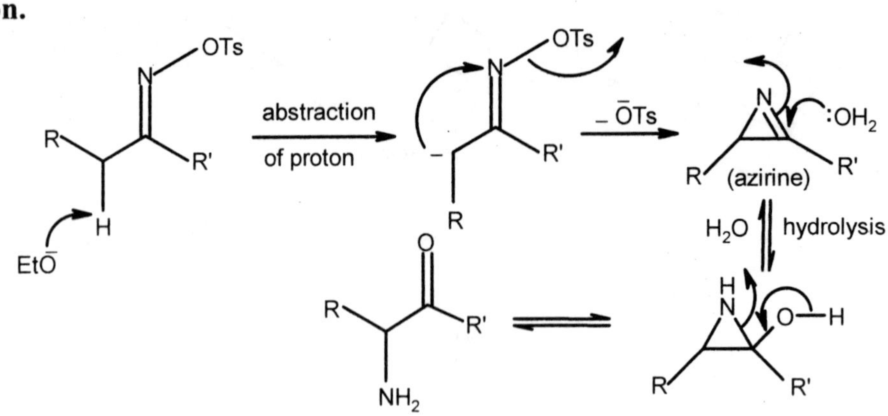

Problem 5.1.21

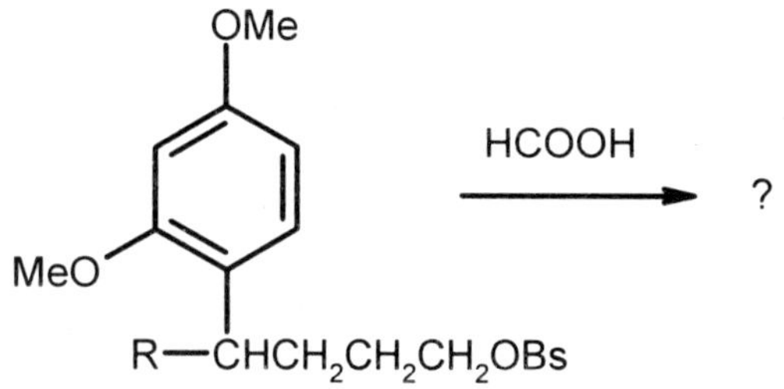

Solution.

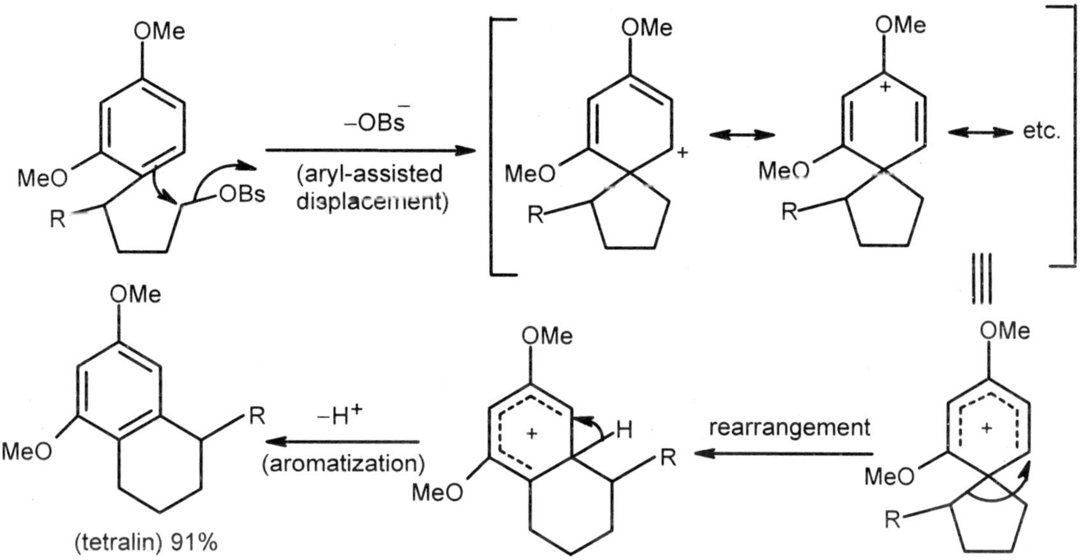

Problem 5.1.22

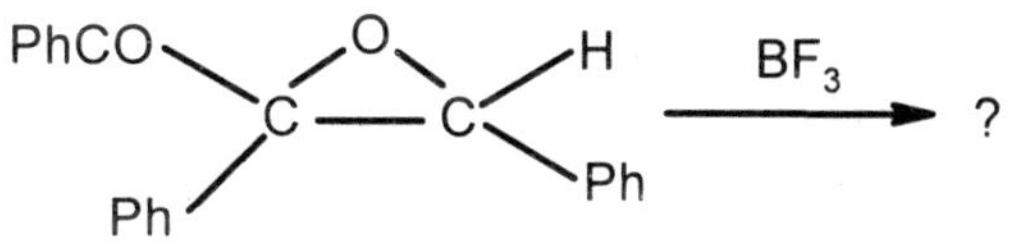

Solution.

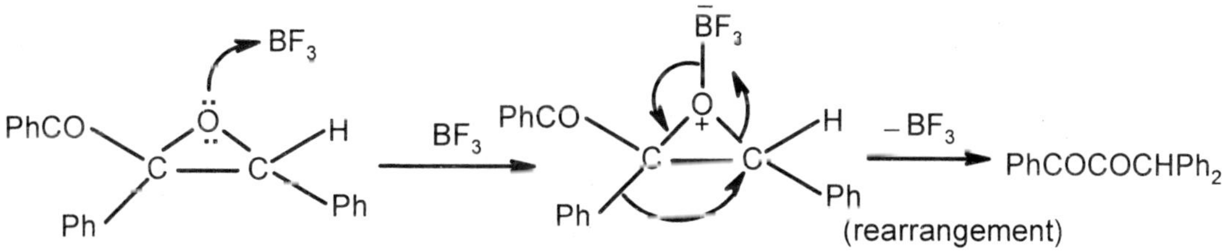

It has been found that the rearrangement takes place intramolecularly, and in most of the cases the process is *concerted*.

Problem 5.1.23

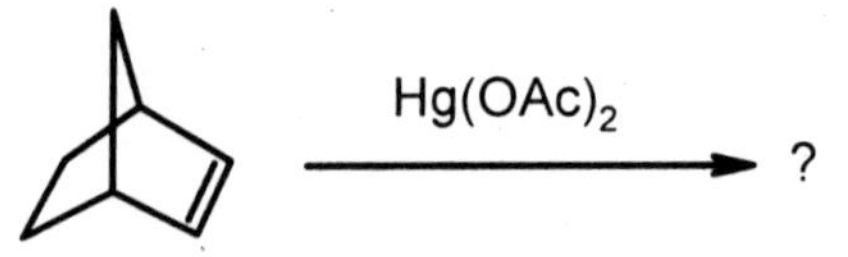

Solution.

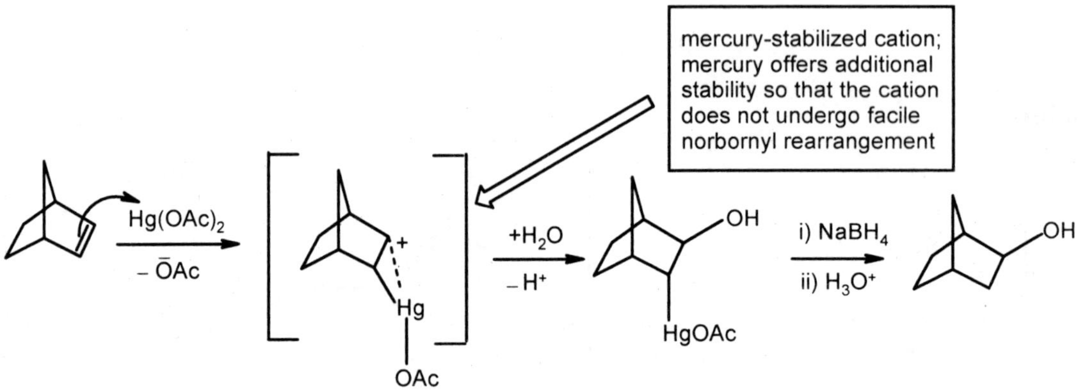

Problem 5.1.24

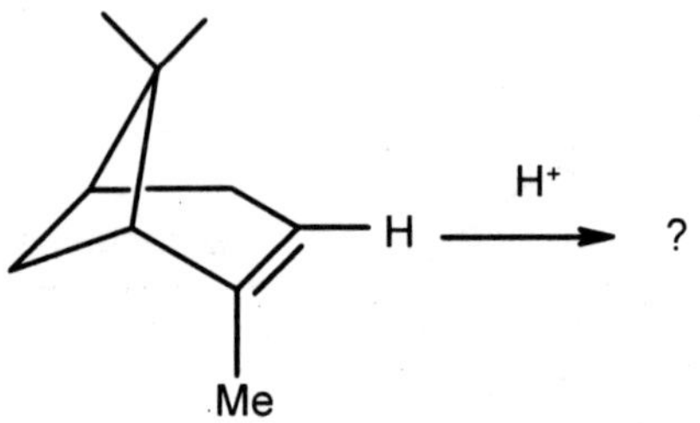

Solution.

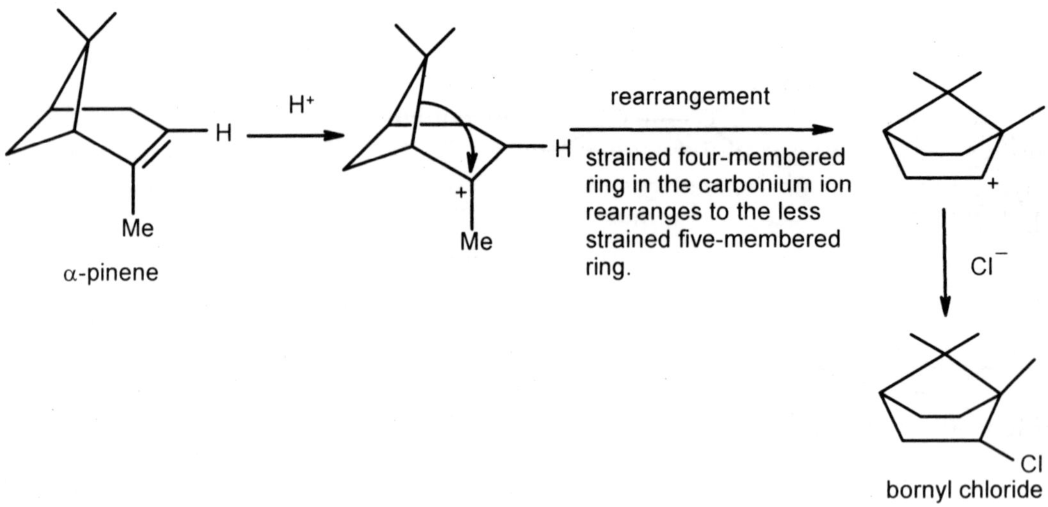

Problem 5.1.25

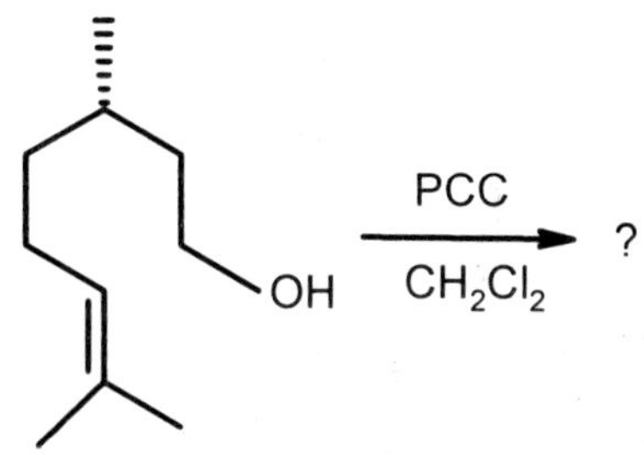

Solution.

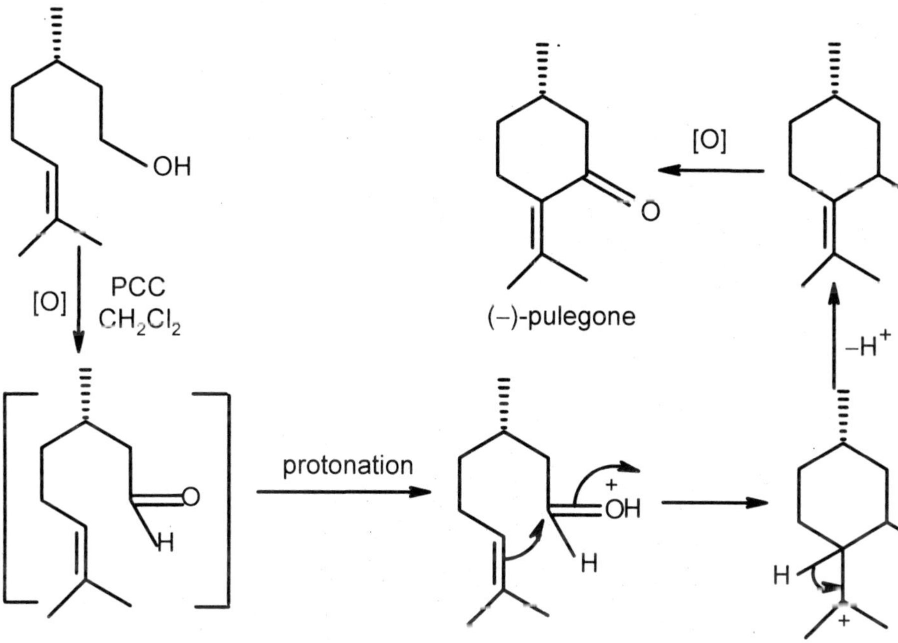

PCC (pyridinium chlorochromate) solution contains minute amount of acid as impurities, and is thus mildly acidic. Hence, the initially formed acid sensitive aldehydic product is protonated as a result of which it cyclizes to the secondary alcohol. This secondary alcohol then undergoes further oxidation under reaction condition to give (–)-pulegone as the final product.

Problem 5.1.26

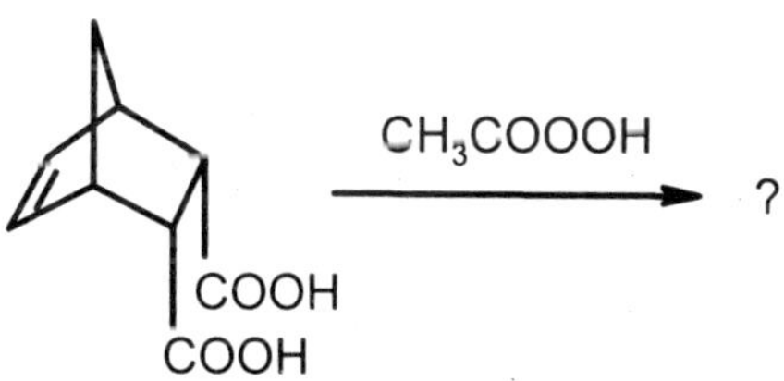

Solution.

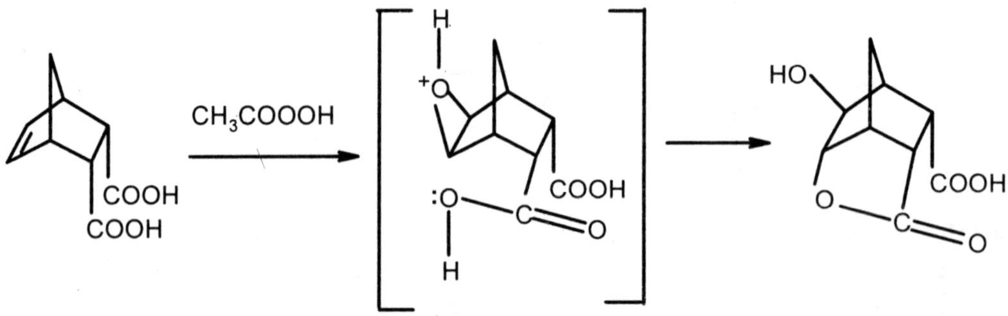

Problem 5.1.27

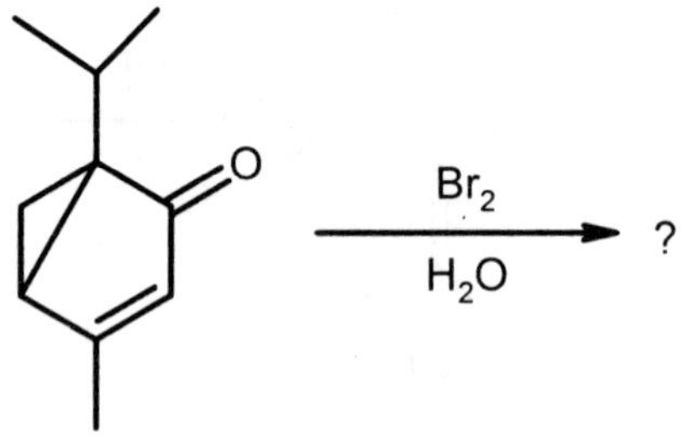

Solution. The reaction may proceed *via* more than one pathway:

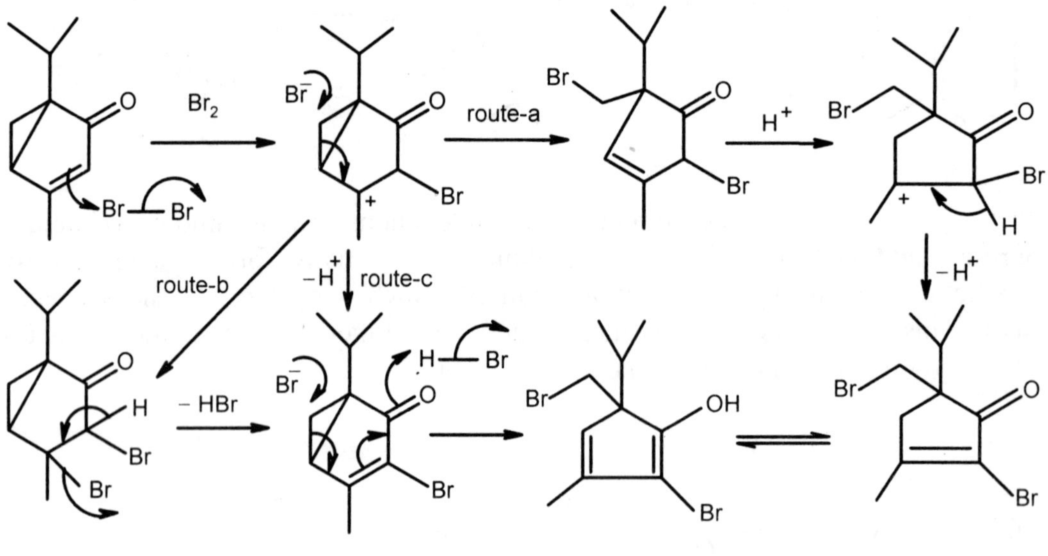

Problem 5.1.28 $(CH_3)_2C(OCH_3)CH_2OBs \xrightarrow{H_2O} ?$

Solution.

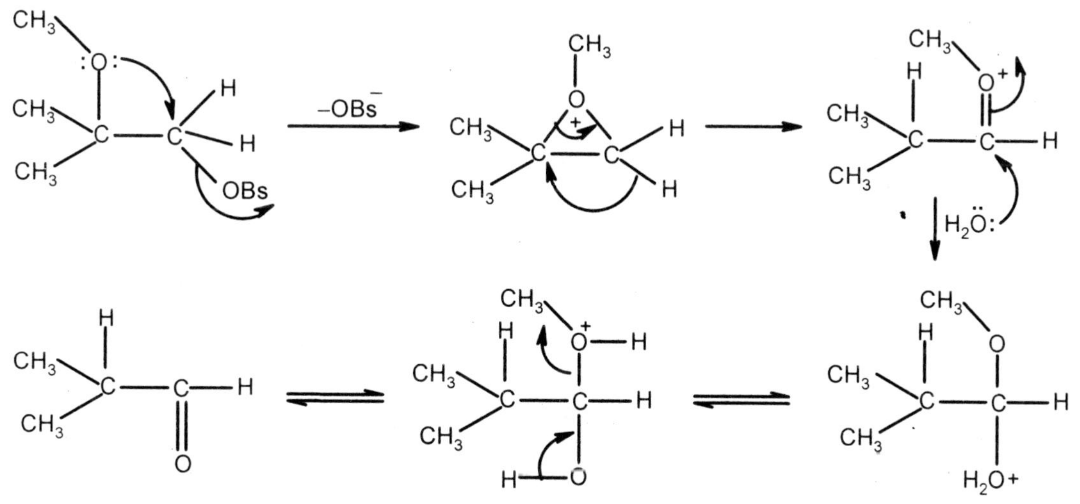

Problem 5.1.29

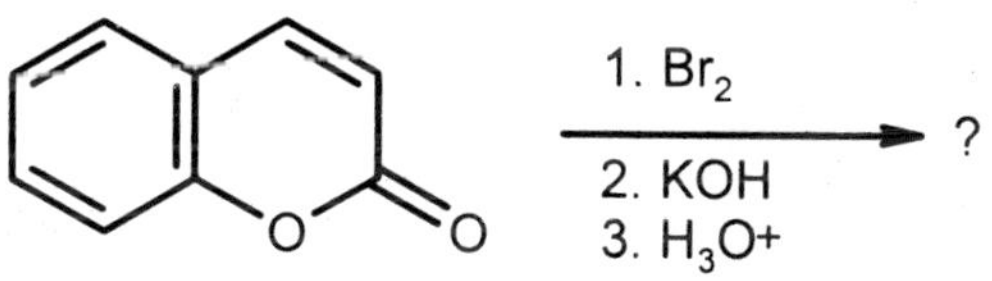

Solution.

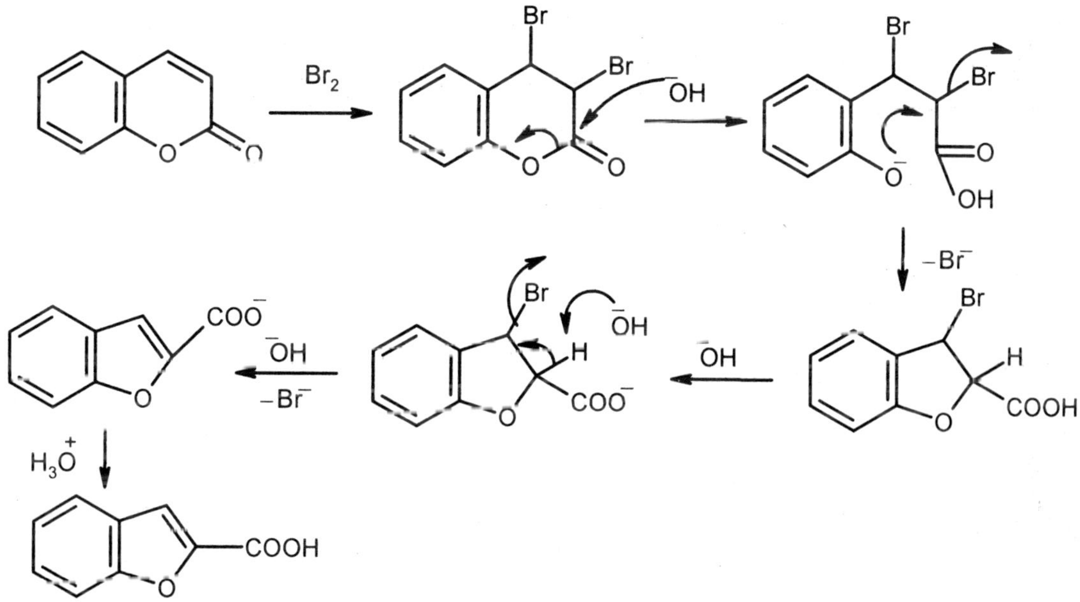

Problem 5.1.30 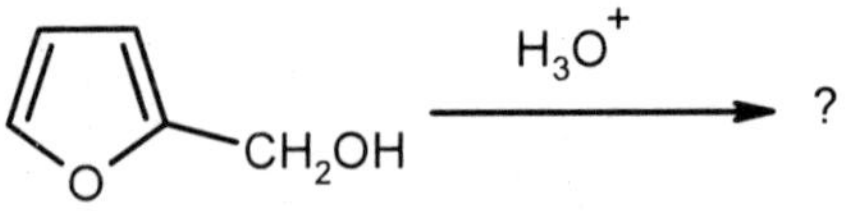

Solution.

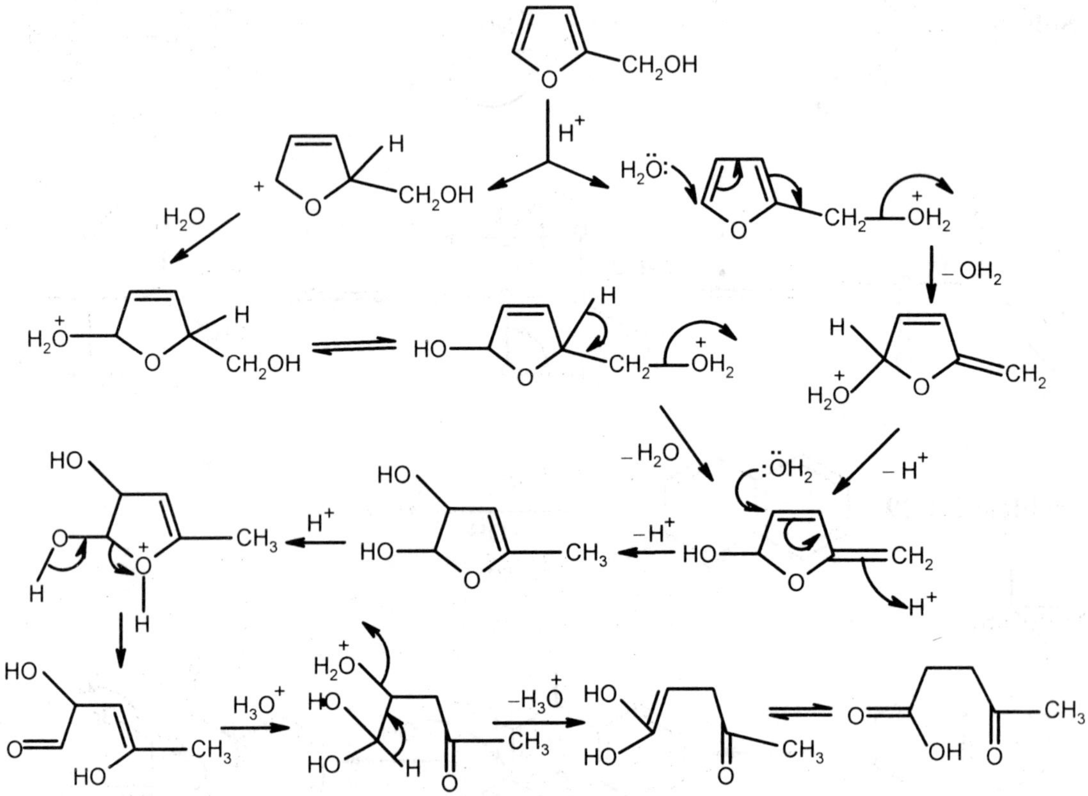

Problem 5.1.31 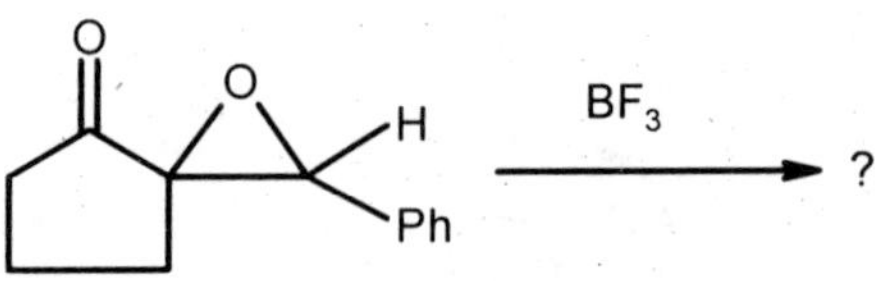

Solution.

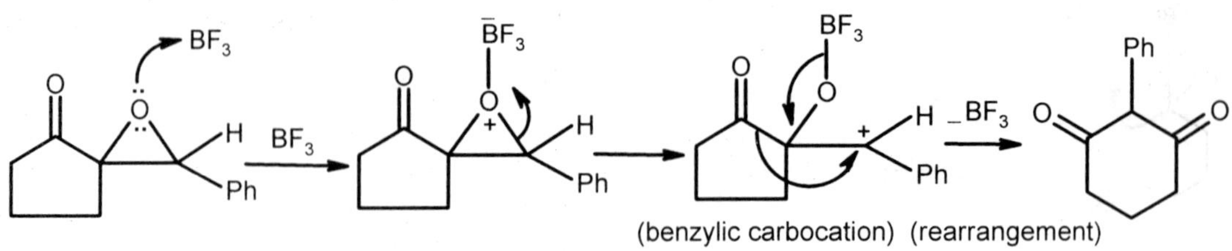

Problem 5.1.32

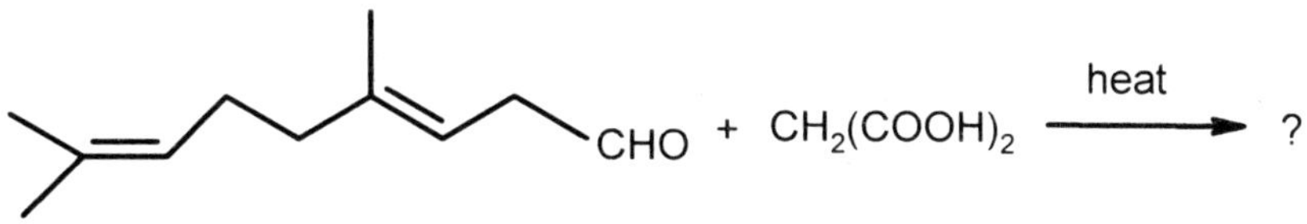

Solution.

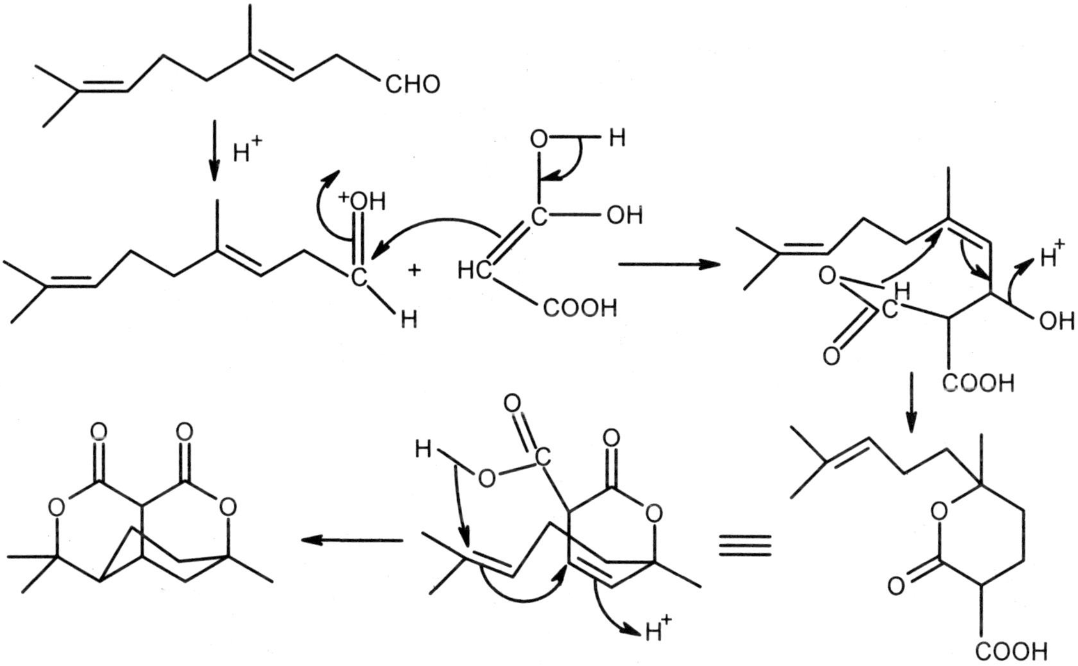

Problem 5.1.33

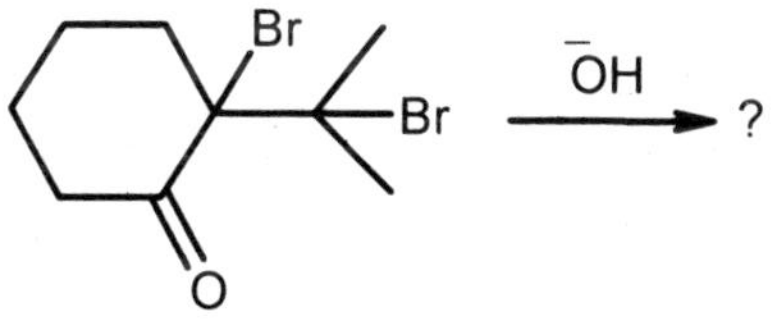

Solution.

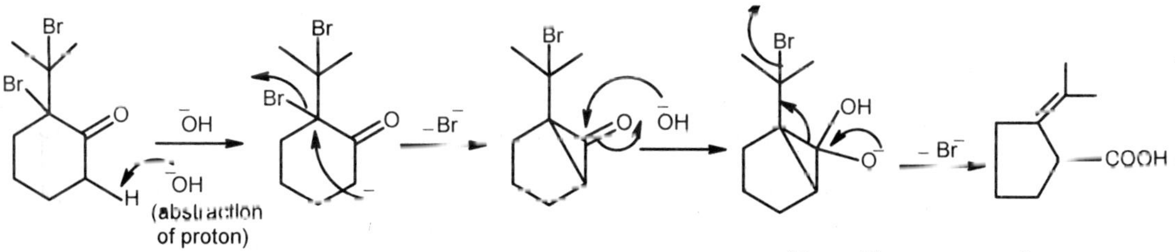

(*Favorskii rearrangement*)

Problem 5.1.34 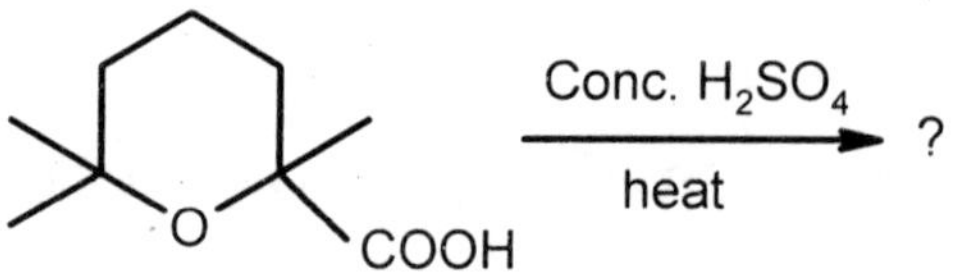

Solution.

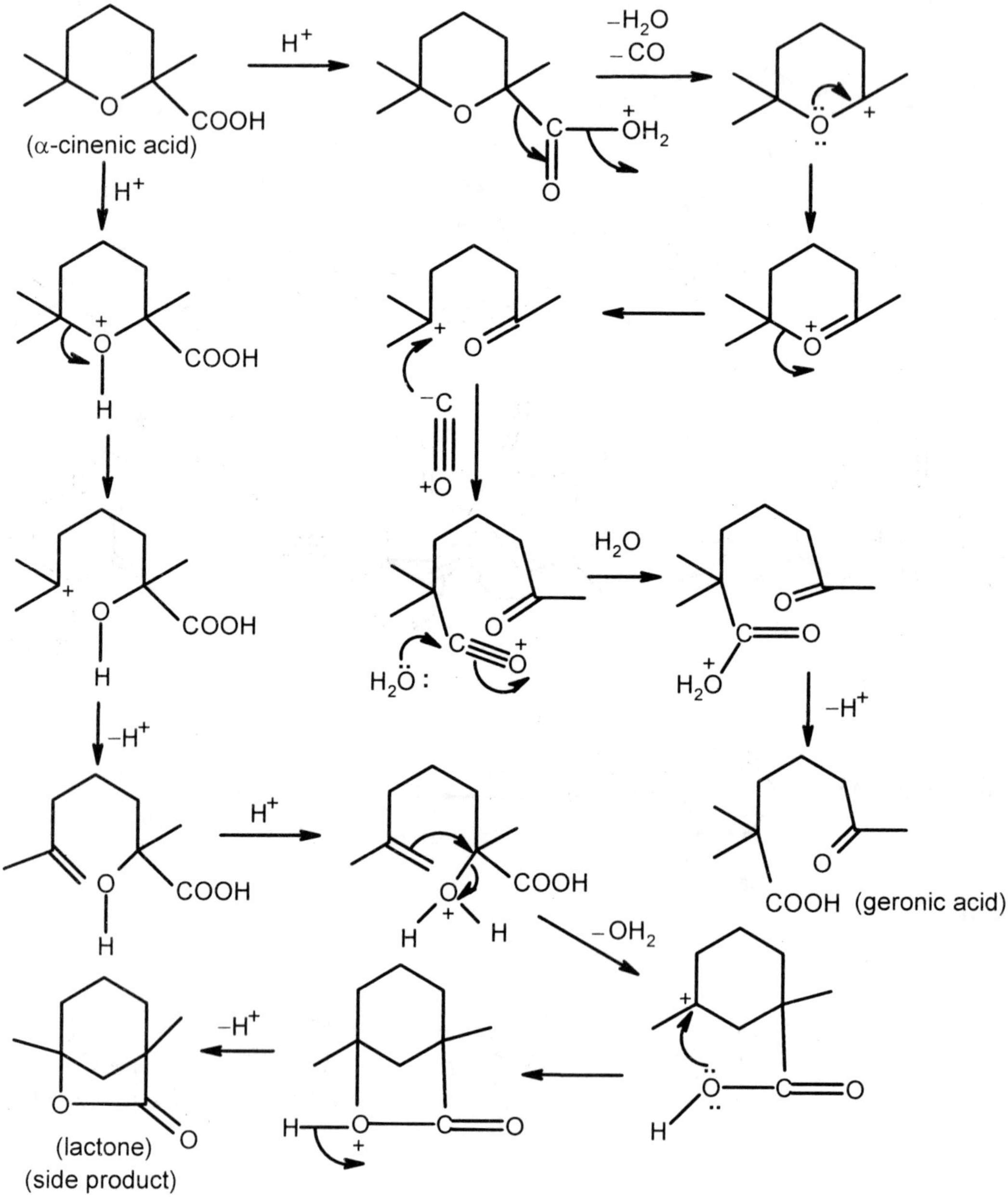

Problem 5.1.35 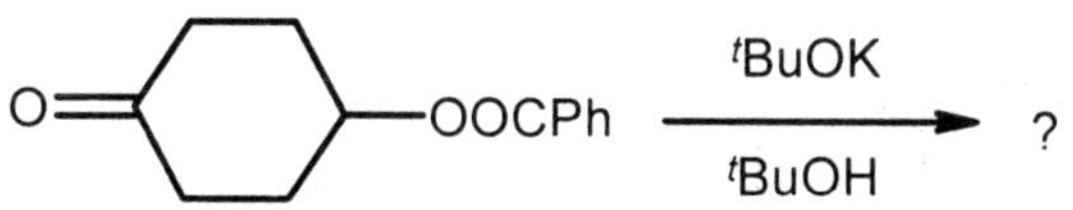

Solution.

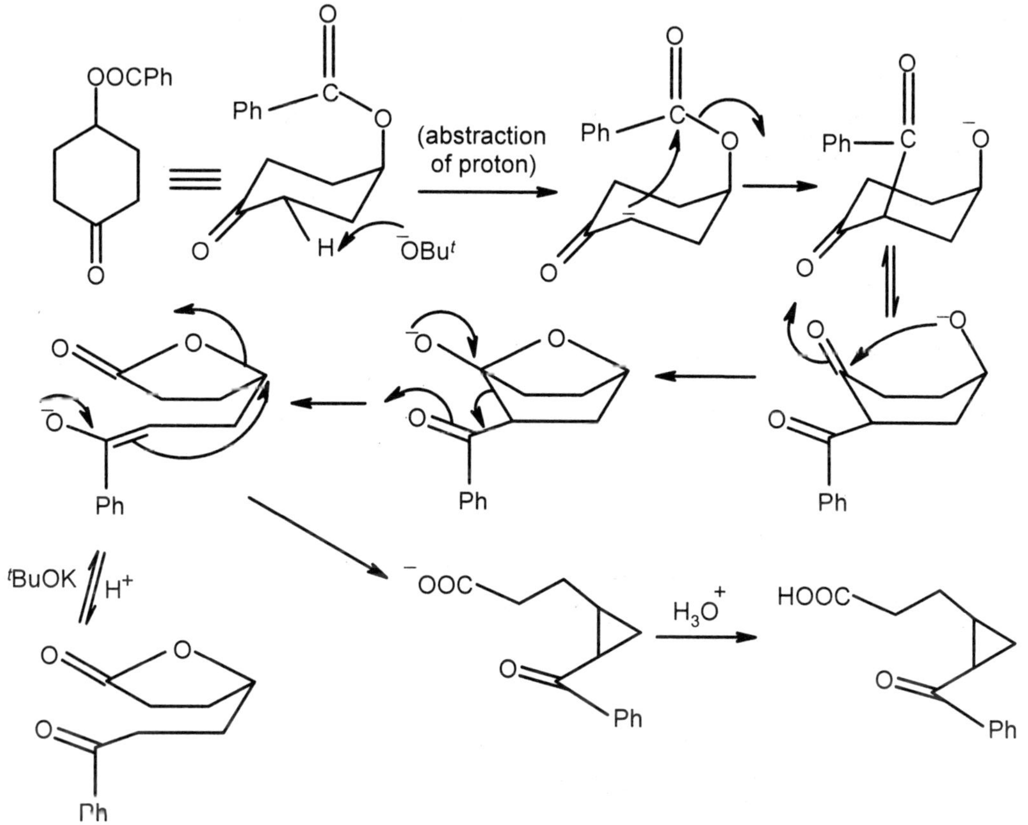

Problem 5.1.36 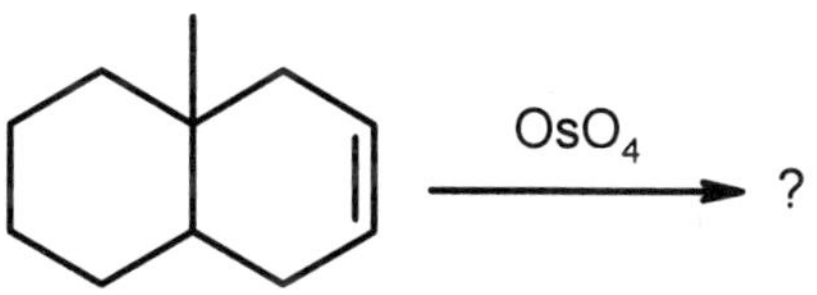

Solution.

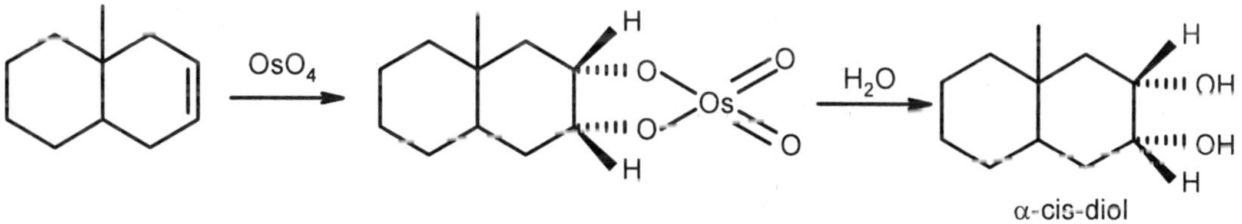

(OsO_4 attacks predominantly from the less hindered back side to give α-cis-diol)

Problem 5.1.37

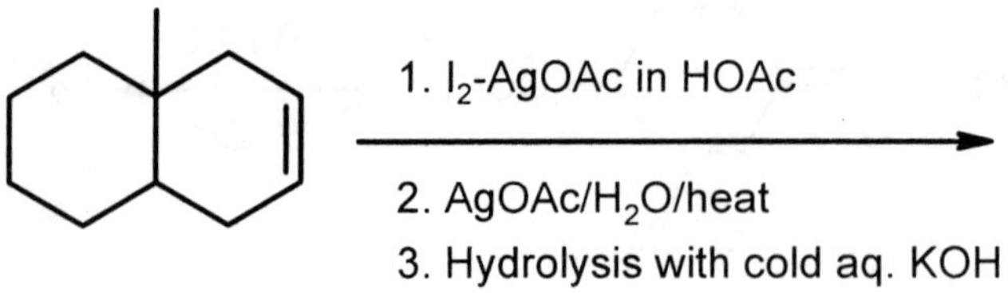

Solution.

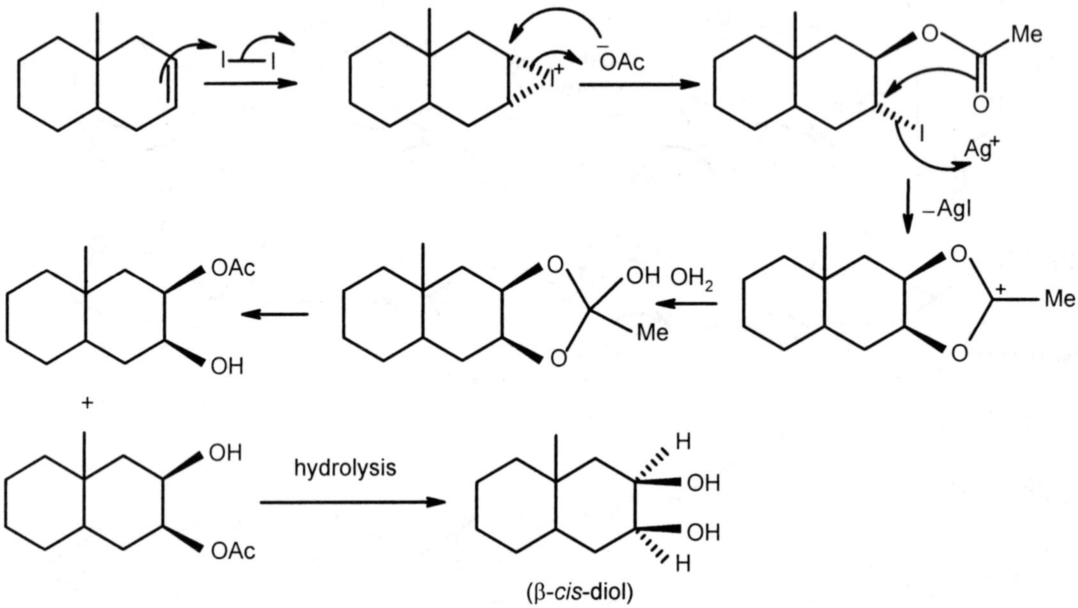

Problem 5.1.38

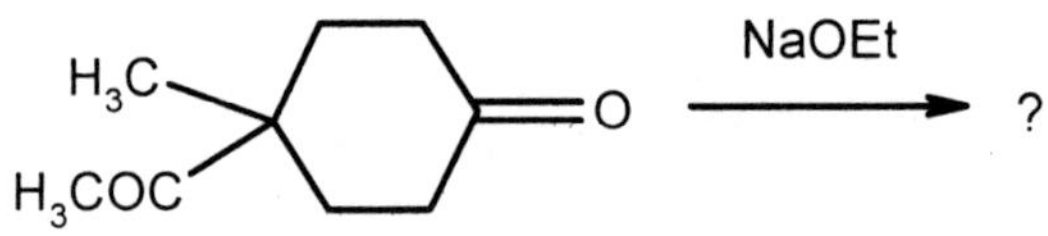

Solution.

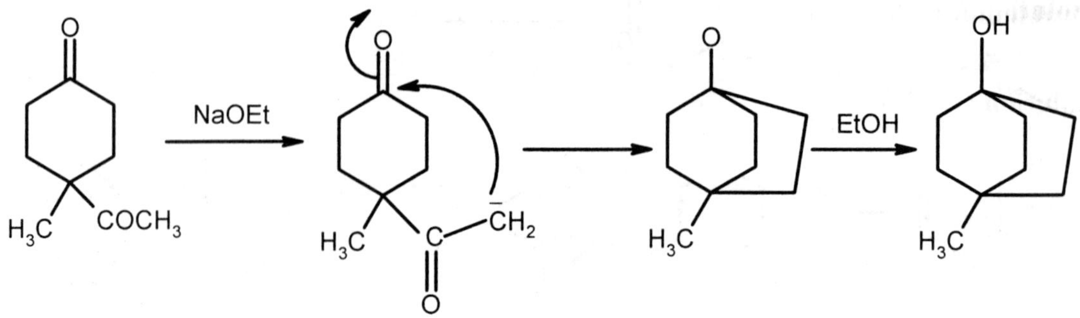

Problem 5.1.39

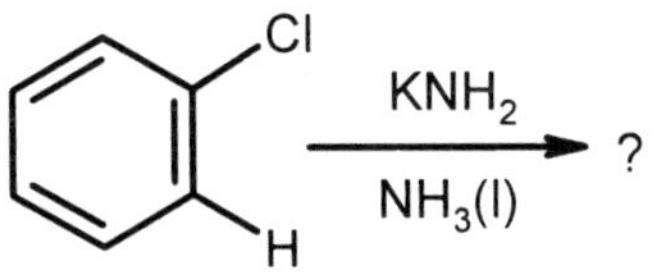

Solution.

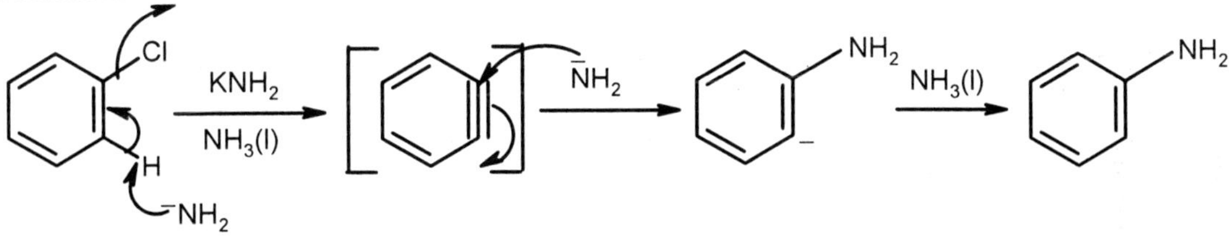

Problem 5.1.40

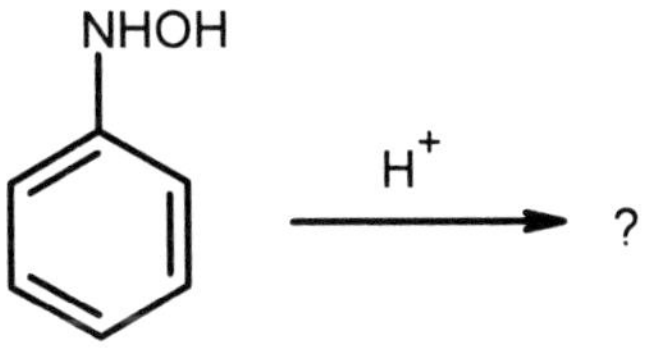

Solution.

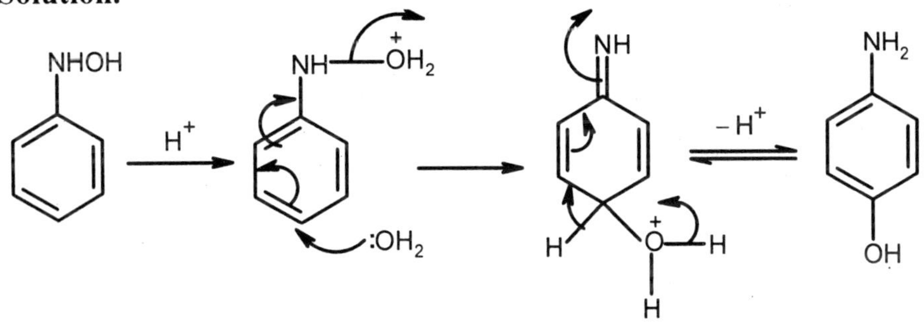

Problem 5.1.41

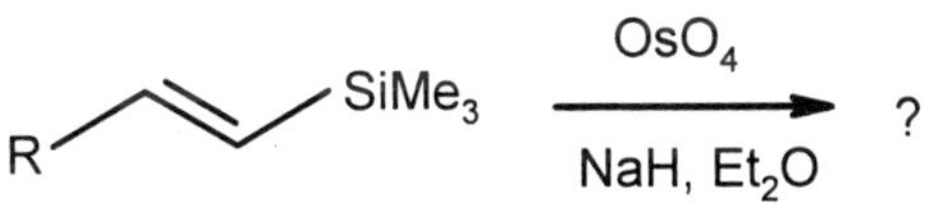

Solution.

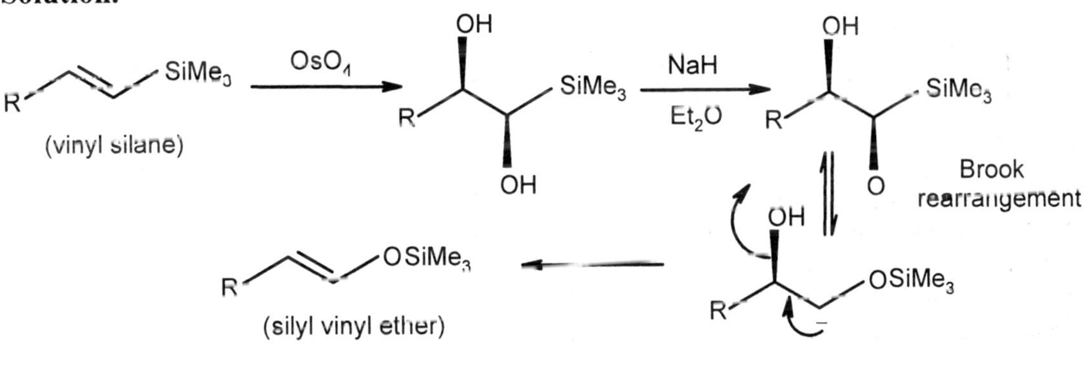

Problem 5.1.42 CH_3COCH_3 + HCHO + CH_3COCH_2COOEt $\xrightarrow{\text{NaOEt/EtOH}}$

Solution.

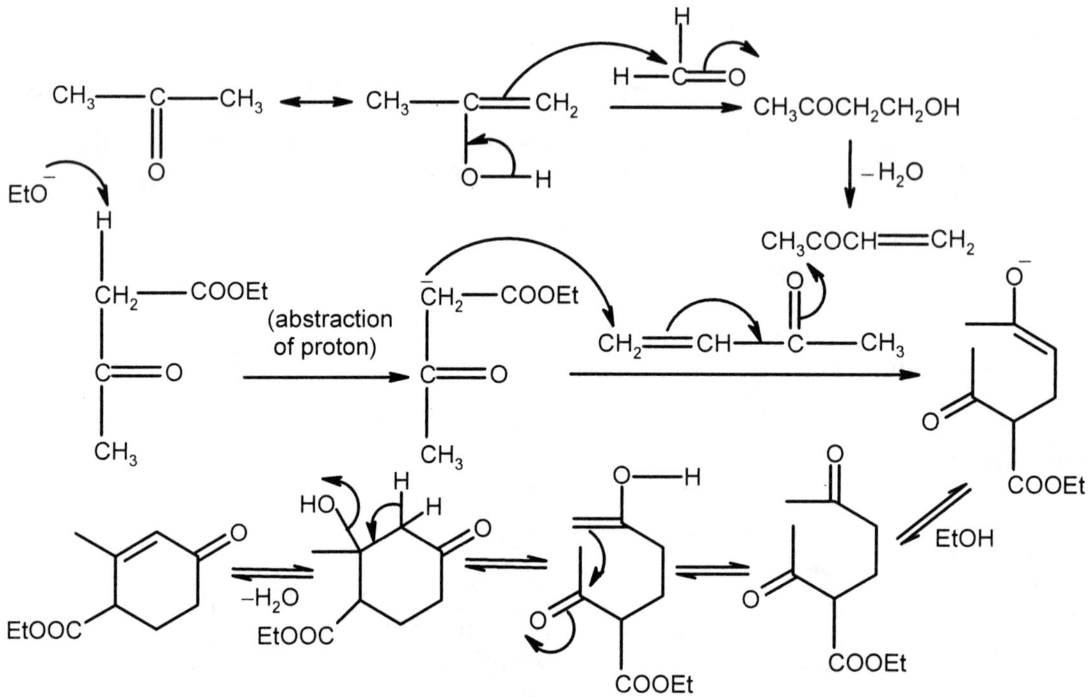

Problem 5.1.43

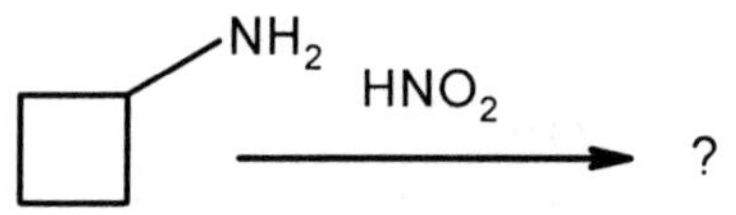

Solution.

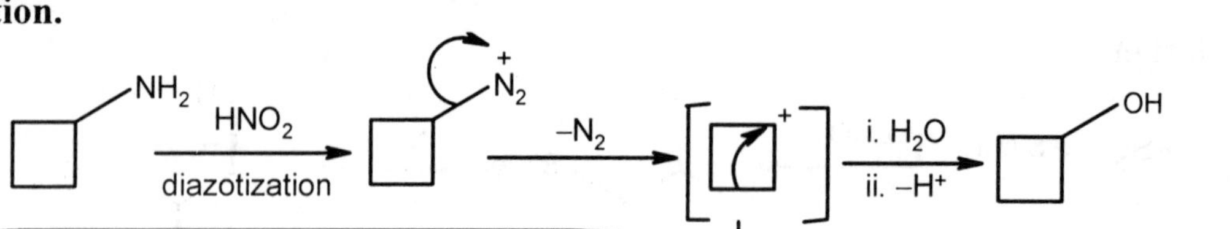

Cyclopropylmethyl cation is expected to relatively more stable because of effective charge delocalization involving cyclopropyl ring C – C bonds and the positive charge due to the fact that C – C bonds in cyclopropane bear appreciable p-character.

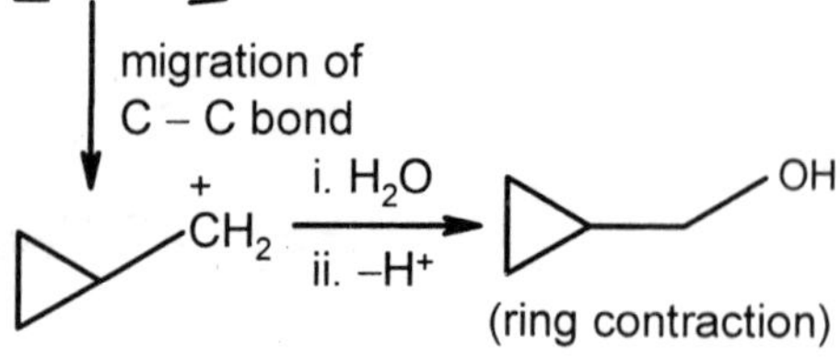

Problem 5.1.44 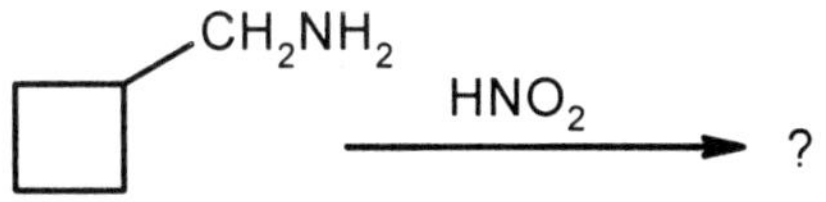

Solution.

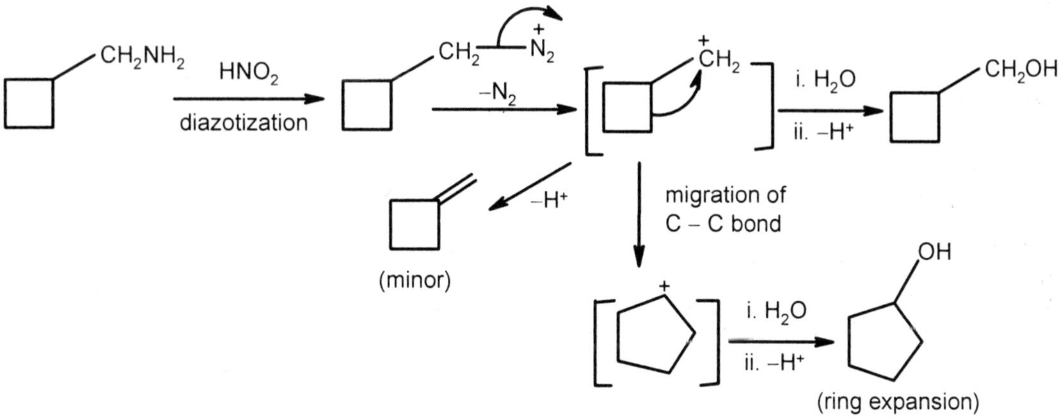

Problem 5.1.45 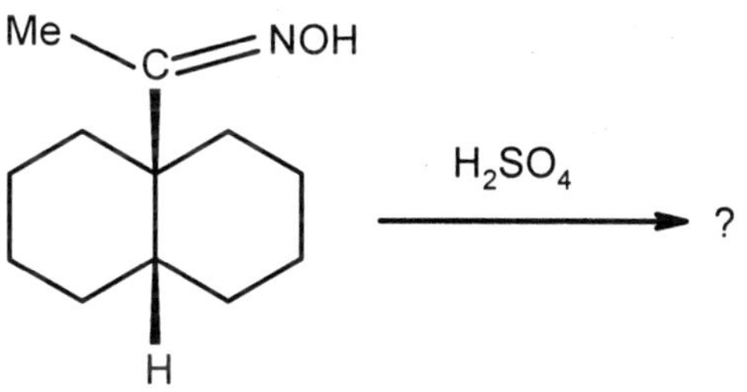

Solution.

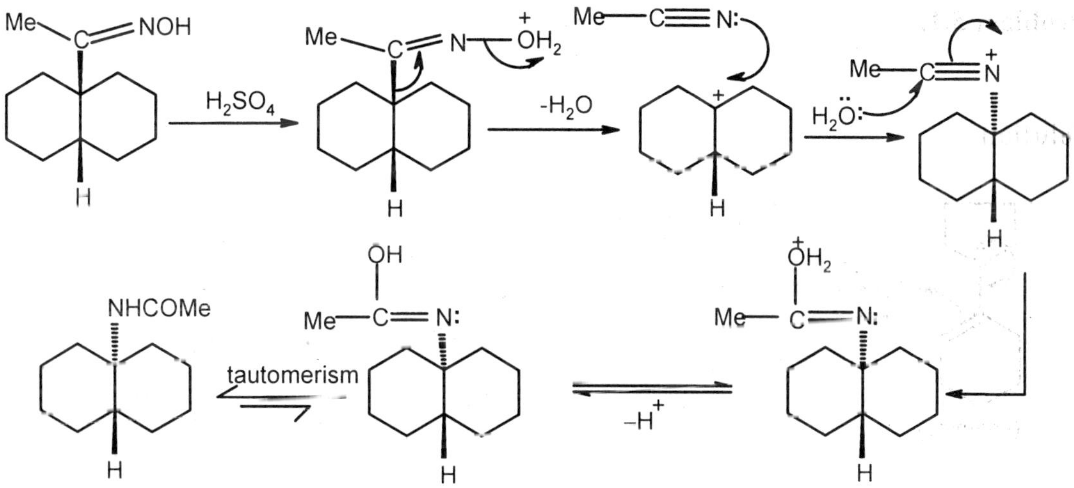

Problem 5.1.46

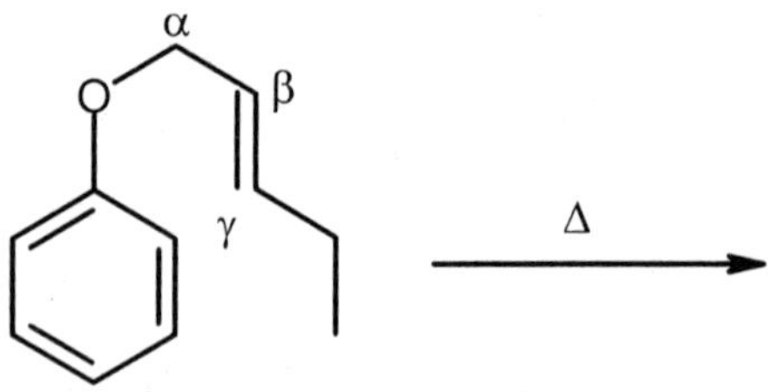

Solution.

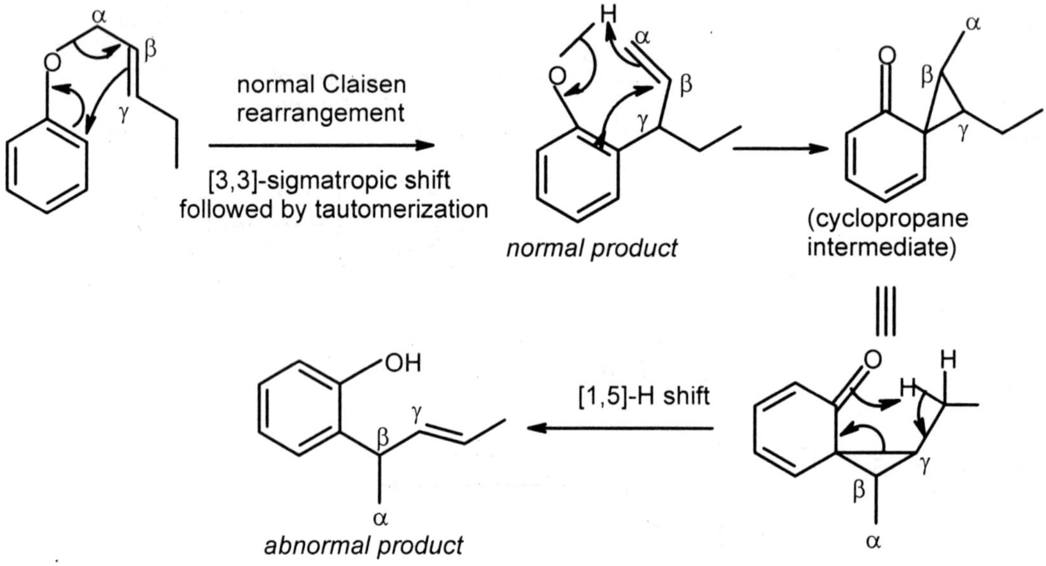

Problem 5.1.47

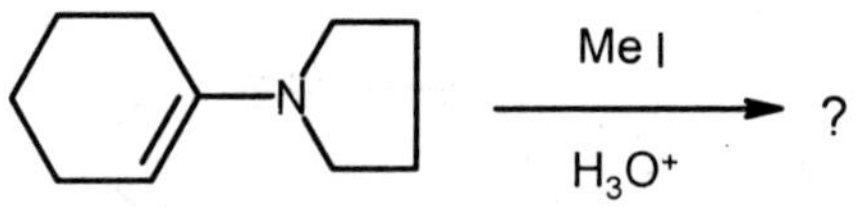

Solution.

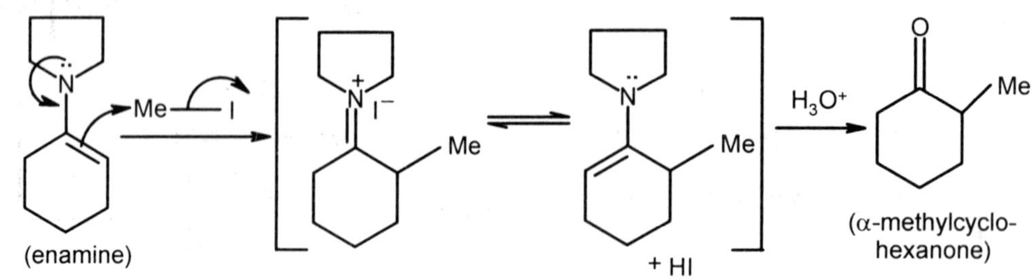

Problem 5.1.48 RCOCHO $\xrightarrow{\bar{O}H}$?

Solution.

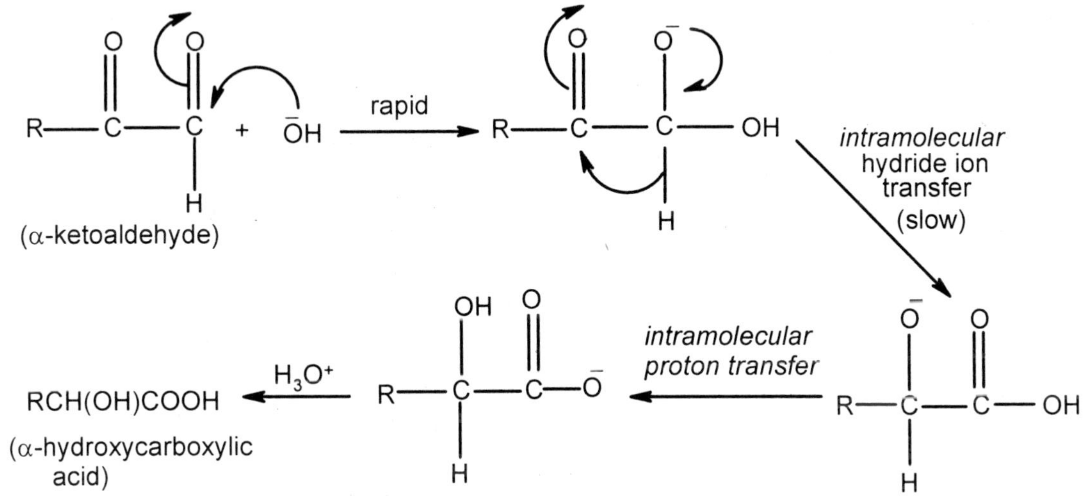

Problem 5.1.49 PhCOCOPh $\xrightarrow{\bar{O}H}$?

Solution. This is benzil-benzilic acid rearrangement reaction.

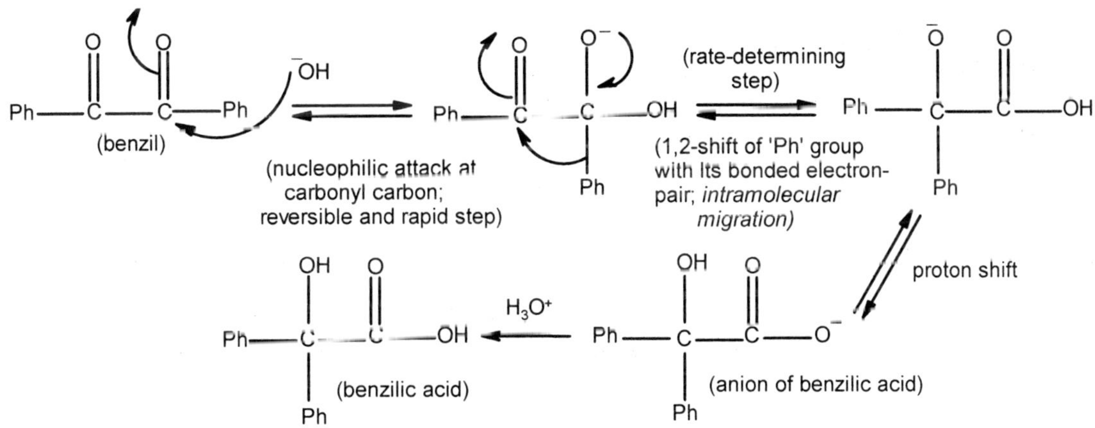

Problem 5.1.50

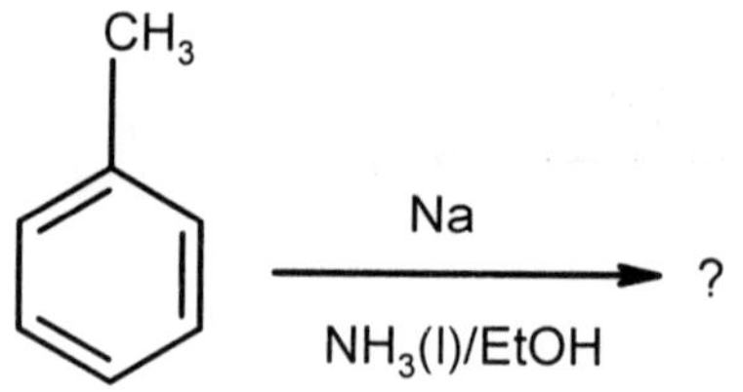

Solution.

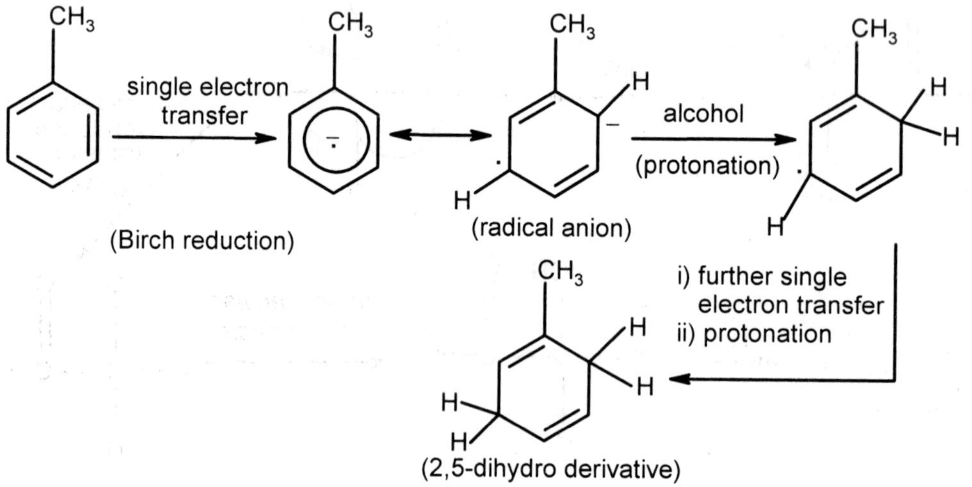

Problem 5.1.51

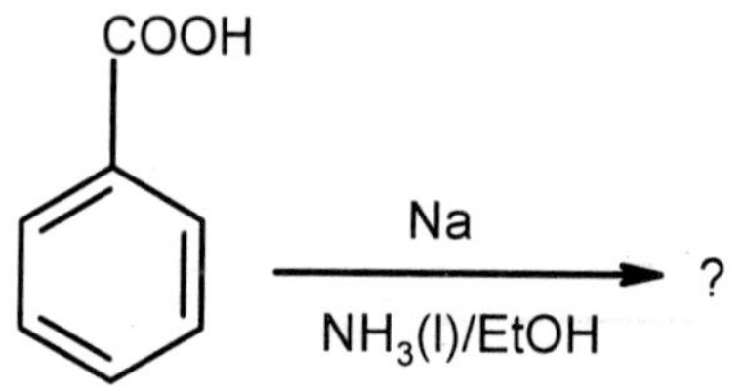

Solution.

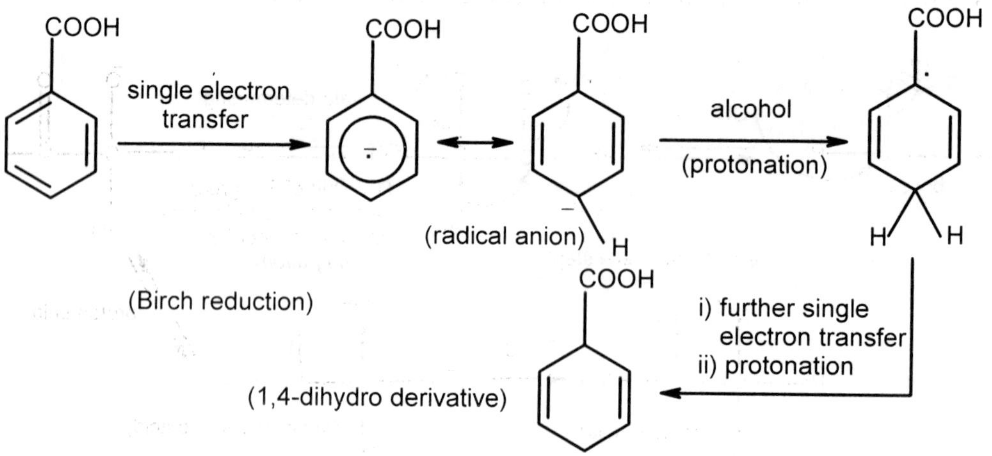

Problem 5.1.52 ArCHO $\xrightarrow{KCN}$?

Solution. This is benzoin condensation reaction.

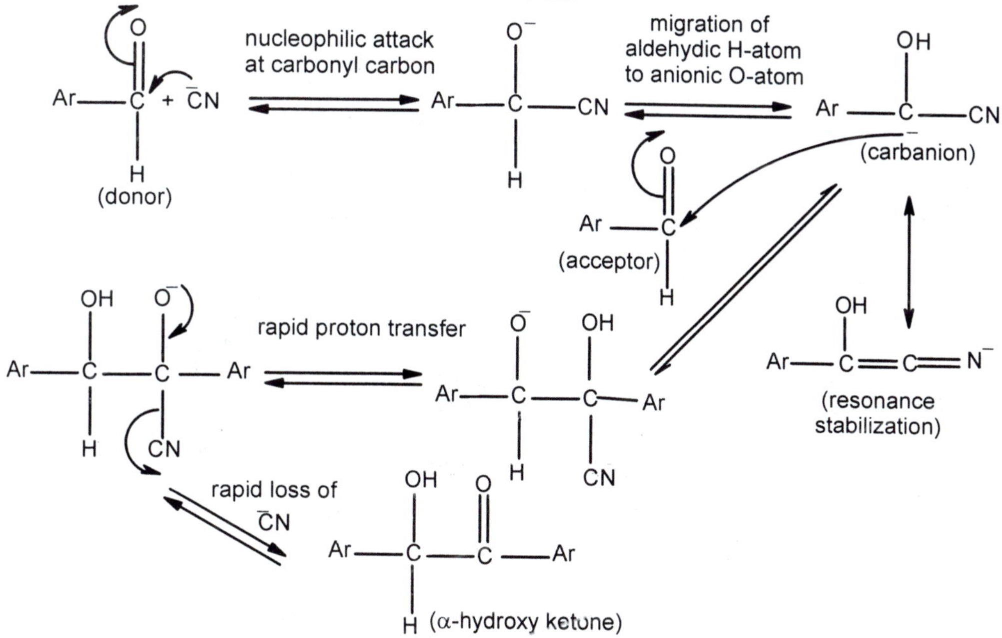

Problem 5.1.53

Solution.

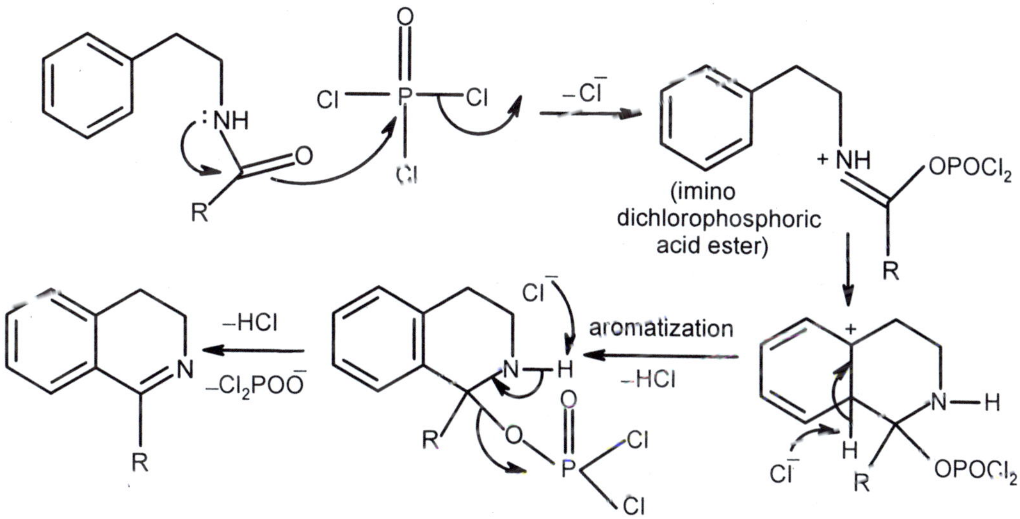

Problem 5.1.54 $EtOOC(CH_2)_4COOEt \xrightarrow[THF]{base} ?$

Solution.

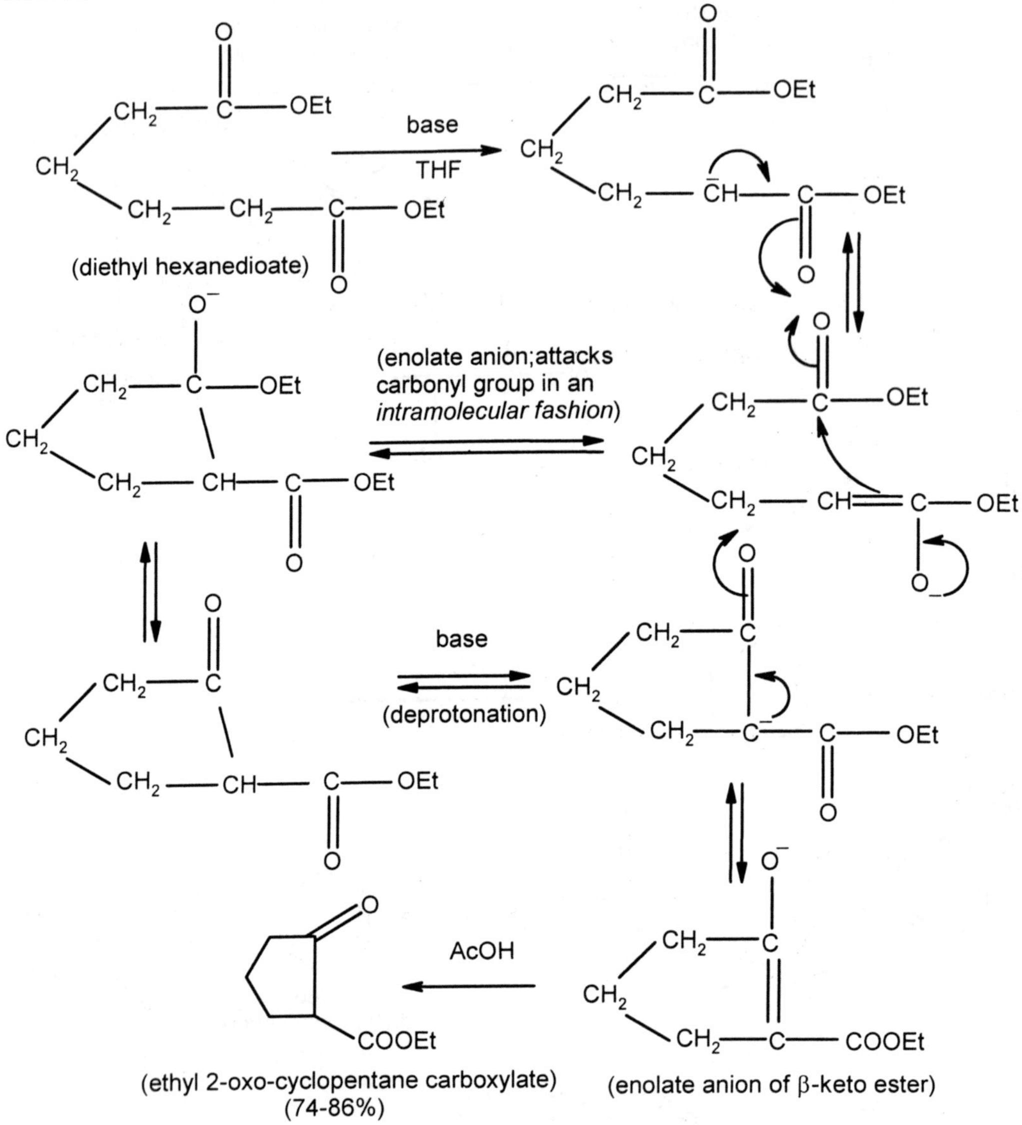

Problem 5.1.55

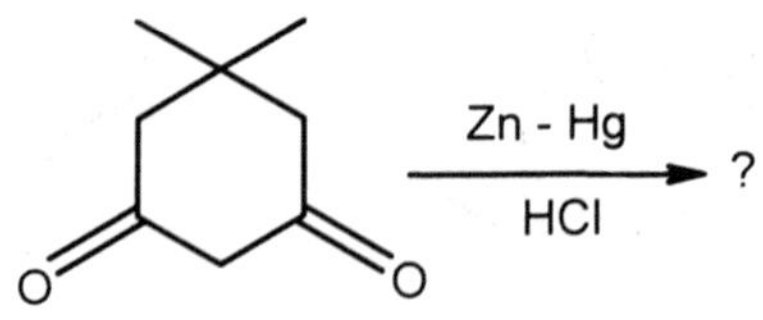

Solution.

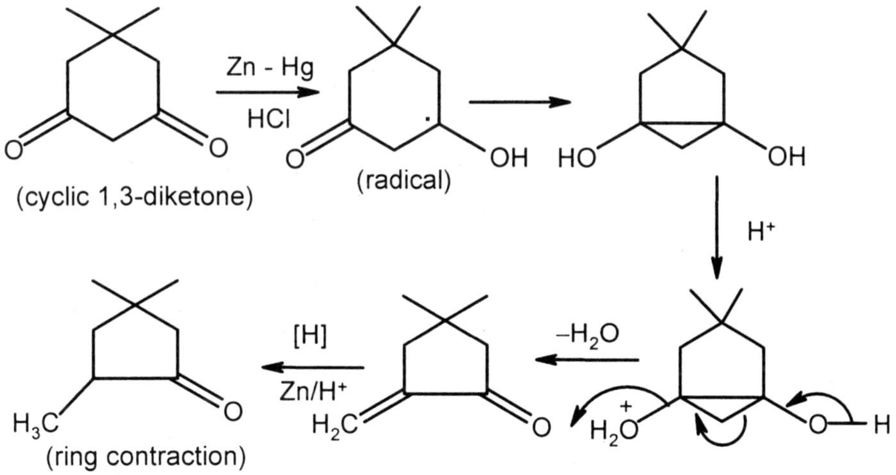

Problem 5.1.56

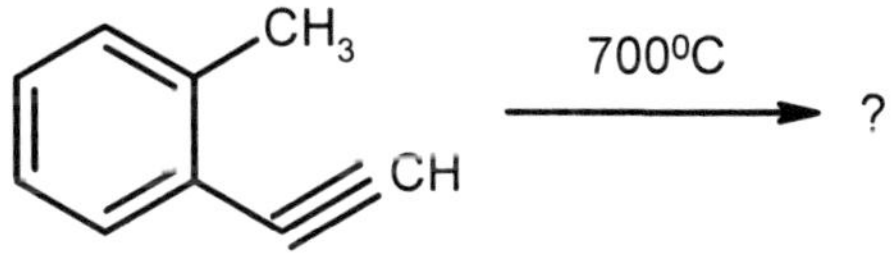

Solution.

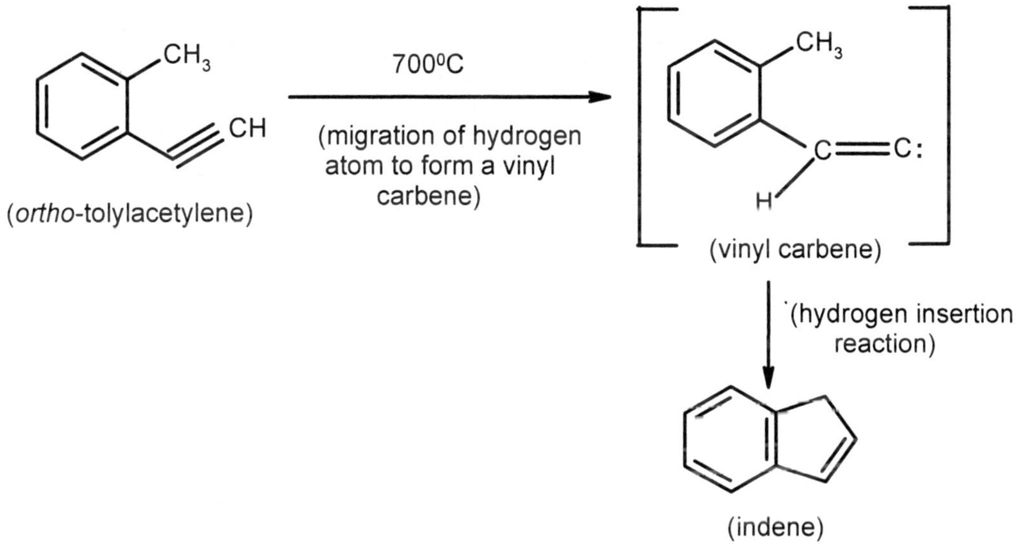

Problem 5.1.57

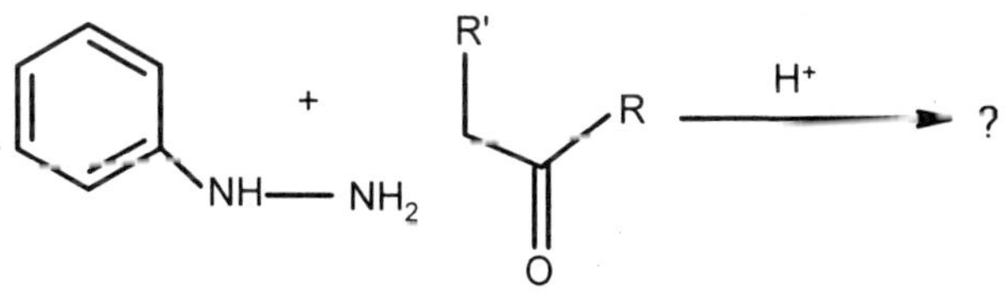

Solution. This method is known as *Fischer indole synthesis*.

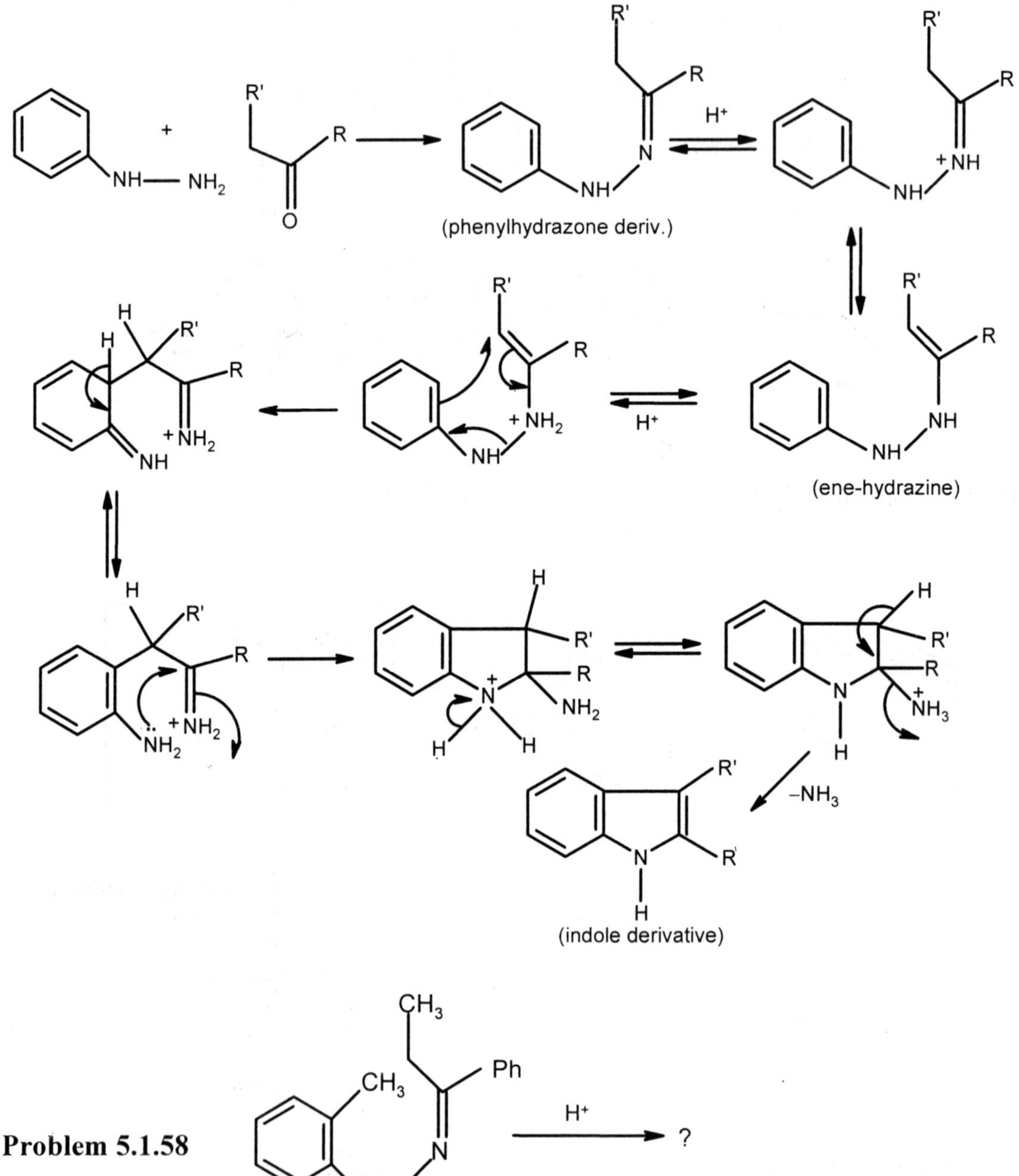

Problem 5.1.58

Solution.

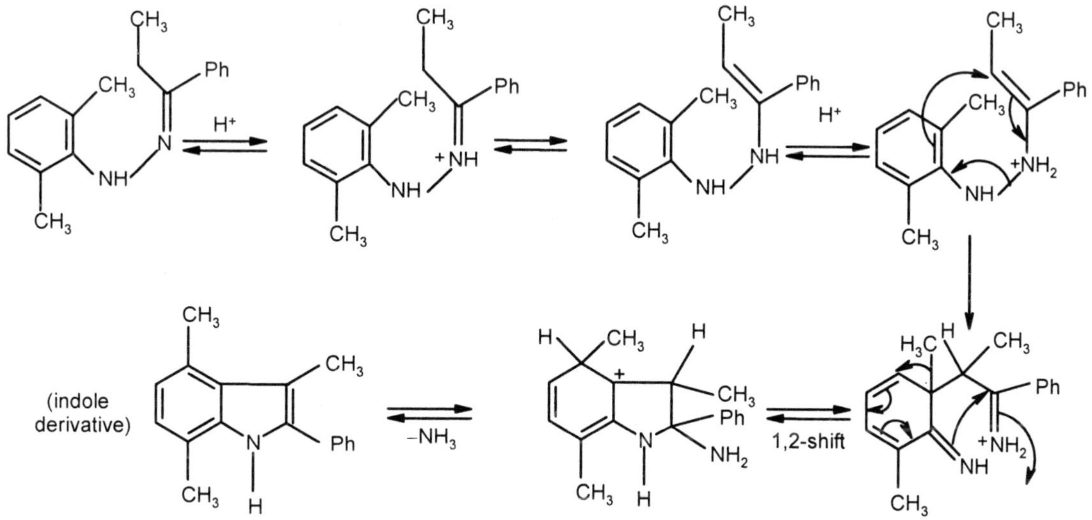

Problem 5.1.59

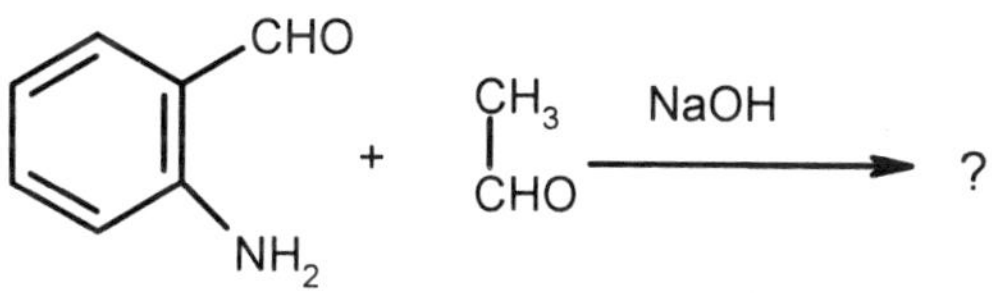

Solution. This method is known as Friedlander synthesis.

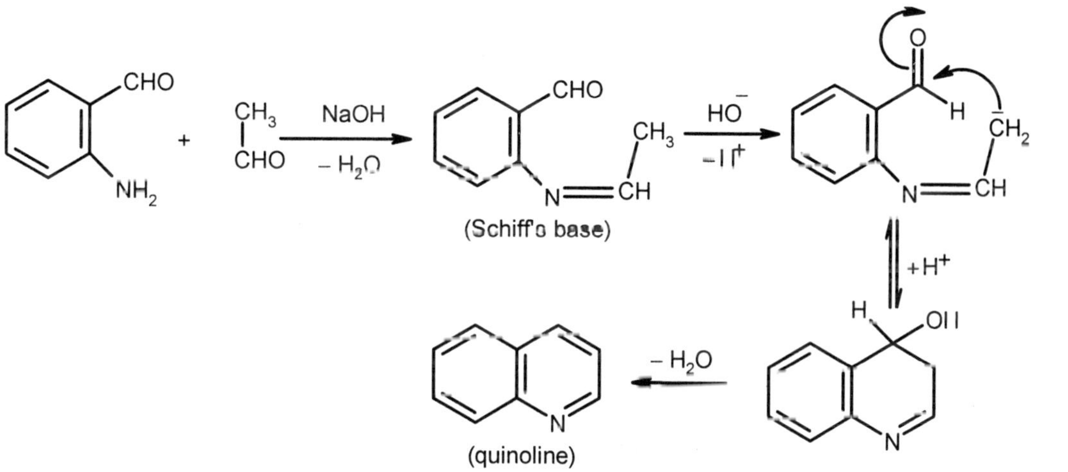

Problem 5.1.60

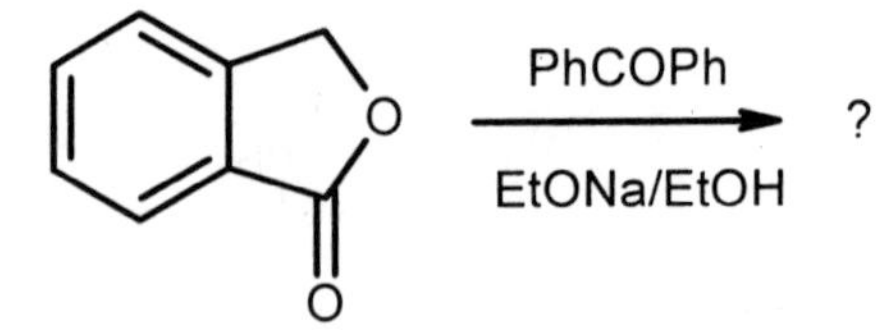

Solution.

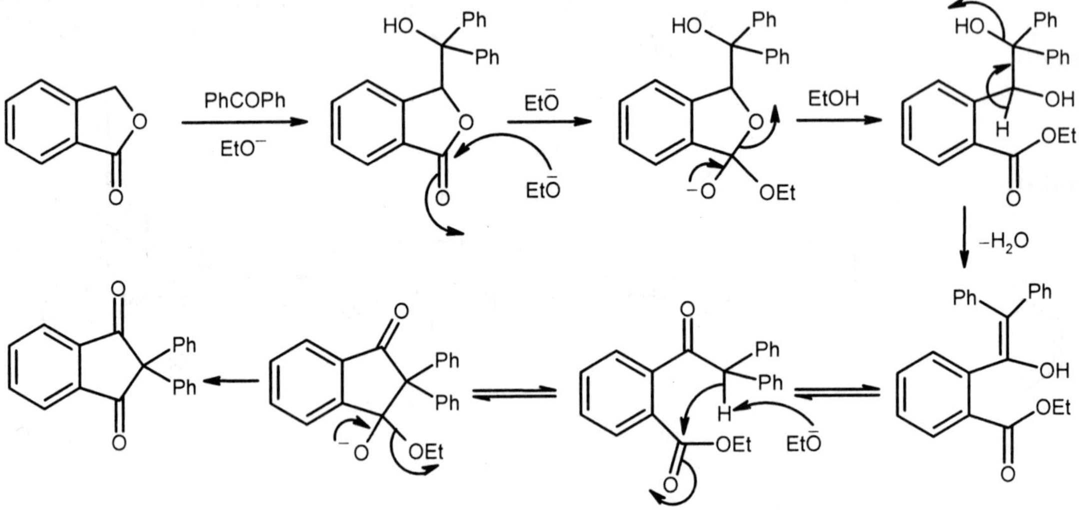

Problem 5.1.61

Solution.

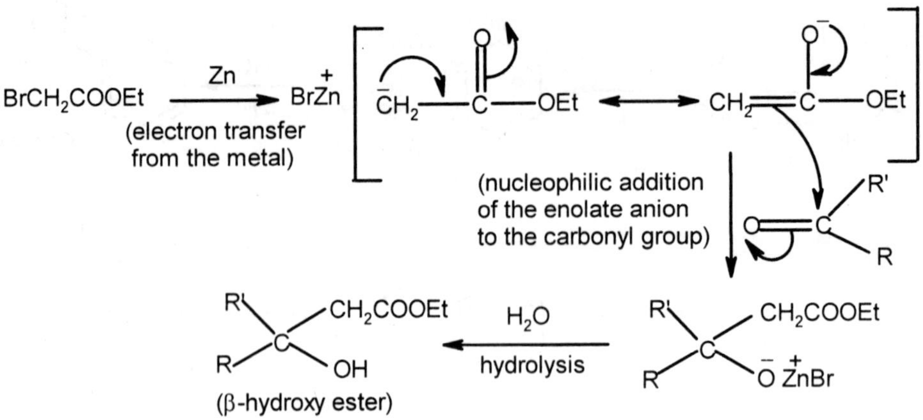

Problem 5.2 Outline the reasonable synthetic path for each of the following conversions:

Problem 5.2.1

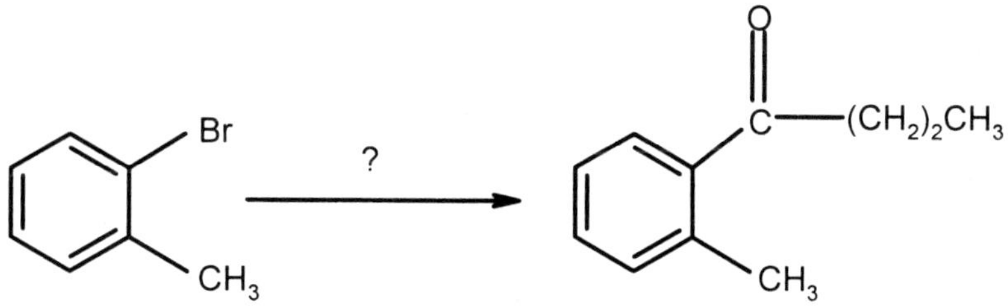

Solution.

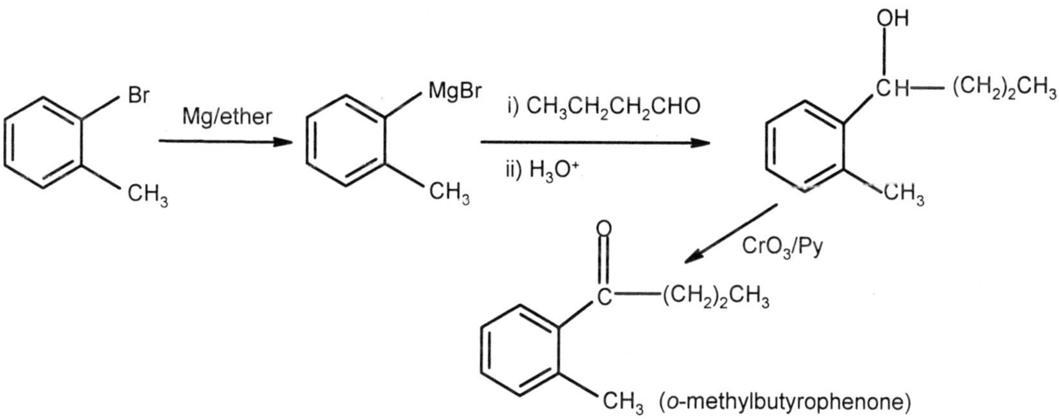

Problem 5.2.2

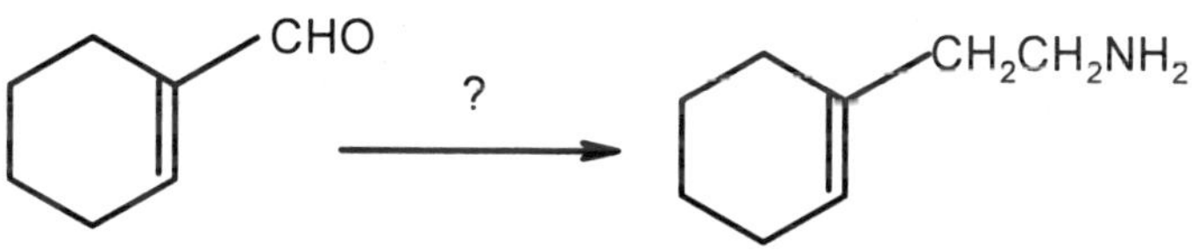

Solution.

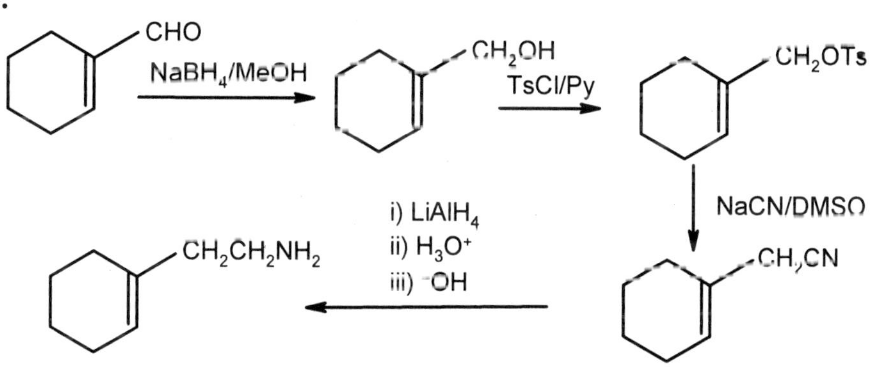

Problem 5.2.3

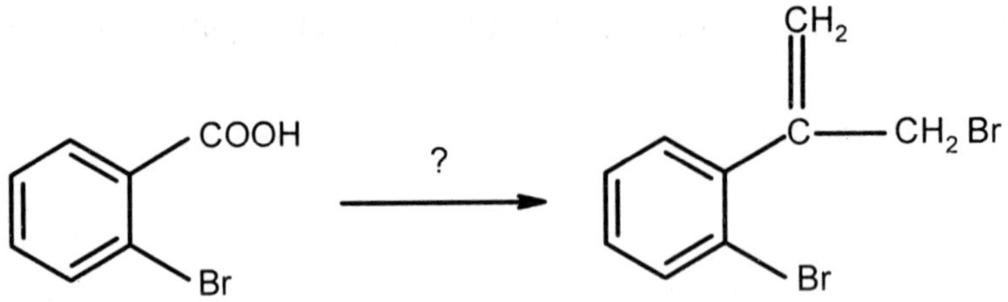

Solution.

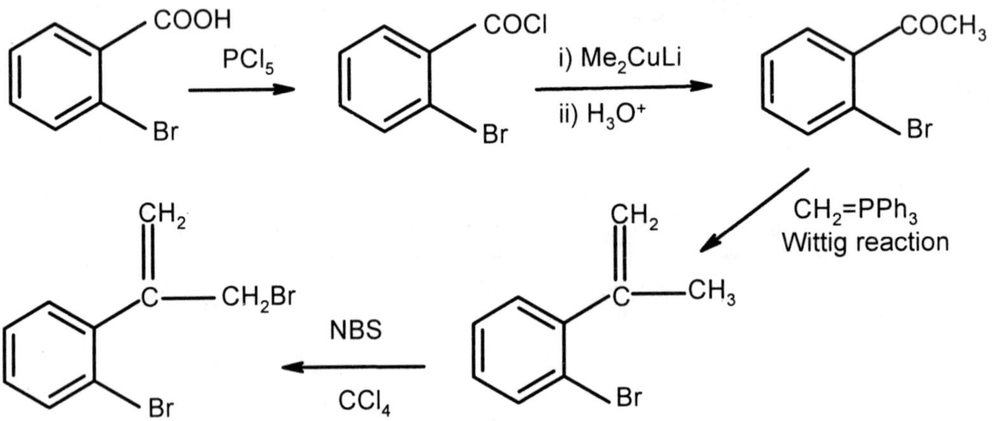

Problem 5.2.4

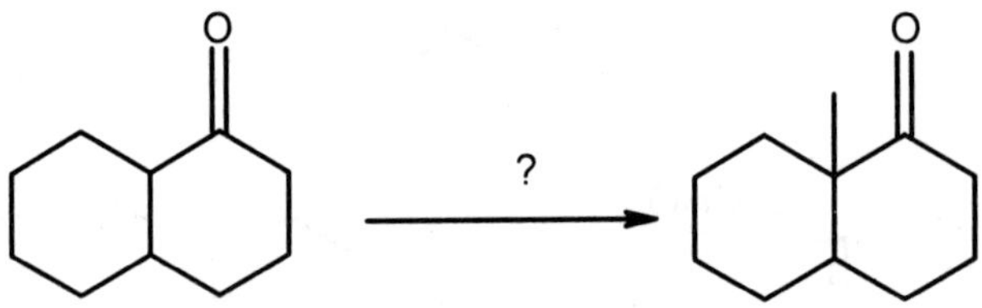

Solution.

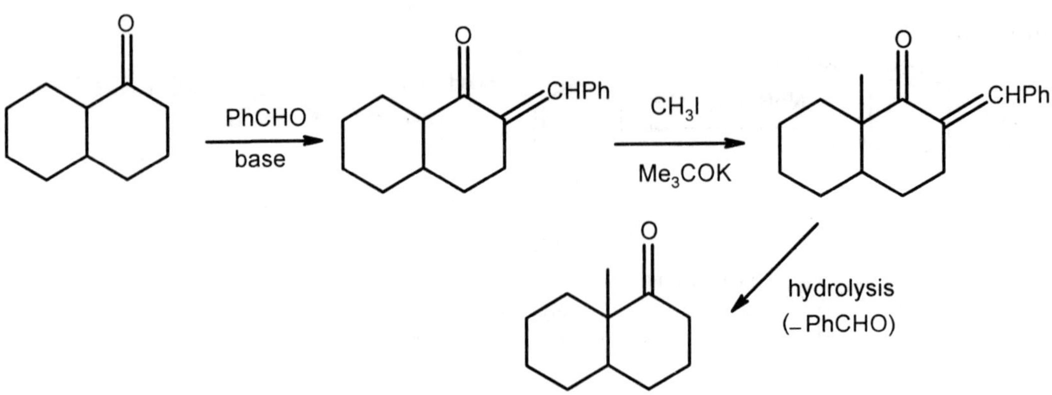

Problem 5.2.5

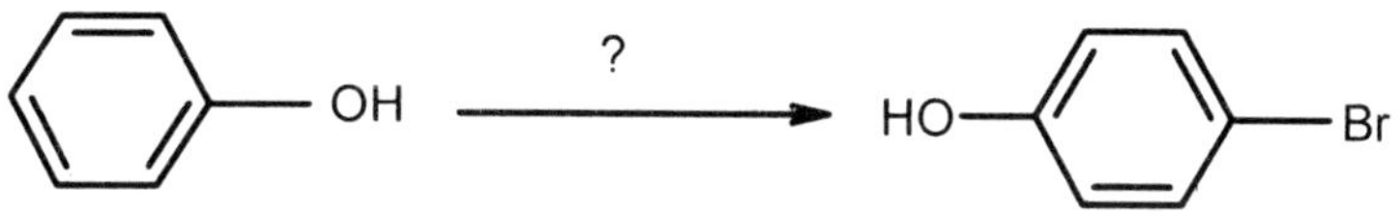

Solution.

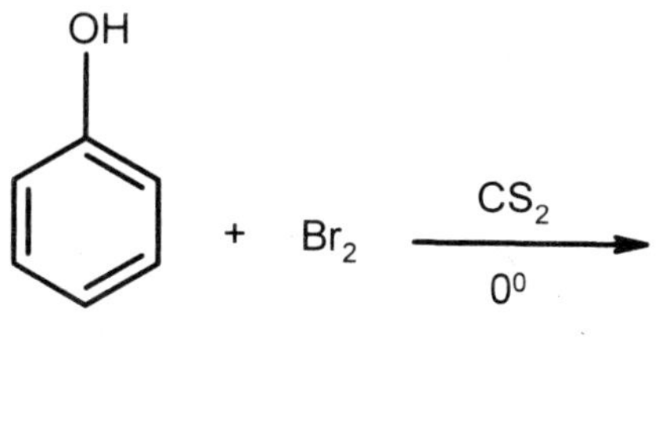

Problem 5.2.6

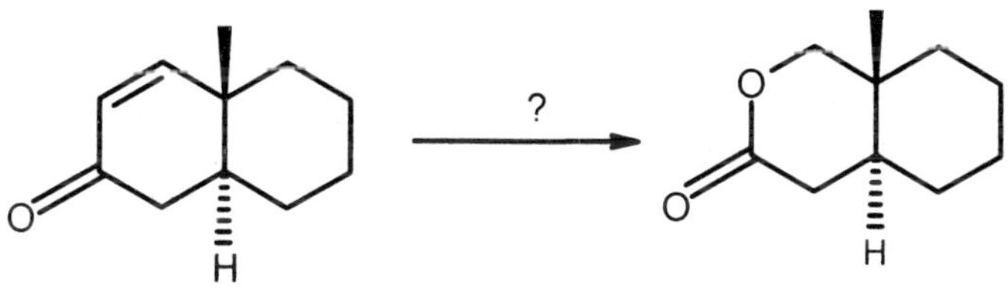

Solution.

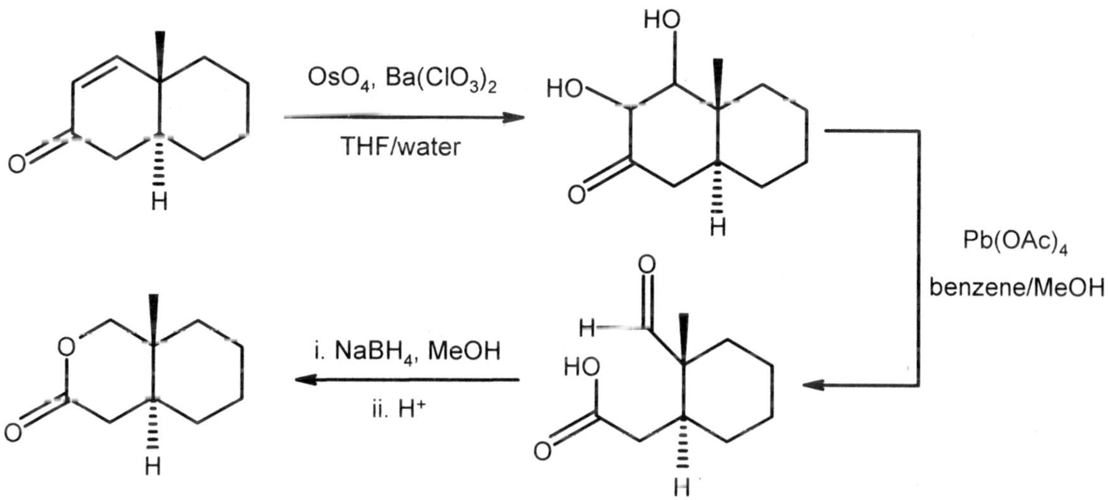

Problem 5.2.7 $CH_2(COOEt)_2$

Solution.

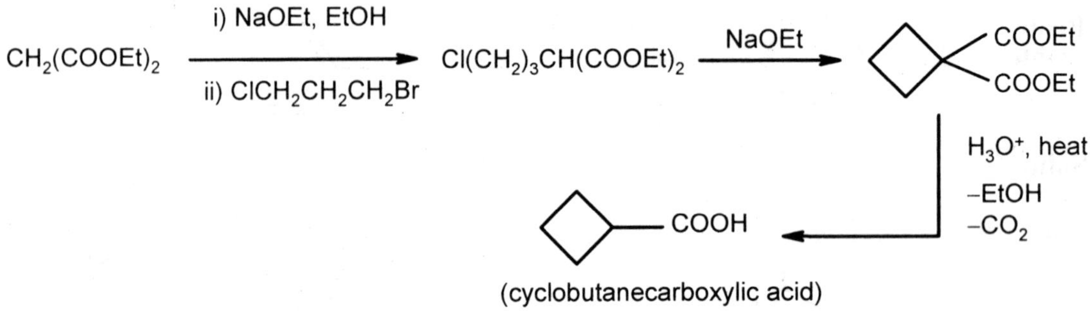

Problem 5.2.8

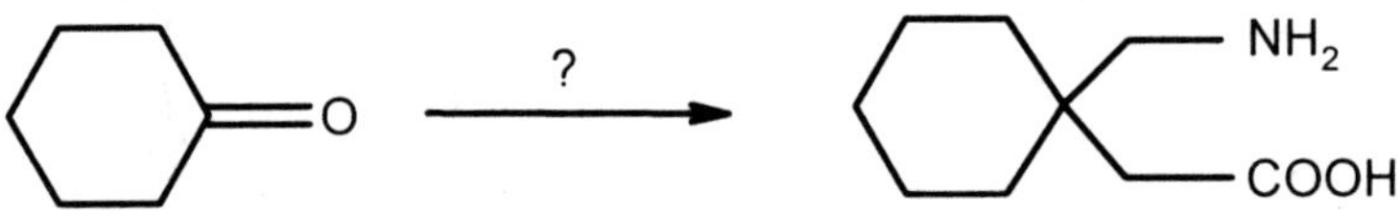

Solution.

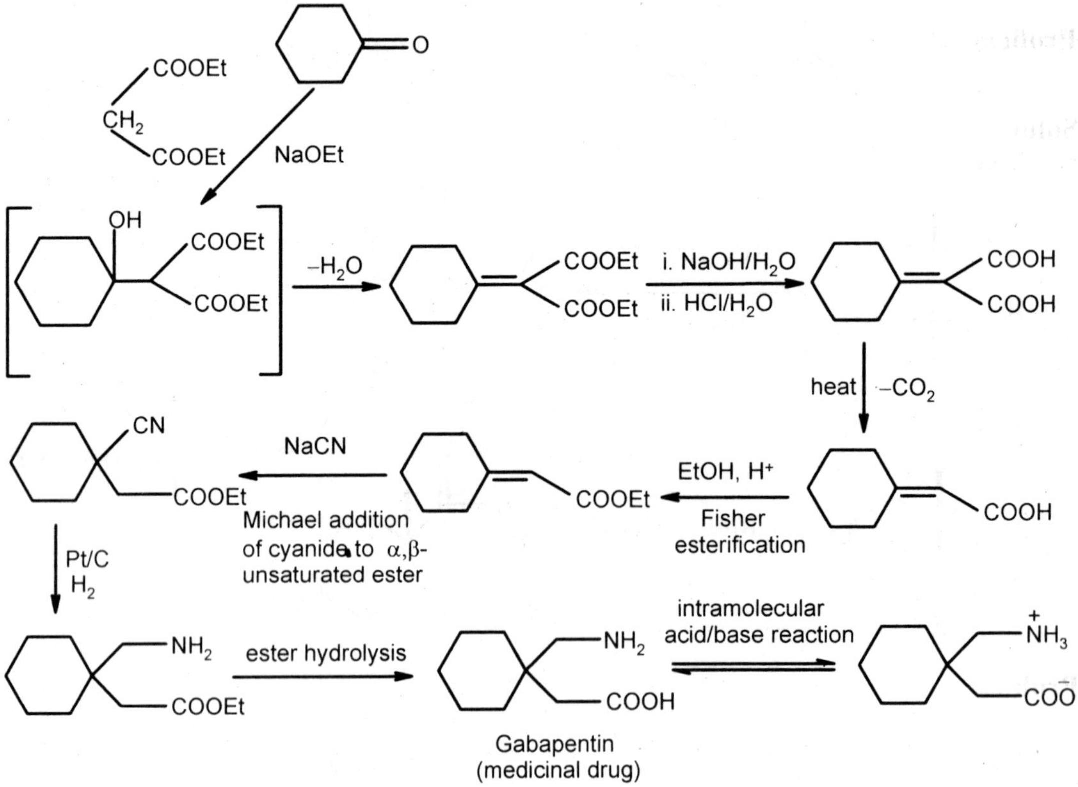

Problem 5.2.9

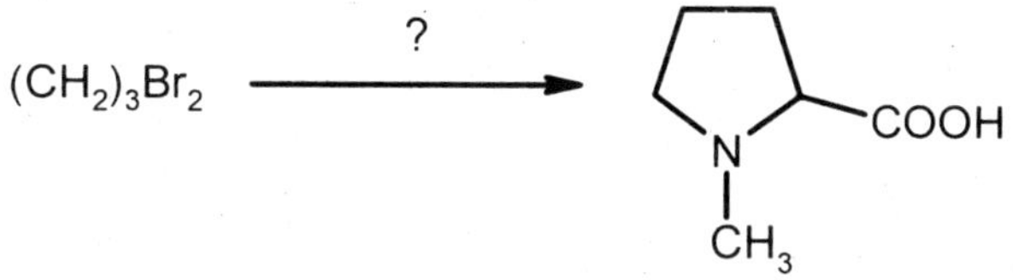

Solution.

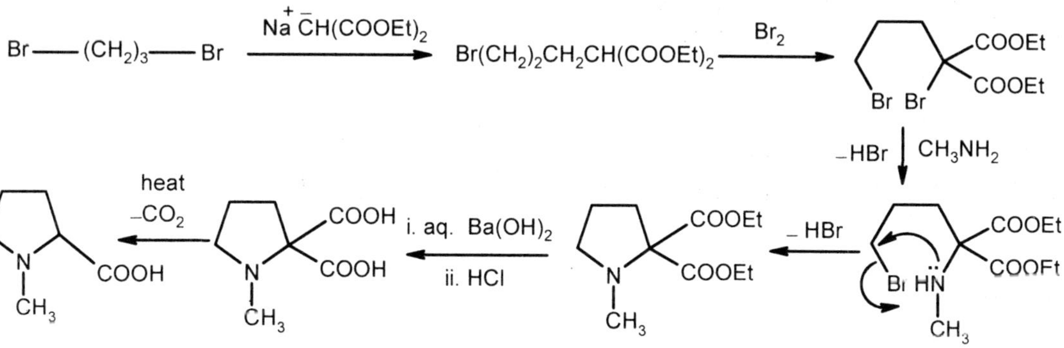

Problem 5.2.10

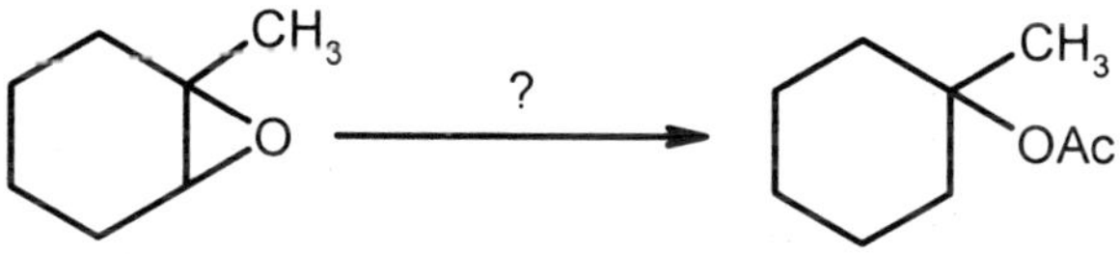

Solution.

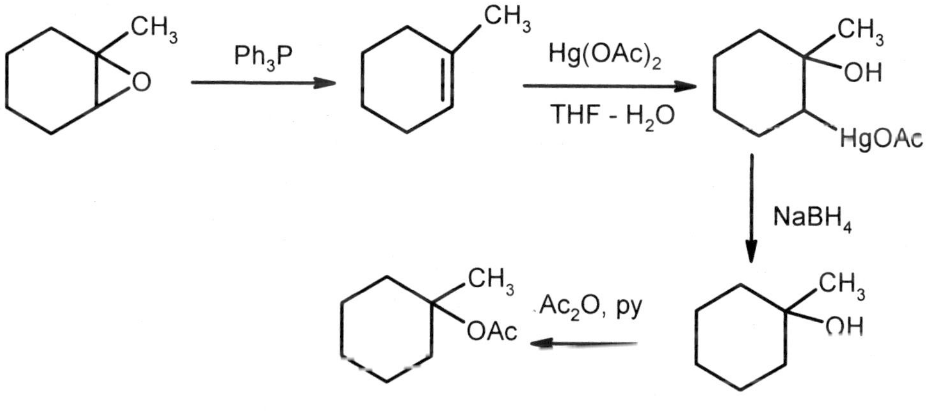

Problem 5.2.11

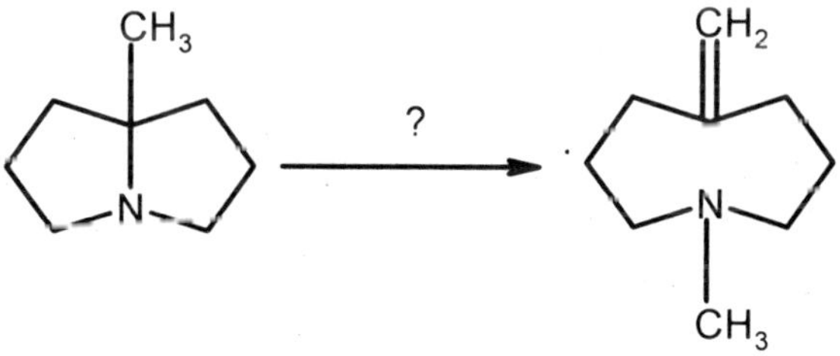

Solution.

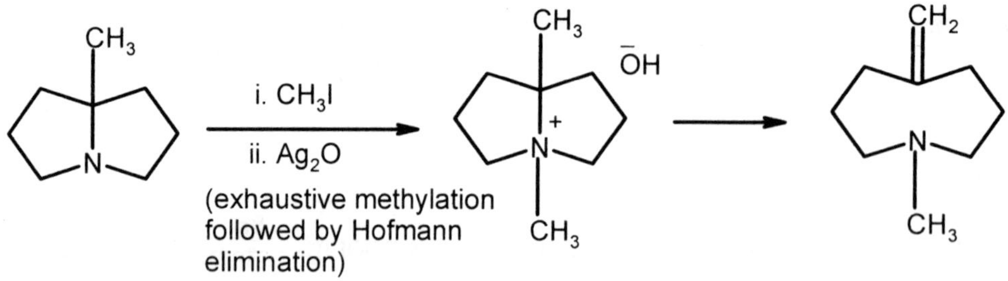

Problem 5.2.12

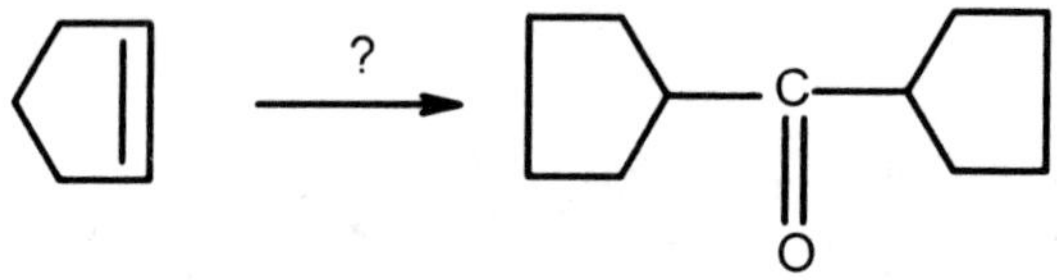

Solution.

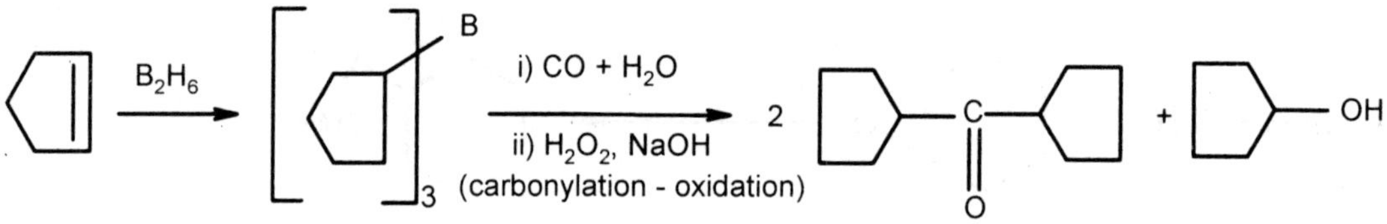

Problem 5.2.13

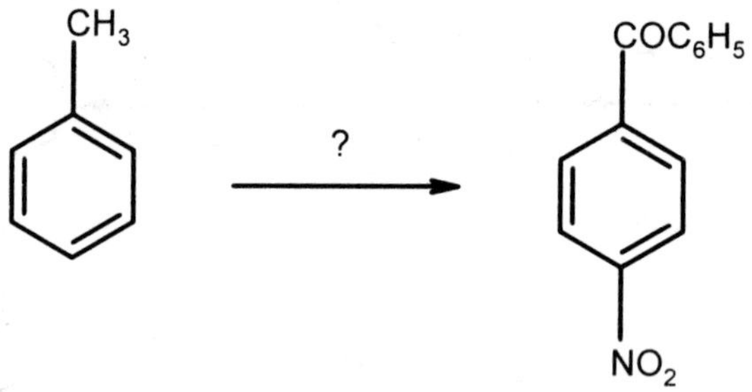

Solution.

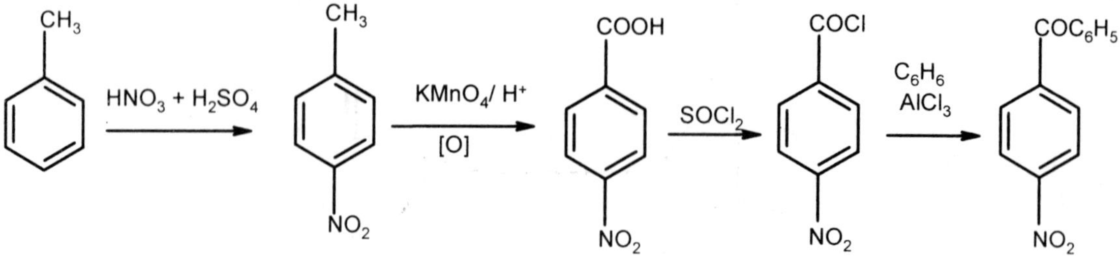

Problem 5.2.14

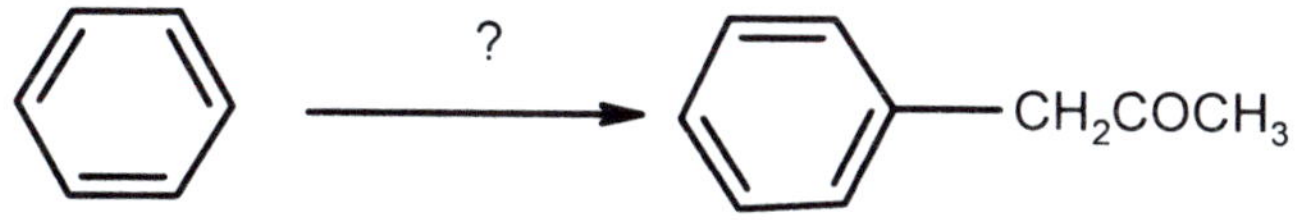

Solution.

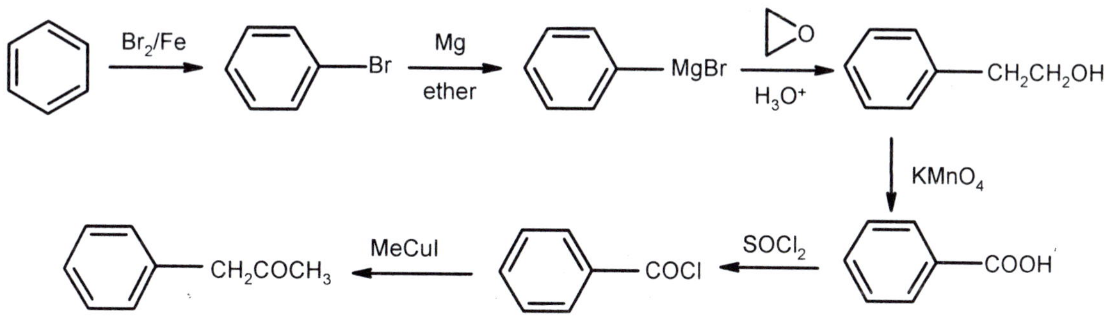

Problem 5.2.15

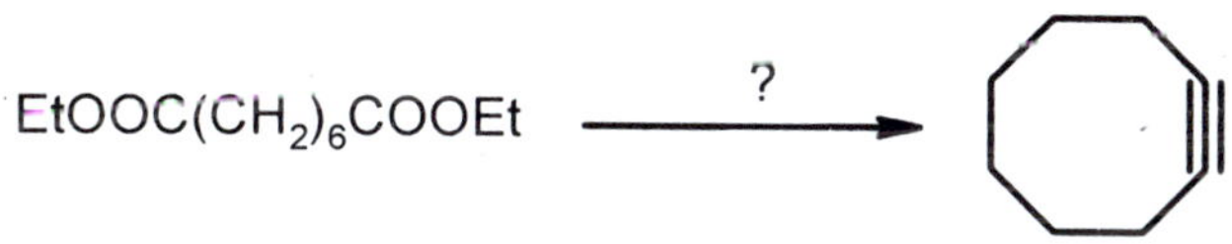

Solution.

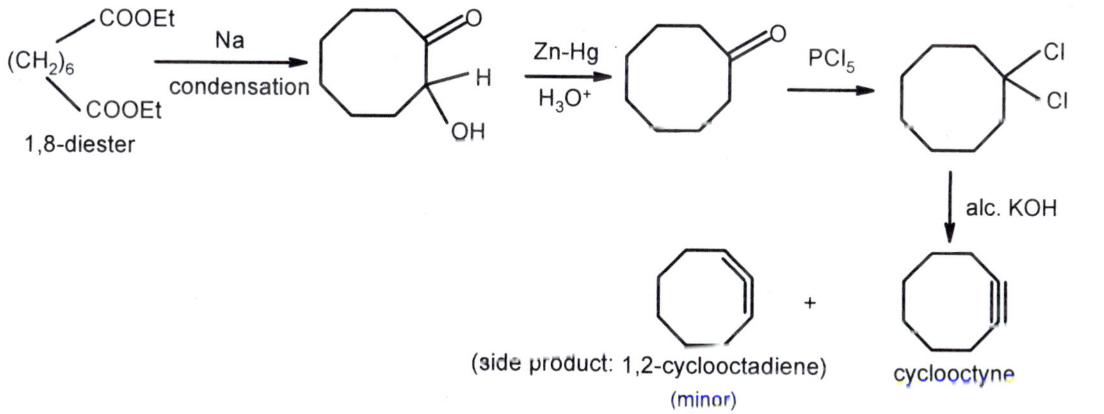

Problem 5.2.16

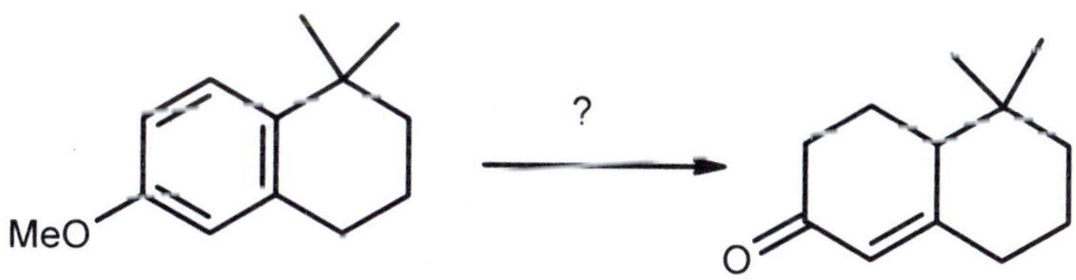

Solution.

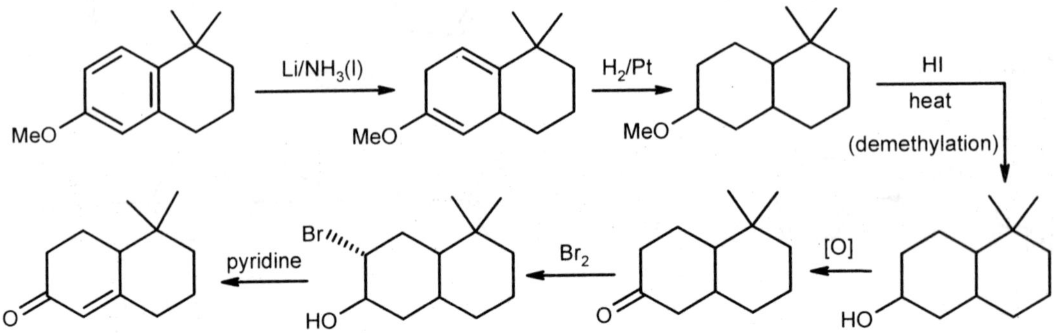

Problem 5.2.17

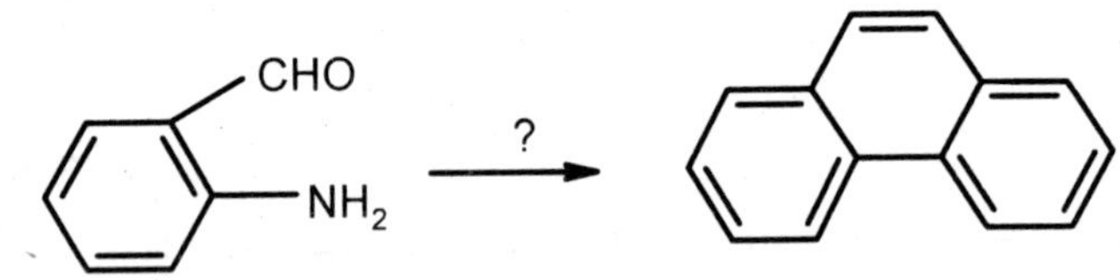

Solution.

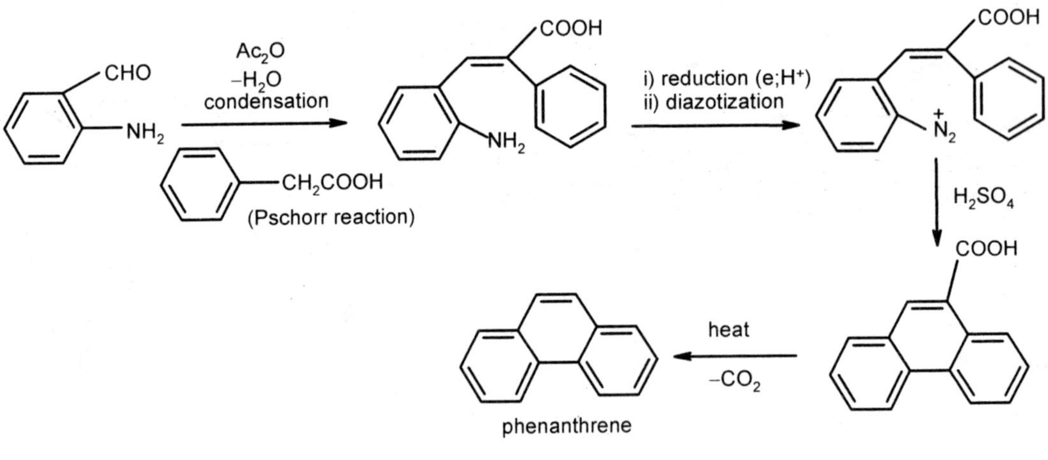

Problem 5.2.18

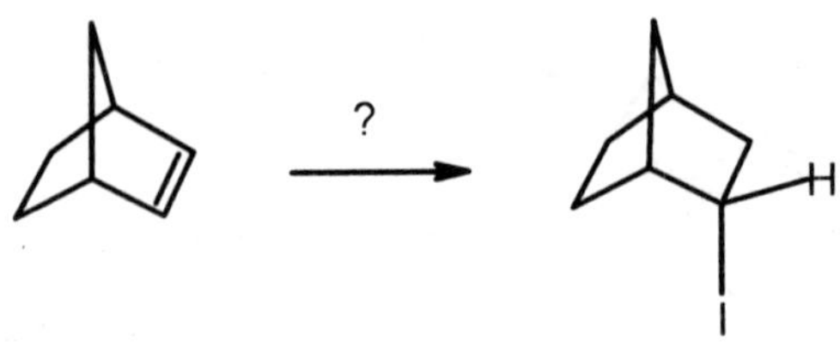

Solution.

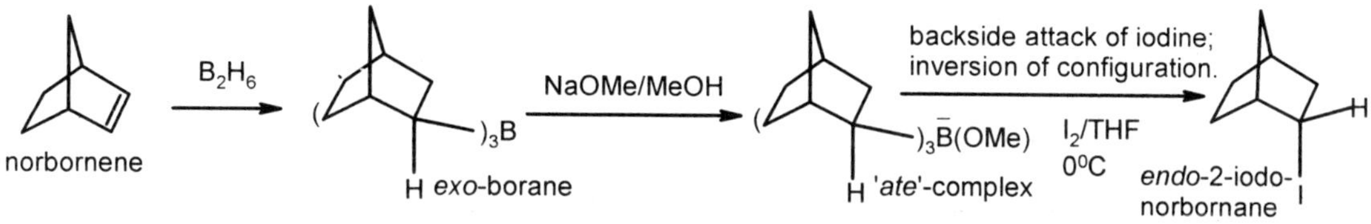

Problem 5.2.19

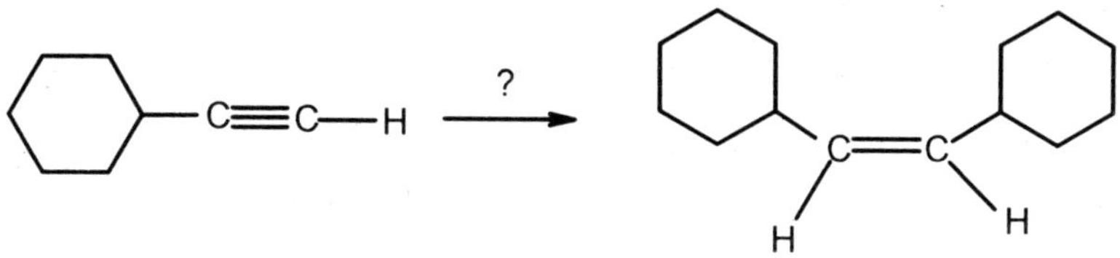

Solution.

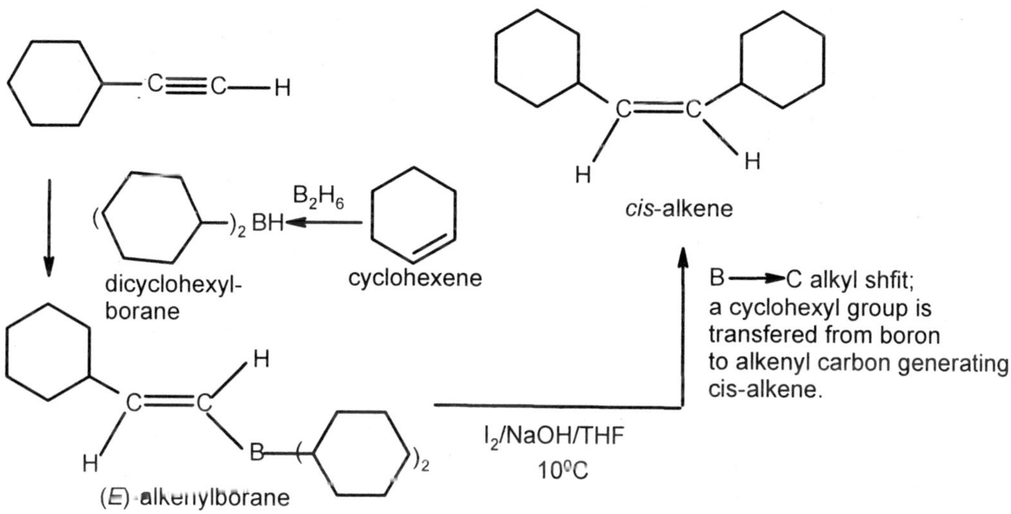

Problem 5.2.20

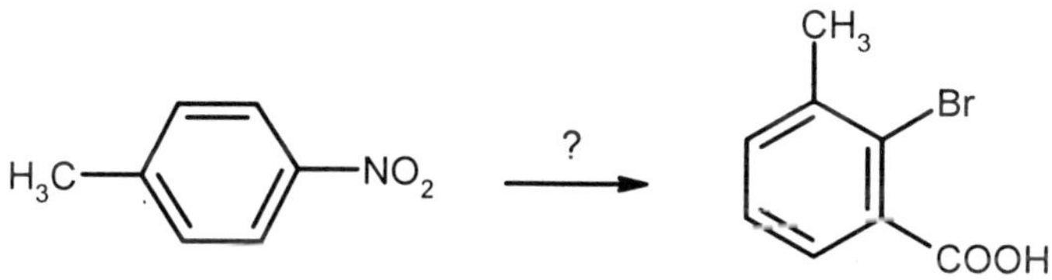

Solution.

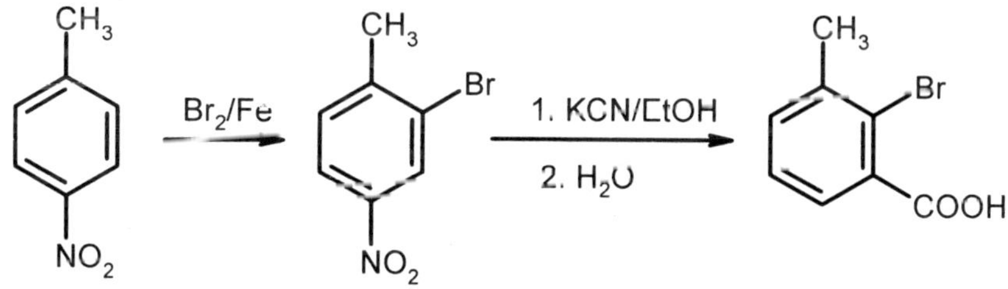

Problem 5.2.21

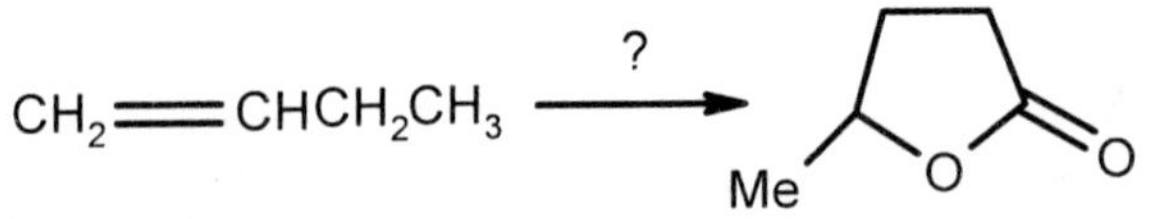

Solution.

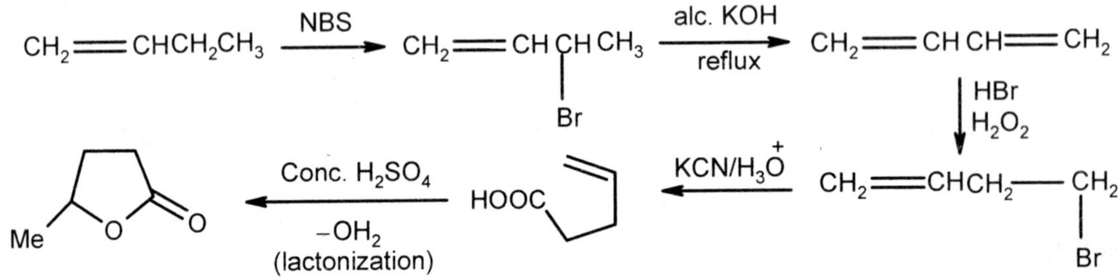

Problem 5.2.22

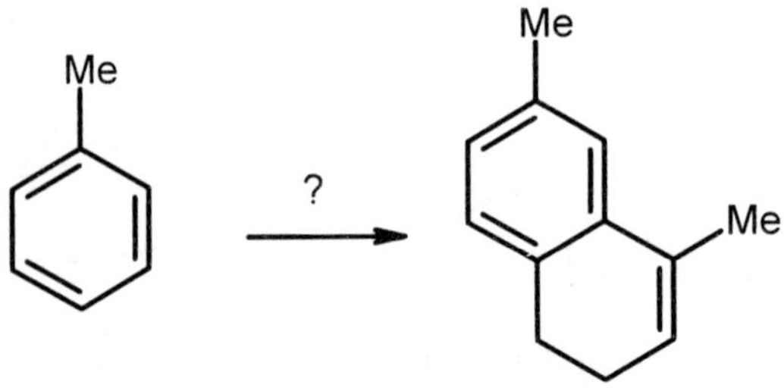

Solution.

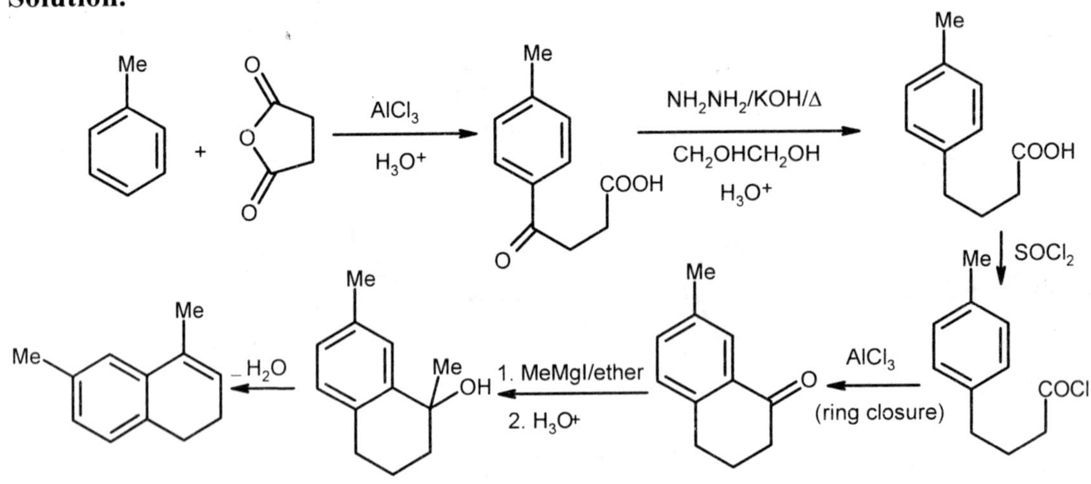

Problem 5.2.23 $CH_3CH_2C \equiv CH \xrightarrow{?} CH_3CH_2CH_2CHO$

Solution.

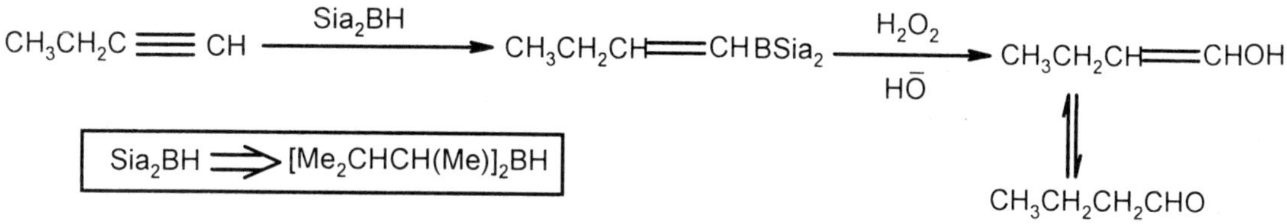

Problem 5.2.24

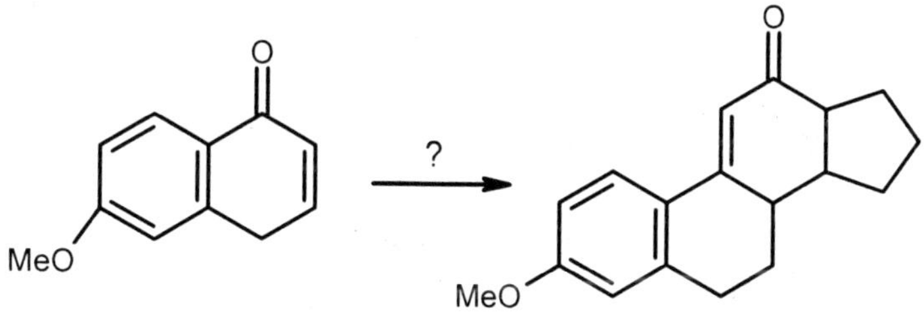

Solution.

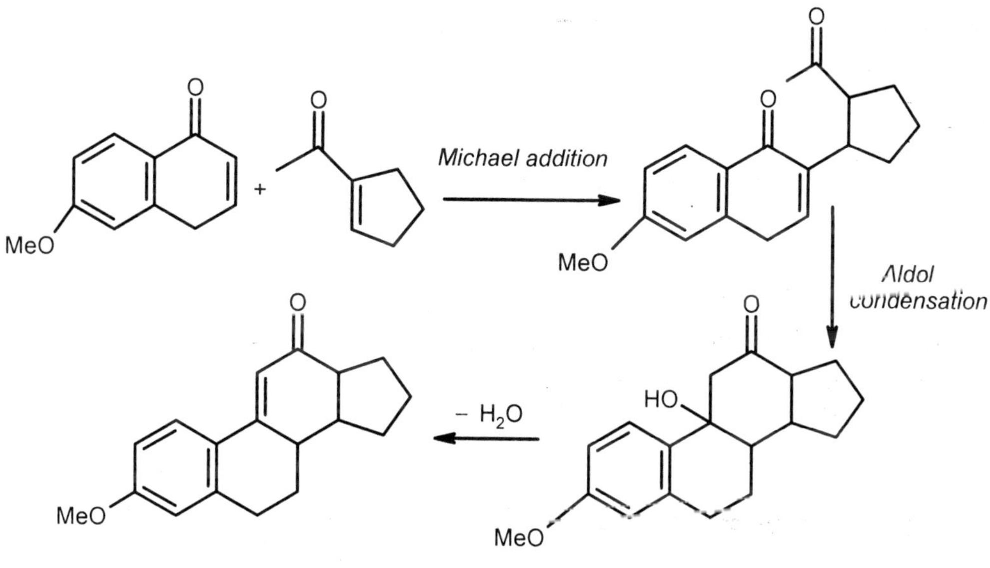

Problem 5.2.25

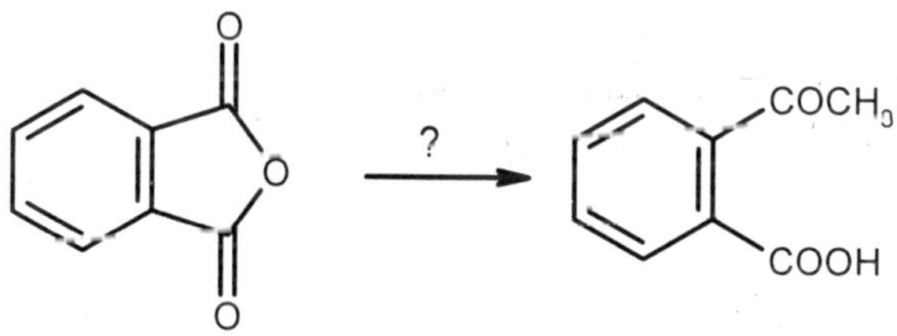

Solution.

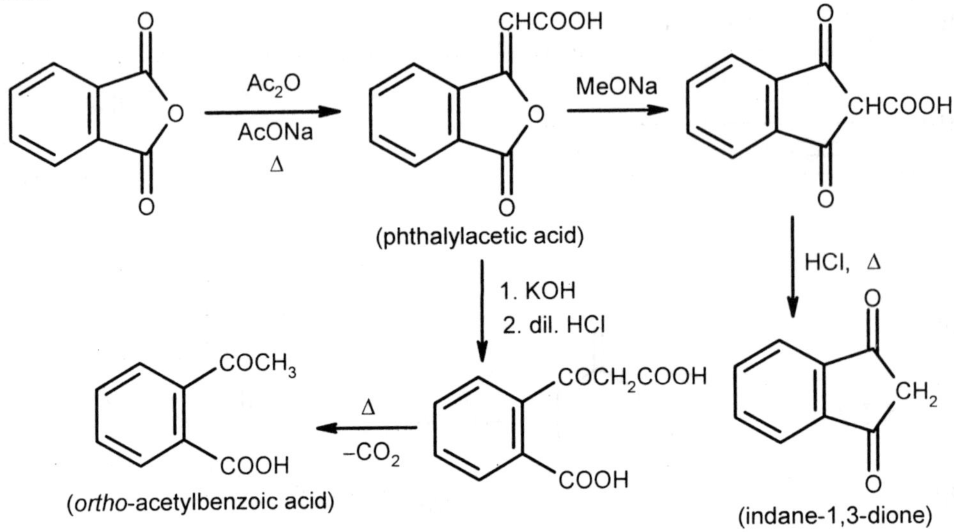

Problem 5.2.26 $Me_2CHCH_2OH \xrightarrow{?}$

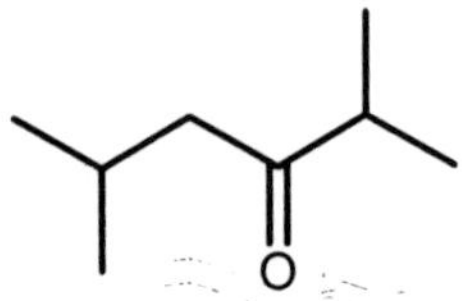

Solution.

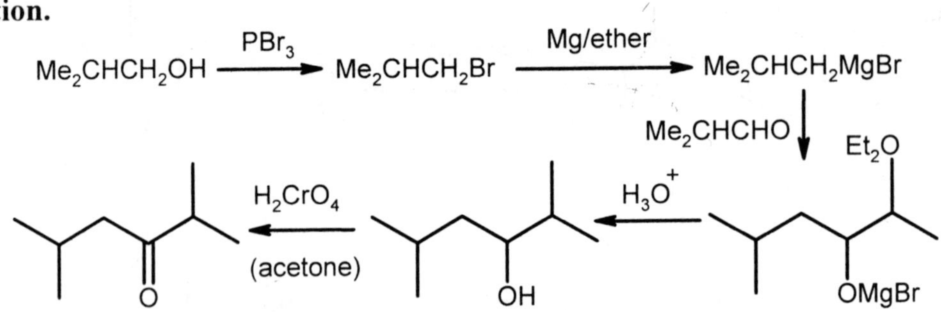

Problem 5.2.27 $CH_3CH{=}CHCH_2CH_2CH_2CN \xrightarrow{?} CH_3CH{=}CHCH_2CH_2CH_2CHO$

Solution.

$$CH_3CH{=}CHCH_2CH_2CH_2CN \xrightarrow[\text{(hexane)}]{(^{i}Bu)_2AlH} CH_3CH{=}CHCH_2CH_2CH_2CH{=}NAl(^{i}Bu)_2$$

$$\xrightarrow[\text{(hydrolysis)}]{H_3O^+} CH_3CH{=}CHCH_2CH_2CH_2CHO$$

Problem 5.2.28 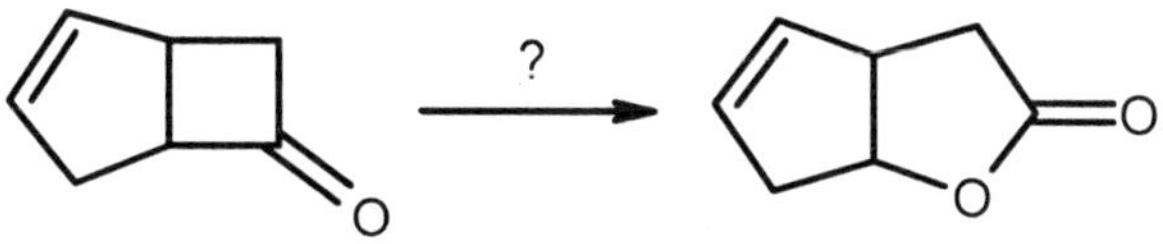

Solution.

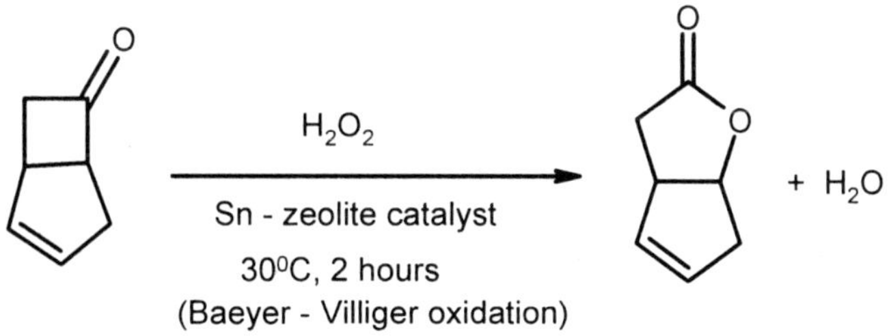

This method is considered to be eco-friendly because oxidation with hydrogen peroxide produces only water as the by-product.

Problem 5.2.29 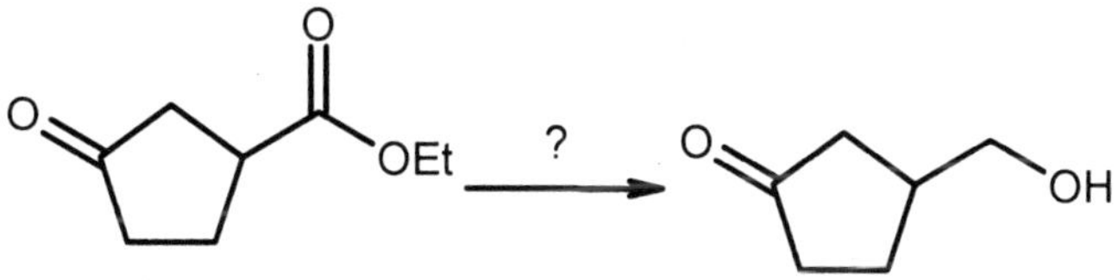

Solution.

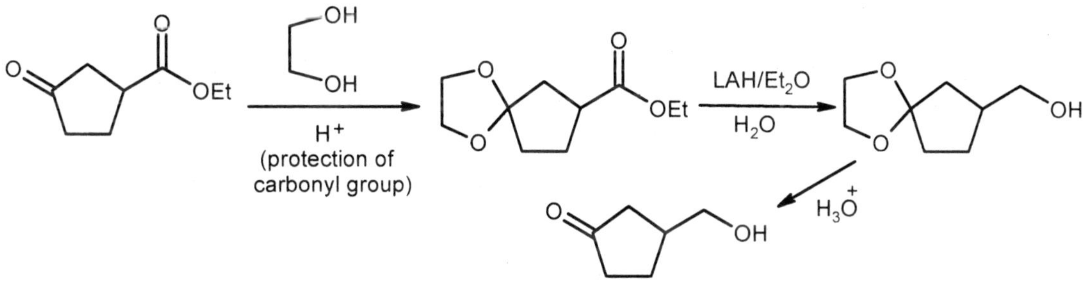

Problem 5.2.30 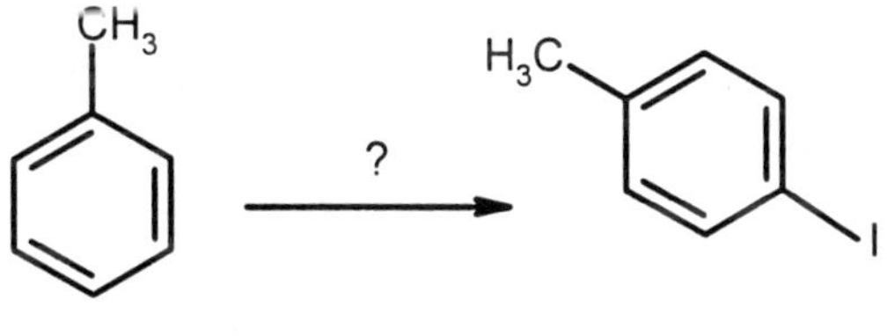

Solution.

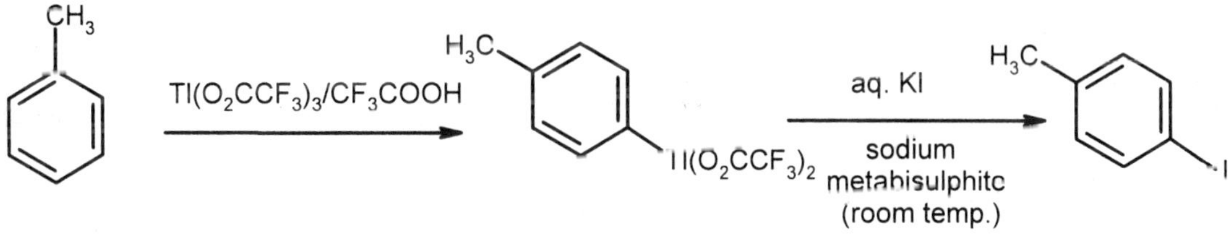

Problem 5.2.31

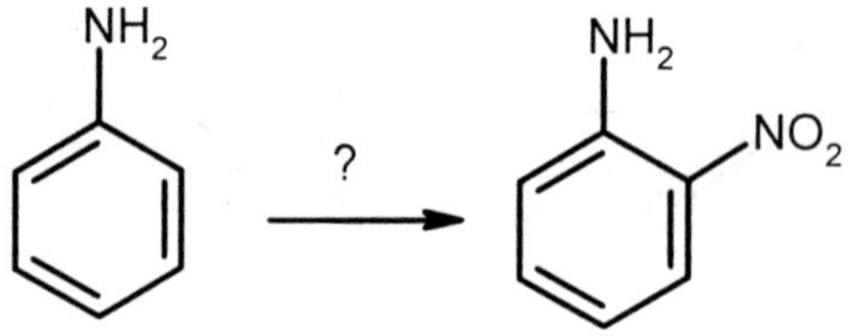

Solution.

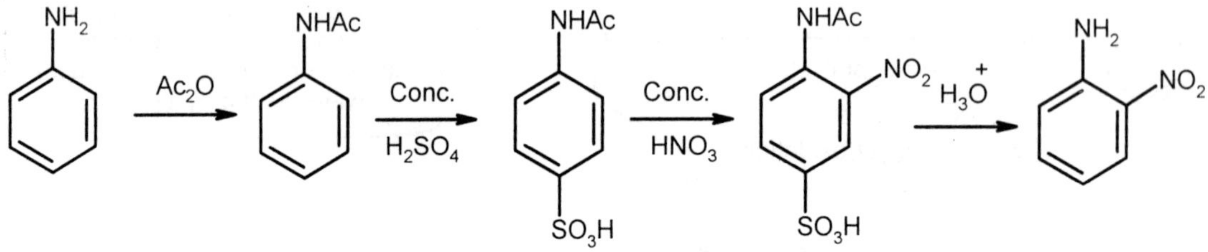

Problem 5.2.32

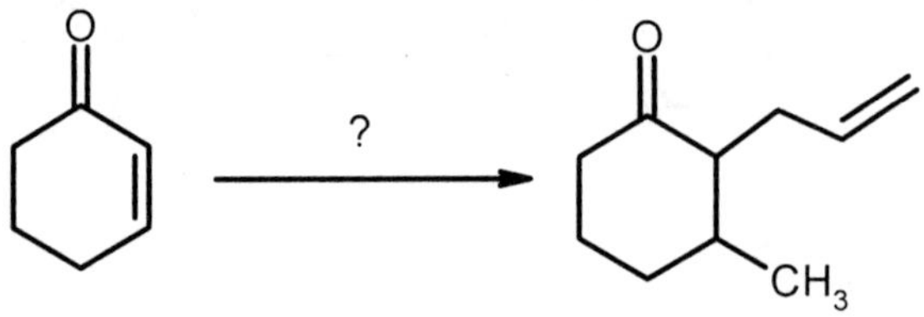

Solution.

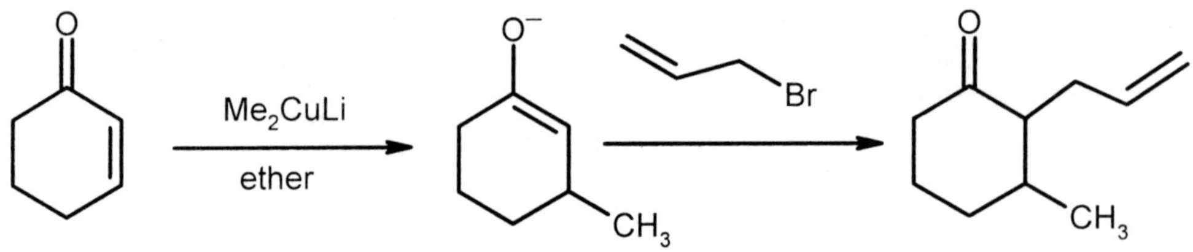

Problem 5.2.33

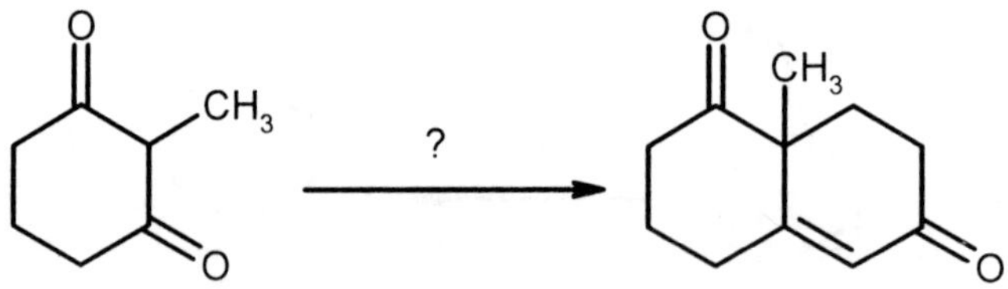

Solution.

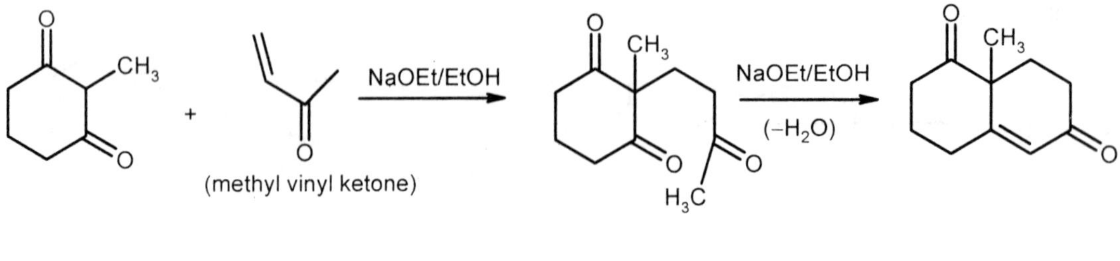

Problem 5.2.34

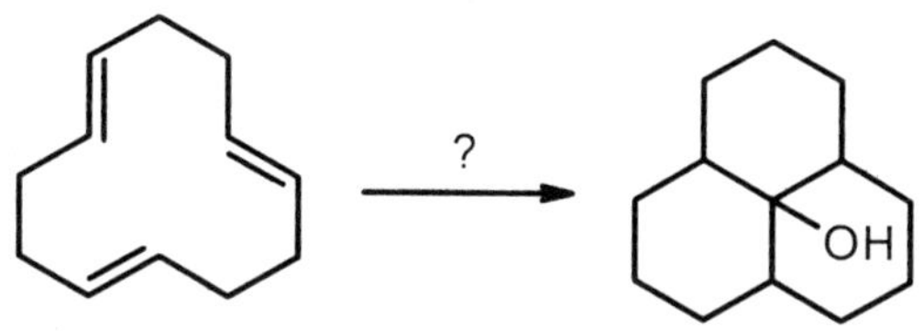

Solution.

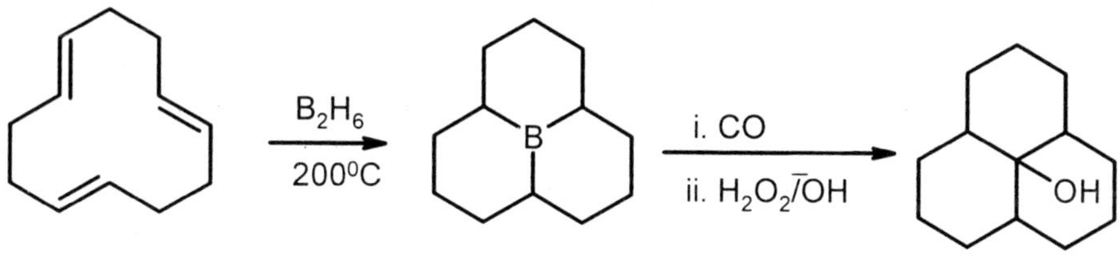

Problem 5.2.35 $CH_3COCH_2COOEt \xrightarrow{?} CH_3COCH_2CH_2COOH$

Solution.

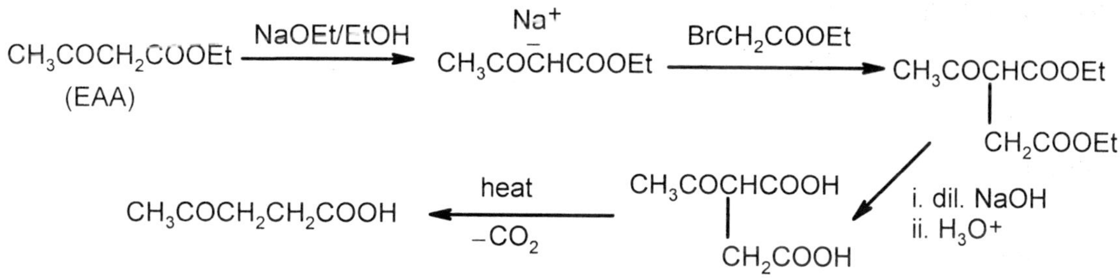

Problem 5.2.36 $H_2C(COOEt)_2 \xrightarrow{?} HOOCCH_2CH_2CH_2COOH$

Solution.

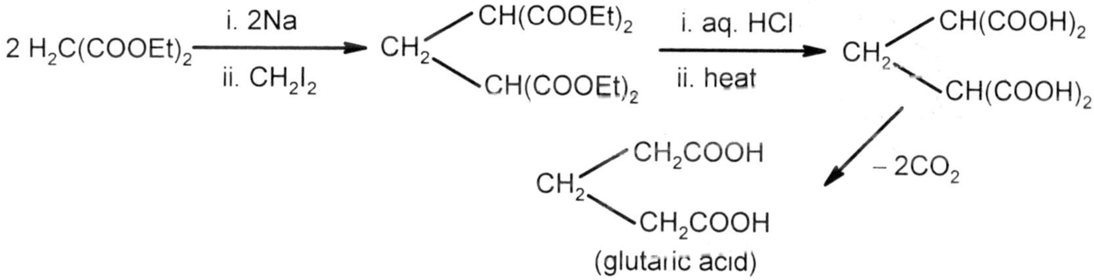

Problem 5.2.37

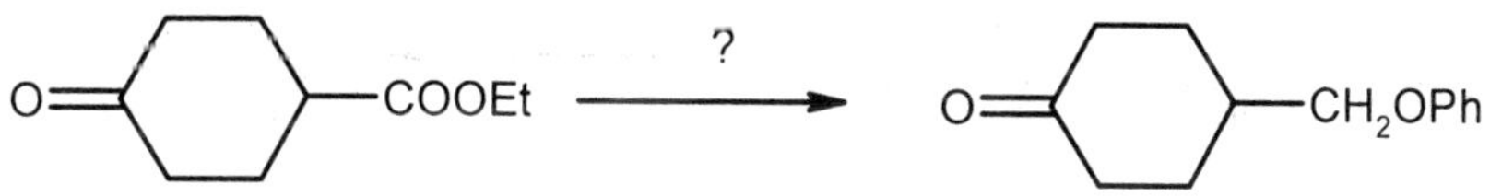

Solution.

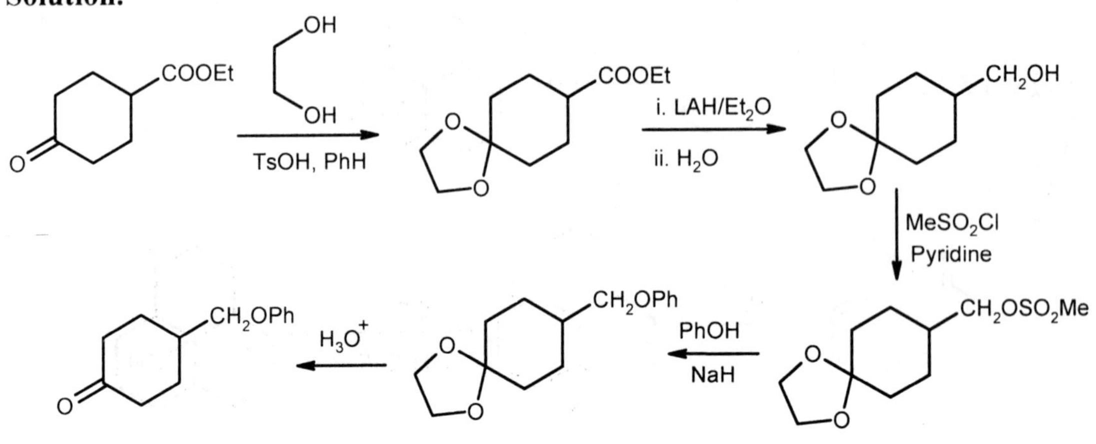

Problem 5.2.38

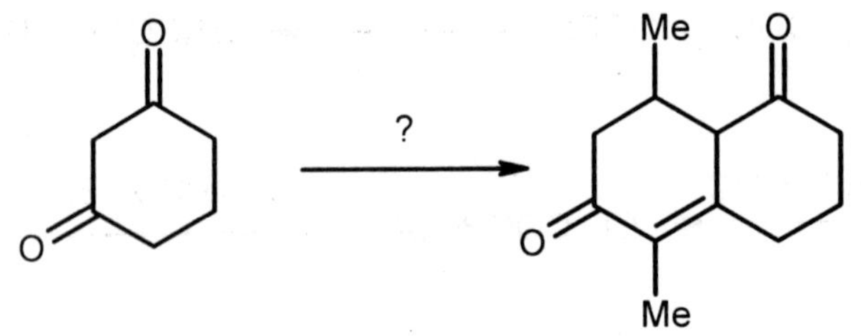

Solution.

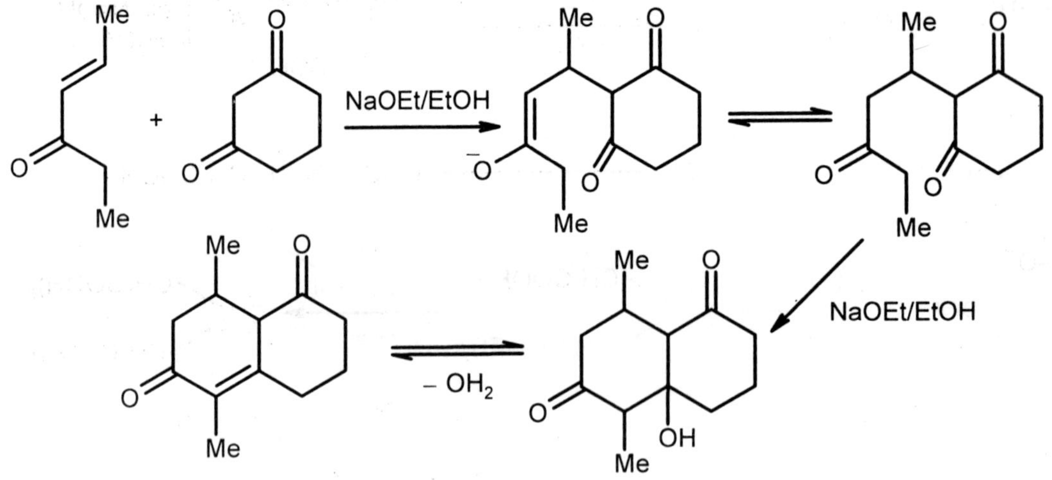

Problem 5.2.39

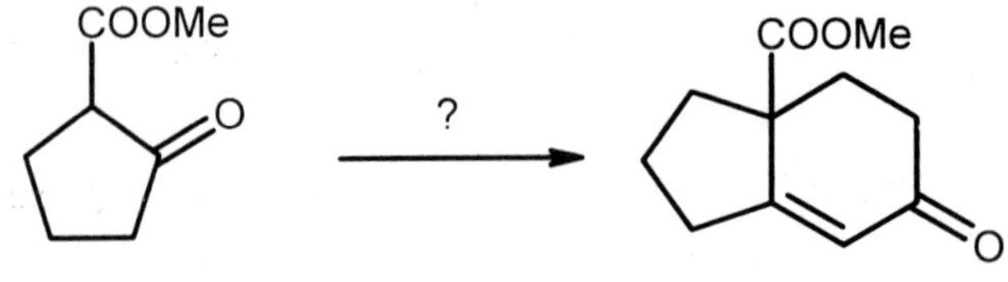

Solution.

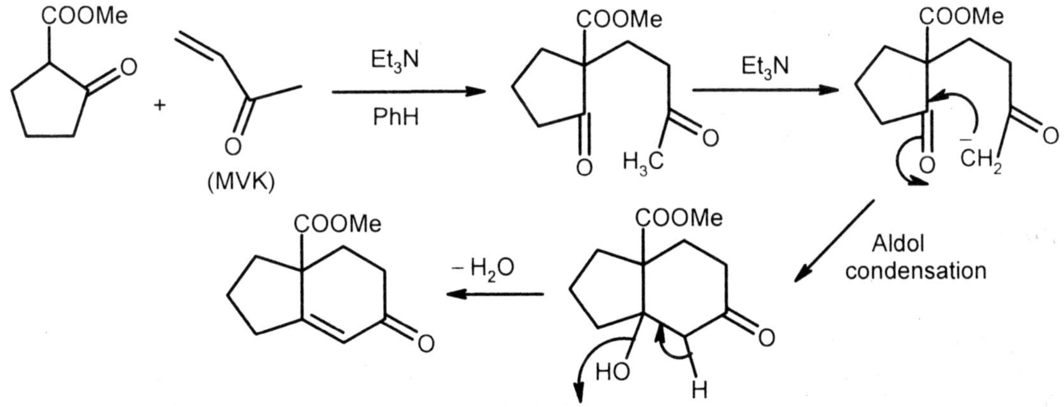

Problem 5.2.40

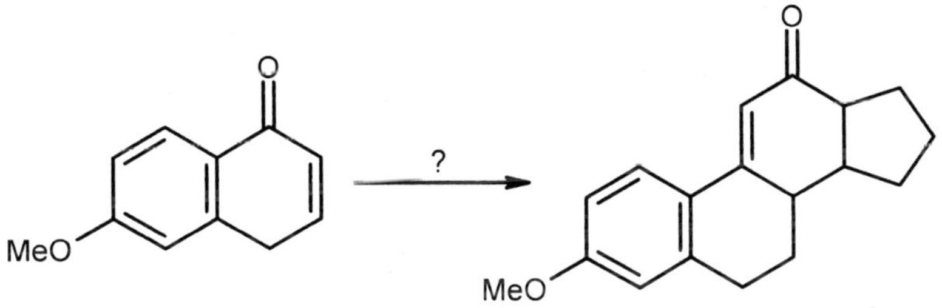

Solution.

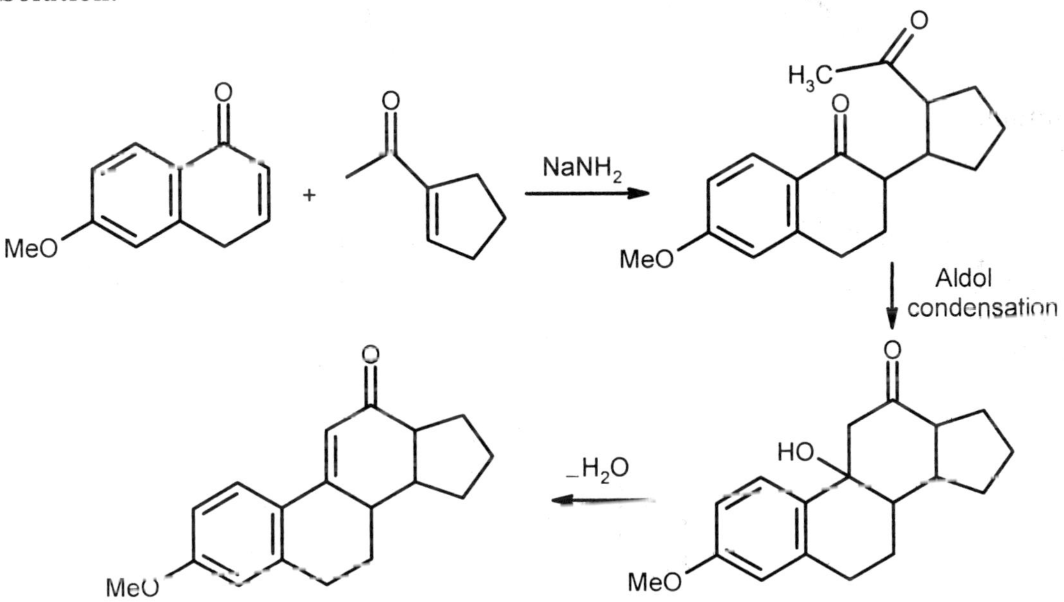

Problem 5.2.41

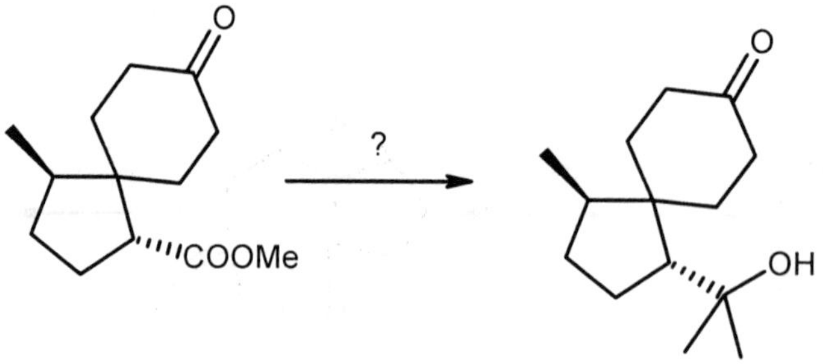

Solution.

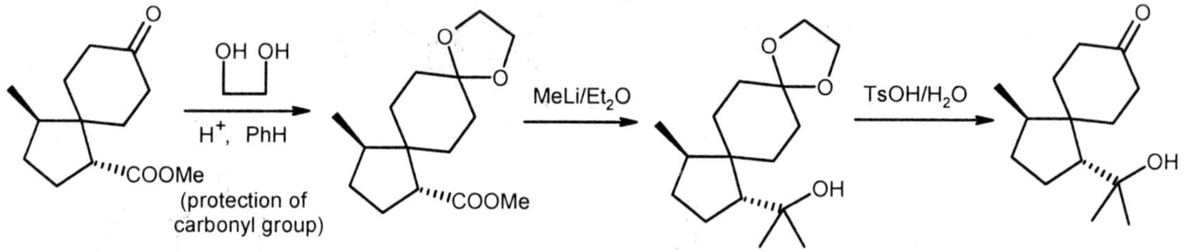

Problem 5.2.42

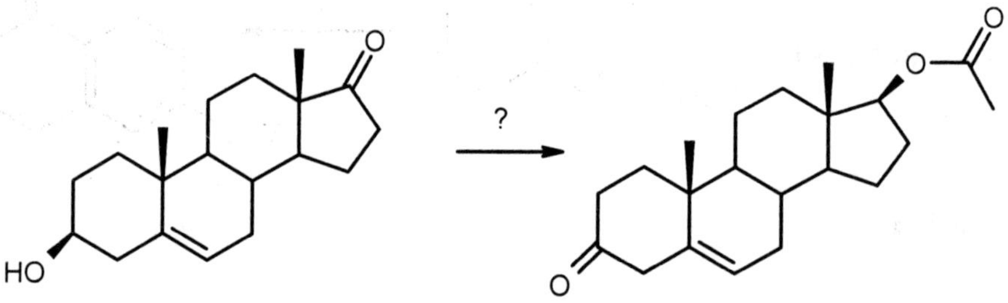

Solution.

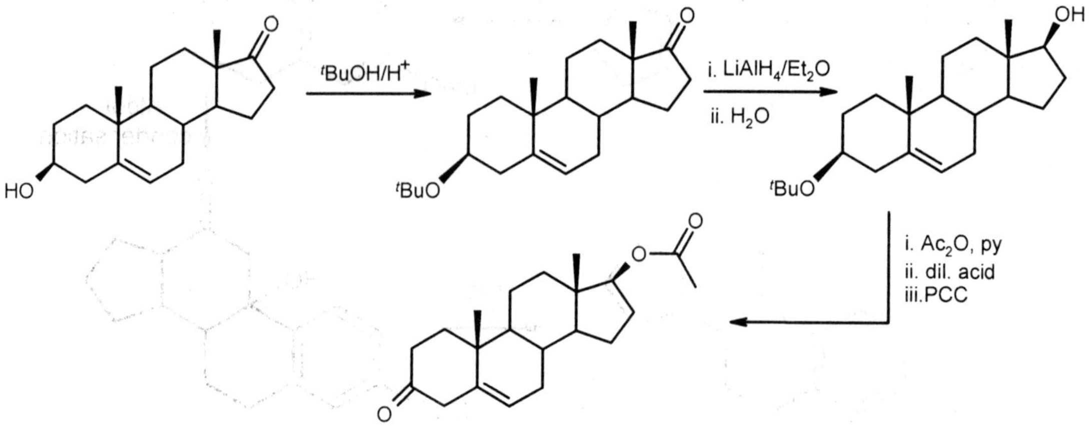

Problem 5.2.43

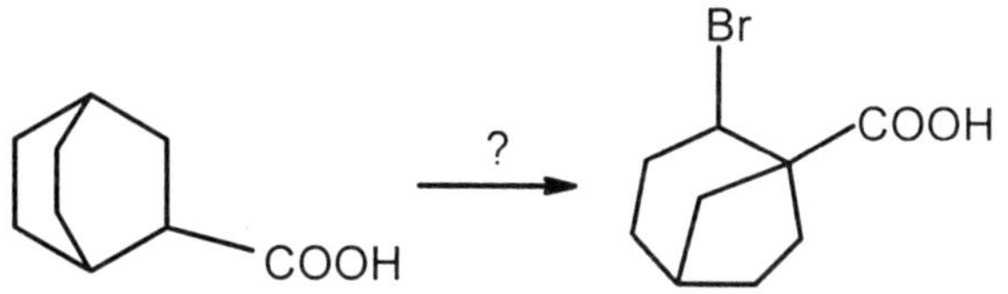

Solution.

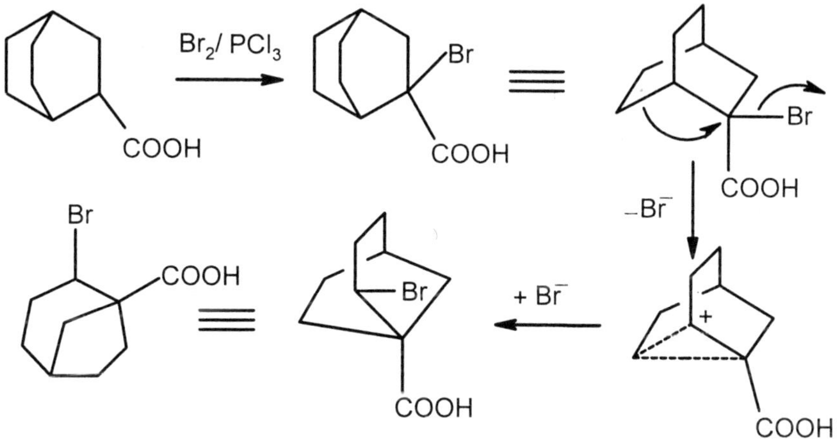

Problem 5.2.44

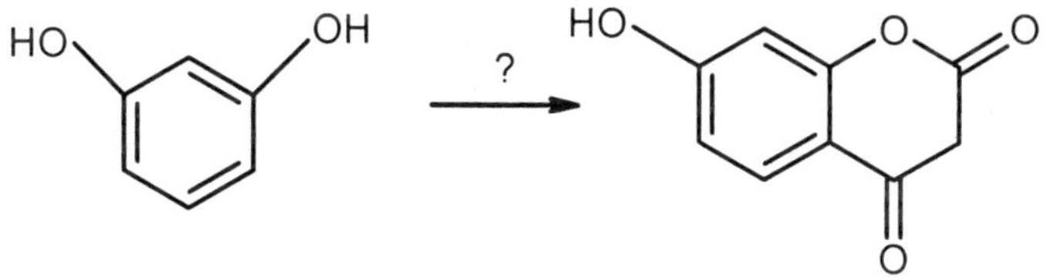

Solution.

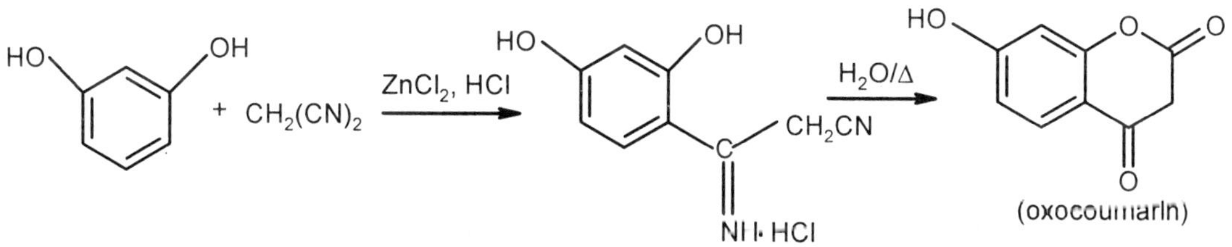

Problem 5.2.45

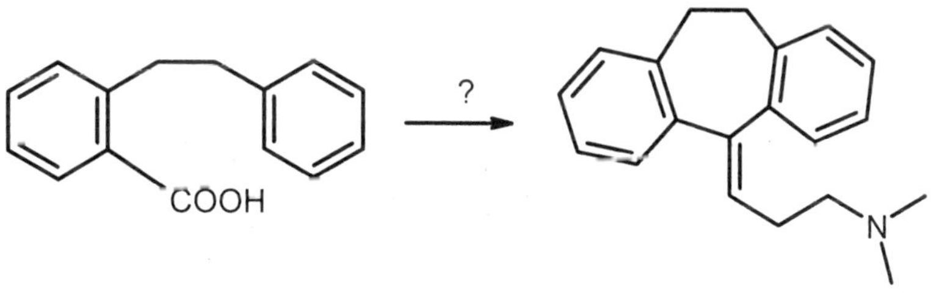

Solution.

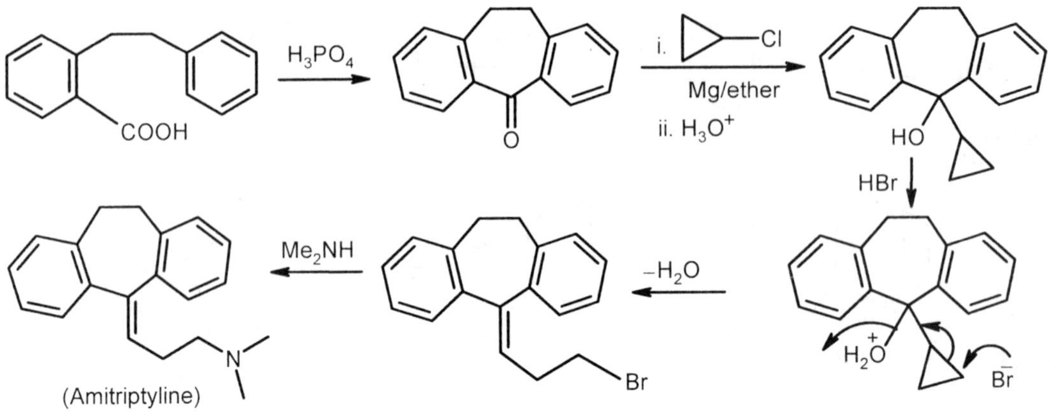

Problem 5.2.46

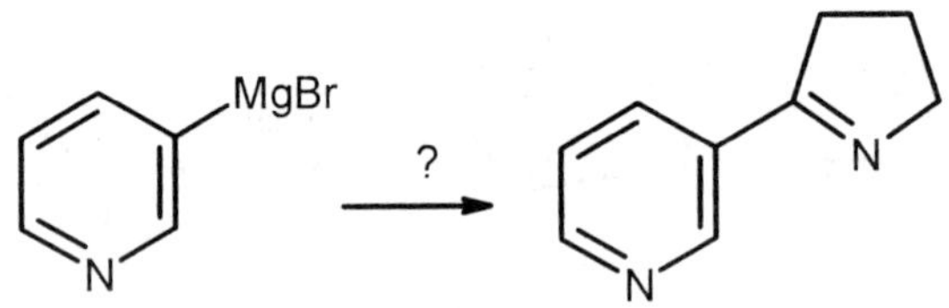

Solution.

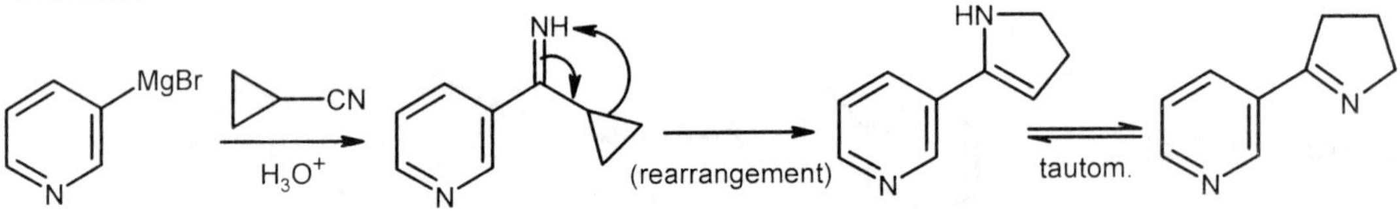

Problem 5.2.47

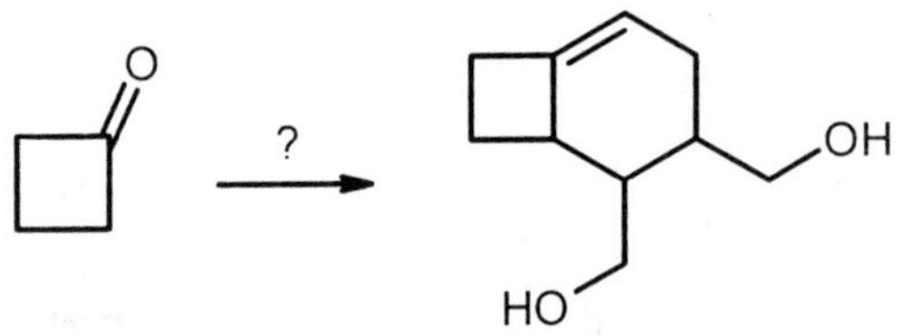

Solution.

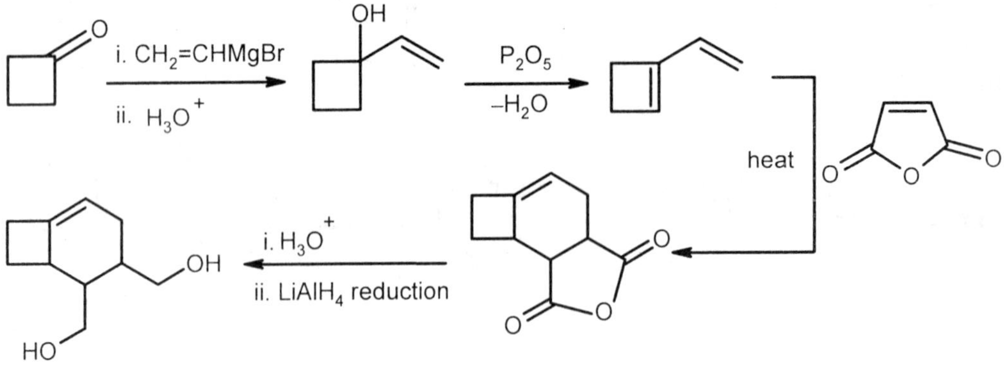

Problem 5.3 Complete the reaction sequence in each of the following cases:

Problem 5.3.1

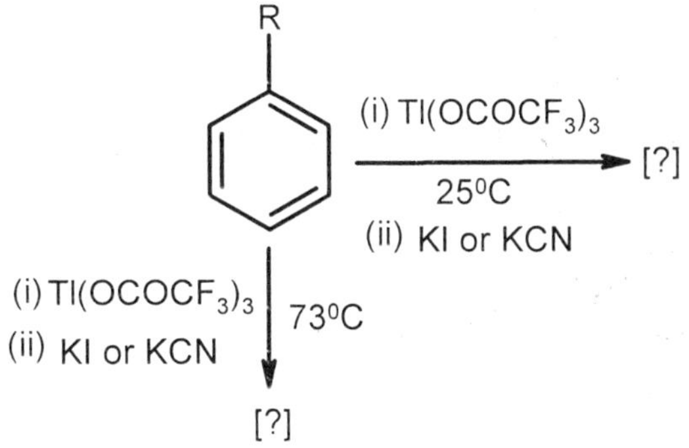

Solution.

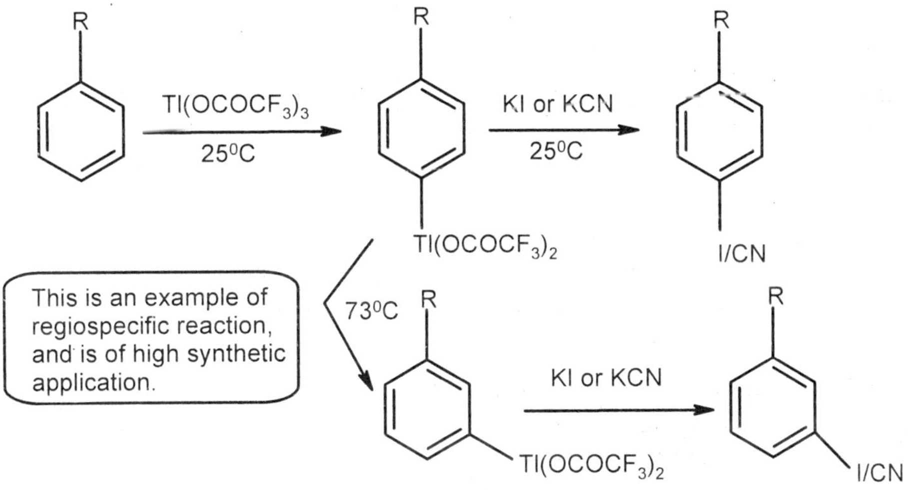

Problem 5.3.2

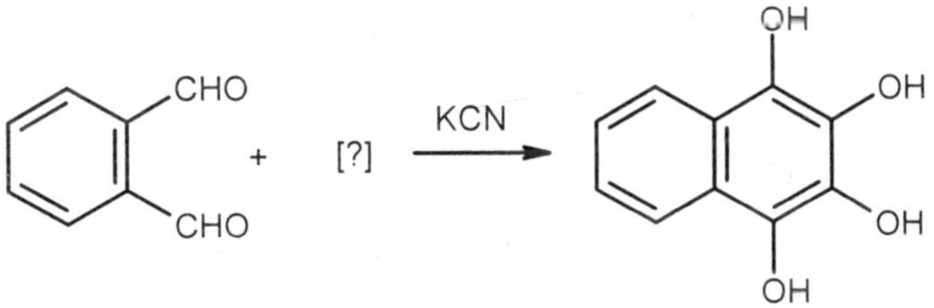

Solution.

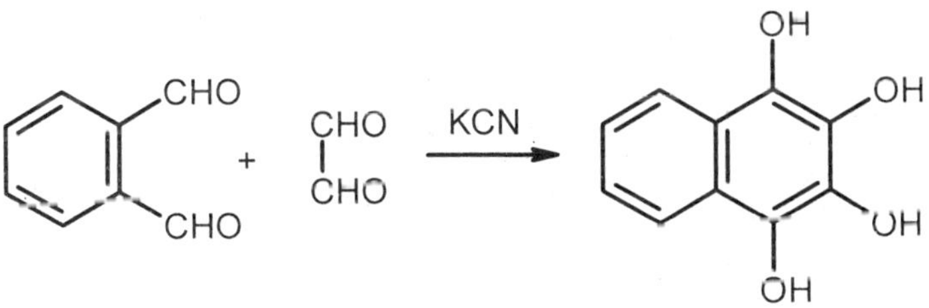

Problem 5.3.3

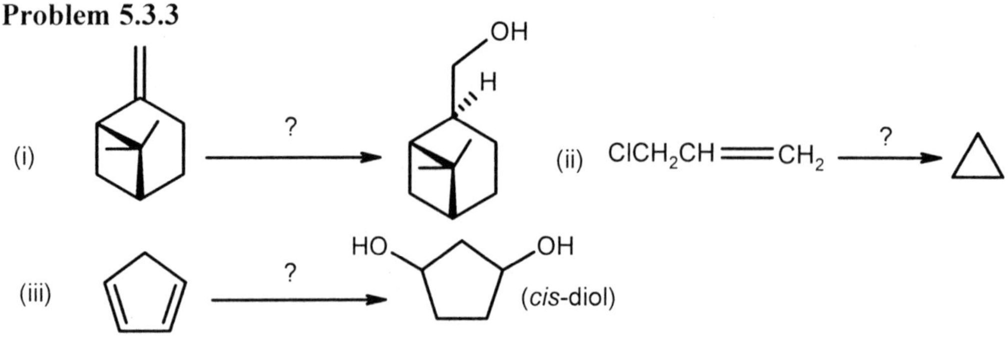

Solutions.

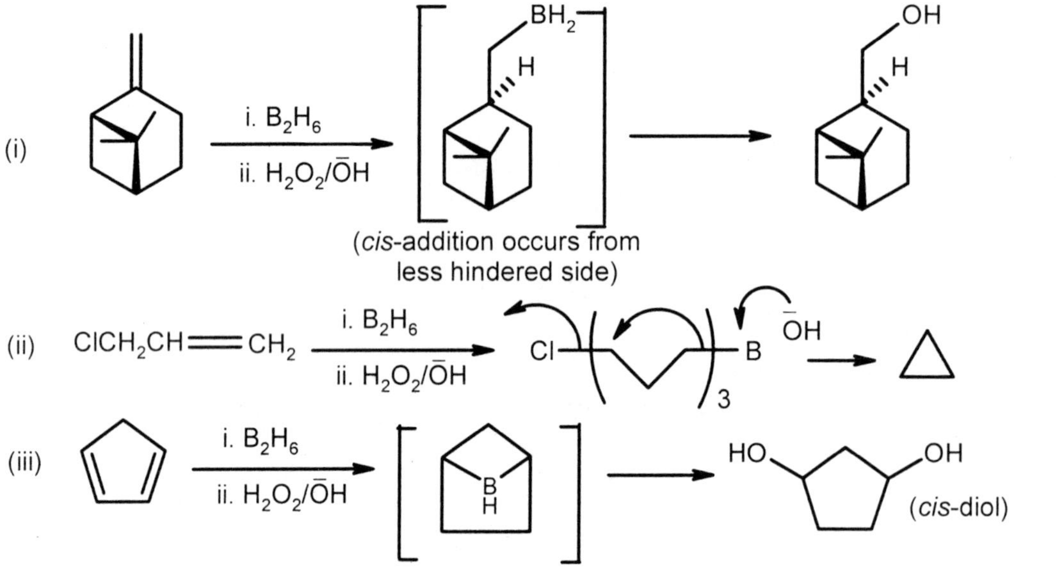

Problem 5.3.4

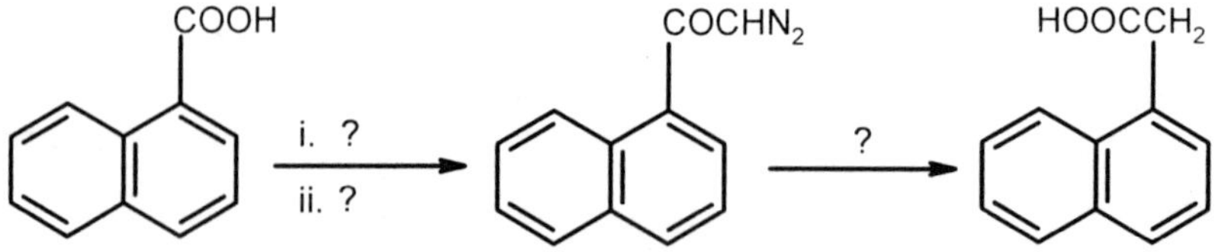

Solution.

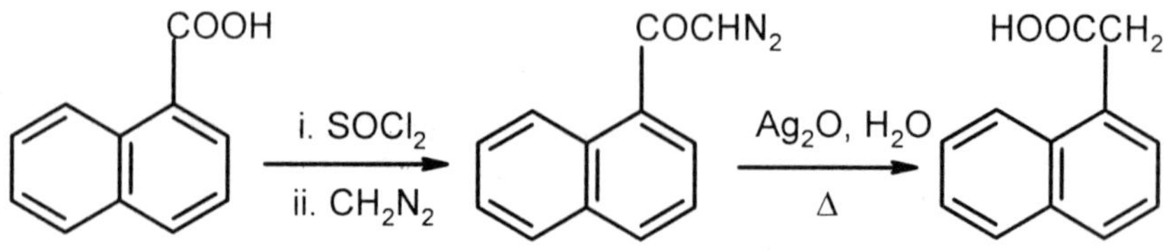

α–Naphthoic acid Diazo-α–acetonaphthone α–Naphthylacetic acid

Problem 5.3.5

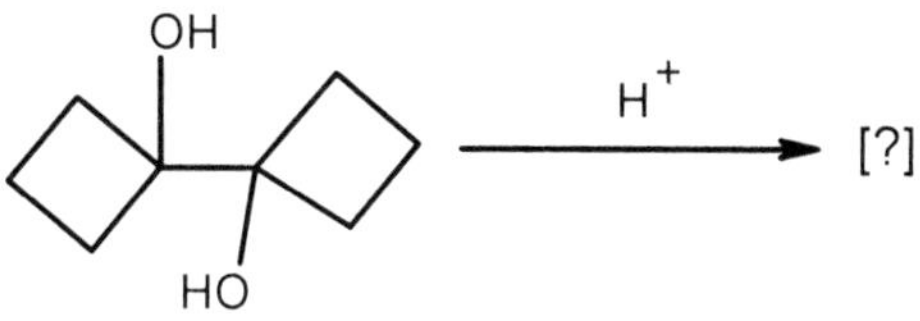

Solution.

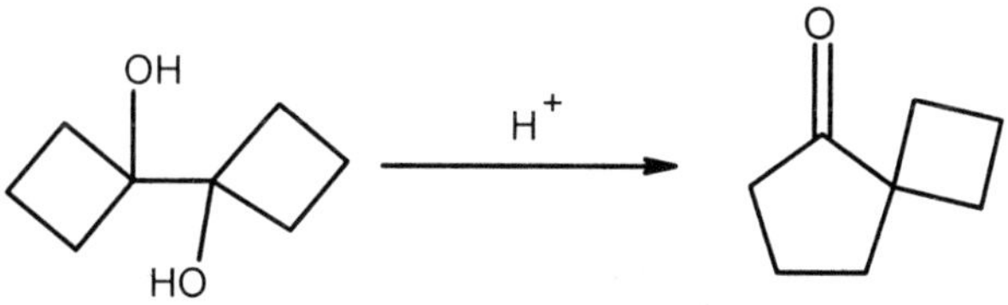

Problem 5.3.6

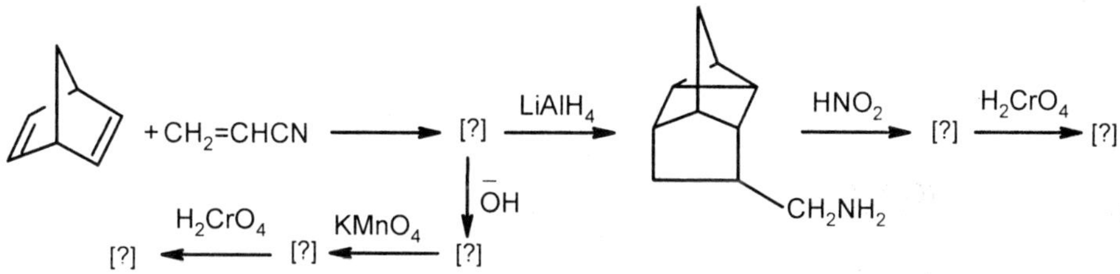

Solution.

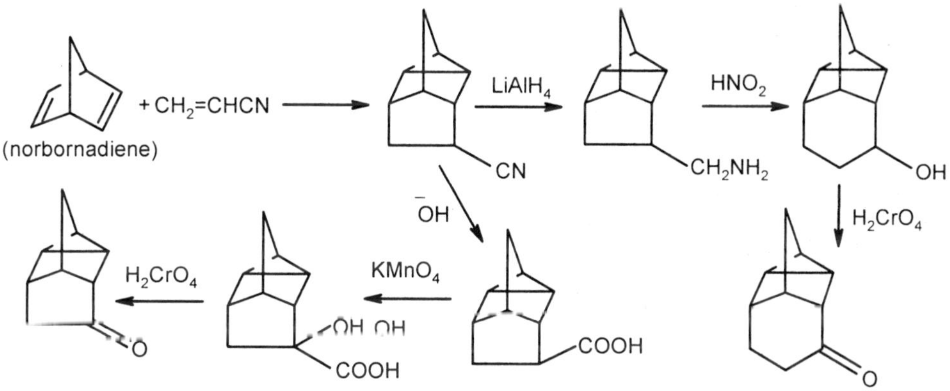

Problem 5.3.7

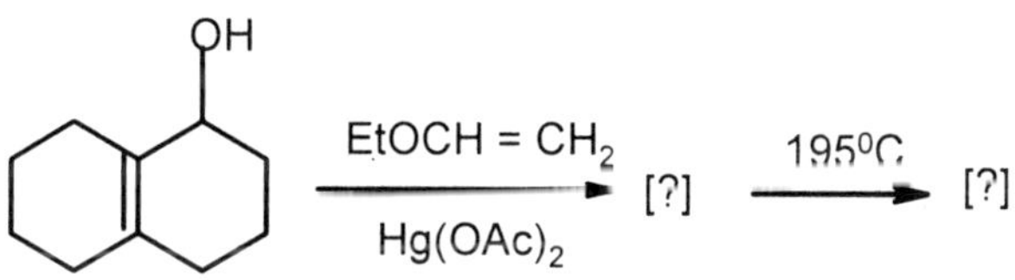

Solution.

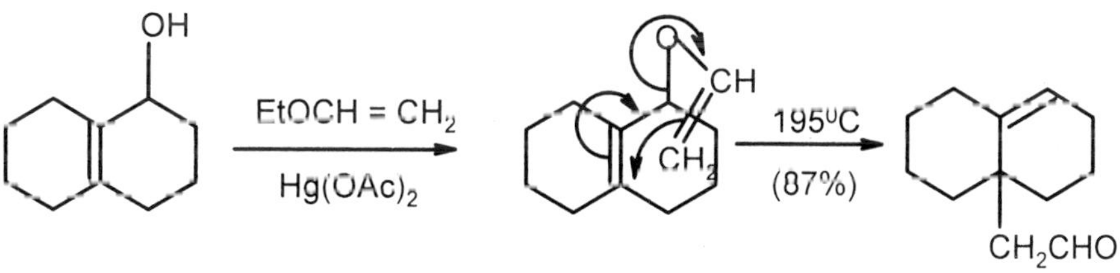

Problem 5.3.8

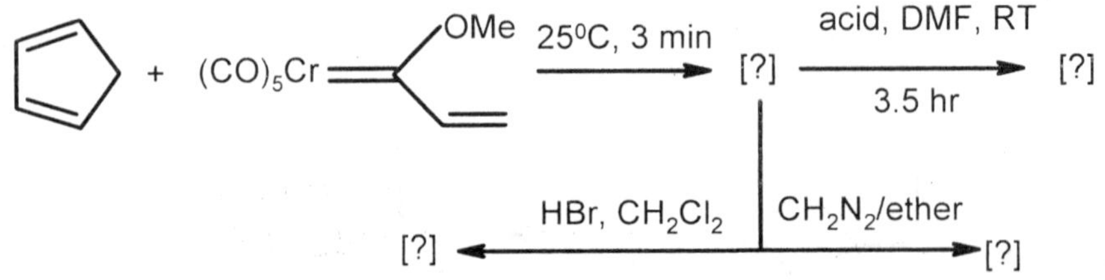

Solution.

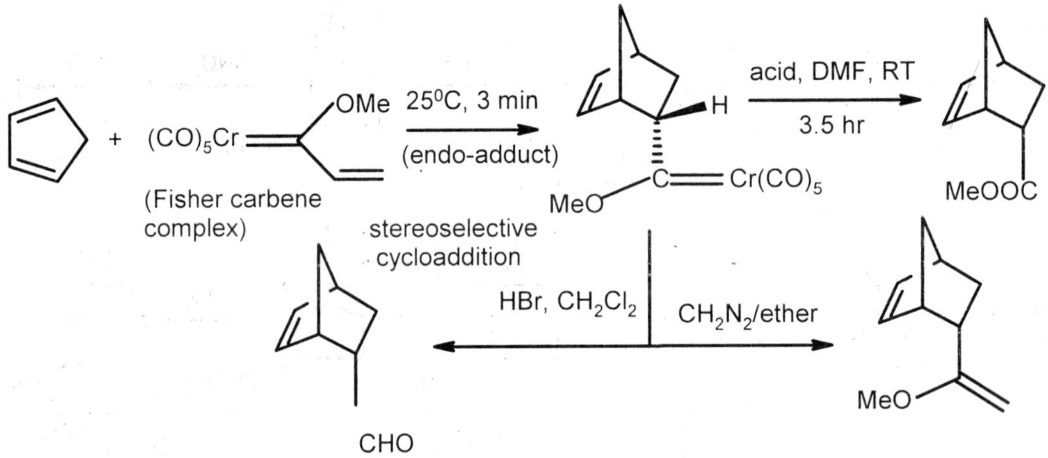

Problem 5.3.9

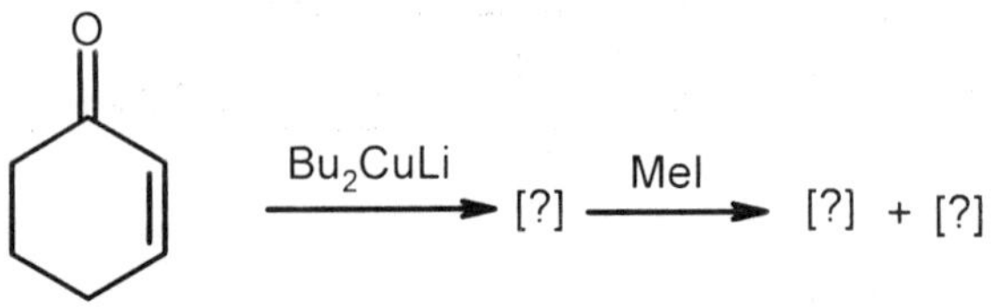

Solution.

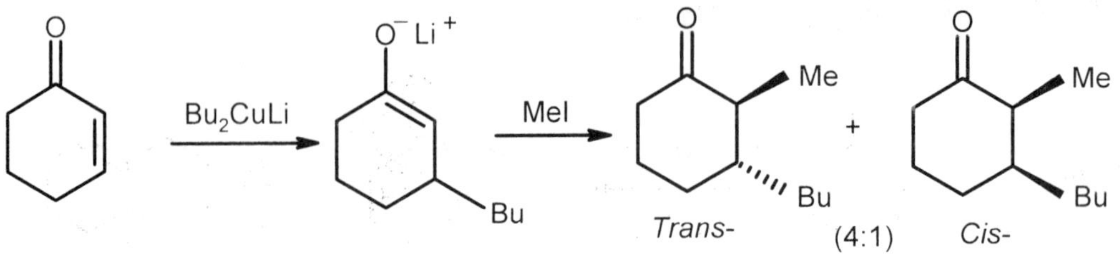

Problem 5.3.10

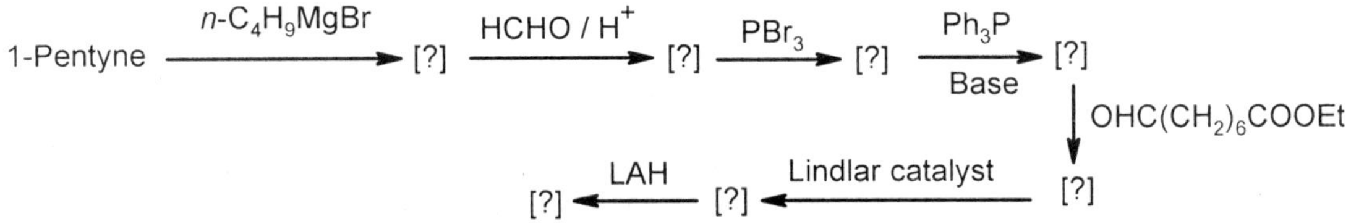

Solution.

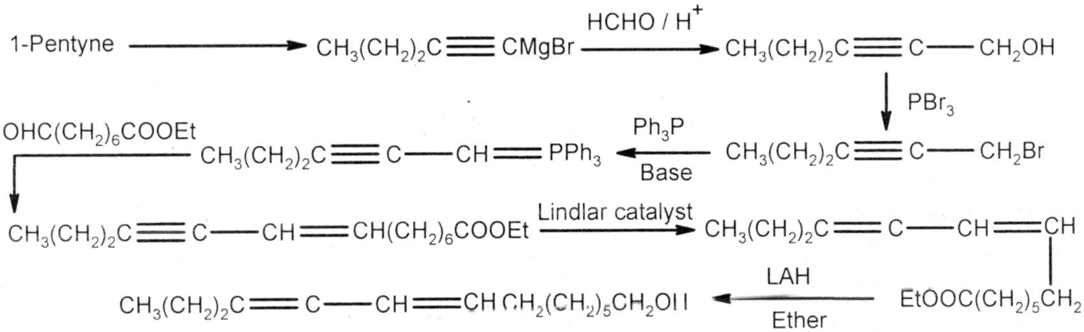

Problem 5.3.11

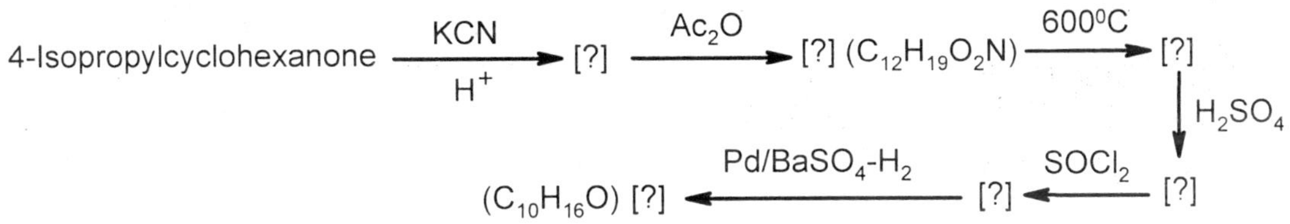

Solution.

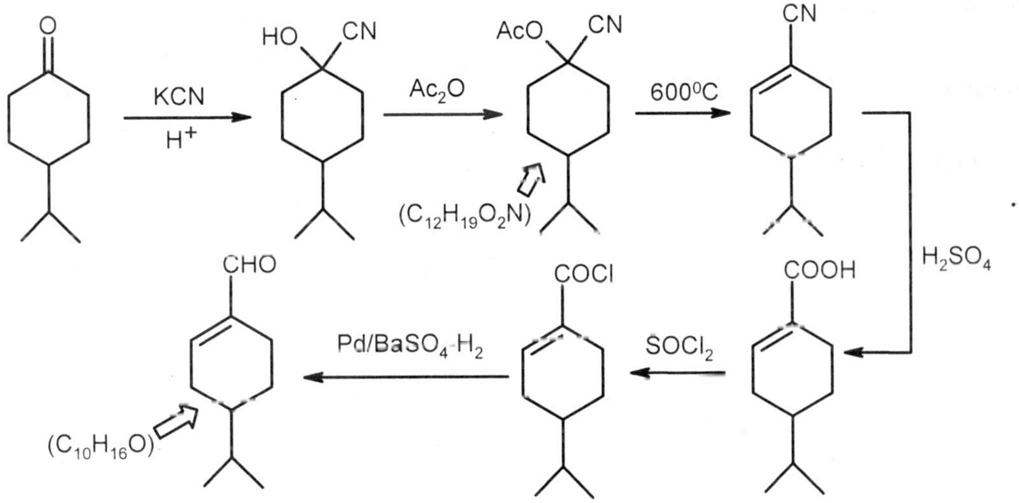

Problem 5.3.12

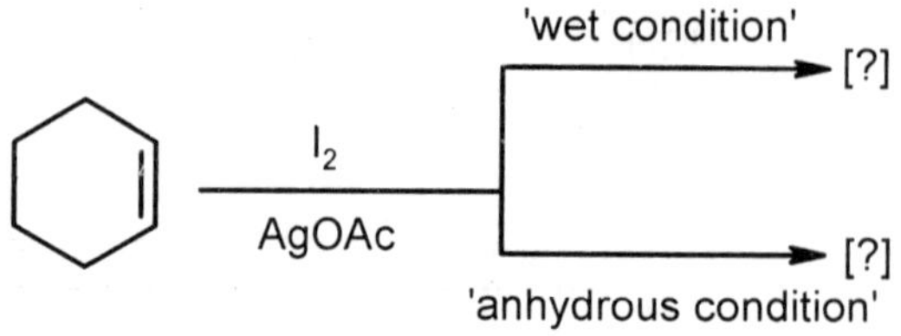

Solution.

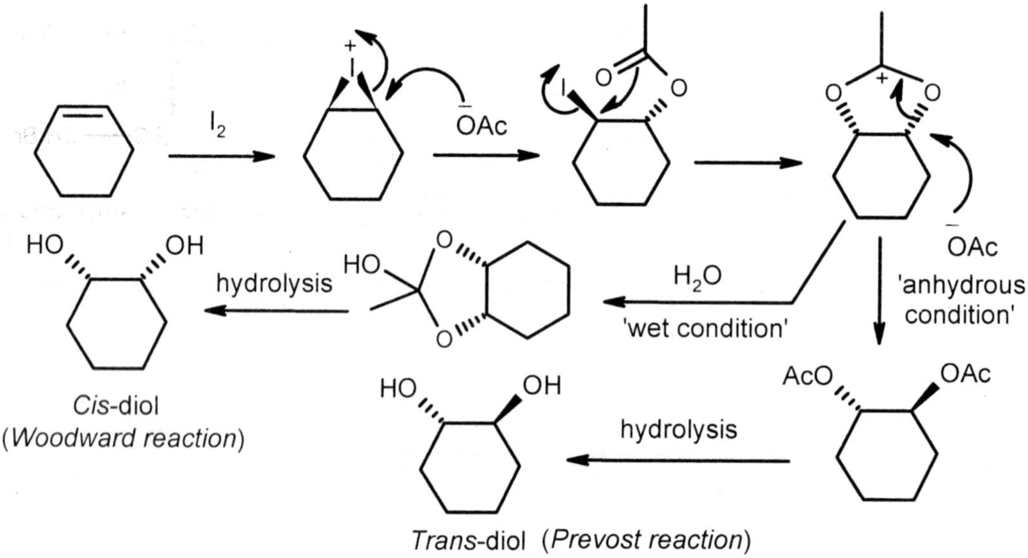

Problem 5.3.13 $Me_2CHLi + CH_2{=}CH_2 \longrightarrow [?] \xrightarrow[\text{ii. } H_2O]{\text{i. } CO_2} [?]$

Solution. $Me_2CHLi + CH_2{=}CH_2 \longrightarrow Me_2CHCH_2CH_2Li \xrightarrow[\text{ii. } H_2O]{\text{i. } CO_2} Me_2CHCH_2CH_2COOH$

Problem 5.3.14

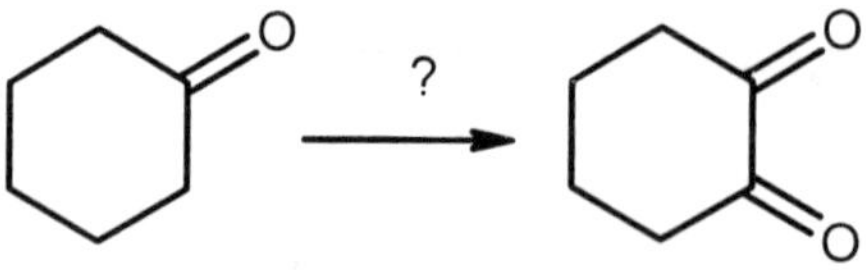

Solution.

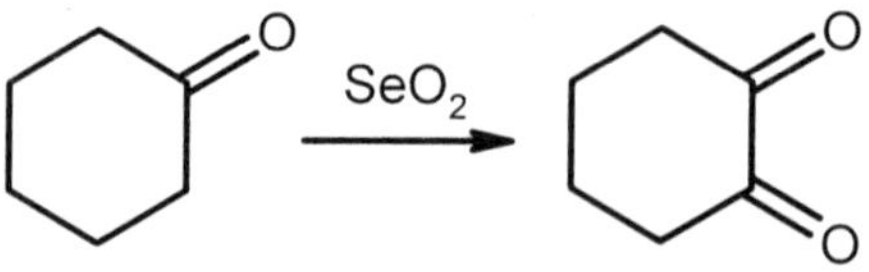

Problem 5.3.15

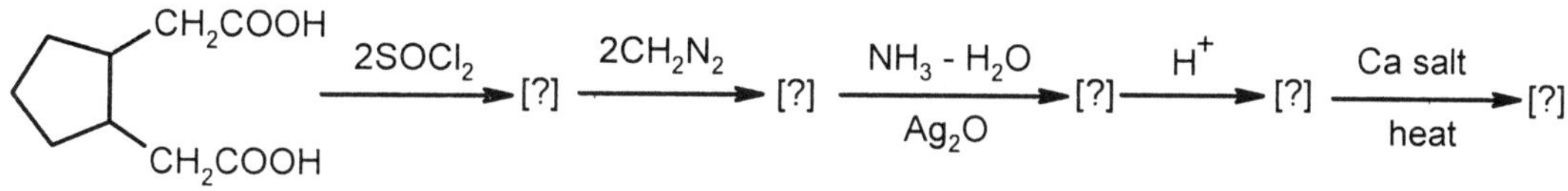

Solution.

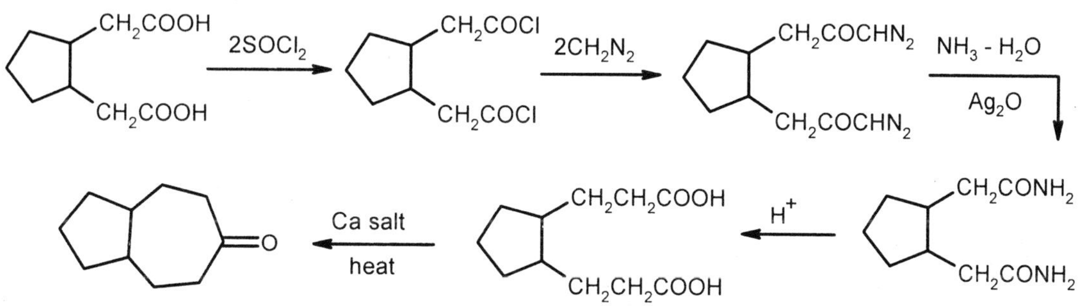

Problem 5.3.16

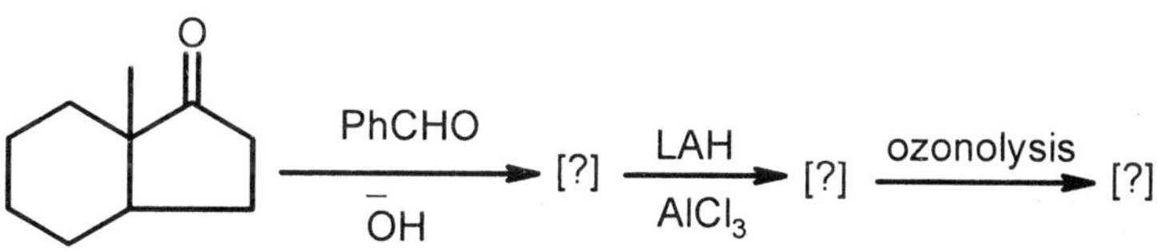

Solution.

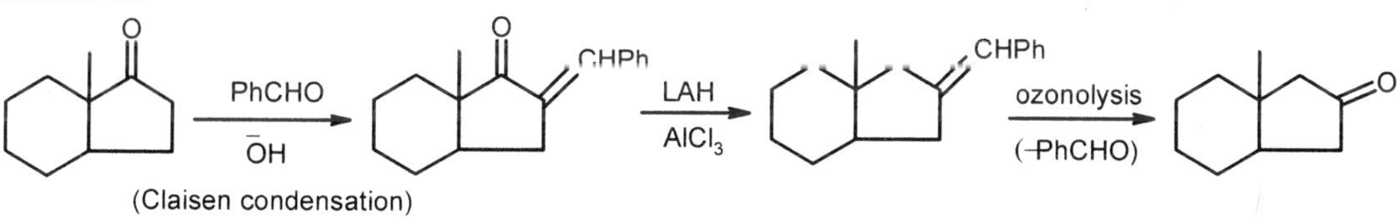

Problem 5.3.17

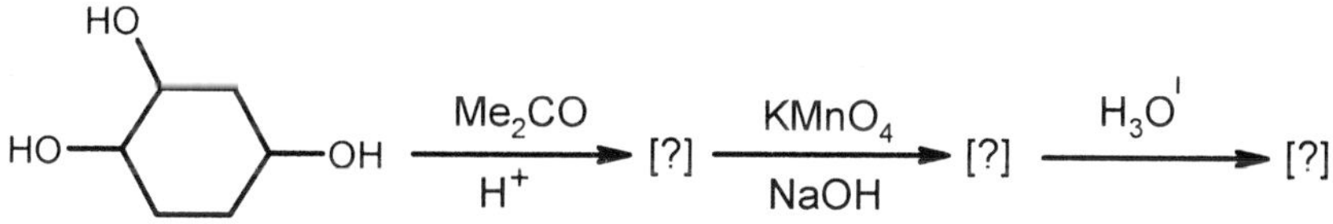

Solution.

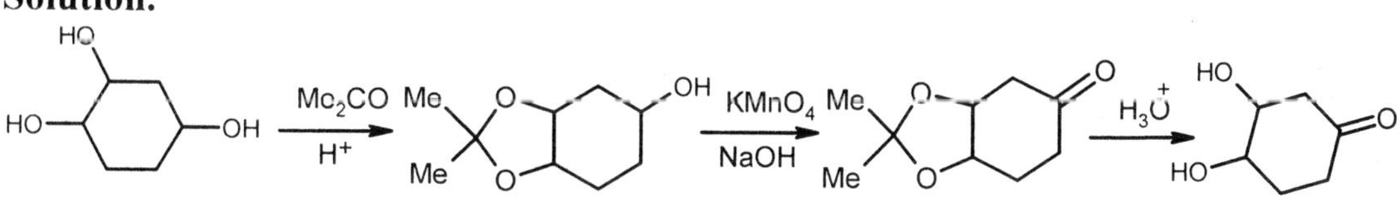

Problem 5.3.18

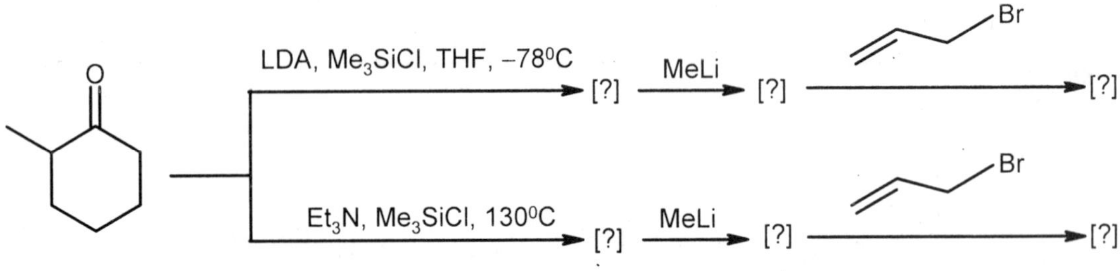

Solution.

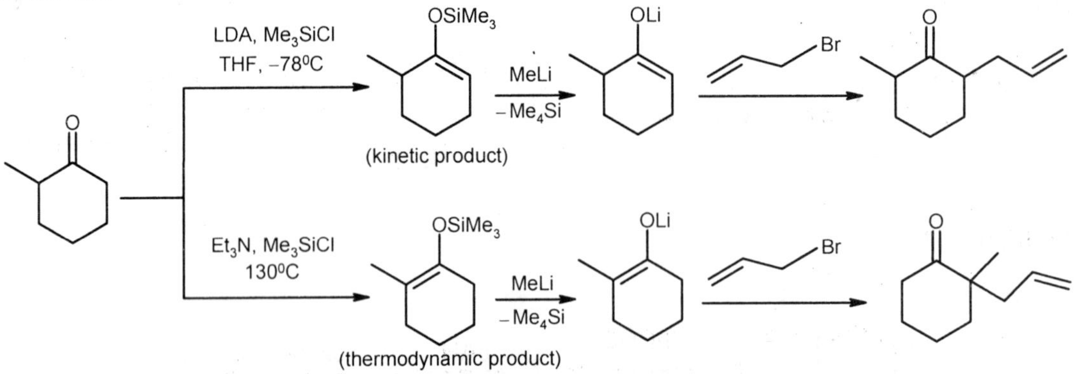

Problem 5.3.19

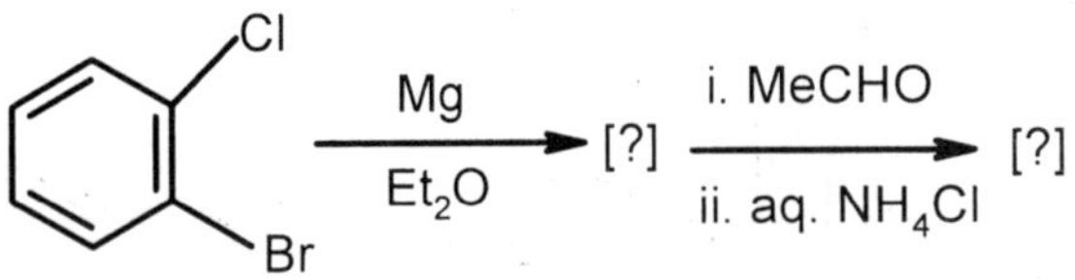

Solution.

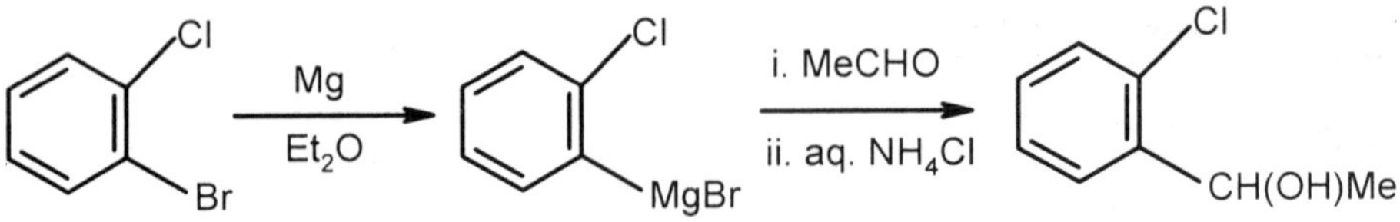

Note: Aryl chlorides do not form Grignard reagent in Et_2O, but in THF; on the hand, aryl bromides and iodides readily react in ether solution to form Grignard reagents.

Problem 5.3.20

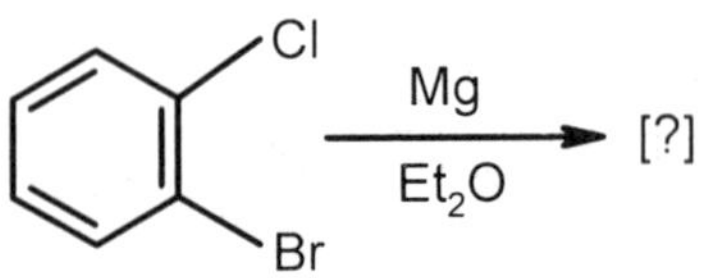

Solution.

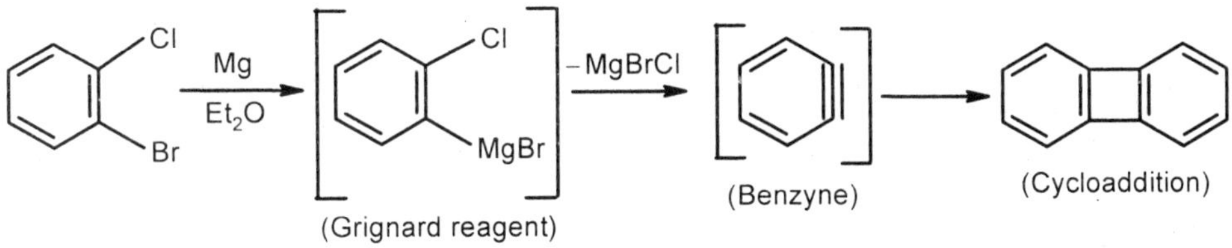

Problem 5.3.21

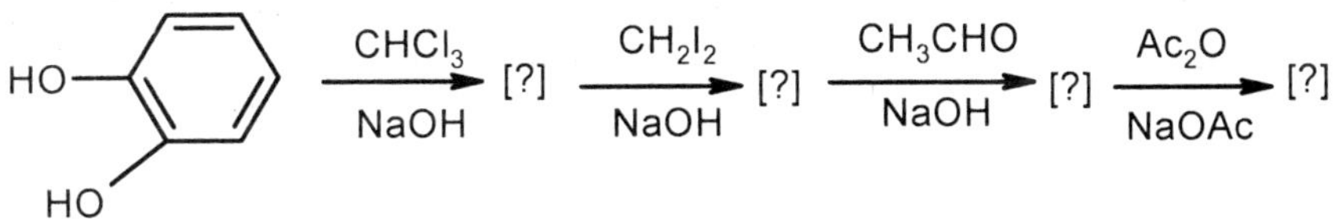

Solution.

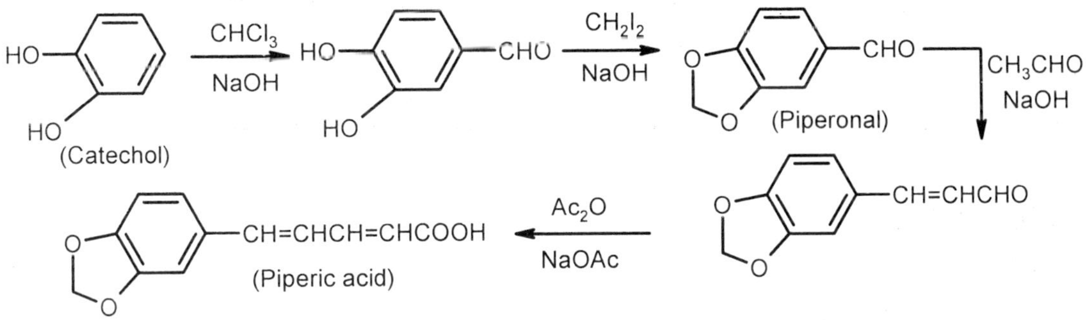

Problem 5.3.22

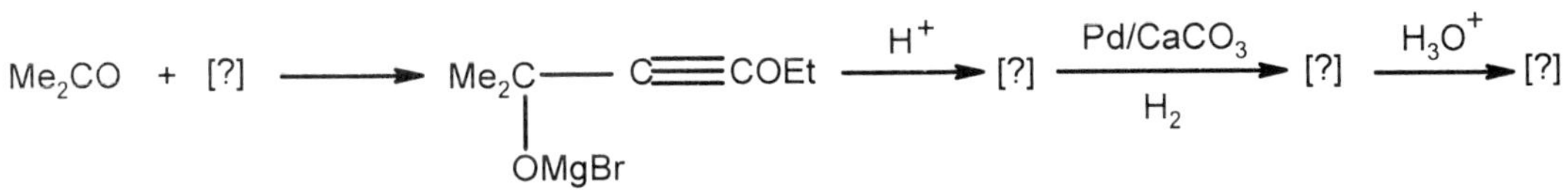

Solution.

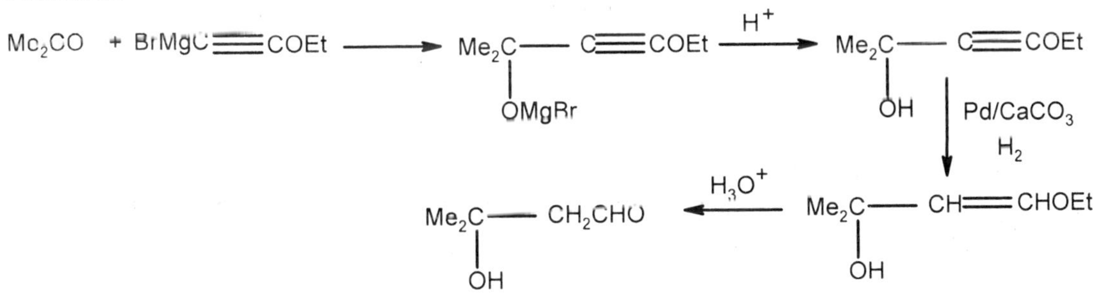

Problem 5.3.23

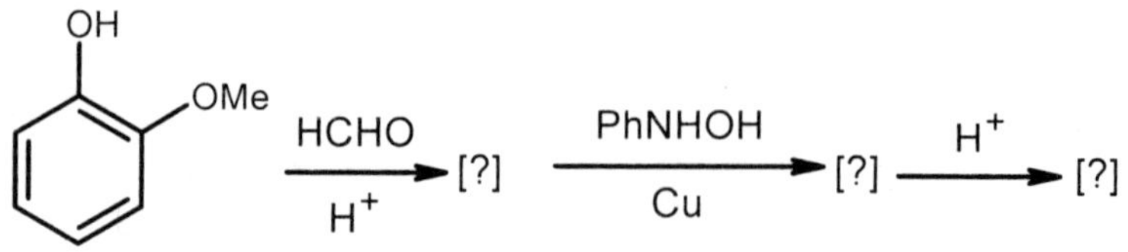

Solution.

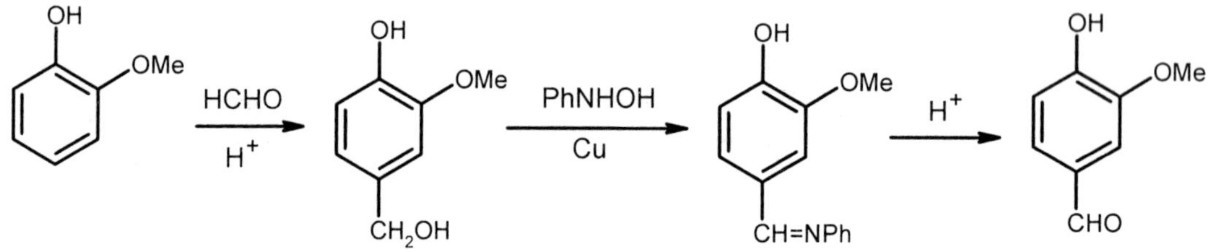

Problem 5.3.24 $PhCOCH_3 \xrightarrow{HNO_2} [?] \xrightarrow{Ac_2O} [?] \xrightarrow{H^+} [?]$

Solution.

$$PhCOCH_3 \xrightarrow{HNO_2} [PhCOCH_2NO] \longrightarrow PhCOCH{=}NOH \xrightarrow[-H_2O]{Ac_2O} PhCOCN \xrightarrow{H^+} PhCOCOOH$$

Problem 5.3.25

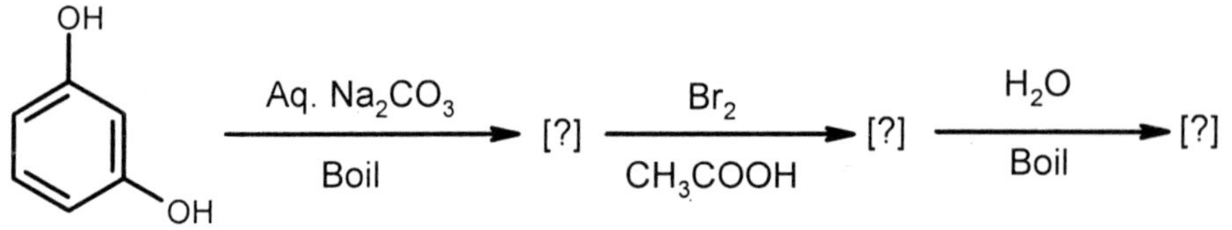

Solution.

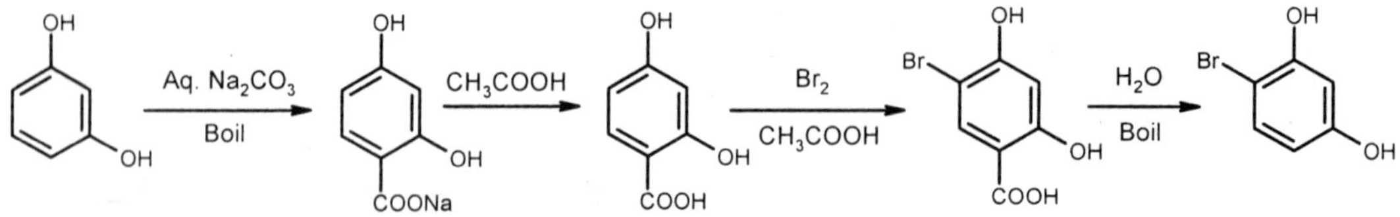

Problem 5.3.26

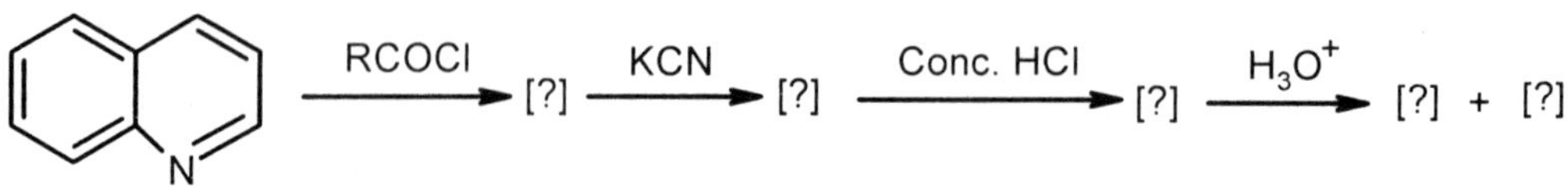

Solution.

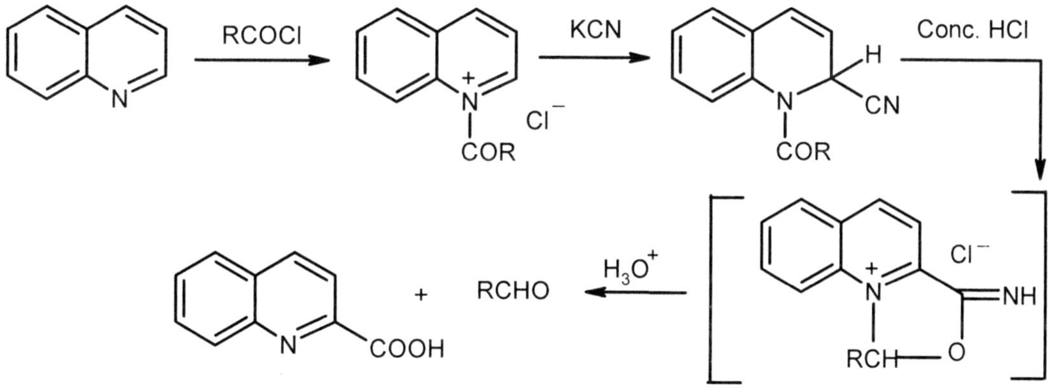

Problem 5.3.27

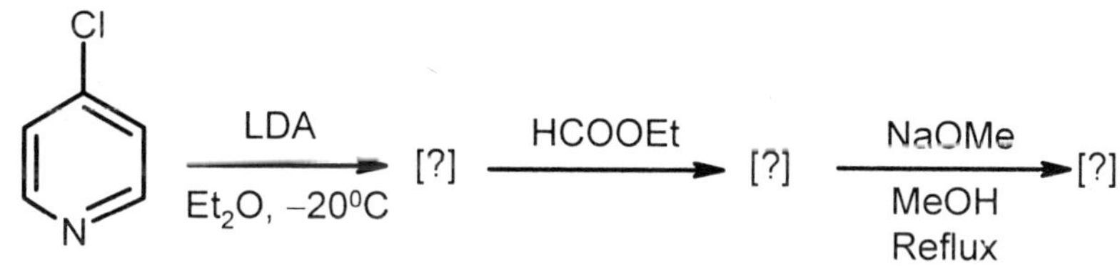

Solution.

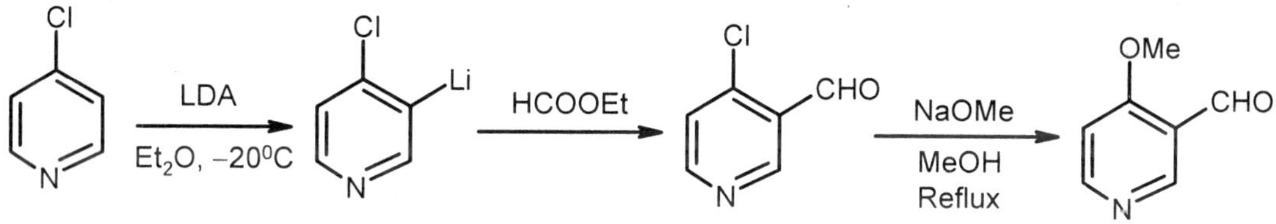

Problem 5.3.28

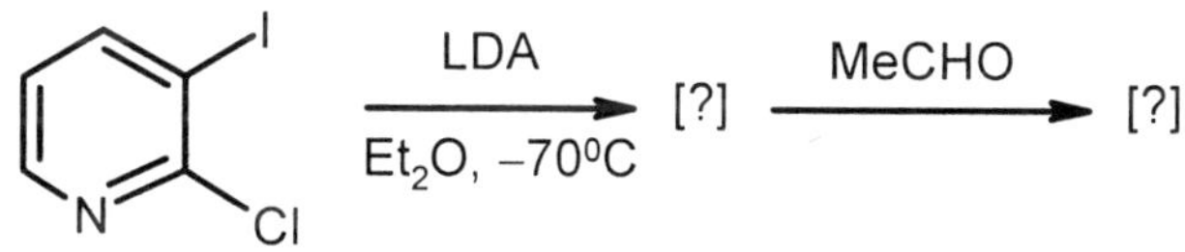

Solution.

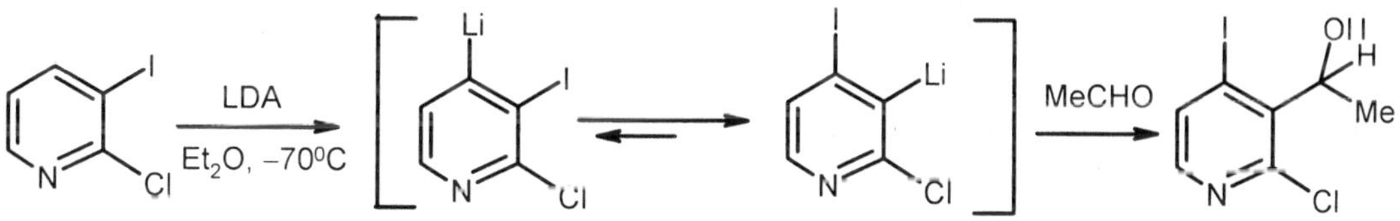

(Isomerization takes place due to the formation of more stable lithio derivative; such isomerization is called as *'halogen dance'*)

Problem 5.3.29

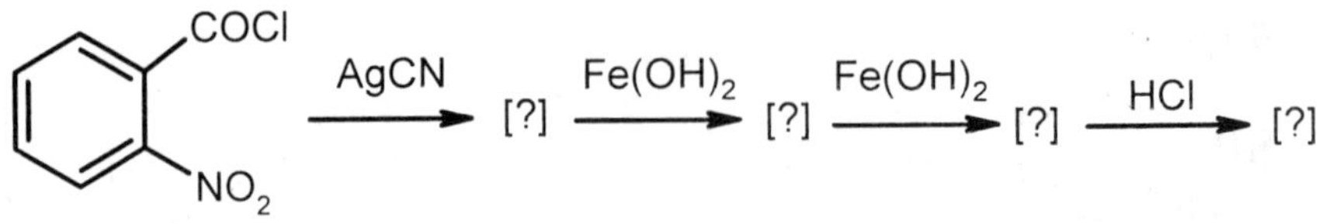

Solution.

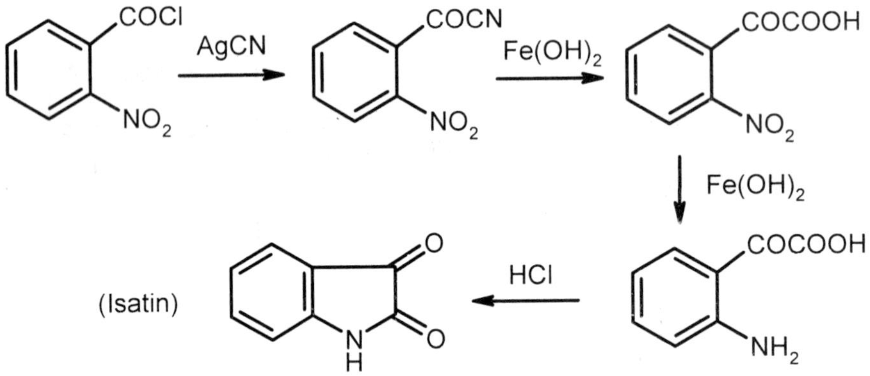

Problem 5.3.30

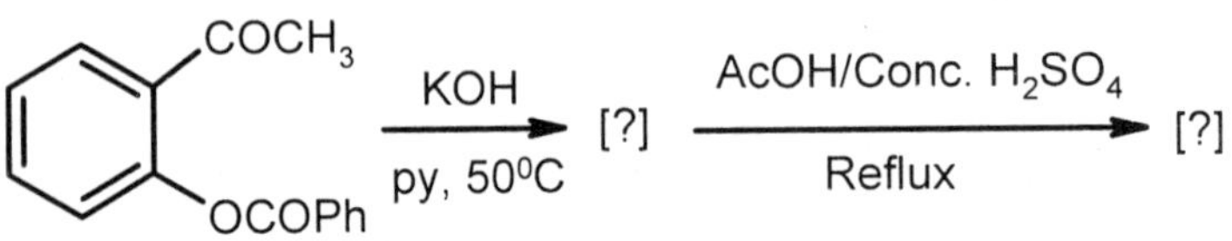

Solution.

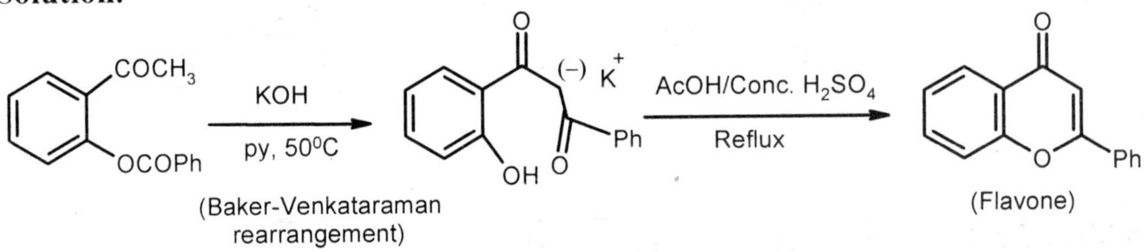

Problem 5.3.31

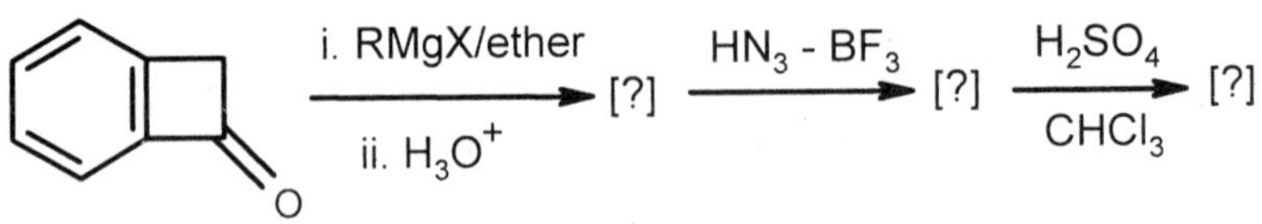

Solution.

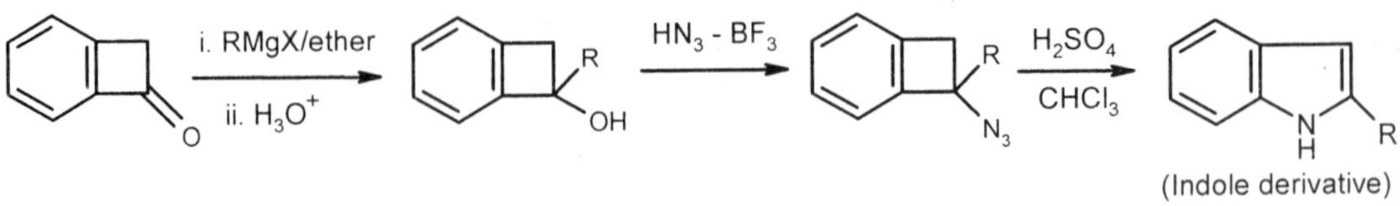

Supplementary problems

SP-5.1 Predict the product(s) in each of the following reactions with the plausible mechanism involved:

SP-5.1.1 $CH_3CH(Br)CH{=}CH_2 \xrightarrow{\text{moist } Ag_2O} [?]$

SP-5.1.2 $(CH_3)_2C(Br)CH_2CH_3 \xrightarrow{^tBuOK} [?]$

SP-5.1.3 $(CH_3)_2CHCH_2OH \xrightarrow{HBr} [?]$

SP-5.1.4 $(CH_3)_2CHCHO \xrightarrow{\text{strong alkali}} [?]$

SP-5.1.5 $Ph_2C(OH)C(OH)Me_2 \xrightarrow{H^+} [?]$

SP-5.1.6 $CH_3{-}C({=}N{-}OH){-}C_2H_5 \xrightarrow{\text{Conc. } H_2SO_4} [?]$

SP-5.1.7 *Cis*-2-Butene + $Br_2 \xrightarrow{MeOH} [?]$

SP-5.1.8 $RCN \xrightarrow{SnCl_2/HCl} [?]$

SP-5.1.9 $CH_3CH_2CH{=}CHCH_3 \xrightarrow{HCl} [?]$

SP-5.1.10 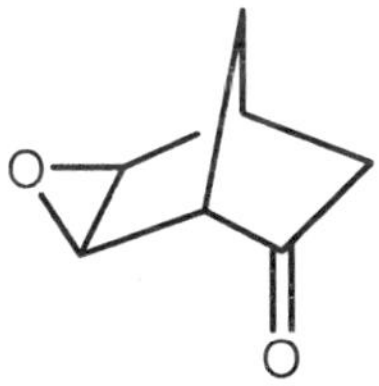 $\xrightarrow{H_3O^+} [?]$

SP-5.1.11 Me_2CHCH_2Me (with $^+NMe_3$ on CH) $\xrightarrow[\text{heat}]{^-OH} [?]$

SP-5.1.12 Anisole $\xrightarrow[DMF]{POCl_3} [?]$

SP-5.1.13 Phenylacetate $\xrightarrow[\text{heat}]{\text{anhydrous } AlCl_3} [?]$

SP-5.1.14 *p*-Chlorotoluene $\xrightarrow[NH_3(l)]{NaNH_2}$ [?]

SP-5.1.15 (*S*)-EtCH(Me)COCl $\xrightarrow[\text{ii. } CHCl_3/\Delta \text{ iii. } H_2O]{\text{i. } NaN_3}$ [?]

SP-5.1.16 (*R*)-EtCH(Me)CH(Me)CONH$_2$ $\xrightarrow{NaOBr/H_2O}$ [?]

SP-5.1.17 MeC$_6$H$_4$NO$_2$(*o*) $\xrightarrow[EtONa]{(COOEt)_2}$ [?]

SP-5.1.18 RMgBr + *CO$_2$ $\longrightarrow$ [?] $\xrightarrow{H^+}$ [?] $\xrightarrow[\text{ii. } CH_2N_2 \text{ (excess)}]{\text{i. } SOCl_2}$ [?] $\xrightarrow{Ag_2O}$ [?] $\xrightarrow{H_2O}$ [?]

SP-5.1.19 α-Naphthol $\xrightarrow{NaHSO_3}$ [?]

SP-5.1.20 HOOCC$\equiv$CCOOH $\xrightarrow{Br_2}$ [?]

SP-5.1.21 PhC$\equiv$CH $\xrightarrow[\Delta]{HgSO_4,\ H_2O}$ [?]

SP-5.1.22 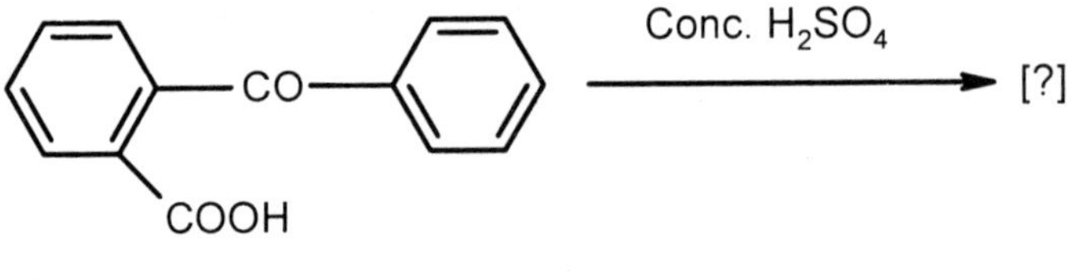

SP-5.1.23 *Cis*- and *trans*-2-aminocyclohexanol (separately treated) $\xrightarrow[HCl]{NaNO_2}$ [?]

SP-5.1.24 CH$_2$=CHCN + CH$_2$(COOEt)$_2$ $\xrightarrow{EtONa}$ [?]

SP-5.1.25 C$_6$H$_5$CHO $\xrightarrow{RCO_3H}$ [?]

SP-5.1.26 CH$_2$=CHCOCH$_3$ $\xrightarrow[HO^-]{H_2O_2}$ [?]

SP-5.1.27 RCH=CH$_2$ $\xrightarrow[\text{ii. } H_2O_2/HO^-]{\text{i. } B_2H_6}$ [?]

SP-5.1.28 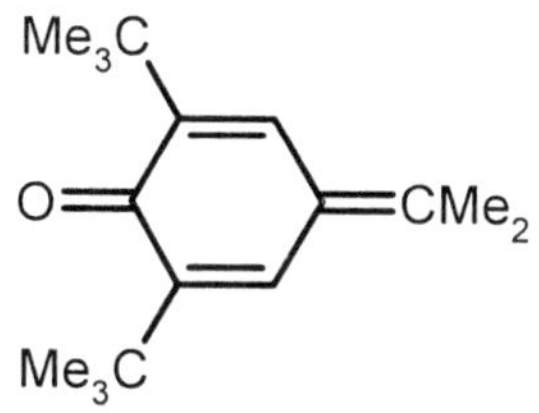 $\xrightarrow{\bar{O}H}$ [?]

SP-5.1.29 *p*-Benzoquinone + 2 $CH_2(CN)COOEt$ $\xrightarrow[\text{boil}]{\text{aq. HCl}}$ [?]

SP-5.1.30 *p*-Methoxytoluene $\xrightarrow[\text{ii. } H_3O^+]{\text{i. Li, } NH_3(l)}$ [?]

SP-5.2 How would you carry out the following transformations?

SP-5.2.1 Benzaldehyde ⟶ Benzilic acid

SP-5.2.2 Nitrobenzene ⟶ Benzidine

SP-5.2.3 Phenol ⟶ Salicyladehyde

SP-5.2.4 Ethylacetoacetate ⟶ 2,5-Dimethylpyrrole

SP-5.2.5 Aniline ⟶ *p*-Bromoaniline

SP-5.2.6 Toluene ⟶ Sacharin

SP-5.2.7 Toluene ⟶ *p*-Aminobenzoic acid

SP-5.2.8 *m*-Dinitrobenzene ⟶ *m*-Nitroaniline

SP-5.2.9 *m*-Dinitrobenzene ⟶ 3-Bromophenol

SP-5.2.10 Benzadehyde ⟶ Diphenylacetic acid

SP-5.2.11 Me_3COMe ⟶ $MeCOOCMe_3$

SP-5.2.12 PhCOMe ⟶ PhCOCH=NOH

SP-5.2.13 $MeCO(CH_2)_2CO_2Et$ ⟶ $MeCH(OH)(CH_2)_2COOEt$

SP-5.2.14 *n* C_4H_9OH ⟶ $CH_3C{\equiv}CH$

SP-5.2.15 PhCOMe ⟶ PhCH(Me)CHO

SP-5.2.16 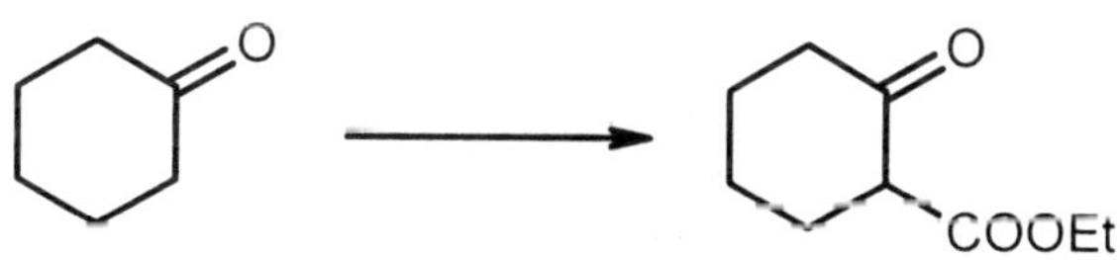

SP-5.2.17 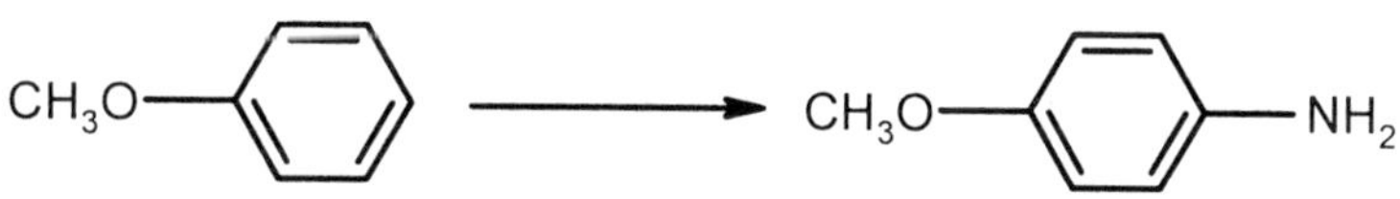

SP-5.2.18 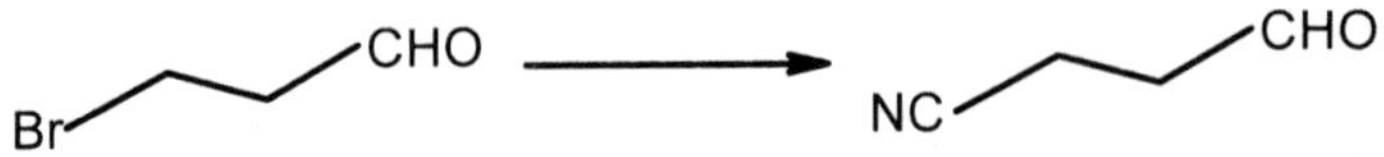

SP-5.2.19 

SP-5.2.20 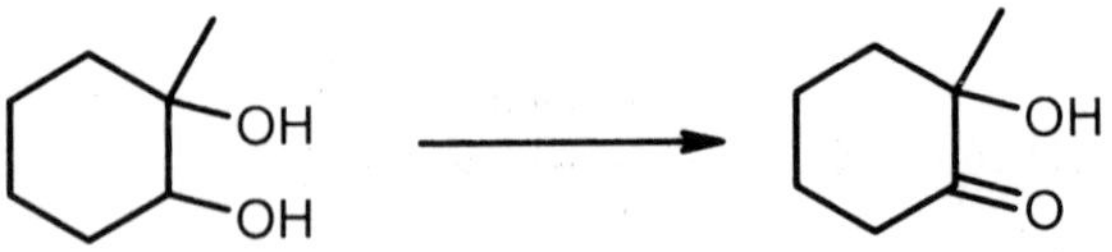

SP-5.2.21 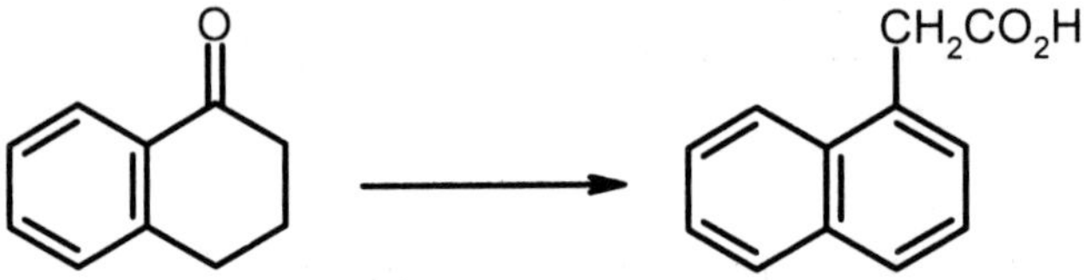

SP-5.2.22 RCHO ⟶ $RCOCH_3$

SP-5.2.23 PhCH=CHCHO ⟶ $PhCH{=}CHCH_2OH$

SP-5.2.24 *p*-Hydroxybenzaldehyde 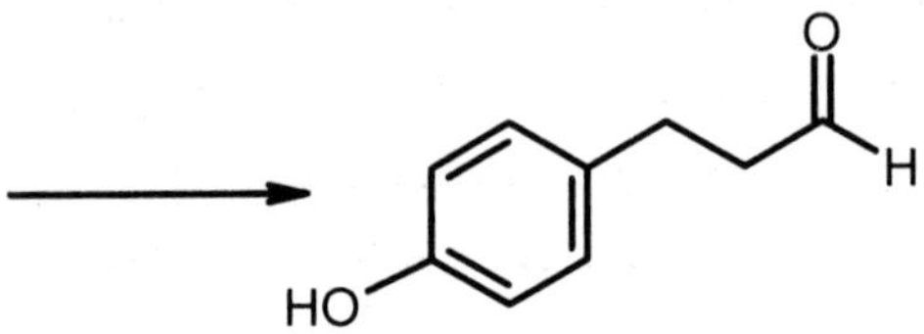

SP-5.2.25 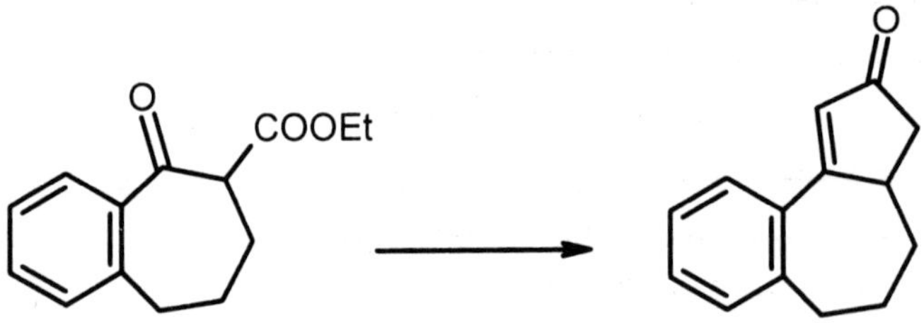

SP-5.3 Complete the reaction sequence in each of the following cases:

SP-5.3.1 RCOCl $\xrightarrow{NH_2OH}$ [?] $\xrightarrow{H^+}$ [?]

SP-5.3.2 1-Bromopropane $\xrightarrow{AgCN}$ [?] (major) + [?] (minor)

[?] (major) $\xrightarrow{[H]}$ [?] $\xrightarrow{NaNO_2/HCl}$ [?]

SP-5.3.3 $Me_2C{=}CHCOCH_3$ $\xrightarrow[Cu^{2+}]{MeMgI}$ [?]

SP-5.3.4 MeCOEt $\xrightarrow[\text{ii. Zn}]{\text{i. BrCH}_2\text{COOEt /Zn}}$ [?] $\xrightarrow{\Delta}$ [?]

SP-5.3.5 $PhCH_2Cl$ $\xrightarrow{Ph_3P}$ [?] $\xrightarrow{PhLi}$ [?] $\xrightarrow{PhCHO}$ [?]

SP-5.3.6 $PhCH_2COOH$ $\xrightarrow[H^+]{EtOH}$ [?] $\xrightarrow[{}^*CH_3I]{NaH}$ [?] $\xrightarrow[\text{iii. resolution}]{\text{i. NaOH , ii. H}}$ (*R*)-compound $\xrightarrow{\text{i. SOCl}_2\text{, ii. PhH, AlCl}_3}$ [?]

SP-5.3.7 $Me_2C{=}CH_2$ $\xrightarrow[\text{pressure}]{H_2SO_4,\ H_2O}$ [?] $\xrightarrow[\text{pressure}]{Me_3COH}$ [?]

SP-5.3.8 $Me_2C(OH)CH_2OHEt$ $\xrightarrow{H^+}$ [?]

SP-5.3.9 Me_2CO $\xrightarrow[\text{THF, }-78^0\text{C}]{\text{LDA}}$ [?] $\xrightarrow{EtCHO}$ [?] $\xrightarrow{H_3O^+}$ [?]

SP-5.3.10 Me_2CO $\xrightarrow[NaNH_2]{C_2H_2}$ [?] $\xrightarrow{H_3O^+}$ [?] $\xrightarrow{H_2/\ Pd}$ [?] $\xrightarrow[400^0]{Al_2O_3}$ [?]

SP-5.3.11 PhCOCH(OH)Ph $\xrightarrow{\text{Tollens' reagent}}$ [?]

SP-5.3.12 $PhCOCHBr_2$ $\xrightarrow{HO^-}$ [?]

SP-5.3.13 CH_3COOEt + [?] $\xrightarrow{\text{NaOEt}}$ $PhCOCH_2CO_2Et$ $\xrightarrow[I_2]{\text{NaOEt}}$ [?] $\xrightarrow[\text{iii. P2O5}]{\text{i. dil. NaOH, ii. H}_3\text{O}^+}$ [?]

SP-5.3.14 *p*-Benzoquinone $\xrightarrow{Ac_2O}$ [?]

SP-5.3.15 C_2H_2 $\xrightarrow[\text{(excess)}]{\text{MeMgI}}$ [?] $\xrightarrow[\text{ii. H}_3\text{O}^+]{\text{i. CO}_2}$ [?] $\xrightarrow[H^+]{MeOH}$ [?] $\xrightarrow[\Delta]{}$ [?]

SP-5.3.16 *Cis*-2-butene $\xrightarrow[h\nu]{CH_2N_2}$ [?]

Cis-2-butene $\xrightarrow[h\nu,\ N_2\ \text{gas}]{CH_2N_2}$ [?]

SP-5.3.16 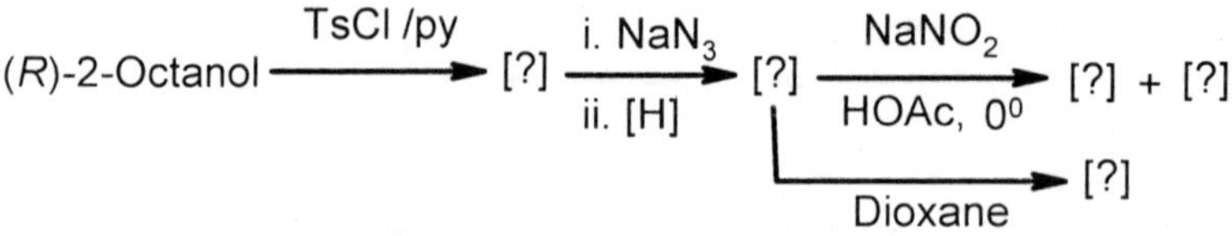

SP-5.3.17 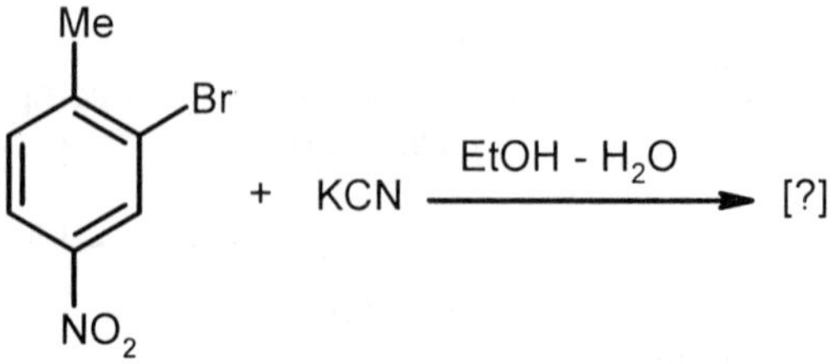

SP-5.3.18 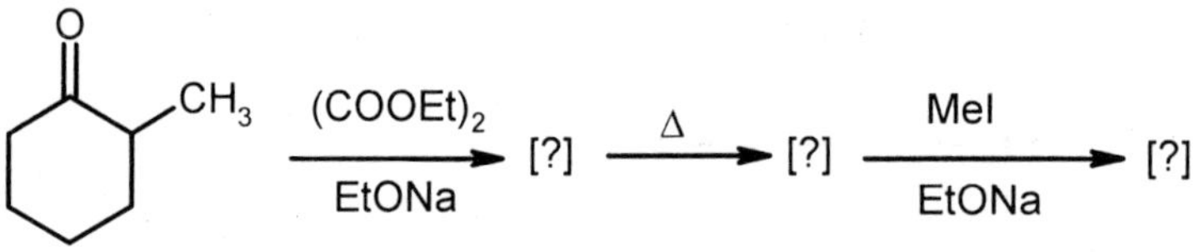

SP-5.3.19 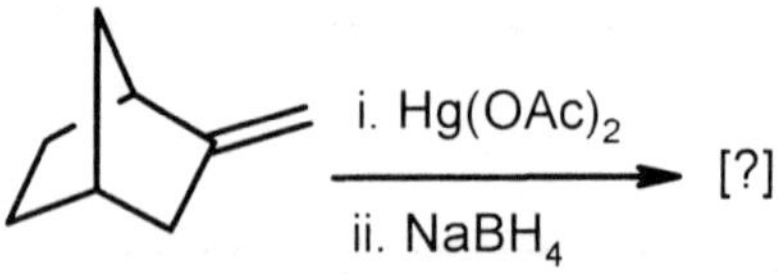

SP-5.3.20 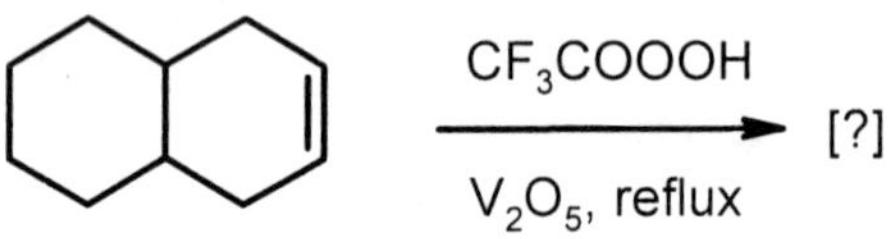

SP-5.3.21 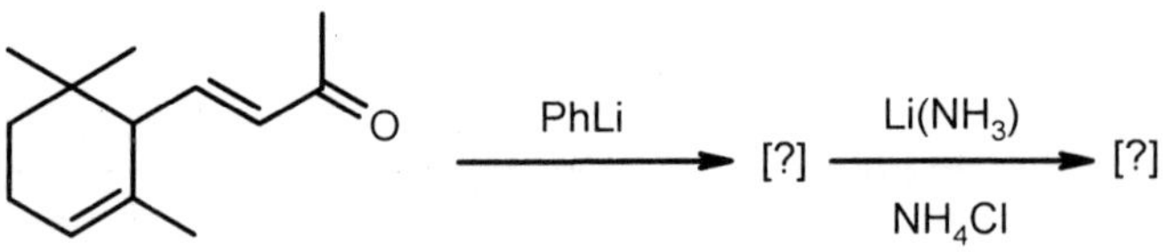

SP-5.3.22 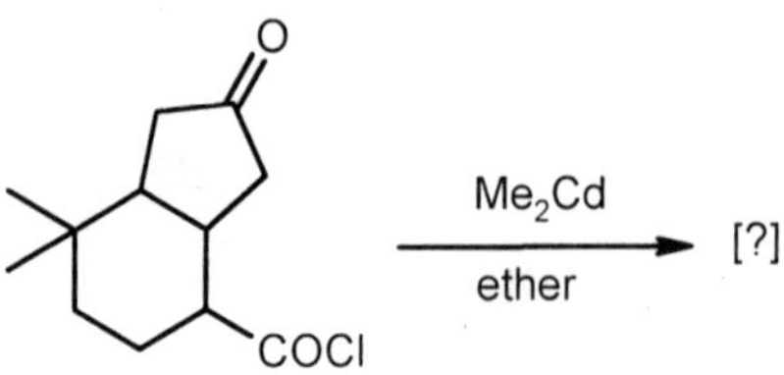

SP-5.3.23 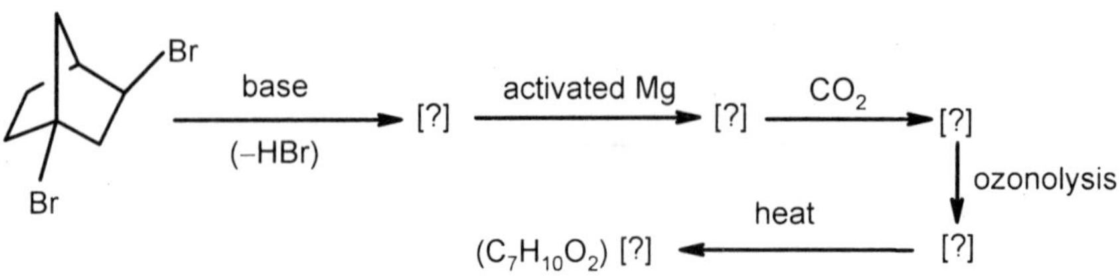

CHAPTER 6

PERICYCLIC REACTIONS AND ORGANIC PHOTOCHEMISTRY

Introduction

Pericyclic reactions have been defined as "those (reactions) in which all first-order changes in bonding relationships take place in concert on a closed curve" (Woodward and Hoffmann, 1970). The term '*pericyclic*' (means around the cycle) arises from the fact that the changes in bonding take place in concerted processes through reorganization of electrons (pair-wise) within a closed loop of interacting orbitals. In concerted reaction, reactant bonds are broken and product bonds are formed at the same time, without intermediates. These reactions do not follow any ionic or radical pathway and remain unaffected by polar solvents, radical initiators or inhibitors and catalysts (exceptions are some acid-catalyzed Diels-Alder reactions). Pericyclic reactions take place thermally or photochemically. The outcomes of pericyclic reactions were first rationalized by Woodward and Hoffmann (1965, 1970) with the help of certain selection rules developed by them on the principle of conservation of orbital symmetry. Besides, some other approaches such as '*aromatic transition state concept*' and '*Frontier Orbital Approach* (FMO)' have also been arrived to explain the outcomes of pericyclic reactions. These reactions may be divided into four broad categories — electrocyclic, cycloaddition, sigmatropic and chelotropic reactions.

All types of reactions that are occurred by absorbing energy from light source are commonly known as *light catalyzed* or *photochemical reactions*. A variety of such types of reactions are known. Organic photochemical reactions find extensive use in preparative and synthetic organic chemistry.

In this chapter, we would restrict our discussion only in representing a number of examples covering various pericyclic and organic photochemical reactions.

Solved Problems

Problem 6.1

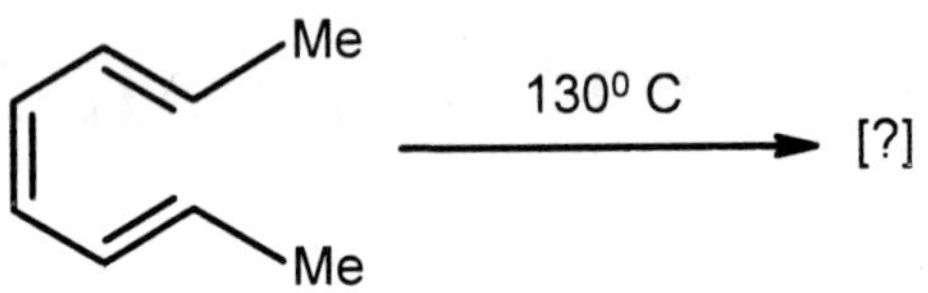

Solution.

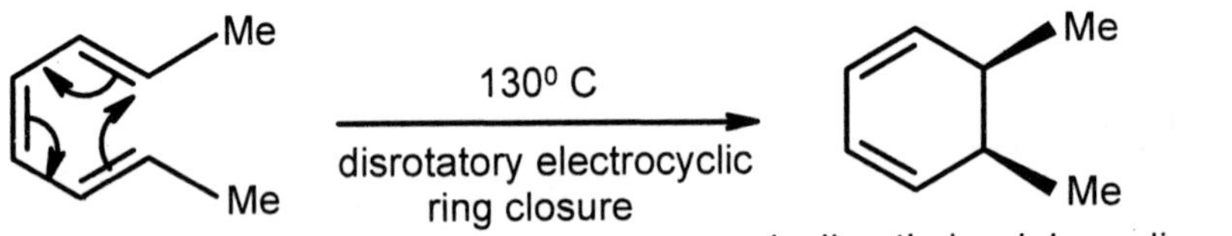

acyclic triene

[*Tet. Lett.*, 377(1965)]

cis-dimethylcyclohexadiene (>90%)

The acyclic triene undergoes thermal electrocyclization reaction using 6πe-electrons following disrotatory path. The cyclized 1,2-dimethylcyclohexadiene product attains *cis*-geometry.

Problem 6.2

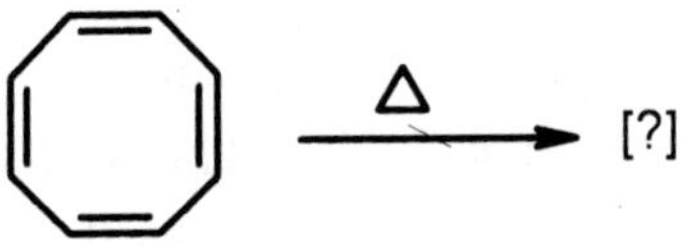

Solution.

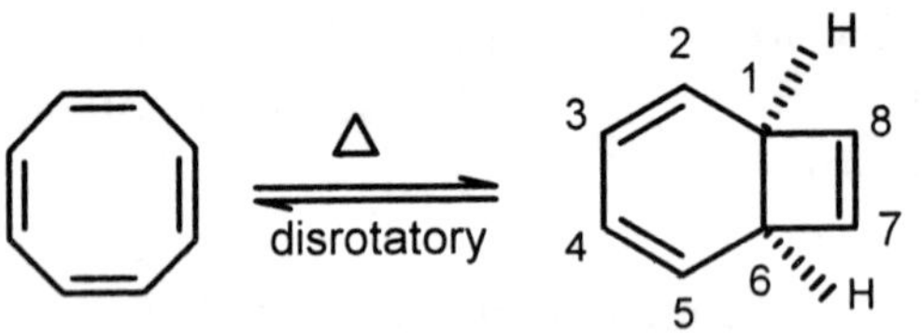

[*Angew. Chem. Int. Edn. Engl.*, **3**, 83(1964)]

cyclo-octatetraene

bicyclo[4,2,0]-octa-2,4,7-triene

This is an example of thermally occurring 6πe-electrocyclic reactions following disrotatory path. The equilibrium favours the backward direction because the bicyclo system is more strained compared to cyclo-octatetraene.

Problem 6.3

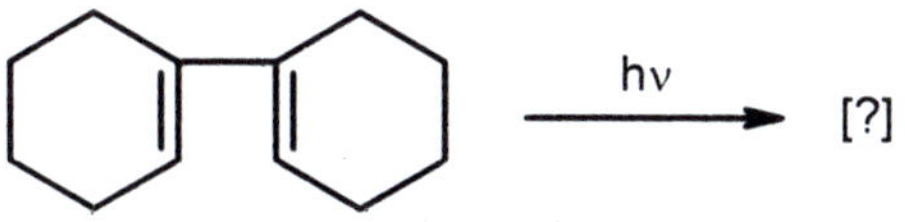

Solution.

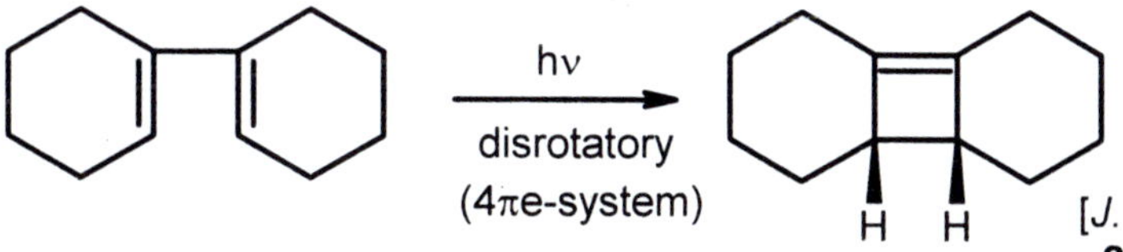

[*J. Am. Chem. Soc.*, **88**, 2742(1966)]

Problem 6.4

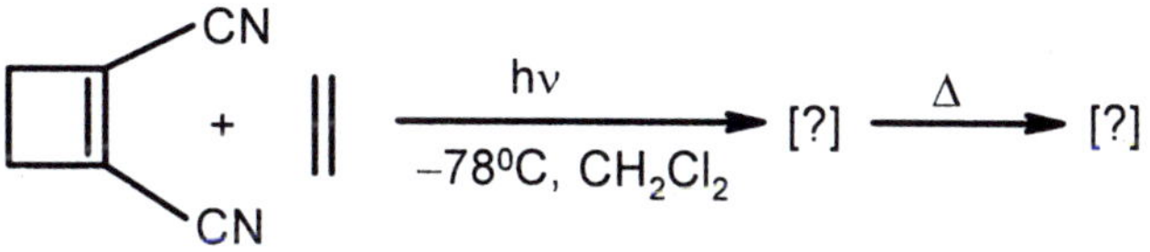

Solution.

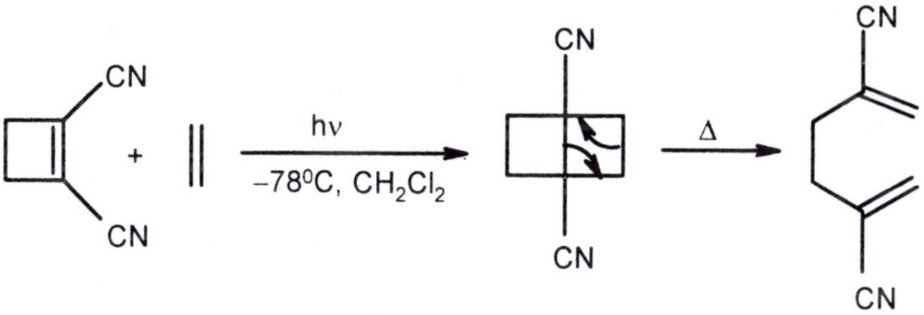

[*Helv. Chim. Acta.*, **57**, 194(1974)]

Problem 6.5

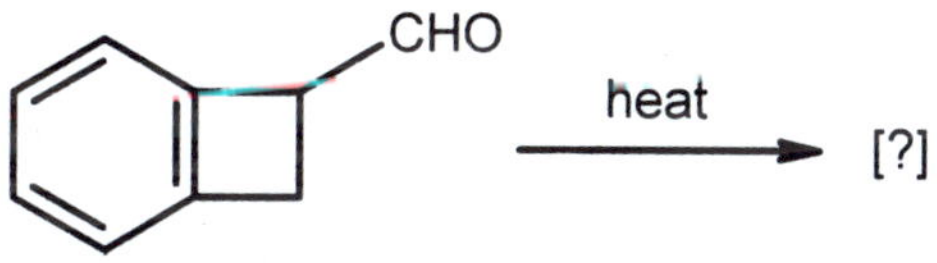

Solution.

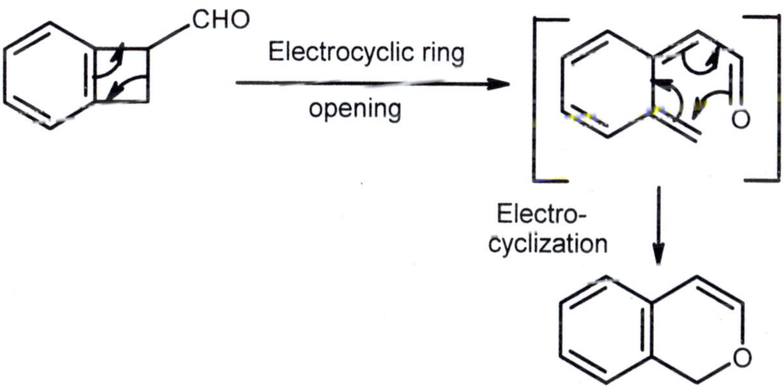

[*J. Org. Chem.*, **42**, 2642(1977)]

Problem 6.6

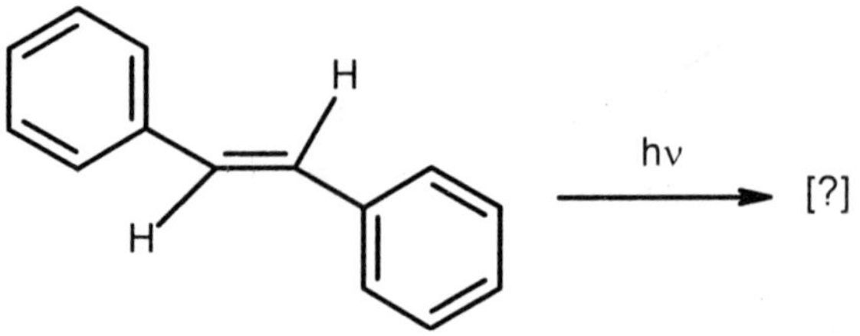

Solution.

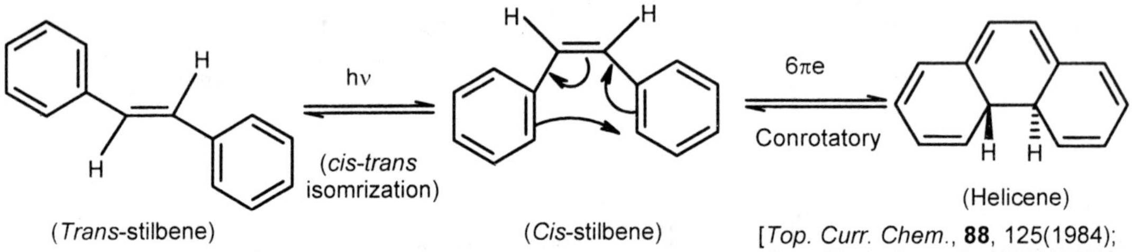

[*Top. Curr. Chem.*, **88**, 125(1984); *Org. Photochem.*, **9**, 1299(1987)]

Problem 6.7

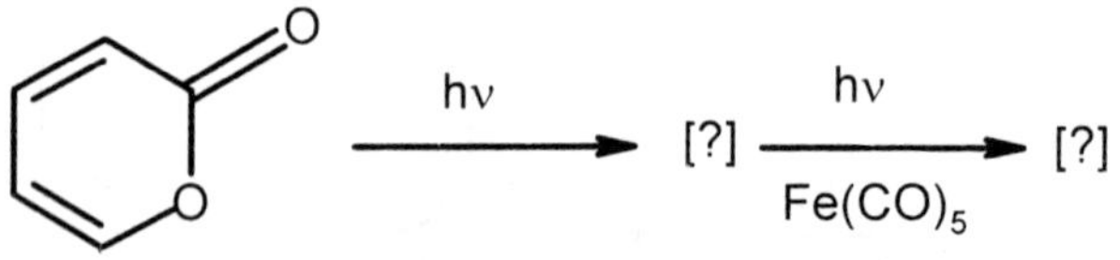

Solution.

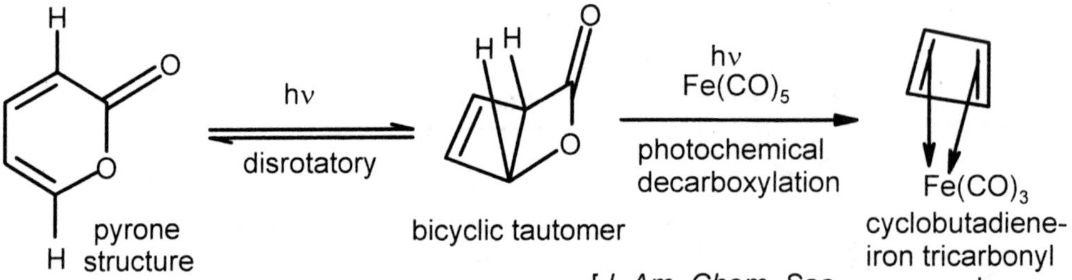

[*J. Am. Chem. Soc.*, **86**, 950(1964)]

Problem 6.8

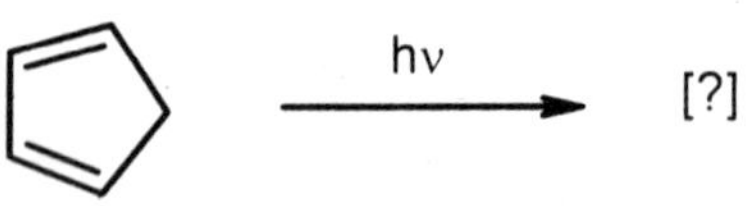

Solution.

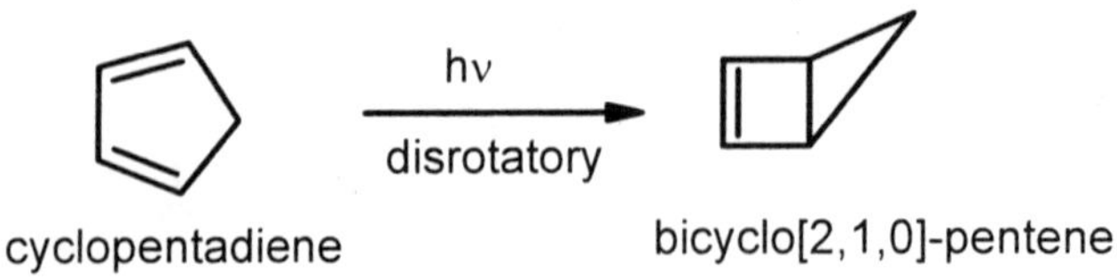

[*J. Am. Chem. Soc.*, **88**, 846(1966)]

Despite of its strained structure bicyclo[2,1,0]-pentene has been found to have a half-life of about two hours at room temperature.

Problem 6.9

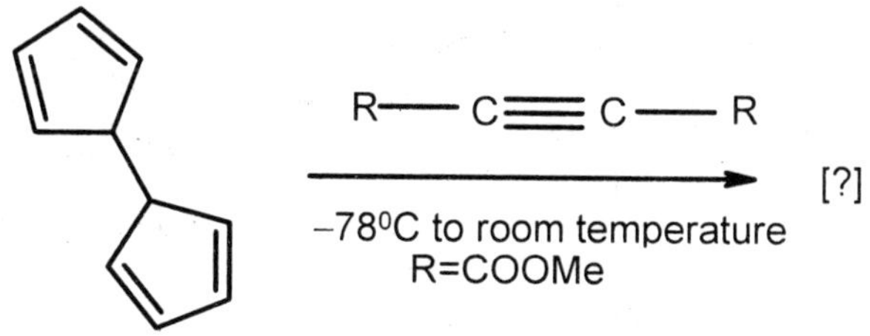

Solution.

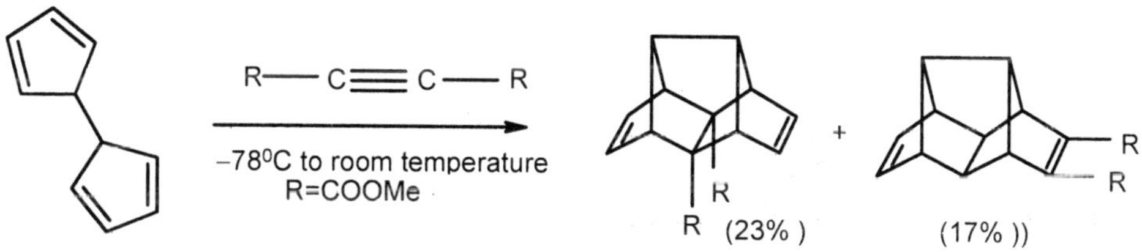

[*J. Am. Chem. Soc.*, **96**, 4671(1974)]

Problem 6.10

Solution.

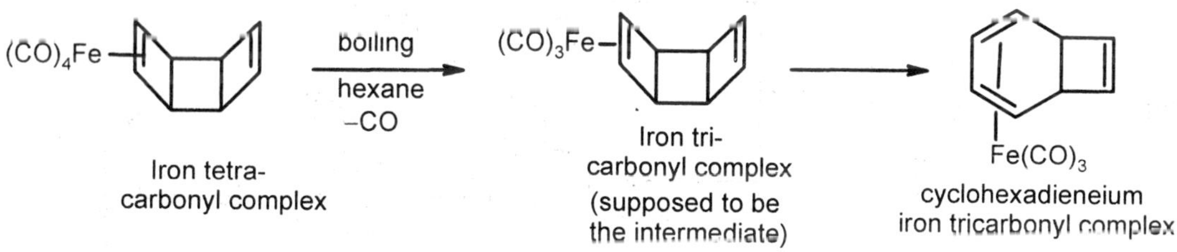

The reaction is an electrocyclic ring opening of a complexed cyclobutene. It is suggested that carbon monoxide liberated in the first step inhibits the reaction [Ref. *J. Am. Chem. Soc.*, **96**, 287(1974)].

Problem 6.11

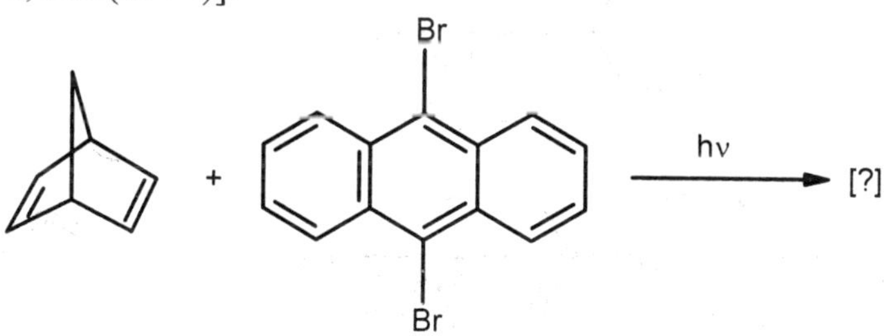

Solution.

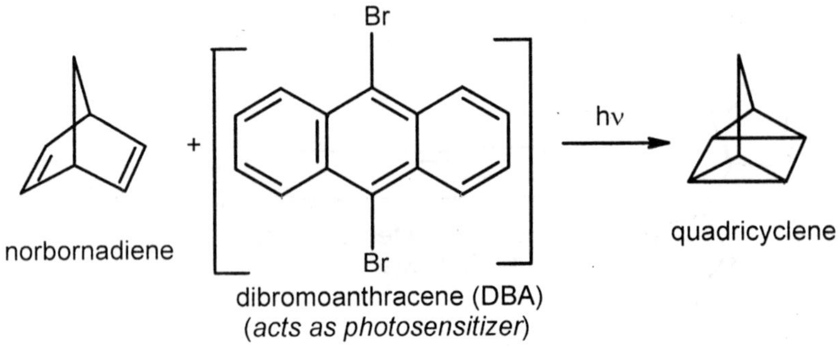

[*Tet.*, **15**, 197(1961); *J. Am. Chem. Soc.*, **83**, 4674(1961)]

Problem 6.12

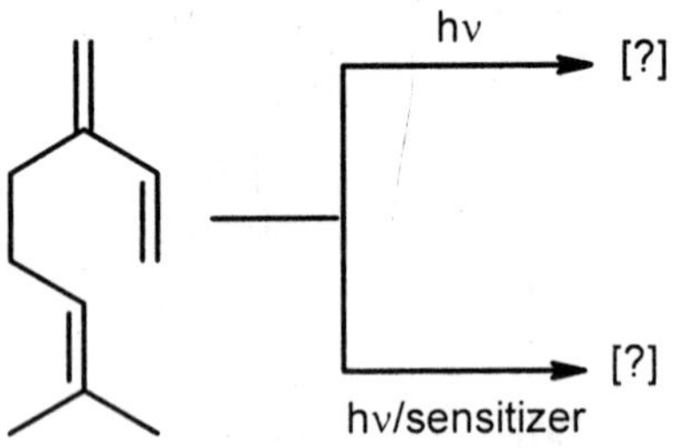

Solution.

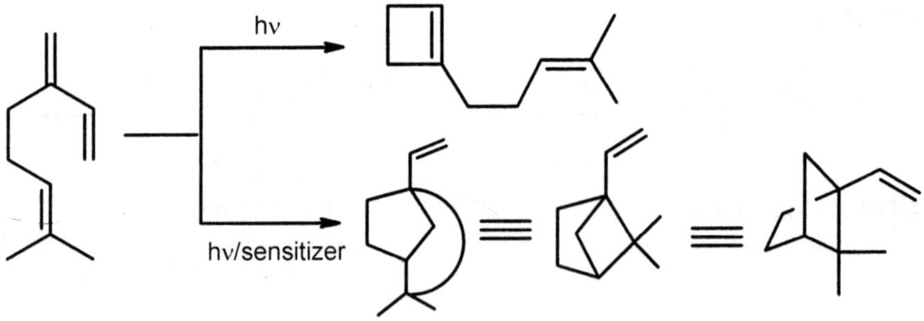

Problem 6.13

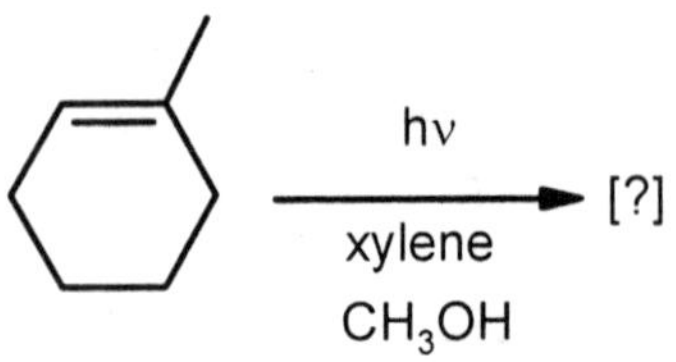

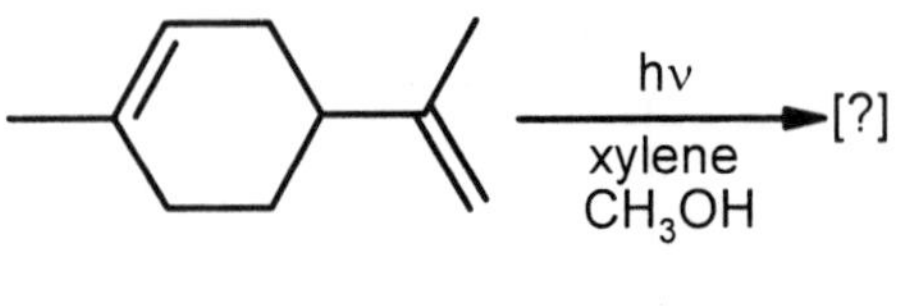

Solution. 1-Methyl cyclohexene (I) in methanol undergoes solvent addition as well as ionic rearrangement in presence of triplet-xylene sensitizer. The reactant is assumed to produce firstly a *trans*-cyclohexene, which then gets protonated by methanol solvent.

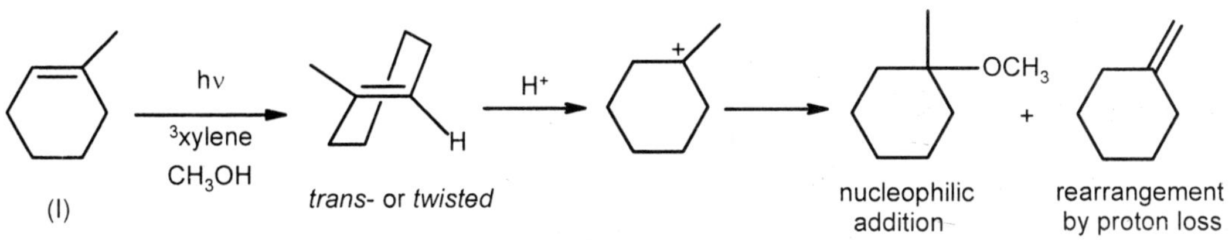

But in case of compound (II) the exocyclic double bond restricts the formation of metastable *trans*- or *twisted* intermediate; thus there results only ionic addition product involving the endocyclic double bond.

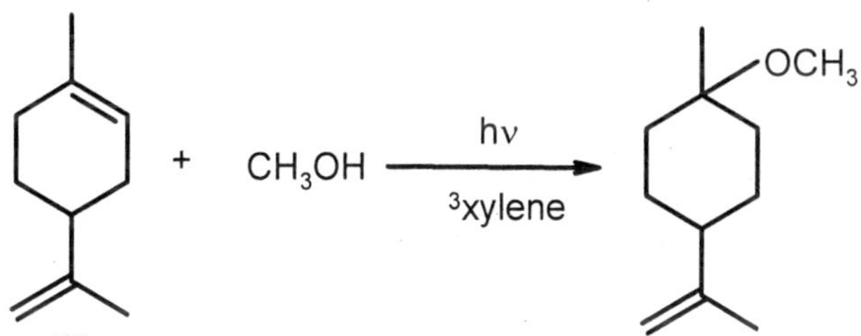

(II)

Problem 6.14

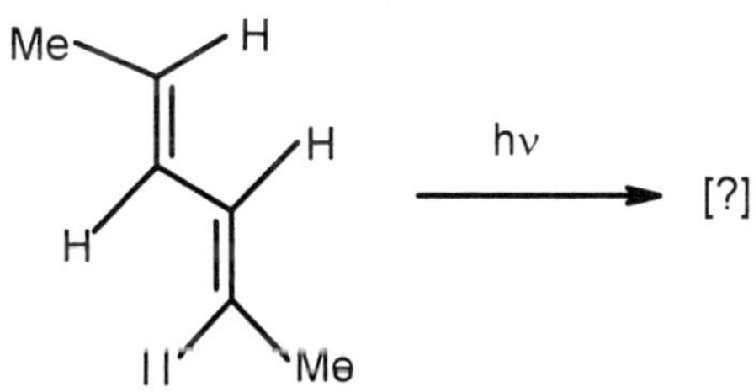

Solution.

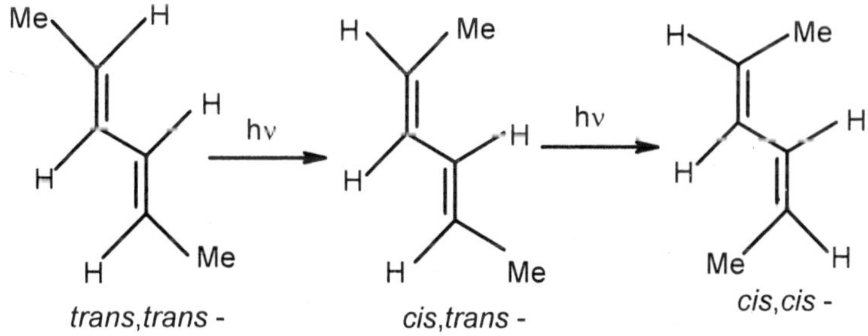

[*Chem. Revs.*, **55**, 625(1955)]

This is an example of photo-induced isomerization; on irradiation *trans,trans*-isomer of hexa-2,4-diene undergoes *cis-trans* isomerization resulting a mixture of geometrical isomers.

Problem 6.15

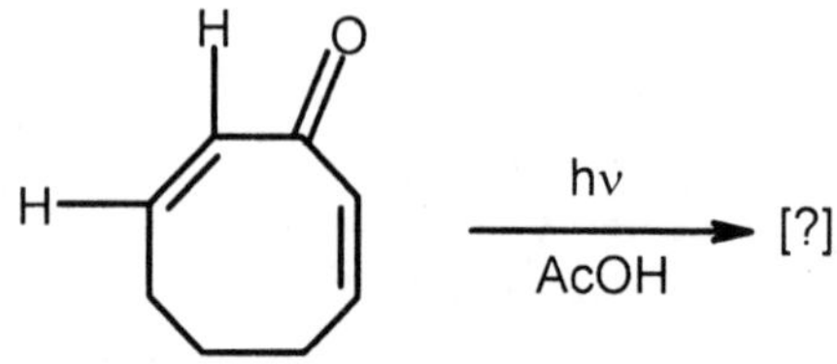

Solution.

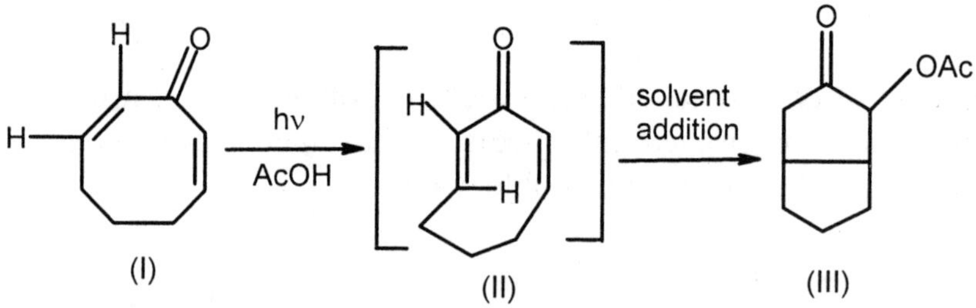

The compound (I) first undergoes a photochemical *cis – trans* isomerization to form highly reactive *trans*-cyclic intermediate (II), which in turn proceeds to the solvent addition product (III) *via* secondary thermal reaction.

Problem 6.16

Solution.

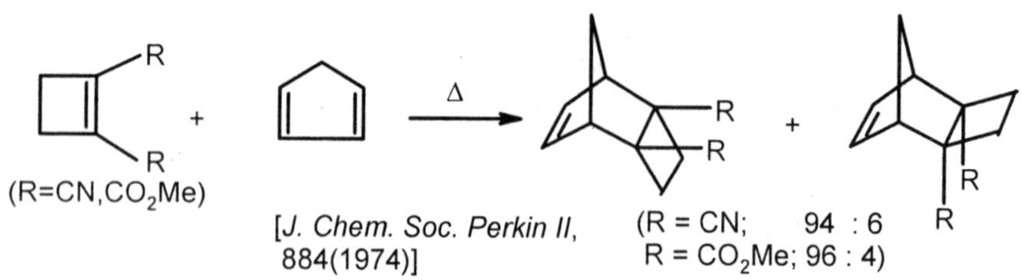

Problem 6.17

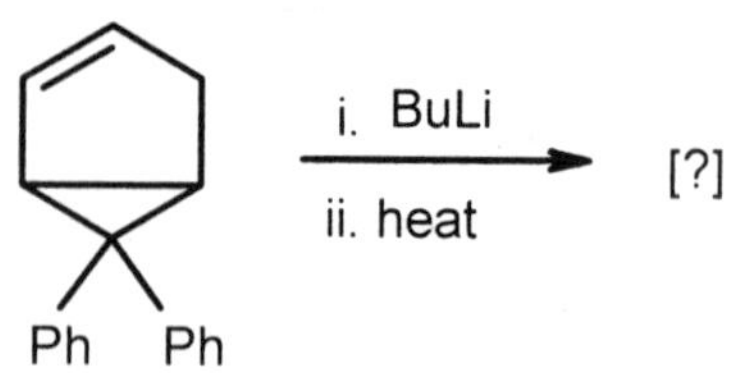

Solution.

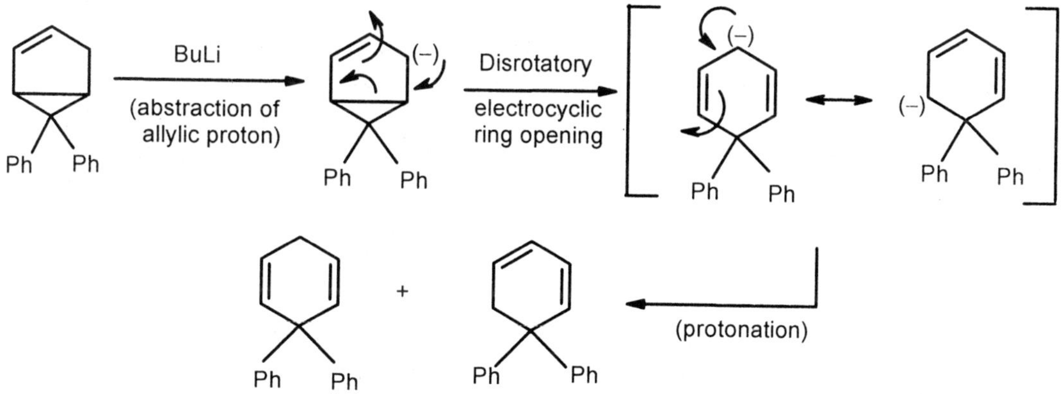

Problem 6.18

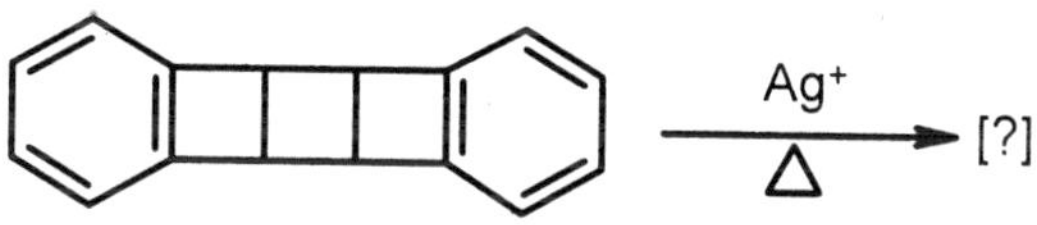

Solution.

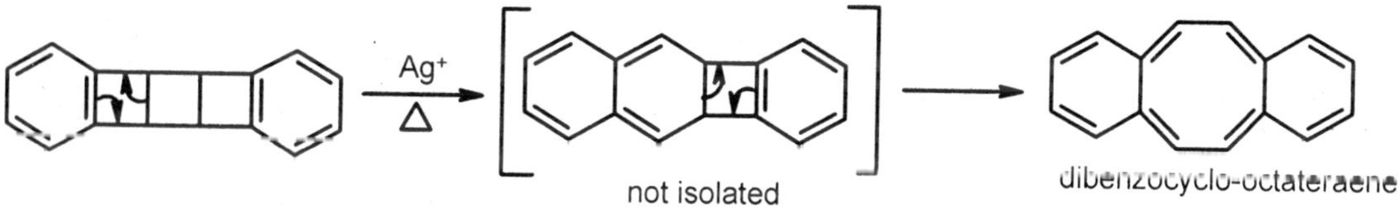

This is an example of metal-catalyzed electrocyclic disrotatory ring opening reaction; activation energy of 96 kJ/mole for the non-catalyzed reaction was found to be about 33 kJ/mole for the catalyzed one [Ref. *Adv. Catalysis*, **20**, 91(1969); *Coord. Chem. Rev.*, **15**, 109(1975)].

Problem 6.19

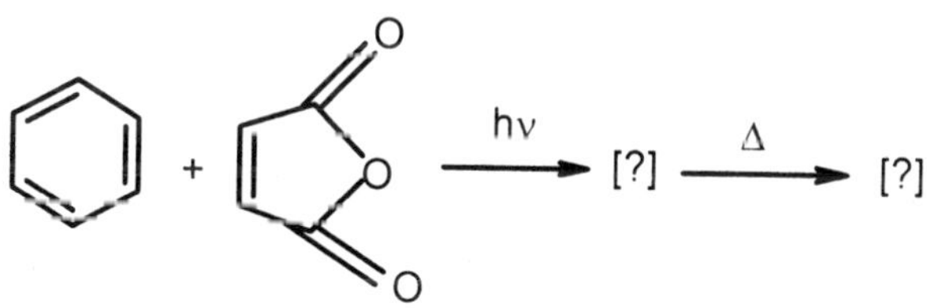

Solution. The adduct (X) is formed during [2+2]-photoaddition reaction of benzene and maleic anhydride followed by a thermal [4+2] cycloaddition.

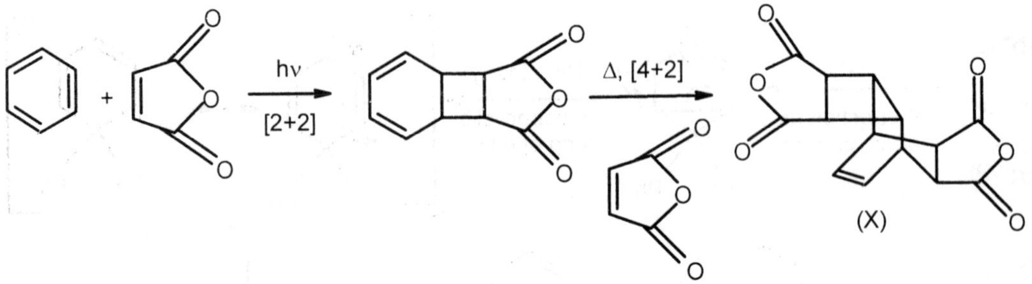

Problem 6.20

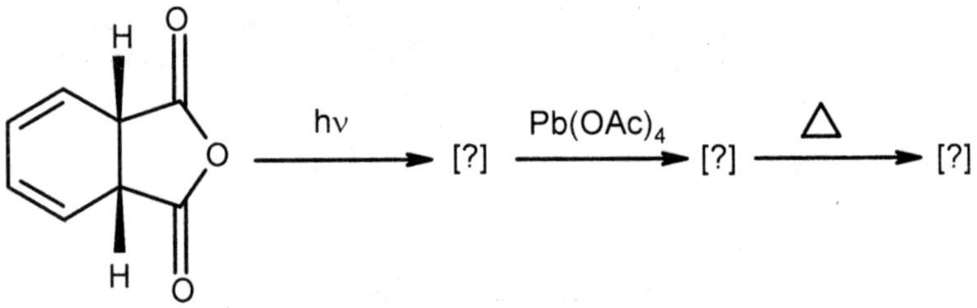

Solution.

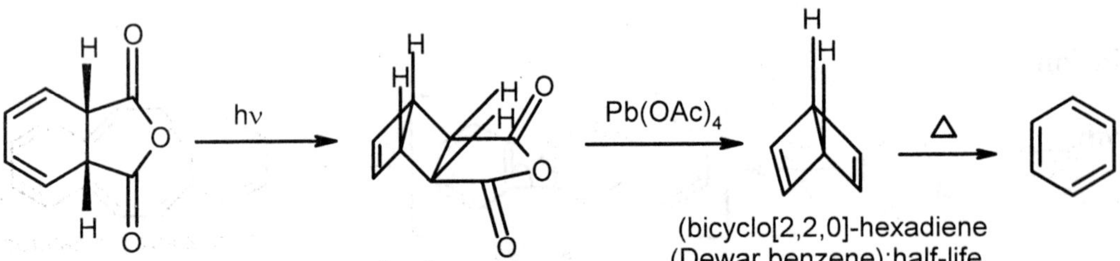

(undergoes photochemical decarboxylation)

(bicyclo[2,2,0]-hexadiene (Dewar benzene);half-life of about twodays in pyridine solution at room temperature)

Problem 6.21

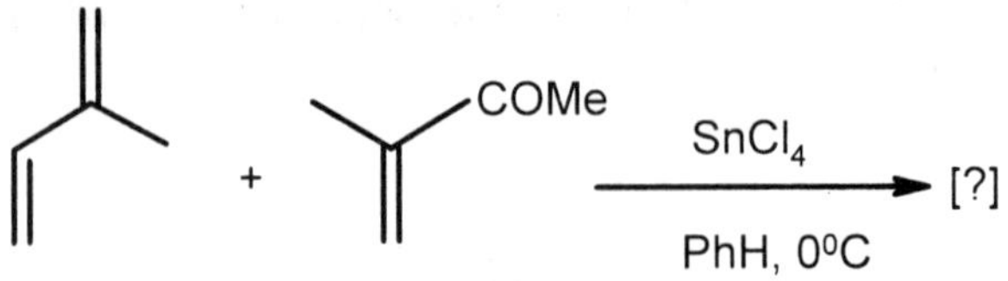

Solution.

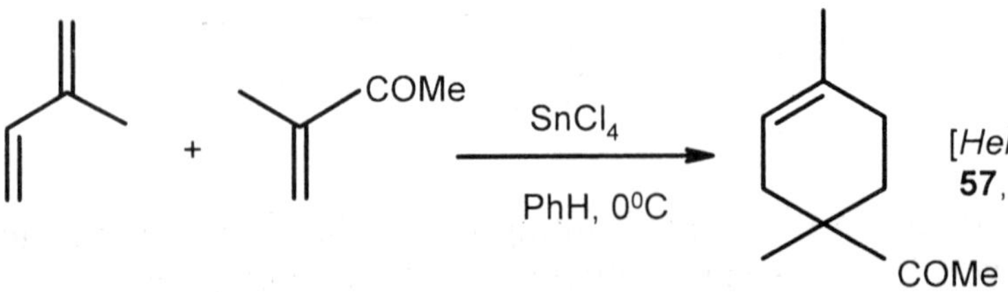

[*Helv. Chim. Acta.*, **57**, 164(1974)]

Problem 6.22

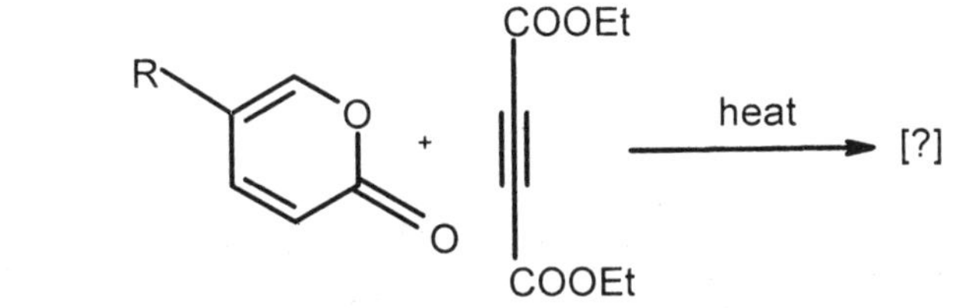

Solution.

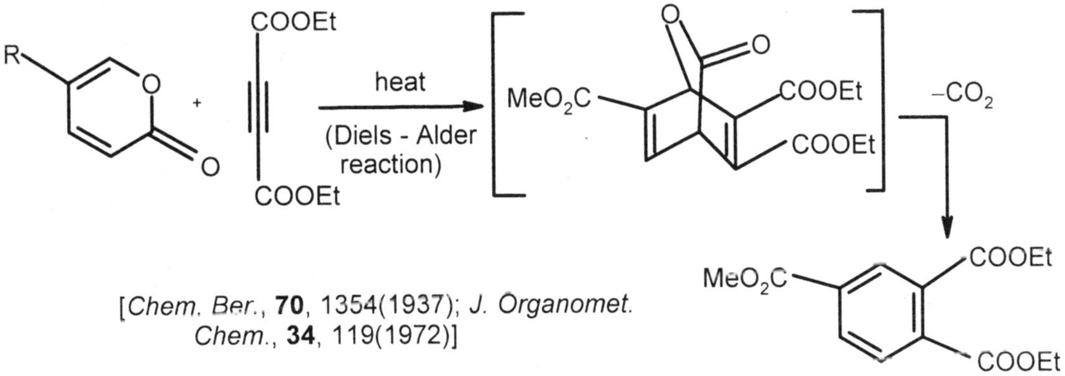

[*Chem. Ber.*, **70**, 1354(1937); *J. Organomet. Chem.*, **34**, 119(1972)]

Problem 6.23

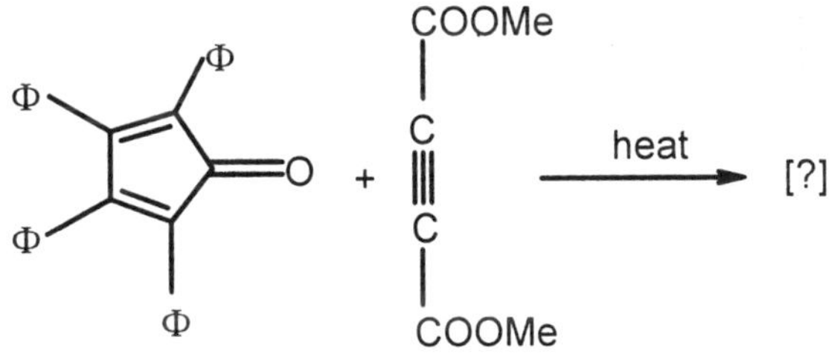

Solution.

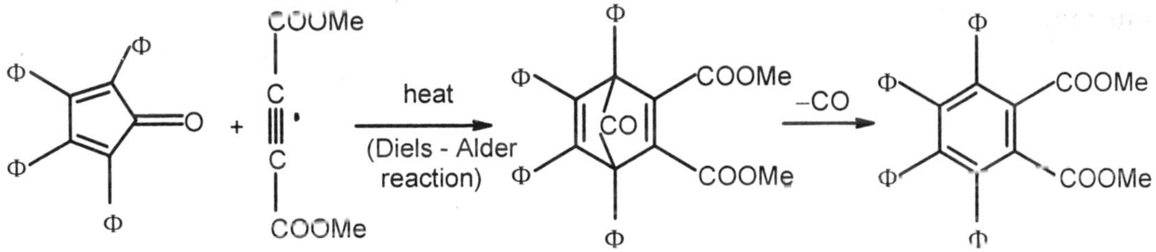

Problem 6.24

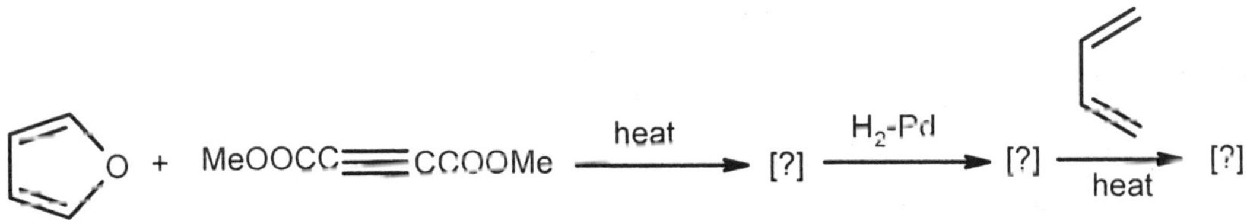

Solution.

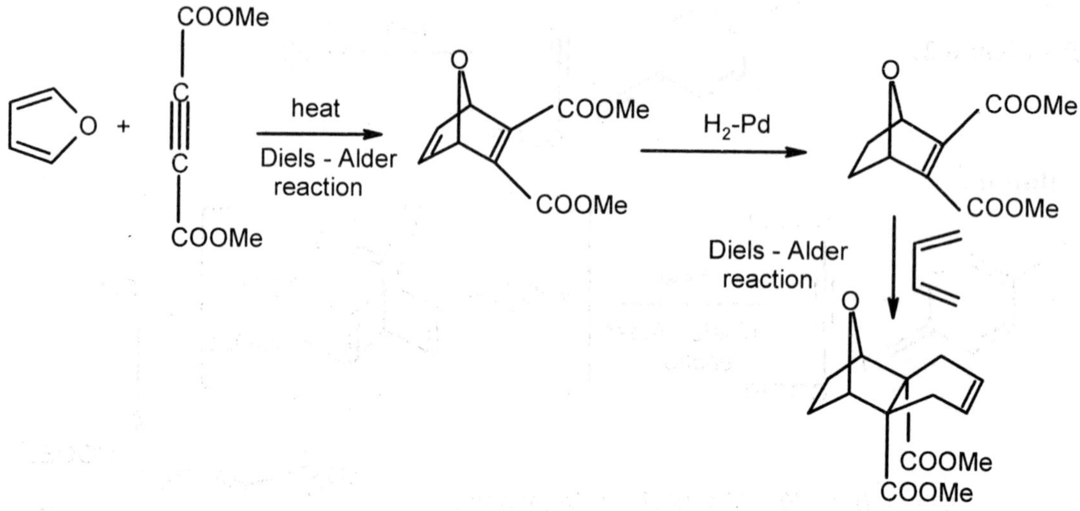

Problem 6.25

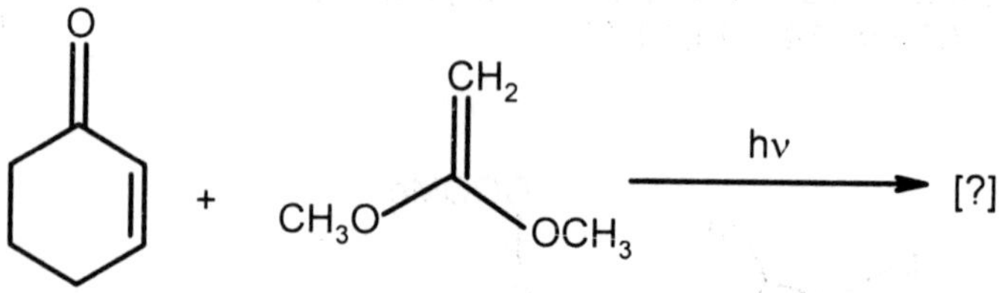

Solution.

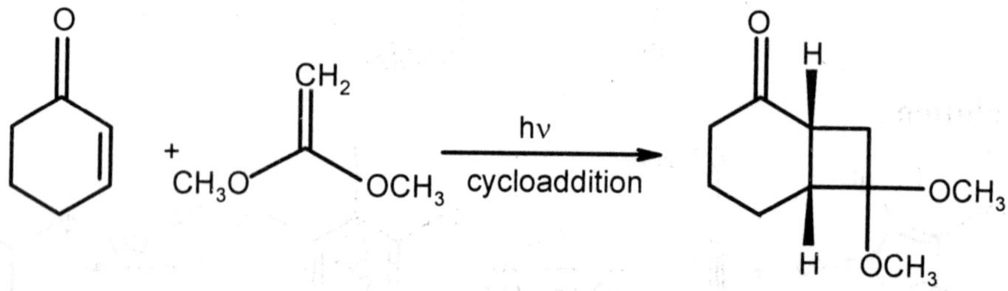

Problem 6.26

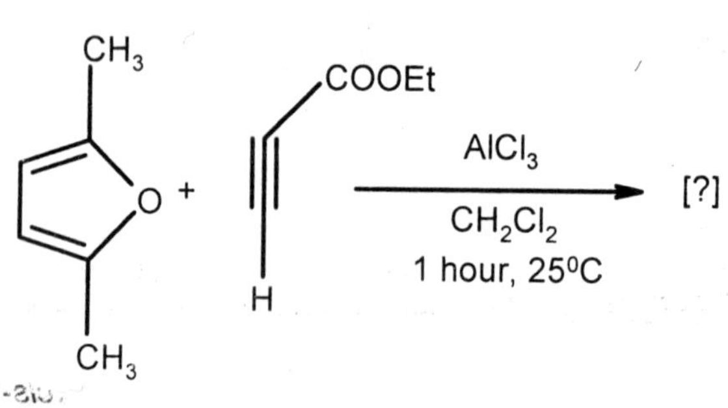

Solution.

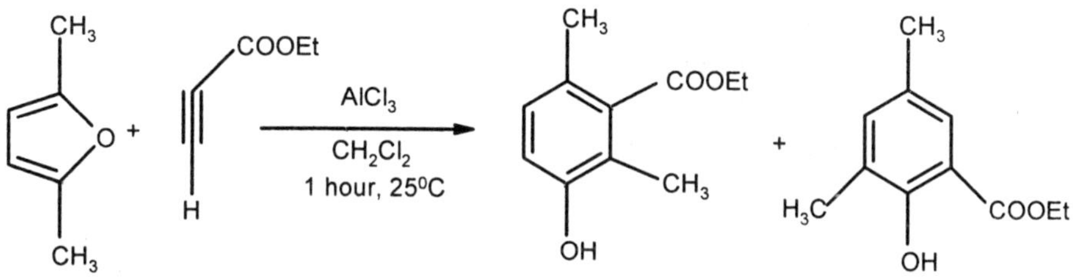

(1 : 6) [*Can. J. Chem.*, **52**, 143(1974)]

Problem 6.27

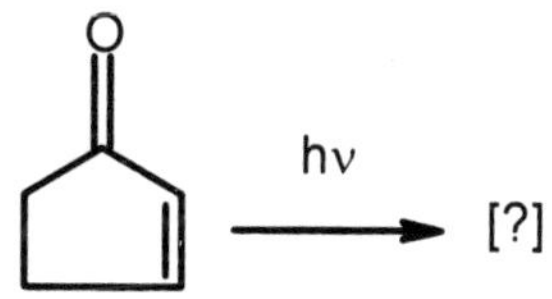

Solution.

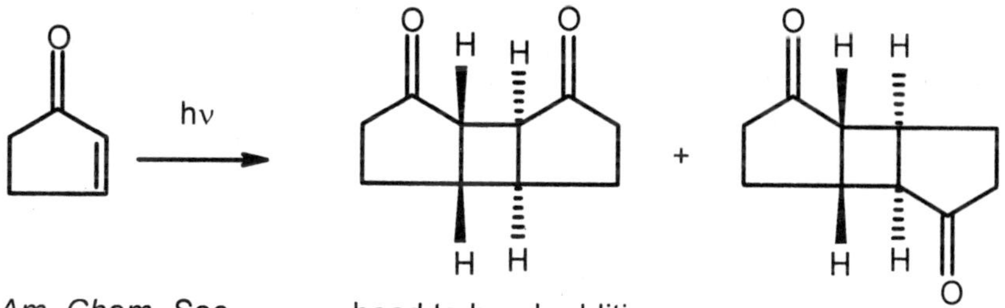

[*J. Am. Chem. Soc.*, **84**, 2344(1962)]

head to head addition (*cis-anti-cis*)

head to tail addition (*cis-anti-cis*)

(equal amount)

Problem 6.28

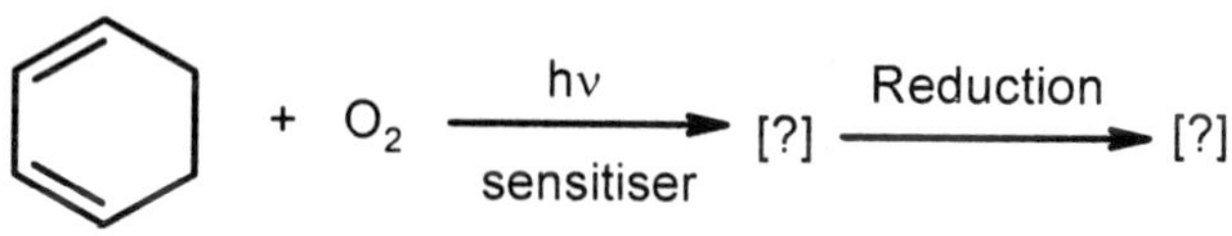

Solution.

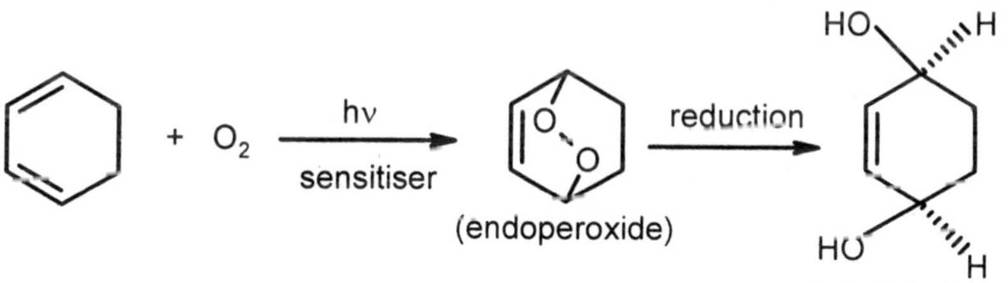

(endoperoxide)

(*cis*-1,4-diol)

Problem 6.29

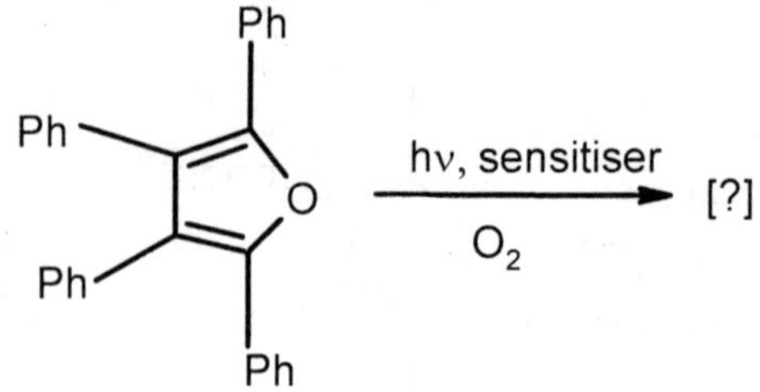

Solution.

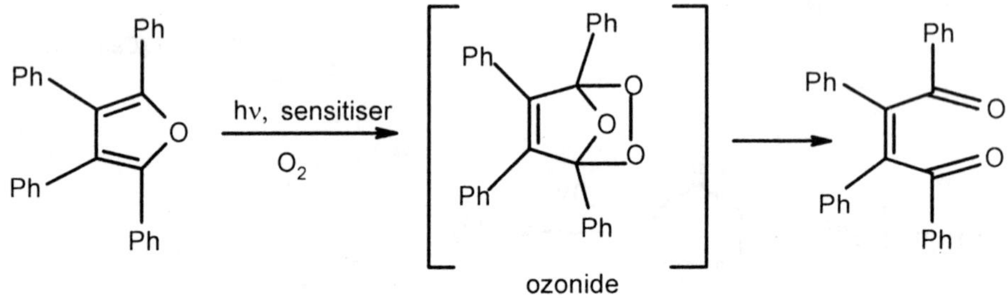

Problem 6.30

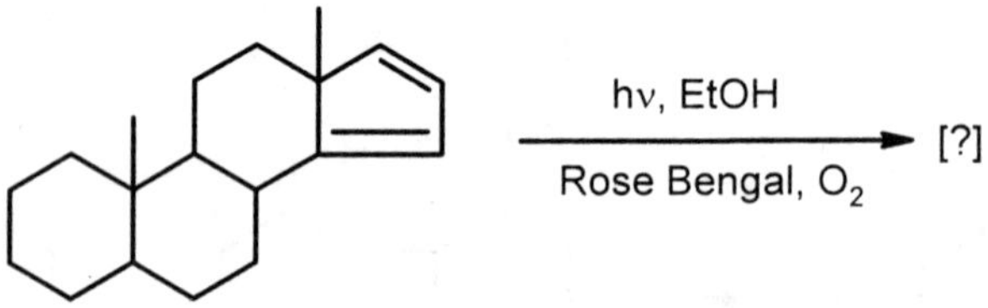

Solution.

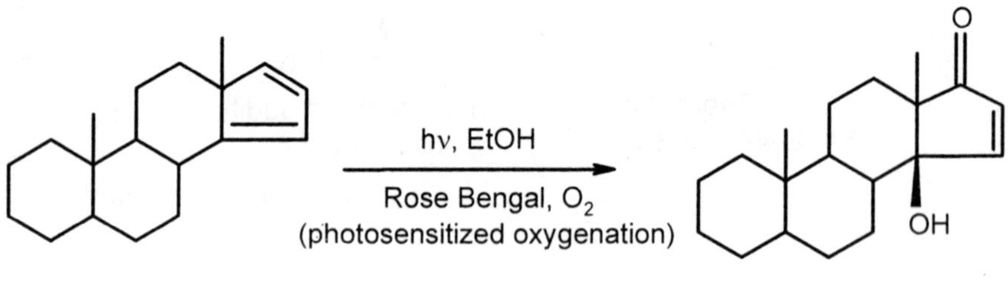

Problem 6.31

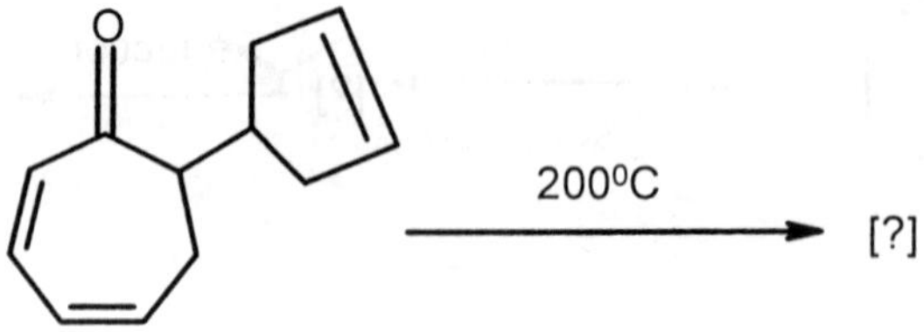

Solution.

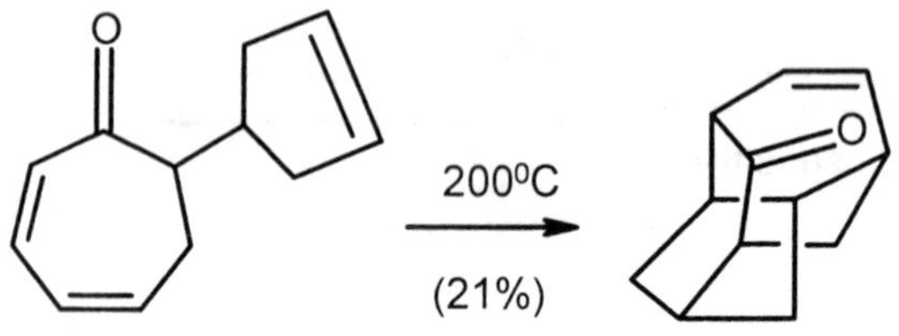

[*J. Am. Chem. Soc.*, **96**, 4668(1974)]

Problem 6.32

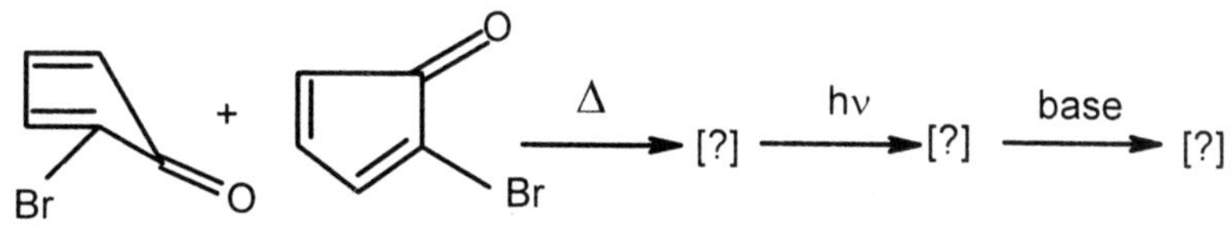

Solution.

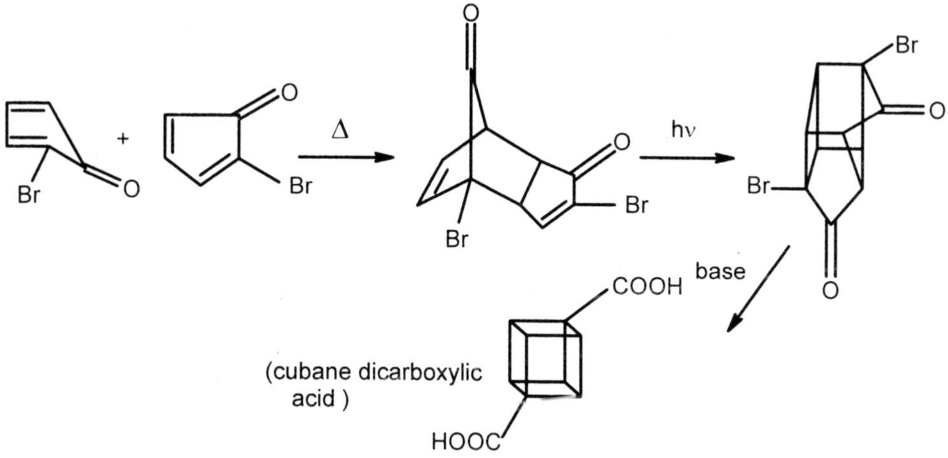

Problem 6.33

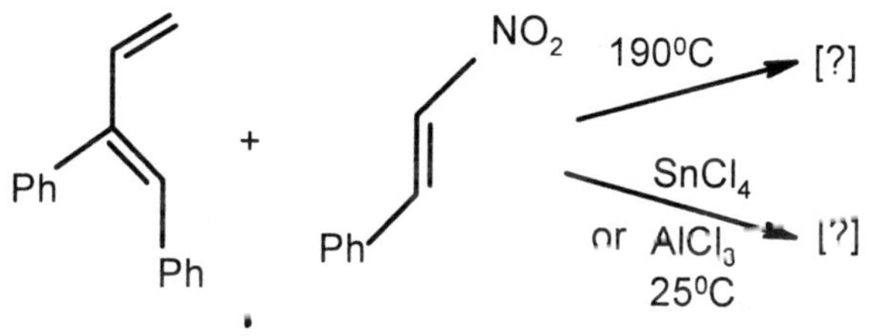

Solution.

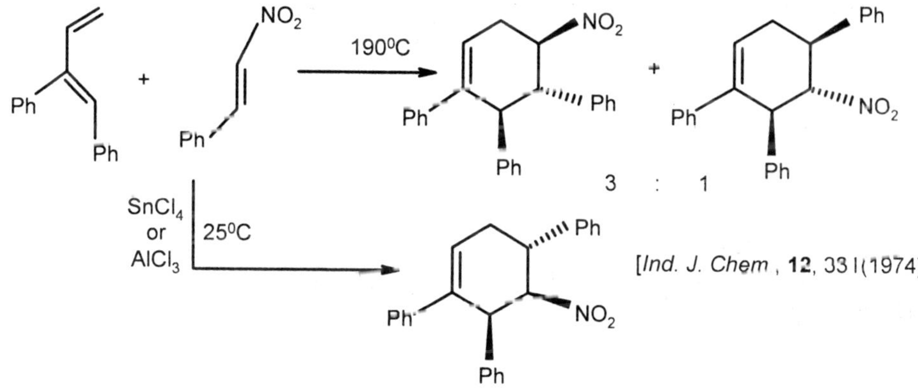

[*Ind. J. Chem* , **12**, 331(1974)]

Problem 6.34

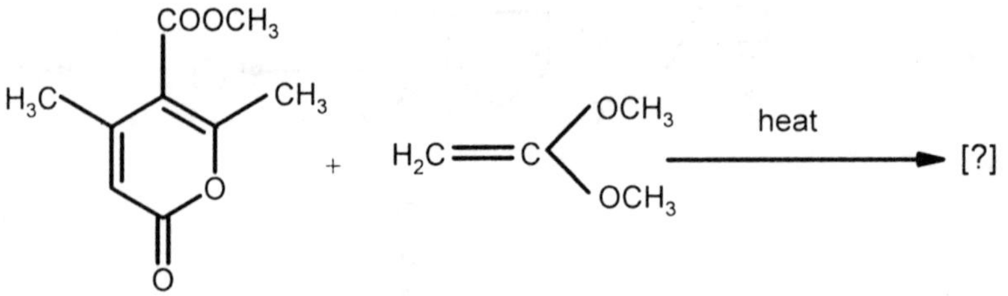

Solution.

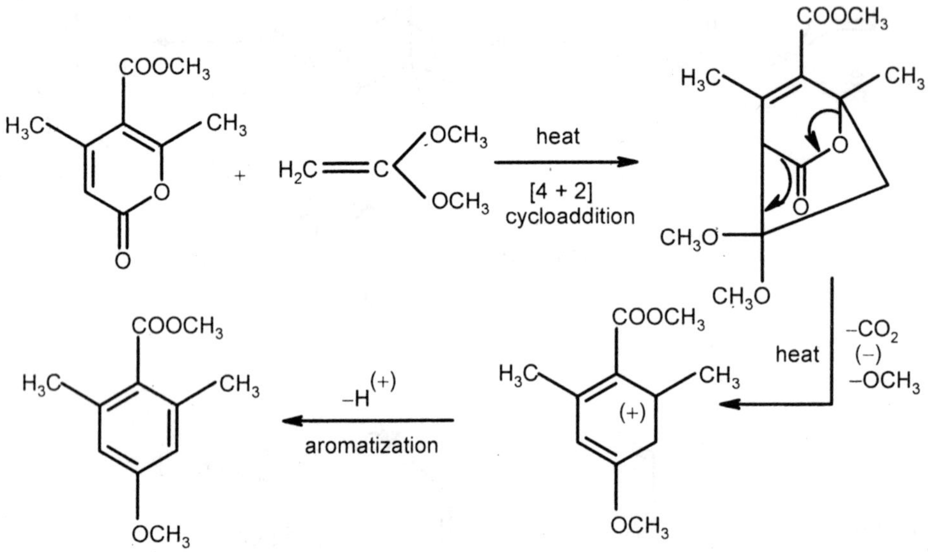

Problem 6.35

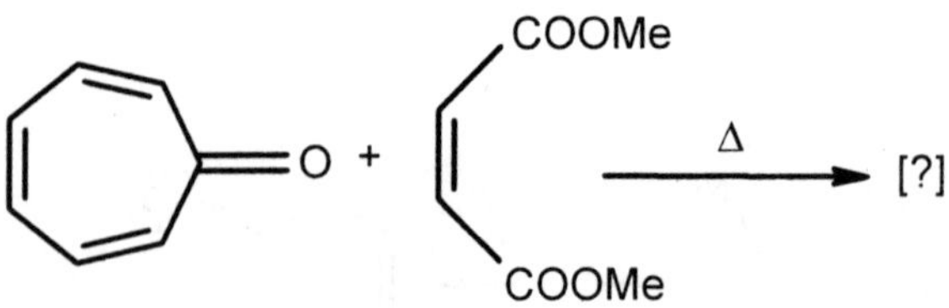

Solution.

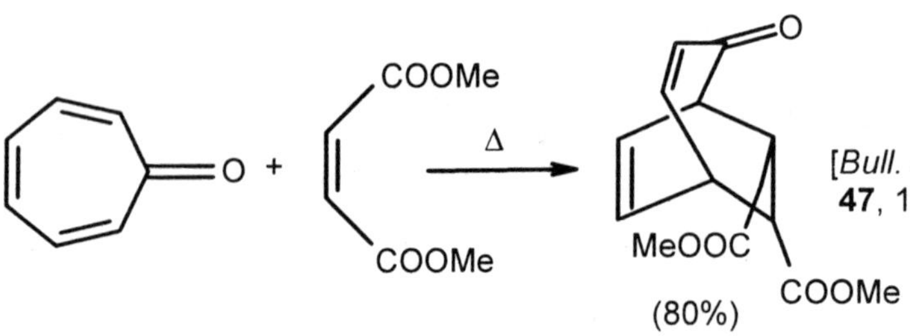

[*Bull. Soc. Chem. Jap.*, **47**, 1041(1974)]

Problem 6.36

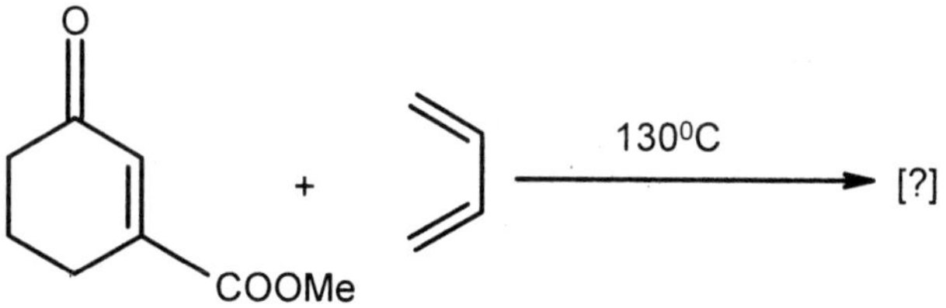

Solution.

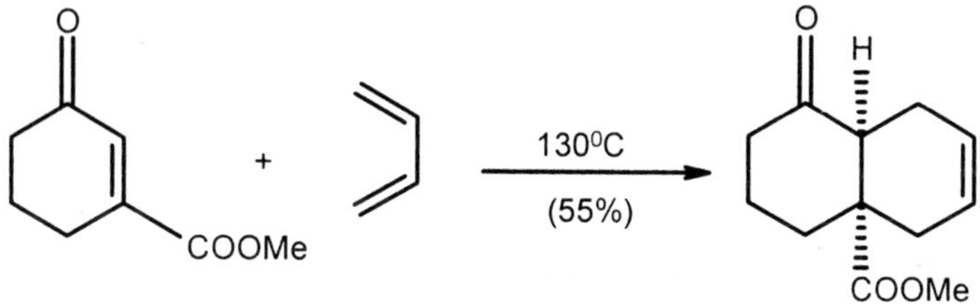

Problem 6.37

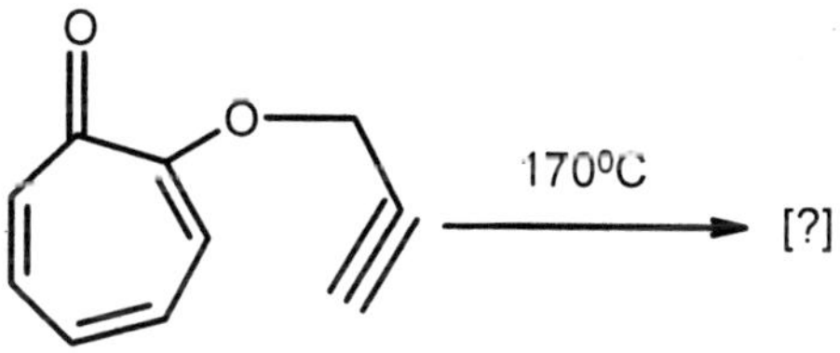

Solution.

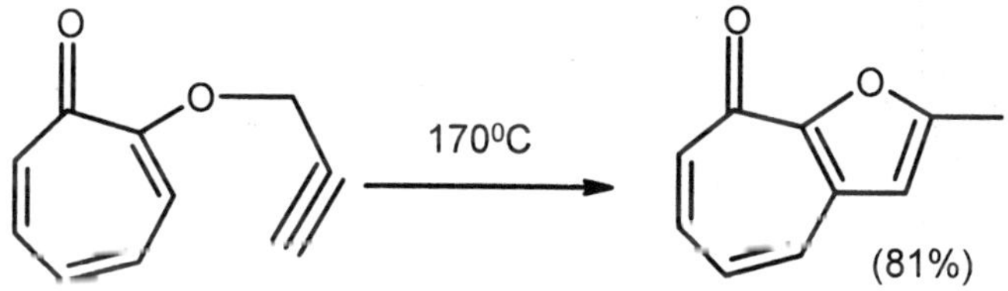

Problem 6.38

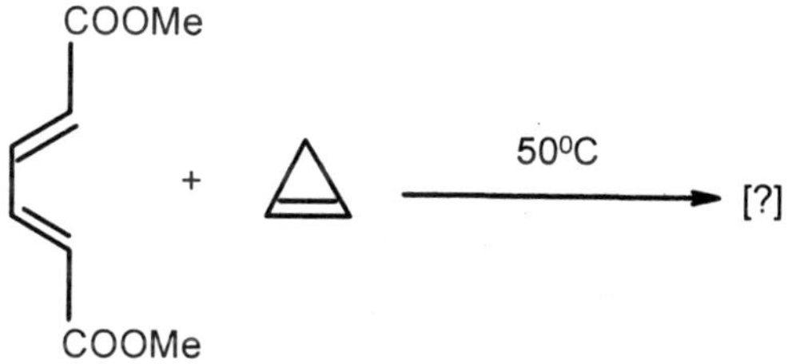

Solution.

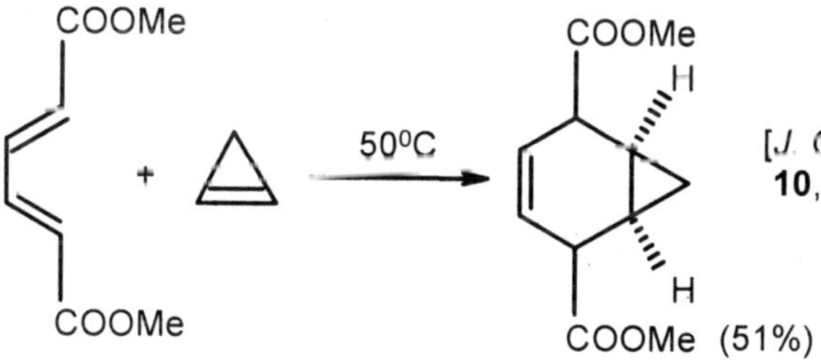

[*J. Org. Chem. U.S.S.R.*, **10**, 1672(1974)]

Problem 6.39

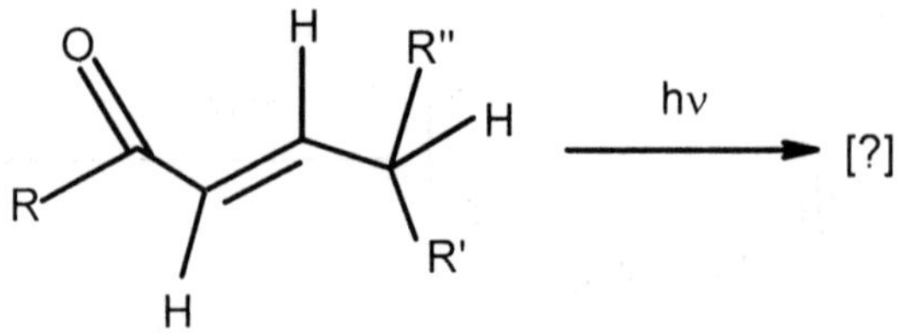

Solution.

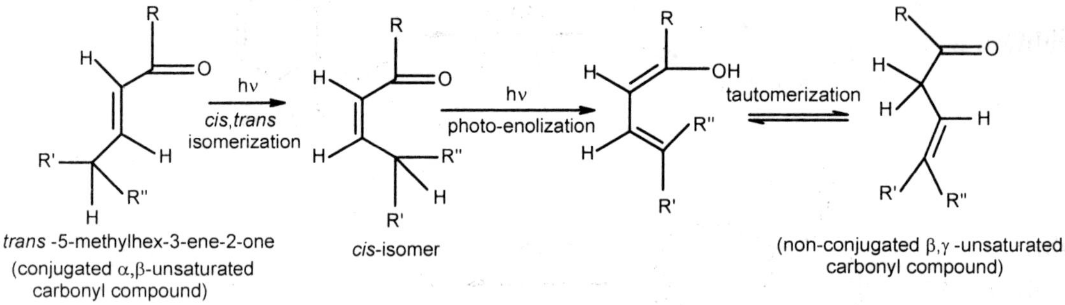

Problem 6.40

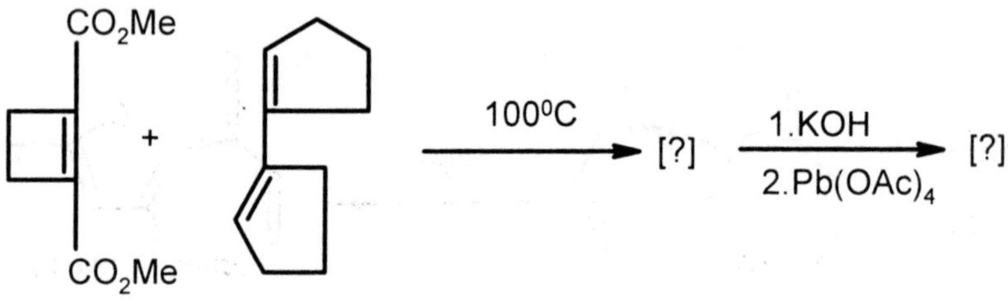

Solution.

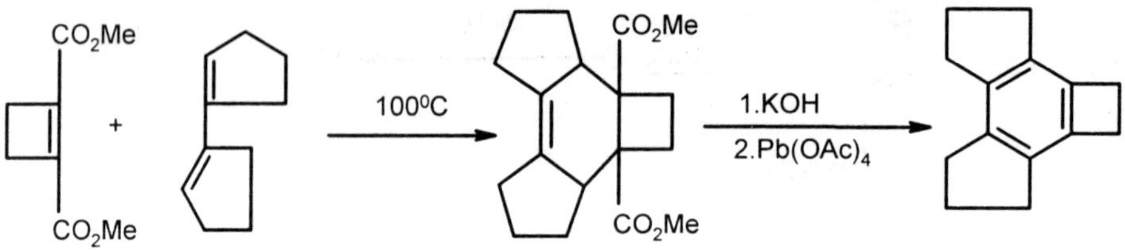

Problem 6.41

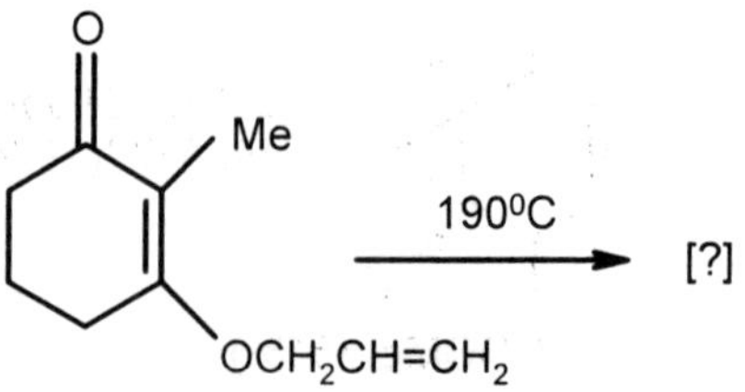

Solution.

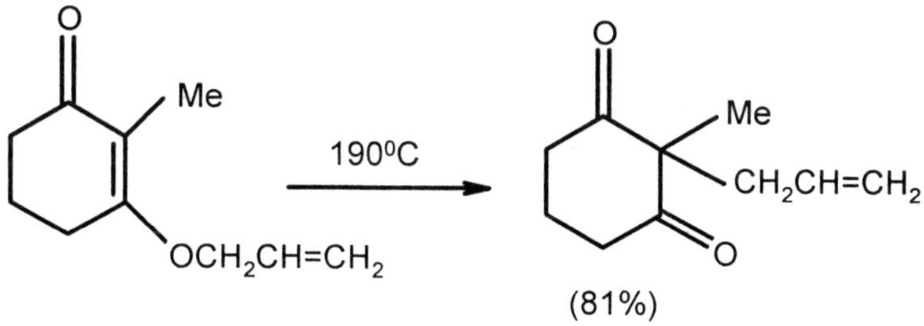

Problem 6.42

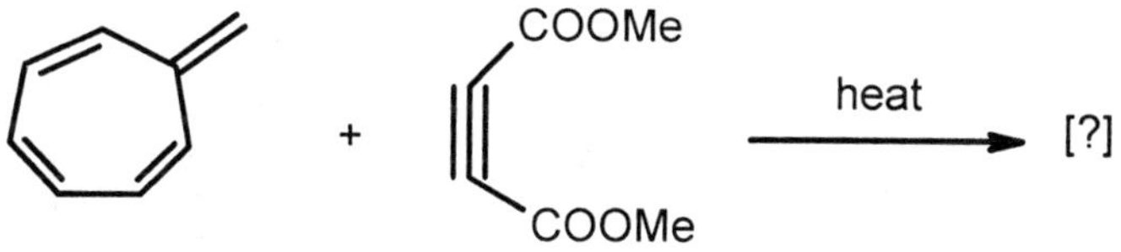

Solution.

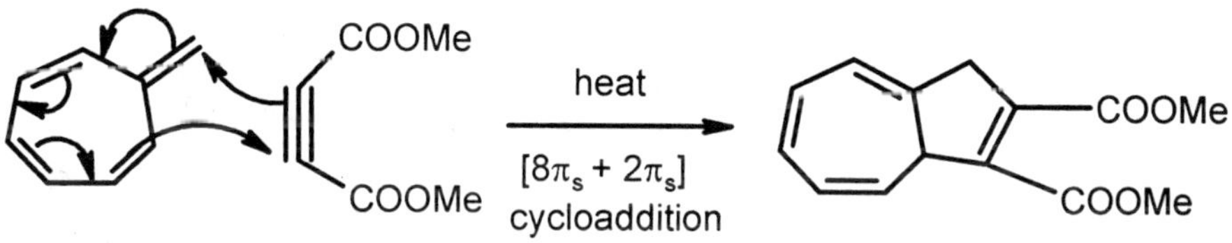

[*Tet.*, **11**, 183(1960)]

Problem 6.43

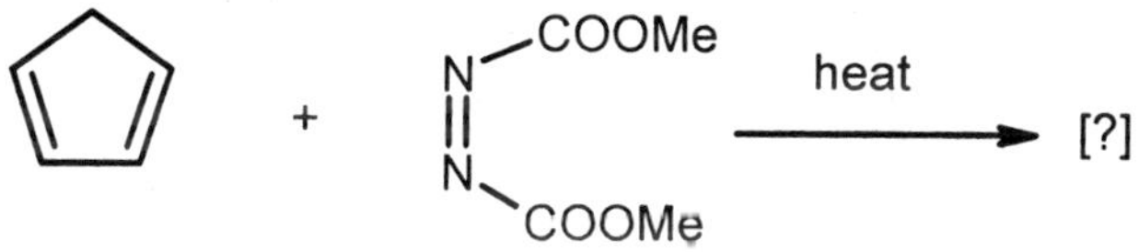

Solution.

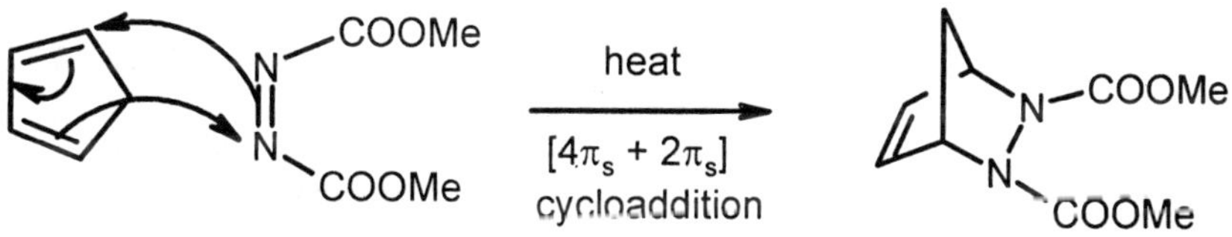

Problem 6.44

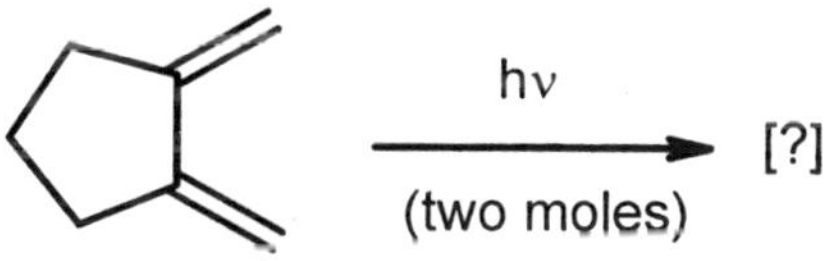

Solution.

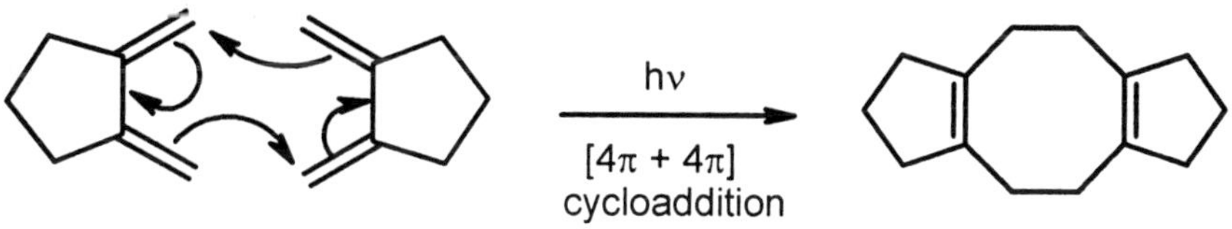

Problem 6.45

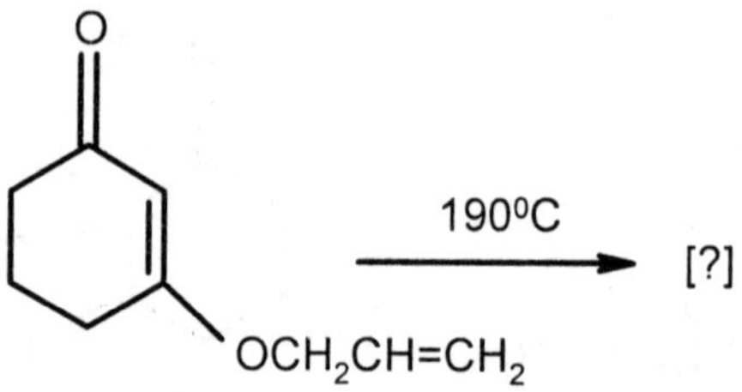

Solution.

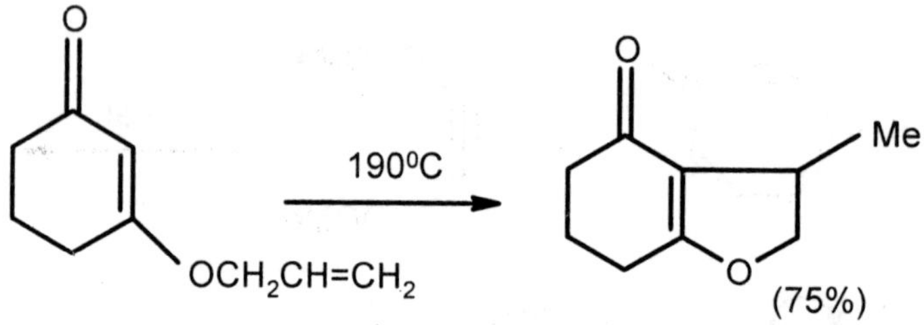

Problem 6.46

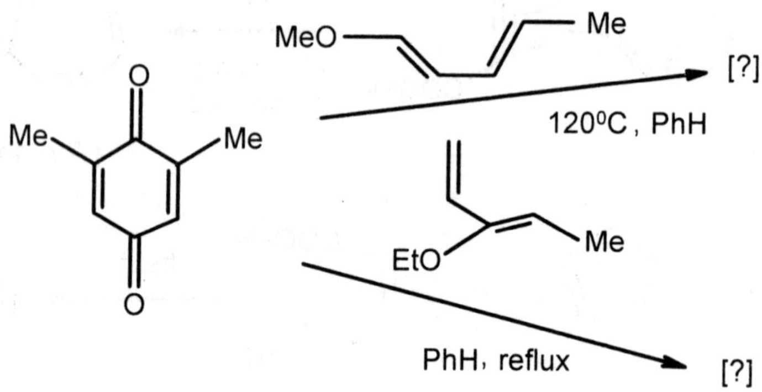

Solution.

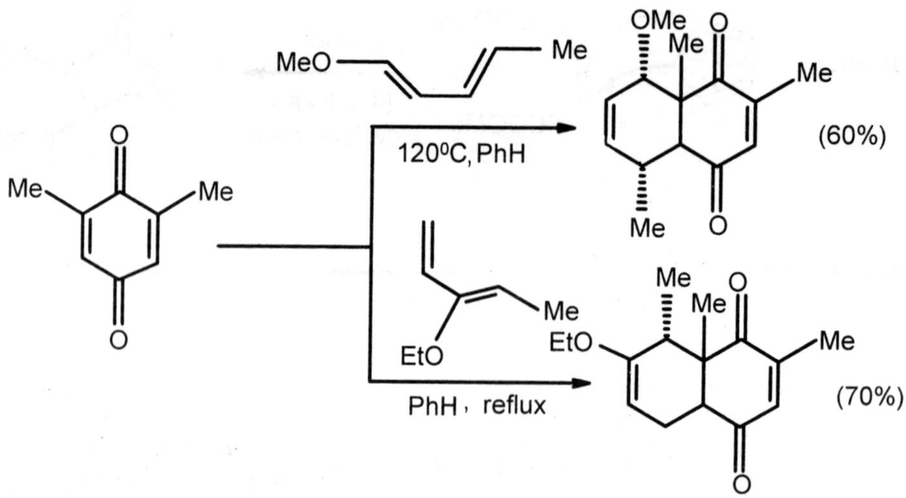

Problem 6.47 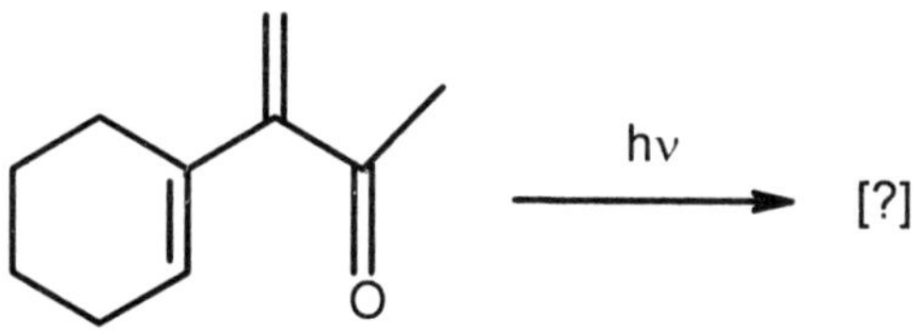

Solution. 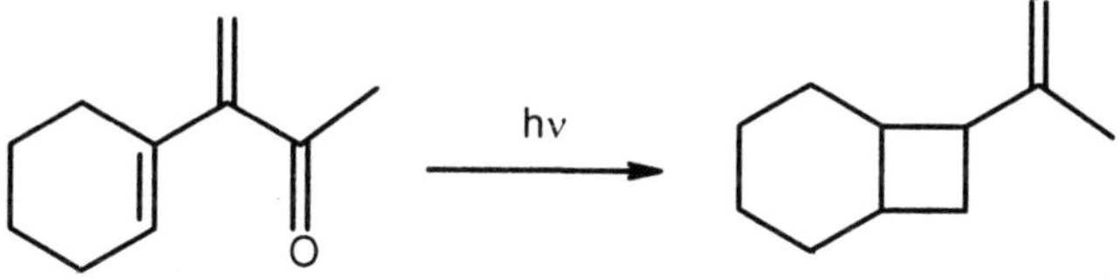

Problem 6.48 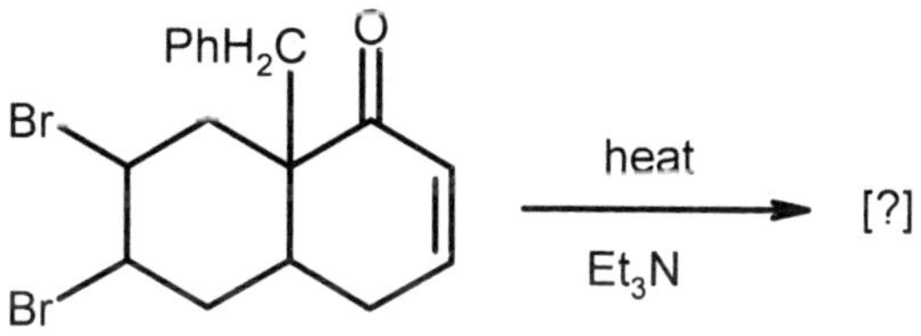

Solution.

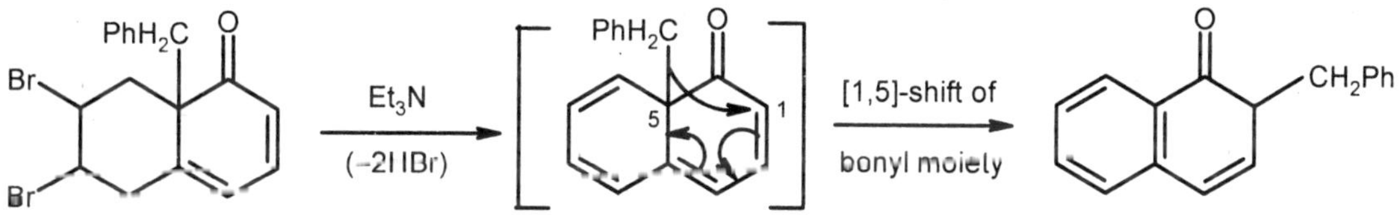

[*J. Org. Chem.*, **52**, 3390(1987)]

Problem 6.49 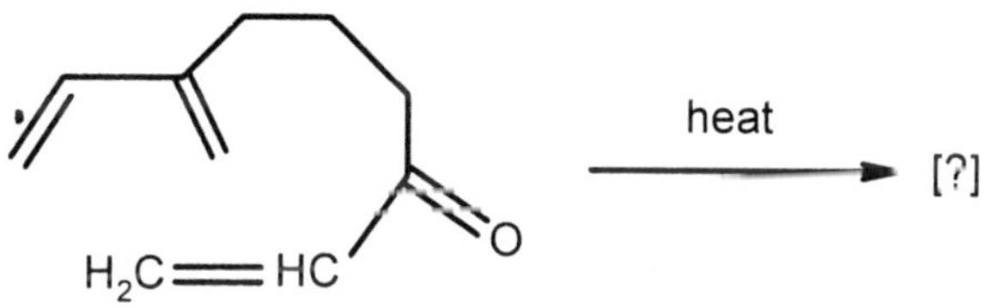

Solution.

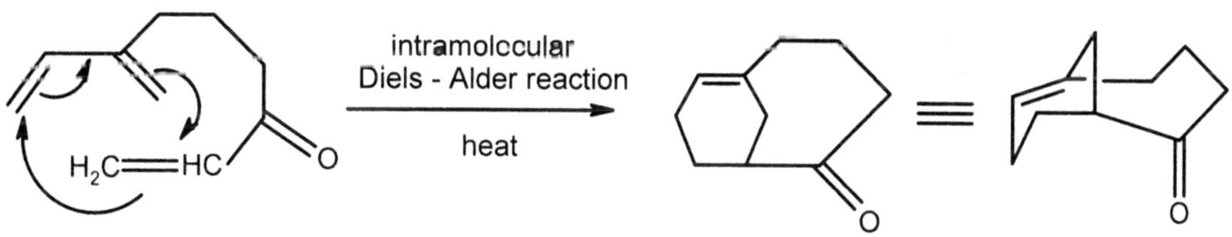

The bridged double bond in the product does not violate Bredt's rule, because it is a part of a large ring.

Problem 6.50

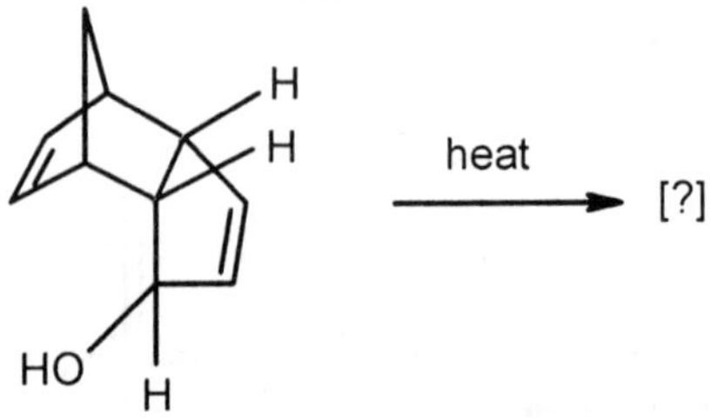

Solution. The *Diels-Alder adduct* (I), when heated at 140^{0}C, undergoes an intramolecular rearrangement to form (II). The formation of (II) could be explained considering the cleavage of bond 'a' followed by recombination of the allylic fragments as shown in (III). In this *retro-Diels-Alder cleavage* bond 'a' cleaves before bond 'b' [Ref. *Tetrahedron*, **5**, 70(1959)].

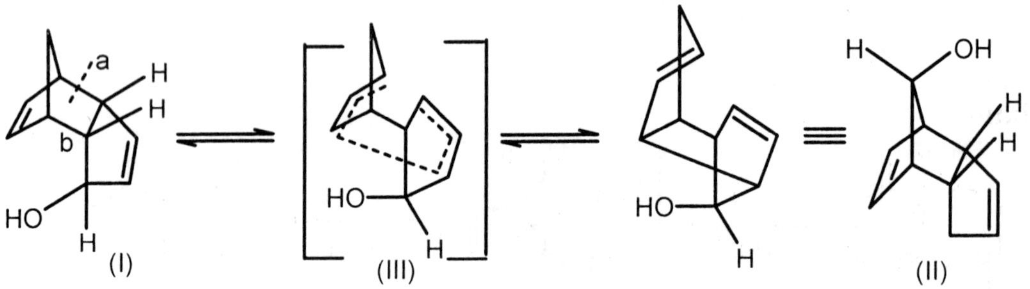

Problem 6.51

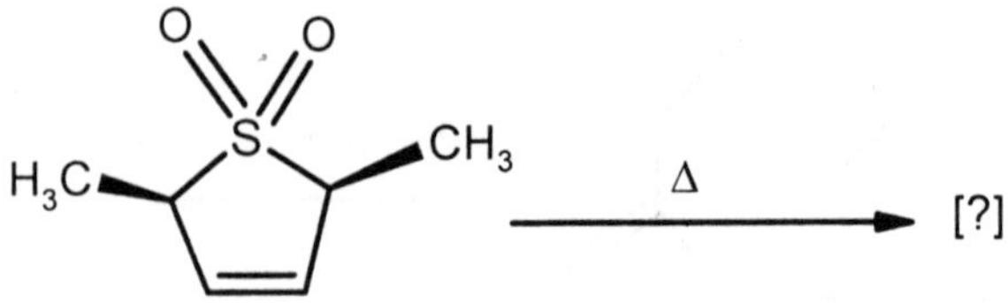

Solution.

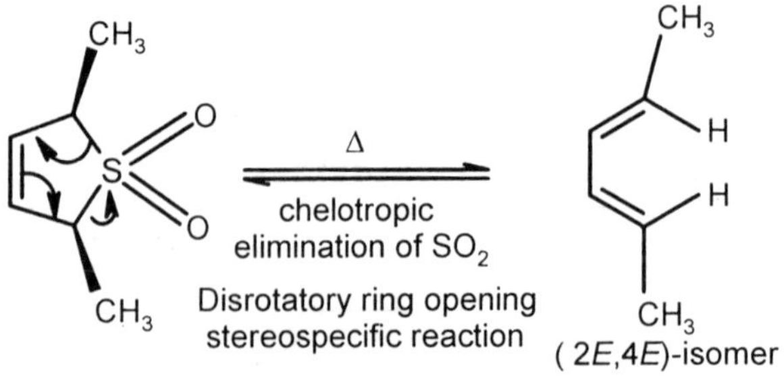

[*J. Am. Chem. Soc.*, **88**, 2857 (1966); *ibid*, **97**, 3666(1975)]

Problem 6.52

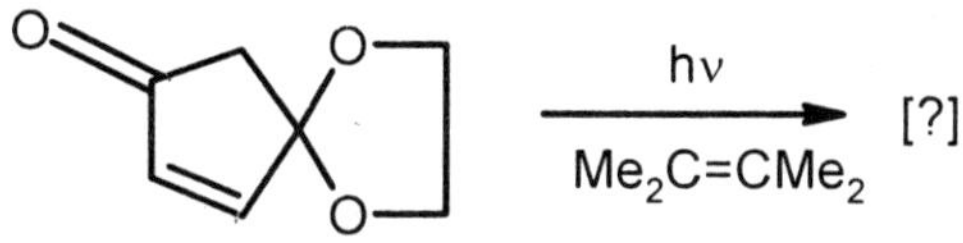

Solution.

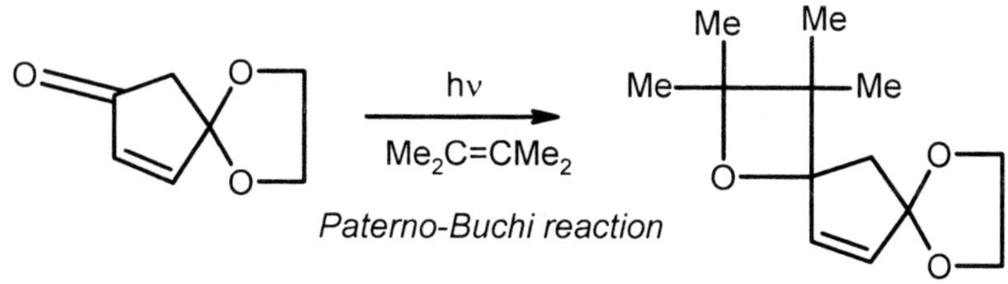

Problem 6.53

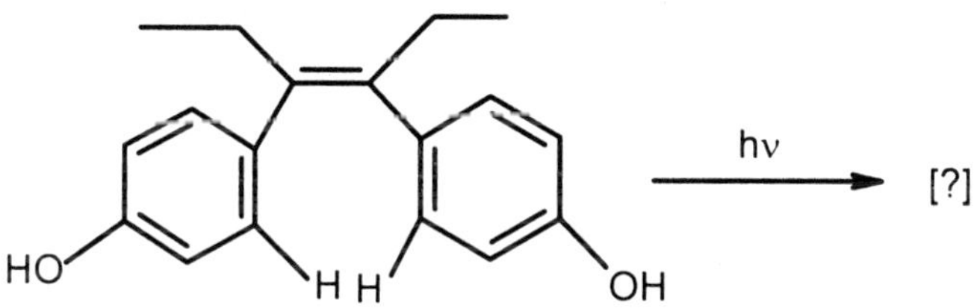

Solution.

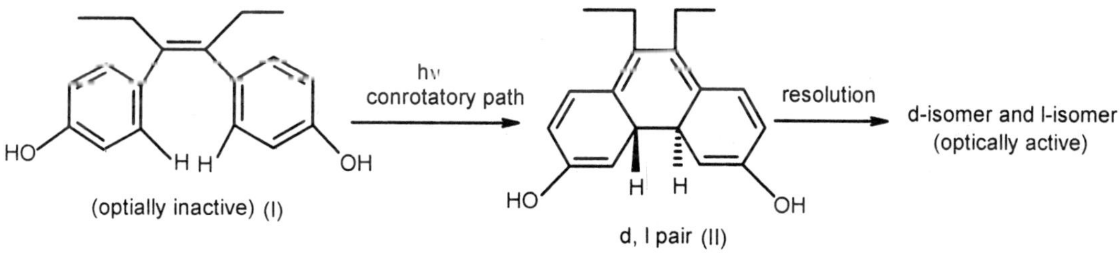

The compound (I, *cis*-derivative) exhibits no optical activity; but on photo-irradiation it undergoes cyclization through *conrotatory* path to produce the *trans*-product (II). It exists as d,*l*-pair. On resolution optically pure isomer can be obtained.

Problem 6.54

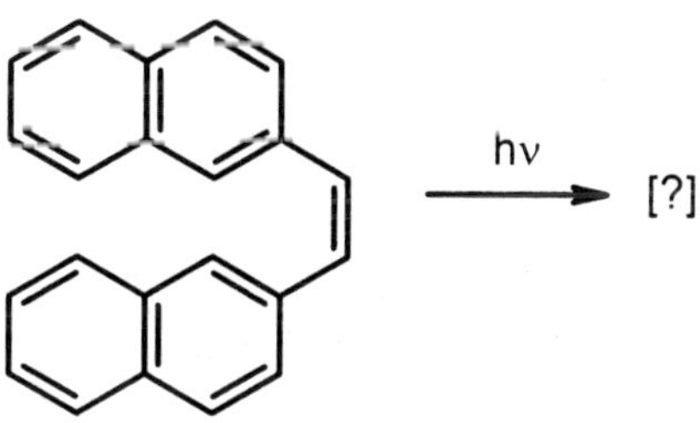

Solution. Compound (I) is optically inactive. Absorption of circularly polarized light initiates photocyclization of (I) leading to the formation of cyclized product — *trans* isomer (II) which shows optical activity. The isomer is chiral due to its helical shape. This is an interesting example of photons acting as *chiral reagents*.

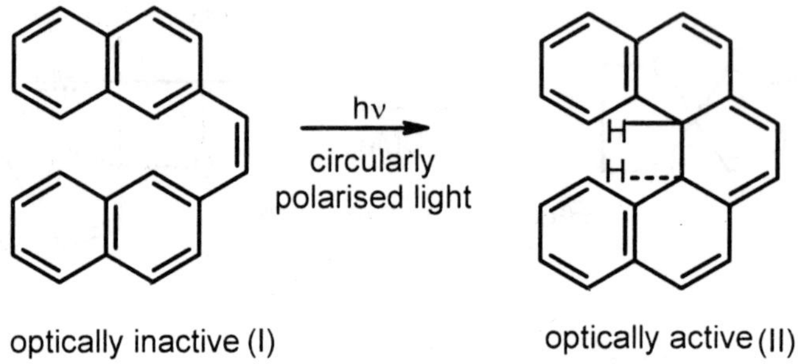

Problem 6.55

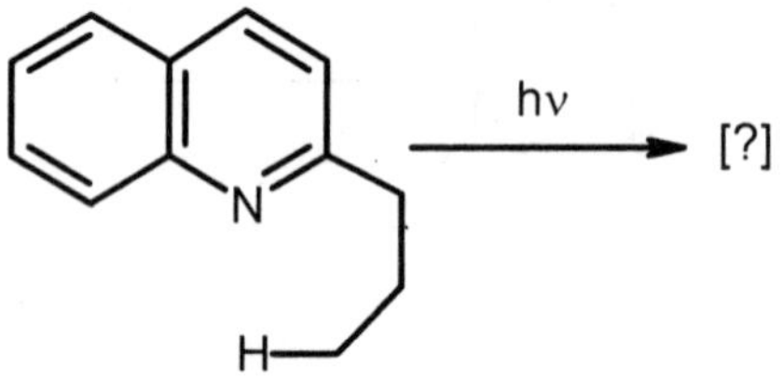

Solution.

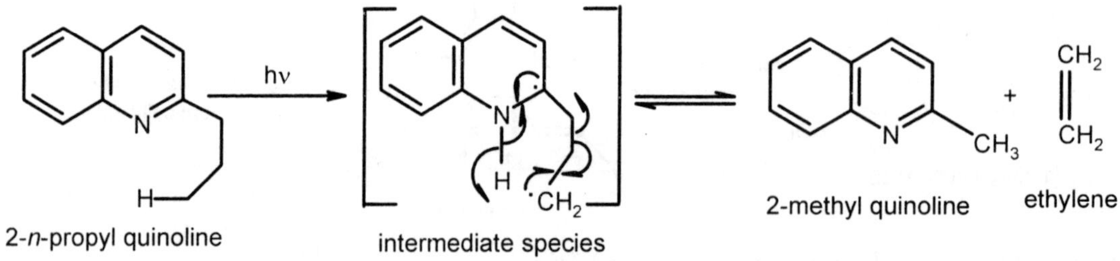

Problem 6.56

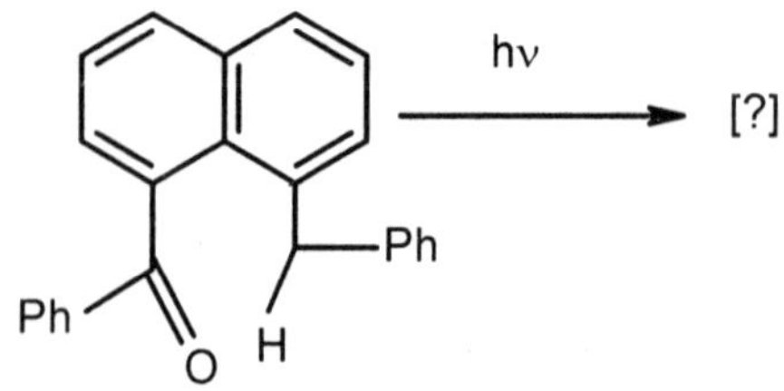

Solution. The naphthyl ketone (I) undergoes photochemical cyclization to yield compound (II); the diradical intermediate is formed *via* a seven-membered cyclic transition state.

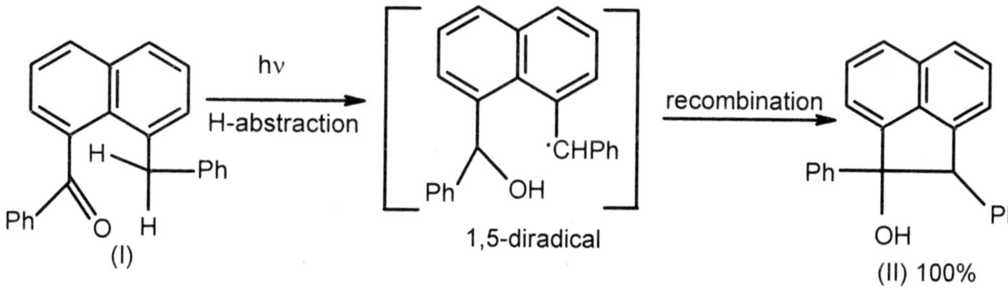

Problem 6.57

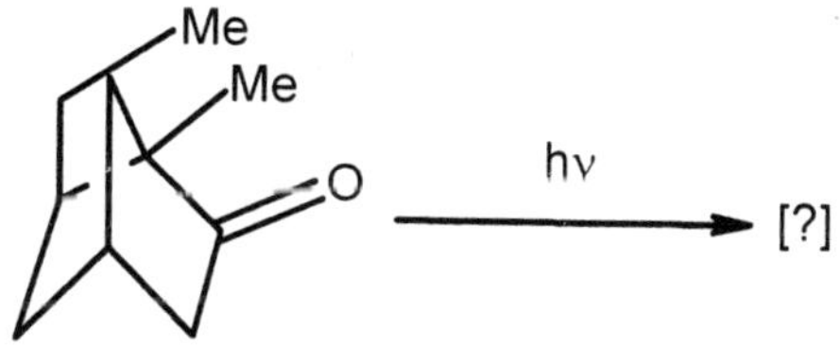

Solution. In protic solvents, cyclic ketones on *Norrish Type I* reaction furnish ketene intermediates, which in turn undergo solvent addition to yield carboxylic acids or their derivatives — as observed in the case of carvone.

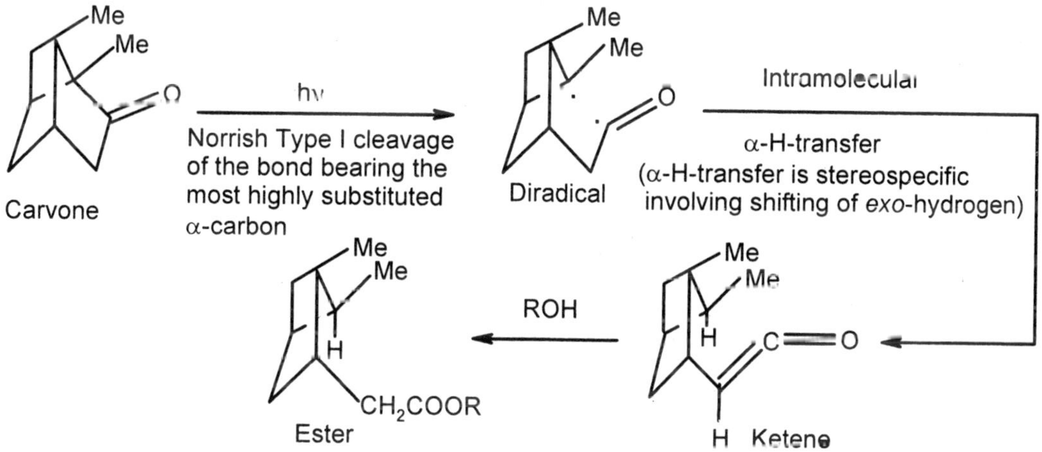

Problem 6.58

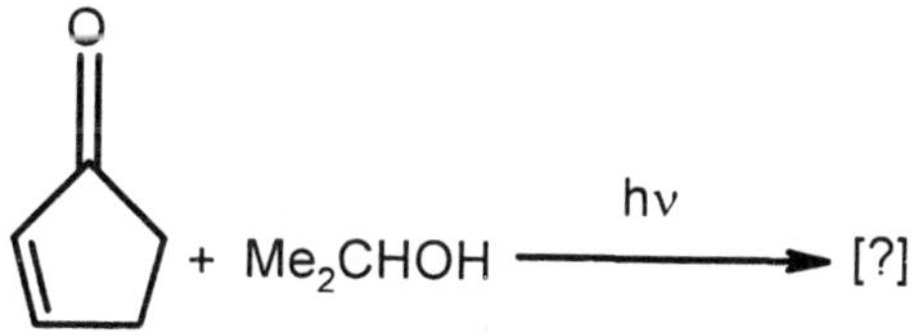

Solution.

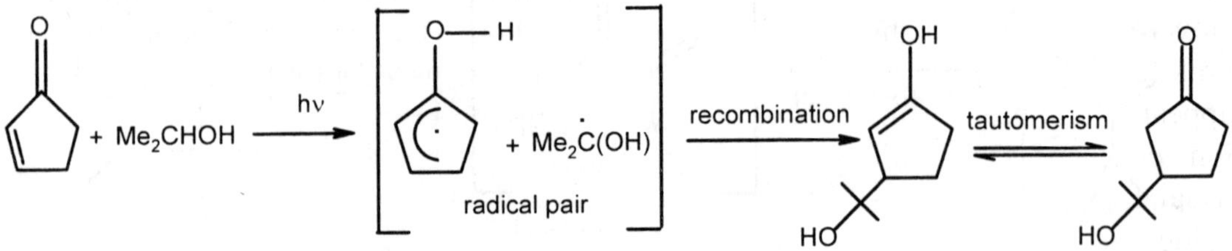

Problem 6.59

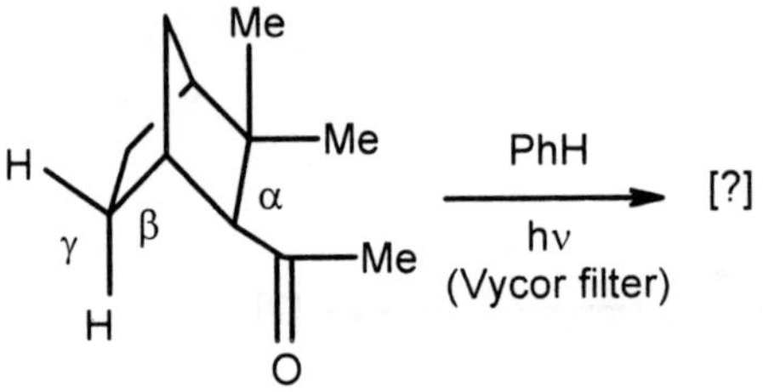

Solution. This is an example of *Norrish Type II cleavage* — photolysis of (I) to form (II) that undergoes ring closing to afford the alcohol (III).

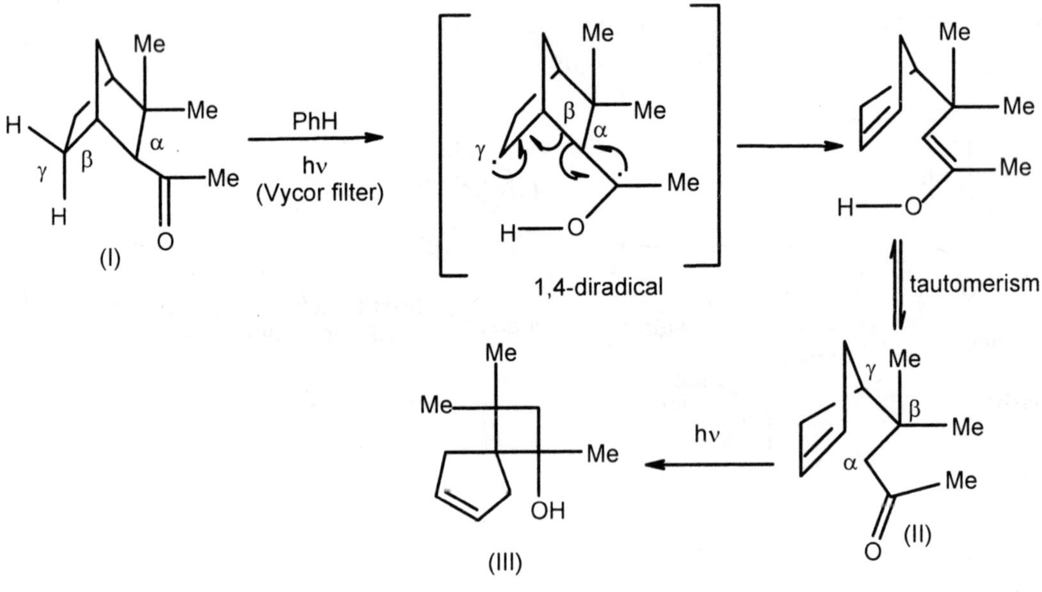

Problem 6.60

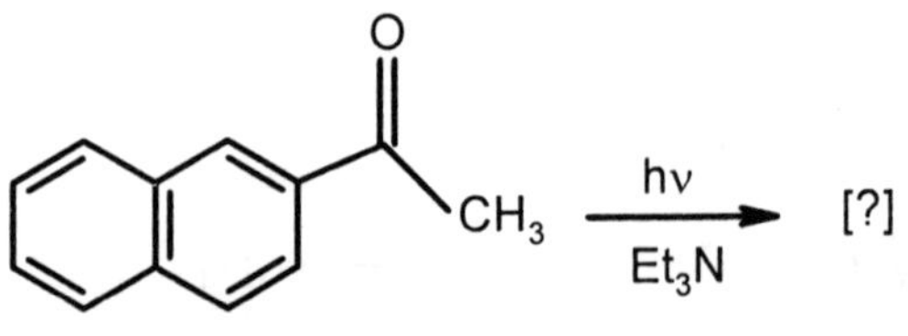

Solution. It is observed that n,π* triplets of naphthyl ketones are generally unreactive towards photoreduction by alcohols and saturated hydrocarbons, presumably due to slows down rate constant of hydrogen abstraction. But these triplets are found to undergo efficient reduction in presence of amines, as exemplified in this case. It appears that an electron transfer, followed by proton transfer from the amine radical ion, produces ketyl radicals, dimerization of which gives the pinacol derivative in usual fashion. The efficiency of the reaction is obviously due to a relatively large value of k_e, the rate constant for electron abstraction.

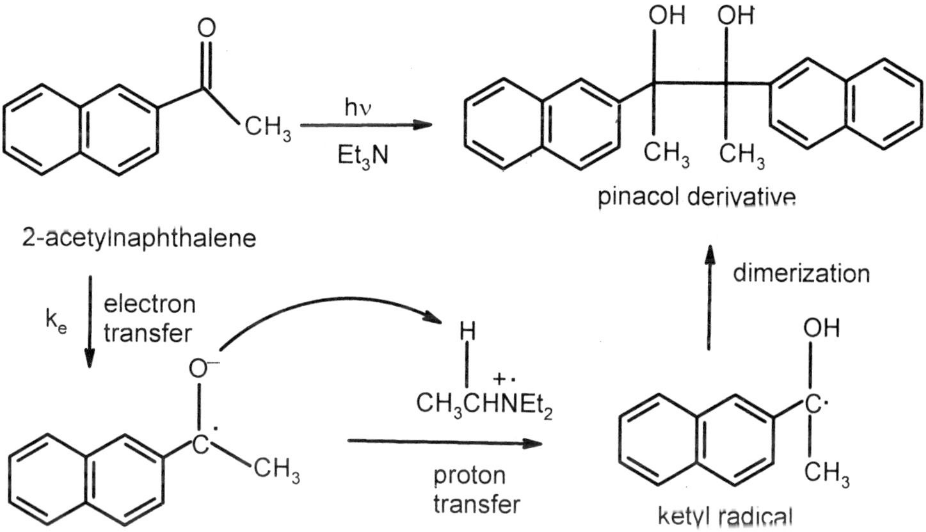

Problem 6.61

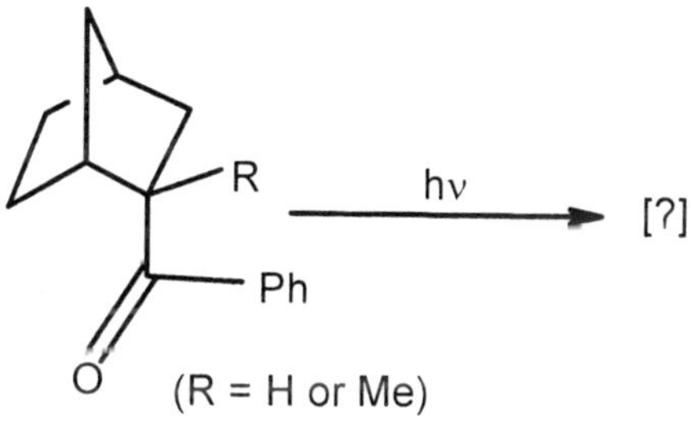

Solution. On photo-irradiation *endo*-2-benzoylnorbornane (Ia, R=H) yields the cleavage product (II), but *endo*-2-benzoyl-2 methylnorbornane (Ib, R=Me) results in formation of the tricyclo[3,2,1,0]-octane (III). The presence of the 2-methyl substituent and the consequent high-energy eclipsing interaction with the phenyl group in the planar transition state required for elimination, results in cyclobutanol formation.

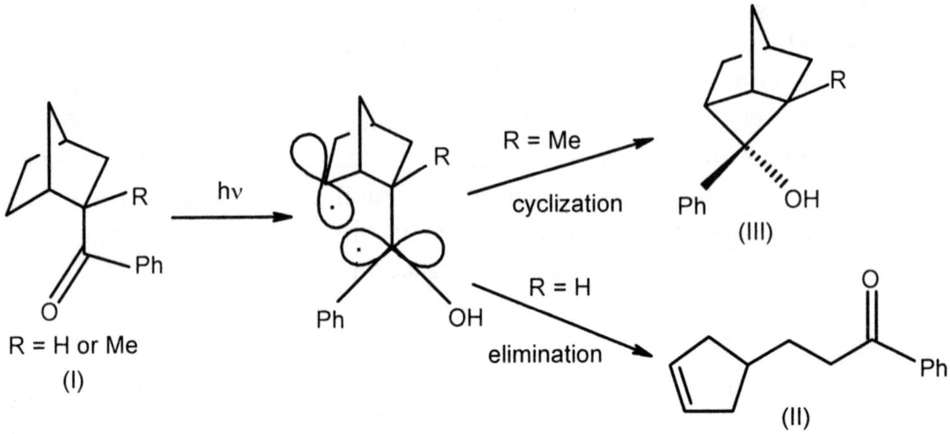

Problem 6.62

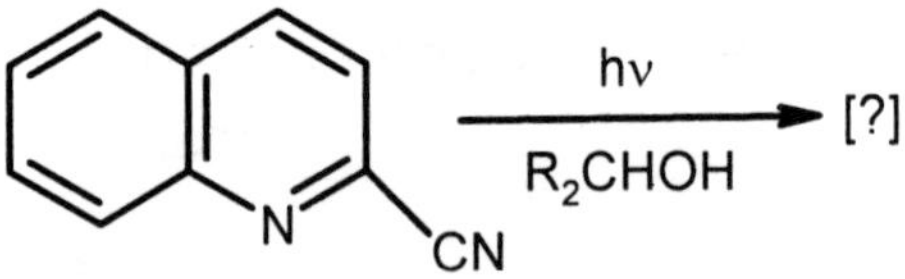

Solution.

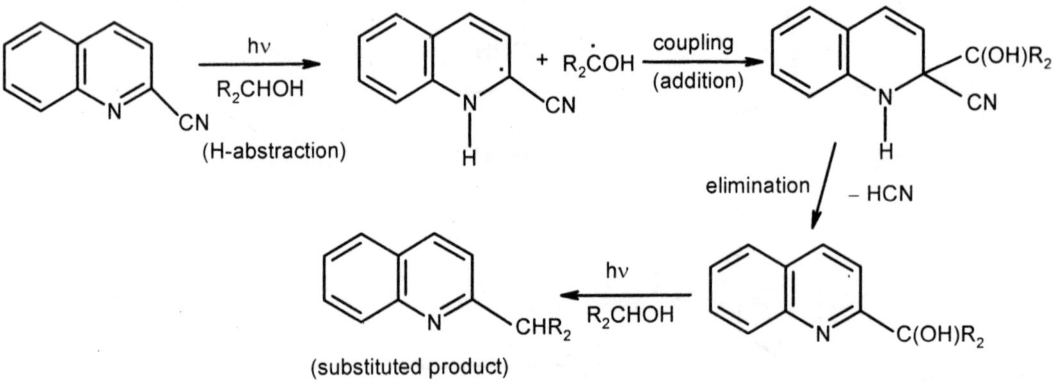

Problem 6.63

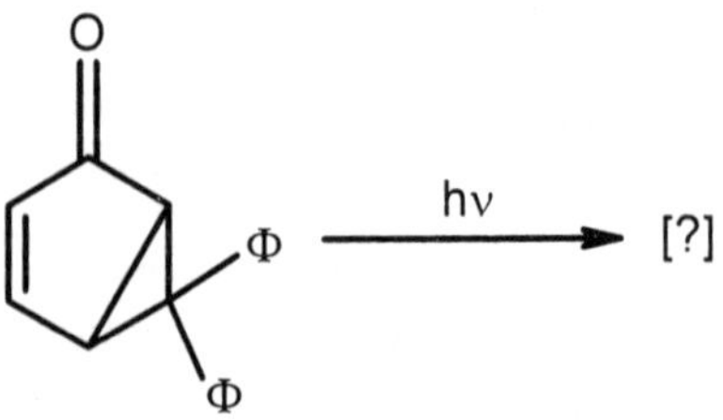

Solution. On photo-irradiation the bicyclic ketone (I) in aqueous dioxane rearranges to 2,3-diphenylphenol (II), 6,6-diphenyl-3,5-hexadienoic acid (III) and minor amounts of 3,4-diphenylphenol.

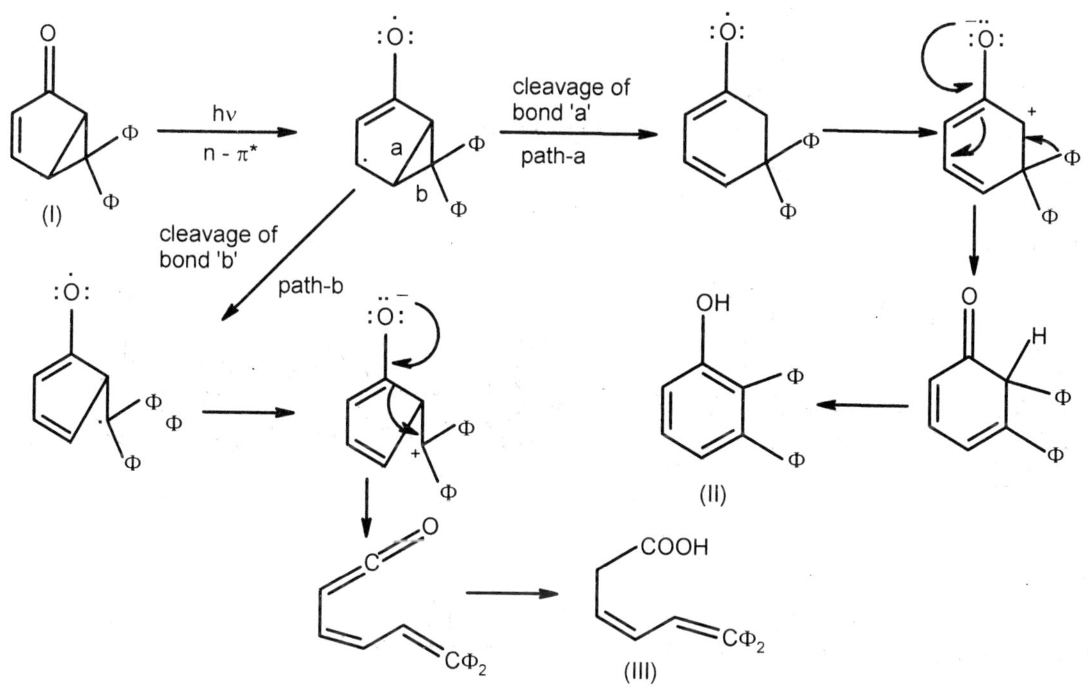

Problem 6.64

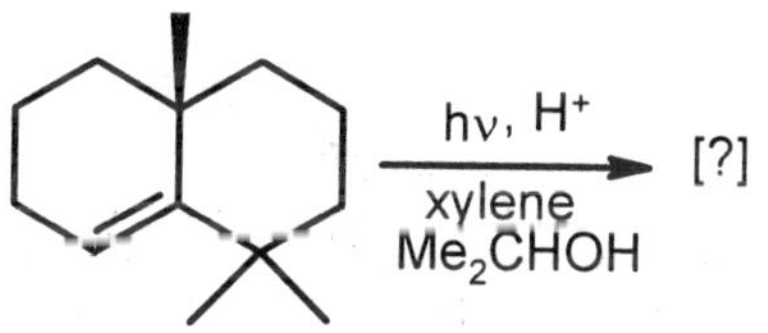

Solution.

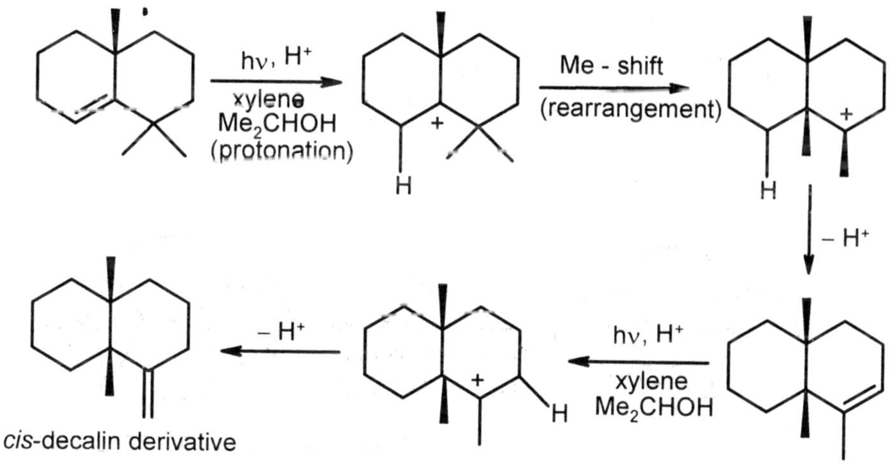

Problem 6.65

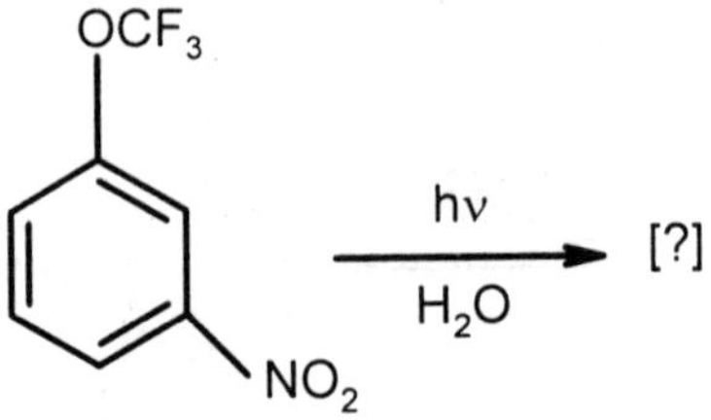

Solution.

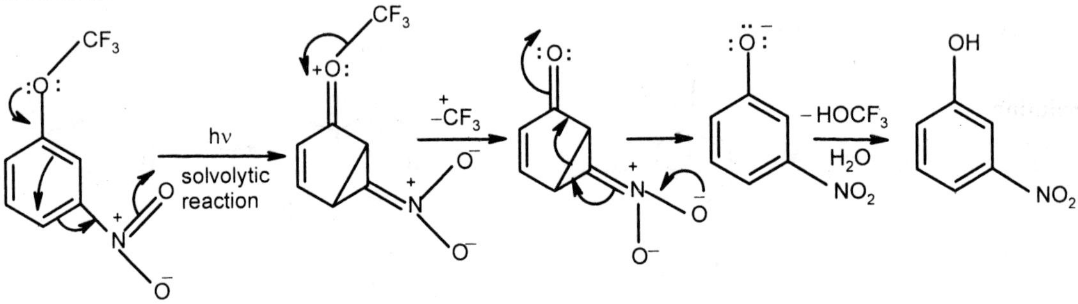

Problem 6.66

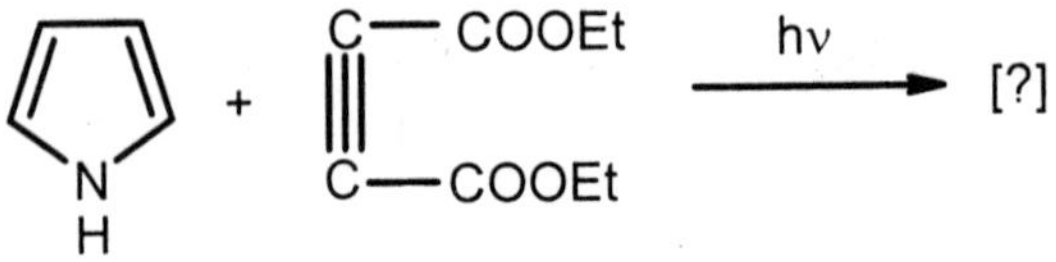

Solution.

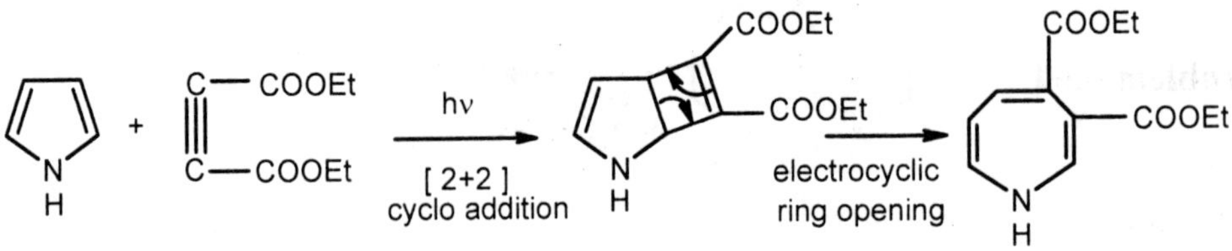

Problem 6.67

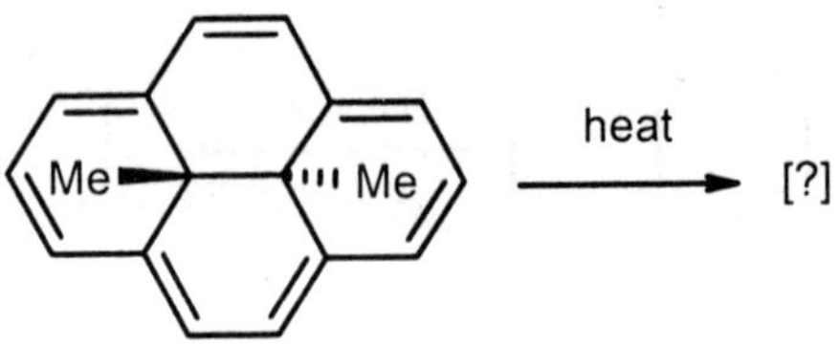

Solution.

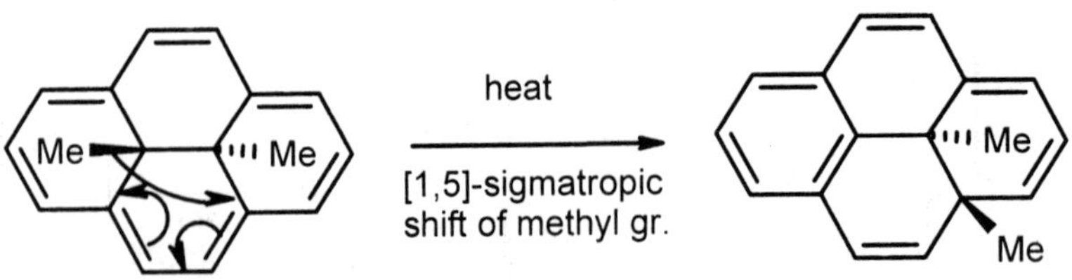

[*J. Am. Chem. Soc.*, **92**, 3669(1970)]

Problem 6.68

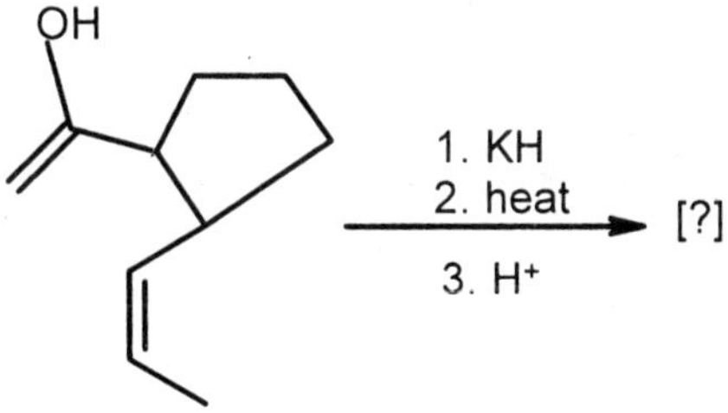

Solution.

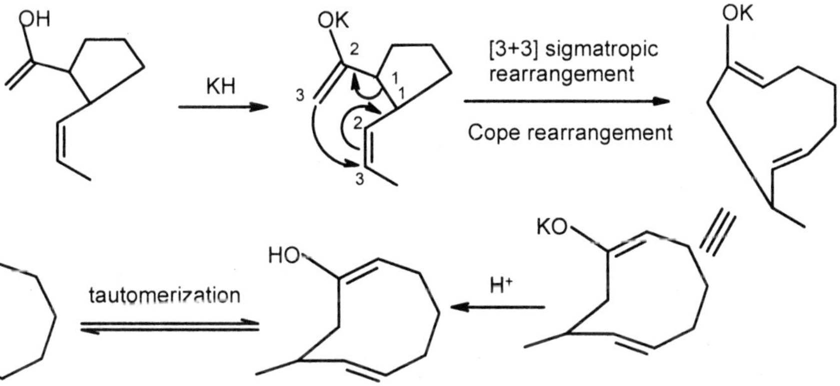

Problem 6.69

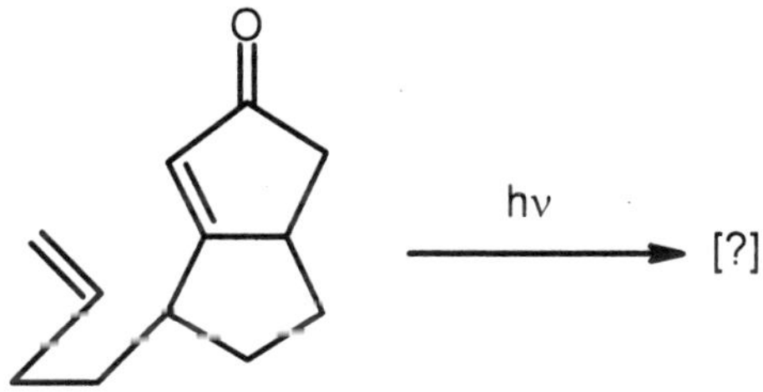

Solution.

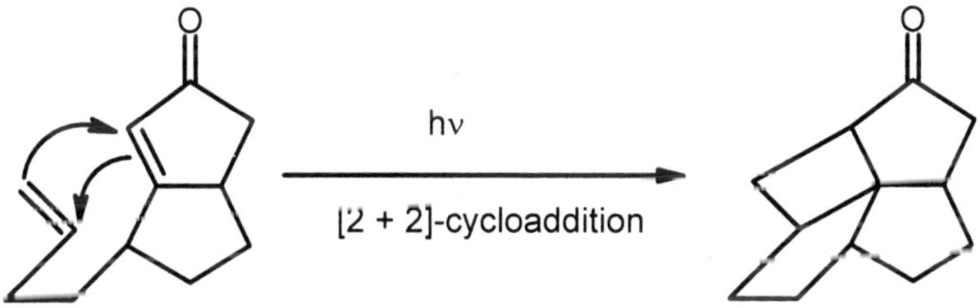

Problem 6.70

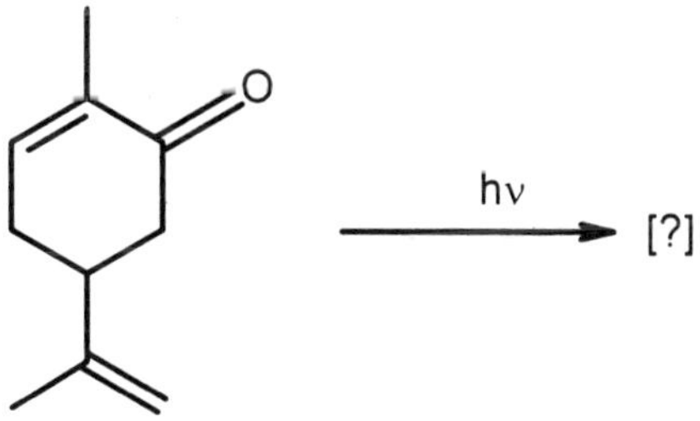

Solution.

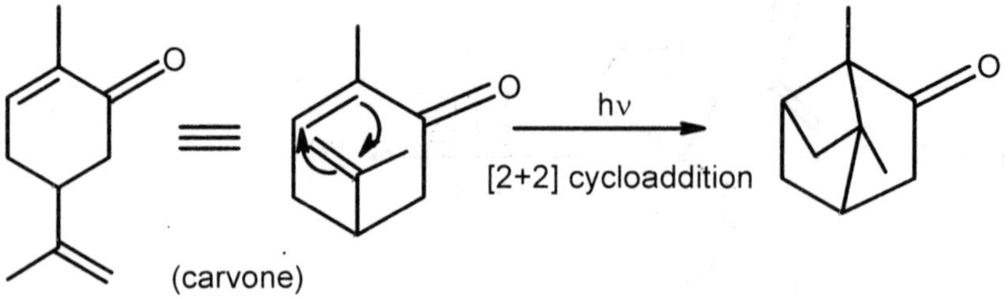

Problem 6.71

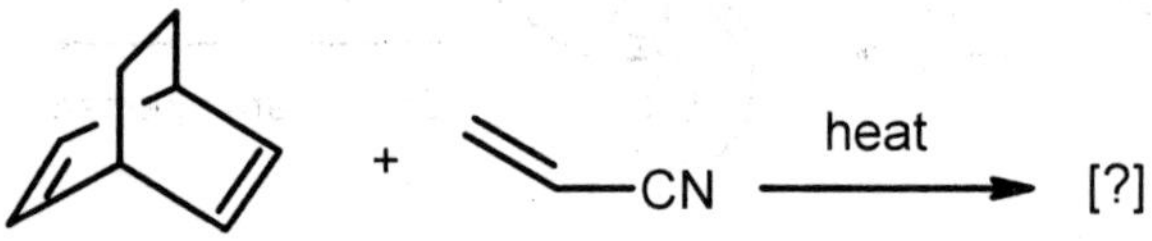

Solution.

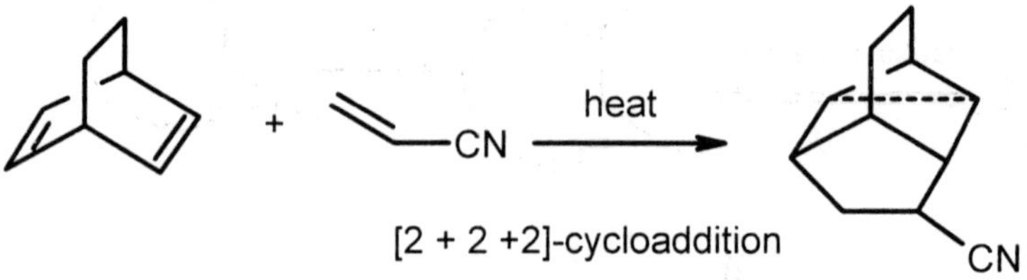

[*J. Org. Chem.*, **25**, 42(1960)]

Problem 6.72

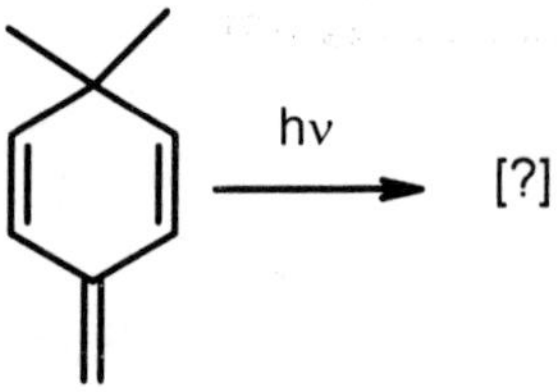

Solution.

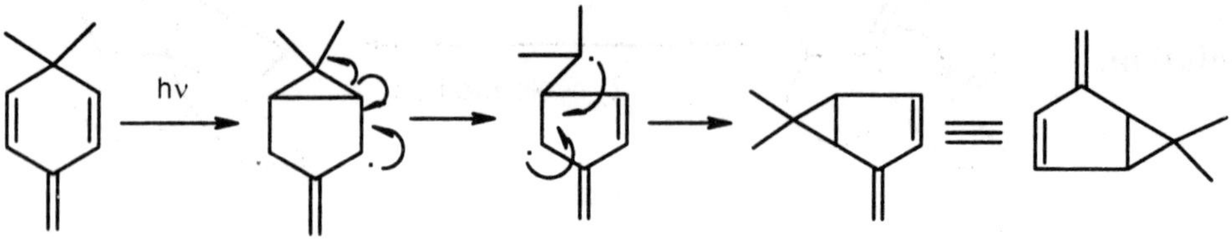

Problem 6.73

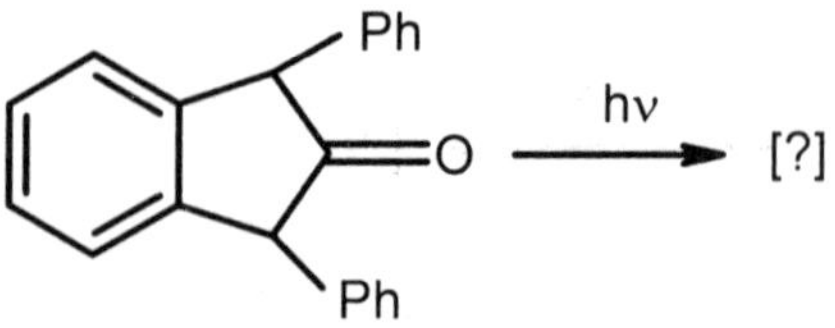

Solution.

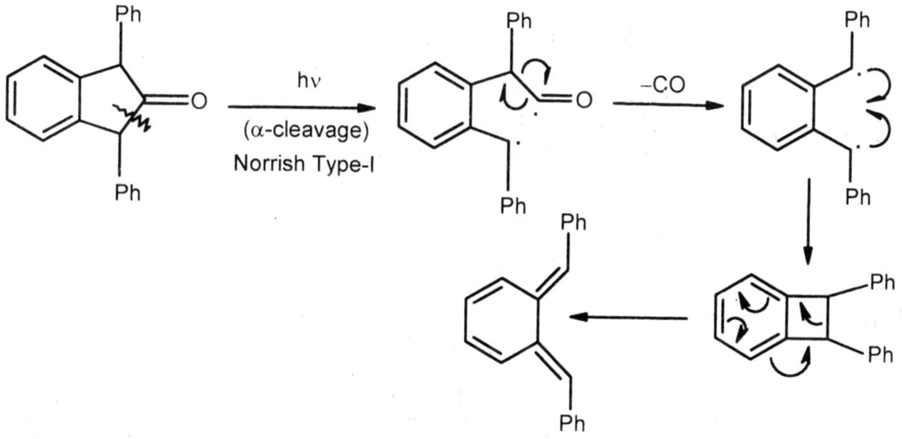

Problem 6.74

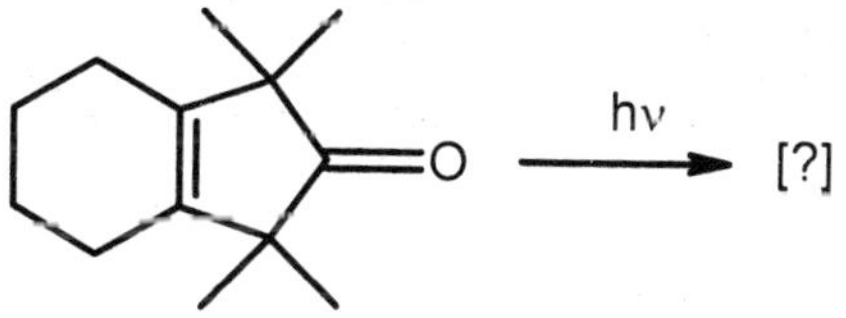

Solution.

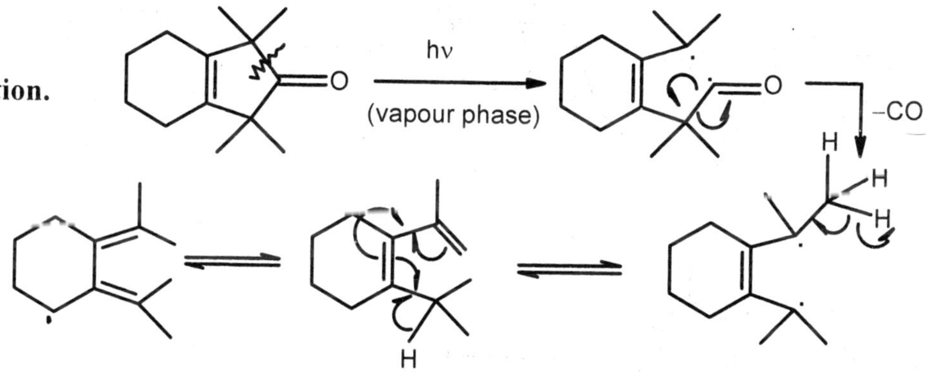

Problem 6.75

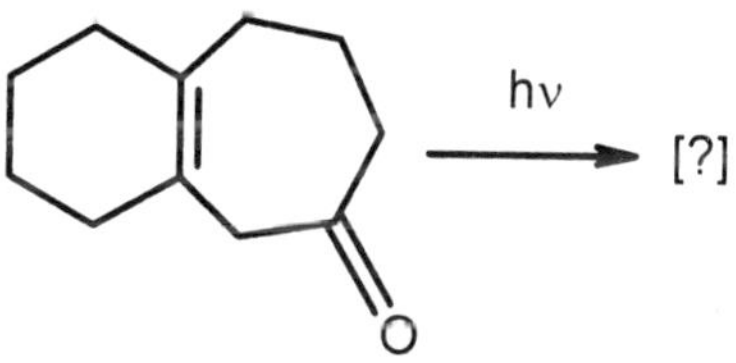

Solution.

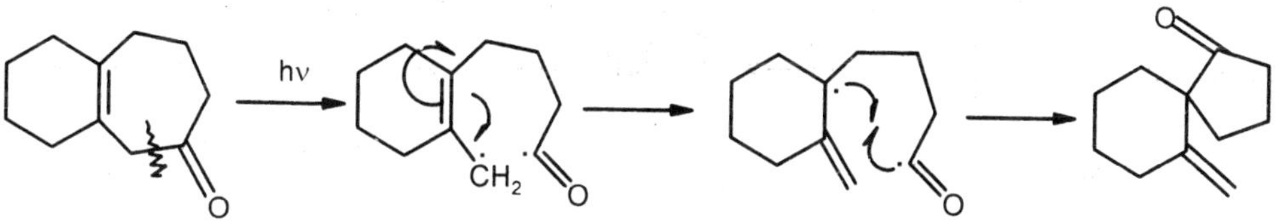

Problem 6.76

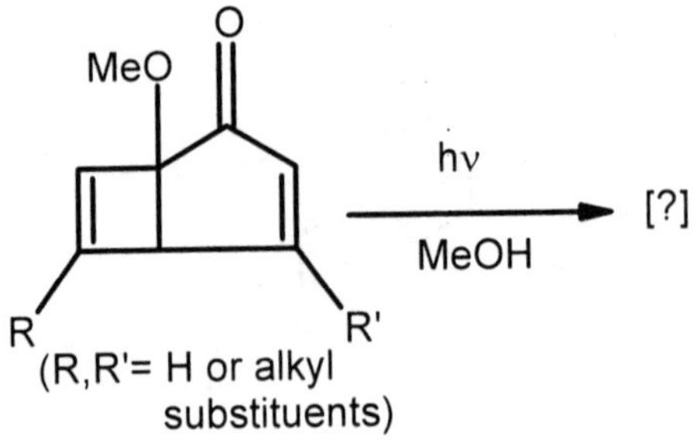

Solution.

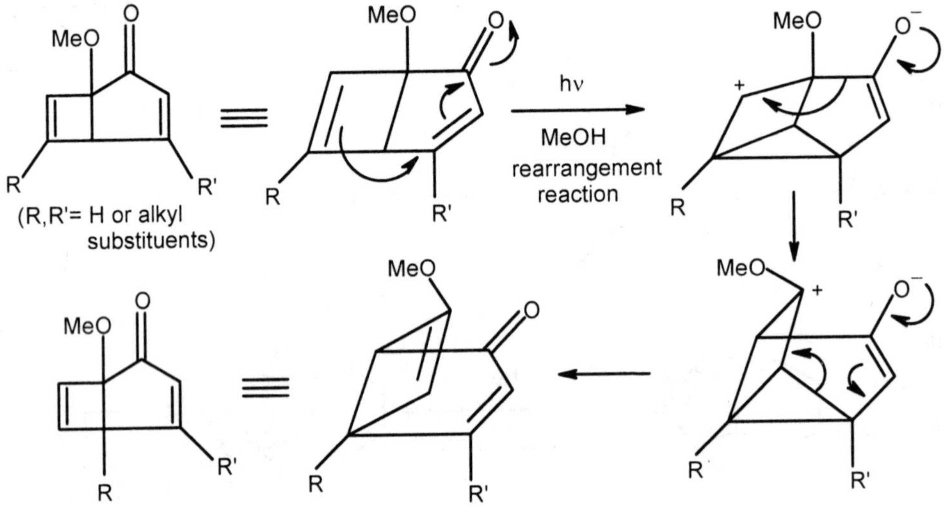

Problem 6.77

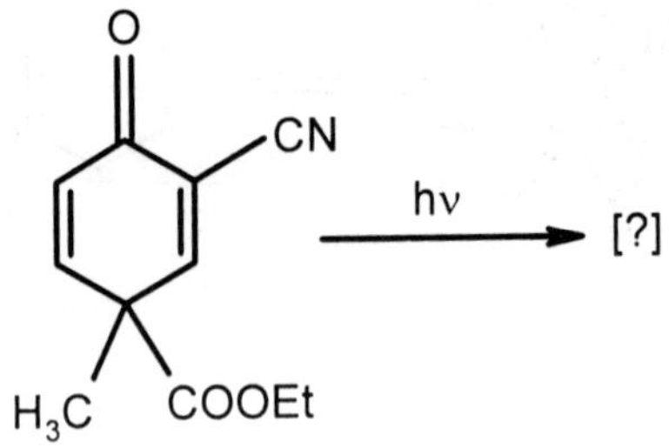

Solution.

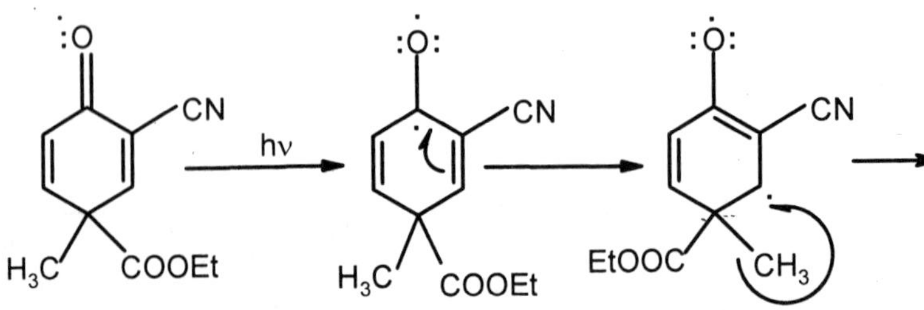

(see at the next page)

(from the previous page)

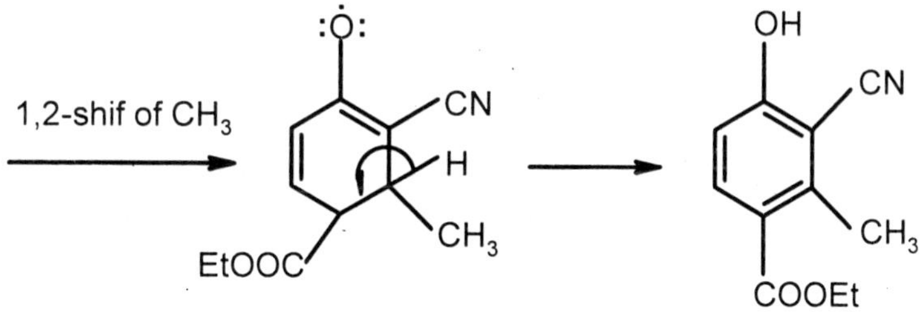

Problem 6.78

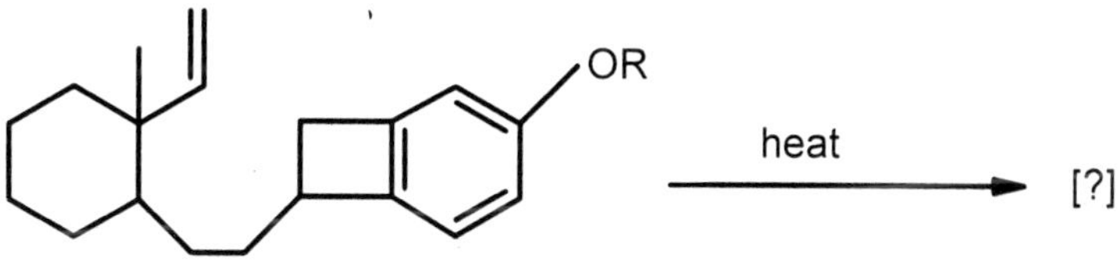

Solution.

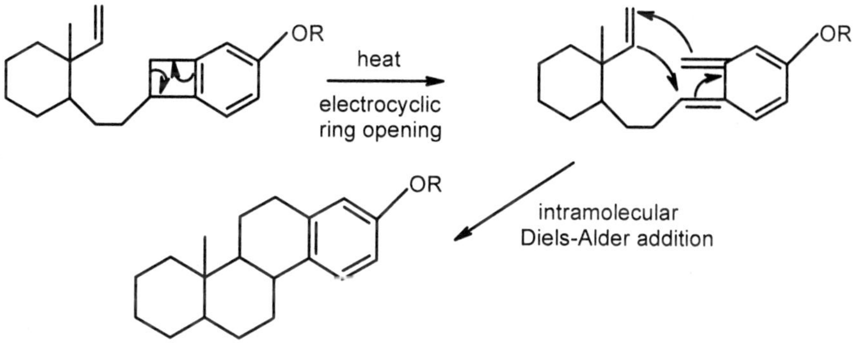

Problem 6.79

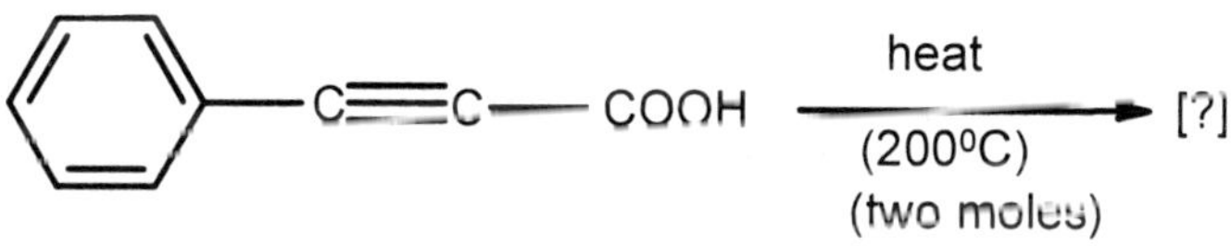

Solution.

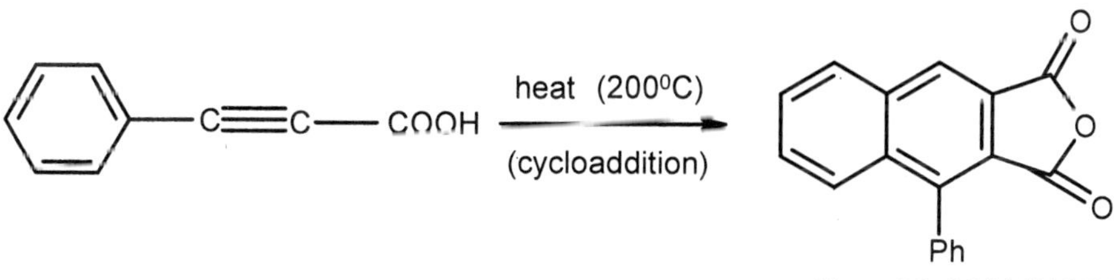

[*Ber.*, **28**, 2511(1895)]

Problem 6.80

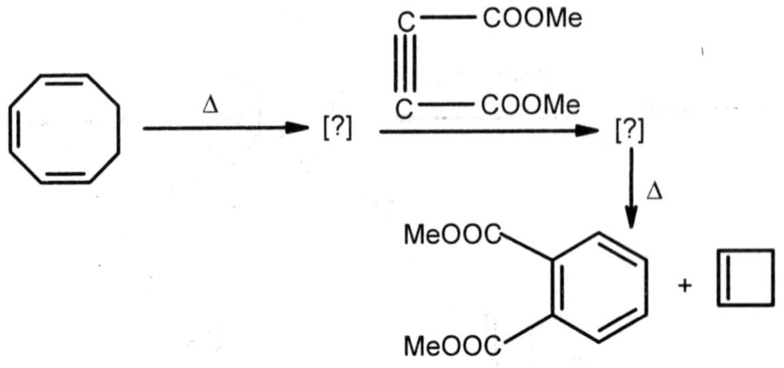

Solution.

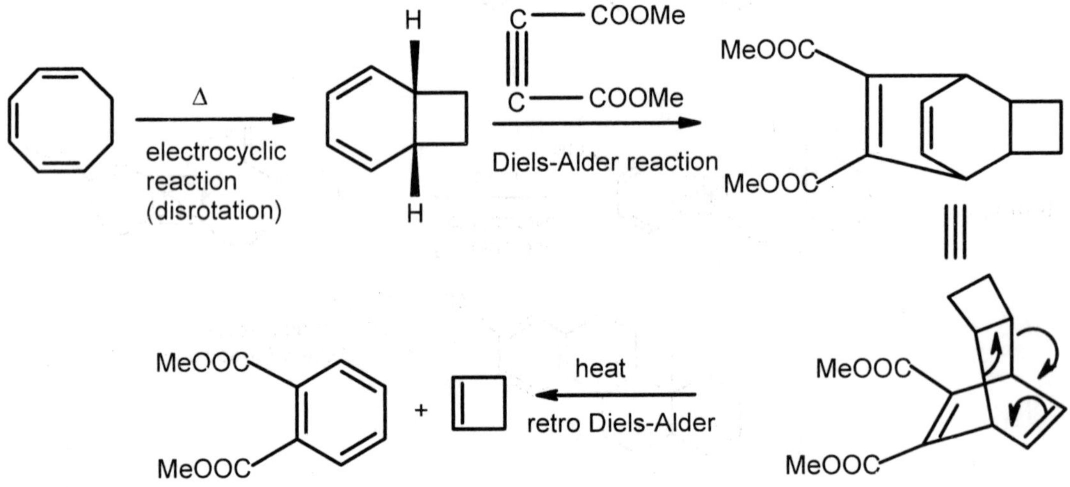

Problem 6.81

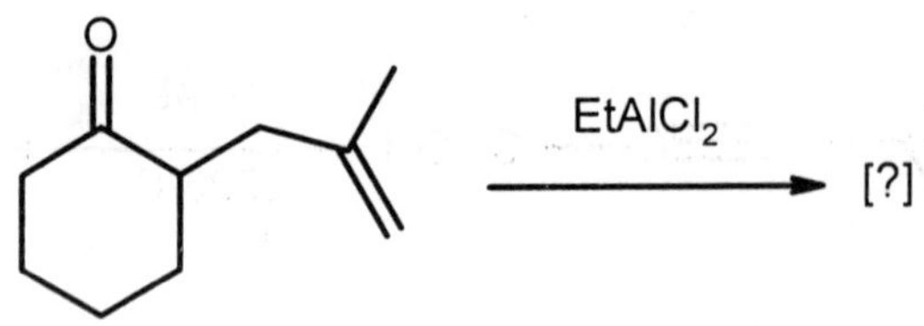

Solution.

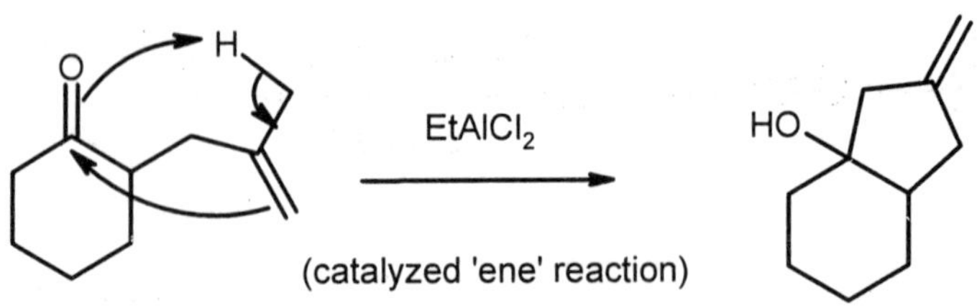

[*J. Org. Chem.*, **49**, 2443(1984)]

Problem 6.82

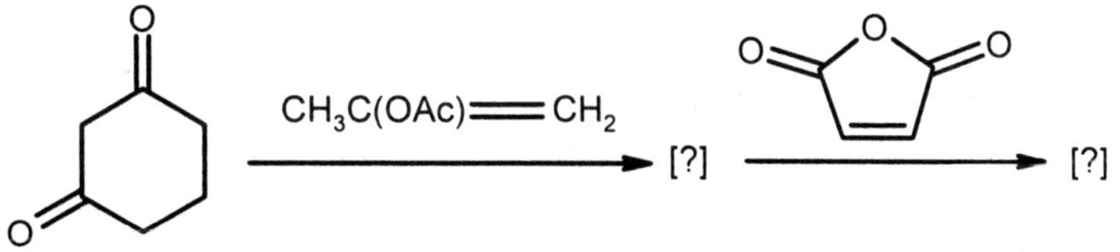

Solution. Cyclohexane-1,3-dione on reaction with isopropenyl acetate generates cyclic 1,3-diacetoxy-1,3-dienes in situ, which then undergoes Diels-Alder reaction with maleic anhydride giving 1-acetoxybicycloalkane dicarboxylic anhyudride derivative as product.

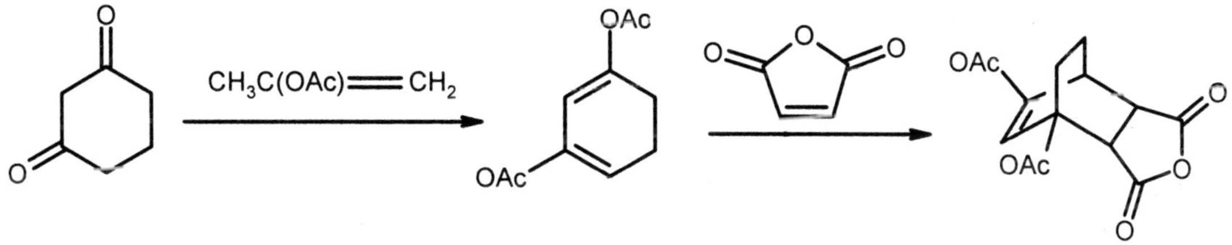

Problem 6.83

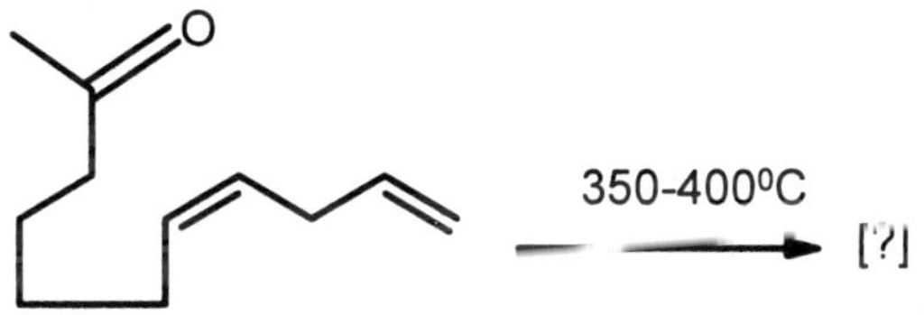

Solution.

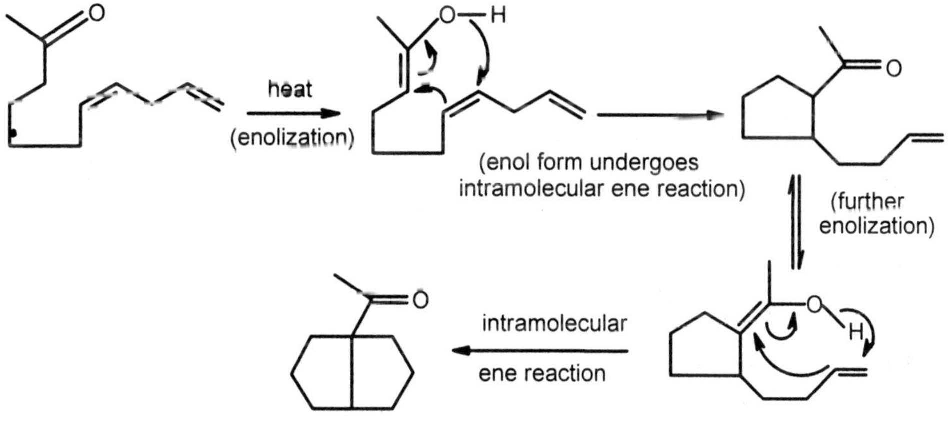

Supplementary problems

Write down the product(s) of the following pericyclic reactions. Also predict the stereochemistry of the product(s) where applicable:

Sp-6.1
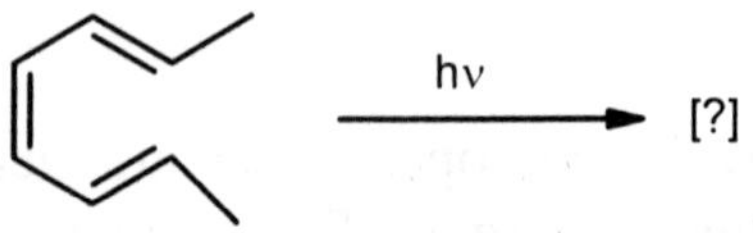

Sp-6.2
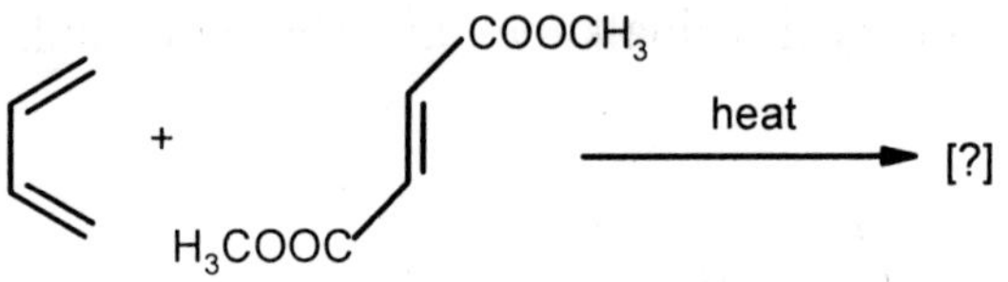

Sp-6.3
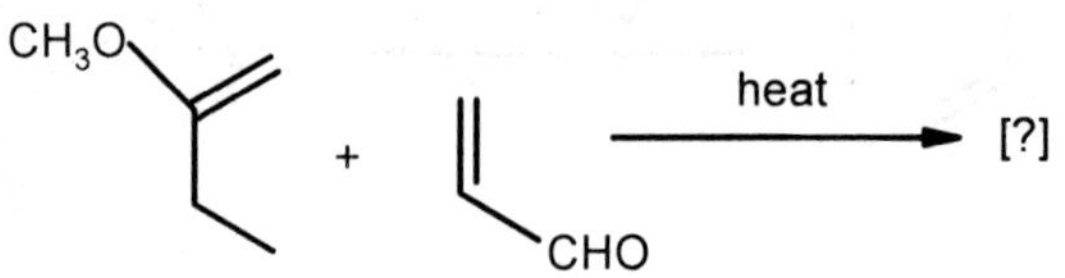

Sp-6.4
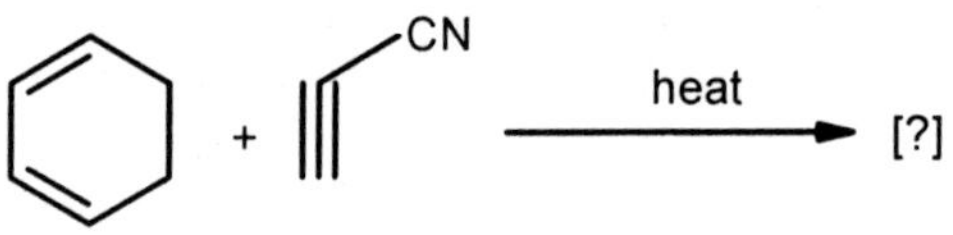

SP-6.5
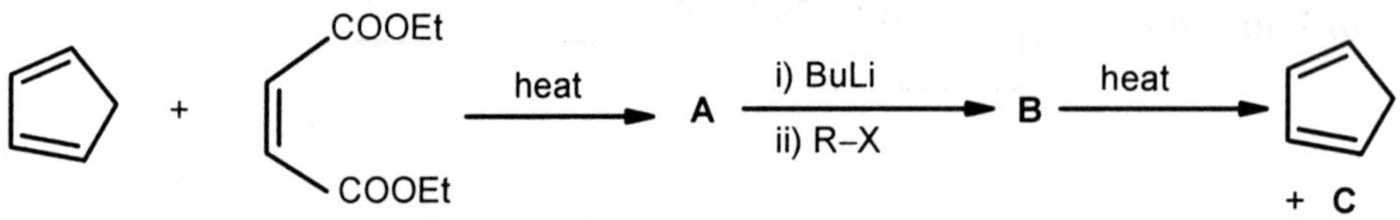

SP-6.6
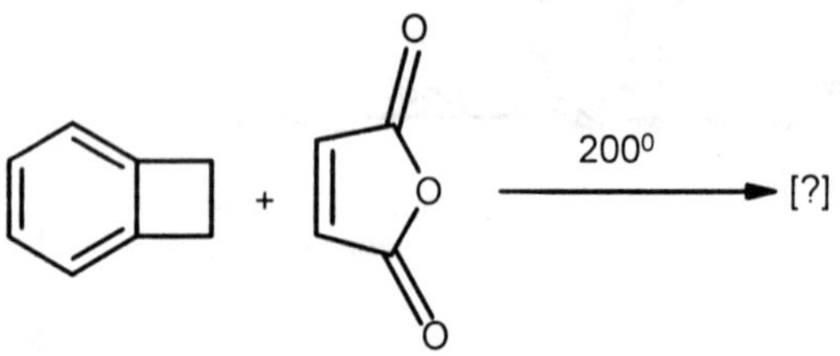

SP-6.7
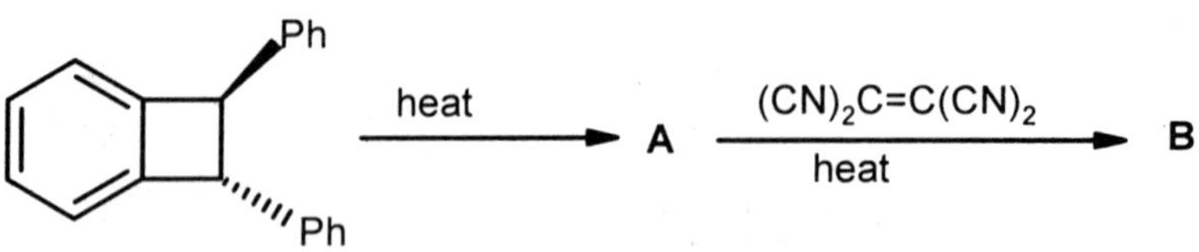

SP-6.8 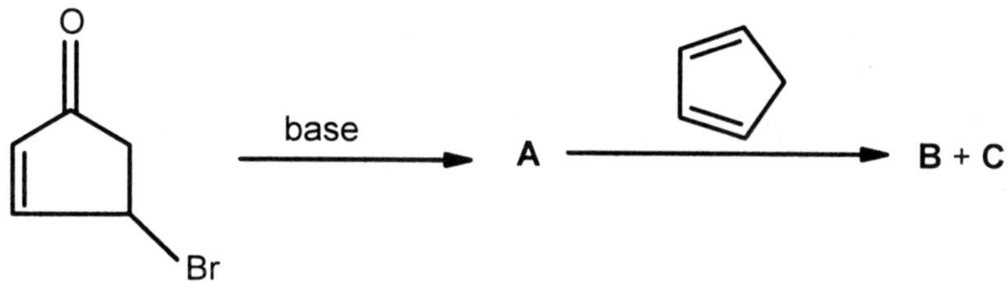

SP-6.9 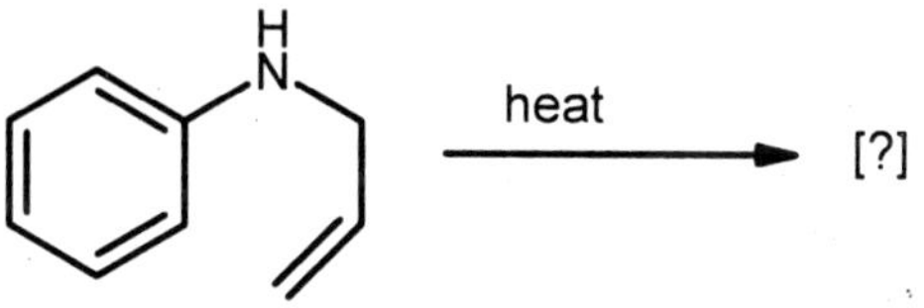

SP-6.10 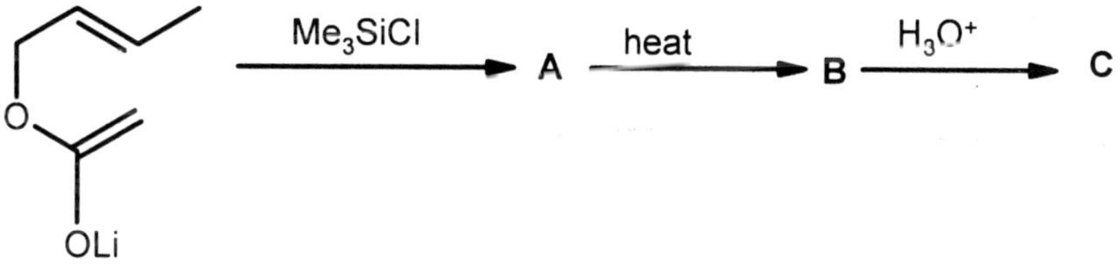

SP-6.11 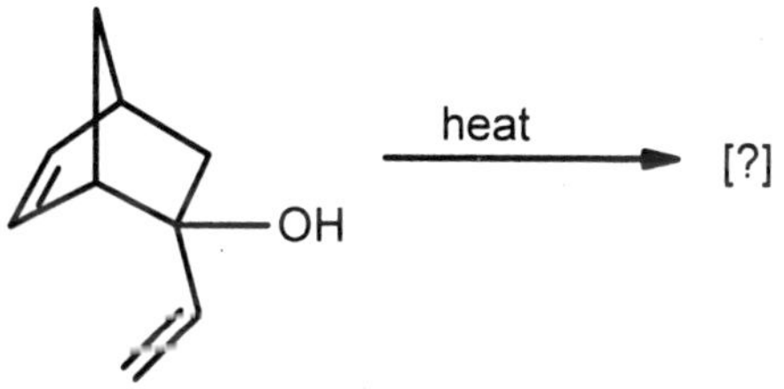

SP-6.12 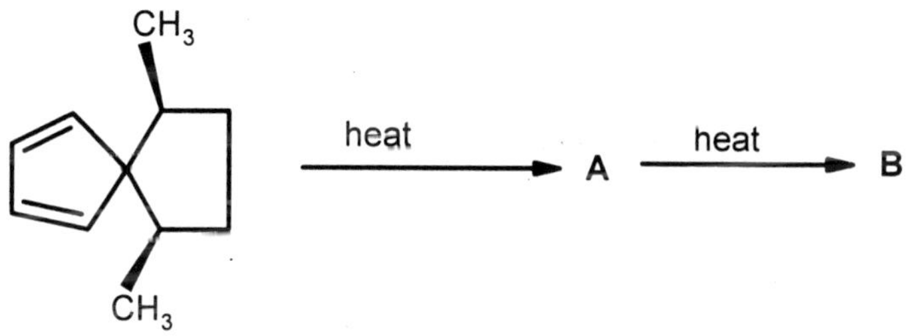

(**Hint.** Suprafacial [1,5]-sigmatropic shift of the carbon with retention of configuration, followed by [1,5]-sigmatropic shift of hydrogen)

SP-6.13 $Me_3C—CO—CH_3 \xrightarrow[\text{(solution phase)}]{h\nu}$ [?]

SP-6.14 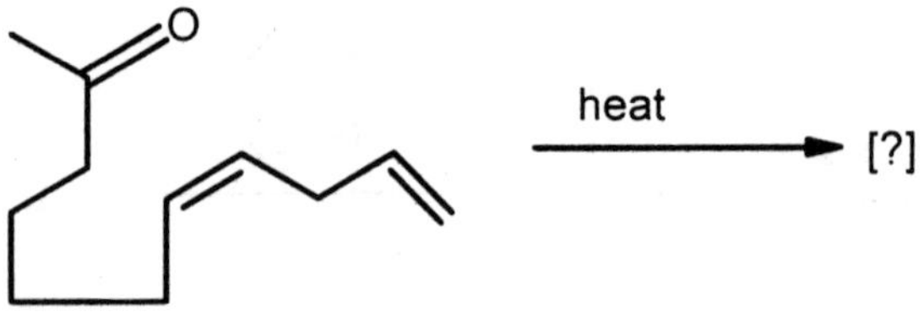

(**Hint.** Enolisation followed by intramolecular ene reaction)

SP-6.15 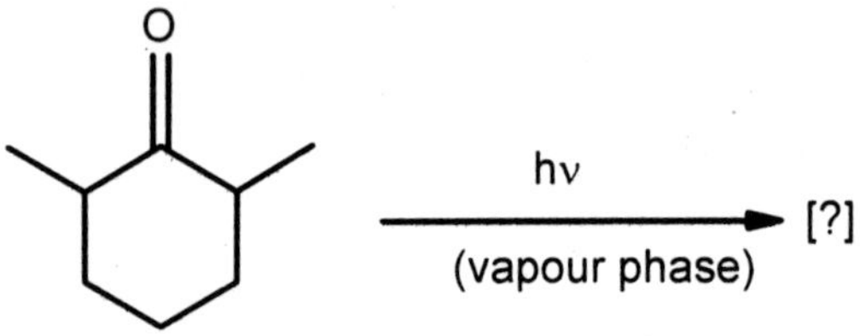

SP-6.16 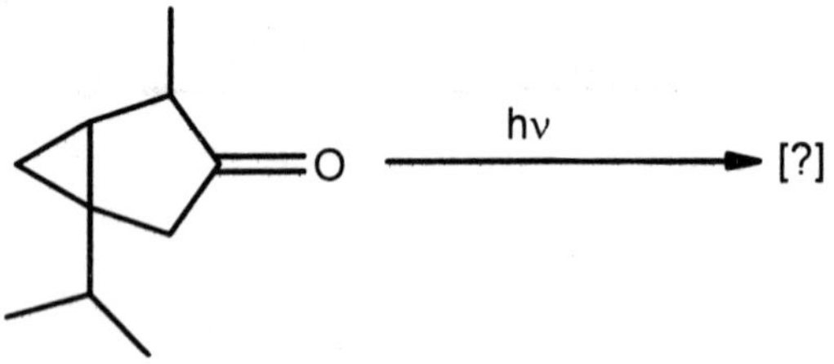

SP-6.17 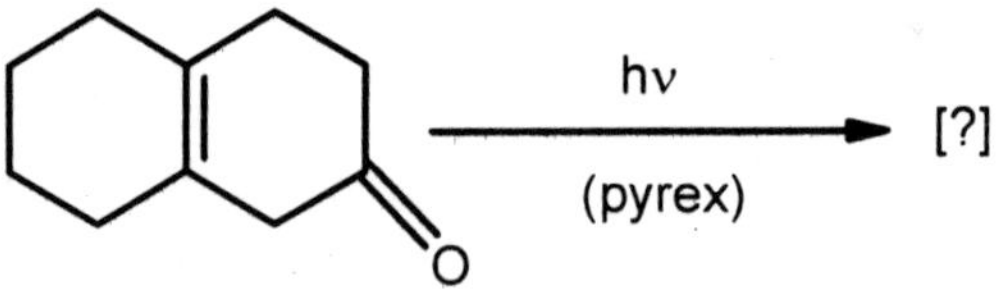

SP-6.18 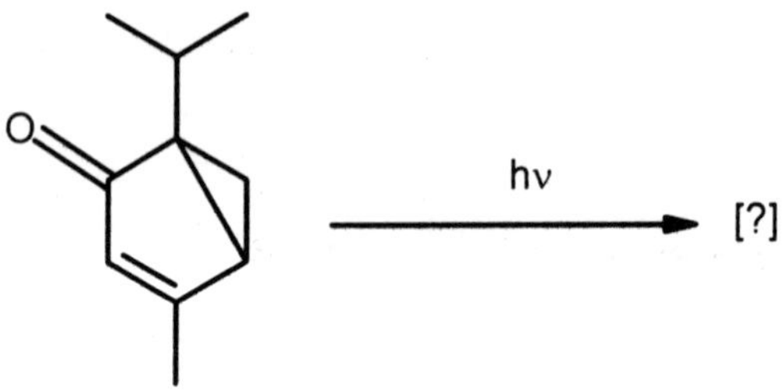

SP-6.19 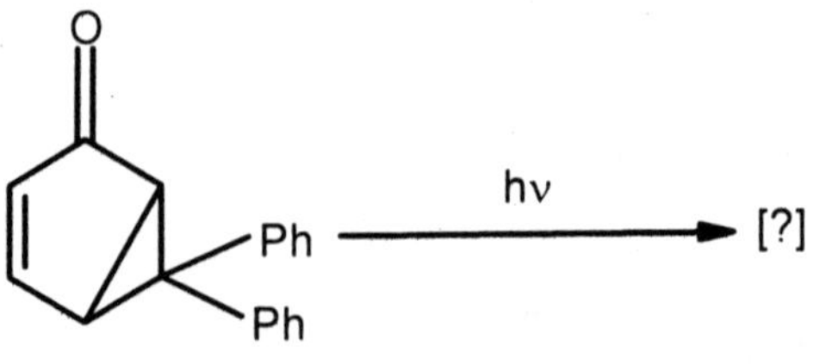

SP-6.20 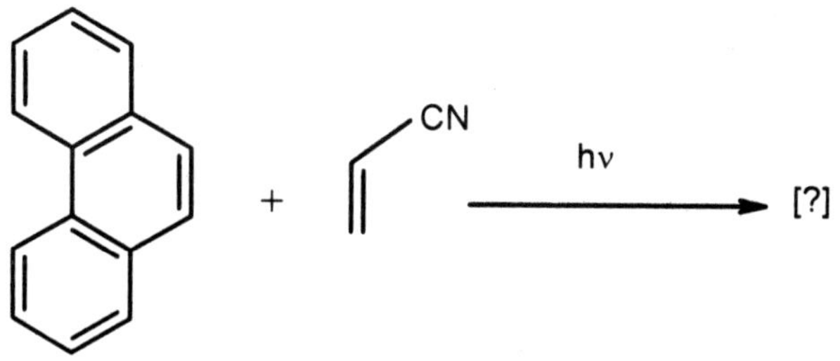

SP-6.21 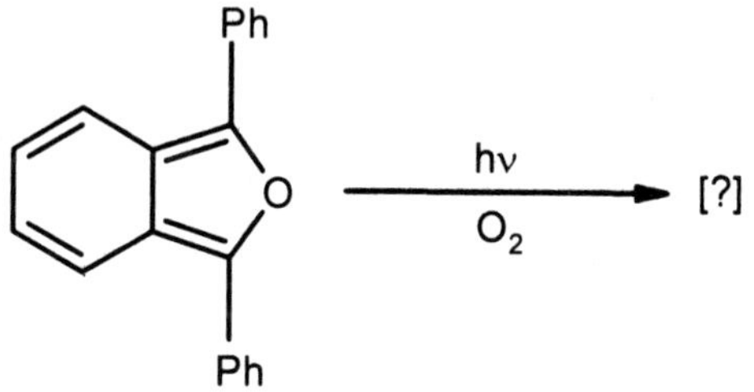

SP-6.22 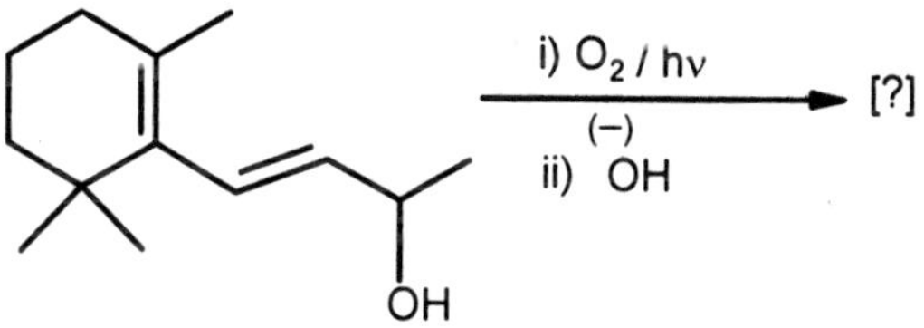

SP-6.23 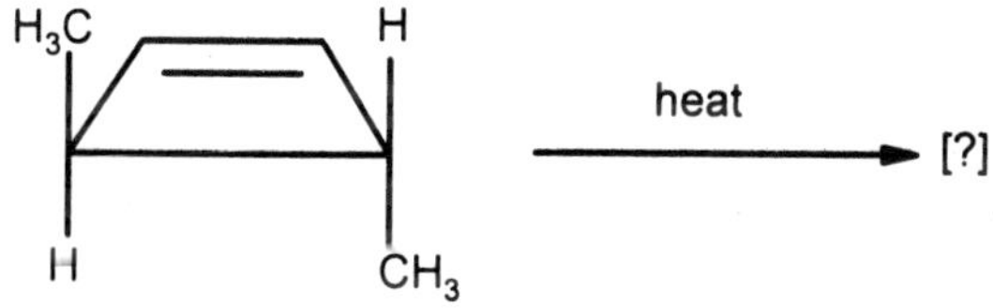

SP-6.24 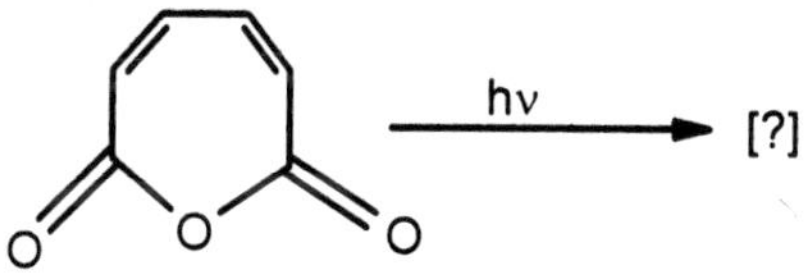

SP-6.25 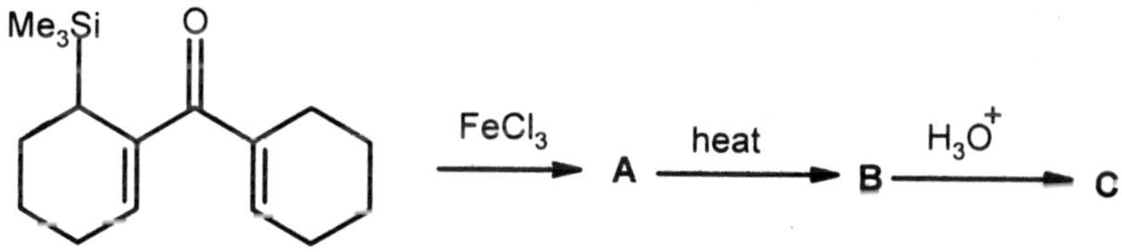

SP-6.26 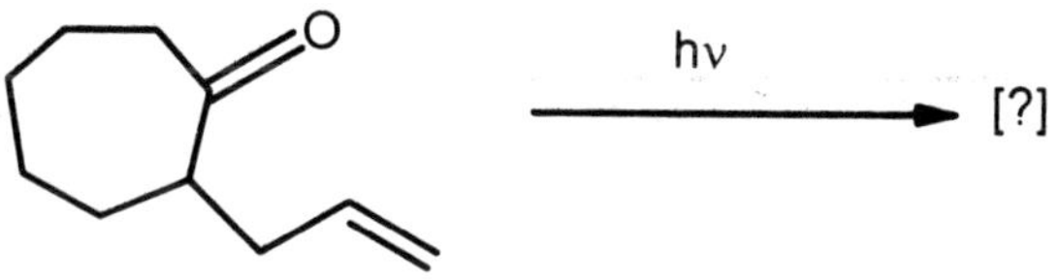

SP-6.27

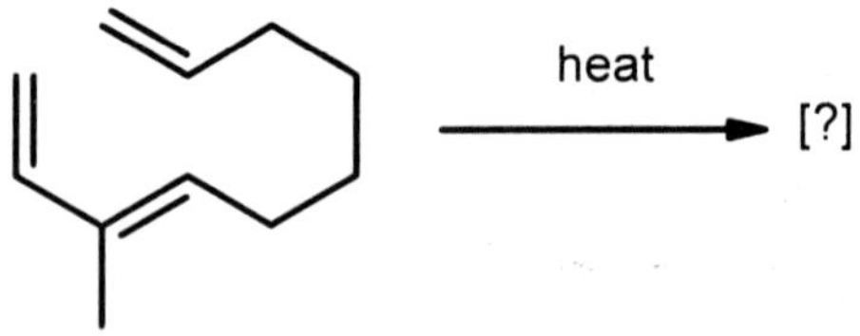

SP-6.28

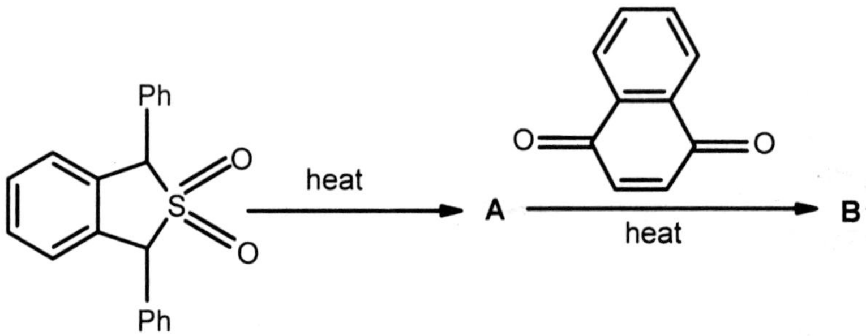

SP-6.29

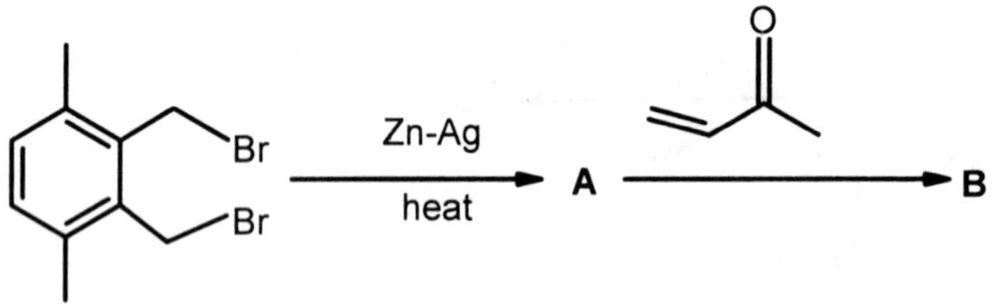

SP-6.30

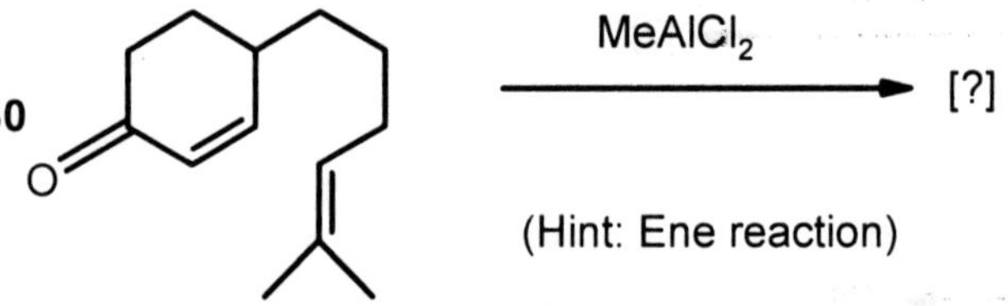

SP-6.31

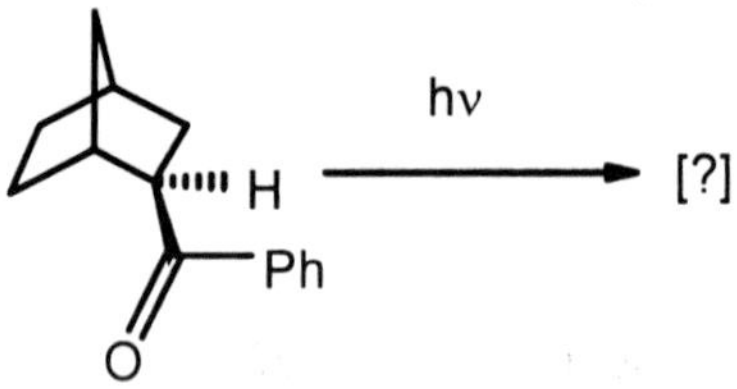

SP-6.32

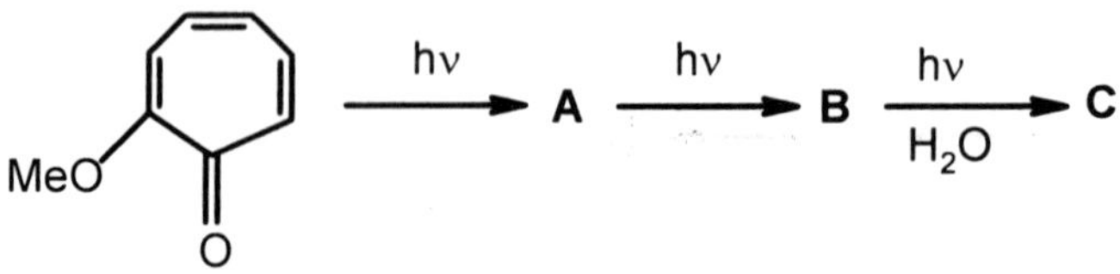

SP-6.33 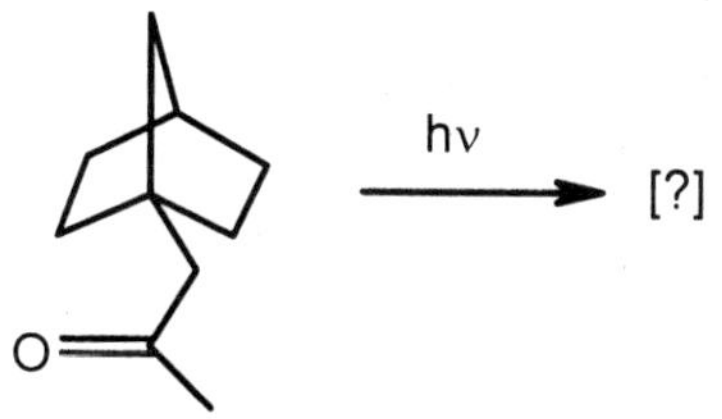

SP-6.34 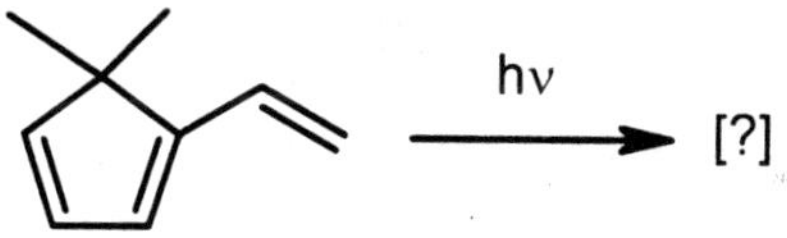

SP-6.35 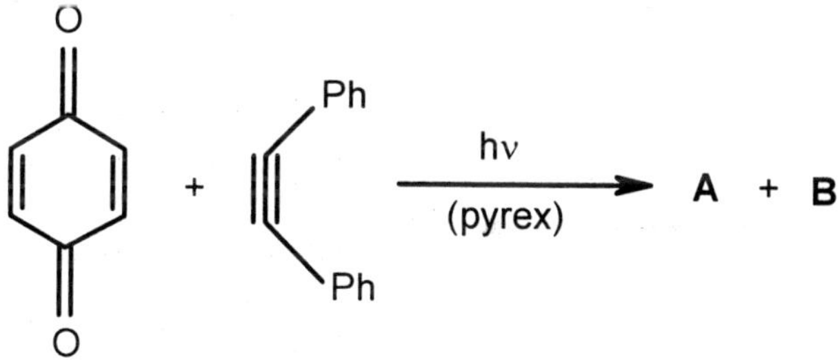

SP-6.36 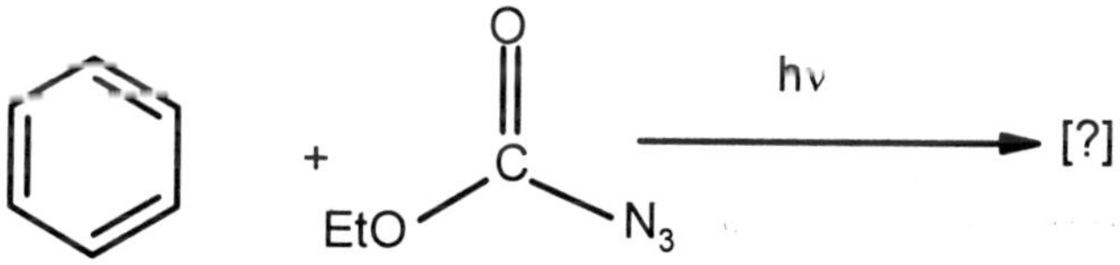

SP-6.37 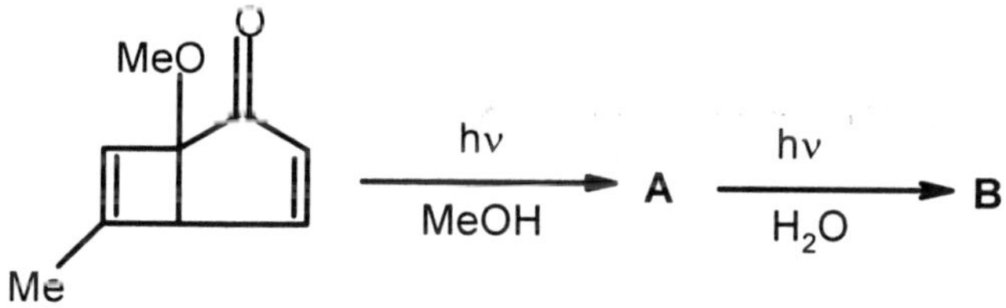

SP-6.38 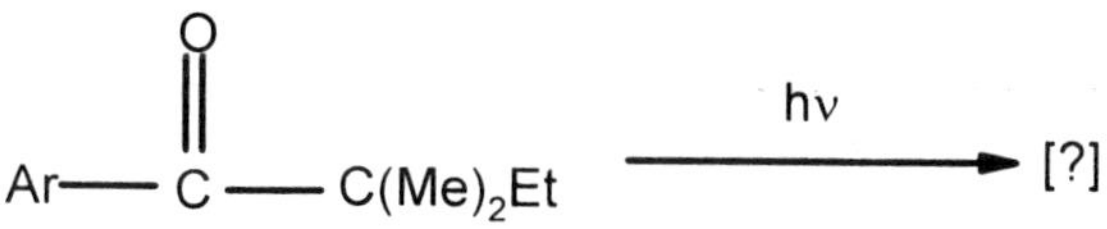

SP-6.39

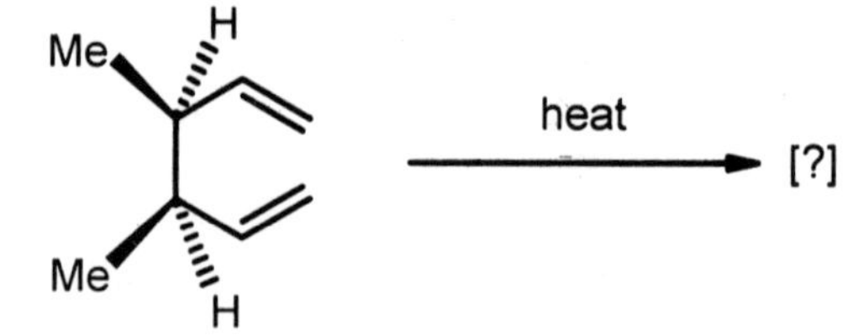

SP-6.40

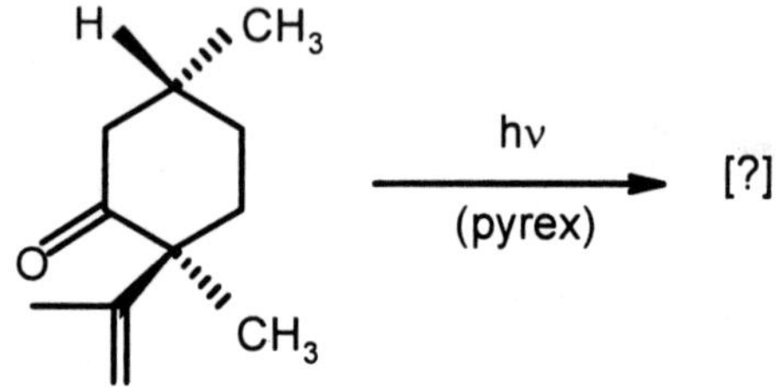

SP-6.41

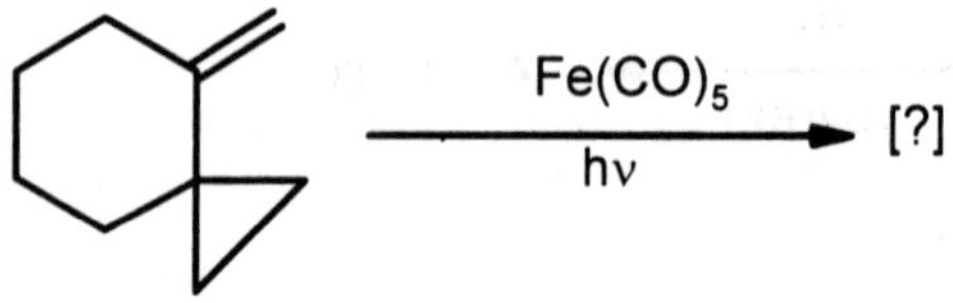

SP-6.42

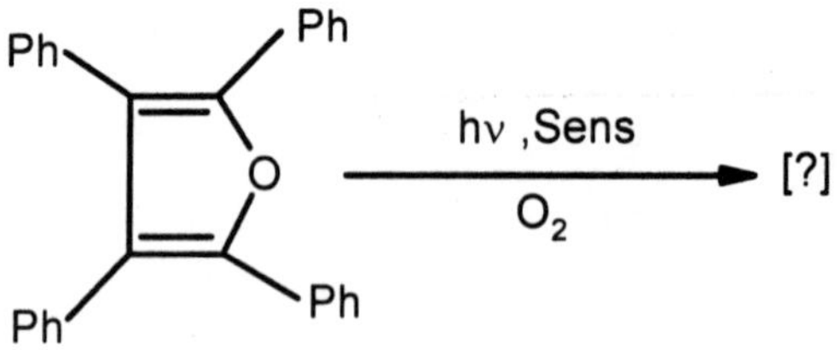

SP-6.43 $\text{MeCO CH}{=}\text{CHCHMe}_2 \xrightarrow[\text{MeOD}]{h\nu}$ [?]

SP-6.44

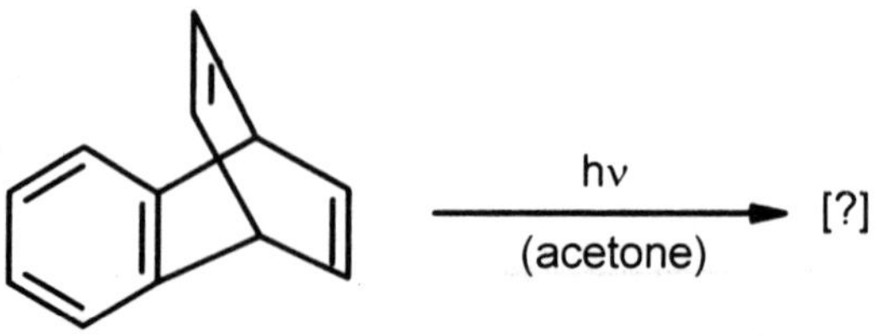

SP-6.45 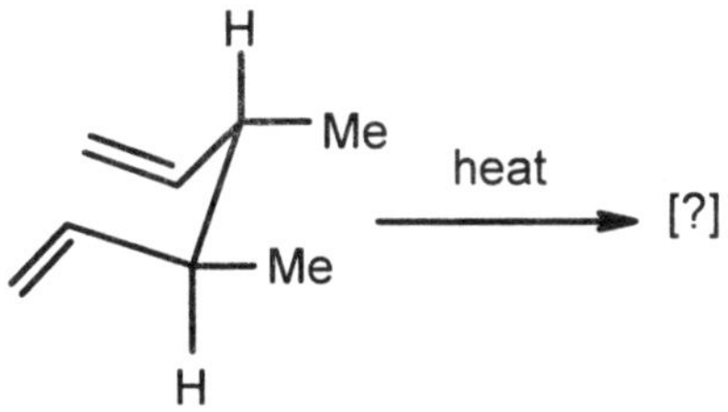

SP-6.46 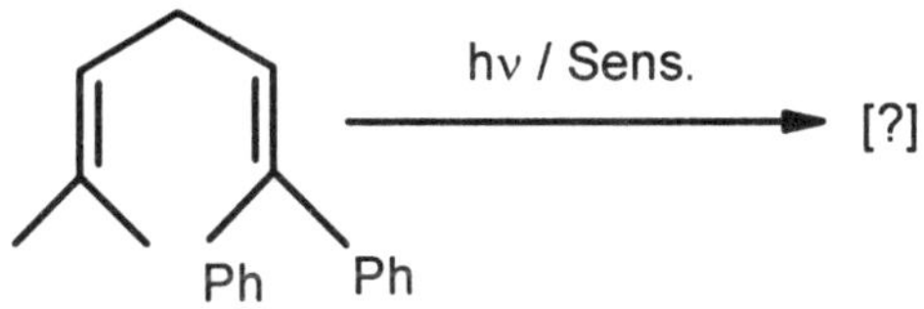

SP-6.47 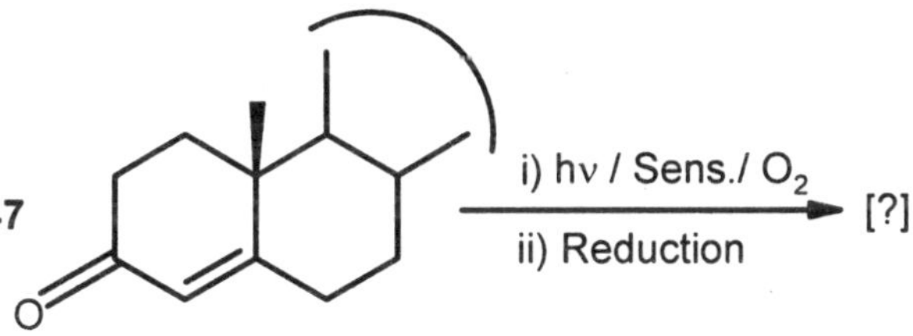

SP-6.48 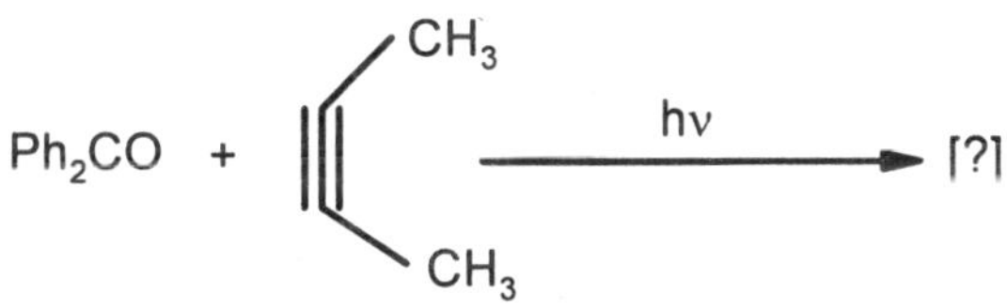

SP-6.49 Ph_2CBr_2 —— i) MeLi; ii) *Cis* -2-butene ——→ [?]

SP-6.50 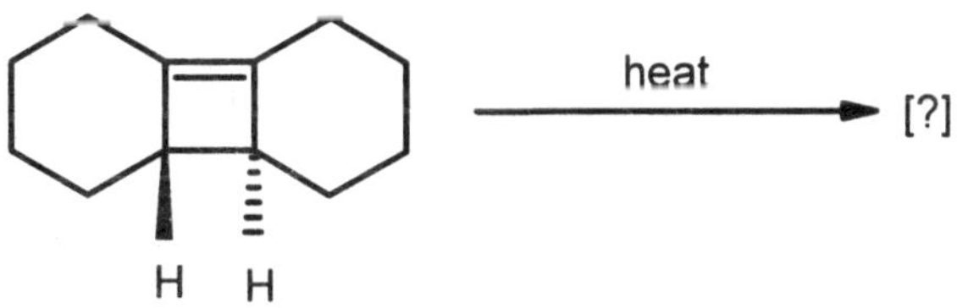

SP-6.51

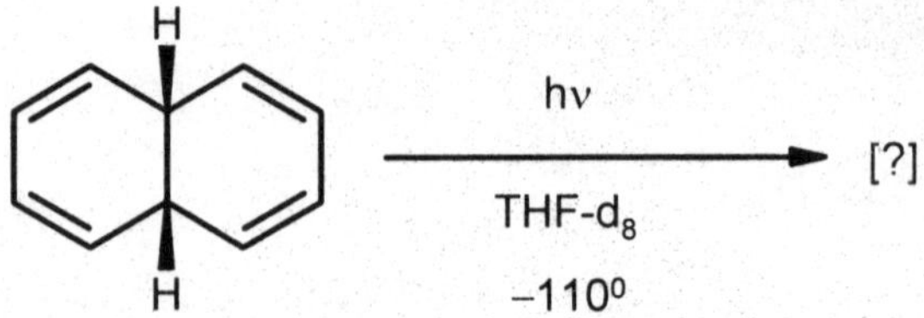

SP-6.52

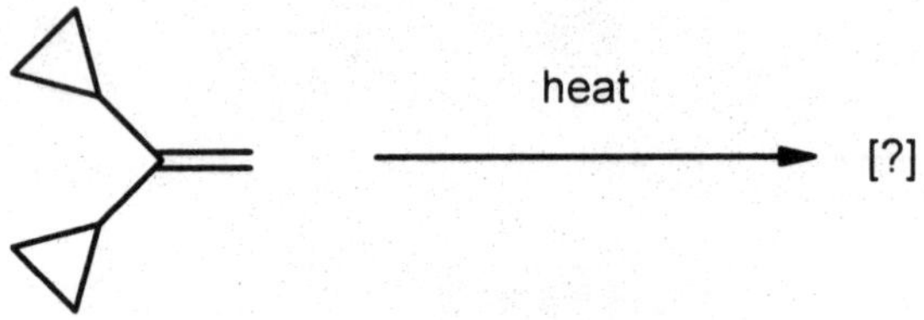

SP-6.53

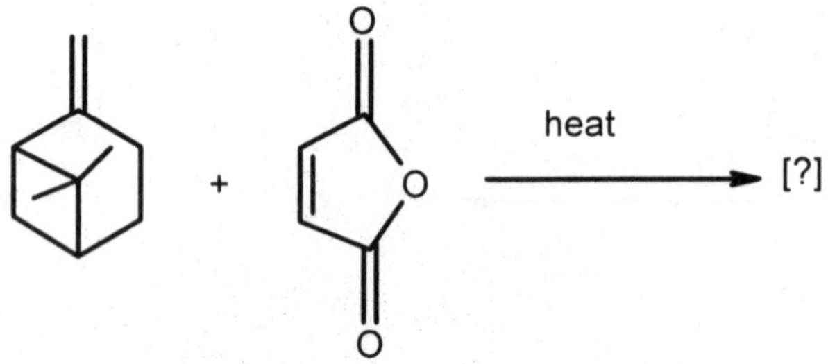

SP-6.54

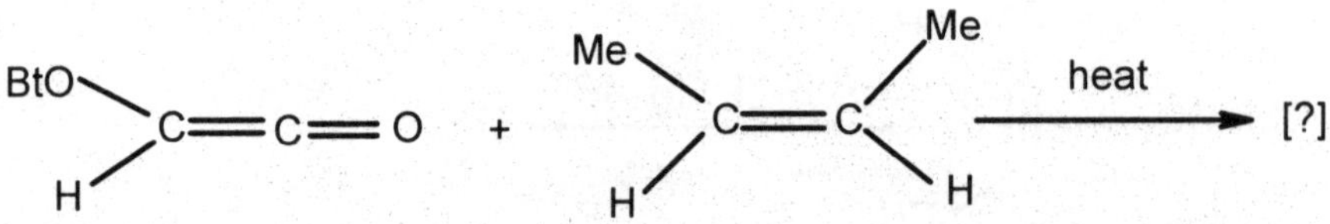

SP-6.55

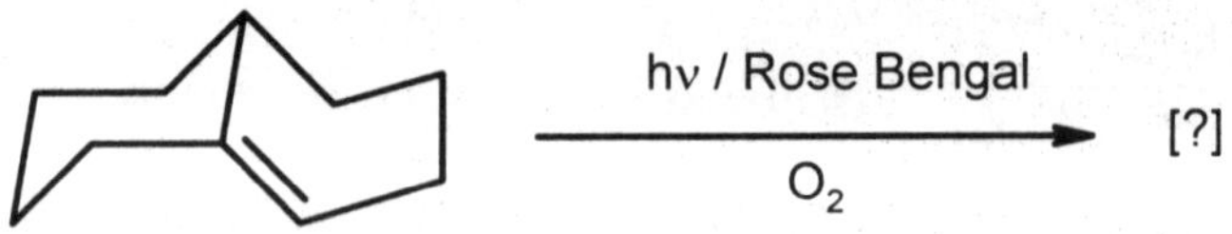

SP-6.56

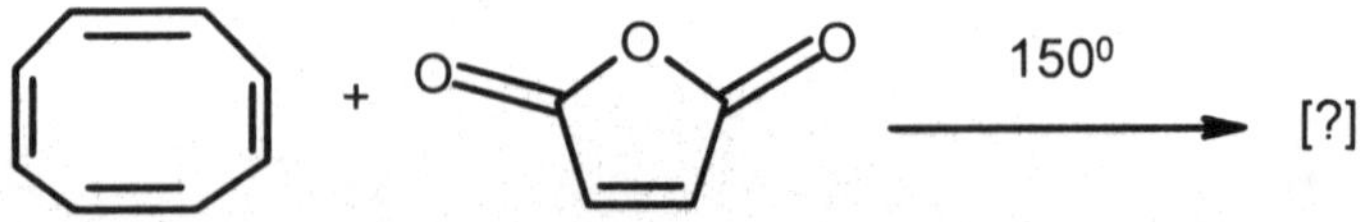

SP-6.57

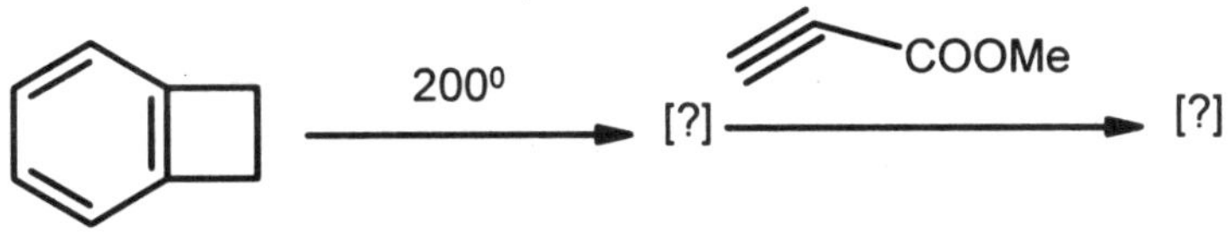

SP-6.58

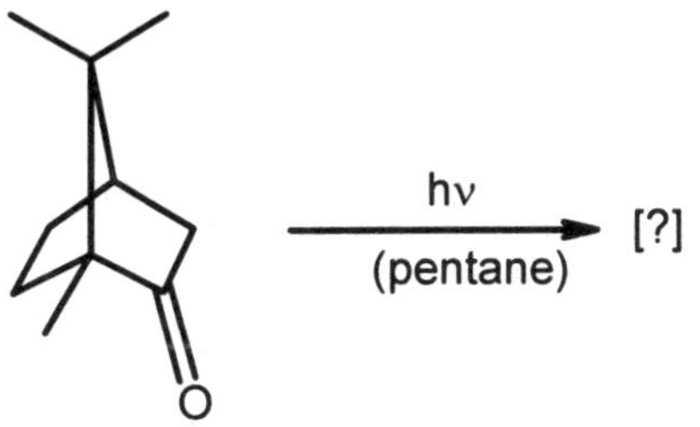

SP-6.59

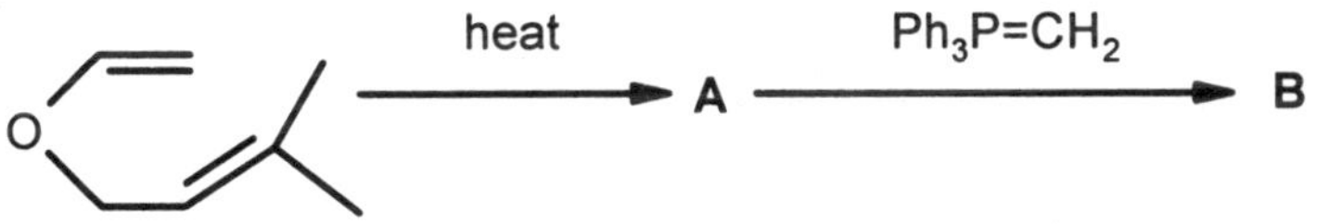

SP-6.60

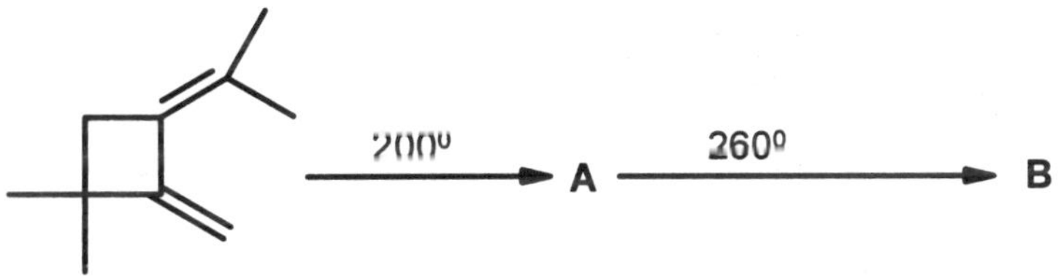

SP-6.61

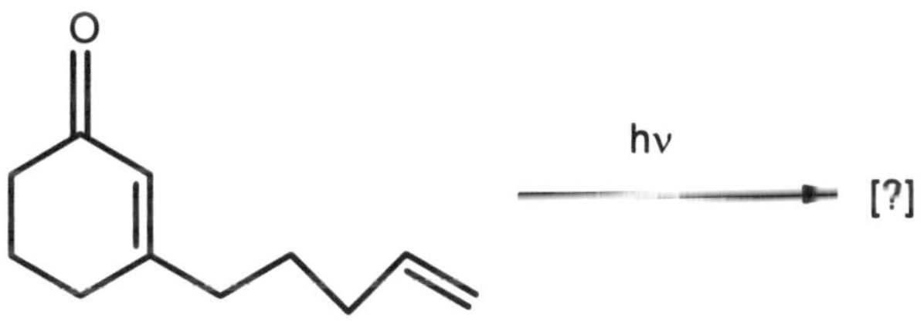

SP-6.62

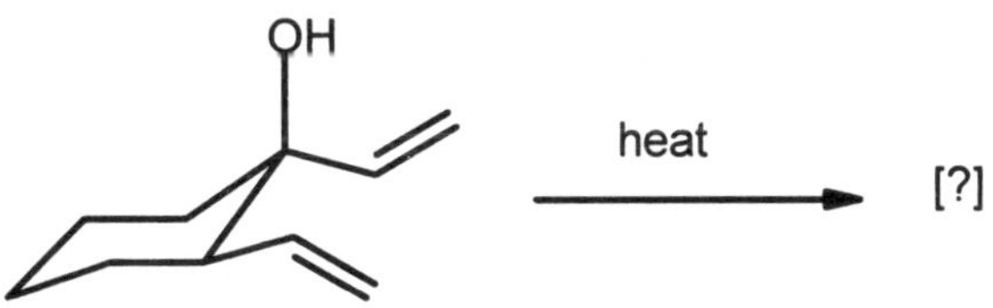

SP-6.63

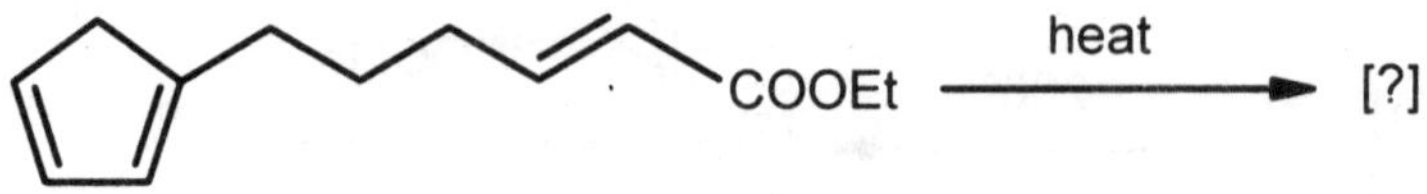

SP-6.64

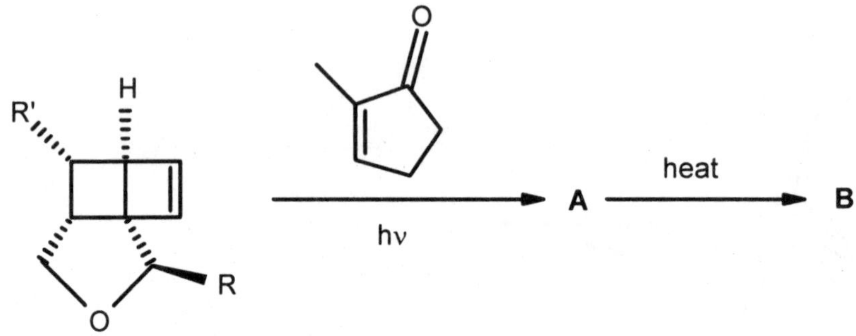

SP-6.65

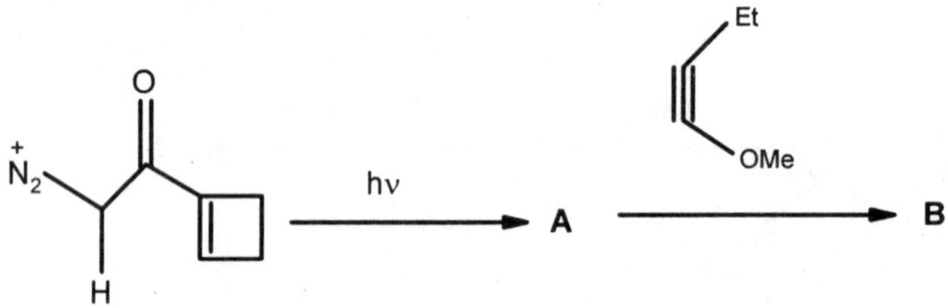

(**Hint.** Removal of N_2 with the formation of ketene intermediate, through rearrangement, which undergoes cycloaddition)

CHAPTER 7

ORGANIC SPECTROSCOPY

Introduction

Spectroscopic method has already established itself as an authentic as well as one of the most significant techniques not only in solving structural problems, but also in analytical and preparative works. Spectroscopic techniques have resulted a lot, and are of ever-increasing contribution to organic chemistry. It has now been possible to identify an organic compound having complex structural feature by utilizing a small amount of sample (as usually obtained from natural sources) with the help of combined spectral techniques. Many recent successes in the structural elucidations of complex natural products would have remained out of reach without these physical methods and their contribution in extending our understanding of the reactivity of organic compounds can hardly be overestimated.

Application of spectroscopic techniques in solving structural problems has become now a routine work to the organic chemists. In the present chapter, we will discuss some of the significant aspects of ultraviolet/visible, infrared, NMR, and mass spectroscopy through solved problems.

Solved Problems

Problem 7.1 What are the possible electronic transitions to be observed in a molecule when irradiated with UV light? Arrange them in order of energetically increasing process.

Solution. On absorption of ultraviolet/visible light of a particular energy, a molecule gets *excited*, and suffers from a change in the electronic level. The change involves the transition of only one electron to the higher energy level unaffecting all other electrons (according to first approximation of Frank–Condon principle). Possible transitions consist of the excitation of bonding (π or σ) and/or non-bonding (n) electron to the next higher energy anti-bonding orbital (π* or σ*). Thus, all possible electronic excitations are: $n \rightarrow \pi^*$, $n \rightarrow \sigma^*$, $\pi \rightarrow \pi^*$ and $\sigma \rightarrow \sigma^*$ transitions.

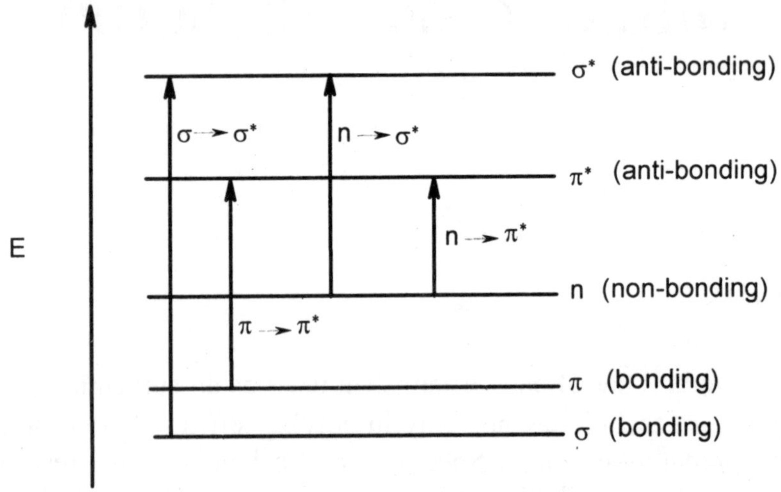

----Electronic excitations----

Hence, the electronic transitions may be arranged in order of increasing ΔE, as:

$$n \rightarrow \pi^* < \pi \rightarrow \pi^* < n \rightarrow \sigma^* < \sigma \rightarrow \sigma^*$$

——→ ΔE increasing ——→

Problem 7.2 Define the terms 'chromophore' and 'auxochrome'.

Solution. ***Chromophore***

The term 'chromophore' is originated from the Greek *chromophoros* meaning *colour carrier*. Though at the beginning the term was applied to the system responsible for imparting colour to a compound, latter it is used in a broader sense that absorbs electromagnetic radiation, irrespective of the fact whether or not a colour is thereby produced. Thus a chromophore, which is a covalently unsaturated group, is responsible for electronic absorption, and offers a characteristic absorption in the ultraviolet or in the visible region.

Some important chromophoric systems are: ethylenic (isolated or conjugated) and acetylenic unsaturations, carbonyls, acids, esters, nitro and nitrile groups, etc. For example, aryl conjugated azo group (Ar–N = N–Ar) is the principal chromophore for providing colour to azo dyes, while nitro group acts as the chromophoric system in the yellow-coloured nitro-compounds. The carbonyl group is also an important chromophoric system: an isolated C=O (as in acetone) absorbs ultraviolet light around 280 nm (characteristic absorption; n → π* transition), and it does not produce any colour.

Auxochrome

An auxochrome that is usually a saturated moiety with non-bonded electrons, when attached with a chromophore, enhances the colour-imparting properties of the chromophore, and thereby brings about a shift of the absorption band towards the red end of the spectrum as well as also altering the intensity of absorption, without being itself a chromophore. Examples are: –OH, –OR, $-NH_2$, –NHR, $-NR_2$, –SH, etc. The effect of an auxochrome is developed due to its ability to extend the conjugation of a chromophore by sharing of non-bonding electrons. Hence, in practice an auxochrome becomes a part of the newly developed chromophore. The new chromophore would show a different value of absorption maximum (at higher wavelength) and as well as extinction coefficient. For example, benzene exhibits an absorption maximum at 255 nm (ε_{max} 203), whereas aniline absorbs at 280 nm (ε_{max} 1430). The amino ($-NH_2$) group behaves as an auxochrome.

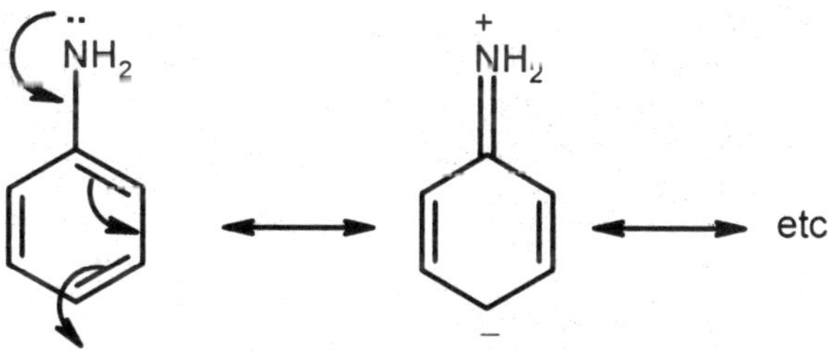

(extended conjugation; availability of π-electrons within the ring enhances)

Problem 7.3. In the following series which one will show relatively maximum value for λ_{max}?

$$CH_3Cl,\ CH_3Br,\ CH_3I$$

Solution. The expected transition is n → σ*. The atomic size increases from Cl→ Br→ I; that is, as one goes from *chloro* to the *bromo* to the *iodo* group, the 'n' electrons are found in higher principal energy levels. On increasing size attractive force of the nucleus towards the outermost shell becomes less. Consequently, 'n' electrons on 'I' suffer least attractive force exerted by the central nucleus among the others in the series, and hence would be excited most feasibly. As a result, absorption occurs at progressively higher λ_{max} since less energy is required.

$$CH_3Cl\ (\lambda_{max}\ 173\ nm) \rightarrow CH_3Br\ (\lambda_{max}\ 204\ nm) \rightarrow CH_3I\ (\lambda_{max}\ 258nm)$$

Problem 7.4 Can you distinguish between *cis*- and *trans*-isomers of stilbene from UV spectral study?

Solution. The geometrical isomers of stilbene can easily be distinguished from their UV spectral studies.

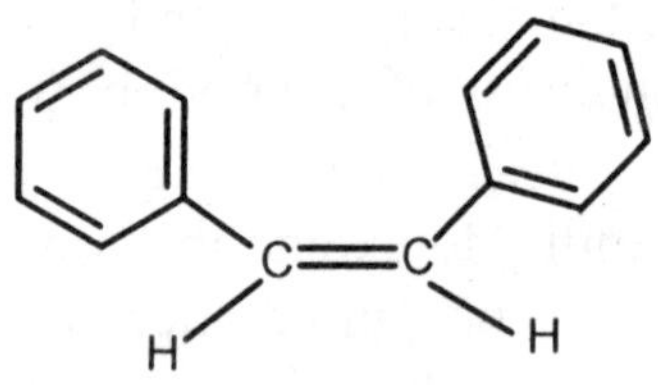

Cis-stilbene

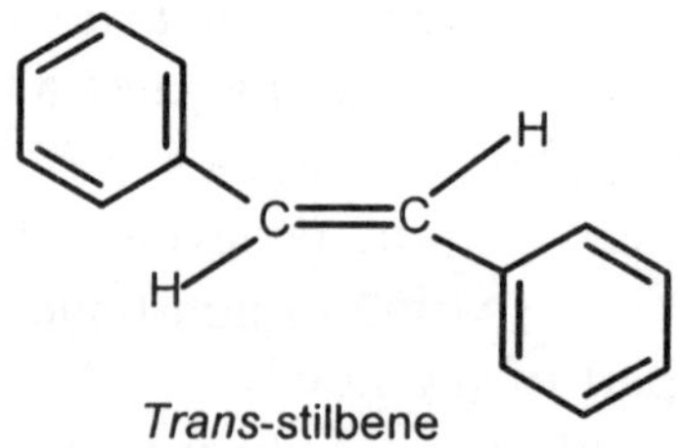

Trans-stilbene

The two bulky phenyl groups are at the same side in the *cis*-isomer; this crowding develops a strong steric hindrance into the molecule, as a result of which a distortion in co-planarity is occurred. Hence, conjugative effect becomes less in comparison to that in the *trans*-isomer where the two bulky substituents are in the opposite direction. That is why, the former absorbs at shorter wavelength (i.e. at higher energy region; λ_{max} 278 nm) and the latter at longer wavelength (i.e. at lower energy region; λ_{max} 295 nm).

Problem 7.5 "In UV spectrum instead of absorption lines, absorption bands are appeared" — do you justify to the observation?

Solution. During the electronic transitions within a molecule subjected under UV irradiation (high energy radiation) the molecule also suffers from both the vibrational and rotational effects. These two additional effects are superimposed on the electronic transitions, and as a result an envelope of transitions occurs as broad bands rather than sharp lines.

Problem 7.6 Mention the types of absorption and intensity shifts in UV spectroscopy.

Solution. Four types of shifts may usually be observed in UV spectroscopy — out of which two are associated with absorption maximum (λ_{max}) and the other two with intensity of absorption (ε_{max}). They are:

(a) Bathochromic shift or Red shift: Absorption maximum is shifted towards longer wavelength due to substitution or solvent effect.
(b) Hypsochromic shift or Blue shift: Absorption maximum is shifted towards shorter wavelength due to substitution or solvent effect.
(c) Hyperchromic shift: Intensity of absorption maximum (ε_{max}) increases.
(d) Hypochromic shift: Intensity of absorption maximum (ε_{max}) decreases.

These four effects can be shown by the following diagram:

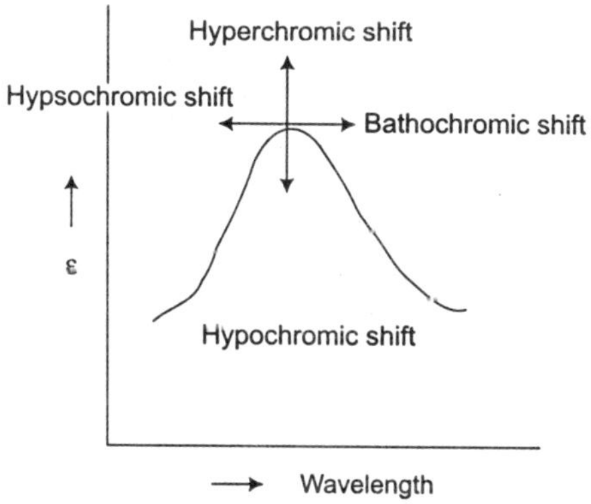

Problem 7.7 How the intensity of absorption of a UV band varies?

Solution. The intensity of absorption of a UV band largely depends upon two factors: (i) the probability of interaction between the radiation energy and the electronic system, and (ii) the difference between the ground and the excited state energy. Intense absorption (higher ε_{max}) occurs when a transition is accompanied by a large change in the transition moment (i.e. dipole moment of transition which is proportional to the change in the electronic charge distribution occurring at the time of excitation).

The exact amount of UV light absorbed at a particular wavelength is expressed as the compound's *molar absorptivity* or *molar extinction coefficient* (ε). This index offers the extent of efficiency of light absorption, and is calculated using *Lambert-Beer's law*

The Lambert-Beer law: $A = \log_{10}(I_0/I)$ I_0 = intensity of incident light I = intensity of transmitted light emerging from sample	$\varepsilon = A/(c \times l)$ A = absorbance c = sample concentration (mole dm^{-3}) l = sample path length (cm)

Problem 7.8 Discuss the solvent effects on UV absorption.

Solution. Polarity of solvents has a profound effect on the position and intensity of a UV absorption band. Both the $\pi \rightarrow \pi^*$ and $n \rightarrow \pi^*$ transitions are influenced by varying polarity of solvents used. Conjugated dienes and aromatic hydrocarbons experience very little solvent shift, while α,β-unsaturated carbonyl compounds show solvent shifts significantly.

(i) *Effect on* $\pi \rightarrow \pi^*$ *transition:* With increasing polarity of solvents $\pi \rightarrow \pi^*$ band moves to longer wavelength (red shift). This is due to the solvation effect. π^* orbital becomes more polar than the corresponding π orbital; dipole-dipole interactions with solvent molecules will, therefore, decrease the energy of the excited state more than that of the ground state, i.e. π^* orbitals are relatively more stabilized through solvation than π orbitals as a result requisite energy of $\pi \rightarrow \pi^*$ transition becomes less with enhancing solvation (red shift). Thus, there occurs a red shift of the order of 10-20 nm in going from hexane as solvent to ethanol.

(ii) *Effect on* $n \rightarrow \pi^*$ *transition:* The $n \rightarrow \pi^*$ transition in carbonyl compounds shows a solvent effect in the opposite direction, i.e. towards shorter wavelength (blue shift) with increasing solvent-polarity. The stabilization of non-bonding orbitals is particularly pronounced with solvents capable of forming hydrogen bonds (like water or ethanol). Stronger is the hydrogen bond to the carbonyl group, greater is the stabilization of non-bonding electrons, and hence energy of $n \rightarrow \pi^*$ transition becomes more in amount (blue shift). For example, λ_{max} of acetone in hexane solution is 278 nm (ε_{max} =1.5), and decreases to 272 nm and 264.5 nm for the solvents ethanol and water, respectively.

Solvent effects on both the $\pi \rightarrow \pi^*$ and $n \rightarrow \pi^*$ transitions may be represented with the help of the following diagram.

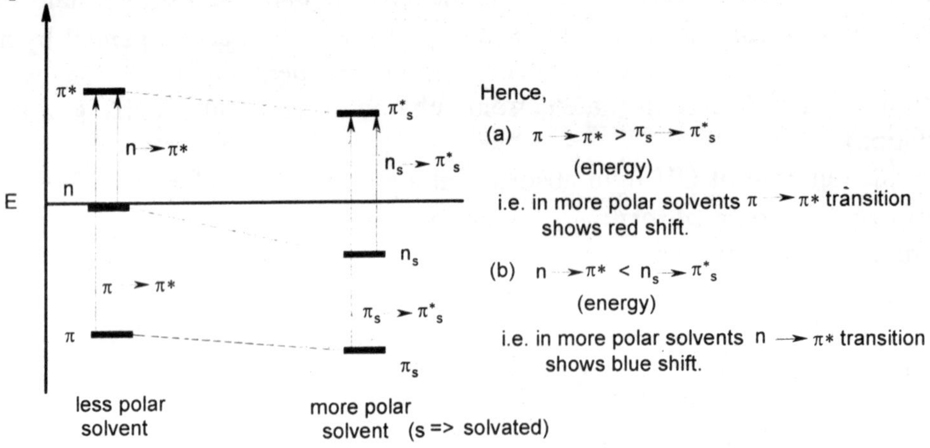

: Effect of solvation in α,β-unsaturated carbonyl compounds

Problem 7.9 Discuss the effects of steric hindrance on UV spectral absorption.

Solution. If coplanarity of π-electron system in a molecule is hindered by means of steric hindrance, extended conjugation becomes disturbed, as a result of which there occurs a marked shift in the absorption maximum and extinction coefficient.

In case of simple biphenyls and 2-substituted biphenyls, it is observed that n→π* transition of 2-substituted derivatives shifts towards shorter wavelength with diminished intensity — because of the fact that π–orbital overlaping in the latter case becomes more distorted due to steric inhibition of the 2-substituted moiety/moieties. Hence, n→π* transition appears at λ_{max} 250 nm (ε_{max} 19,000) in case of simple biphenyl (I), while the value reduces to λ_{max} 237 nm (ε_{max} 10,250) in 2-methyl-biphenyl (II). The medium steric hindrance to coplanarity about a single bond, thus, decreases the intensity of absorption.

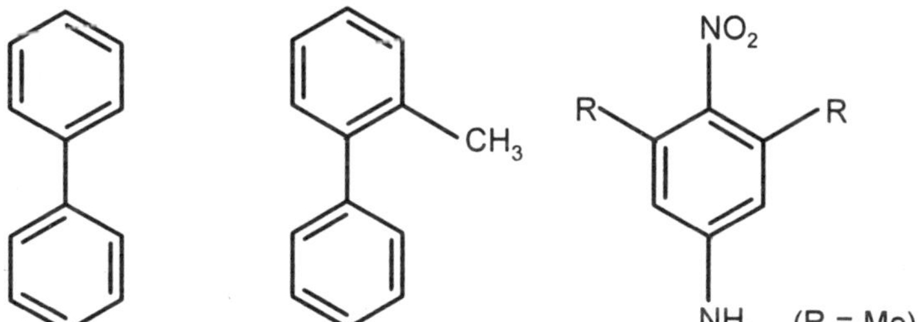

(I) (II) (III) (R = Me) (R = H)

Further, a blue shift is observed in case of 2,4,6-trimethylacetophenone that absorbs at λ_{max} 242 nm (ε_{max} 3200) as compared to *p*-methylacetophenone at λ_{max} 252 nm (ε_{max} 15,000). Steric hindrance may sometimes give rise to red shift also instead of exhibiting blue shift. Let us consider the case of (III) — the λ_{max} value of the 3,5-dimethyl-*p*-nitroaniline (R=Me) is at 385 nm (ε_{max} 4840) exhibiting a slight red shift and marked reduction in intensity from that of the parent compound (R=H) at 375 nm (ε_{max} 16,000).

When an alkene chromophore is capable of showing geometrical isomerism (*cis-trans* isomerism), it is normally observed that the *trans*-isomer absorbs at relatively longer wavelength (and higher intensity) than the *cis* isomer. The fact is due to the resultant of more effective π-orbital overlapping possible in the *trans*-isomer. This phenomenon may be best illustrated in the case of *cis* and *trans*- isomers of cinnamic acid and also those of stilbene.

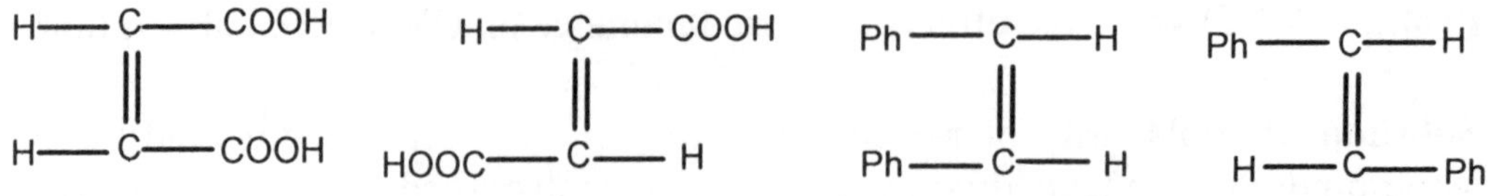

	Cis-cinnamic acid	*Trans*-cinnamic acid	*Cis*-stilbene	*Trans*-stilbene
λ_{max}(nm)	268	272	283	295
ε_{max}	10,7000	15,900	12,300	25,000

Thus, UV absorption largely depends upon steric factors.

Problem 7.10 The UV spectrum of the α,β-unsaturated ketone, $Me_2C=CHCOMe$, shows absorption bands at 230 nm and 321 nm in isoocatane. Assign them in terms of electronic transitions. How do the absorption bands change their positions with the change of the solvent from isooctane to water? Give reasons for your answer.

Solution. The π→π* transition corresponds to absorption band at 230 nm (called K-band), whereas the n→π* transition of lone-pair electrons of carbonyl oxygen to the anti-bonding π-orbital corresponds to absorption band at 321 nm (called R-band).

On increasing polarity of solvent (isooctane to water) the π→π* absorption band moves to longer wavelength (red shift); but the n→π* transition shows the solvent effect in the opposite direction, i.e. towards shorter wavelength (blue shift) with increasing solvent-polarity. For proper explanation, see the answer of the **problem 7.8.**

Problem 7.11 How could you distinguish between the K-bands of conjugated di- and polylene systems with those of enone system?

Solution. The K-bands (π→π* transitions) of conjugated di- and polylene systems can be distinguished from those of enone system by observing the effect of altering solvent polarity. The hydrocarbon double bonds are non-polar; and hence the π→π* transitions of diene and polyene systems are almost unchangeable on altering solvent polarity. On the contrary, the corresponding transitions of enone system undergo a bathochromic shift, usually accompanied by increasing intensity, as the polarity of the solvent is increased. The red shift is presumably resulted from a reduction in the energy level of the excited state accompanying dipole-dipole interaction (and may also hydrogen bonding) with polar solvent(s).

Problem 7.12 Saturated acyclic ketones usually record three absorption bands in their UV spectra at around 160 nm,190 nm and 280 nm. Assign them in terms of electronic transitions. Predict with proper reasoning which one will record the most intense absorption?

Solution. Assignments: $\pi \rightarrow \pi^*$ transition at λ_{max} 160 nm (K-band)
$n \rightarrow \sigma^*$ transition at λ_{max} 190 nm
$n \rightarrow \pi^*$ transition at λ_{max} 280 nm (R-band)

The absorption band at λ_{max} 280 nm (R-band) appeared due to $n \rightarrow \pi^*$ transition is of lowest intensity (weak, $\varepsilon_{max} < 30$), because this absorption band results from the forbidden transition of a loosely held n-electrons to the π^* orbital, the lowest unoccupied orbital of the carbonyl group, although the wavelength is long (lower energy). The most intense band having high extinction coefficient would be at λ_{max} 160 nm developed due to $\pi \rightarrow \pi^*$ transition.

Problem 7.13 1,3-Butadiene absorbs at longer wavelength when it is irradiated with UV radiation, in comparison to ethylene. Give reason.

Solution.

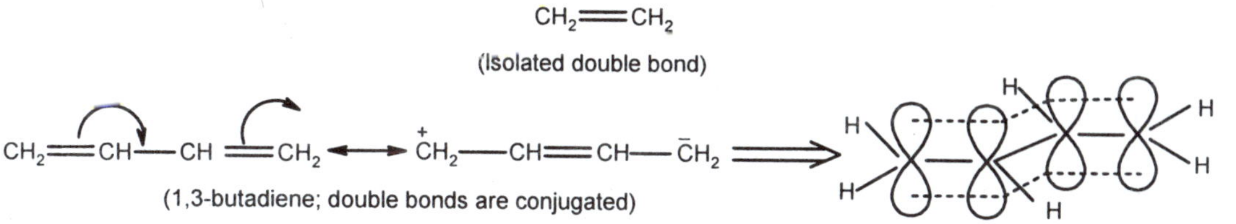

In case of ethylene the unsaturation is described by π orbitals of different energy — one is bonding and the other anti-bonding. But in conjugated diene compounds (π–π conjugated system), such as 1,3-butadiene, two new energy levels are generated due to an effective overlap of the π orbitals (figure below).

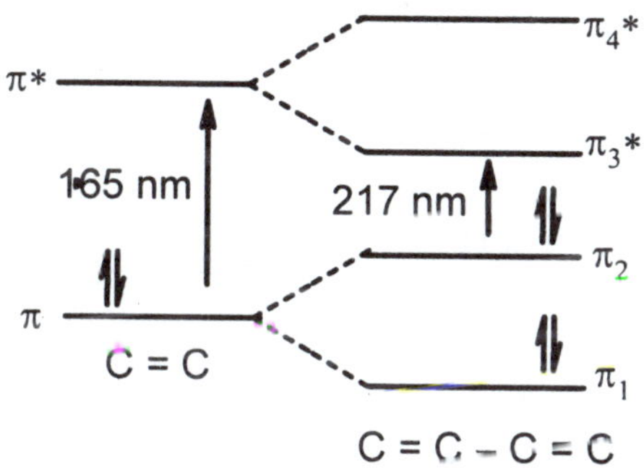

Thus, the $\pi_2 \rightarrow \pi_3^*$ transition of butadiene occurs relatively at longer wavelength in comparison to the $\pi \rightarrow \pi^*$ transition of ethylene. Other $\pi - \pi^*$ transitions also are possible in the conjugated system, but their intensities depend on the "allowedness" of the transitions.

Problem 7.14 How do you determine the strength of hydrogen bonding on the basis of UV spectroscopic technique?

Solution. It is possible to determine the strength of hydrogen bonding by using UV spectral technique. It is done on the basis of shifts in UV absorption (blue shift) of a suitable substrate in polar solvents. Acetone may be considered as a good substrate in this regard. Acetone absorbs at 279 nm and 264.5 nm, respectively in hexane (nonpolar solvent) and water (polar solvent). The calculated energies associated with these wavelengths are approximately 121 kcal/mole and 126 kcal/mole, respectively. It is obvious that in water medium, acetone molecules must be hydrogen-bonded with the solvent molecules. Thus more energy (lower wave-length) is required for electronic excitation of acetone in water medium in comparison to that in hexane. Hence, the difference in energy (i.e. ~5 kcal/mole) may be considered to be equivalent to the strength of the hydrogen bond, and is found to be in good agreement with the known energy value of hydrogen bond

Problem 7.15 The position of UV absorption maxima of aniline in aqueous solution are different from those of benzene but are almost identical with those in a solution of pH=1. Explain.

Solution. In aniline, the lone-pair of electrons on nitrogen of $-NH_2$ group takes part in delocalization with the π–electron system of benzene ring. This enhanced conjugation lowers the $\pi \rightarrow \pi^*$ transition energy of benzene nucleus, as a result of which bathochromic shifts of the absorption maxima are noticed.

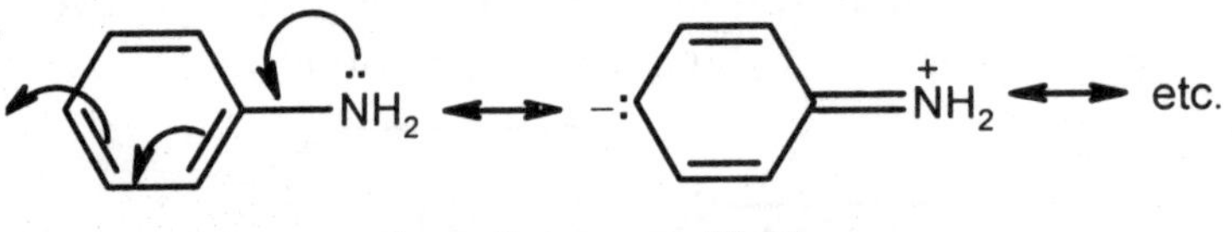

(in ordinary condition)

But in strong acidic solution (pH=1) the $-NH_2$ group of aniline molecule becomes protonated giving rise to the anilinium cation; the non-bonding electron pair is no longer available for interaction with π-electrons of the ring, and a spectrum appears almost identical to that of benzene.

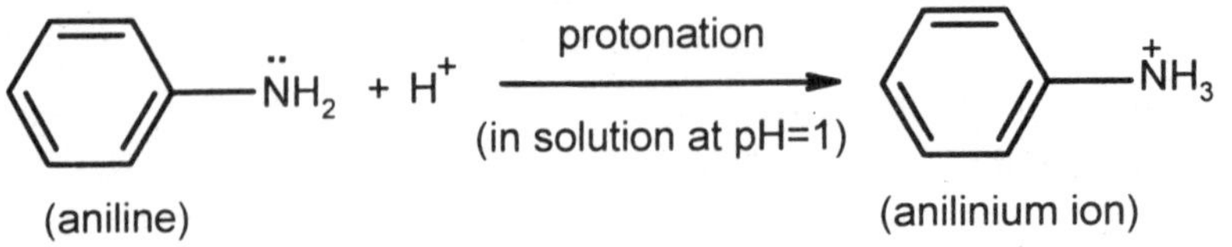

Problem 7.16 Compare the UV absorption spectra of the following compounds: *N,N*-dimethylaniline, 2,*N,N*-trimethylaniline and 2,6,*N,N*-tetramethylaniline.

Solution.

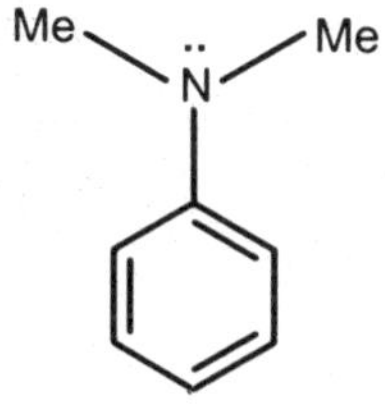

(I)

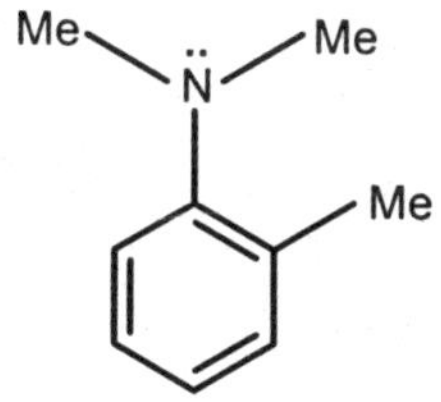

(II)

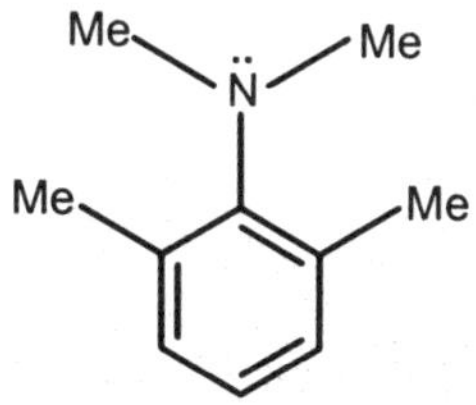

(III)

	(I)	(II)	(III)
λ_{max} (nm)	176	185	194
	200	207	211
	250	247	259
	296	288	

Extended conjugation of the lone-pair of electrons on 'N' atom within $-NMe_2$ moiety with the ring π-electrons prevails in (I). Presence of bulky methyl group at the *ortho*-position of $-NMe_2$ in (II) creates some steric inhibition to the parallel orientation of the π–orbital of the non-bonding electrons with the π-orbitals of the ring; hence the extended conjugation is less. Such steric inhibition becomes maximum in case of molecule (III), as a result of which $-NMe_2$ moiety goes out of the plane, and there is no extended conjugation at all. Thus the absorption band at λ_{max} 250 nm in (I), appeared due to electron transfer, becomes low intense in (II) (intensity reduces about to half), while this particular band is eliminated in (III).

Problem 7.17 "Azobenzene is deep orange-red while hydrazobenzene is a colourless compound" – explain.

Solution.

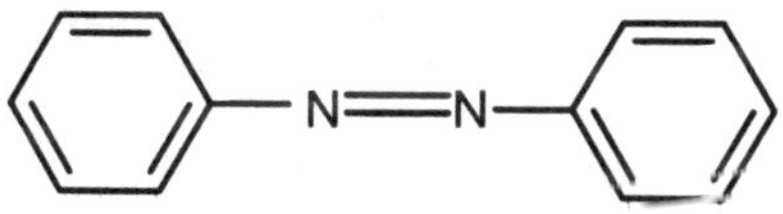

(azobenzene)

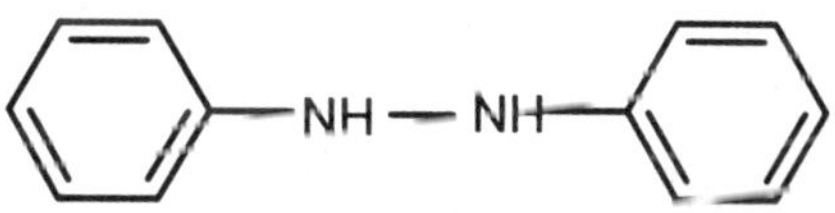

(hydrazobenzene)

Azobenzene has a chromophoric system (–N=N–), which is conjugated with both the benzene rings. That's why it absorbs at a longer wavelength, 440 nm for n→π* transition giving rise to the colour and the π→π* bands also appear at 320 and 230 nm. On the other hand, hydrazobenzene should have a spectrum similar to that of aniline for which the longest band is at 180 nm, since the two lone-pair orbitals on nitrogen are approximately at right angles to each other.

Problem 7.18 Does alkyl substitution or ring residue play any role on the absorption maximum of a chromophore?

Solution. An alkyl substitution or a ring residue, if properly placed with respect to the chromophoric moiety, offers a bathochromic shift of the corresponding absorption maximum. This is due to the resultant of hyperconjugation in which the σ-electrons of the alkyl group are mobile enough to interact with the chromophoric group, thereby, extending the conjugation.

It may be cited that B-band of benzene appears at λ_{max} 256 nm (ε_{max} 200),, while that for toluene it appears at λ_{max} 261 nm (ε_{max} 300). This bathochromic shift is attributed to hyperconjugation in which the σ-electrons of the methyl C–H bond participate in resonance with the ring. Hyperconjugative effect of methyl group is more pronounced than other alkyl groups.

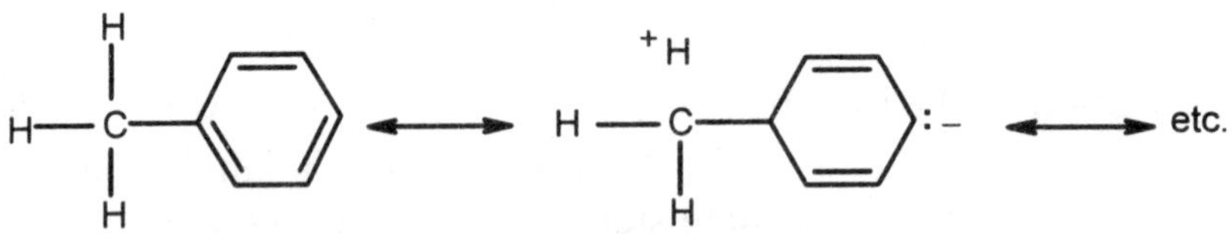

(hyperconjugation)

Problem 7.19 "Biphenyl exhibits a very intense absorption band at 252 nm, but the absorption band exhibited by its 2,2'-dimethyl derivative is of low intensity at ~270 nm'' — explain.

Solution.

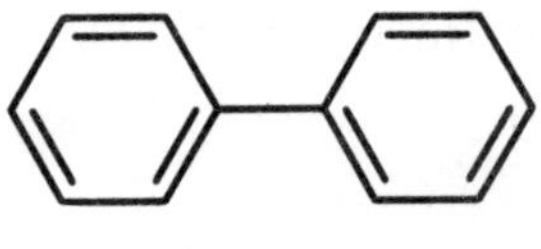

(biphenyl)

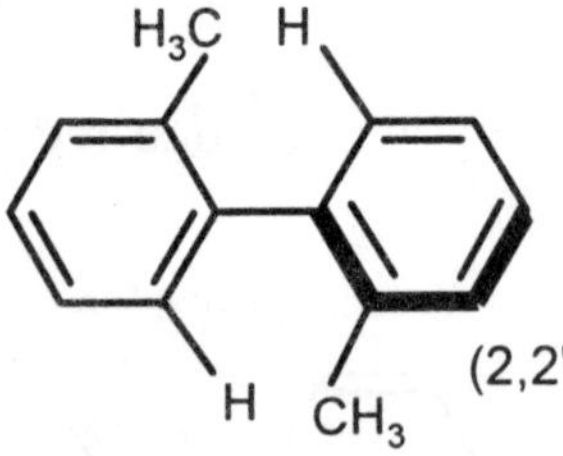

(2,2'-dimethylbiphenyl)

Since the angle of twist in biphenyl is small, conjugation of the π-electron systems between the two rings is not much affected; due to this extended conjugation as prevailed in a biphenyl molecule the intensity of the absorption band (at 252 nm) becomes intense (ε_{max}=19000). On the contrary, non-bonded steric repulsion between the *ortho*-substituents in 2,2'-dimethylbiphenyl molecule is comparatively much higher, as result of which a twisted conformation is preferred for this molecule. For twisted conformation extent of conjugation would obviously be much lower, which is reflected through the appearance of low-intensed absorption band, λ_{max} 270nm (ε_{max}=800) for the molecule.

Problem 7.20 Acetylacetone in aqueous solution records a UV absorption band at $\lambda_{max}(H_2O)$ 274 nm (ε_{max} 2050), intensity of which differs markedly when recorded in isooctane [λ_{max} (isooctane) 272 nm, ε_{max} 12,000]. Justify the observation.

Solution. Acetylacetone exists as an equilibrium mixture of two tautomeric forms — keto and enol.

(β-dikeonic form)

(enolic form stabilized by intramolecular H-bonding)

$\lambda_{max}(H_2O)$ 274 nm; ε_{max} 2050
λ_{max} (isooctane) 272 nm; ε_{max} 12,000

The UV spectra of β-diketones depend on the degree of enolization. Once again, the absorption is directly dependent on the concentration of the enolic tautomer. In aqueous solution the enol form of acetylacetone exists to the extent of only 15%, while in non-polar solvents such as isooctane (and also in vapour phase) the proportion enhances to the extent of about 90–92%. That's why, the absorption band of the compound in isooctane (a non-polar solvent) is more intense than that one observed in water (a polar solvent).

Problem 7.21 In alcoholic solution the UV spectra of *p*-hydroxy acetophenone and *p*-methoxy acetophenone are virtually identical. What method you will follow so that the two compounds can be distinguished by UV spectroscopy?

Solution. It is possible to distinguish the compounds by using UV spectroscopy. In *p*-hydroxy acetophenone the hydroxy group is free, while in the other it is blocked. Under this circumstance if we measure UV absorption behaviour of both the compounds in presence of sodium acetate solution in alcohol, an appreciable amount of bathochromic shift is noticed in case of *p*-hydroxy acetophenone while the absorption maximum for the other remains practically unaltered. In presence of sodium acetate (offers alkaline medium) the free auxochromic moiety (–OH) in *p*-hydroxy acetophenone becomes capable (through anionic oxygen) to extend conjugation more effectively, thereby shifting the absorption maximum towards higher wavelength.

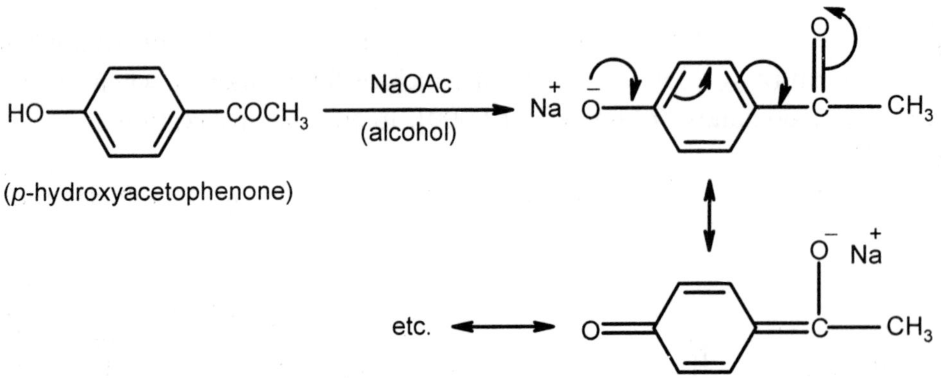

Problem 7.22 "Solution of iodine in hexane becomes brown when poured into benzene" — offer explanation.

Solution. The appearance of brown colour when iodine is added to benzene is due to the formation of a charge-transfer benzene-iodine complex.

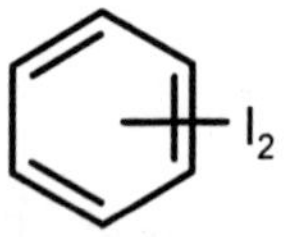

(benzene-I_2 complex)

It is assumed that benzene π-cloud interacts with the outer electrons of the iodine molecule to form the charge-transfer complex. While both benzene and molecular iodine in hexane have UV-VIS absorptions of their own (benzene: λ_{max} 255 nm; iodine: λ_{max} 500 nm), the charge-transfer complex displays an intense additional absorption band at λ_{max} 290 nm (ε_{max}=150000).

Probl

Problem 7.23 Calculate λ_{max} values for the following compounds according to the existing emperical rules.

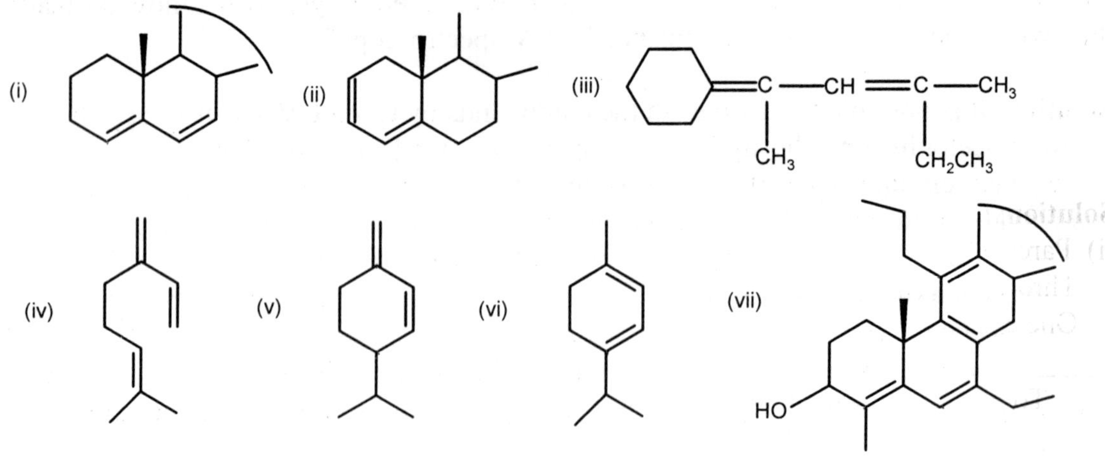

(viii) 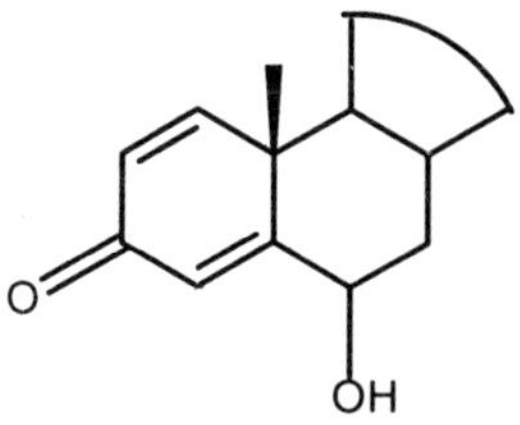

(ix) 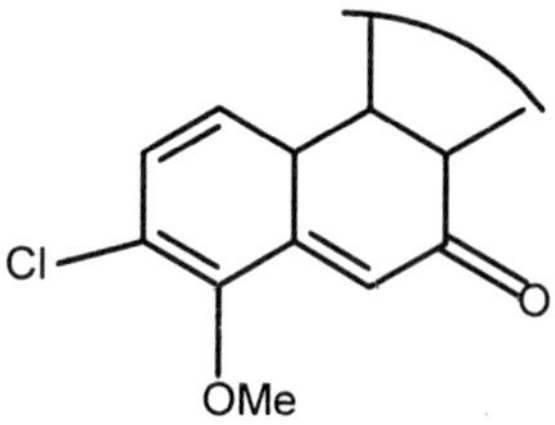

(x) 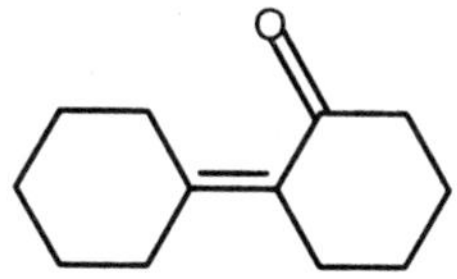

(xi) 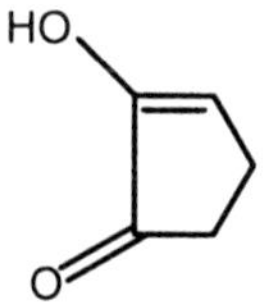

(xii) 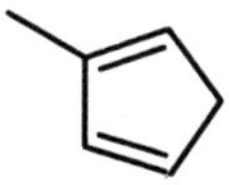

(xiii) 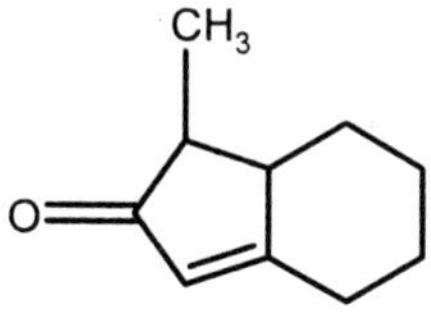

(xiv) 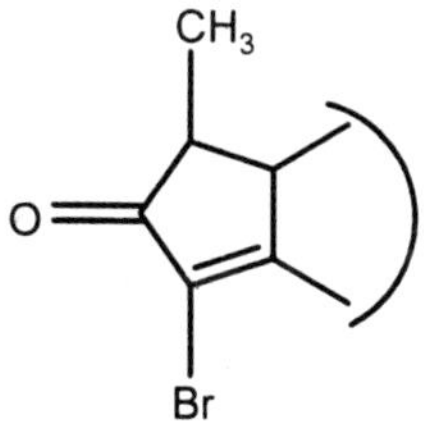

(xv) 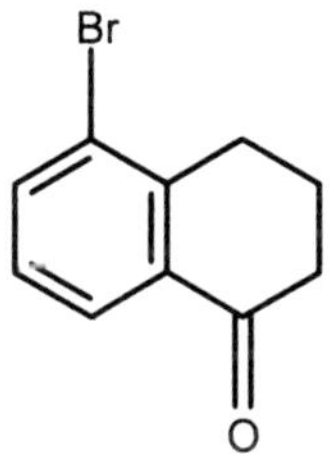

(xvi) 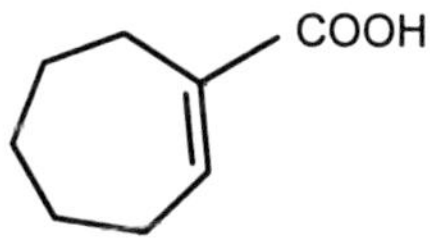

(xvii) 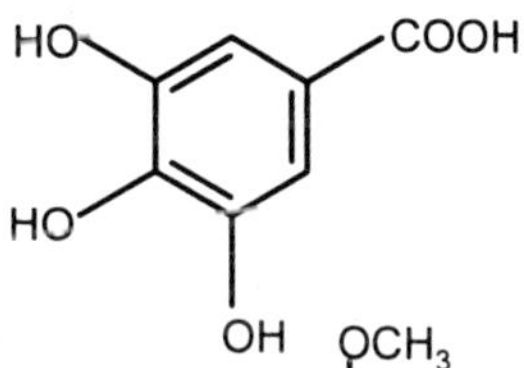

(xviii) 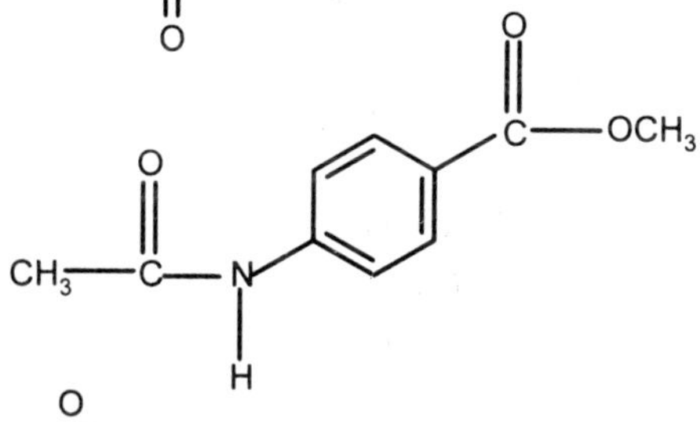

(xix) 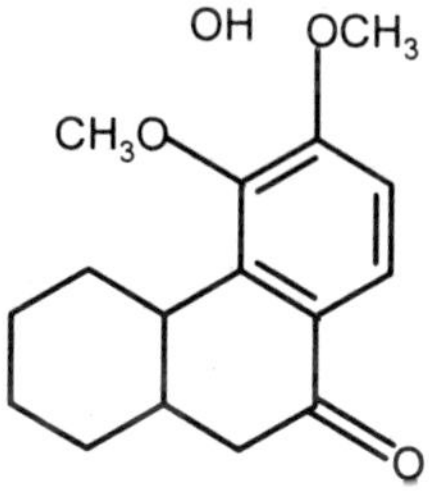

(xx) 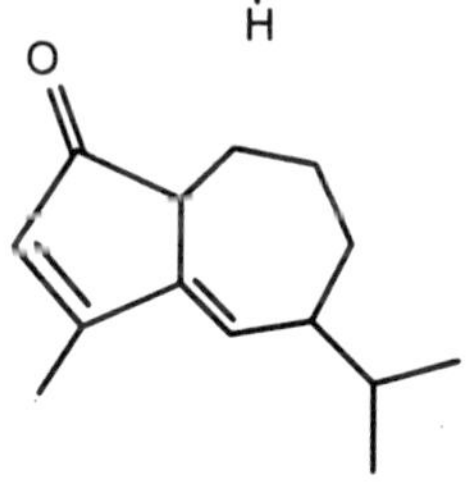

(xxi) 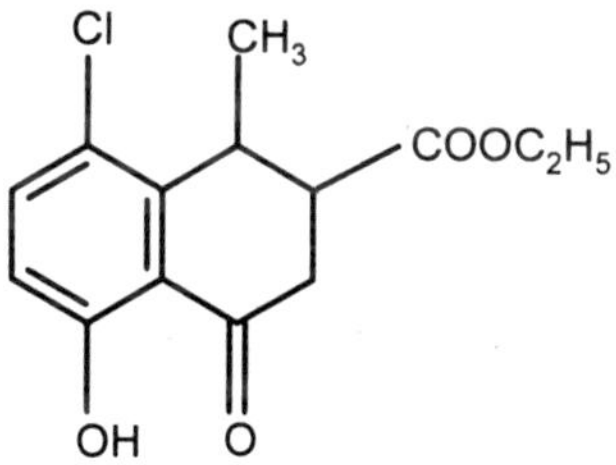

Solutions:

(i)

Parent value for heteroannular system		214 nm
Three-ring residues	3 x 5 =	15 nm
One exocyclic double bond		5 nm
Total (λ_{max} calculated)		234 nm

(ii) Parent value for homoannular system		253 nm
Three-ring residues	3 x 5 =	15 nm
One exocyclic double bond		5 nm
Total (λ_{max} calculated)		273 nm

(iii) Parent value for conjugated diene		217 nm
Three-alkyl residues	3 x 5 =	15 nm
Two-ring residues	2 x 5 =	10 nm
One exocyclic double bond		5 nm
Total (λ_{max} calculated)		247 nm

(iv) Parent value for acyclic diene system		214 nm
One-ring residues		5 nm
Total (λ_{max} calculated)		219nm

(v) Parent value for heteroannular system		214 nm
Two-ring residues	2x 5 =	10 nm
One exocyclic double bond		5 nm
Total (λ_{max} calculated)		229 nm

(vi) Parent value for homoannular system		253 nm
Four-ring residues	4 x 5 =	20 nm
Total (λ_{max} calculated)		273 nm

(vii) Parent value for homoannular system		253 nm
Eight-ring residues	8 x 5 =	40 nm
One exocyclic double bond		5 nm
Extended conjugation	2 x 30 =	60 nm
Total (λ_{max} calculated)		358 nm

(viii) Parent value for α,β-unsaturated carbonyl		215 nm
Two β-substituents	2 x 12 =	24 nm
One exocyclic double bond		5 nm
γ-hydroxy group		30 nm
Total (λ_{max} calculated)		274 nm

(ix) Parent value for α,β-unsaturated carbonyl		215 nm
One β-substituent		12 nm
One ω-substituent		18 nm
Two extended conjugation	2 x 30 =	60 nm
Homoannular diene component		39 nm
One exocyclic double bond		5 nm
γ-methoxy group		17 nm
δ-chloro		12 nm
Total (λ_{max} calculated)		378 nm

(x) Parent value for α,β-unsaturated carbonyl	215 nm
One α-substituent	10 nm
Two β-substituent	24 nm
One exocyclic double bond	5 nm
Total (λ_{max} calculated)	254 nm

(xi) Base value for five membered α,β-unsaturated carbonyl	202 nm
One β-substituent	12 nm
α-hydroxy group	35 nm
Total (λ_{max} calculated)	249 nm

(xii) Base value for 1,3-cyclopentadiene	239 nm
Two ring residues	10 nm
One alkyl residue	5 nm
Total (λ_{max} calculated)	254 nm

(xiii) Parent value for five-membered enone		202 nm
Two β-ring residues	2 x 12 =	24 nm
One exocyclic double bond		5 nm
Total (λ_{max} calculated)		231 nm

(xiv) Parent value for five-membered enone		202 nm
α–Bromo		25 nm
Two β-ring residues	2 x 12 =	24 nm
One exocyclic double bond		5 nm
Total (λ_{max} calculated)		256 nm

(xv) Parent chromophore	246 nm
ortho–Ring residue	03 nm
meta–Bromo	02 nm
Total (λ_{max} calculated)	251 nm

(xvi) Parent chromophore (having α,β–Dialkyl groups)	217 nm
Double bond in a seven-membered ring	05 nm
Total (λ_{max} calculated)	222 nm

(xvii) Parent chromophore		230 nm
Two *meta*–OH	2 x 7 =	14 nm
One *para*–OH		25 nm
Total (λ_{max} calculated)		269 nm

(xviii) Parent chromophore (benzoyl derivative; ester)	230 nm
para– $NHCOCH_3$	45 nm
Total (λ_{max} calculated)	275 nm

(xix) Parent chromophore (benzoyl derivative)		246nm
ortho– Ring residue		03 nm
meta– OCH_3		7 nm
para– OCH_3		25 nm
Total (λ_{max} calculated)		281 nm

(xx) Parent value for five-membered enone		202 nm
Double bond extending conjugation		30 nm
One β-ring residue		12 nm
One γ– and one δ–ring residue	2 x 18 =	36 nm
One exocyclic double bond		5 nm
Total (λ_{max} calculated)		285 nm

(xxi) Parent chromophore (benzoyl derivative)		246 nm
ortho– Ring residue		03 nm
meta– Cl		00 nm
ortho– OH		07 nm
Total (λ_{max} calculated)		256 nm

Problem 7.24 Outline the basic principle behind IR spectroscopy.

Solution. IR spectroscopy is an important physical tool to the organic chemists for getting sufficient information about the structural pattern of an organic compound. Spectrum recorded between 4000–400 cm^{-1} is of greatest practical use to them. It is observed that even a very simple molecule can give an extremely complex spectrum; and also that no two compounds, except enantiomers, can have exactly the same IR spectrum. The organic chemist exploits this advantage when he matches the spectrum of an unknown compound against that of an authentic sample. IR spectrum is very much characteristic of a molecule as a whole; but it is also true that certain groups of atoms offer bands at or near the same frequency regardless of the structure of the rest of the molecule. These characteristic bands permit one to gather useful structural information by simple inspection and reference to generalized charts of characteristic group frequencies.

When infrared radiation is passed through a sample of an organic compound, some of the frequencies are absorbed, while others are transmitted without being absorbed. IR spectrum plots such absorbance or transmittance of IR radiation against frequency. The frequency or wavelength of absorption depends upon the relative masses of the atoms, the force constants of the bonds, and the geometry of the atoms. Infrared radiation causes excitation of the ***quantized molecular vibration states***. Only those molecular vibrations, ***which are responsible for carrying out a change in the dipole moment of the molecules***, are recorded in the IR spectrum. The change in dipole moment is attributed to a change in charge distribution during specific molecular vibrations — the alternating electric field being produced thereby interacts with the oscillating electric field of the electromagnetic IR beam. Such vibrational spectra appear as bands rather than lines, because a single vibrational energy change is accompanied by a number of rotational energy changes. These vibrational–rotational bands appearing between 4000–400 cm^{-1} are of general interest.

Molecular vibrations or fundamental vibrations involve no change in the centre of gravity of the molecule. Molecular vibrations are of two types:

(a) ***Stretching:*** A stretching vibration is the movement of atoms with respect to a particular atom within the molecule along the bond axis such that the interatomic distance is being increased or decreased. Stretching modes (***Figure-a***) may be of two types — ***symmetric stretching and asymmetric stretching***. In symmetric stretch, the movement of the atoms would be in the same direction, whereas in asymmetric stretch one atom approaches the central atom while the other departs from it.

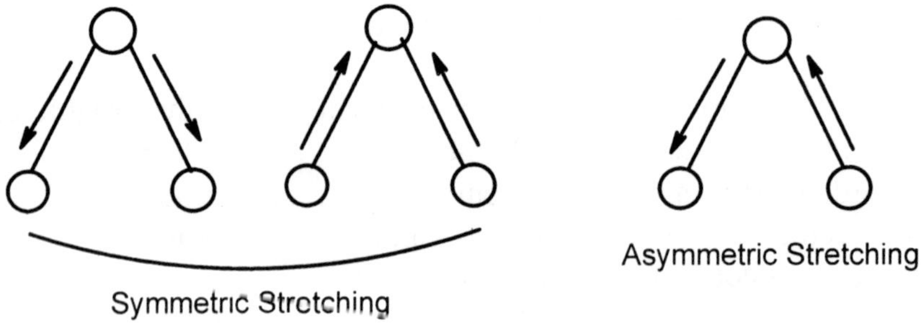

Figure-a: Stretching Vibration ----- symmetric and asymmetric

(b) ***Bending:*** A bending vibration involves a change in bond angle between bonds with a common atom. Bending vibrations (***Figure-b***) are of four modes: ***scissoring***, ***rocking***, ***wagging***, and ***twisting***.

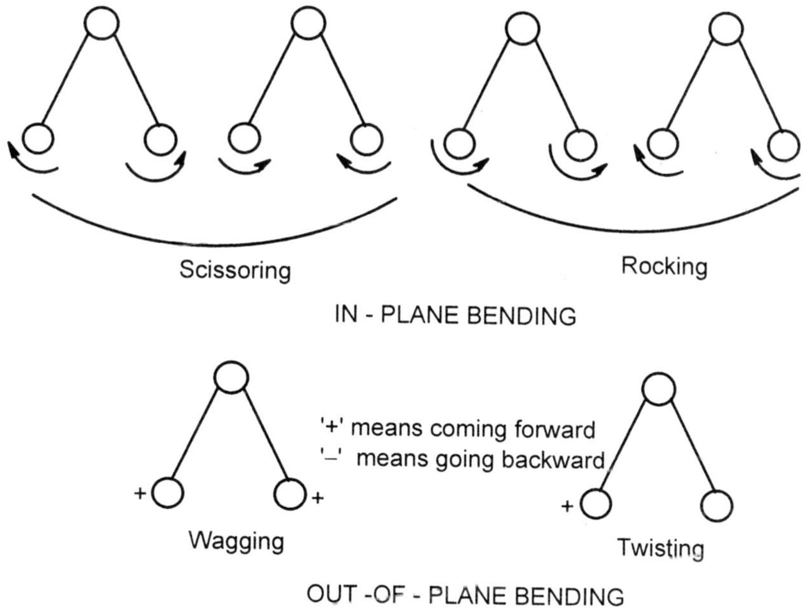

Figure-b: In - Plane and Out - of - Plane Bending Vibrations

It should be noted in this connection that less amount of energy is required for bending vibrations in comparison to that for stretching vibrations, and hence IR absorption bands for bending would appear at higher wavelength or lower wave-numbers than those for stretching vibrations.

Problem 7.25. How hydrogen bonding affects IR frequency in an organic compound? Would you able to distinguish between intramolecular and intermolecular hydrogen bonding by applying IR spectroscopic technique?

Solution. Hydrogen bonding has a considerable impact on IR absorption frequencies of organic compounds where such bonding persists. The usual proton donor groups in organic molecules are hydroxyl, carboxyl, amine or amide groups, whereas common proton acceptor atoms are oxygen, nitrogen, and the halogens. Let us represent a hydrogen bond as:

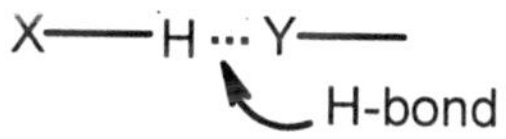

The hydrogen bond can be regarded as a resonance hybrid of (I) and (II):

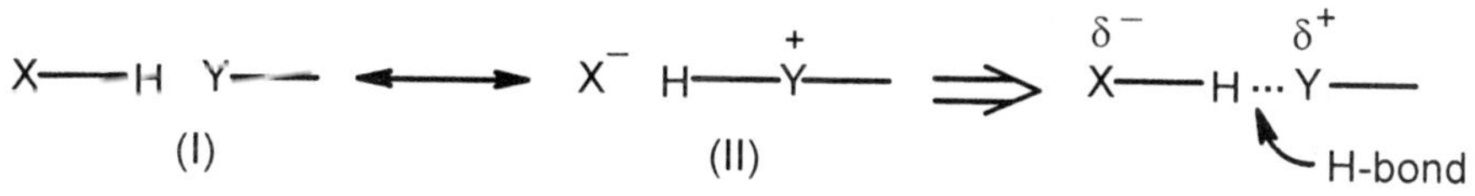

Thus, hydrogen bonding involves a lengthening of the original X–H bond making it weak — the phenomenon is responsible for reduction in the force constant of X–H bond as well as that of the acceptor group. Consequently, the X–H stretching band moves to lower frequency (i.e. longer wavelength) usually with increased intensity and band widening. The stretching frequency of the acceptor group is also reduced but to a lesser extent. Hydrogen bonding affects bending vibration of hydrogen also; it is observed that bending vibration of hydrogen bonded X–H bond shifts to a shorter wavelength but the effect is less pronounced in comparison to stretching frequency. Stronger the hydrogen bond, the longer is the X–H bond — hence shifting of the stretching frequency is in lower range (i.e. higher wavelength) resulting a broader and more intense absorption band; the value of the stretching vibration is thus in turn a measure of the strength of a hydrogen bond.

Hydrogen bonding may be of two types — intramolecular and intermolecular hydrogen bonding. Distinction between them can easily be done using this spectroscopic technique. This can be achieved by the effect of dilution — the effect of concentration on intermolecular hydrogen bonding is markedly different than that on the other. On dilution with non-associative and inert solvents, the absorption band that results from intermolecular hydrogen bonding generally disappears at a very low concentration (less than ~0.01M in nonpolar solvents like CCl_4, CS_2, etc.). On dilution the intermolecular hydrogen bond gradually breaks up resulting a gradual decrease in intensity of the absorption band for X–H, and gradual increase in intensity of the absorption band for the free X–H moiety. On the contrary, intramolecular hydrogen bonding remains unaffected and thereby also the absorption band, since such bonding arises due to internal effect and persists at very low concentrations. Let us illustrate the whole matter with the help of following examples.

Illustrations

In IR spectra of pure alcohols or phenols (called neat spectra), a broad O–H stretching vibration is observed in the range from 3400-3300 cm^{-1} due to intermolecular hydrogen bonding. As the parent compound is diluted with CCl_4, a sharp 'free' (non-hydrogen-bonded) O–H stretching band appears at about 3600 cm^{-1}. On further dilution intensity of the broad intermolecular hydrogen-bonded band is reduced considerably leaving the free O–H stretching absorption band as the major one.

Presence of intramolecular hydrogen bonding in *ortho*-carbonyl-substituted phenols is responsible for the appearance of the O–H IR stretching vibration band at a lower frequency. Hydrogen bonding in enols and chelates is particularly strong and the observed O–H stretching frequencies may be very low (down 2800 cm^{-1}). For example, the O–H band is centred at about 3200 cm^{-1} in the neat (pure state) spectrum of methyl salicylate, while O–H bands from normal phenols are centred at about 3350 cm^{-1}. The intramolecularly hydrogen-bonded band does not change its position significantly even at high dilution, because the internal bonding is not altered by a change in concentration, and there appears no peak for free O–H group at a very diluted solution.

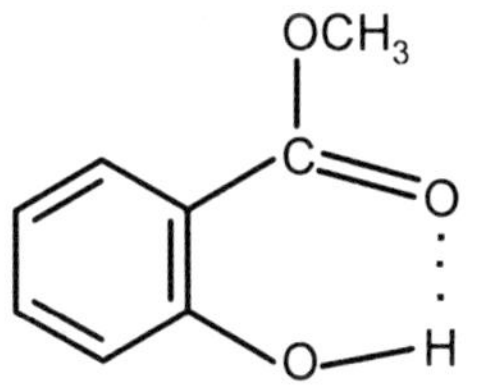

(chelate)
(methyl salicylate)

Strong intramolecular hydrogen bonding occurs in *o*-hydroxyacetophenone also; the resulting absorption at 3077 cm^{-1} is broad, and independent of concentration. But the *para*-isomer, which exhibits a broad band at 3100 cm^{-1} for strongly intermolecular hydrogen-bonded hydroxyl in the spectrum of a neat sample, gives a sharp 'free' hydroxyl peak at 3600 cm^{-1} in dilute CCl_4 solution.

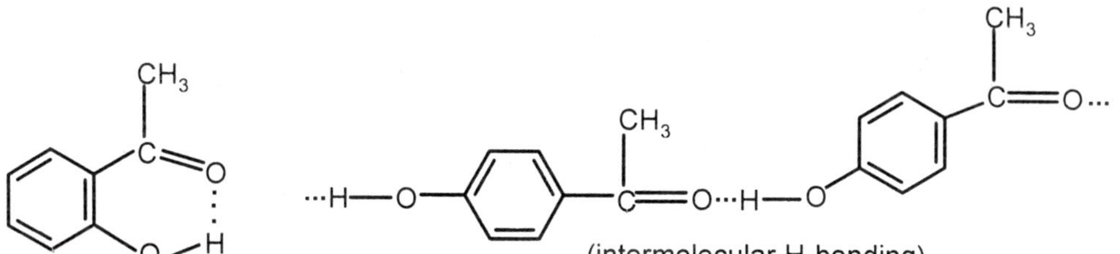

(intermolecular H-bonding)
(*p*-hydroxyacetophenone)

Similarly, IR absorption frequencies for carboxylic hydroxyls, carbonyl groups, N–H bonds, enols and non-chelated alcoholic O–H groups are significantly influenced by hydrogen bonding.

A carboxylic acid exists in monomeric form in very dilute solution, and absorbs at about 1760 cm^{-1} for carbonyl stretching vibration. However, acids in concentrated solution or in the form of neat liquid or in the solid state tend to demerize through intermolecular hydrogen bonding. This dimerization weakens the C=O bond and lowers the stretching force constant resulting a lowering of the carbonyl frequency to about 1710 cm^{-1}.

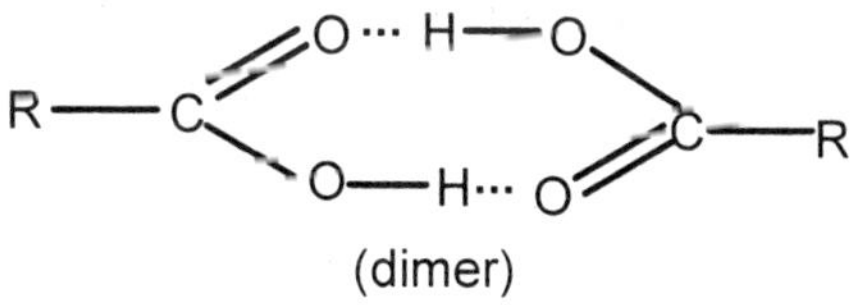

(dimer)

Problem 7.26 The carbonyl IR stretching frequencies of the following cyclic ketones are observed as:

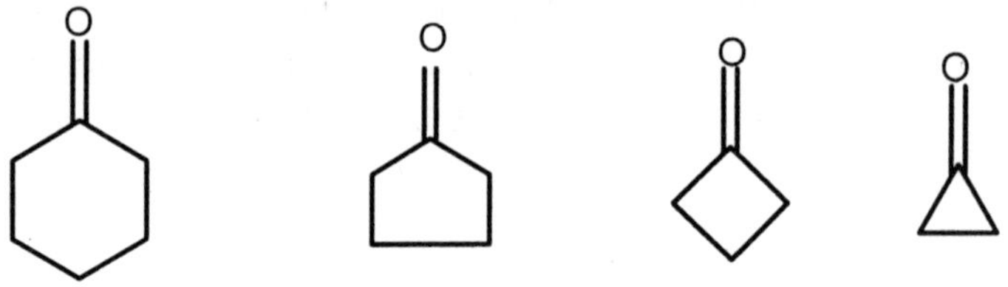

1715 cm^{-1} 1745 cm^{-1} 1780 cm^{-1} 1813 cm^{-1}

Offer explanation for the above observation.

Solution. It is noticed that frequency of C=O absorption increases with decrease in ring-size. As the ring-size decreases, more strain will be introduced inside the ring. Cyclic ketones can achieve the perfect 120^0 angle at the carbonyl group only if the ring is at least six-membered.

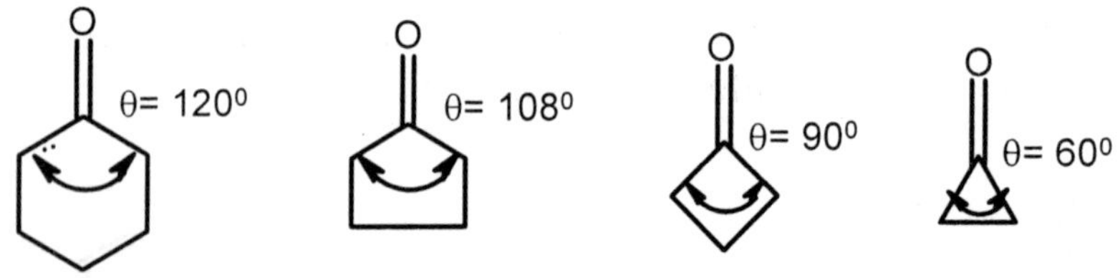

0⁰ of strain 12⁰ of strain 30⁰ of strain 60⁰ of strain

1715 cm^{-1} 1745 cm^{-1} 1780 cm^{-1} 1813 cm^{-1}

Thus, imposed ring-strain is responsible for raising the frequency of a carbonyl group. So, it may be argued that as the strain is increased, the C=O bond outside of the ring becomes more strengthened and shifted leading to enhancement of the corresponding force constant. This phenomenon can be explained in the following way:

In a six-membered ring, the sp^2 orbitals forming the σ framework around the carbonyl group can overlap perfectly with the sp^3 orbitals on neighbouring carbon atoms because the orbital angle and the bond angle are the same. So six-membered ketones show the normal frequency found for open-chain compounds.

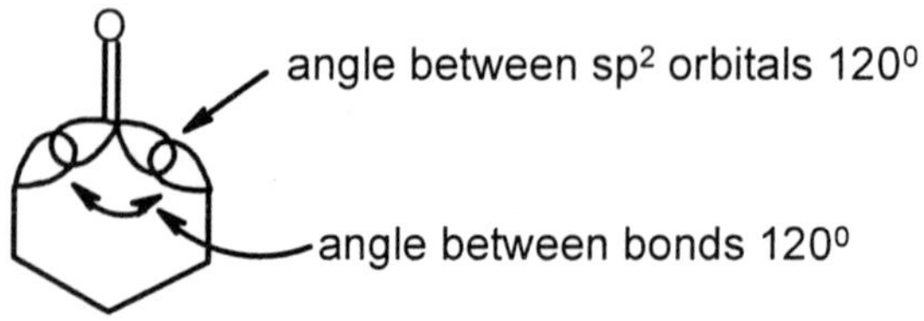

But the situation differs in smaller ring systems. Here, angles between bonds within the rings are less than that of the angle between sp^2 orbitals (120^0). In case of four-membered ring, the angle difference is 30^0. To cope with this situation the small rings introduce higher s character outwardly. It is suggested that *some* more p character goes into the ring bonds and same amount of extra s character goes into the σ-bond to oxygen.

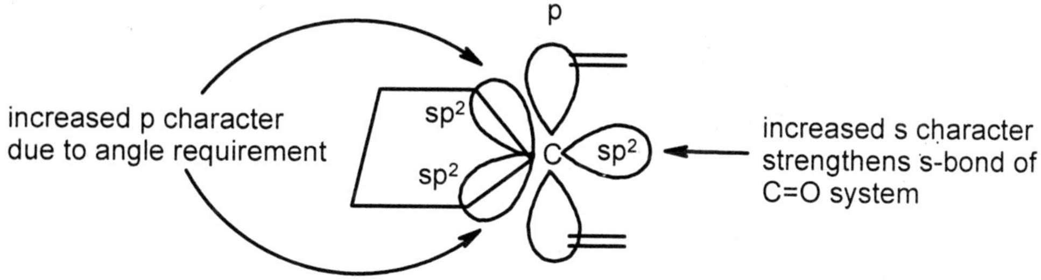

As the s character of the sp^2-orbital forming σ–bond with oxygen is increased, the C=O bond becomes more shorten requiring more energy for its stretching vibration that ultimately shifts the frequency to higher value as observed in the above series.

Problem 7.27 Offer explanation for the observed variation in C=O stretching frequencies of the following carbonyl compounds:

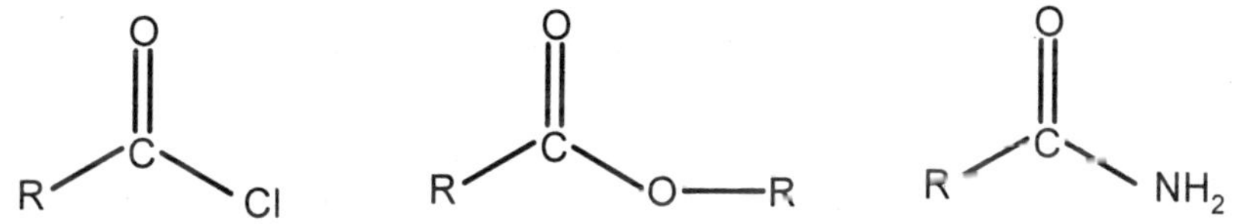

C=O stertch	1815 cm^{-1}	1735 – 1745 cm^{-1}	1650 – 1680 cm^{-1}

Solution. When carbonyl carbon is directly attached with an electronegative substituent 'X' having lone pair(s) of electrons, then two effects are to be considered — one is conjugative effect and the other is electron-withdrawing inductive effect.

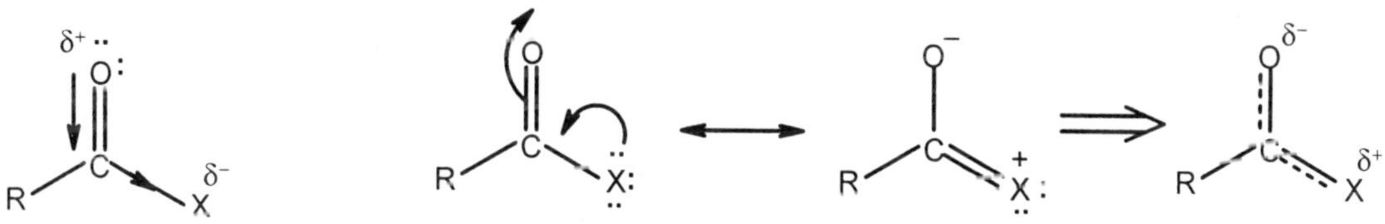

(–Inductive effect of X) (delocalization of lone pair electrons on X with C=O double bond)

The inductive effect as exerted by 'X' results in shortening and strengthening the C=O bond, which will ultimately lead to increase in the observed carbonyl stretching frequency. On the contrary, conjugation lengthens and weakens the C=O bond leading to decrease in the observed frequency.

Extent of inductive effect would be greater and more dominant where electronegativity of 'X' is larger. Thus inductive effect would dominate in acid chlorides, while in esters it is slightly dominating. Conversely, conjugative effect dominates over the inductive effect in amides. This effect becomes more effective when lone-pair orbital is relatively high in energy. Lesser is the electronegativity of 'X', the higher is the energy of its lone-pair orbitals leading to greater overlapping that lowers the amounts of π-bonding, and decrease in carbonyl stretching frequency occurs.

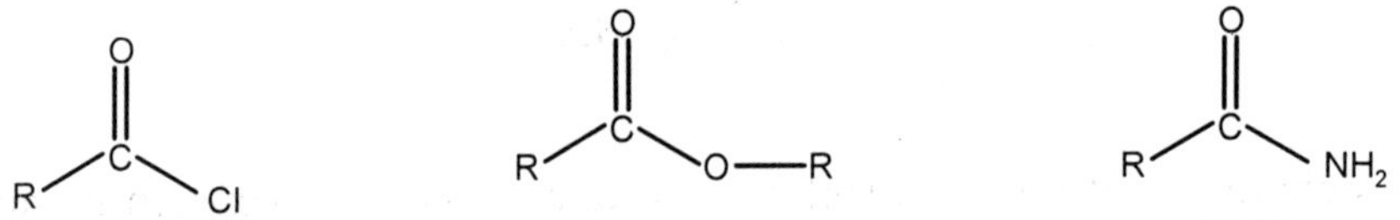

(acid chloride) (inductive effect dominates) | (ester) (inductive effect slightly dominates) | (amide) (conjugation effect strongly dominates)

C=O stertch 1815 cm^{-1} | 1735–1745 cm^{-1} | 1650 - 1680 cm^{-1}

Problem 7.28 *Axial*-2-chloro-1-decalone absorbs at 1725 cm^{-1} for the C=O stretching vibration, while the *equatorial* isomer absorbs at 1750 cm^{-1}. Why?

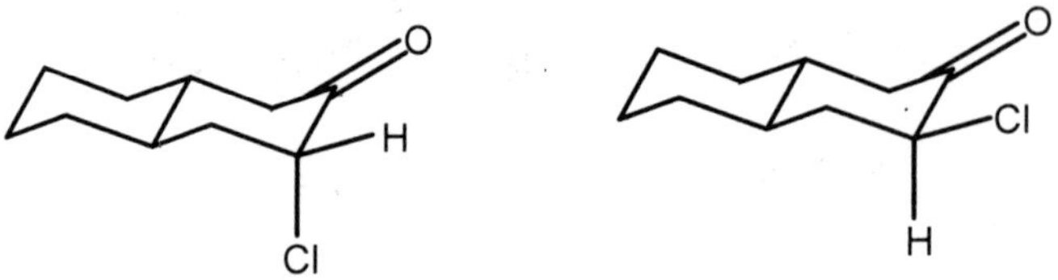

C=O stertch ~1725 cm^{-1} ~1750 cm^{-1}

Solution. When an α-carbon with respect to carbonyl function is attached to a strong electron-withdrawing substituent like chlorine, the carbonyl band shifts to a higher frequency; this is because of the fact that the electron-withdrawing effect ultimately shortens the carbonyl π-bond, thereby, increasing the corresponding force constant.

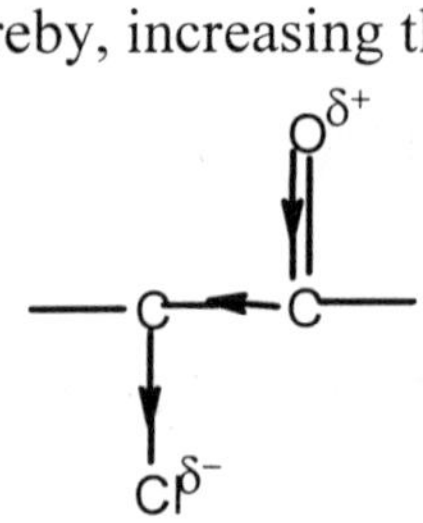

In the *equatorial* isomer of 2-chloro-1-decalone, the chlorine atom is in close proximity of the carbonyl oxygen, but in the *axial* conformation it is far away from the group. The *axial* halogen atom attains a staggered position relative to the carbonyl oxygen; but when it is *equatorially* disposed, the almost parallel dipole of the C–X and that of C=O bonds results in the rise of carbonyl absorption frequency. This is because of the fact that in the *equatorial* conformation the halogen atom next to the carbonyl repels strongly the non-bonded electrons on the oxygen making them to strengthen the carbonyl C=O bond so that the absorption frequency is increased.

Problem 7.29 IR spectrum of benzalacetone in CS_2 at room temperature exhibits two separate peaks for C=O stretching vibration. Explain.

Solution.

$PhCH{=}CHCOCH_3$ (benzalacetone)

A C=C double bond conjugated with carbonyl group lowers the IR carbonyl frequency because of lowering of the corresponding force constant as a result of delocalization—

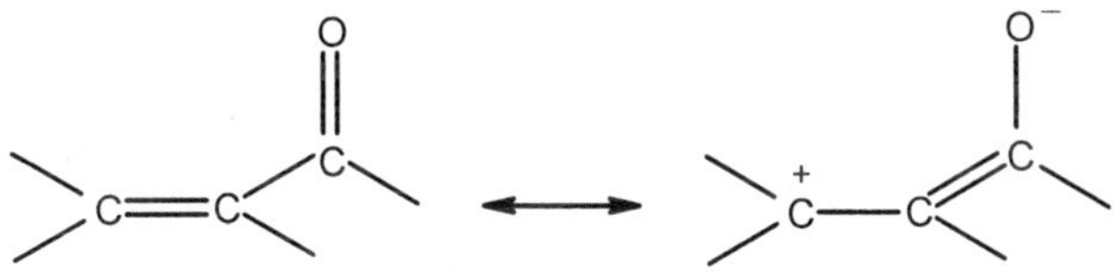

(C=O bond attains more single bond character)

Thus, IR carbonyl frequency of such system usually appears below the normal value for ordinary carbonyl compounds. However, stereochemical orientation of the conjugated carbonyl system has an impact on the IR absorption of α,β-unsaturated ketones. Here benzalacetone may exist both in *s-cis* and *s-trans* conformations at room temperature.

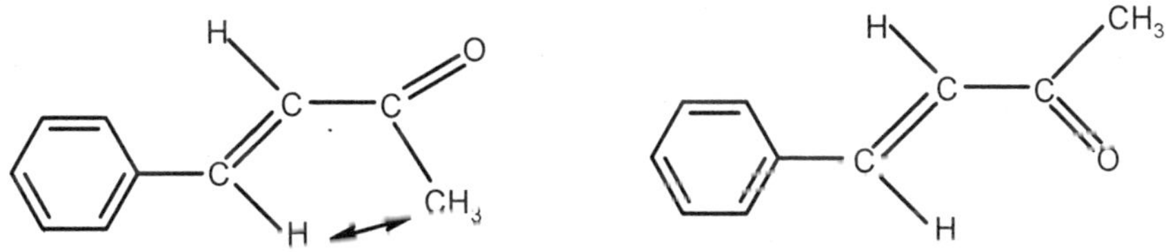

s-trans *s-cis*

C=O stretch 1674 cm^{-1} 1699 cm^{-1}

Steric effects that reduce the coplanarity of the conjugated system reduce the effect of conjugation. In the absence of steric hindrance, a conjugated system will tend toward a planar conformation. Hence, the *s-cis* conformation absorbs (having of steric interaction) at a frequency higher than the *s-trans* conformation (lacking of such steric interaction).

Problem 7.30 Under ordinary condition acetylacetone exists in tautomeric equilibrium. Comment on the IR spectral behaviour of the compound.

Solution. β-Diketones (i.e. 1,3-diketones) often exhibit tautomerization that yields an equilibrium mixture of enol and keto tautomers of which the enolic form predominates. Thus, acetylacetone (as liquid) at 40^{0}C is found to exist up to the extent of 64% in the enolic form.

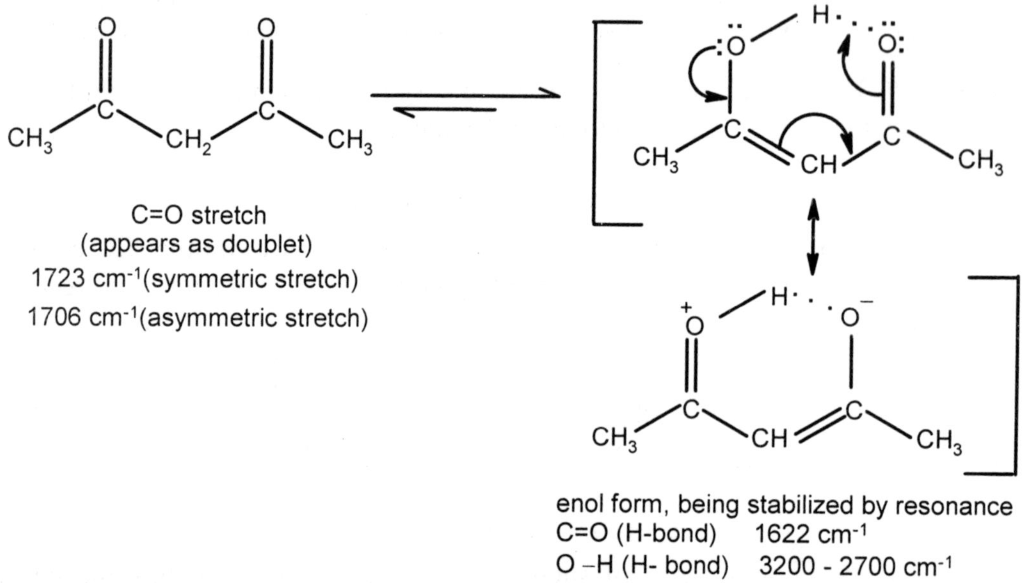

The enolic form does not show the normal absorption of conjugated ketones — the absorption peak becomes substantially shifted to lower frequency region and appears as a broad and more intensified band. This is because of the combined effect of both internal hydrogen bonding and conjugations shown in the above scheme — both of them contribute to the lowering of carbonyl frequency (at 1622 cm^{-1}) in enol form. A weak, broad O–H stretching absorption is also observed for the enolic form at 3200 – 2700 cm^{-1}.

Since the keto form is also present, a doublet for the symmetric and asymmetric stretching frequencies is observed for the two carbonyls. Relative intensities of the enol and keto carbonyl absorptions depend on the relative proportions of the tautomers present at equilibrium.

Problem 7.31 Explain why:

(a) In the substituted phenols the O–H stretching is at 3608 cm^{-1} in (I), and at 3605 and 3643 cm^{-1} in (II) whereas the stretching frequency is at 3643 cm^{-1} in (III).

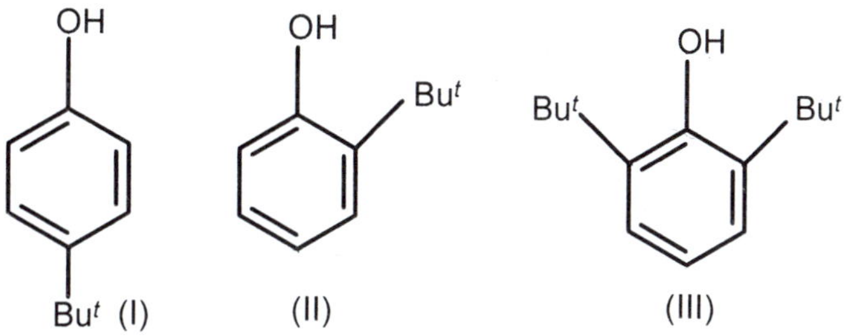

(I) (II) (III)

(b) *Trans*-1-methylcyclohexane-1,2-diol has O–H infrared absorption bands at 3618 and 3597 cm^{-1} whereas *trans*-1-isopropylcyclohexane-1,2-diol has a single peak at 3625 cm^{-1}.

(c) 2-Hydroxy-3-nitroaetophenone shows two carbonyl-stretching frequencies at 1692 and 1658 cm^{-1}.

(d) In both KBr pellet and $CHCl_3$ solution *o*-nitrophenol exhibits its O–H stretching band at 3200 cm^{-1} whereas the position varies in case of the *p*-isomer in the two media (3300 and 3520 cm^{-1}, respectively)

(e) *Cis*-cinnamic acid absorbs at a higher frequency than the *trans*-isomer.

Solutions.

(a) *Para-t*-butyl phenol molecules are associated (through intermolecular hydrogen bonding) together, and so the associated OH group in (I) exhibits a single absorption band at 3608 cm^{-1} in its IR spectrum. But in the *ortho*-isomer (II), a steric inhibition prevails as a result of which molecular association is disturbed; the fact has revealed from the appearance of two separate hydroxyl stretching bands at 3605 and 3643 cm^{-1} — attributed respectively to associated hydroxyl and free hydroxyl. The steric crowding offered by the large *t*-butyl moiety is found to be extreme in 2,6-di-*t*-butyl phenol (III) where there is no formation of hydrogen bond, i.e. no molecular association at all, and that's why the molecule exhibits a single absorption peak at 3643 cm^{-1} for free hydroxyl O–H stretching vibration.

(b) *Trans*-1-methylcyclohexane-1,2-diol exists at a equilibrium mixture of the following two conformers (I) and (II):

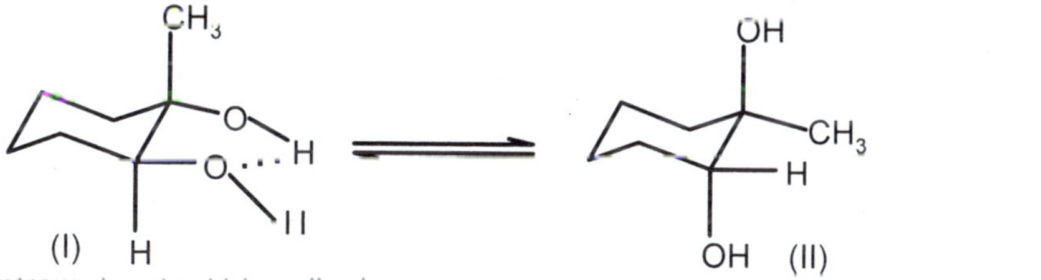

(I) (II)

(Intramolecular H-bonding) (Intramolecular H-bonding is absent)

(*trans*-1-methylcyclohexane-1,2-diol)

In conformer (I) the equatorially oriented hydroxyl functions are internally hydrogen bonded, and hence the O–H stretching vibration appears a lower frequency at 3597 cm^{-1} as compared to the absorption band appeared at 3618 cm^{-1} for the axially oriented hydroxyl groups in conformer (II) where they are not internally hydrogen bonded. Thus, two separate infrared absorption bands are obtained in the IR spectrum of *trans*-1-methylcyclohexane-1,2-diol for O–H stretching vibration.

On the contrary, *trans*-1-isopropylcyclohexane-1,2-diol exhibits only one such band at 3625 cm^{-1}. This is because of the fact that the molecule exists predominantly in conformation (IV) since it is thermodynamically more stable(the bulky isopropyl moiety attains equatorial orientation). In this conformation the axially oriented hydroxyls can't form internal hydrogen bond and the O–H infrared absorption band shifts to relatively higher frequency (at 3625 cm^{-1}).

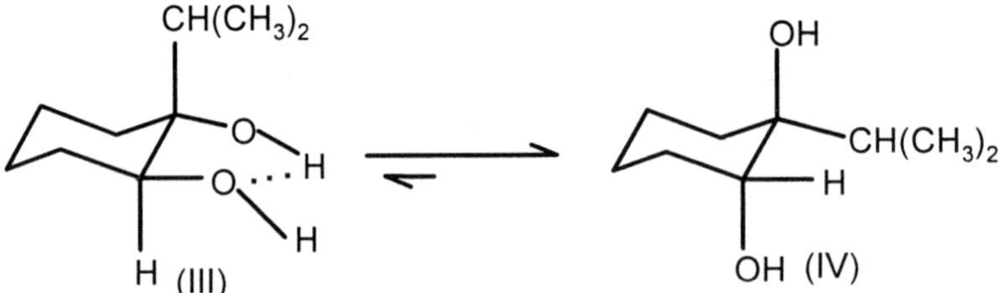

(III) (little or no existence)

(IV) (thermodynamically stable; predominant conformation)

(*trans*-1-isopropylcyclohexane-1,2-diol) (intramolecular H-bonding is absent)

(c) 2-Hydroxy-3-nitroacetophenone remains in the following equilibrium mixture:

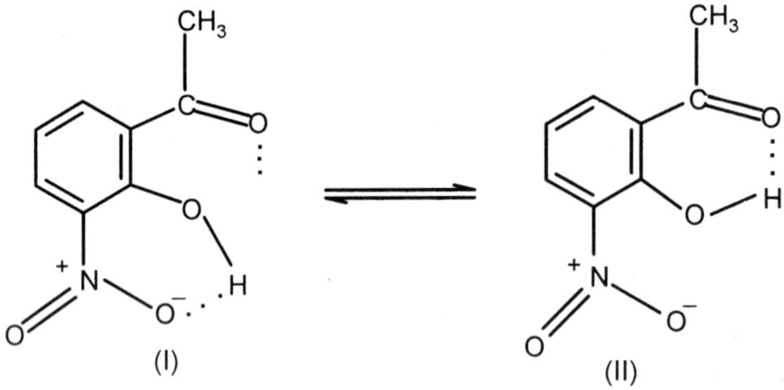

(I) (II)

The phenomenon is a consequence of competing hydrogen bond formation. In form (I) the free carbonyl shows the normal conjugated C=O stretch at 1692 cm^{-1}, while the corresponding stretching frequency shifts to lower value (1658 cm^{-1}) in form (II) where the conjugated carbonyl is also internally hydrogen bonded (i,e. chelated) that lowers the stretching force constant of the C=O bond.

(d)

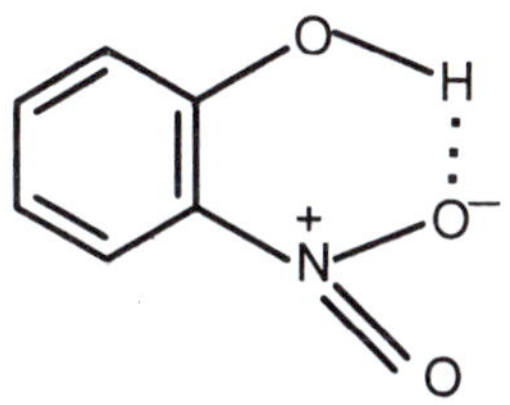

(*o*-nitrophenol
internal hydrogen bonding)

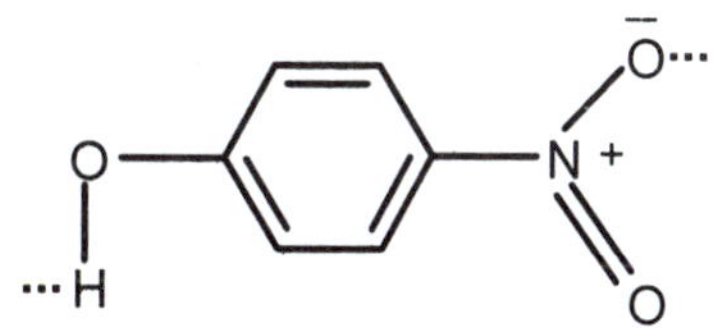

(*p*-nitrophenol
intermolacular hydrogen bonding)

Since the O–H bond in *o*-nitrophenol is internally hydrogen bonded (chelation), it remains unaffected by alteration of medium; therefore the O–H stretching frequency appears at the same position (3200 cm^{-1}) when measured in two different media (using KBr pellet and $CHCl_3$ solution).

But the situation is different in case of the *para*-isomer; here instead of internal hydrogen bonding, intermaolecular hydrogen bonding is present as a result of which the molecules become associated. The extent as well as strength of intermolecular hydrogen bonding is affected by dilution effect; thus strength of the hydrogen bond becomes more in the solid state than that in chloroform solution. As a consequence, *para*-nitrophenol exhibits its O–H stretching frequency at two different positions when measured in KBr pellet (3300 cm^{-1}) and in chloroform solution (3520 cm^{-1}).

(e)

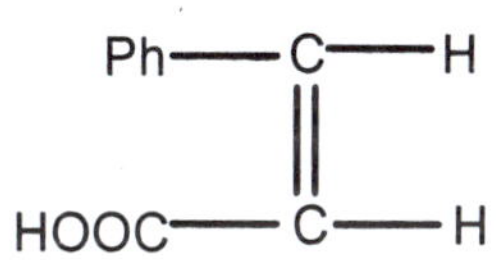

cis-cinnamic acid

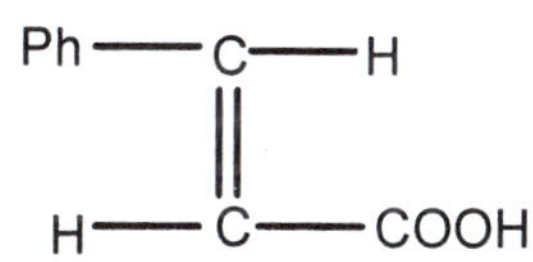

trans-cinnamic acid

C=O stretch 1705 cm^{-1} 1680 cm^{-1}

In the *cis*-isomer the two bulky substituents (phenyl and carboxylic acid groups) are at the same side of C=C double bond. Due to repulsive interactions, the C=O unit of –COOH group goes out of the plane of double bond. Thus, conjugation is reduced as a result of which C=O stretching vibration shifts to higher frequency range. But the conjugative effect is more pronounced in the *trans*-isomer, where the two bulky groups are on the opposite side of the double bond. The double bond character of carboxylic carbonyl group in *trans*-cinnamic acid is reduced leading to the decrease in the corresponding force constant. Thus, it absorbs at relatively lower frequency at 1680 cm^{-1} for C=O stretching vibration.

Problem 7.32 How can you differentiate the following pairs by applying IR spectroscopy?

a) Methylbenzoate and *o*-hydroxyacetophenone
b) δ-Valerolactone and δ-valerolactum
c) Salicylic acid and *p*-hydroxybenzoic acid
d) Vinylacetate and methylacrylate
e) Acetophenone and *p*-tolualdehyde
f) Methylbenzoate and phenylacetate
g) *Cis* and *trans*- isomers of 4-*t*-butylcyclohexanol
h) *Gauche* and *anti*-conformers of 1,2-dichloroethane
i)

j) CH_3CH_2CHO and CH_3COCH_3
k) CH_3CH_2COCl and $ClCH_2CH_2OCH_3$
l) *Para*-aminoacetophenone and acetanilide
m) *Para*-(aminomethyl)benzoic acid and *para*-aminophenylacetic acid

Solutions. ***Hints only***

a)

(methylbenzoate)

(*o*-hydroxyacetophenone)

The carbonyl stretching frequency (νC=O) in the ester molecule, methylbenzoate usually appears at 1724 cm^{-1}; negative inductive effect exerted by the methoxyl group is responsible for enhancing the absorption frequency.

But in *o*-hydroxyacetophenone, one should expect two characteristic bands for carbonyl stretching and O–H stretching vibration. Both the groups are bound by intramolecular hydrogen bond; thus the effect of internal hydrogen bonding would be reflected in their corresponding IR absorptions — the carbonyl and hydroxyl stretching frequencies are shifted towards much lower values; the absorption bands appear respectively at 1634 cm^{-1} and ~3400 cm^{-1}.

b) Carbonyl stretching frequency in δ-valerolacetone appears at a normal value for esters, i.e. at 1735 cm^{-1}. But in δ-valerolactum the conjugative effect of nitrogen lone-pair appears to be a predominant factor that weakens the C=O bond (the respective force constant is decreased); as a consequence carbonyl stretching frequency in the cyclic lactum appears at a much lower value, 1670 cm^{-1}.

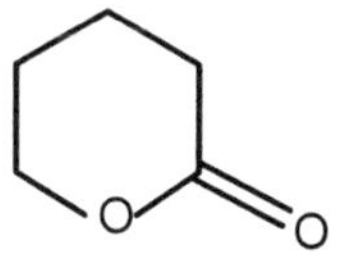

(δ-valerolacetone)

νC=O 1735 cm^{-1}

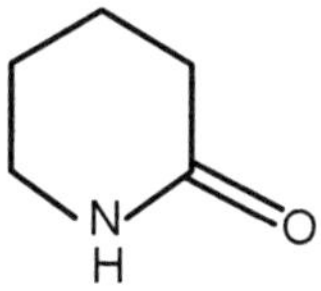

(δ-valerolactum)

1670 cm^{-1}

c)

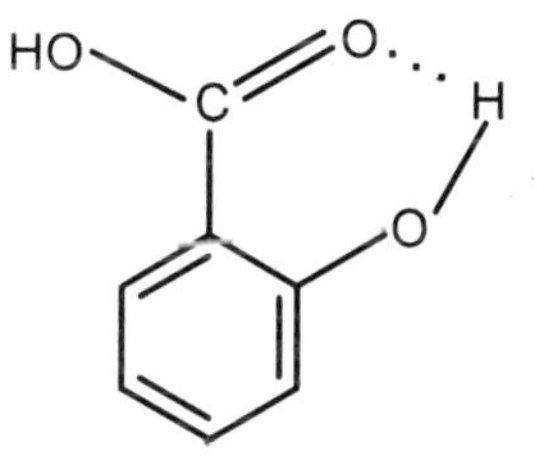

(*o*-hydroxybenzoic acid)
salicylic acid

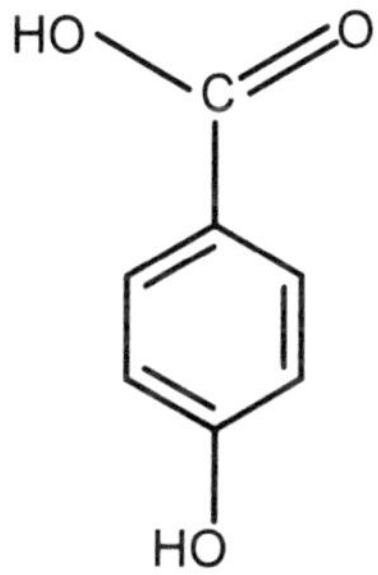

(*p*-hydroxybenzoic caid)

The *ortho*-phenolic OH group in salicylic acid is involved in intramolecular hydrogen bonding and hence the value of O–H-stretching bonded absorption near 3400-3200 cm^{-1} remains almost the same at all dilutions whereas the phenolic OH group in *p*-hydroxybenzoic acid is involved in intermolecular hydrogen bonding; consequently, dilution of the solution of the latter compound will convert the low frequency, broad, and intensed OH (bonded) absorption band (appeared at high concentration) to a high frequency, sharp, and weak OH (free) band of unbonded hydroxy group.

d)

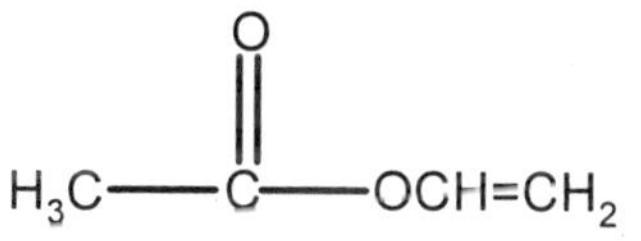

vinylacetate

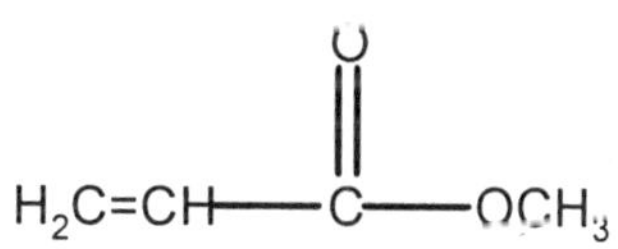

methylacrylate

C=O stretch 1776 – 1762 cm^{-1} 1740–1715 cm^{-1}

Conjugation of a single-bonded oxygen atom with C=C increases the carbonyl stretching frequency by an extent of ~30 cm^{-1} above the normal ester carbonyl absorption; this situation is observed in vinylacetate. Electron-withdrawing effect of the vinyloxyl moiety as a whole strengthens the C=O bond leading to the shifting of its stretching frequency at 1776 cm^{-1}. On the other hand methylacrylate, where the carbonyl unit is conjugated with α,β–unsaturated double bond, exhibits its carbonyl stretching vibration at a lower frequency region, 1740–1715 cm^{-1}. For both the molecules vinylic double bond (C=C) stretch appears around 1640 cm^{-1}.

e)

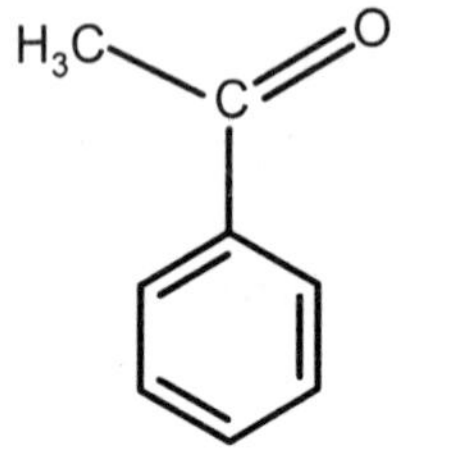

(acetophenone)

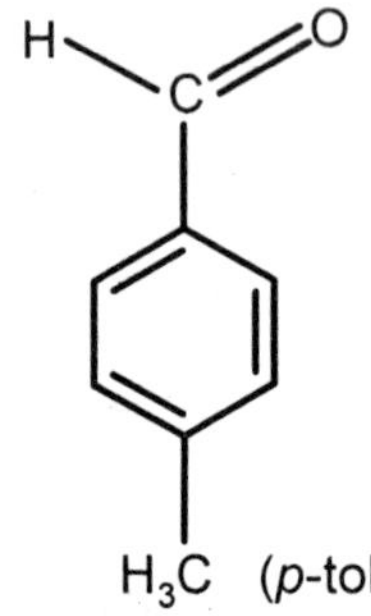

(*p*-tolualdehyde)

Aldehydes generally absorb 10-15 cm^{-1} above the corresponding ketones. The C=O stretching vibration in *p*-tolualdehyde appears at ~1700 cm^{-1}, while that in acetophenone appears at around 1685cm^{-1}. But the distinguishing factor between an aldehyde and a ketone is that the aldehydic C–H bond offers a weak stretching vibration in the region 2830-2695 cm^{-1}, which often appears as a doublet due to Fermi resonance. Thus, *p*-tolualdehyde shows its aldehydic C–H stretch at 2827 and 2736 cm^{-1}, but acetophenone does not.

f)

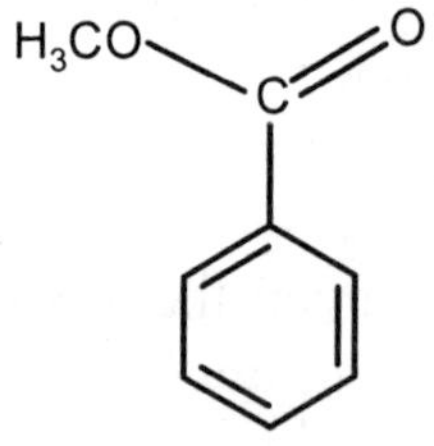

(methylbenzoate)

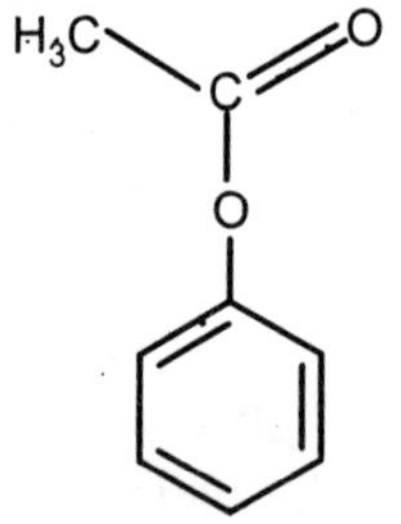

(phenylacetate)

In methylbenzoate the carbonyl unit is in conjugation with the phenyl ring. But this unit in phenylacetate is not conjugated with the phenyl ring; rather it is suffered from strong electron-withdrawing effect of the phenyloxyl group. Consequently, the C=O bond becomes strengthened in the latter case and becomes weakened in the former. Thus it is expected that carbonyl stretching frequency would be lower in methylbenzoate and relatively higher in phenylacetate. The respective observed values are ~1720 cm^{-1} and 1770 cm^{-1}.

g)

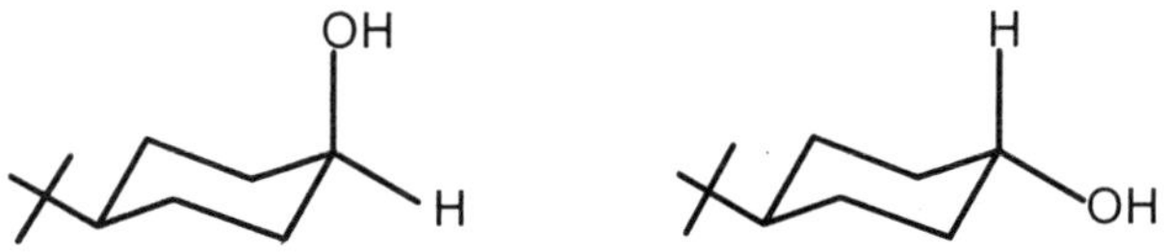

cis-isomer (*axial*-alcohol) | *trans*-isomer (*equatorial*-alcohol)

C–O stretch ~950 cm^{-1} ~1060 cm^{-1}

Axial and *equatorial* substituents on cyclohexane ring exhibit variation in C–X stretching vibration frequencies. An *equatorial* substituent usually absorbs at a higher frequency than does the *axially* disposed same substituent. The higher frequency of the C–X bond results due to less steric interaction faced with adjacent hydrogen atoms. Thus, *trans*-4-*t*-butylcyclohexanol (*equatorial*-alcohol) exhibits C–O stretching frequency at ~1060 cm^{-1} while the *cis*-isomer (*axial*-alcohol) at ~950 cm^{-1}. [X = electronegative atom].

h)

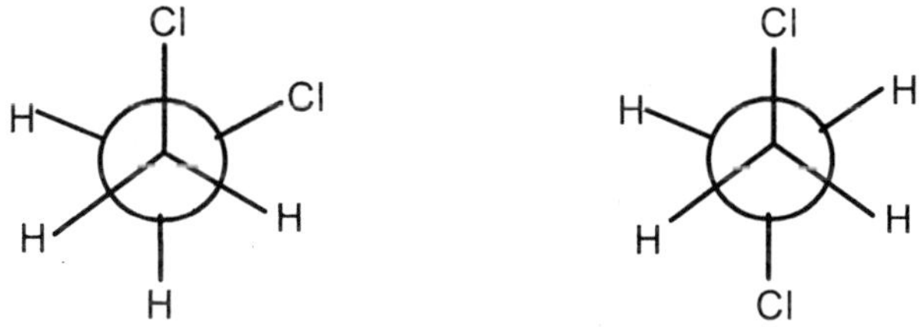

gauche 1235 cm^{-1} (CH_2 rocking) | *anti* 1291 cm^{-1} (CH_2 rocking)

Both the *gauche* and *anti* conformations of 1,2-dichloroethane are present together at ordinary condition. Their existence can easily be detected by means of IR spectroscopy. In its IR spectrum 1,2-dichloroethane exhibits bands at 1291 and 1235 cm^{-1} attributed to CH_2 rocking vibrations respectively in the *anti* and *gauche* conformations; their relative intensities depend upon temperature. The *anti* form predominates at lower temperature, while the *gauche* form becomes predominant at relatively higher temperature. The band at 1291 cm^{-1} becomes thus less intensed as the temperature is raised whereas the band at 1235 cm^{-1} becomes more intensed.

i)

νC=O 1800cm^{-1} 1750 cm^{-1}

Unsaturation α to C=O group in cyclic lactones reduces C=O stretching frequency; but if unsaturation is present α to an –O– unit the corresponding frequency is increased as observed in the present situation. Besides, when the double bond is adjacent to the –O-, a strong C=C absorption is observed in the region of 1685-1660 cm^{-1}.

j) Ketones absorbs at a lower frequency than aldehydes because of their additional alkyl substituent, which is electron-donating and supplies electrons to the C=O bond. This effect weakens the C=O bond in the ketone leading to the lowering its force constant, and thereby also the absorption frequency. Thus, the carbonyl stretching frequency of acetone is observed around 1715 cm^{-1} and that of the aldehyde around 1725 cm^{-1}. Besides, in an aldehyde the aldehydic C–H bond (–CHO) offers a weak stretching vibration in the region 2830-2695 cm^{-1}, which often appears as a doublet due to Fermi resonance, as also observed in case the IR spectrum of 1-propanal.

k) An intense band around 1800 cm^{-1} would appear in the IR spectrum of the acid chloride due to C=O stretch. But such band will be absent in case of the ether ($ClCH_2CH_2OCH_3$); instead it exhibits an intense C–O stretching absorption in the range of 1300–1000 cm^{-1}.

l) The spectrum of *p*-aminoacetophenone exhibits two N-H stretching absorptions in the range of 3500-3300 cm^{-1} whereas multiple N-H stretching bands in the range from 3300-3120 cm^{-1} will appear in the spectrum of acetanilide.

m)

$$H_2NCH_2C_6H_4COOH \Longrightarrow H_3\overset{+}{N}CH_2C_6H_4CO\overset{-}{O}$$

p-(aminomethyl)benzoic acid — zwitter ion (predominant)

$$H_2NC_6H_4CH_2COOH \Longrightarrow H_3\overset{+}{N}C_6H_4CH_2CO\overset{-}{O}$$

p-aminophenylacetic acid — zwitter ion (little or no)

Aryl amino groups are relatively less basic in comparison to alkyl amino groups; thus amino acids with the amino group directly attached to an aryl ring, as in the case of *p*-(aminomethyl)benzoic acid, exist predominantly with the free amino and carboxyl functions. But amino acids with the amino group directly attached to an alkyl carbon exits predominantly as zwitterions. Hence, *p*-(aminomethyl)benzoic acid shows separate absorption bands characteristic of both NH_2 and COOH groups; the N – H stretch at 3300 cm^{-1}, broad OH stretch from 3500-2500 cm^{-1} and C=O stretch near 1700 cm^{-1}. On the other hand, *p*-aminophenylacetic acid exhibits the characteristic C=O stretching vibration at ~1600 cm^{-1}.

Problem 7.33 What do you mean to say NMR spectroscopy?

Solution. Nuclear Magnetic Resonance (NMR) spectroscopy is an important technique to study molecules by means of interacting the nuclei of these molecules, placed in a strong magnetic field, with radiofrequency electromagnetic radiation.

Under a strong magnetic field, the NMR active nucleus (termed as '*spinning magnet*') attains a particular ***precessional motion*** with definite ***precessional frequency***. According to Quantum Mechanics, a rotating particle can absorb electromagnetic radiation only when it exhibits a uniform periodic motion with a frequency that exactly matches the frequency of the absorbed radiation. When these two frequencies exactly match, the electromagnetic fields can "*constructively*" interfere with the oscillations of the particle. Such system is then said to be "*in resonance*", and then absorption of radiofrequency energy occurs, thus causing a change in spin-state. The terminology "Nuclear Magnetic Resonance" is originated from this fact, *i.e.* creating a perfect 'resonance' between the precessional frequency of the nuclear magnet (spinning nucleus) and the frequency of electromagnetic radiation as passed through, at a point where they have been exactly matched and radiofrequency energy absorption takes place — as a result of which NMR signal comes out. NMR signals, arising out of particular nuclei residing in different electronic environments in a specific molecule, when plotted onto a chart paper NMR spectrum of that molecule is said to be obtained.

Problem 7.34 What type of nuclei become NMR active?

Solution. Certain nuclei behave as if they are spinning; in fact, any atomic nucleus that possesses either *odd* mass or *odd* atomic number, or both will have a quantized spin angular momentum and a magnetic moment. To exhibit NMR phenomenon nuclei must have their spin quantum number '*I*' greater than *zero*. Following table offers few examples:

Table: ***Few nuclei showing their spin quantum No. and level of spin states***

Nucleus	Atomic Number	Mass Number	Nuclear spin quantum No. (*I*)	No. of spin states (2*I*+1)	NMR activity
$^{1}_{1}H$	odd	odd	1/2	2	active
$^{2}_{1}H$	odd	even	1	3	active
$^{12}_{6}C$	even	even	0	0	inactive
$^{13}_{6}C$	even	odd	1/2	2	active

Continued...

Table continued

Nucleus	Atomic Number	Mass Number	Nuclear spin quantum No. (*I*)	No. of spin states (2*I*+1)	NMR activity
$^{14}_{7}N$	odd	even	1	3	active
$^{15}_{7}N$	odd	odd	1/2	2	active
$^{16}_{8}O$	even	even	0	0	inactive
$^{17}_{8}O$	even	odd	5/2	6	active
$^{10}_{5}B$	odd	even	3	7	active
$^{11}_{5}B$	odd	odd	3/2	4	active
$^{19}_{9}F$	odd	odd	1/2	2	active
$^{31}_{15}P$	odd	odd	1/2	2	active
$^{34}_{16}S$	even	even	0	0	inactive
$^{35}_{17}Cl$	odd	odd	3/2	4	active

Problem 7.35 Outline the basic principle of NMR spectroscopy.

Solution. Any moving charge generates a magnetic field of its own. A nucleus, being a charged particle, will have a magnetic moment (μ) to be generated by its charge and spin. Spinning of nuclear charge of certain nuclei, whose spin quantum numbers are greater than *zero*, on the nuclear axis generates a magnetic dipole along the axis. This spin angular momentum of the spinning charge can be described in terms of spin quantum number (*I*). The intrinsic magnitude of the generated dipole is expressed in terms of nuclear magnetic moment (μ):

$$\mu = \frac{\gamma h[I(I+1)]^{1/2}}{2\pi}$$

where h = Planck's constant
γ = Magnetogyric ratio

Under ordinary condition (no external field) all the spin states of a given nuclear magnet are degenerate (are of equivalent energy) and oriented in random fashion. But under the influence of an external magnetic field, the degeneracy is lost and orientations become quantized — the magnetic nucleus can adopt only certain discrete orientations out of (2I+1) possible orientations. For example, nuclei with spin quantum No. ½ (*viz.* ^{1}H, ^{13}C, ^{19}F) orient themselves only in two allowed directions; each of these orientations corresponds to a certain energy level.

Let us consider the case of a proton; it can adopt only two orientations with respect to an external magnetic field — either *aligned with* the field (spin = +1/2; lower energy state; α-state) or *opposed to* the field (spin = –1/2; higher energy state; β-state). We can also describe these orientations respectively as *parallel* with or *antiparallel* with the applied field.

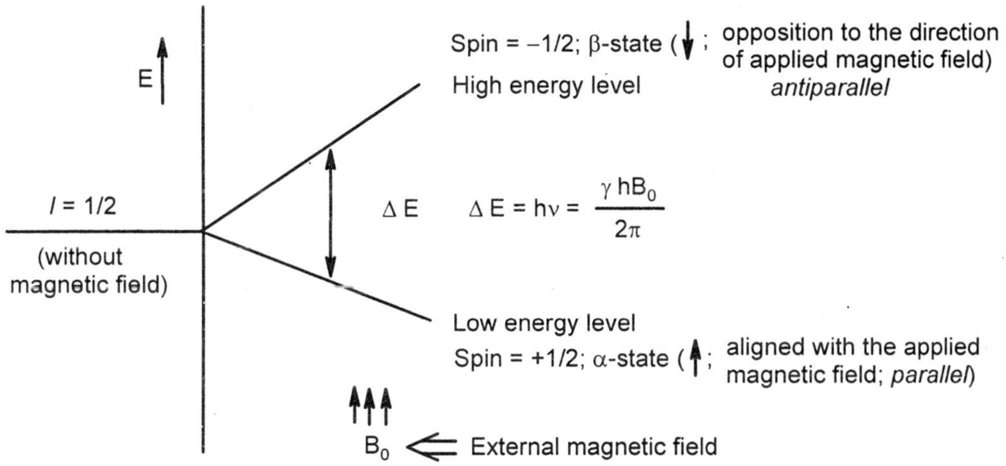

The energy difference (ΔE) between these two spin states (α- and β-states) is directly proportional to the strength of the external magnetic field. As the proton is a spinning magnet, under the influence of an external magnetic field it begins to precess about its own axis of spin with an angular frequency, which is termed as "***Larmor frequency***"; degeneracy of spin states is lost and the nucleus exhibiting the characteristic precessional motion (*i.e.* wobbling motion) attains two principal orientations — either *aligned* with the field or *opposed* to the field. The precessional frequency is directly proportional to the strength of the applied field — the stronger the applied field, the higher is the rate of precession. As an example, if a proton is placed at an external magnetic field of 1.4 Tesla (i.e. 14,000 Gauss), it will precess at 60 million times per second, and the corresponding Larmor frequency or precessional frequency would be ν = 60 MHz; similarly, under an external magnetic field of 2.3 Tesla, 'ν' would be 100 MHz, and so on.

The precession of the charged nucleus generates an oscillating electric field of the same frequency. If a radiofrequency radiation is now applied in a direction perpendicular to the external magnetic field and at a frequency that exactly matches the precessional frequency of the spinning nucleus, absorption of energy will occur; that is, at a ***resonance point*** where the frequency of the oscillating electric field component of the incoming radiation just matches the frequency of the electric field generated by the precessing nucleus, the two fields can couple, and energy can be transferred from the incoming radiation to the nucleus resulting a sudden 'flip' of the nucleus from its lower energy level (α-spin state) to the higher energy level (β-spin state). At this time of perfect resonance NMR signal comes out.

If populations of the upper and lower energy states become equal, then there would not exist any energy difference between the two states of the nucleus in its parallel and antiparallel orientations out of which no NMR signal be observed; no net absorption of energy takes place at the state of equality. However, at equilibrium there is a slight excess (called the 'Boltzmann excess') of nuclei in the lower energy state (α-state) as compared to the upper energy state (β-state) and this difference in the populations in the two levels is responsible for the NMR signal. The ratio is expressed as:

$$\frac{N_\beta}{N_\alpha} = \exp\left(\frac{-\Delta E}{kt}\right) \quad \text{. . . . . Boltzmann Equation}$$

At the resonance point, energy absorption from the supplied radiofrequency electromagnetic radiation continues till the number of nuclei in lower and higher spin states become equal. As soon as the equality develops, energy absorption stops. This state, where development of equal number of nuclei in higher and lower level occurs, is termed as ***saturation state***. Since the process of energy absorption is a continuous one, there must be a difference in number of nuclei between the states. For this reason, it is assumed that nuclei in the higher state release energy and step down to the lower energy level, thereby increasing the number of nuclei in the lower state. The stepping down of nuclei from higher to lower energy level by releasing corresponding amount of energy is termed as a ***process of relaxation***. Relaxation process may be of two types — ***spin-spin relaxation*** and ***spin-lattice relaxation***, out of which the latter one is of importance in NMR technique.

It may be pointed out that it is impossible to imagine a bare nucleus. What we can imagine is an atom, *i.e.* nucleus surrounded by electrons. Once again, electrons are magnetic in character and so they will have magnetic moment. This magnetic moment of electrons will definitely depend upon its density. Greater the density, greater is the magnetic moment of the electrons. Hence, it is obvious that when we are measuring the magnetic moment of the nuclear magnet, we are also measuring the magnetic moment of the surrounding electrons — because the applied magnetic field associated with radiofrequency radiation will not directly fall on the nucleus. It must have to pass through the electronic magnetic filed surrounding the nucleus; this effect is known as ***shielding effect***. Further, extra nuclear electrons are intimately associated with chemical bond. Thus determination of the magnetic moment of the nuclear magnet actually amounts to the determination of the strength (*i.e.* nature) of the chemical bond. In this respect, *i.e.* to determine the nature of chemical bonds, the principle of NMR is applied in Chemistry.

Most of the organic compounds invariably contain hydrogen (^{1}H) besides carbon; other atoms that may also be present are N, O, S, P, halogens, etc. In organic compounds, H-atoms are associated with these different atoms in different electron-environments. So, determination of magnetic moment of different protons will ultimately give rise the nature of different chemical bonds present in the particular molecule. For this reason, commonly used NMR is **PMR** — Proton Magnetic Resonance Spectroscopy, generally designated as ^{1}H-NMR spectroscopy.

Thus, an organic molecule may be regarded as an assembly of hydrogen atoms with different electronic environments. When such molecule is placed in a static magnetic field and an applied revolving magnetic field with radiofrequency is designed, protons with different electronic environments will absorb energy at different magnetic field. If the applied field is gradually increased starting from minimum to maximum at different field strength, resonances at different respective points will be created. As soon as one resonance point appears, energy absorption takes place and signal is produced. Different signals are recorded and are plotted onto a chart-paper as a graph. The positions of different signals in the graph-paper indicate the type of protons present in the molecule — the idea from which the concept regarding its structural pattern can be understood. Since different types of protons are recorded in a graph as signals, the method is termed as '***spectroscopy***'. This chart-paper having recorded signals is regarded as NMR spectrum. The position of the different signals in the chart-paper appearing as peaks is indicated with respect to the position of the signal of protons of a known compound. This known compound is termed as 'Reference Standard'.

Problem 7.36 Give a brief account on 'Reference standards' used in ^{1}H NMR spectral technique.

Solution. The position of proton peaks of a reference compound is regarded as the starting point, i.e. zero. The proton(s) signals of unknown compounds are normally placed at the left hand side of the position of the peak of the reference compound. This means electronic environment surrounding the protons of any unknown organic molecule is supposed to be less electron-dense than that of the protons of the standard.

If the reference standard is used externally, confusion regarding the exact position of a particular proton(s) signal of an unknown compound under different condition may arise. In order to avoid such confusion regarding the exact placement of different proton peaks of any compound, reference compound is used *internally* along with the unknown sample. For this reason, reference is usually termed as ***internal reference standard.***

Internal reference compound must have to satisfy the following criteria:

(a) Since proton-peaks of unknown compounds are placed at the left-hand direction of the corresponding proton-peaks of reference, the electronic environment surrounding the protons of reference compound should have to be more dense than that of mostly available organic compounds.

(b) As it is used internally, it should be miscible with any organic compound under experiment, and also should be chemically inert.

Tetramethylsilane (Me_4Si), abbreviated as TMS, is a compound that satisfies all the criteria required for reference compound, and that's why is used normally as internal reference standard. The protons of its methyl groups are more shielded than those of most other known compounds. Although compounds with more shielded proton than that in TMS are known today, yet TMS is being used as a good reference substance since in most cases it mark one end of the range. TMS does not involve in intermolecular association with sample, and due to its high volatility it can easily be removed when sample is to be recovered.

Recently a substitute of TMS, termed as DSS [sodium 2,2-dimethyl-2-silapentane-5-sulphonate *i.e.* Sodium salt of 3-(trimethylsilyl)propane sulphonate — $Me_3SiCH_2CH_2CH_2SO_3Na$], is also in use as internal reference in aqueous solution for running PMR spectra of water soluble compounds in D_2O solvent.

Problem 7.37 Predict ^{1}H-NMR spectral behaviour of the following organic molecules considering the use of TMS as internal standard in each case:
(i) *p*-Chloroaniline; (ii) 2,4-Dichloroaniline; (iii) 2-Bromo-4-nitrotoluene; (iv) (*E*)-Ethyl 2-butenoate; (v) *Trans*-Crotonic acid; (vi) Vinyl acetate; (vii) *m*-Nitrobenzoic acid;(viii)Coniochaetone A;(ix) Coniochaetone B; (x) (*E*)-$C_6H_5CH{=}CHCOOCH_2CH_3$

Solutions.

(i)

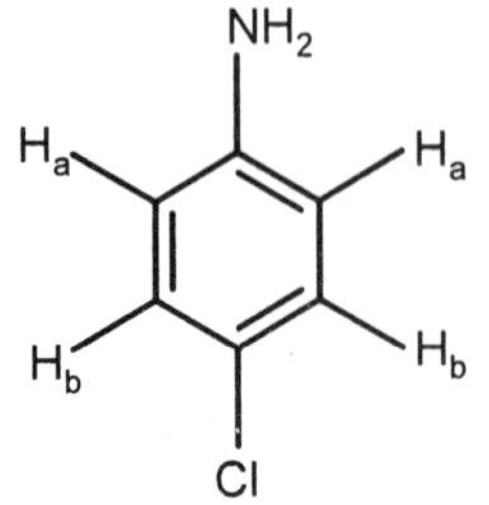

Signals

H_a: δ 6.5 (2H, d, *J* = 9 Hz) ; H_b: δ 7.1 (2H, d, *J* = 9 Hz);
N$\underline{H}_2$: δ 3.6 (2H, broad singlet)

(ii)

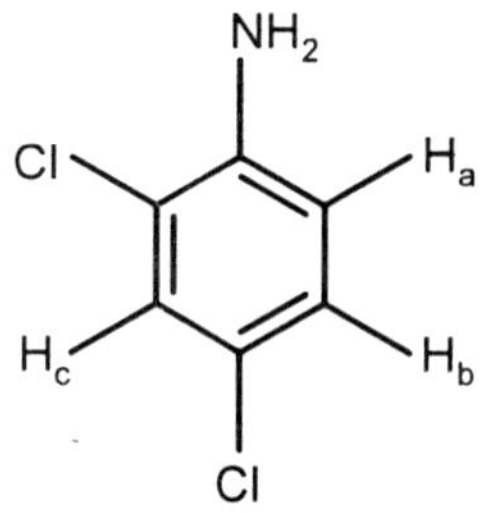

Signals

H_a: δ 6.6 (1H, d, J = 9 Hz); H_b: δ 6.95 (1H, dd, J = 3 & 9 Hz); H_c: δ 7.2 (1H, d, J = 3 Hz); $N\underline{H}_2$: δ 3.9 (2H, broad singlet)

(iii)

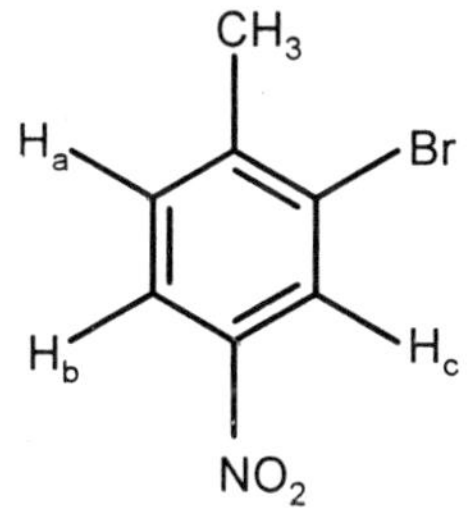

Signals

H_a: δ 7.00 (1H, d, J = 8 Hz) ; H_b: δ 7.6 (1H, dd, J = 2 & 8 Hz); H_c: δ 7.85 (1H, d, J = 2 Hz); Ar-$C\underline{H}_3$: δ 2.40 (3H, singlet)

(iv)

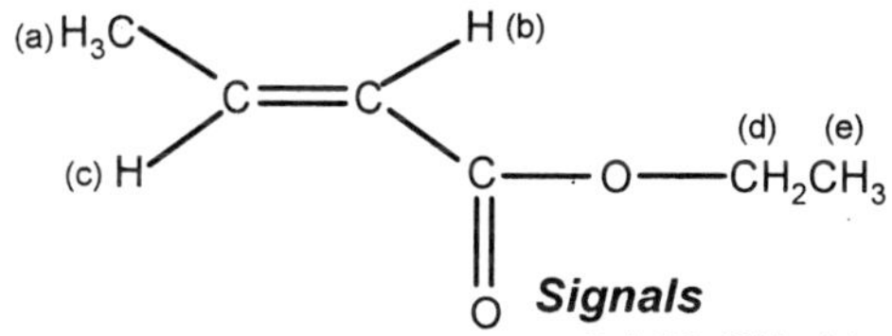

Signals

a: δ 1.88 (3H, dd, J = 1.7 & 6.8 Hz) ; b: δ 5.81 (1H, dq, J = 1.7 & 16 Hz); c: δ 6.95 (1H, dq, J = 6.8 & 16 Hz); d: δ 4.13 (2H, q, J = 7 Hz); e: δ 1.24 (3H, t, J = 7 Hz)

(v)

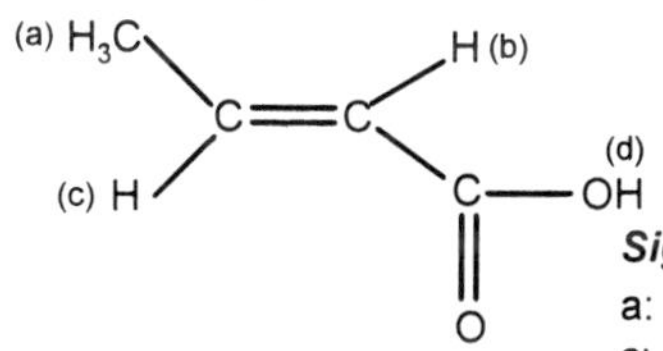

Signals

a: δ 1.92 (3H, dd, J = 1.6 & 6.9 Hz) ; b: δ 5.86 (1H, dq, J = 1.6 & 15.6 Hz); c: δ 7.10 (1H, dq, J = 6.9 & 15.6 Hz); d: δ 12.2 (1H, s)

(vi)

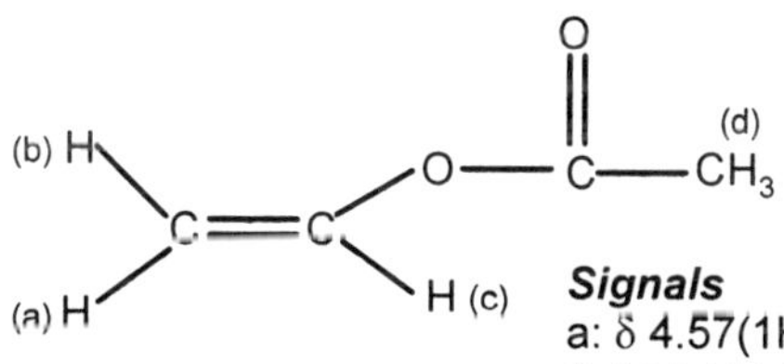

Signals

a: δ 4.57(1H, dd, J = 1.47 & 6.25 Hz) ; b: δ 4.88 (1H, dd, J = 1.47 & 14.34 Hz);c: δ 7.27 (1H, dd, J = 1.47 & 13.97 Hz); d: δ 2.2 (3H, s)

(vii)

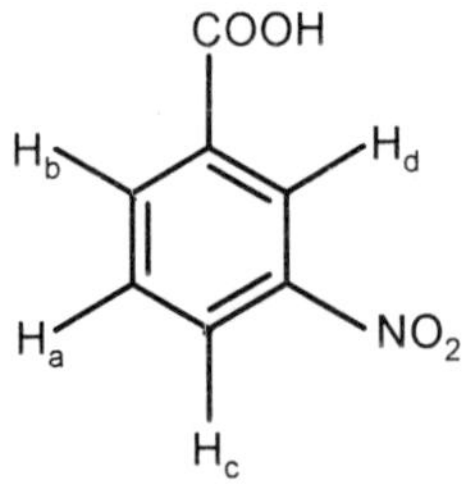

Signals

H_a: δ 7.22 (1H, dd, J = 7.7 & 8.1 Hz) ; H_b: δ 8.45 (1H, ddd or dt, J = 1.5 & 7.7 Hz); H_c: δ 8.50 (1H, ddd, *ortho* and two *meta*-coupled); H_d: δ 8.96 (1H, dd, two *meta*-coupled)

(viii)

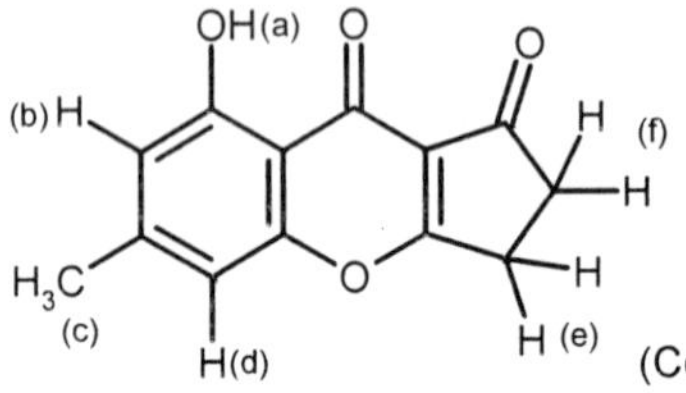

Signals

a: δ 12.21(1H, s); b: δ 6.69 (1H, s); c: δ 2.41 (3H, s); d: δ 6.77 (1H, s); e: δ 3.07 (2H, m); f: δ 2.70 (2H, m)

(Coniochaetone A)

(ix)

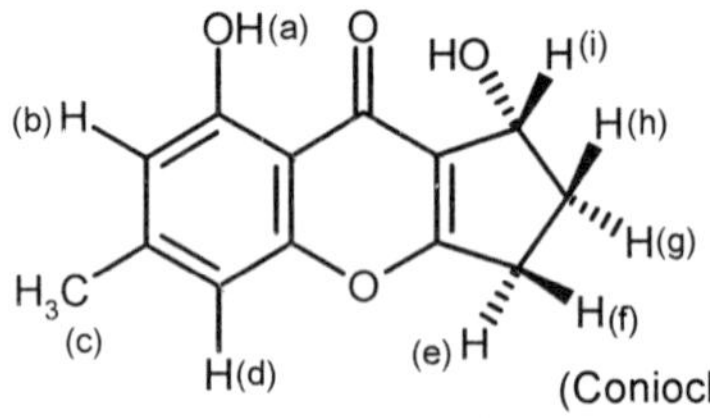

Signals

a: δ 12.25(1H, s); b: δ 6.70 (1H, s); c: δ 2.38 (3H, s); d: δ 6.62 (1H, s); e: δ 2.81 (1H, ddd, J = 5.1, 9.3, 18 Hz); f: δ 3.10 (2H, dq, J = 1.4, 54.1, 9.4, 18 Hz); g; δ 2.03(1H, m); h: δ 2.49 (1H, m); i: δ 5.43 (1H, ddd, J = 1.4, 3.3, 7.6 Hz)

(Coniochaetone B)

(x)

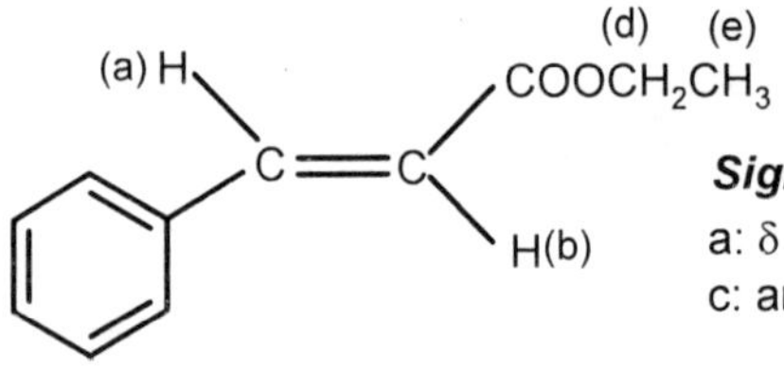

Signals

a: δ 7.7(1H, d, J = 17 Hz); b: δ 6.4 (1H, d, J = 17 Hz); c: around δ 7.4(5H, m); d: δ 4.3 (2H, q); e: δ 1.3 (1H, t)

Problem 7.38 What do you mean by 'Larmor frequency'? What is its importance in an NMR experiment?

Solution. The frequency with which a nucleus precesses around the applied magnetic field is called its 'Larmor frequency' or 'Larmor precessional frequency'.

Larmor frequency depends on the strength of the applied magnetic field (B_0) and on the magnetogyric ratio of the nucleus. When radiofrequency (B_1) is applied in a direction perpendicular to the external magnetic field (B_0), and at the moment when the value of B_1 exactly matches the Larmor frequency, absorption of energy occurs causing the nucleus to flip to a higher-energy orientation. This is responsible for NMR signals. With the increase in magnetic field (B_0), *i.e.* increase in Larmor frequency (as 'ν' α 'B_0'), the population excess of the lower energy level over the upper energy level increases, leading to a corresponding enhancement in the sensitivity of the NMR experiment.

Problem 7.39 What will happen if the radiofrequency pulse is applied for an unusually long time?

Solution. By absorbing energy from the radiofrequency radiation passed, nuclei flip to higher energy state (β); by relaxation process, they revert to the lower energy state (α) and an equilibrium is established with a slight Boltzmann excess in the lower α-state. This slight excess of population in the lower energy state will be disturbed on continuous irradiation and a state of saturation will be reached, since the nuclei in the β-state will not be able to dissipate (waste out) the extra energy via relaxation processes.

Problem 7.40 Cite example of two molecules that bear protons more shielded than those in TMS.

Solution. Cyclic compounds having π-electrons within the ring, if produce induced ring current upon exposure to external magnetic field, show ***anisotropic effect***. Under this situation, if a proton is placed in the centre of the ring rather than on its periphery, it would be found to be shielded. Thus protons falling within the circulating zone become shielded, while those falling outside the circulating zone become deshielded. For examples:

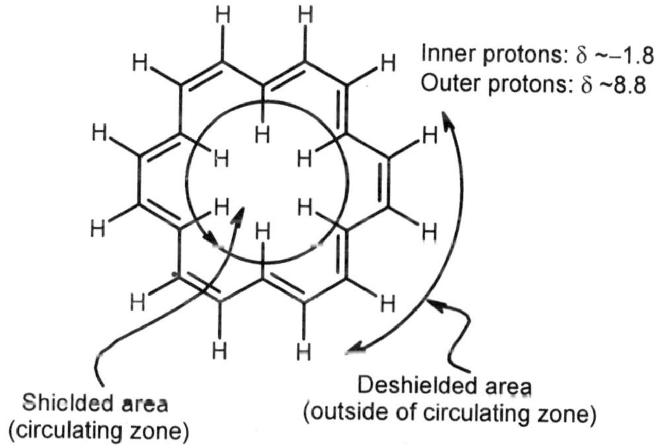

[18]-Annulene

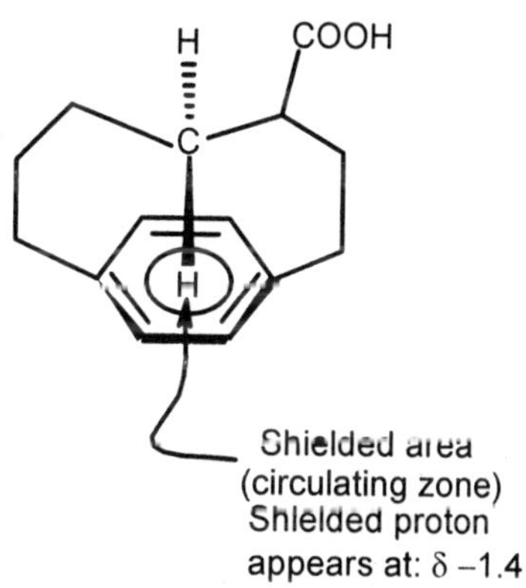

(Cyclophane compound)

Problem 7.41 Assign the structure of a compound, $C_{10}H_{12}O$, whose mass spectrum shows *m/z* values of 15, 43, 65, 57, 91, 105, and 148.

Solution. The mass peak at *m/z* 148 is the molecular ion peak. The peak at *m/z* 91 is likely to be $C_7H_7^+$, whose precursor is the stable benzyl cation, $C_6H_6CH_2^+$. The molecule bears an acetyl group (CH_3CO) as evidenced from the appearance of mass fragments peaks at *m/z* 43, 105 (148 – CH_3CO), and 15 (CH_3CO – CO). The benzyl and acetyl groups are linked through a methylene (CH_2) unit as also evidenced from the appearance of mass fragments peaks at *m/z* 57 (CH_3COCH_2), and 91(148 – CH_3COCH_2). Thus, the structure of the compound is $CH_3COCH_2CH_2C_6H_5$.

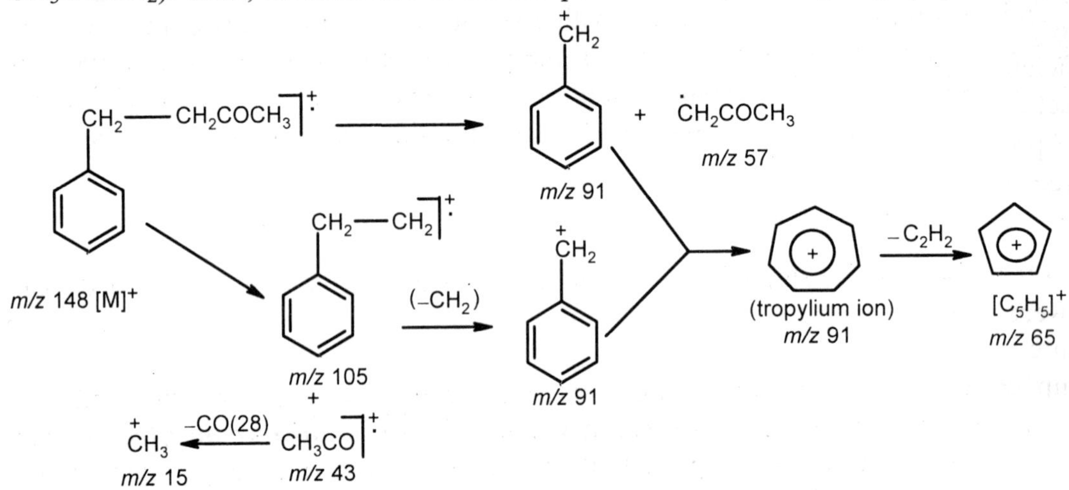

Problem 7.42 An organic compound having molecular formula $C_6H_{11}BrO_2$ exhibits ^{1}H-NMR signals at: δ 4.1 (2H, q, J = 7.5 Hz), 4.0 (2H, t, J = 7.5 Hz), 1.5 –2.2 (4H, m), 1.25 (3H, t, J = 7.5 Hz). Predict the structure of the compound.

Solution. ***Hint*** Probable structure:

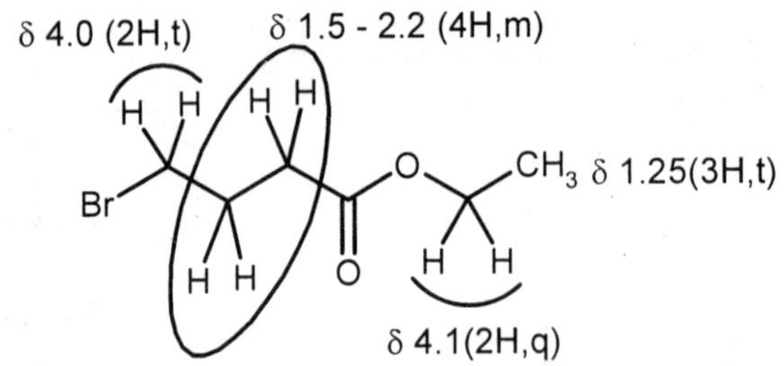

Problem 7.43 An organic compound of molecular formula, $C_6H_{12}O$ showed two peaks in the PMR spectrum at δ 1.0 (9H, s) and δ 2.0 (3H, s). One of the intense IR bands of this compound appears at 1715 cm^{-1}, but no absorption at 2720 cm^{-1}. Deduce the structure of the compound.

Solution.

Molecular formula: $C_6H_{12}O$

Level of Unsaturation (LU): C + 1 – H/2 – X/2 + N/2

= 6 + 1 – 6 = 1 (signifies the presence of one unsaturation or one ring)

The molecule showed IR absorption band at 1715 cm^{-1} characteristic for carbonyl group which may be either an aldehyde or a ketone; but since there is no characteristic signal at 2720 cm^{-1}, presence of aldehydic function is rejected. The presence of the keto group (>C=O) thus accounts for the 'level of unsaturation' (or otherly hydrogen deficiency index) as well as oxygen atom of the molecular formula. Hence, the part structure may be given as: [>C=O]C_5H_{12}

Appearance of three proton singlet at δ 2.0 in the PMR spectrum of the molecule suggests the presence of a methyl group on a carbonyl carbon, *i.e.*

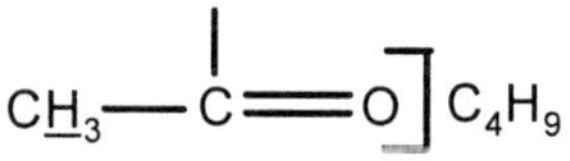

Remaining nine equivalent protons resonating at δ 1.0 (9H, s) typical of alkane nature must be attributed to the presence of a tertiary butyl group. Thus, the organic compound is 3,3-dimethyl-2-butanone.

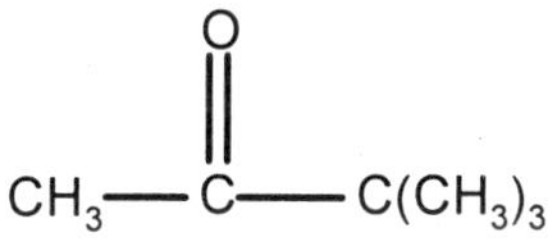

Problem 7.44 A sweet smelling organic compound 'A' (m.f. $C_8H_{16}O_2$) on reaction with LAH furnishes a single primary alcohol 'B'. But on reaction with excess of MeMgBr, the compound 'A' yields two alcohols 'B' and 'C'. In the ^{1}H-NMR spectrum, compound 'B' exhibits signals at δ 3.8 (2H, d, J = 7 Hz), 1.8 (1H, m), 1.6 (1H, brs, exchangeable with D_2O) and 0.9 (6H, d, J = 7.2 Hz). Identify the structures of A, B and C, and explain the reactions involved.

Solution.

Molecular formula of 'A': $C_8H_{16}O_2$

Level of Unsaturation (LU): C + 1 – H/2 – X/2 + N/2

= 8 + 1 – 8 = 1 (signifies the presence of one double bond or one ring)

The physical property as well as the outcomes of chemical reactions of 'A' with LAH and MeMgBr (excess) reveals that 'A' is an ester. The carbonylic unsaturation satisfies the value of 'LU'. Thus the part structure of 'A' can be represented as:

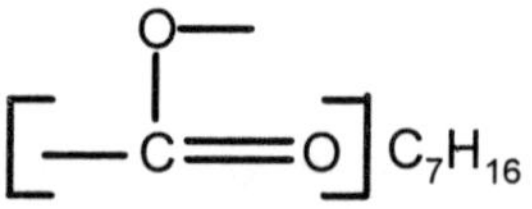

One mole of an ester on reaction with LAH gives two moles of primary alcohols; hence 'A' (say RCOOR') reacting with LAH gives two moles of identical 1^0-alcohol (B), but furnishes two different alcohols 'B' and 'C' on reaction with excess of MeMgBr.

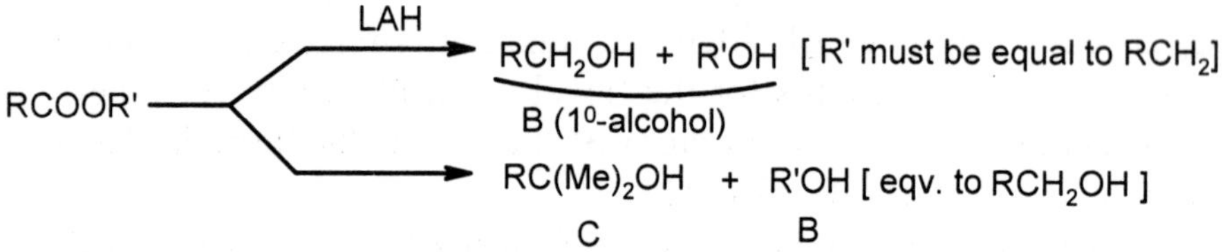

Hence, the part structure of A, B, and C may be given as:

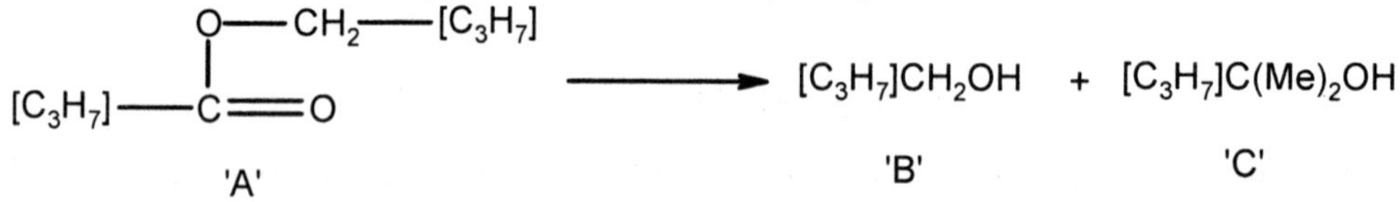

From the scrutiny of the ^{1}H-NMR data for 'B', its structure is assigned as 2-methyl-1- propanol.

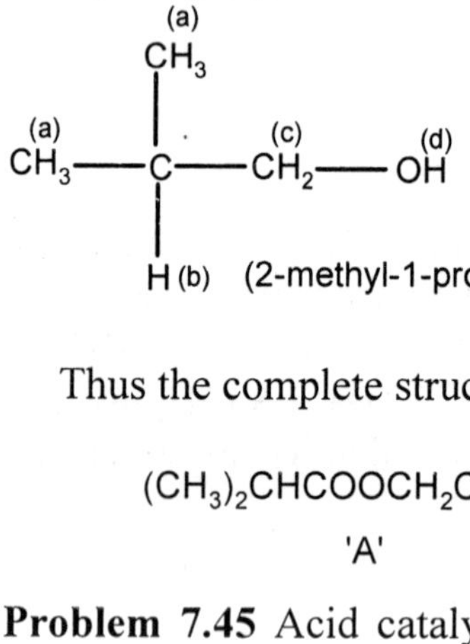

(2-methyl-1-propanol)

^{1}H-NMR data at:
a: δ 0.9(6H, d, J =7.2 Hz) ; b: δ 1.8 (1H, m);c: δ 3.8 (2H, d, J = 7Hz); d: δ 1.6 (1H, brs, exchangeable with D_2O)

Thus the complete structure of 'A' would be 2-methylpropyl 2-methylpropanoate:

$(CH_3)_2CHCOOCH_2CH(CH_3)_2$ 'A' $(CH_3)_2CHCH_2OH$ 'B' $(CH_3)_2CHC(Me)_2OH$ 'C'

Problem 7.45 Acid catalyzed dehydration of a tertiary alcohol 'A' ($C_6H_{14}O$) gives one major compound 'B' and one minor compound 'C', both having the same molecular formula of C_6H_{12}. Spectral data of these two compounds are as follows:

Compound 'B':

IR: 1660, 3080 cm^{-1}

^{1}H-NMR: δ 0.91 (3H, t, J = 7 Hz), 1.60 (3H, s), 1.70 (3H, s), 1.98 (2H, q, J = 7 Hz), 5.08 (1H, t, J = 7 Hz)

Compound ‘C’:

IR: 1640, 3090 cm^{-1}

^{1}H-NMR: δ 0.92 (3H, t, $J = 7$ Hz), 1.40 (2H, sextet, $J = 7$ Hz), 1.74 (3H, s), 2.02 (2H, t, $J = 7$ Hz), 4.78 (2H, s)

Deduce the structures of ‘A’, ‘B’ and ‘C’.

Solution.

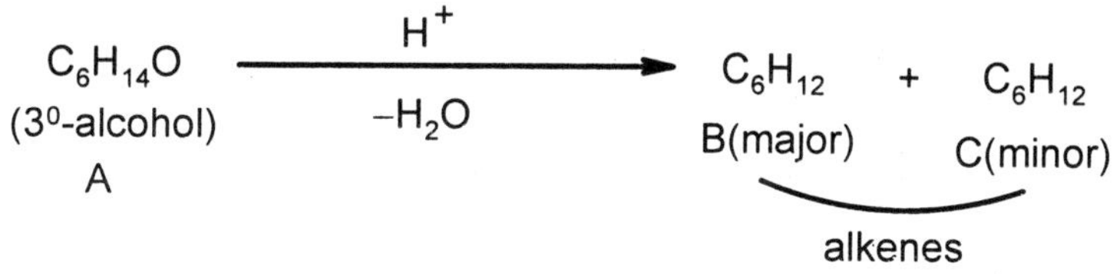

Partial structures of A, B, and C may be given as:

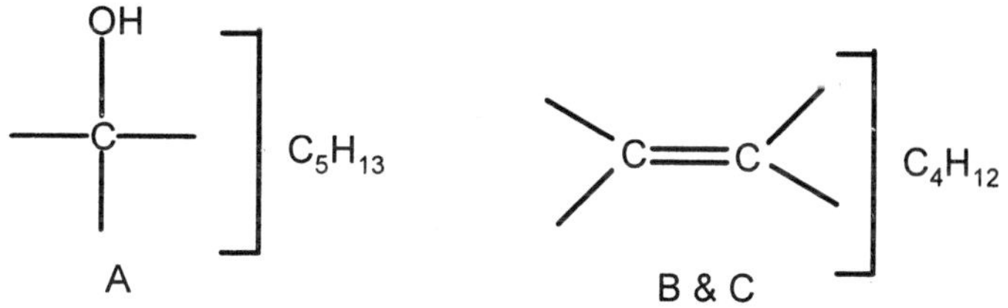

From the IR and PMR spectral analysis the structures of ‘B’ and ‘C’ can be formulated as:

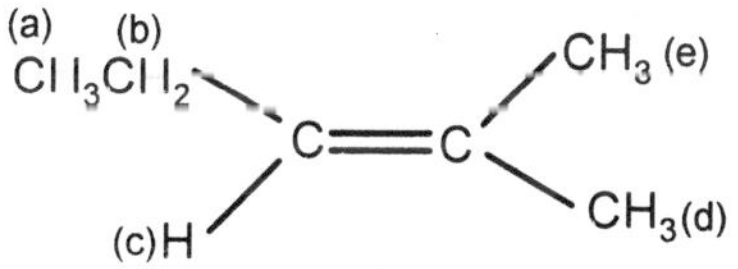

'B'

(IR absorption bands at 1660 and 3080 cm^{-1} are atttributed to the trisubstituted double bond)

^{1}H-NMR data at:

a: δ 0.91(3H, t, $J = 7$ Hz) ; b: δ 1.98 (2H, q, $J = 7$ Hz); c: δ 5.08 (1H, t, $J = 7$Hz);d: δ 1.60 (3H, s); e: δ 1.70 (3H, s)

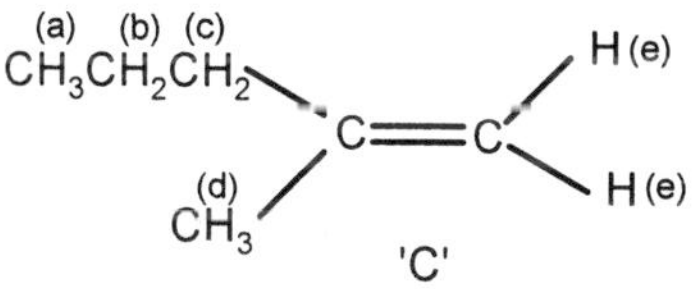

'C'

(IR absorption bands at 1640 and 3090 cm^{-1} are atttributed to the disubstituted double bond)

^{1}H-NMR data at:

a: δ 0.92(3H, t, $J = 7$ Hz) ; b: δ 1.40 (2H, sextet, $J = 7$ Hz); c: δ 2.02 (2H, t, $J = 7$Hz);d: δ 1.74 (3H, s); e: δ 4.78 (2H, s)

Thus, ‘A’ must be assigned as 2-methyl-2-pentanol:

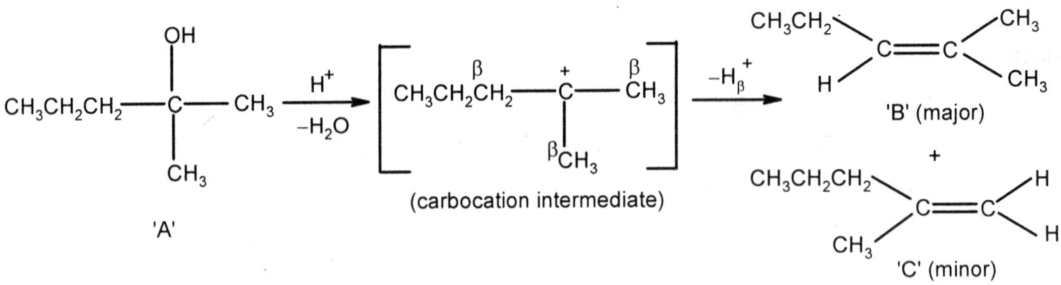

Problem 7.46 A cyclic ketone 'A' (m.f. C_5H_8O) on reaction with $NaOCH_3$ gives a product 'B' which has the following spectral properties:
IR: 1740, 1160 cm^{-1}; ^{1}H-NMR: δ 0.92 – 1.22 (7H, m), 2.20 (2H, t), 3.60 (3H, s); Mass: *m/z* 116, 85, 59, 31.
Deduce the structures of 'A' and 'B'.

Solution.

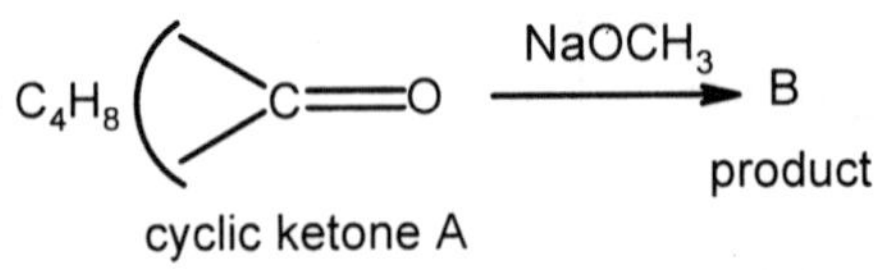

[B must be an ester as evidenced from the appearance of IR absorption bands at 1740 cm^{-1} for ester carbonyl and at 1160 cm^{-1} for C – O stretching; thus B is an open chain molecule]

Thus,

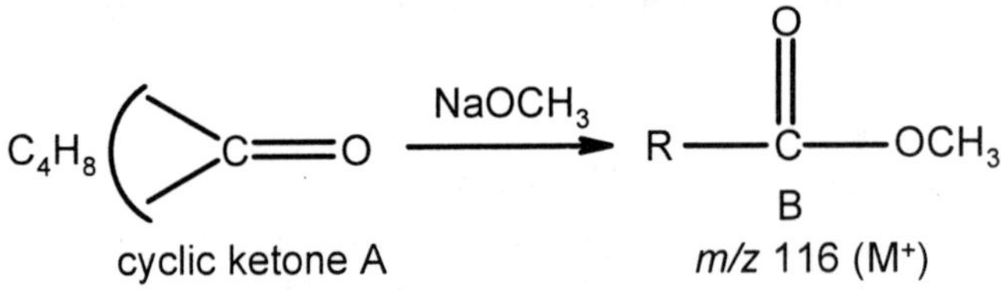

From ^{1}H-NMR and mass spectral analysis, it may be argued that 'R' would be *n*-$C_3H_7CH_2$ moiety. Thus, 'B' can be formulated as methyl pentanoate that satisfies IR, ^{1}H-NMR and mass spectral properties for it.

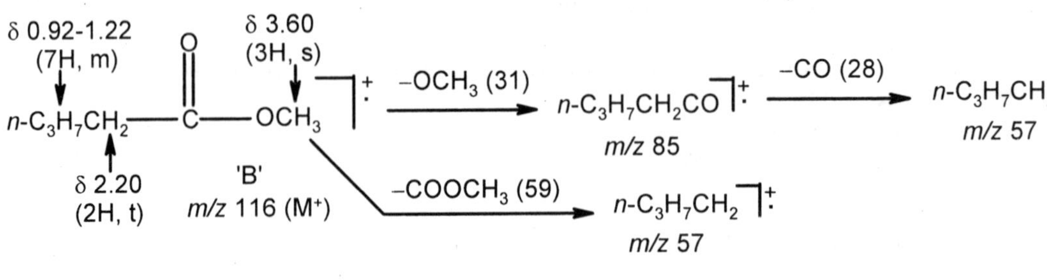

Methyl pentanoate (B) is formed from cyclopentanone (A) on reaction with sodium methoxide, as shown below:

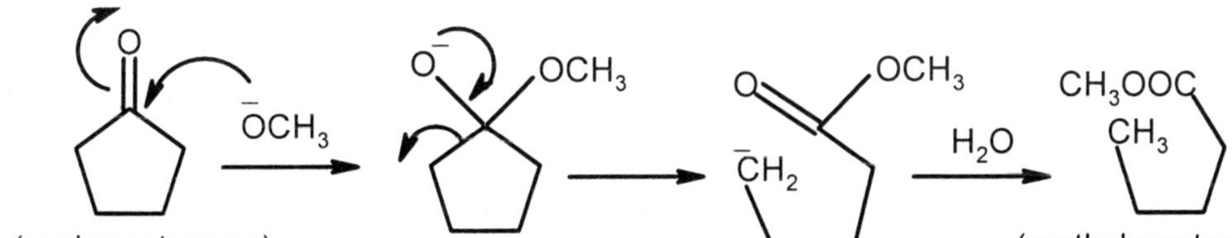

(cyclopentanone)
'A'

(methyl pentanoate)
'B'

Problem 7.47 Reaction of an organic compound 'A' (m.f. $C_7H_7NO_3$) with a mixture of conc. H_2SO_4 and conc. HNO_3 gave two isomeric compounds 'B' (major) and 'C' (minor). The compound 'B' exhibited the following spectral behaviour:

IR: 3040, 1610, 1510, 1350, 1030 cm^{-1}

^{1}H-NMR: δ 8.1 (2H, d, J = 8 Hz), 6.8 (2H, d, J = 8 Hz), 3.9 (2H, s)

Deduce the structures of 'A', 'B' and 'C', and explain the reaction involved.

Solution. ***Hint***

IR and PMR spectral data of 'B' suggest that it is an aromatic compound. On the basis of molecular formula, nature of chemical reaction undergone, and spectral properties, structural formulations of the compounds (A – C) can be given as:

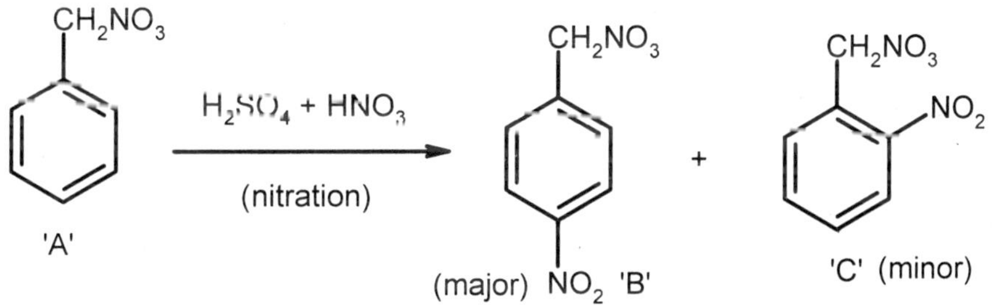

'A'

(major) NO_2 'B'

'C' (minor)

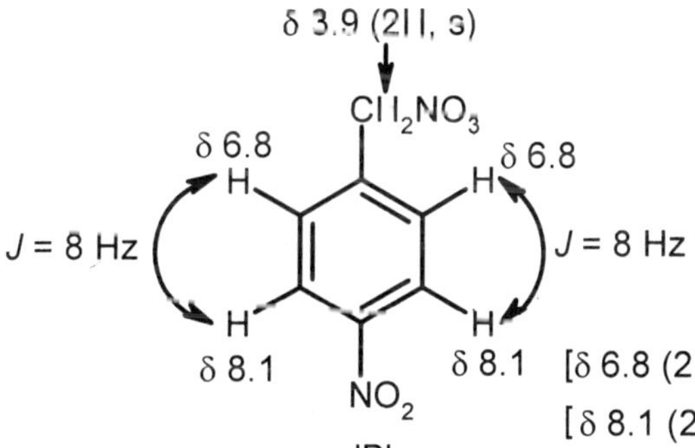

'B'

[δ 6.8 (2H, d, J = 8 Hz)]

[δ 8.1 (2H, d, J = 8 Hz)]

Problem 7.48 A ketone 'A'of molecular formula, C_8H_8O, upon treatment with $NaBH_4$ gives 'B' (m.f. $C_8H_{10}O$). The IR spectrum of 'B' has a strong band at 3600 cm^{-1}. Reaction of 'B' with methanesulfonyl chloride in the presence of a base furnishes 'C' that on treatment with NaI in acetone produces a derivative, which results in producing a hydrocarbon 'D' on reaction with $Na(CN)BH_3$. The hydrocarbon 'D' exhibits the PMR data as: δ 0.9 (3H, t), 2.6 (2H, q), 7.1 (5H, brs).

Deduce the structures of 'A' to 'D', and represent all the reactions involved.

Solution. ***Hint***

From the PMR data, the hydrocarbon 'D' can be formulated as ethylbenzene:

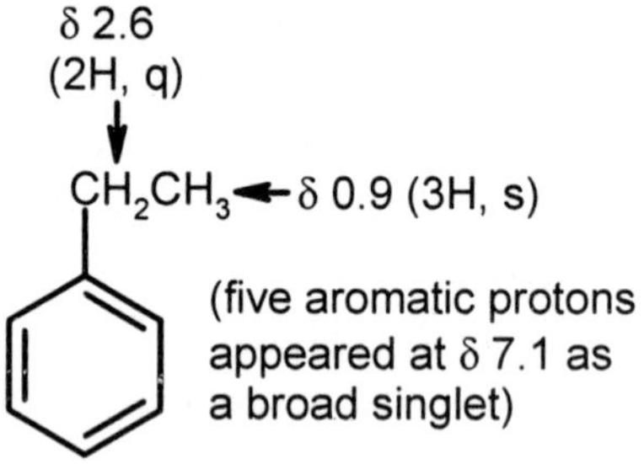

(ethyl benzene) 'D'

Hence, the reaction scheme along with the structures of the compounds (A–D) is shown below:

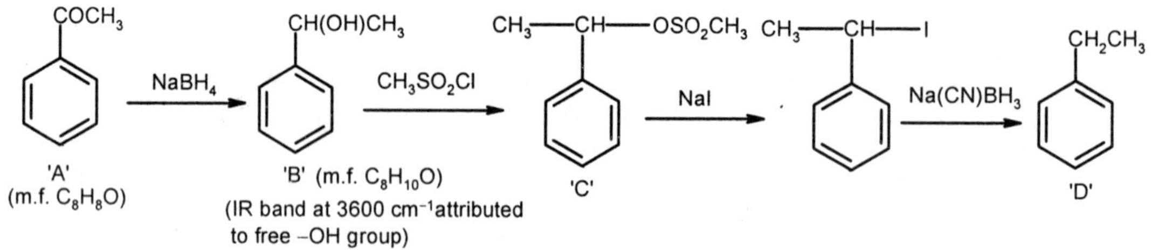

Problem 7.49 An organic compound with molecular formula $C_9H_{10}O_2$ showed the following spectral characteristics:

UV: 270 nm

IR: 1680 cm^{-1}

^{1}H-NMR: δ 7.6 (2H, d, $J = 8$ Hz), 6.9 (2H, d, $J = 8$ Hz), 3.9 (3H, s), 2.0 (3H, s).

Deduce the structures of the compound.

Solution. ***Hint*** Molecular formula: $C_9H_{10}O_2$

Level of Unsaturation (LU): C + 1 – H/2 – X/2 + N/2

= 9 + 1 – 5 = 5 (signifies five unsaturations/rings)

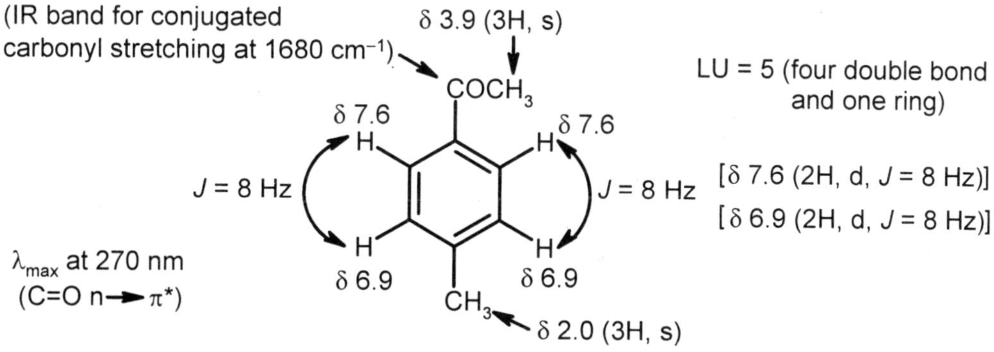

(*p*-methylacetophenone)

Problem 7.50 An organic compound with molecular formula $C_6H_{10}O$ showed the following spectral characteristics:

IR: 3600 – 3500, 3300, 2800, 2140 cm^{-1}

^{1}H-NMR: δ 1.0 (3H, t), 1.52 (2H, q), 1.4 (3H, s), 2.48 (1H, s), 3.3 (1H, brs)

Mass spectrum: *m/z* 98 $(M)^+$, 83, 82, 69 (base peak)

Deduce the structures of the compound.

Solution. ***Hint*** Molecular formula: $C_6H_{10}O$

Level of Unsaturation (LU): C + 1 – H/2 – X/2 + N/2

= 6 + 1 – 5 = 2 (signifies two unsaturations/rings)

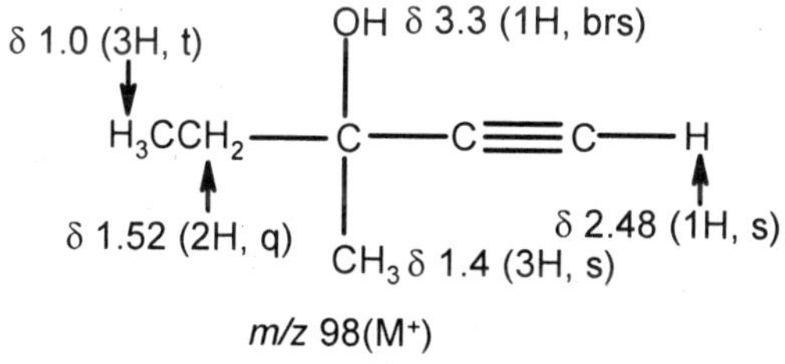

LU = 2 [signifies one triple bond]

IR bands (cm^{-1}): 3600-3500 (free OH); 3300 (≡C—H str.); 2140 (—C≡C— str.); 2800 (C—H str.)

Mass: *m/z* 98 (M^+); 83 $[M - CH_3]^+$; 82 $[M - CH_3 - H]^+$, 69 $[M - C_2H_5]^+$

Problem 7.51 An organic compound with molecular formula $C_8H_8O_2$ showed the following spectral characteristics:

UV: λ_{max} 229 and 257 nm; IR: 1710 cm^{-1}

^{1}H-NMR: δ 7.2 (5H, s), 3.5 (2H, s); Mass spectrum: *m/z* 136 $(M)^+$, 91 (base peak), 45.

Deduce the structures of the compound.

Solution. ***Hint***

Molecular formula: $C_8H_8O_2$

Level of Unsaturation (LU): C + 1 – H/2 – X/2 + N/2

= 8 + 1 – 4 = 5 (signifies five unsaturations/rings)

From PMR data (signal at δ 7.2 for five protons) it is evident that the compound bears an aromatic nucleus. UV absorptions at λ_{max} 229 (n→π*) and 257 (π→π*) nm suggest the presence of a keto group as well as the aryl moiety. The keto group is supposed to be the part of a carboxylic acid functionality as evidenced from the appearance of IR absorption peak at 1710 cm^{-1}. The PMR signal at δ 7.2 appeared as singlet for five protons is indicative of a phenyl ring, and a singlet at δ 3.5 for two protons is suggestive of the presence of a methylene group in between the phenyl group and carboxylic acid function. Thus, the proposed structure would be:

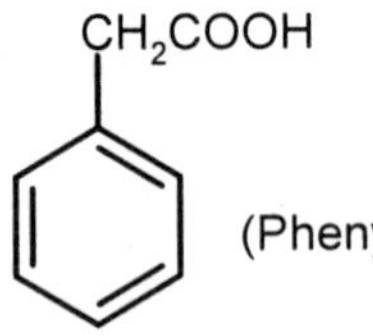

(Phenylacetic acid)

This proposed structure clearly supports the mass fragmented ion-peaks:

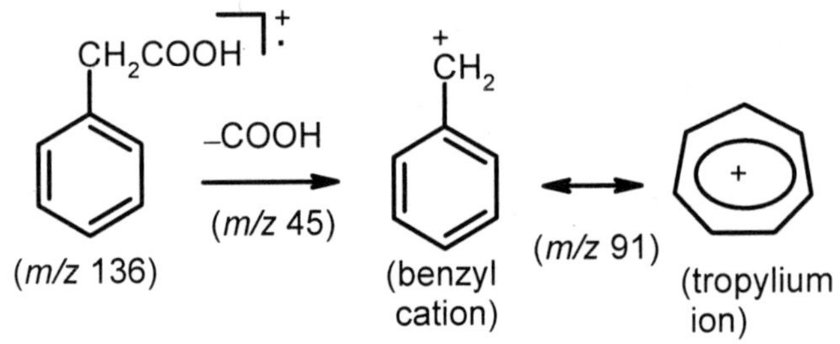

Problem 7.52 In an experiment performed to oxidize 1,4-butanediol to prepare succindialdehyde, the product obtained had no aldehyde properties. It had a strong band at 1780 cm^{-1} and PMR signals at δ 1.2 (2H, m), 2.5 (2H, m), 4.0 (2H, m). Mass spectrum showed M+ at *m/z* 86. Identify the product and account for its formation.

Solution. ***Hint***

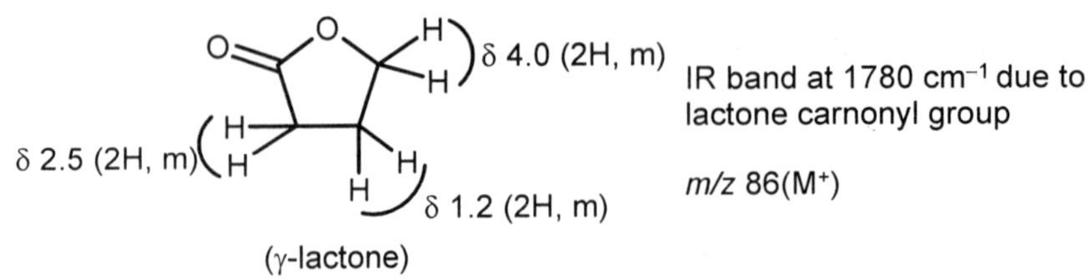

(γ-lactone)

Problem 7.53 A compound (m.f. C_7H_8O) shows a broad and intense band at 3300 cm^{-1} in its IR spectrum; its PMR spectrum consists of three sharp singlets at 3.68, 4.43 and 7.21 ppm with the peak area ratio 1:2:5. What would be its structure?

Solution. ***Hint***

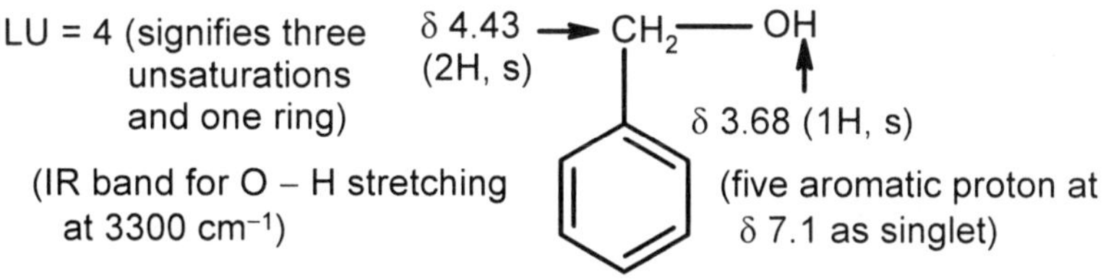

Benzyl alcohol (C_7H_8O)

Problem 7.54 An organic compound with molecular formula $C_{14}H_{10}O_2$ showed the following spectral characteristics:

UV: 260 nm

IR: 3080, 3040, 1690, 1680, 1600, 1590, 1450, 1215, 720, 690 cm^{-1}

^{1}H-NMR: δ 8.0 (4H, m), 7.8 (6H, m)

Mass spectrum: *m/z* 210 $(M)^+$, 182, 152, 105 (base peak), 77, 51.

Deduce the structures of the compound.

Solution. ***Hint***

Molecular formula: $C_{14}H_{10}O_2$

Level of Unsaturation (LU): C + 1 – H/2 – X/2 + N/2

= 14 + 1 – 5 = 10 (signifies ten unsaturations/rings)

The spectral observations clearly suggest that the compound bears aromatic moiety. IR absorptions at 3080 and 3040 cm^{-1} are due to aromatic C–H stretching vibrations that is substantiated by the appearance of the IR bands at 1066, 1590 and 1450 cm^{-1}. Carbonyl stretching frequencies appeared at 1690 and 1680 cm^{-1} for two such groups are indicative that they are attached with aromatic nucleus (i.e. conjugated). UV absorption at λ_{max} 260 ($\pi \rightarrow \pi^*$) nm supports the benzenoid nature of aromatic moiety. IR absorptions at 720 and 690 cm^{-1} are indicative of monosubstituted benzene ring while the absorption peak at 1215 cm^{-1} is due to C–C–C stretching and bending vibrations of C–CO–C moiety in aromatic ketone. From these spectral observations as well as 10 DBE (double bond equivalent/level of unsaturation), the probable structure may be assumed to be benzil.

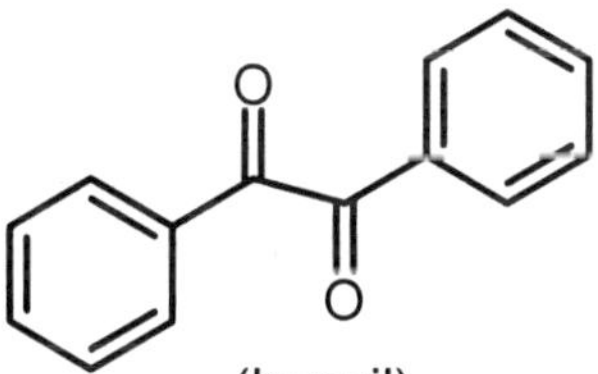

(benzil)

The four *ortho*-protons of two phenyl rings appeared at δ 8.0 while the remaining six protons of both the rings appeared at δ 7.8, both the peaks as multiplets in its PMR spectrum. The structure also successfully explains the mass fragment-ion peaks as shown below:

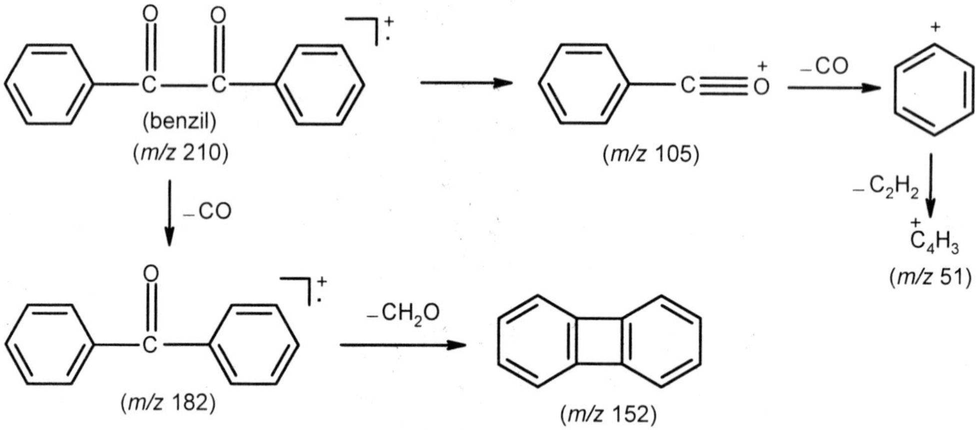

Problem 7.55 Predict the IR and PMR spectral properties of ethyl 3-aminobenzoate.

Solution. The IR spectrum of the compound would record a significant absorption bands at 1708 cm^{-1} for conjugated ester carbonyl, and also at around 3400 cm^{-1} as a doublet for the primary amino group. The *meta*-disubstitution of the benzene ring would be attributed to the appearance of aromatic out-of-plane bending bands at around 680, 760, and 880 cm^{-1}. The PMR spectral properties are shown in the figure below:

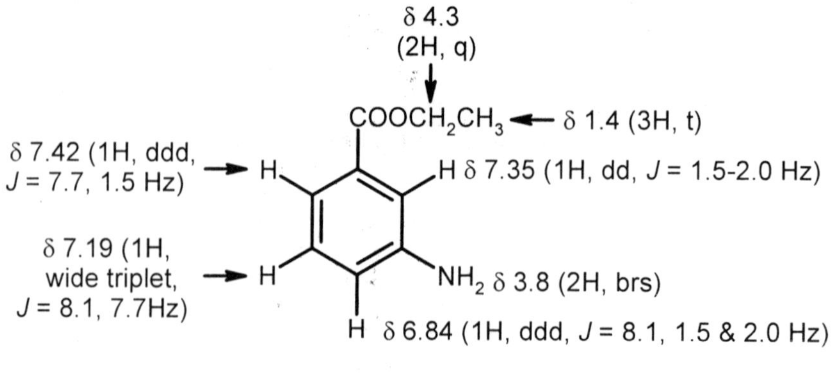

(ethyl 3-aminobenzoate)

Supplementary problems

SP-7.1 When *p*-aminophenol is dissolved in water, λ_{max} is at longer wavelength than in acid solution. Explain.

SP-7.2 Yellow colour of *p*-nitrophenol dissolved in water deepens in intensity and moves to longer wavelength on addition of NaOH to the solution. Explain.

SP-7.3 The wavelength as well as the extinction coefficient increases with the increase in conjugation in the compound. Justify the statement.

SP-7.4 The position of absorption of acetone shifts in different solvents: 279 nm (hexane); 272 nm (ethanol), and 264.5 nm (water). Explain.

SP-7.5 The UV spectrum of acetone shows absorption maxima at 166, 189, and 279 nm. What type of transition is responsible for each of these bands?

SP-7.6 How will you distinguish an aliphatic aldehyde from an aliphatic ketone by applying IR spectroscopy?

SP-7.7 Explain why does maleic acid absorb at a higher frequency as compared to fumaric acid in an IR spectrum?

SP-7.8 Write a short note on finger print region.

SP-7.9 Justify the following observations:

(i) The νC=O frequency of absorption for β-lactone is higher as compared to γ-lactone.

(ii) A compound in its vapour phase absorbs for stretching of a particular bond at a higher wave-number as compared to that when it is in the solid state.

(iii) 2-hydroxy-3-nitroacetophenone exhibits two carbonyl stretching frequencies at 1692 and 1658 cm^{-1}.

(iv) The O–H infrared absorption frequency for 1,1,2,2-tetra-*t*-butylethane-1,2-diol is much smaller than that for ethylene glycol.

(v) 2,3-Dimethyl-2-butene shows no IR absorption for C=C stretching, whereas isobutylene does.

SP-7.10 Construct a structure consistent with each of the following spectra:

(i) C_4H_9Br [^{1}H-NMR data: δ 1.04 (6H, d), 1.95 (1H, m), 3.33 (2H, d)]

(ii) $C_{10}H_{14}$ [^{1}H-NMR data: δ 1.30 (9H, s), 7.28 (5H, s)]
(iii) $C_9H_{11}Br$ [^{1}H-NMR data: δ 2.15 (2H, q), 2.75(2H, t), 3.38 (2H, t), 7.22 (5H, s)]
(iv) C_7H_8O [^{1}H-NMR data: δ 3.7 (brs), 4.45(s), 7.25 (s) with peak area ratio of 1:2:5]
(v) $C_4H_8Br_2$ [^{1}H-NMR data: δ 1.90 (3H, s), 3.9 (1H, s)]
(vi) C_8H_9Br [^{1}H-NMR data: δ 2.0 (3H, d), 5.15 (1H, q), 7.35 (5H, m)]
(vii) $C_5H_9BrO_2$ [^{1}H-NMR data: δ 1.28 (3H, t), 2.86(2H, t), 3.58 (2H, t), 4.19 (2H, q)]
(viii) $C_{10}H_{13}NO_2$ [^{1}H-NMR data: δ 1.5 (3H, t), 2.2 (3H, s), 4.2 (2H, q), 7.1 (2H, dd); 7.7 (2H, dd); 8.2 (1H, brs)] [Hint. *N*-(*p*-ethoxyphenyl)acetamide]

SP-7.11 Explain how the peaks at *m/z* 115, 101 and 73 arise in the mass spectrum of 3-methyl-3-heptanol.

SP-7.12 The mass spectrum of 3-butyn-2-ol shows a large peak at *m/z* 55. Draw the structure of the fragment and explain why it is particularly stable.

SP-7.13 How can you use the information of *ortho* effect to explain the formation of the ion peak at *m/z* 149 in diethyl phthalate?

SP-7.14 Ethyl butanoate in its mass spectrum exhibits two characteristic peaks due to odd-electron ions at *m/z* 88 and 60, and an abundant ion at *m/z* 71. Explain the fragmentation pattern.

SP-7.15 An organic compound of molecular formula, $C_9H_{12}O_2$ showed three peaks in the PMR spectrum at δ ~7.2 (5H, brs), 3.6 (2H, q), 1.96 (3H, t). One of the intense IR bands of this compound appears at 1740 cm^{-1}. Deduce the structure of the compound.

SP-7.16 In the ^{1}H-NMR spectrum of *p*-nitroethylbenzene, the four chemical shifts appear at approximately 7.8, 7.1, 2.8 and 1.1 ppm. Assign the resonances to the appropriate protons, and indicate the expected splitting pattern in each case.

SP-7.17 A carboxylic acid 'A' (m.f. $C_{11}H_{14}O_2$) upon treatment with $SOCl_2$ followed by treatment of the product (B) with anhydrous $AlCl_3$ in CS_2 gives 'C' (m.f $C_{11}H_{12}O$). The IR spectrum of 'C' has a strong absorption band at 1690 cm^{-1}, and its PMR spectrum has three signals in the intensity ratio (low to high field) 2:1:3. On heating 'C' with hydrazine and KOH in ethylene glycol, a hydrocarbon 'D' is obtained, with the following PMR signals at: δ 1.22 (6H, s), 1.85 (2H, t, *J* = 7 Hz), 7.02 (4H, s). Deduce the structures of 'A' to 'D' along with interpretation of the spectral data, and represent all the reactions involved.

SP-7.18 An organic compound showed the following spectral characteristics:
IR: 1730, 1540 cm^{-1}
^{1}H-NMR: δ 1.6 (3H, d, $J = 7$ Hz), 2.4 (2H, t), 4.2 (2H, t), 5.1 – 5.6 (2H, m), 7.3 – 8.1 (4H, symm. multiplet)
Mass spectrum: *m/z* 235 $(M)^+$, 207, 167, 150. Deduce the structure of the compound.

SP-7.19 (i) Treatment of neopentyl chloride with sodamide yields a hydrocarbon of formula C_5H_{10}, which rapidly decolourizes bromine in carbon tetrachloride; but it is not oxidized by cold, dilute neutral permanganate. Its PMR spectrum shows absorption at δ 0.20 and δ 1.05 with peak area ratio 2:3. What is the probable structure of the hydrocarbon? Rationalize its formation.

(ii) In aqueous acid 2,5-dimethylfuran gives a product 'A' in high yield. The mass spectrum of 'A' shows a molecular ion at m/z 114(100%), with a M + 1 peak at 115 (6.7%) and a M + 2 peak at 116 (0.6%). The IR spectrum of 'A' exhibits an intense absorption at 1715 cm^{-1}. The PMR spectrum shows two singlets at δ 2.6 and δ 2.1 with peak area ratio 2:3. What is the structure of 'A'? (Given: Natural abundance of $^{13}C = 1.08\%$, $^2H = 0.016\%$).

SP-7.20 Compound 'A' reacts with HIO_4 to give a single compound 'B' that shows a strong IR absorption band at 1720 cm^{-1}. Compound 'A', when treated with conc. H_2SO_4, gives another compound 'C'. Compound 'C' (m.f. $C_6H_{12}O$) exhibits only two singlets at δ 1.10 and δ 2.05, with an area ratio of 3:1 in its PMR spectrum. Deduce the structures of 'A', 'B' and 'C'.

SP-7.21 A water-soluble liquid (b.p. 60^0C) shows a strong IR absorption band at 1720 cm^{-1}; its PMR spectrum consists of three signals at δ 9.5 (1H, d), 2.6(1H, m), 1.1 (6H, d). Propose a structure consistent with these data.

SP-7.22 A compound of molecular formula C_6H_{12} shows only one unsplit peak in the ^{1}H-NMR spectrum at δ 1.77. What is its structure?

SP-7.23 The IR spectrum of compound 'A' (m.f. C_4H_7N) exhibited a moderate band at 2250 cm^{-1}. Its ^{1}H-NMR spectrum records proton resonances at δ 1.33 (6H, d) and 2.72 (1H, m). What is the structure of 'A'?

SP-7.24 Hydrocarbon 'A' (m.f. C_6H_6) gave a PMR spectrum with two signals, δ 6.55 and δ 3.84, with an area ratio of 2:1. On warming with pyridine for 3 hrs. 'A' was quantitatively converted to benzene. Mild hydrogenation of 'A' gave 'B' (m.f. C_6H_{12}), which showed no absorption in the 1650 cm^{-1} region of the infrared, and exhibited only one broad peak in the PMR spectrum at δ 2.34. Assign structures to 'A' and 'B'.

SP-7.25 A compound 'A' (m.f. C_7H_4BrClO), on treatment with diazomethane forms compound 'B' that is converted into another compound 'C' (m.f. $C_8H_7BrO_2$) upon heating with moist silver oxide. The ^{1}H-NMR spectral data of 'C' are: δ 2.7 (s), 7.0(m), 12.0 (s) with peak area ratio of 2:4:1. What are the structures of compounds 'A', 'B', and 'C'?

SP-7.26 A colourless basic liquid 'A' (m.f. C_7H_9N) reacts immediately with $NaNO_2$ in cold dil. sulphuric acid to give neutral yellow oil (B). Reaction of 'A' with acetic anhydride in pyridine solution furnishes a neutral white solid (C), the ^{1}H-NMR spectrum of which shows a singlet at δ 1.9, a singlet at δ 3.5, and a multiplet at δ 7.3 with intensities of 1:1:1.67, respectively. Assign structures to 'A', 'B', and 'C'.

SP-7.27 A compound of molecular formula C_6H_{10} absorbs only one mole of hydrogen on catalytic hydrogenation. Its ^{1}H-NMR spectrum records signals at δ 4.82 (2H), 2.22 (4H), 1.65 (4H). What is the structure?

SP-7.28 The strongest bands in the IR spectrum of an organic compound (m.f. $C_4H_{10}O_2S$) were 1150 and 1330 cm^{-1}. The ^{1}H-NMR spectrum of the compound records a triplet at δ 1.32 (3H), and a quartet at δ 3.03 (2H). What is the structure of the compound?

SP-7.29 Deduce structures for the products of the following reactions, on the basis of spectroscopic properties as supplied:

(i)

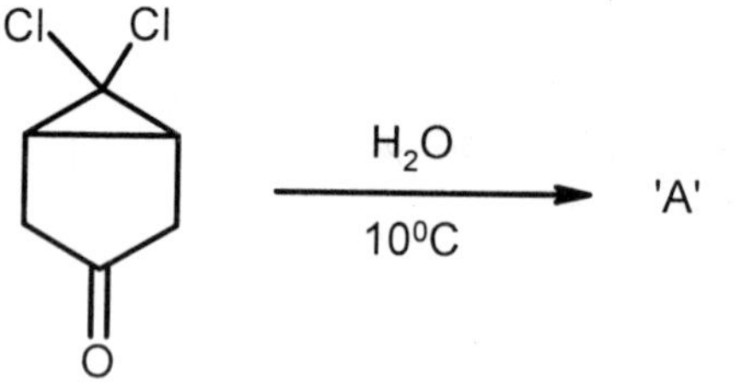

$[M]^+$ 128 and 130

IR: 3500, 1600 and 1500 cm^{-1}

PMR: δ 7.1 (2H, d, *J* = 7 Hz); 6.8 (2H, d, *J* = 7 Hz); 5.4 (1H, brs)

(ii)

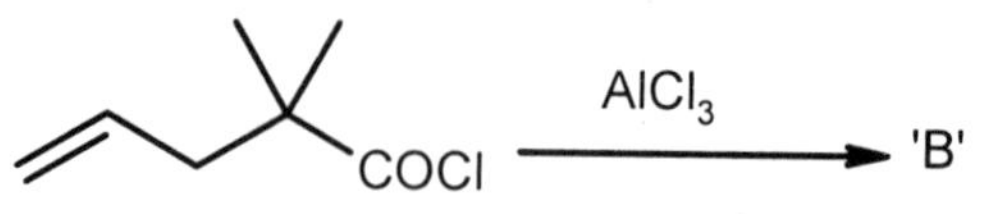

$[M]^+$ 110

IR: 17200 cm^{-1}

PMR: δ 7.7 (1H, ddd, *J* = 6 and 3 Hz); 6.2 (6H, ddd, *J* = 6 and 2 Hz); 2.2 (1H, dd, *J* = 3 and 2 Hz); 1.1 (6H, s)

SP-7.30 An organic compound (m.f. C_3H_7NO) showed the following spectral characteristics:

UV: 238 nm
IR: 3428(m), 2941 – 2857 (w), 1681 (s), and 1452 (w) cm^{-1}
^{1}H-NMR: δ 1.9 (3H, s), 2.7 (3H, s), 8.13 (1H, s)
Deduce the structure of the compound.

SP-7.31 An organic compound (molecular weight 116) showed the following spectral characteristics:
UV: 283 nm
IR: 3000– 2500, 1715, and 1342 cm^{-1}
^{1}H-NMR: δ 2.2 (3H, s), 2.25 (2H, t), 2.6 (2H, t), 11.1 (1H, s)
Deduce the structure of the compound. [***Hint*** m.f. $C_5H_8O_3$]

SP-7.32 An organic compound (m.f. $C_9H_5NO_4$) showed the following spectral characteristics:
IR: 3000– 2500, 2225, 1715, 1605, 1518, 1344, and 900 – 700 cm^{-1}
^{1}H-NMR: δ 7.25–7.4 (4H, unsymm. multiplet), 11.1 (1H, s)
Deduce the structure of the compound.

SP-7.33 An organic compound (molecular weight 156) showed the following spectral characteristics:
UV: 205 nm
IR: 3415, 3060, 2246, 1742, 1674, and 1182 cm^{-1}
^{1}H-NMR: δ 2.1(s), 4.38 (s), 5.51 (s), 6.76 (s) in the ratio of 22:23:7.6:7.5
Deduce the structure of the compound.

SP-7.34 An organic compound (m.f. $C_6H_{15}N$) showed the following spectral characteristics:
UV: Transparent above 200 nm
IR: 2950– 2840, 1280 cm^{-1}
^{1}H-NMR: δ 1.15 (t, J = 6.9), 2.6 (q, J = 6.9 Hz) in the ratio of 10.5 : 6.7
Deduce the structure of the compound.

Problem 7.35 'Pulegone' on condensation with ethyl acetoacetate in alcoholic sodium ethoxide forms 'pulegone acetone'. Predict which one between I and II would be the correct structure for this product on the basis of the given spectral data for the compound?

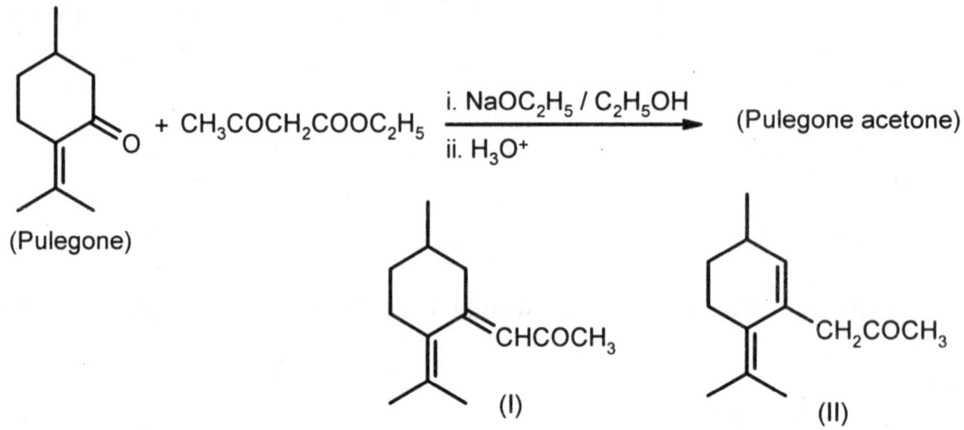

Spectral data (Pulegone acetone):

IR: 1675, 1640 cm^{-1}

^{1}H-NMR: δ 0.97 (3H, d), 1.1 (6H, s), 1.2 – 2.0 (8H, m), 2.3 (2H, s), 6.1 (1H, s)

Given that ozonation of pulegone acetone yields an acid, $C_{13}H_{20}O_4$.

Problem 7.36 An organic compound (molecular weight 87) showed the following spectral characteristics:

UV: 220 nm

IR: 3500, 2960, 1682, 1610, 1405, 1372, 700, 650 cm^{-1}

^{1}H-NMR: δ 1.0 (6H, d), 2.1 (1H, m), 8.1 (2H, br.s)

Deduce the structure of the compound. [***Hint*** m.f. C_4H_9NO]

C H A P T E R 8

HETEROCYCLES

Introduction

Heterocyclic organic compounds comprise of cyclic structures in which one or more atoms are non-carbon — they may be nitrogen, oxygen, sulphur, phosphorous, boron, etc. However, 'N', 'O', and 'S'are the most common heteroatoms, and such heterocyles have been studied extensively. Heterocylcic compounds may be classified into *aliphatic* and *aromatic*.

Heterocyclic compounds are of much importance; they occur widely in nature at a variety of structural skeletons. Besides, a variety of such compounds are prepared on a commercial scale by the dye and drug industries. A large number of heterocyclic compounds are essential to life; various compounds like alkaloids, antibiotics, essential amino acids, the vitamins, the hormones, and haemoglobin along with many synthetic drugs contain heterocyclic ring systems. A good knowledge on heterocyclic chemistry is very much necessary to the organic chemists in general.

Solved Problems

Problem 8.1 Sketch the reaction scheme for each of the following 'Name Reactions' in heterocyclic chemistry.

i. Bischler-Napieralski Synthesis
ii. Fischer-Indole Synthesis
iii. Friedlander Synthesis

iv. Hantzsch Synthesis of Pyridines
v. Hantzsch Synthesis of Pyrroles
vi. Paal-Knorr Synthesis of Pyrrroles
vii. Pomeranz-Fritsch Synthesis
viii. Skraup Synthesis

Solutions.

i. ***Bischler-Napieralski Synthesis***

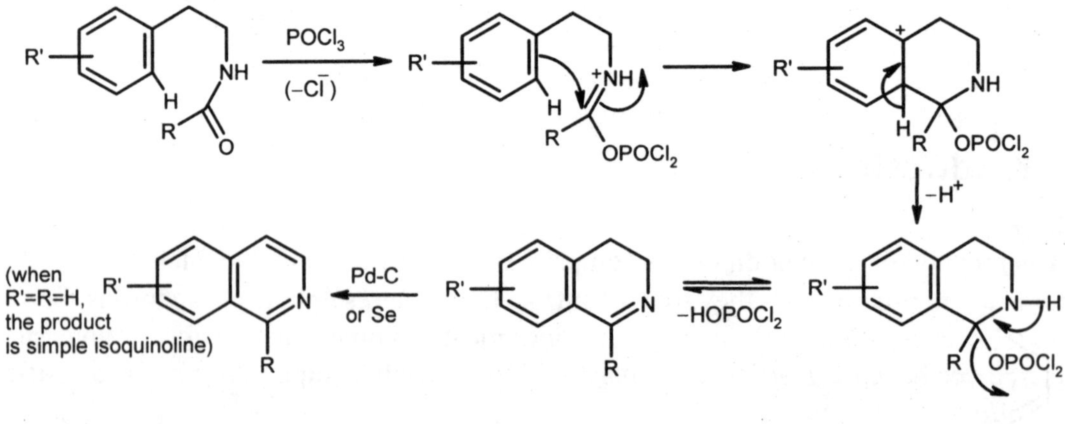

ii. ***Fischer-Indole synthesis***

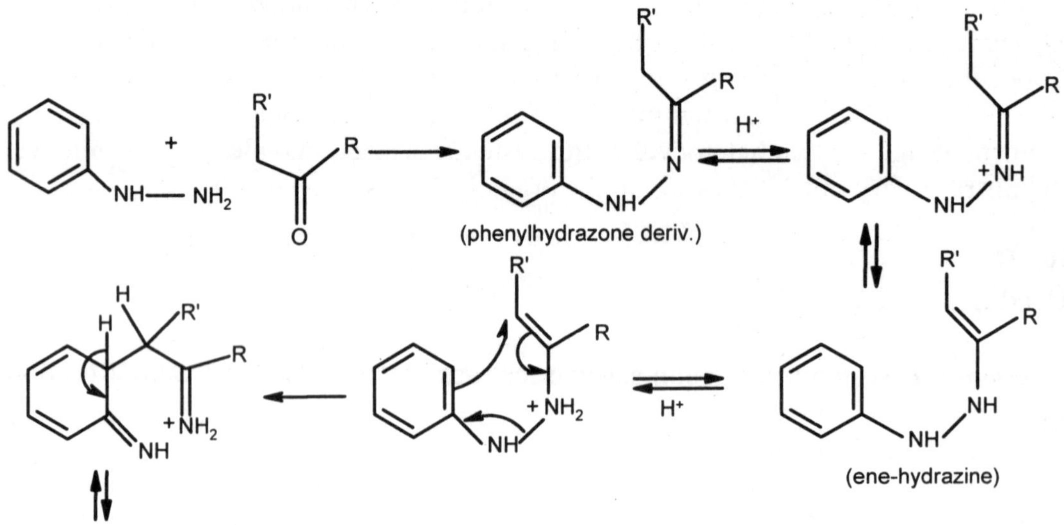

(see at the next page)

(from the previous page)

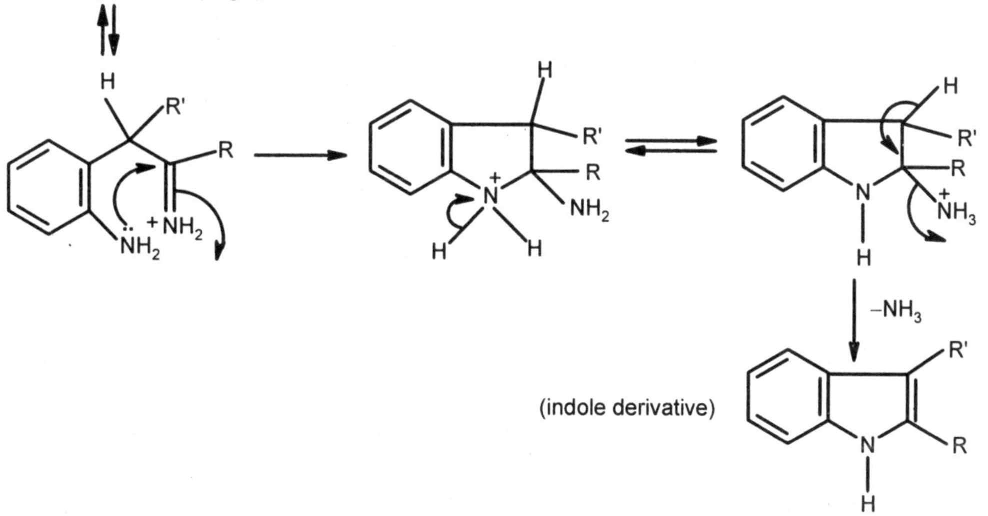

iii. ***Friedlander synthesis***

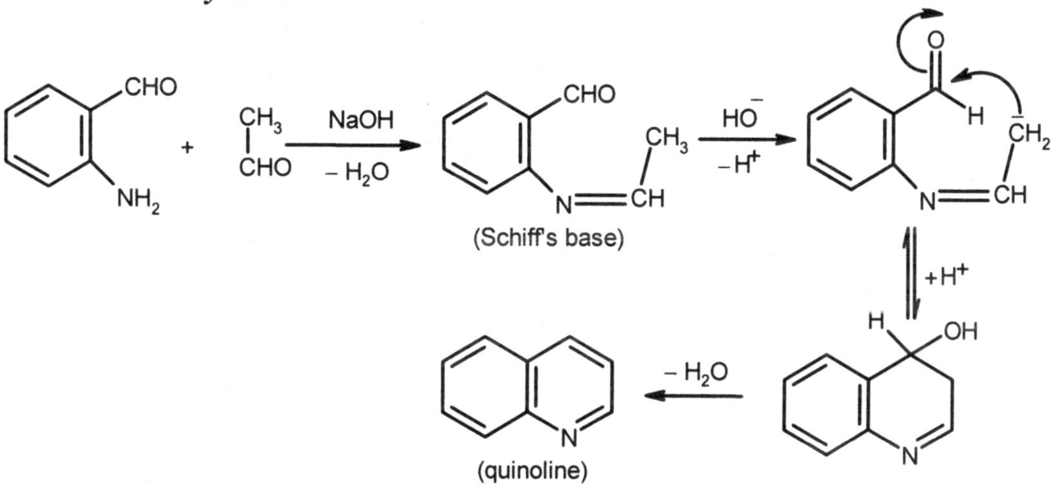

iv. ***Hantzsch synthesis of pyridines***

Condensation reaction of an aldehyde with two moles of a β-dicarbonyl compound/β-ketoester and ammonia followed by oxidation with nitrous acid.

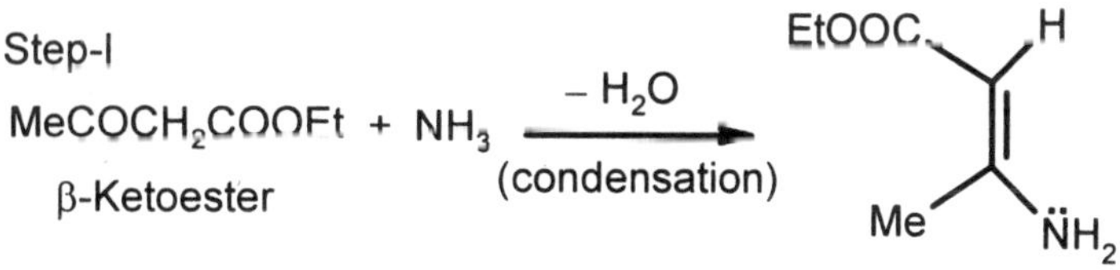

Step-II

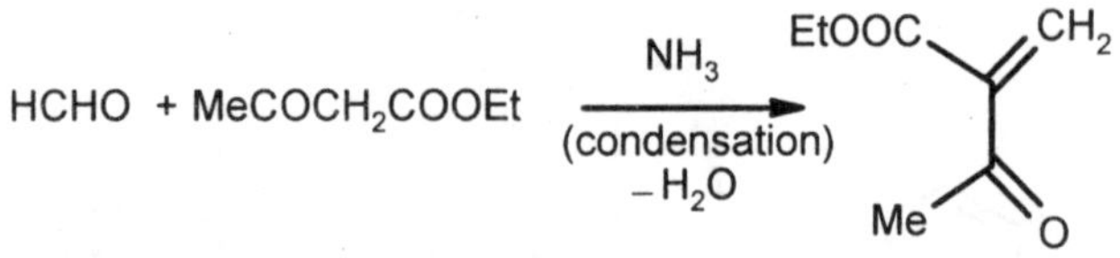

Step-III

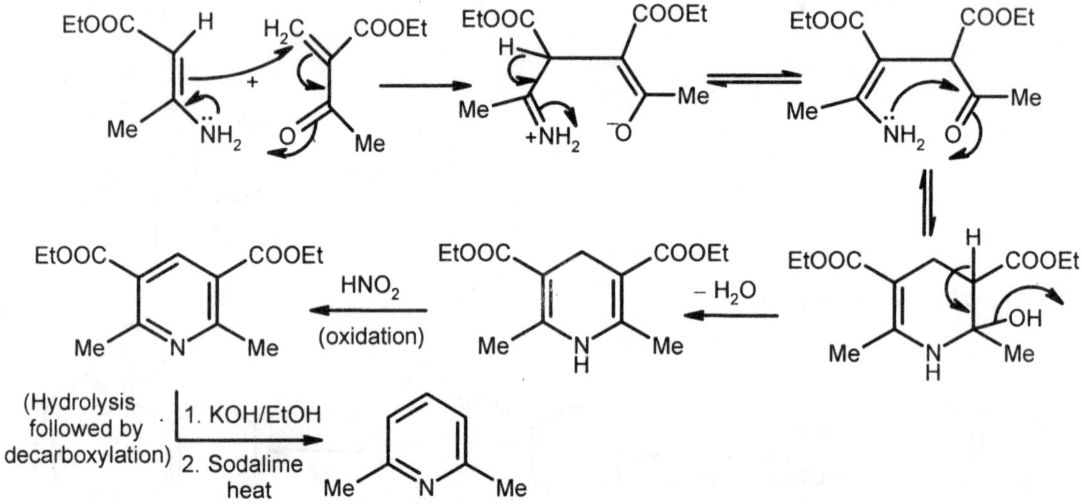

v. ***Hantzsch synthesis of pyrroles***

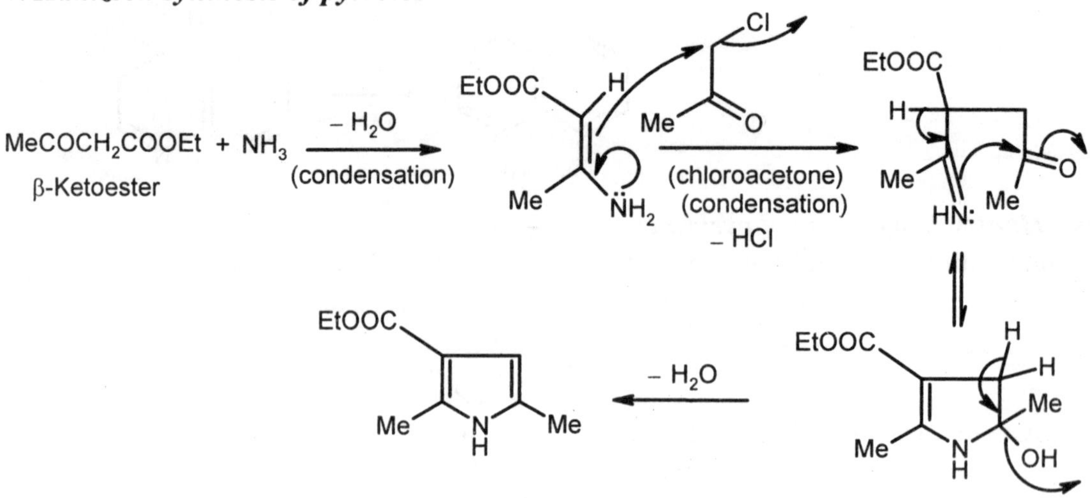

vi. ***Paal-Knorr synthesis of pyrroles***

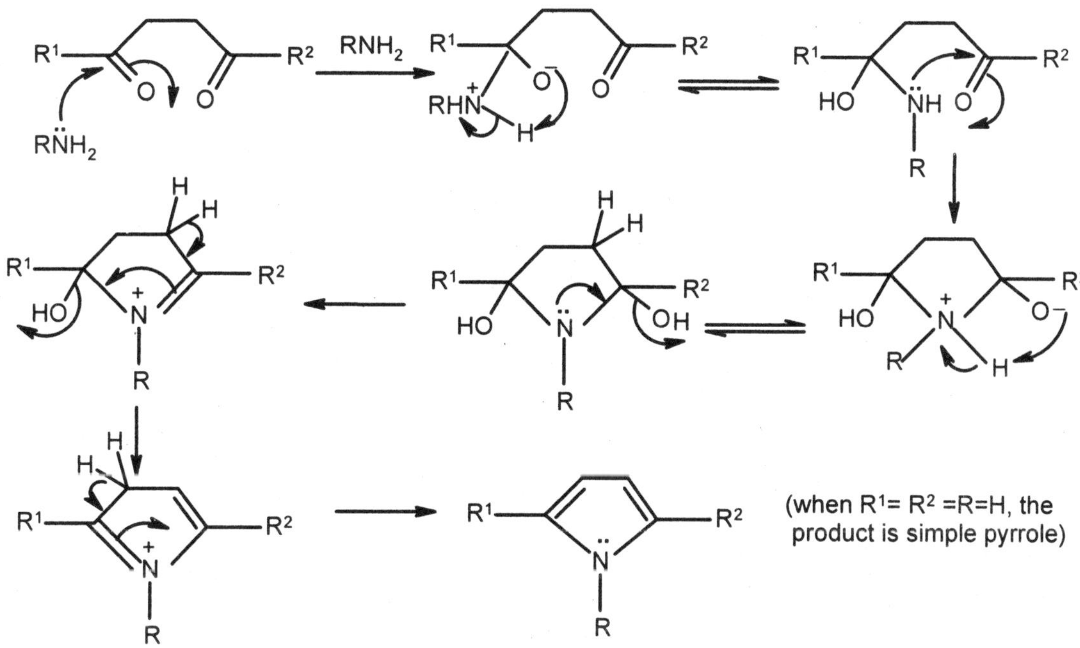

vii. ***Pomeranz-Fritsch Synthesis***

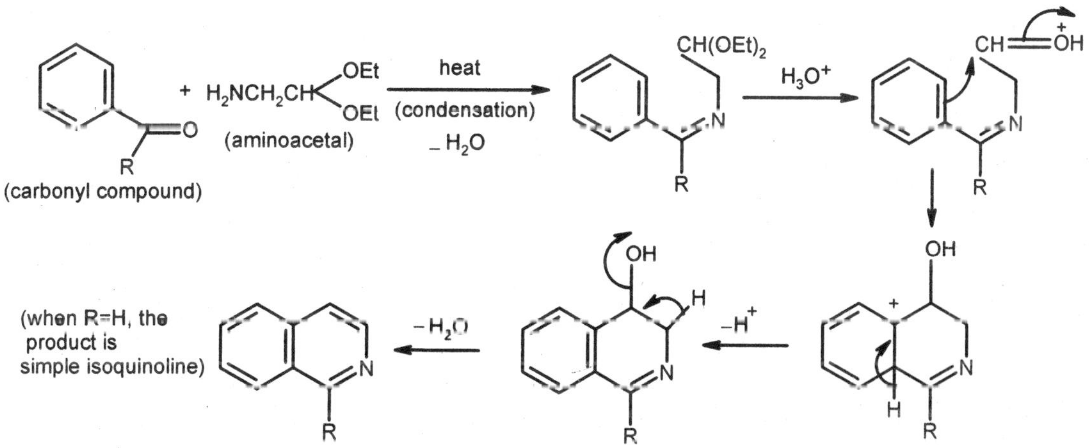

viii. ***Skraup Synthesis:*** This method involves acid-induced condensation of an α,β-unsaturated carbonyl compound with a primary aromatic amine having a free *ortho*-position, in the presence of an oxidizing agent. In simple case, quinoline is produced on heating a mixture of aniline, glycerol, conc. sulphuric acid and nitrobenzene (acts as oxidizing agent). Ferrous sulphate is also sometimes used due its moderating effect (makes the reaction less violent) by acting as oxygen carrier.

Glycerol is taken as a precursor for acrolein (acraldehyde), because it can't be used directly since it undergoes polymerization under the experimental conditions.

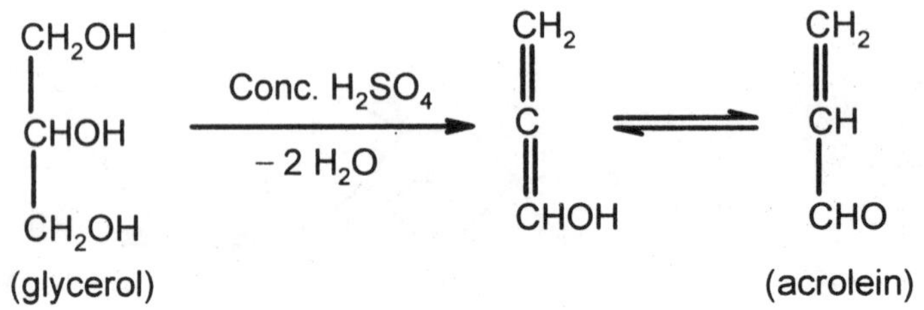

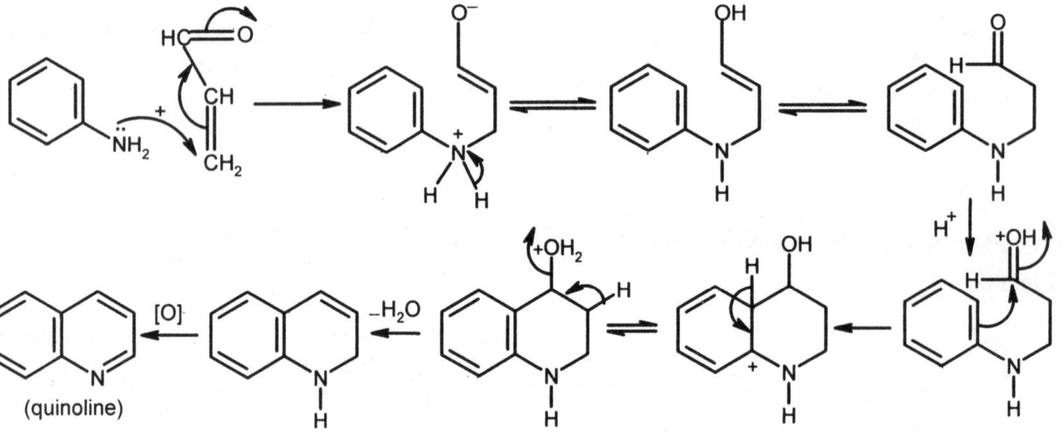

Problem 8.2 Account for each of the following facts:

Problem 8.2.1 Pyrrole becomes protonated on the carbon atom adjacent the nitrogen atom rather than on nitrogen atom.

Solution. The nonbonding electrons on the nitrogen atom in pyrrole are part of the aromatic sextet; hence the pair of nonbonding electrons is much less available for protonation. The α-carbon adjacent to the nitrogen atom becomes relatively electron-dense due to resonance effect, and protonatoion occurs on it — the conjugate acid as formed attains stability to some extent as every atom in this structure has an octet.

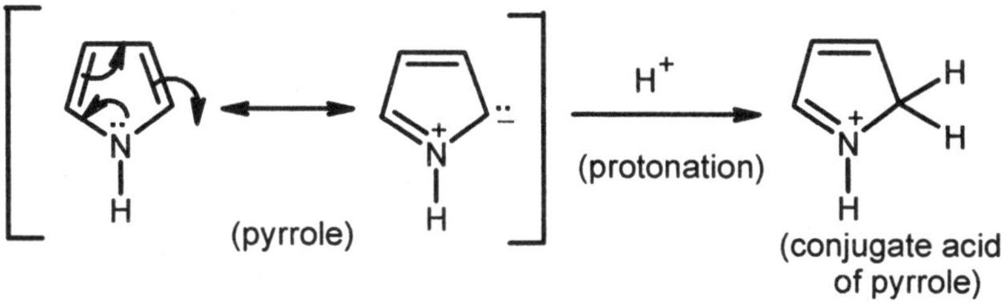

Problem 8.2.2 Pyrrole undergoes electrophilic substitution preferably at the 2-position.

Solution. Pyrrole usually prefers to undergo electrophilic substitution at 2-position rather than 3-position; however, if both the α-positions are blocked, then substitution occurs at the other position(s). The preference for electrophilic attack at 2-position can be visualized from the following resonance contributing structures:

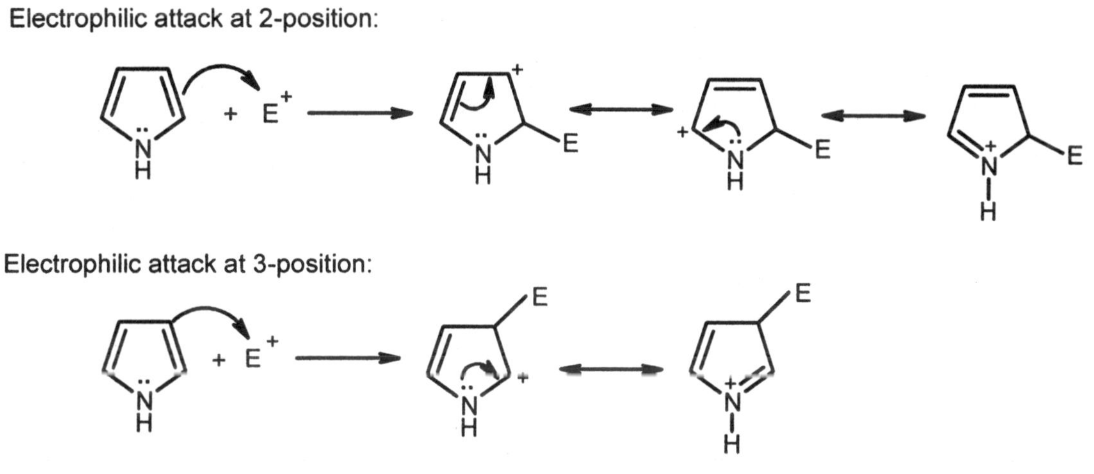

It is evident that there are three resonance contributors for the intermediate ('*Wheland intermediate*') formed due to substitution at the 2-position whereas there are only two such forms for the similar intermediate arisen out of substituition at the 3-position.

Problem 8.2.3 Imidazole is a stronger base than pyrrole.

Solution. The second nitrogen atom in imidazole molecule bears a pair of nonbonding electrons that is not part of the aromatic sextet. Protonation, thus, occurs on this nitrogen atom, and the conjugate acid is stabilized by delocalization of the positive charge to both the nitrogen atoms giving rise to two equivalent resonance contributors.

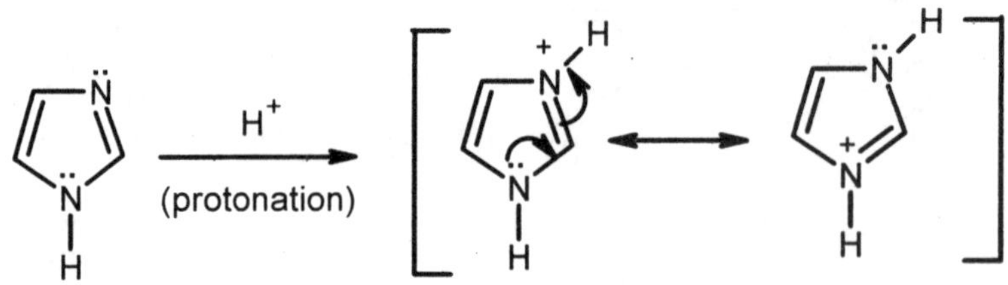

(conjugate acid of imidazole)
stabilized by resonance

Problem 8.2.4 Thiophene is more aromatic compared to furan and pyrrole.

Solution. Sulphur is less electronegative compared to nitrogen and oxygen — thus nonbonding electrons on sulphur in thiophene are readily available for participation in delocalization within the ring. Besides, sulphur atom bears 3d orbitals that are utilized in resonance process (oxygen and nitrogen cannot) giving rise to greater number of canonical forms, thus making thiophene molecule much more aromatic in comparison to furan and pyrrole. Estimation of the ring current in thiophene has shown it to be more aromatic than furan and pyrrole.

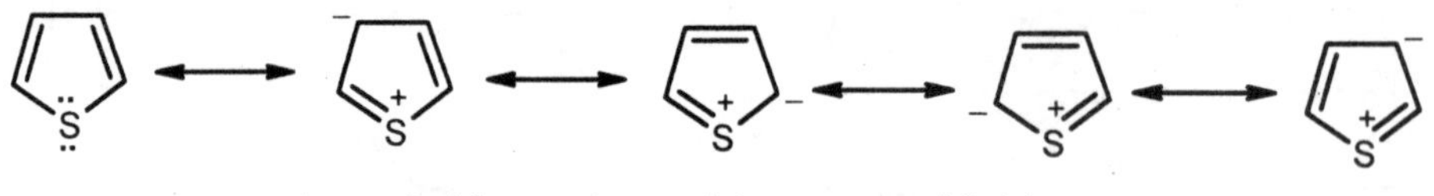
(canonical forms where sulphur uses p-orbitals)

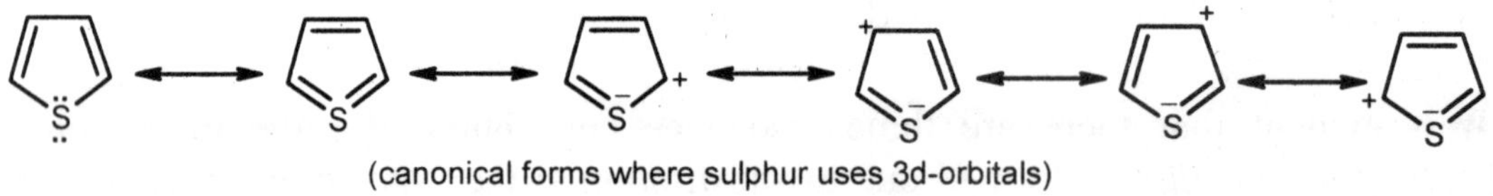
(canonical forms where sulphur uses 3d-orbitals)

Electronegativity of oxygen is high as compared to nitrogen and sulphur; thus in case of furan, the high electronegativity of oxygen reduces the ease of delocalization that reduces the resonance stabilization of furan and makes it least aromatic. Pyrrole has intermediate aromaticity among the series. The resonance energies for thiophene, pyrrole and furan are 29, 22, and 16 kcal/mole, respectively.

Thiophene > Pyrrole > Furan
(aromaticity order)

Problem 8.2.5 Pyridine undergoes electrophilic substitution preferably at the 3-position.

Solution. Pyridine usually prefers to undergo electrophilic substitution at 3-position rather than 4-position. The preference for electrophilic attack at 3-position can be understood by considering the following resonance contributing structures:

Electrophilic attack at 3-position:

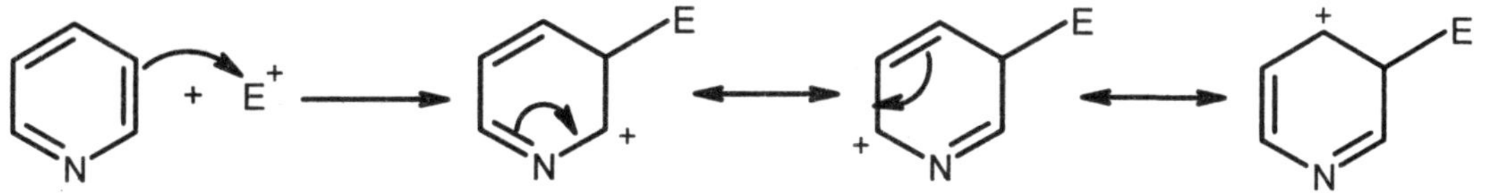

Electrophilic attack at 4-position:

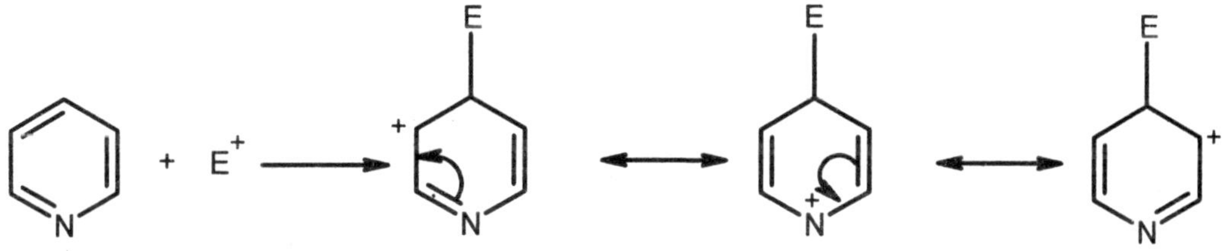

From the scrutiny of the above resonance structures, it is evident that none of the canonical forms for 3-substituted intermediate bears positive charge on ring-nitrogen, whereas in case of 4-substituted intermediate one of such resonance contributors bears positive charge on electronegative nitrogen atom. Thus, canonical forms in the former case are as a whole more contributing making the electrophilic attack at 3-position of pyridine more preferable, which is also consistent with electron density calculations.

Problem 8.2.6 4-Chloro-3-nitropyridine hydrolyses readily even in warm water.

Solution. The strong electron-withdrawing effect of the nitro functionality present at the *ortho*-position of the departing chloro group makes the hydrolysis facilitated as evident from the following mechanistic scheme.

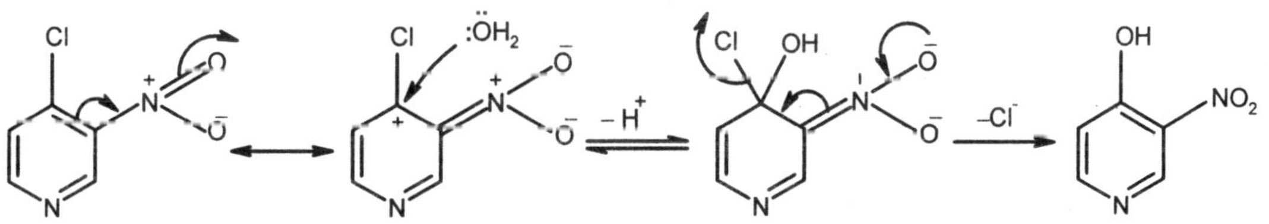

Problem 8.2.7 Pyridine-*N*-oxide is more reactive than pyridine toward both electrophilic and nucleophilic substitutions.

Solution. The 4-position in Pyridine-*N*-oxide may either bear positive charge (II) or negative charge (III) as evident from the resonance contributing structures of the molecule.

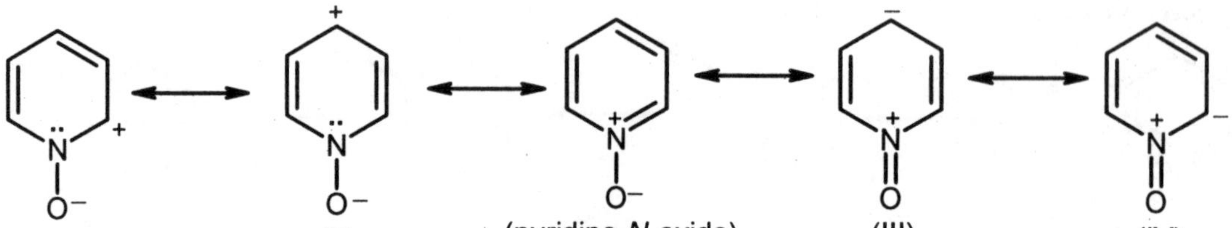

(I) (II) (pyridine-*N*-oxide) (III) (IV)

Therefore, both the incoming electrophile (E^+) and nucleophile (Nu^-) can attack at this position, and the high reactivity is attributed to the stability of the intermediates as formed (all the atoms in both the intermediates attain complete octets).

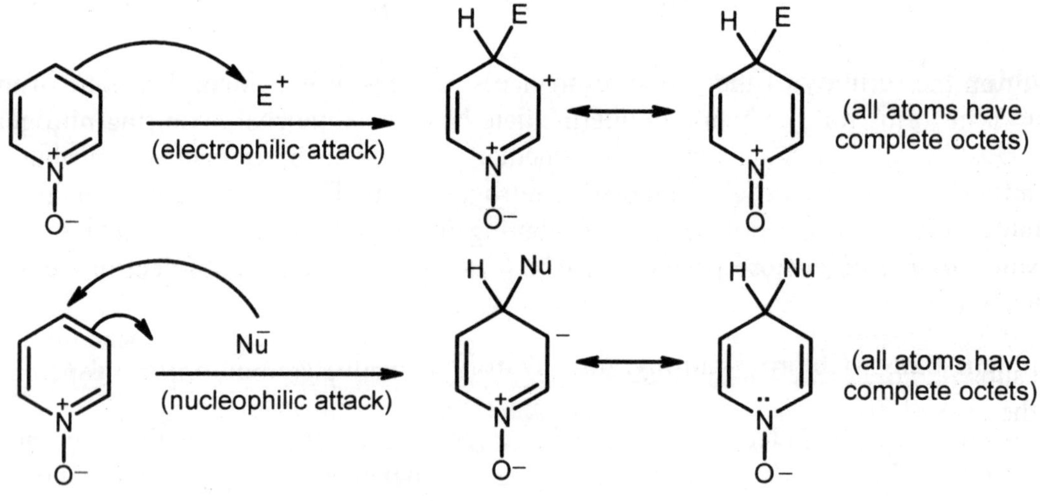

Problem 8.2.8 Pyridine is usually nitrated at 4-position through pyridine-*N*-oxide.

Solution. Direct nitration of pyridine yields 3-nitropyridine, not the 4-nitroderivative. So, nitration at 4-position of pyridine is carried out indirectly through pyridine-*N*-oxide —electrophilic substitution in pyridine-*N*-oxide occurs efficiently at the 4-position. The adduct as obtained can easily be converted into the pyridine derivative.

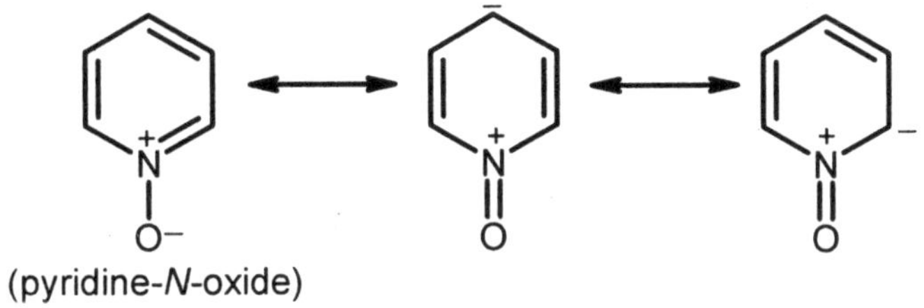

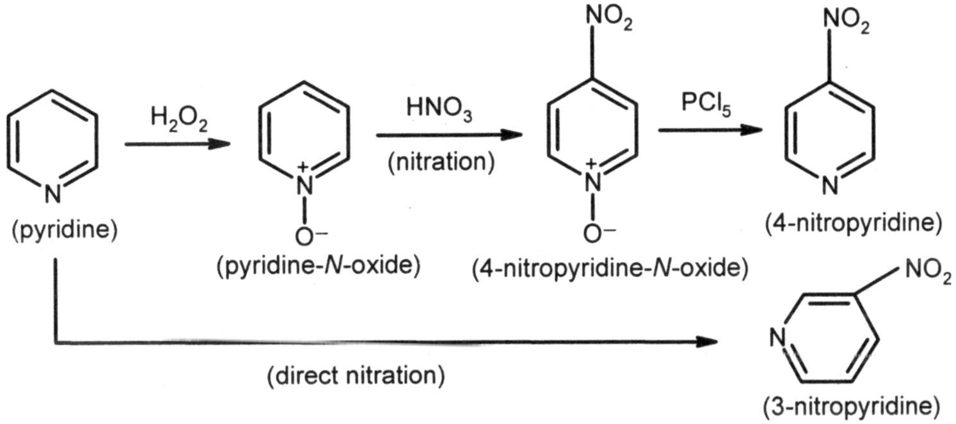

Problem 8.2.9 While nitration of pyridine-*N*-oxide occurs at 4-position, sulphonation takes place at 3-position though both are carried out in the presence of sulphuric acid.

Solution. In contrast to nitration, sulphonation reaction takes place only under severe condition (oleum/$HgSO_4$/~240^{0}C) because pyridine-*N*-oxide is highly reluctant towards sulphonation reaction. Strikingly, sulphonation occurs at the 3-position instead of usually expected 4-position (i.e. the substitution position for nitro group); this drastic change in orientation is attributed to the fact that in sulphonating mixture the substrate molecule firstly forms the species (I) where electron-donating capability of the oxygen atom is reduced so largely that (I) becomes *meta*-directing.

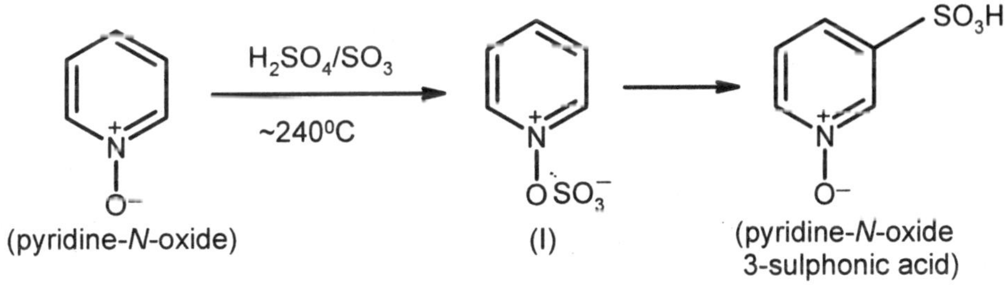

Problem 8.2.10 Electrophilic substitution in indole takes place chiefly at 3-position.

Solution.

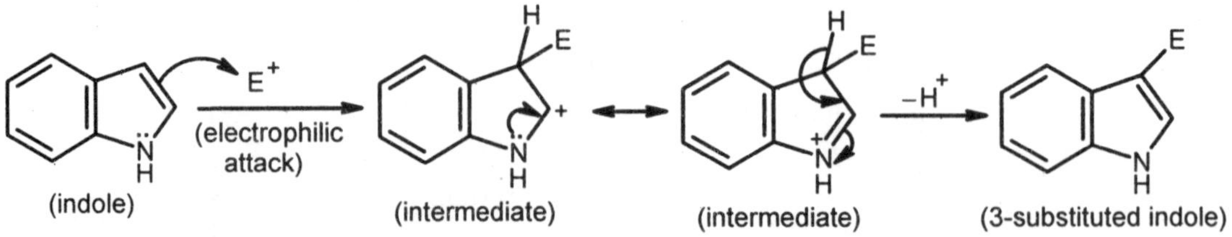

Electrophilic attack at the 3-position of indole gives the most stable cationic intermediate leaving the aromaticity of the six-membered ring intact. Moreover, the positive charge is accommodated by nitrogen. Hence, electrophilic attack at the 3-position of indole is facilitated.

Problem 8.2.11 Piperidine cannot be used instead of pyridine in Schotten-Baumann procedure.

Solution. In Schotten-Baumann acylation procedure, pyridine is widely used as base; pyridine itself does not acylated. But piperidine being a 2^0-amine would itself also be acylated in the reaction condition. That's why piperdine cannot be used in place of pyridine here.

Problem 8.2.12 Isomeric pyridinecarboxylic acids undergoes decarboxylation at a reactivity order of 2 > 3 > 4, whereas the order of reactivity in case of isomeric pyridineacetic acids is 2 or 4 > 3.

Solution.

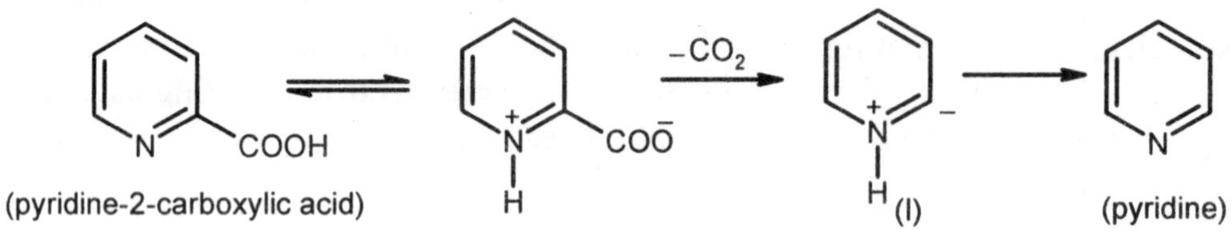

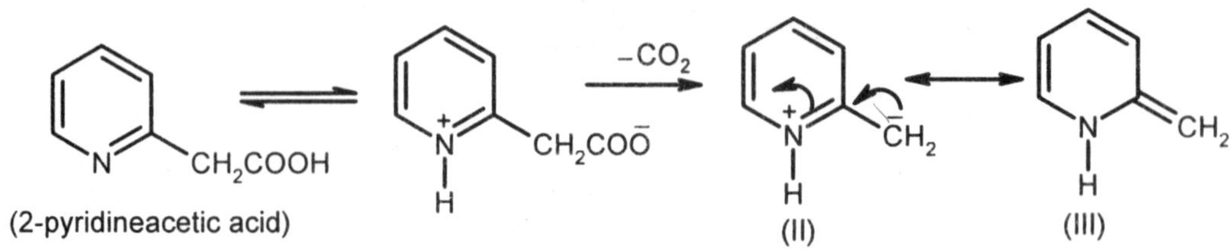

The carbanionic intermediate (I) and its isomers are stabilized by the inductive effect of the positive charge on nitrogen; since inductive effect becomes weaker with distance, it is more pronounced in (I) and would be least in the *para*-isomer. Hence, the order of stability of carbanionic intermediates becomes 2 > 3 > 4, which is the reactivity-order also.

On the other hand, benzylic-like carbanions are formed during the decarboxylation of the isomers of pyridineacetic acid; carbanion intermediate (II) formed on decarboxylation of 2-pyridineacetic acid is stabilized by resonance (as shown in figure; structures II & III). Benzylic-like carbanion from 4-pyridineacetic acid also gets stabilized by conjugative effect, but such resonance is not possible for the intermediate arising from 3-pyridineacetic acid. Consequently, the stability as well as the reactivity order in case of isomeric pyridineacetic acids would be 2 or 4 > 3.

Problem 8.2.13 Hydroxyfurans practically do not exist.

Solution. The chemistry of 2-hydroxyfurans is regarded as the chemistry of unsaturated lactones; this is because of the fact that 2-hydroxyfurans practically do not exist — actually exist, if at all, at undetectably low concentrations in tautomeric equilibria involving 2(5*H*)-furanone and 2(3*H*)-furanone.

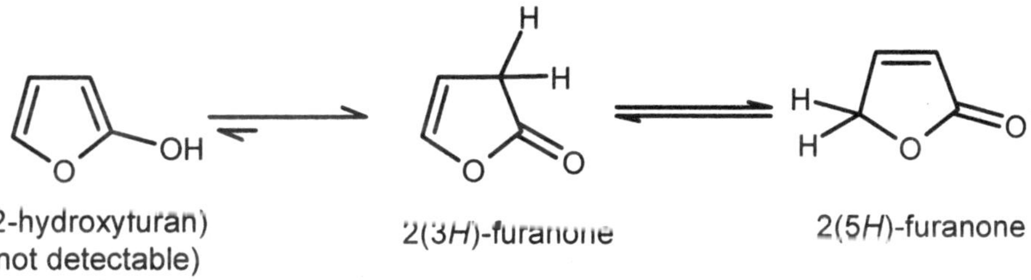

For example, the angelica lactones exist in equilibrium (on treatment with an organic base) — the β-isomer is more stable.

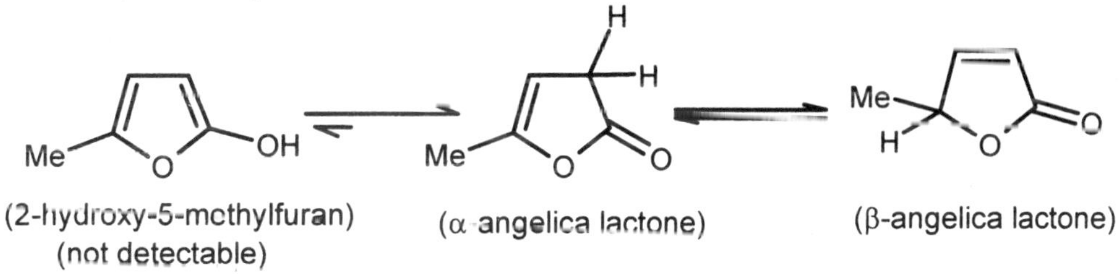

Problem 8.2.14 2,6-Di-*tert*-butyl pyridine does not undergo reaction at all with iodomethane, even at high pressure.

Solution. The presence of substituents and their orientations has a profound impact on the reactivity of heterocycles. Substituents adjacent to the nitrogen atom in pyridine direct the ease of its reactivity with alkyl halides. Introduction of steric interference at α-positions results in slowing the rate of *N*-alkylation; it has been experimentally found that introduction of one methyl group at α-position of pyridine nucleus slows down the rate by about three fold, while 2,6-dimethyl substitution slows the rate between 12–40 times. In 2,6-di-*tert*-butyl pyridine, the two large groups at the two α-positions in the molecule create much steric interference such that the reagent (methyl iodide) cannot come in contact with the nitrogen atom concerned, even at high pressure — the highly reactive methylfluorosulphonate can *N*-methylate the substrate only under high pressure.

Problem 8.2.15 The relative rates of reaction of iodomethane separately with isoquinoline, quinoline, and 8-methylquinoline are 69, 8 and 0.008, respectively.

Solution. *Peri*-substitutions also play a vital role on the reactivity of heterocyles; greater is extent of steric inhibitions caused by *peri*-substituents, slower is the rate of a particular reaction with the heteroatoms in the substrate molecules. In the present case, there is no *peri*-substituent in isoquinoline, whereas quinoline bears a hydrogen atom (small in size) and 8-methylquinoline bears a methyl group (relatively bulky in size) at their respective *peri*-positions. Thus, the reactivity order with iodomethane would be isoquinoline > quinoline > 8-methylquinoline, which has been observed to be 69:8: 0.008.

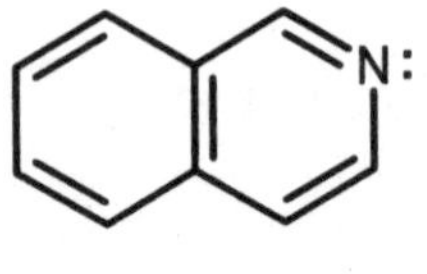

(isoquinoline)

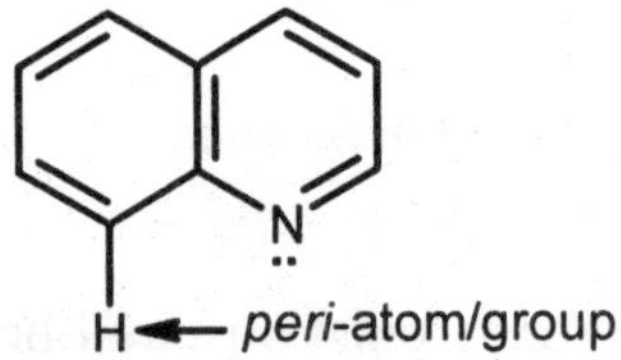

(quinoline)

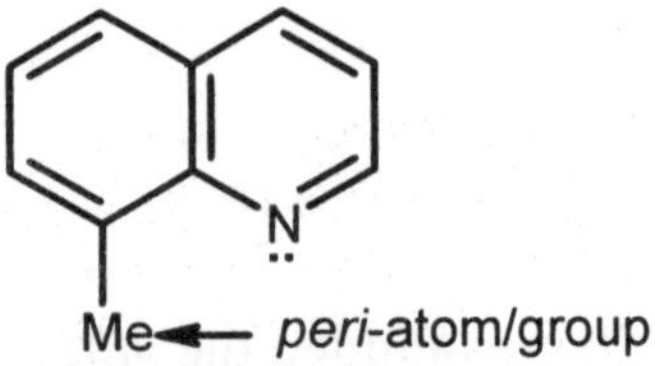

(8-methylquinoline)

Problem 8.2.16 The reaction of either 3- or 4-bromopyridine with secondary amines in the presence of sodamide/sodium *t*-butoxide yields the same 1:1 mixture of 3- and 4-dialkylamino-pyridines.

Solution. The nucleophilic substitution reaction is supposed to proceed *via* an elimination process with the intermediacy of 3,4-didehydropyridine (3,4-pyridyne). 3-Bromo-pyridine and 4-bromopyridine, both form the same intermediate that ultimately yields 1:1 mixture of 3- and 4-dialkylamino-pyridines.

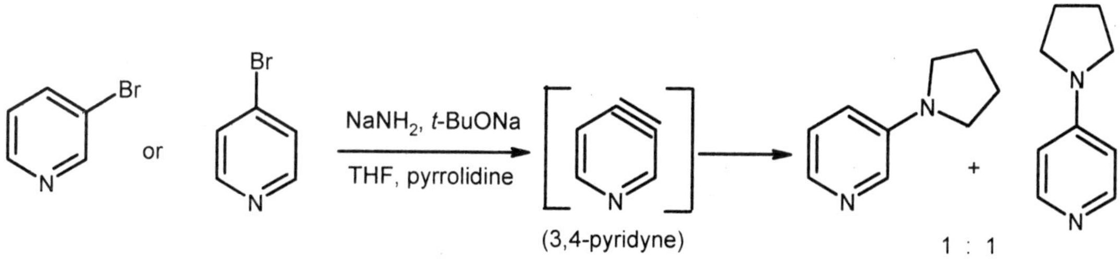

That no 2-aminated pyridine is produced reveals a greater difficulty in generating 2,3-pyridyne.

Problem 8.3 Outline the reasonable synthetic path for each of the following conversions:

Problem 8.3.1

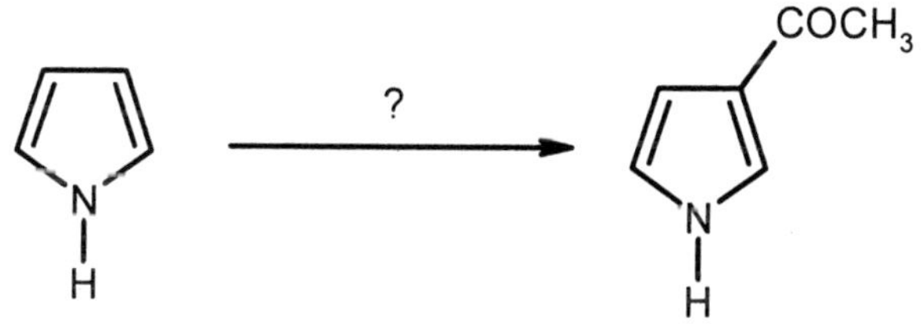

Solution.

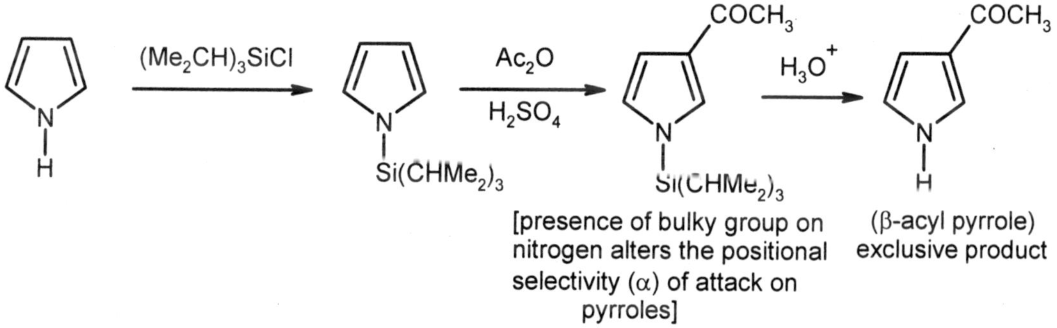

Problem 8.3.2

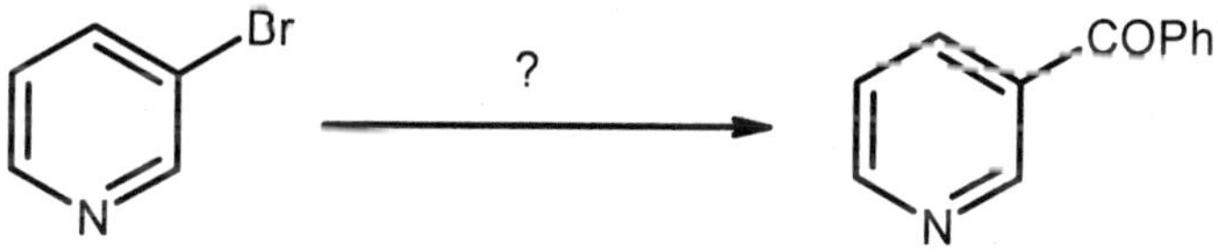

Solution.

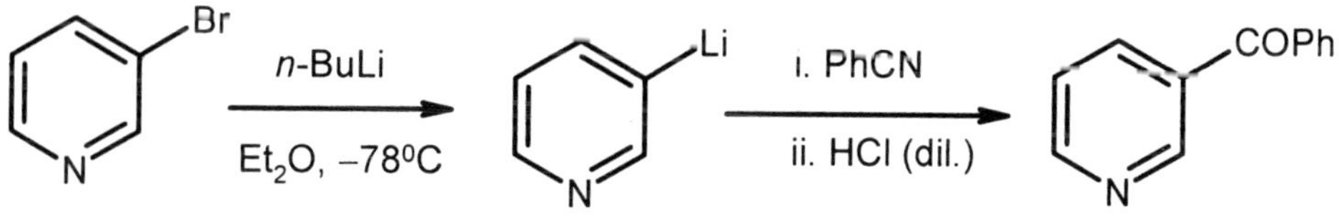

Problem 8.3.3

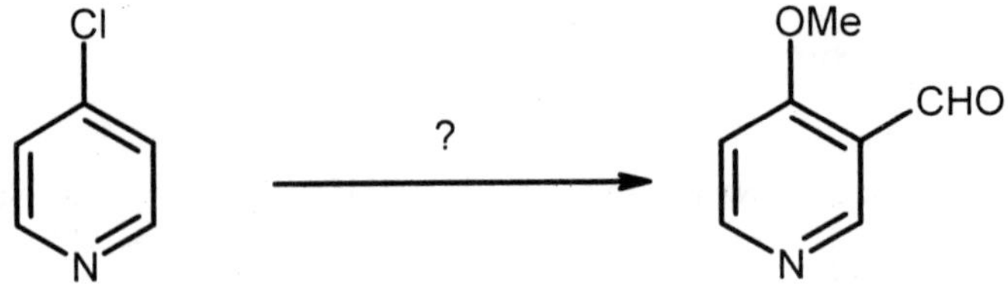

Solution.

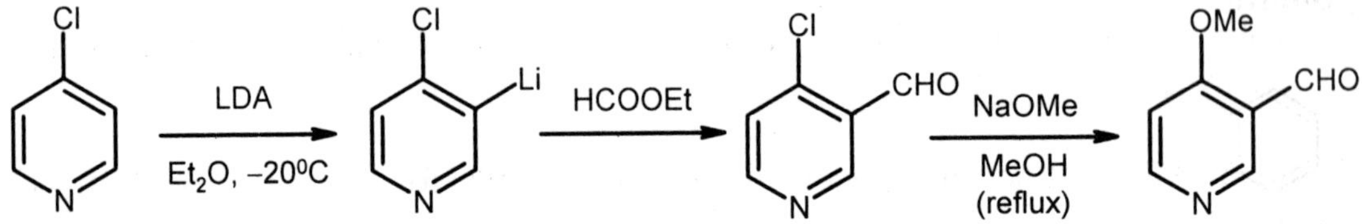

Problem 8.3.4

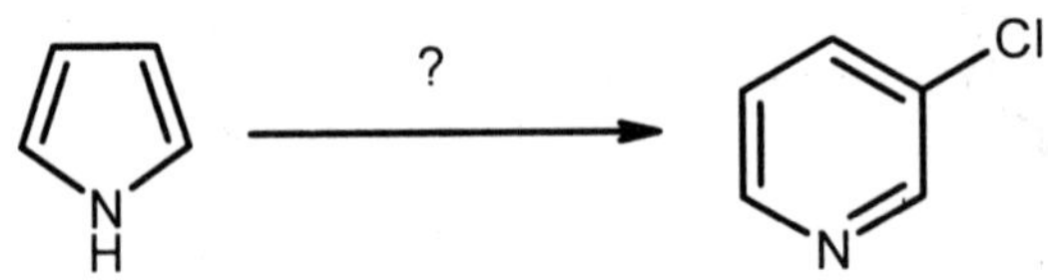

Solution.

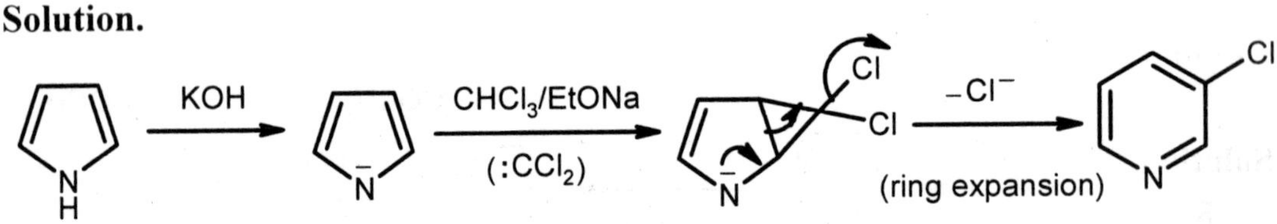

Problem 8.3.5

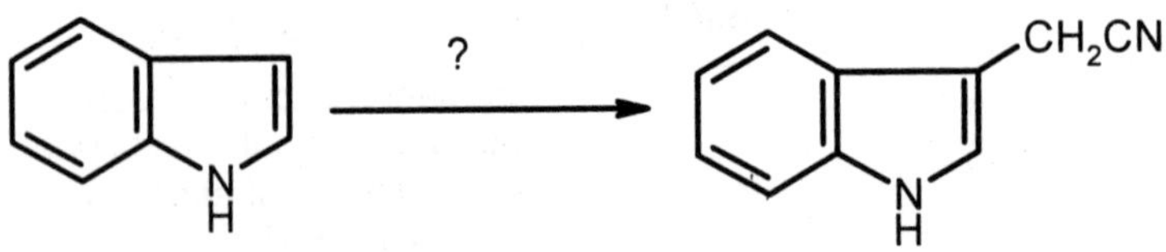

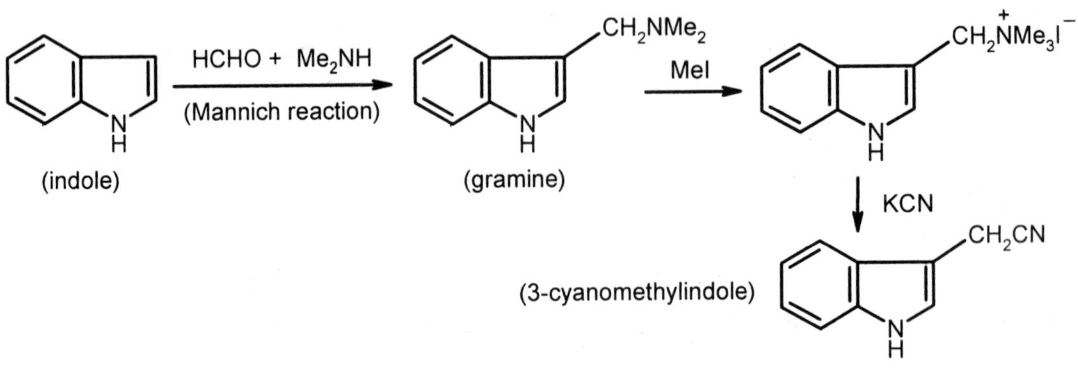

Problem 8.3.6

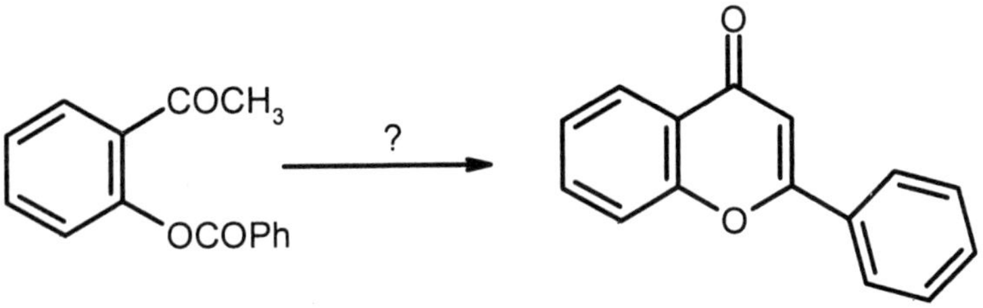

Solution.

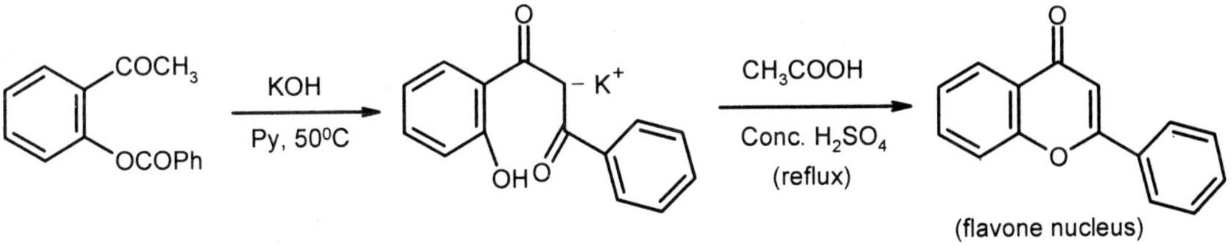

Problem 8.3.7

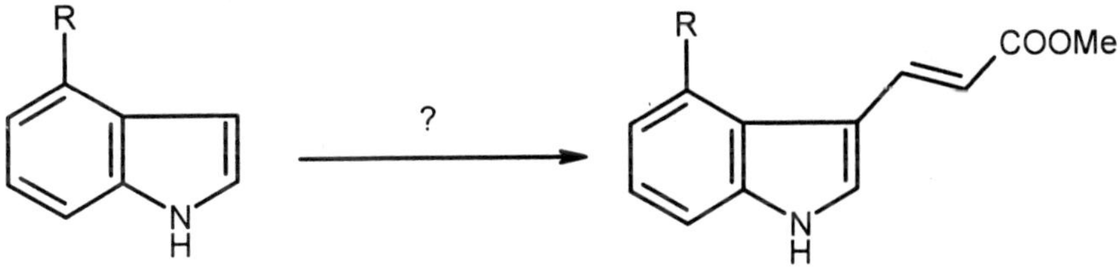

Solution.

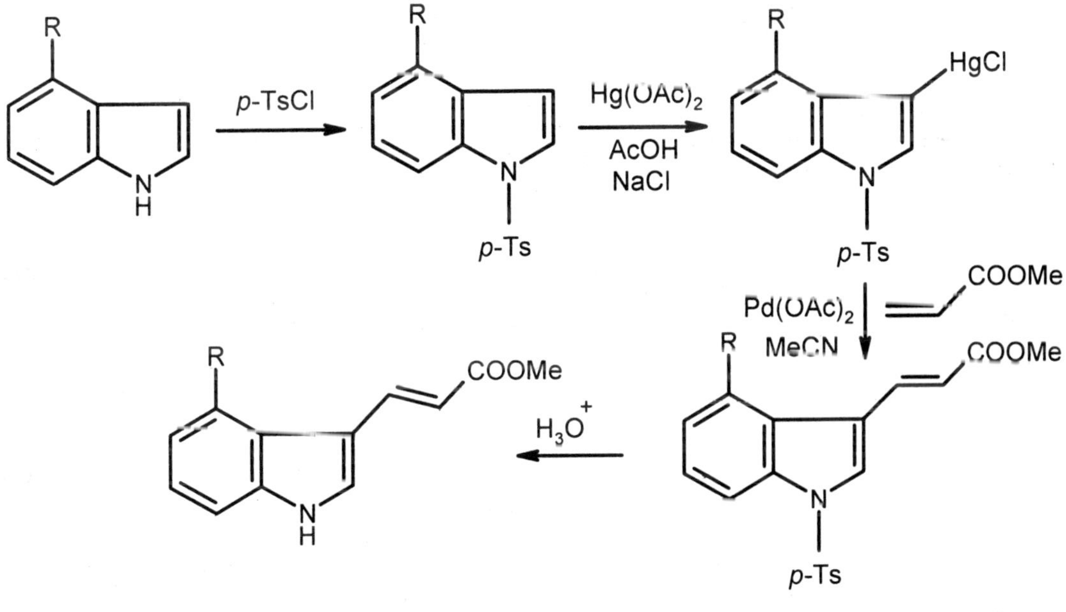

Problem 8.3.8

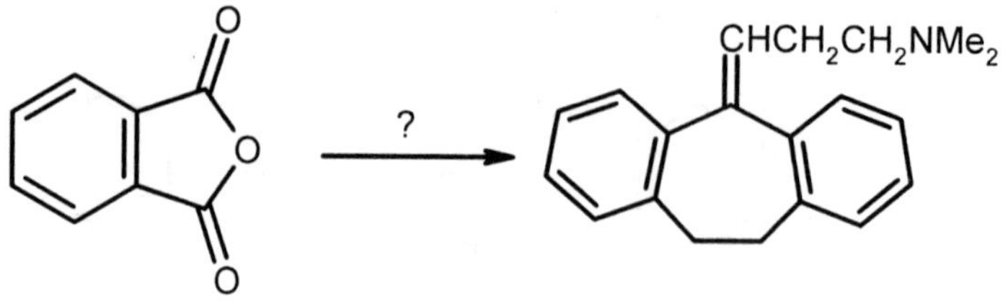

Solution.

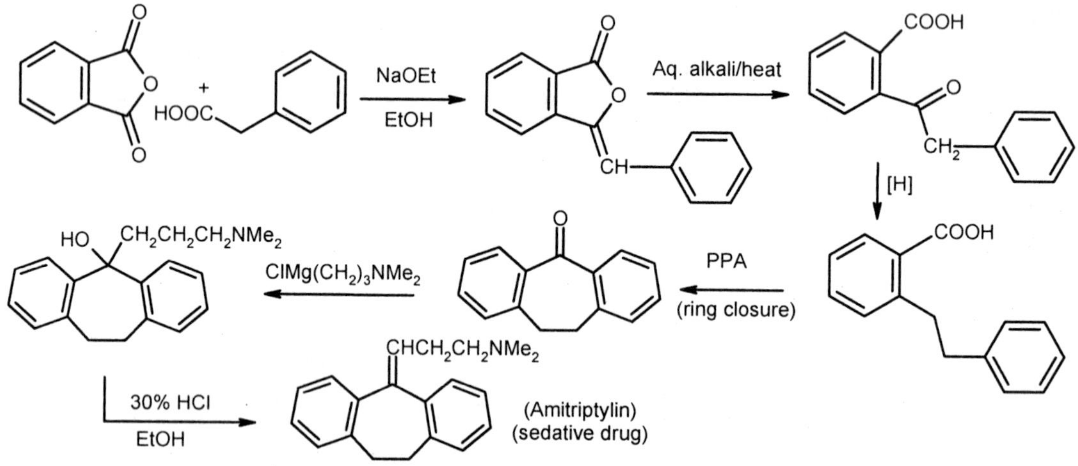

Problem 8.3.9

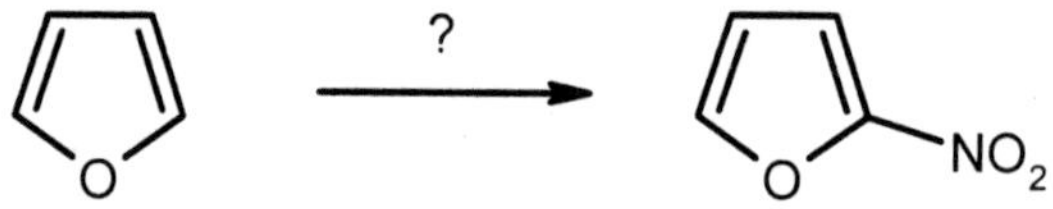

Solution.

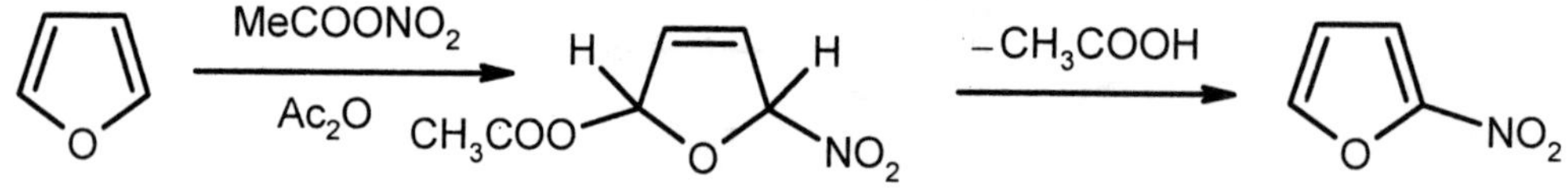

Problem 8.3.10

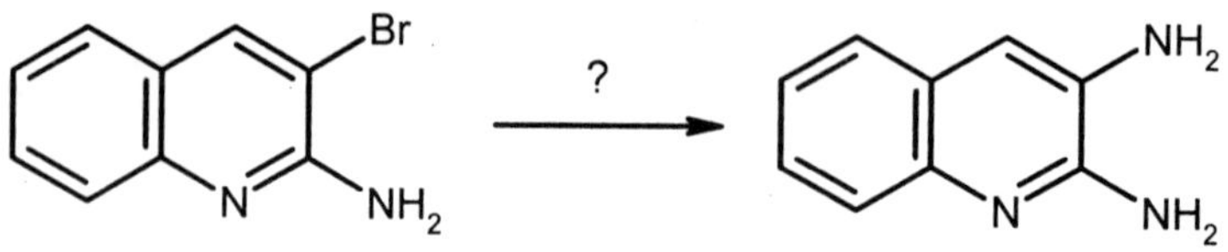

Solution.

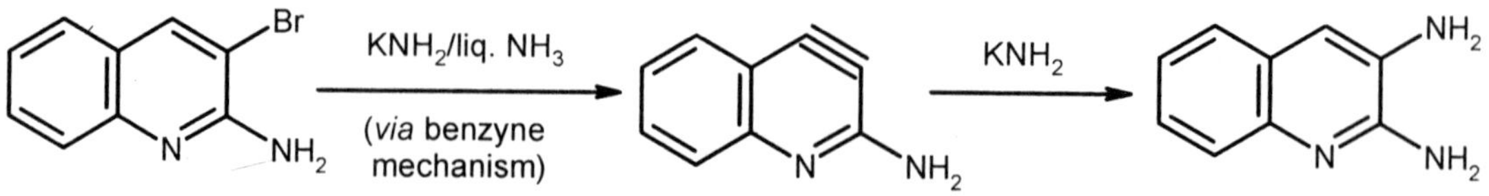

Problem 8.4 Complete the reaction sequence in each of the following cases:

Problem 8.4.1

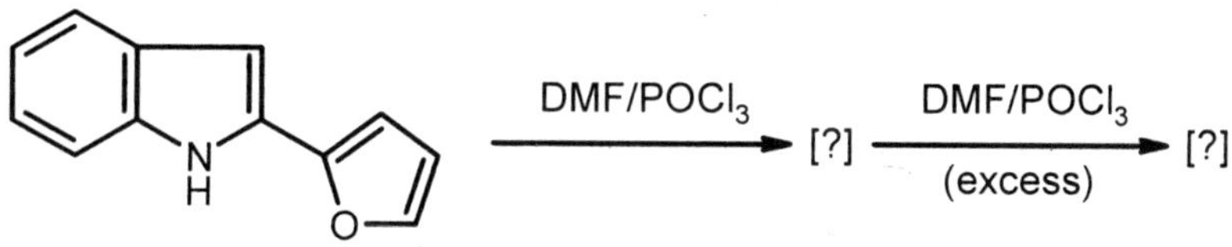

Solution.

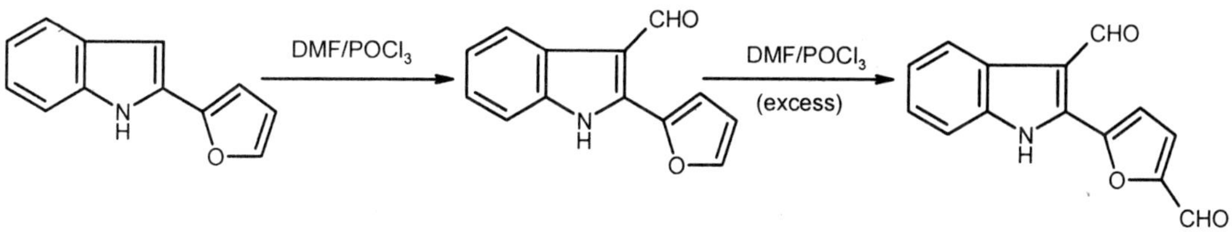

Problem 8.4.2

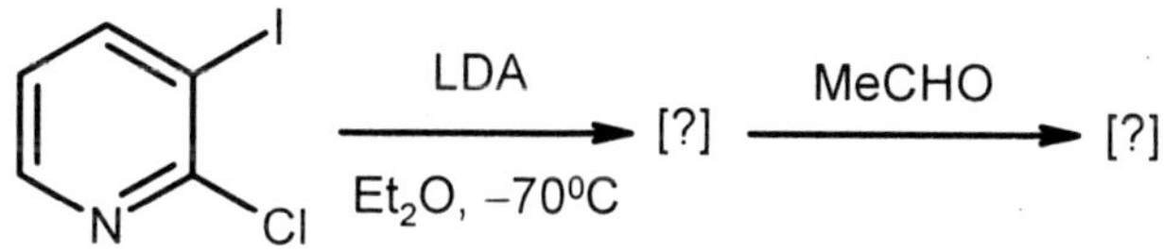

Solution.

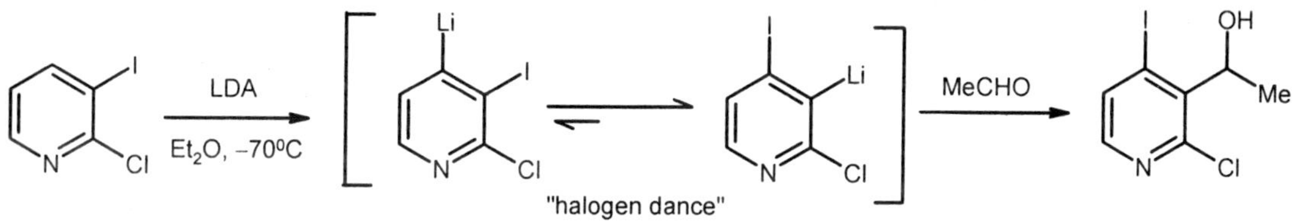

Problem 8.4.3

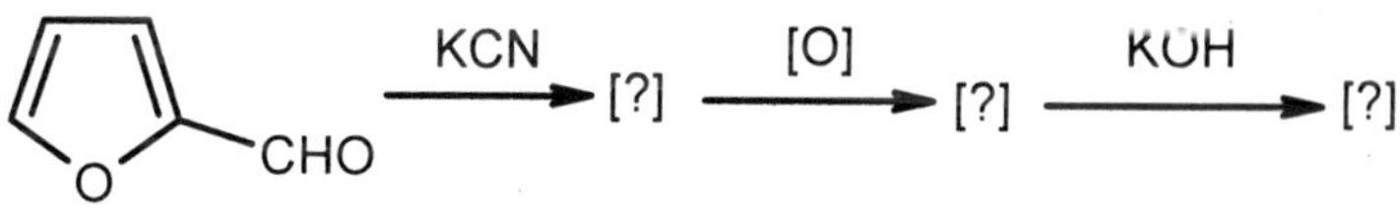

Solution.

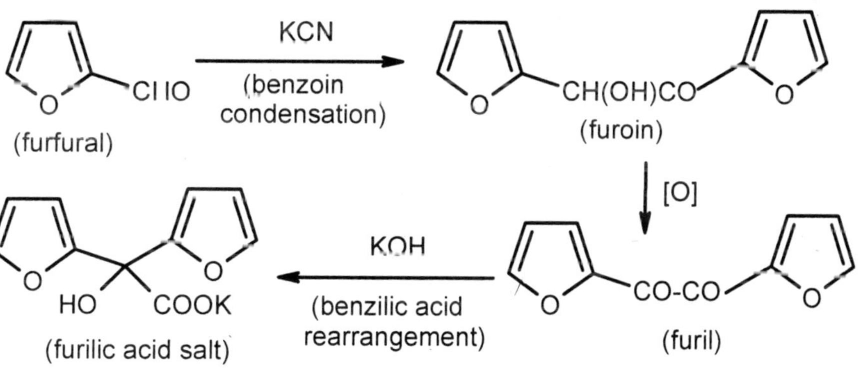

Problem 8.4.4

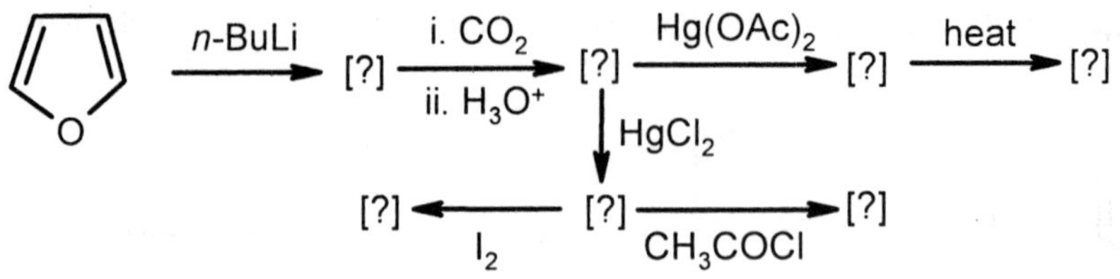

Solution.

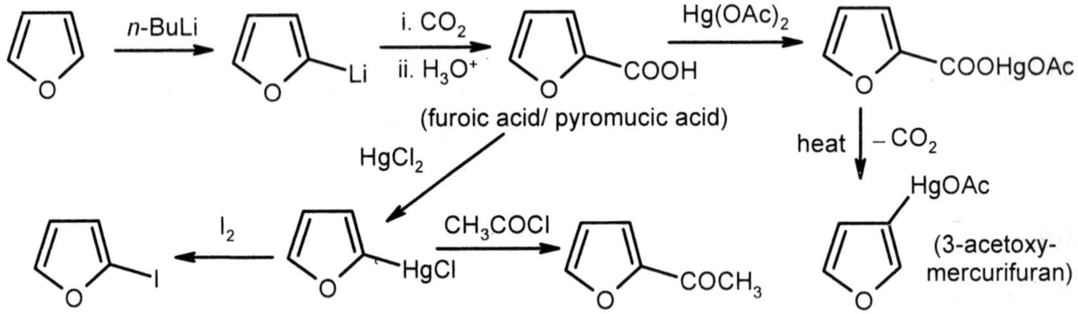

Problem 8.4.5

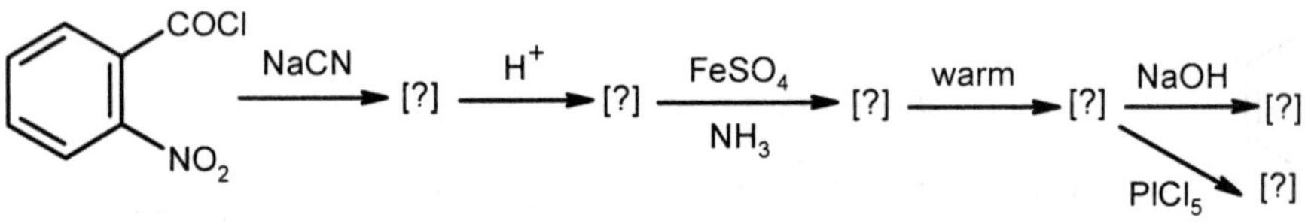

Solution.

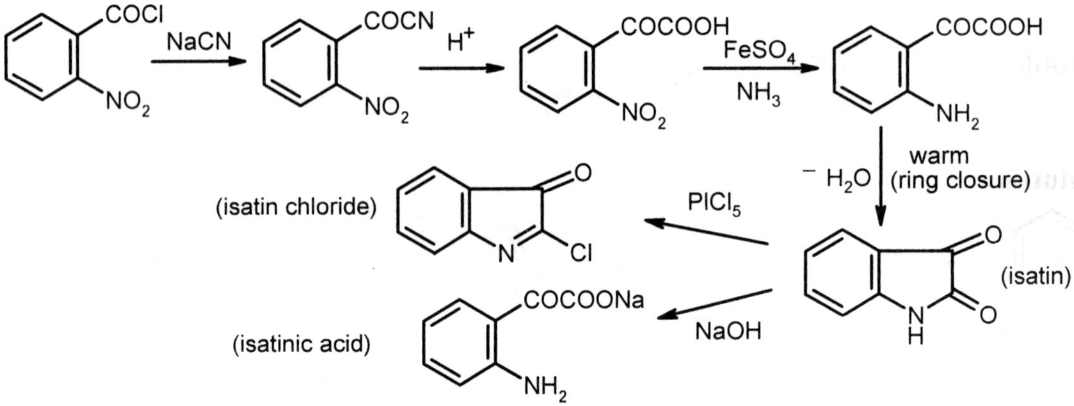

Problem 8.4.6

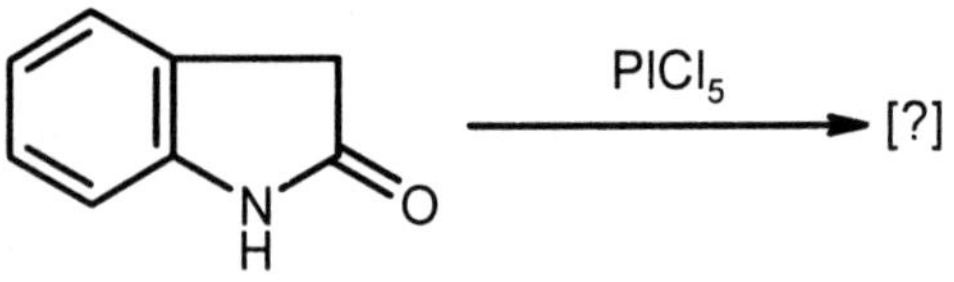

Solution.

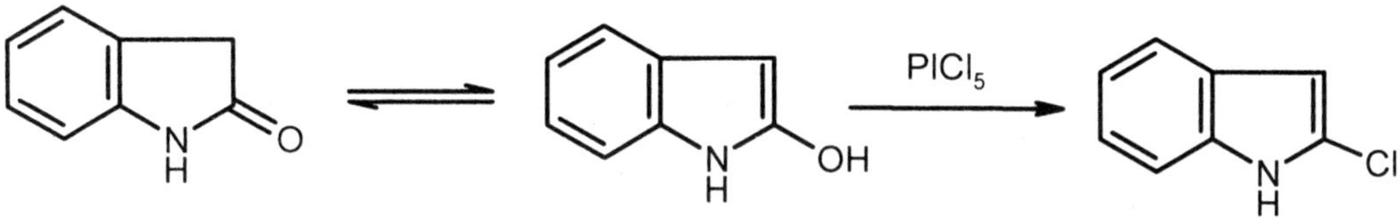

Problem 8.4.7

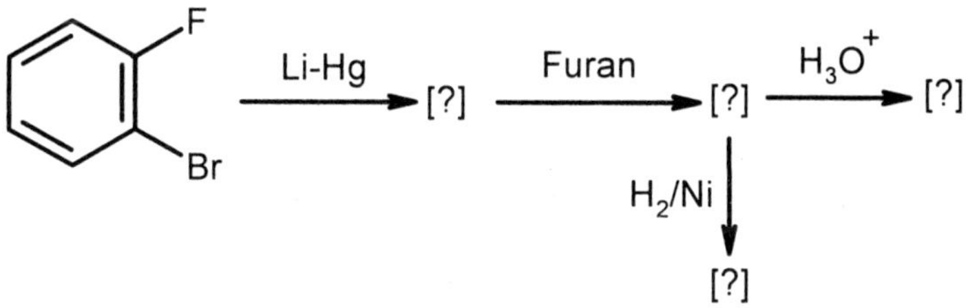

Solution.

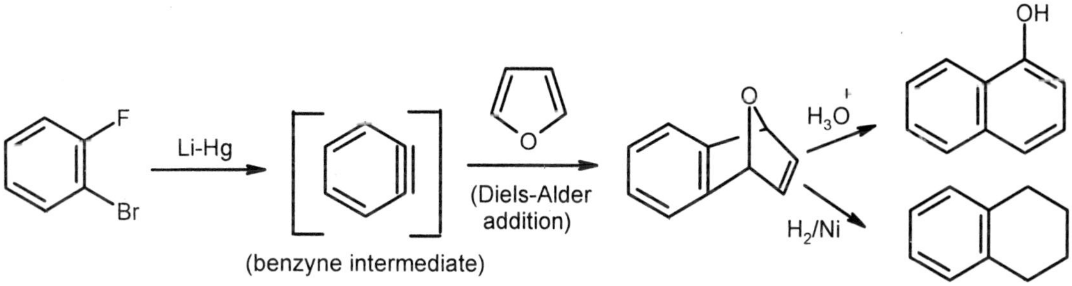

Problem 8.4.8

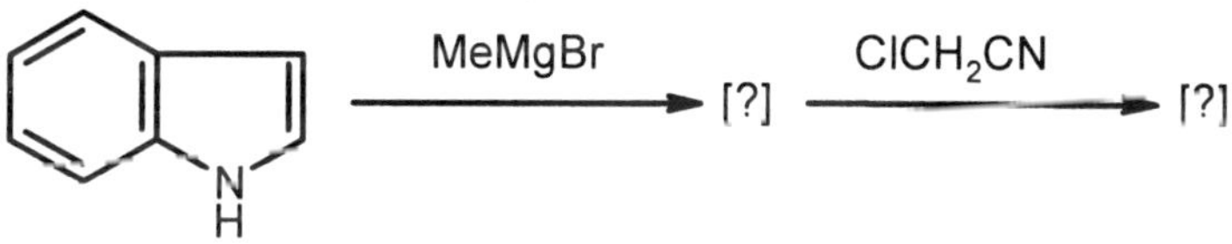

Solution.

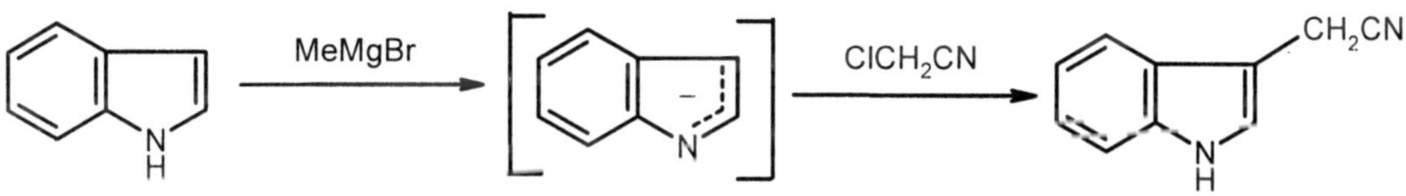

Problem 8.4.9

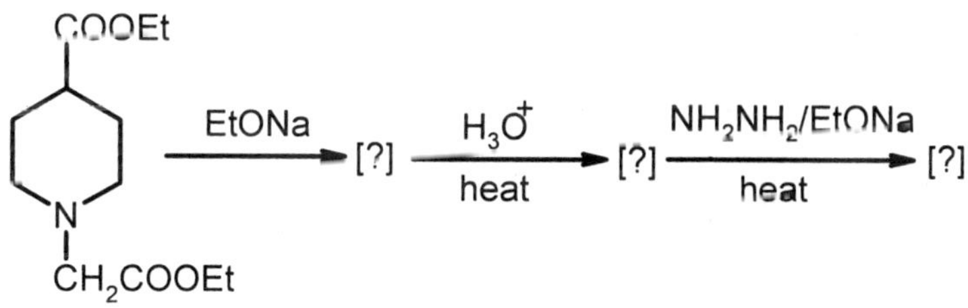

Solution.

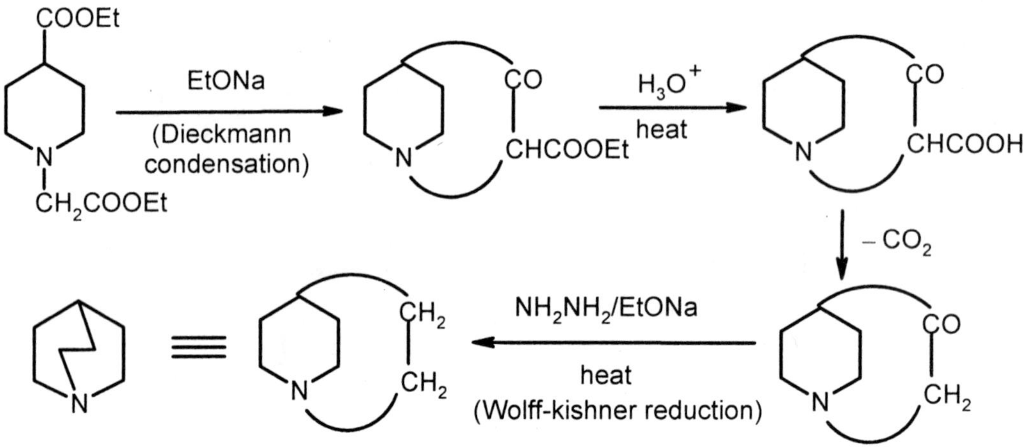

Problem 8.4.10

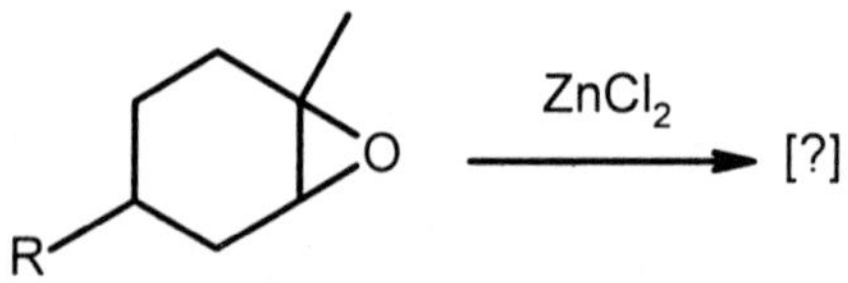

Solution.

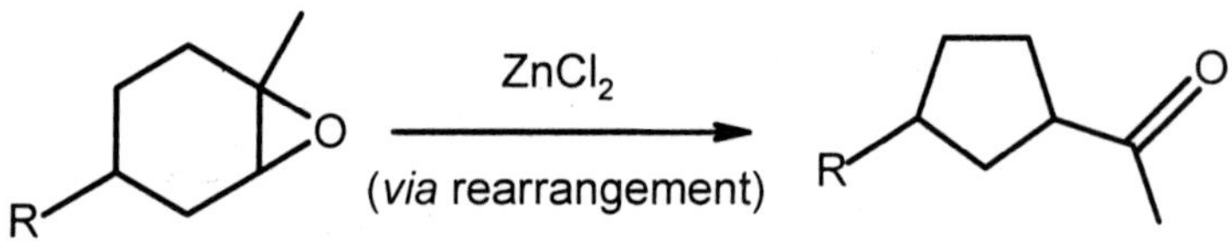

Problem 8.4.11

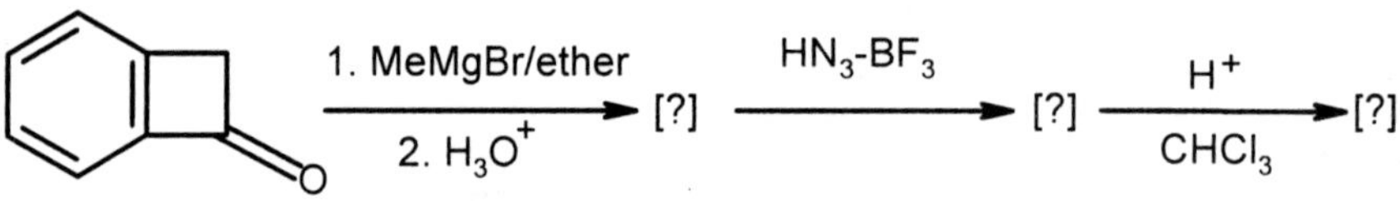

Solution.

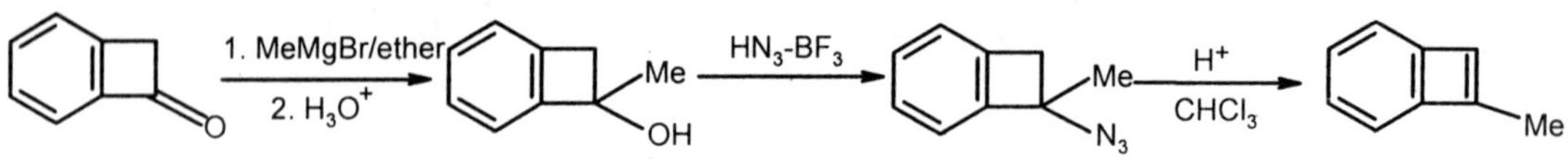

Problem 8.4.12

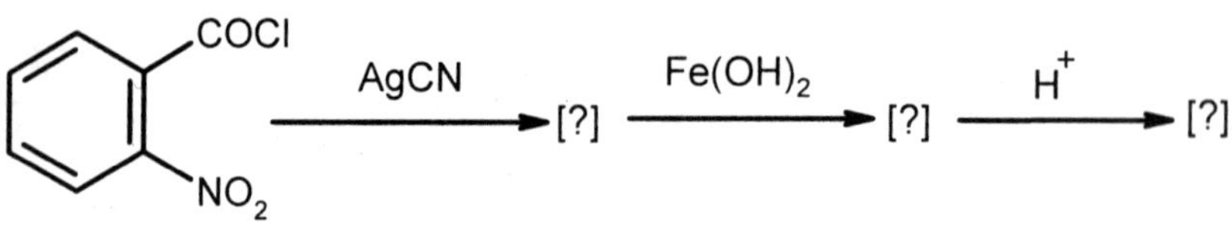

Solution.

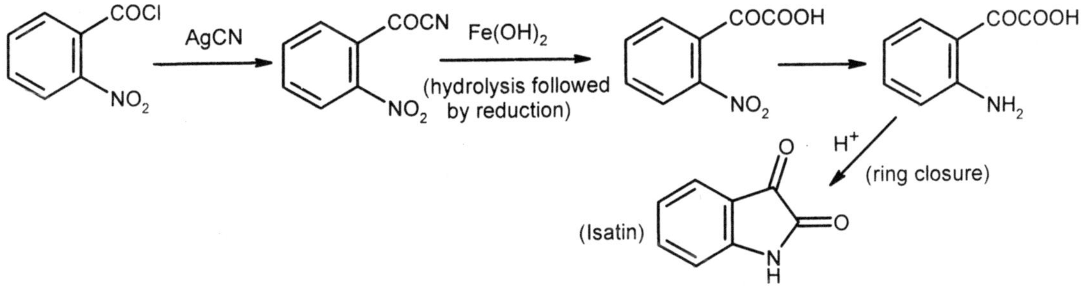

Problem 8.4.13

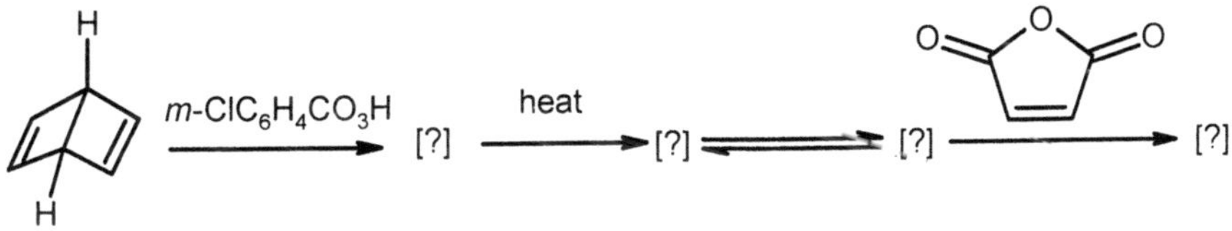

Solution.

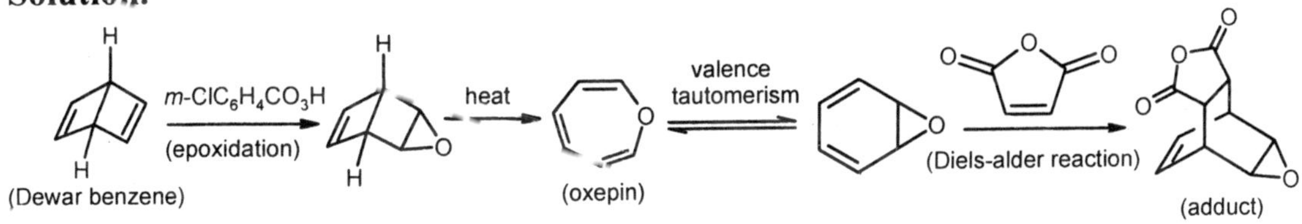

Problem 8.4.14

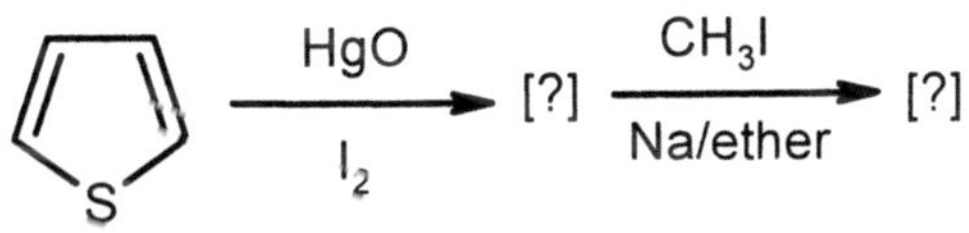

Solution.

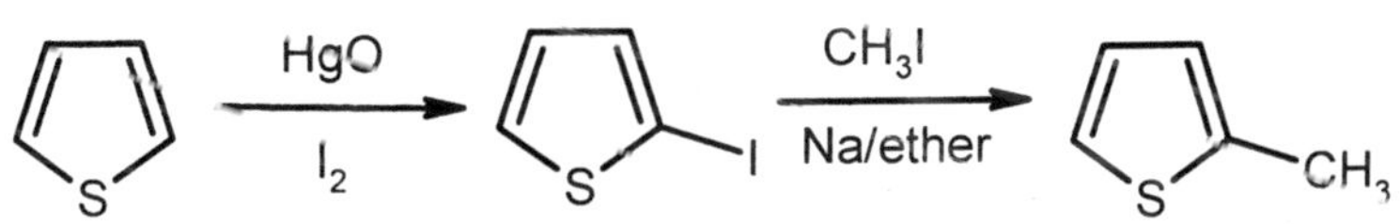

Problem 8.4.15

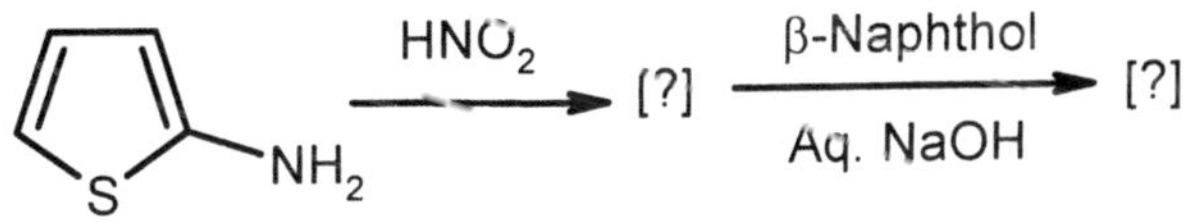

Solution.

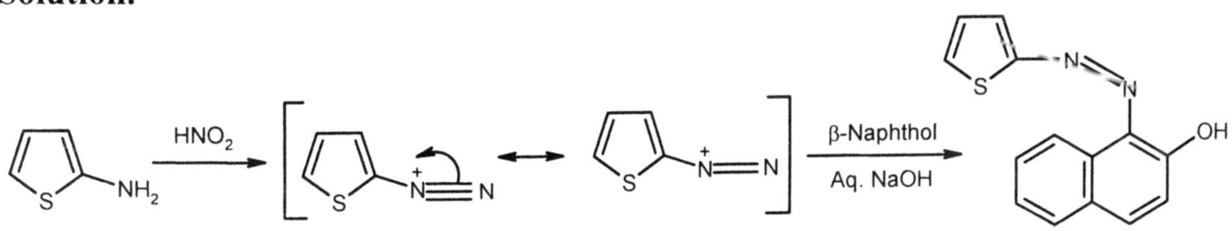

Problem 8.5 How would you introduce an amino group *ortho* to a nitro group in a heterocyclic nucleus?

Solution. Vacarious nucleophilic substitution (VNS) enables one to introduce an amino group *ortho* to a nitro group in a heterocyclic nucleus by the reaction with methoxyamine as illustrated below:

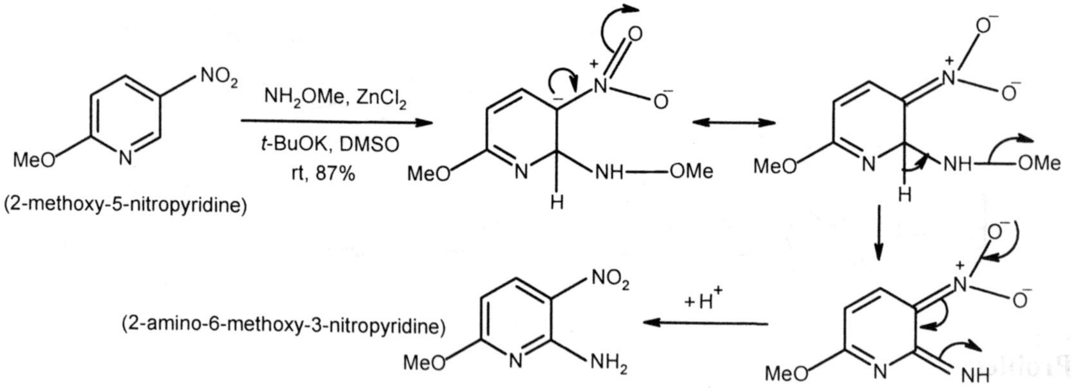

Problem 8.6 Suggest a mechanism for each of the following reactions:

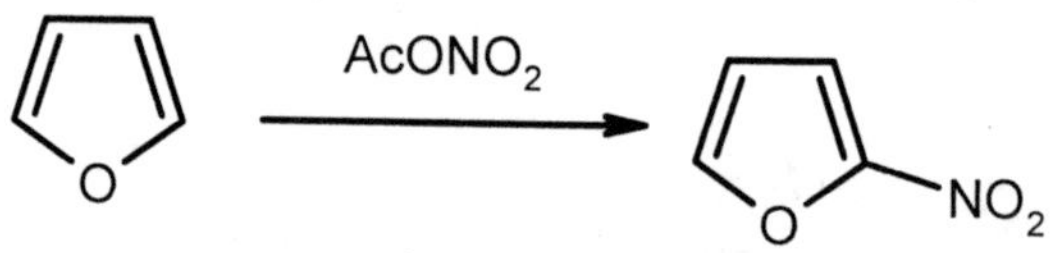

Solution.

$$AcONO_2 \rightleftharpoons Ac\bar{O} + \overset{+}{N}O_2$$

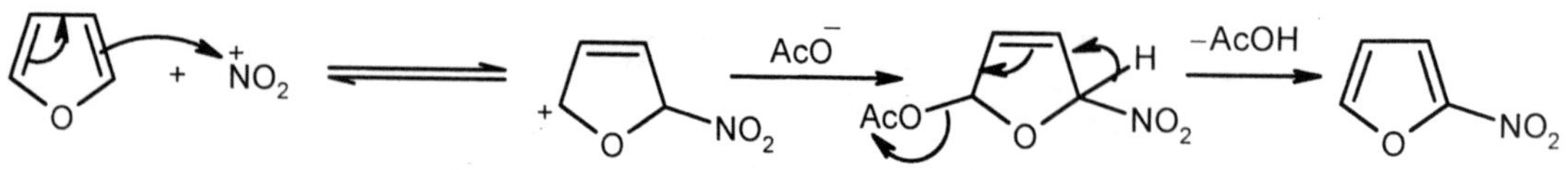

(2,5-adduct)

[nitration of furan with acetylnitrate, thus, proceeds *via* a 2,5-addition intermediate with the reagent]

Problem 8.7 Suggest mechanism for the following transformation:

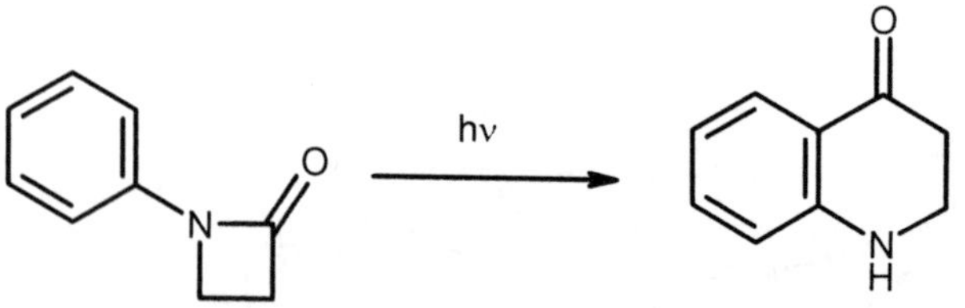

Solution.

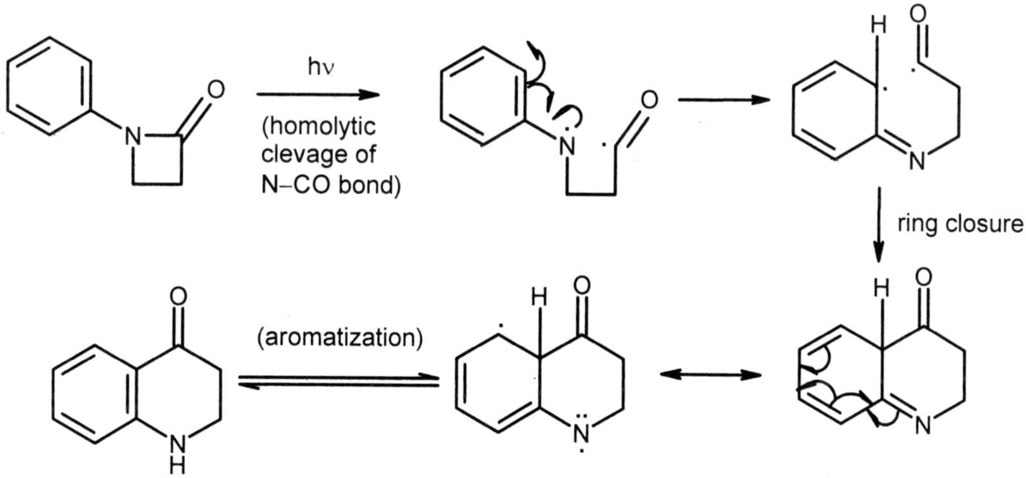

Problem 8.8 Suggest mechanism for the following transformation:

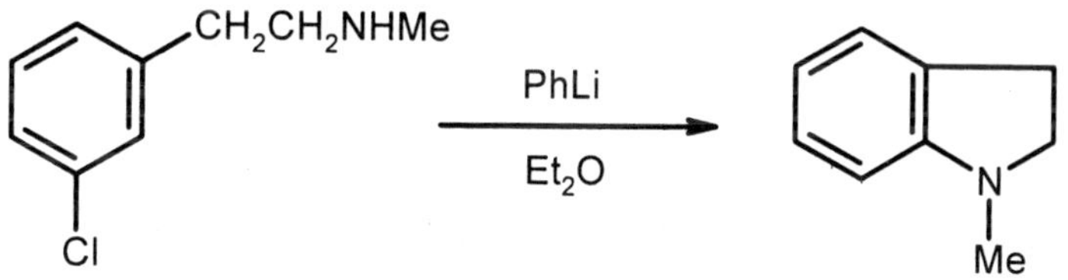

Solution.

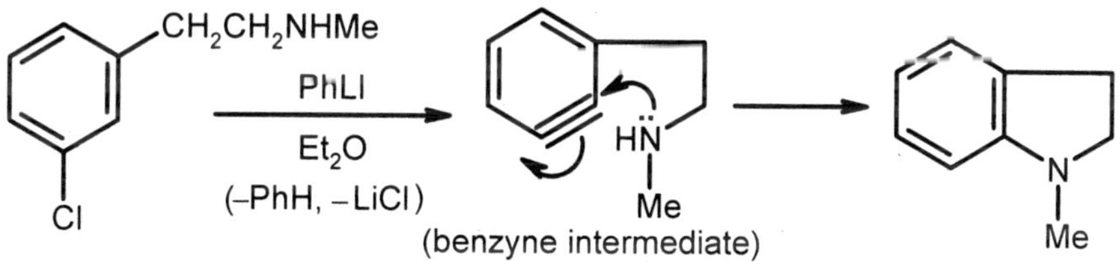

Problem 8.9 Cyclohexene epoxide on reaction with a Grignard reagent yields a rearranged cyclopentanol derivative in addition to usual cyclohexyl product:

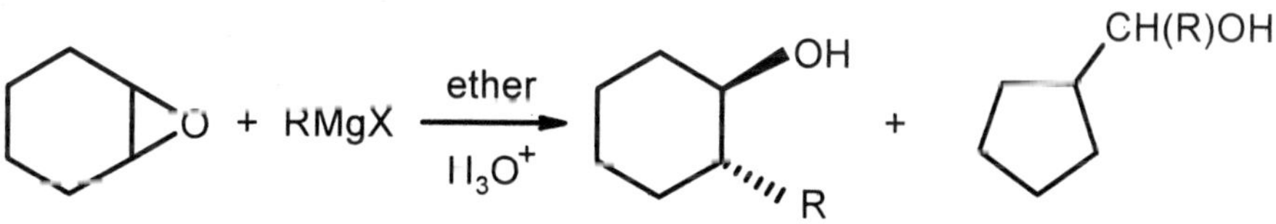

Suggest a plausible mechanism for the reaction.

Solution.

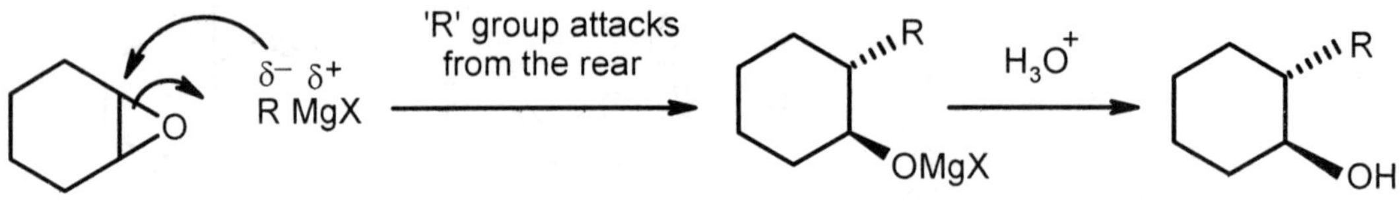

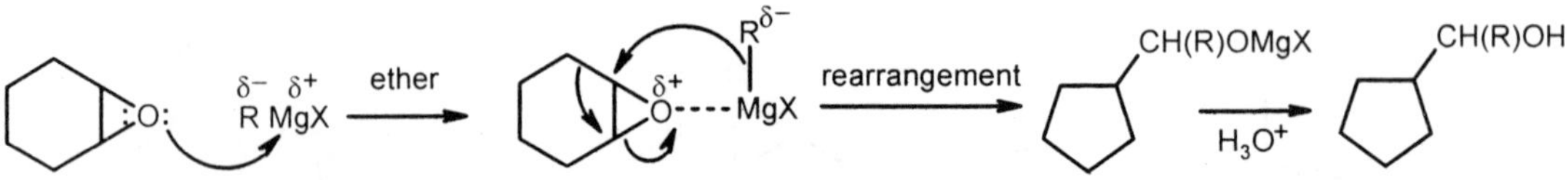

Problem 8.10 Furan on treatment with bromine in dioxane gives chiefly the 2-bromo substituted product, while on treatment with bromine in methanol yields mainly 2,5-dihydro-2,5-dimethoxyfuran. Give the reaction course with plausible mechanism.

Solution. In polar aprotic solvent like dioxane, furan undergoes simply the electrophilic substitution reaction with bromine yielding 2-bromofuran.

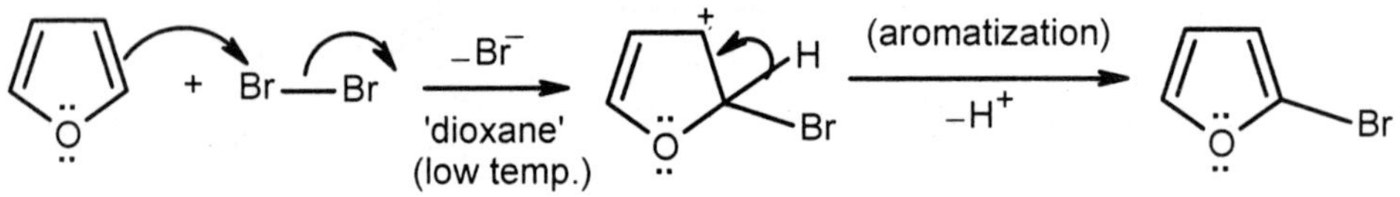

On the other hand, in presence of methanol (a protic polar solvent) furan undergoes addition-elimination mechanism giving ultimately a 1,4-adduct at low temperature.

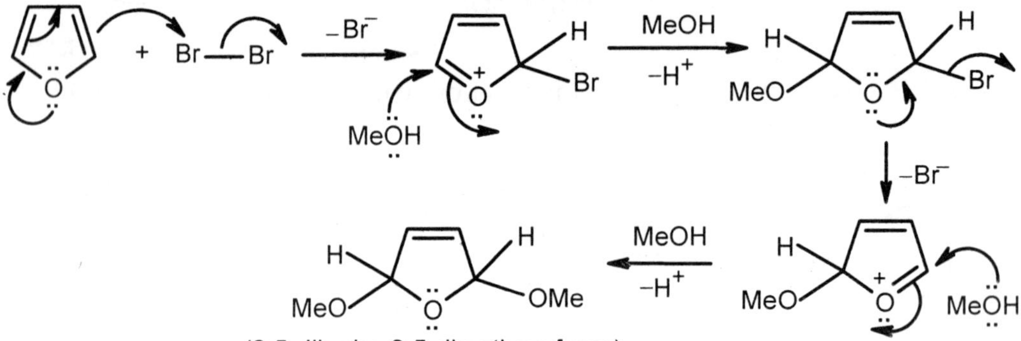

(2,5-dihydro-2,5-dimethoxyfuran)

Problem 8.11 Furan undergoes polymerization in conc. H_2SO_4, but undergoes hydrolytic cleavage to succinaldehyde in dilute H_2SO_4. Justify the statement with proper mechanistic scheme.

Solution. In acidic condition furan first forms first an electrophilic species (I) that is subsequently transformed into (II):

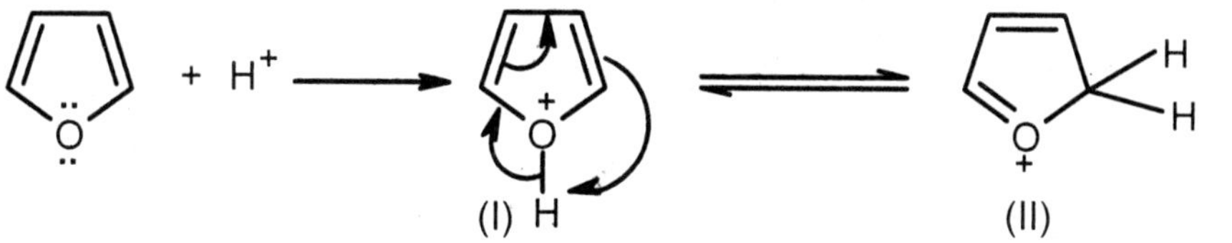

In conc. H_2SO_4 medium, the species (II) undergoes polymerization with another furan molecules as shown below:

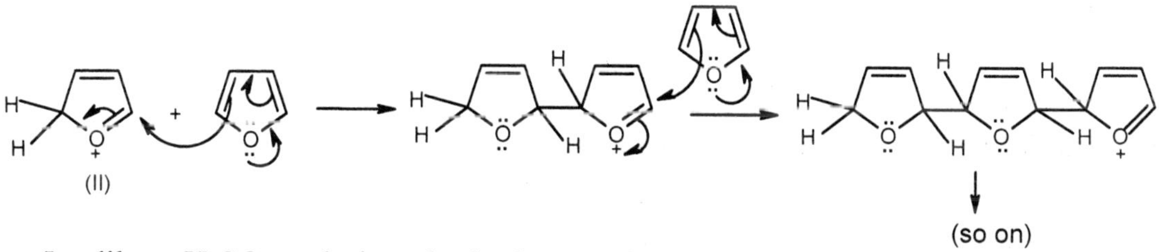

In dilute H_2SO_4 solution, both the species (I) and (II) are attacked by water molecules at methylene carbon to form succinaldehyde:

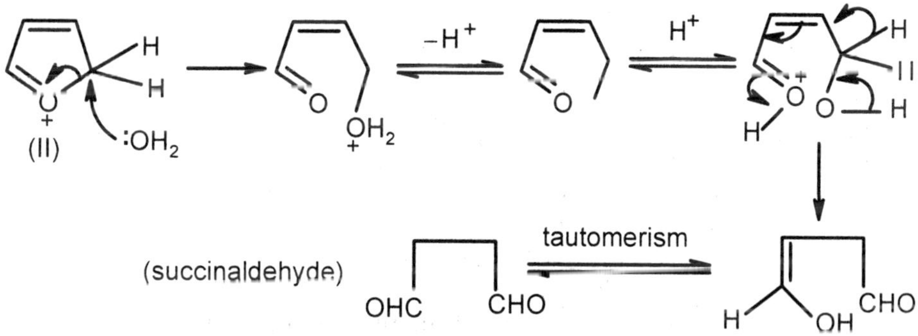

Similarly,

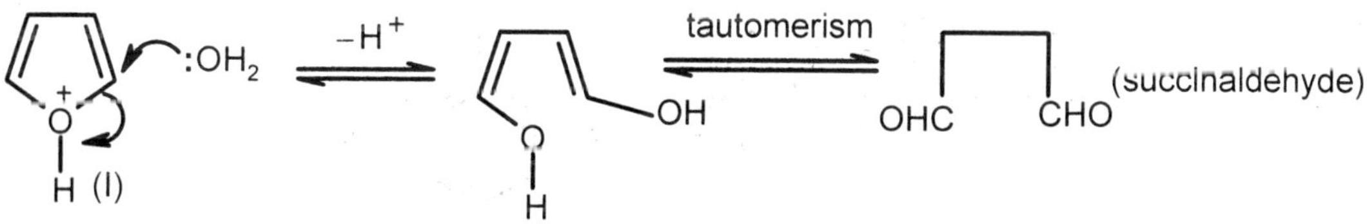

Problem 8.12 How do you prove that quinoline bears benzene nucleus in its molecule?

Solution. From the following reaction sequence it can be proved that benzene nucleus is a part of quinoline molecule:

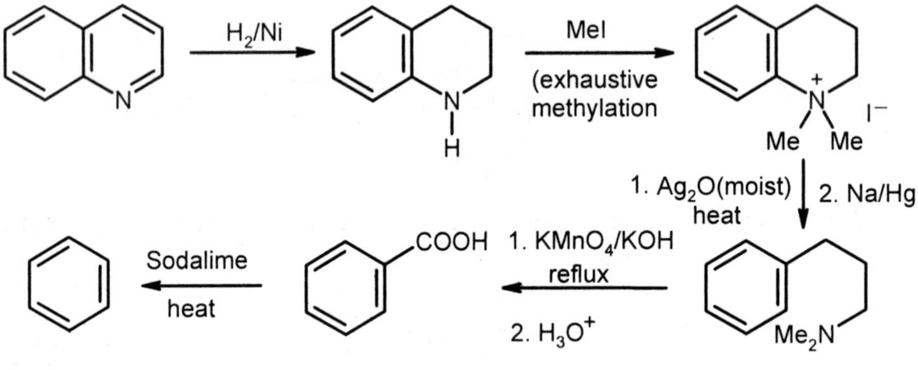

Problem 8.13 Suggest a plausible mechanistic scheme for the following conversion:

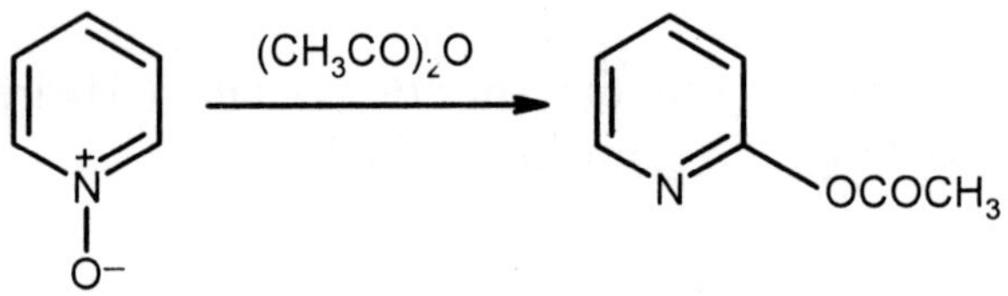

Solution.

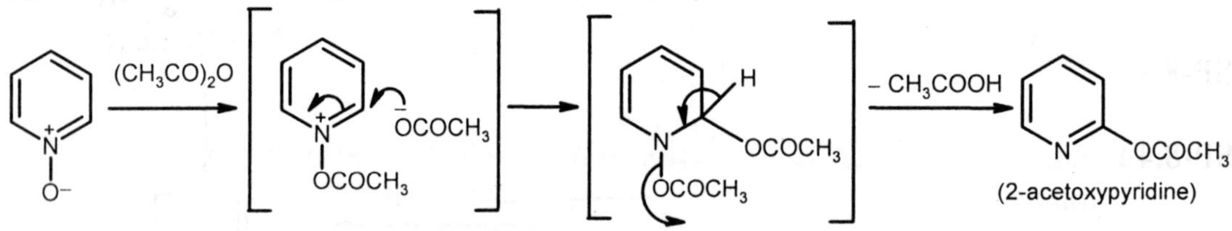

Problem 8.14 Account for the formation of the product in the following reaction:

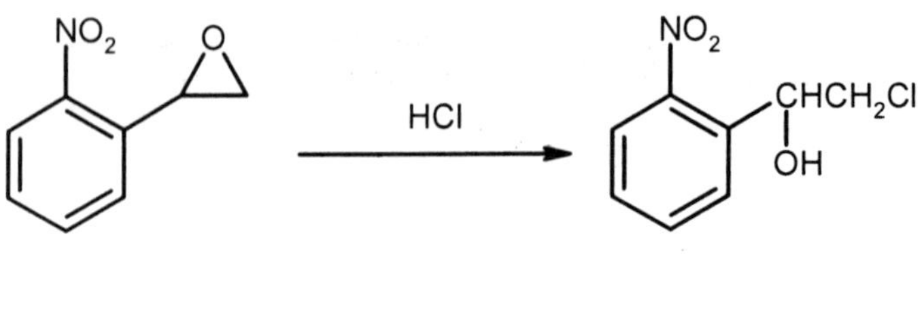

Solution. In the first step there occurs protonation on the epoxy oxygen atom; in the second step ring opening takes place. Direction of ring-opening is markedly influenced by the electronic nature of substituent on oxirane ring — in this case the presence of strong electron-withdrawing *ortho*-nitrophenyl moiety would destabilize the incipient benzyl carbocation, thus, making the ring-opening from the non-substituted end more preferable.

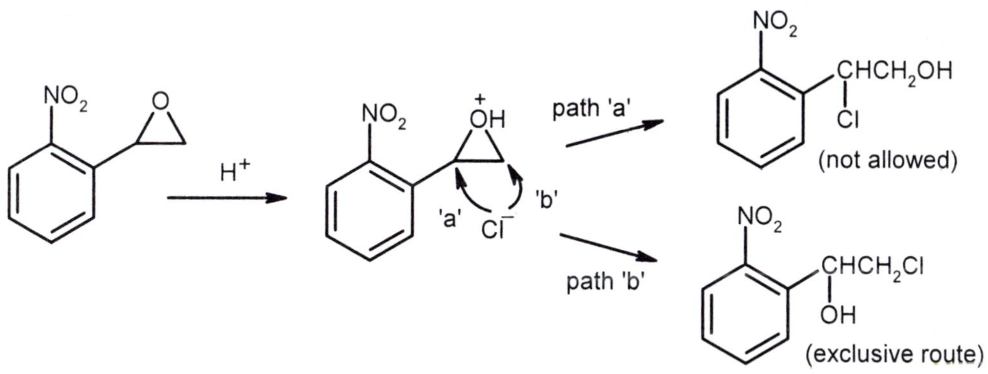

Supplementary problems

SP-8.1 Pyrrole has a higher boiling point than furan and also than thiophene. Comment.

SP-8.2 Pyrrole 2-aldehyde is reluctant to undergo Perkin condensation while furfural does. Explain.

SP-8.3 2-Nitrofuran is less reactive than furan itself towards acids. Explain.

SP-8.4 Electrophilic substitution in 4-phenylpyridine takes place in the benzene ring Explain.

SP-8.5 Imidazole undergoes electrophilic substitution reaction under only drastic conditions. Comment.

SP-8.6 Electrophilic attack on pyrazole is hindered in acidic medium. Offer explanation.

SP-8.7 Nitration of furan with acetyl nitrate proceeds *via* a 2,5-addition intermediate with the reagent. Show the reaction scheme.

SP-8.8 Complete the reaction sequence:

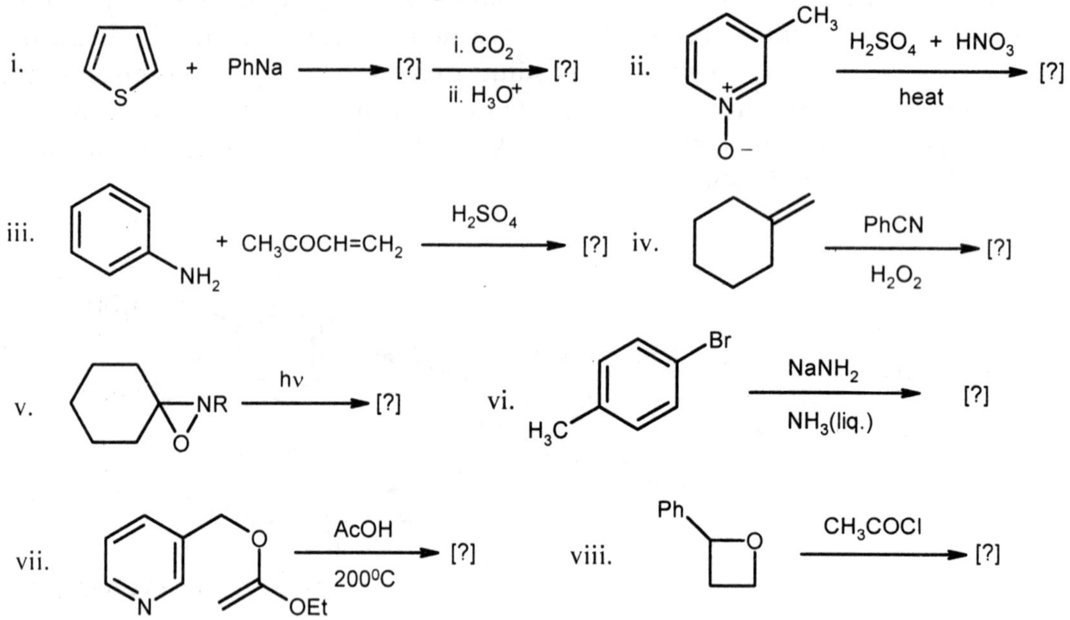

SP-8.9 Imidazole is a stronger acid than pyrrole and a stronger base than pyridine. Explain.

SP-8.10 Acraldehyde itself is not used in the Skraup synthesis of quinoline. Comment.

SP-8.11 Nitration of pyridine *N*-oxide occurs at 4-position while sulphonation at 3-position though both the reactions are carried out in the presence of sulphuric acid. Justify the phenomenon.

SP-8.12 Pyridine does not polymerize in the same fashion as pyrrole does in the presence of strong acid. Explain.

SP-8.13 A hydroxy derivative of pyrimidine is less water-soluble than pyrimidine itself. Offer explanation.

SP-8.14 Compare the basicity of 7-aminoquinoline with that of 8-aminoquinoline.

SP-8.15 Pyrimidine is less basic than pyridine. Comment.

CHAPTER 9

CARBOHYDRATES

Introduction

Carbohydrates are found everywhere in nature. They are essential components of living organisms, and are the key substances to life. Green plants synthesize carbohydrates by means of photosynthesis; such chemical constituents behave alike chemical intermediaries to store solar energy that is used later by living organisms to run their lives — carbohydrates on breaking down through metabolic processes in cells supply the major source of energy required by the living organisms.

Carbohydrates may be defined as polyhydroxy aldehydes or ketones or those compounds that on hydrolysis produce substances having such functionalities; this definition is not fully satisfactory, but closer to structural reality and more suggestive of chemical reactivity of carbohydrates.

Simple carbohydrates are also known as saccharides (Latin *saccharum*, Greek *sakcharon*) or sugars (Sanskrit: sweet sand; '*su*' means sweet, '*gar*' means sand). The names of most of the sugars end with the term '***-ose***'; as for examples: glucose (also known as dextrose), a principal sugar in blood; sucrose, the table sugar; lactose, the milk sugar; fructose, a sugar in fruits and honey.

Carbohydrate chemistry is very rich, and is of high importance to a wide range of scientists including chemistry and biology. In this present discussion an attempt has been taken to introduce the beginners with the essence of carbohydrate chemistry to boost up their interest for further reading.

Solved Problems

Problem 9.1 How do you classify carbohydrates?

Solution. Carbohydrates are classified into four major categories: monosaccharides, disaccharides, oligosaccharides, and polysaccharides.

(i) Monosaccharides: Monosaccharides are simple sugars that cannot be hydrolyzed to simpler molecules. Depending upon the number of carbon atoms in continuous chain, they are termed as *triose* (three-carbons), *tetrose* (four-carbons), *pentose* (five-carbons), *hexose* (six-carbons), and so on. Further, saccharides containing an aldehyde group (–CHO) are called *aldoses*, while those with a keto group (>CO) are referred to as *ketoses*. For examples: glucose (an aldose), fructose (a ketose).

(ii) Disaccharides: Disaccharides are sugars that yield two same or different monosaccharides on hydrolysis. Examples are: *maltose* (composed of two glucose units), *sucrose* (composed of one glucose and one fructose units), *lactose* (composed of one glucose and one galactose units).

(iii) Oligosaccharides: These carbohydrates are formed as a result of joining three to eight monosaccharide units. A few examples are: *maltotriose* (on hydrolysis it gives three molecules of glucose), *raffinose* (it is also a trisaccharide that on hydrolysis yields three different molecules of monosaccharides, namely, glucose, galactose, and fructose).

(iv) Polysaccharides: Polysaccharides are carbohydrates that yield a large number of monosaccharides (ranging from more than eight to over one thousand) on hydrolysis. The most widely spread polysaccharides have the general formula $(C_6H_{10}O_5)_n$. These are the naturally occurring polymers (known as *biopolymers*) of carbohydrates. The most common examples are *starch* and *cellulose*, both of which are the biopolymers of glucose.

$$\text{Starch} \xrightarrow[\text{(hydrolytic cleavage)}]{H_3\overset{+}{O}\text{/heat}} > 1000 \text{ glucose units}$$

$$\text{Cellulose} \xrightarrow[\text{(hydrolytic cleavage)}]{H_3\overset{+}{O}\text{/heat}} > 1000 \text{ glucose units}$$

Problem 9.2 What do mean to say *epimers* and *anomers*?

Solution. Stereoisomers, which differ in configuration at only one centre among a number of asymmetric centers present in the molecules, are called *epimers*. For examples:

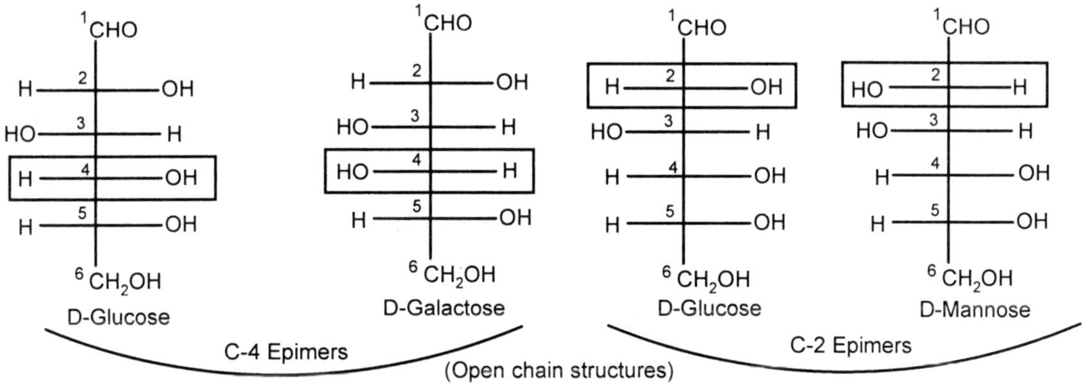

In carbohydrates chemistry, C-1 epimers are known as *anomers*, and the asymmetric carbon atoms are termed as *anomeric carbons*. Such type of epimerism is developed in cyclic hemiacetal structures, as in the case of D-glucose. In ring structure, D-glucose can exist in two different stereochemical forms, known as *anomers* — one is α-form and the other is β-form.

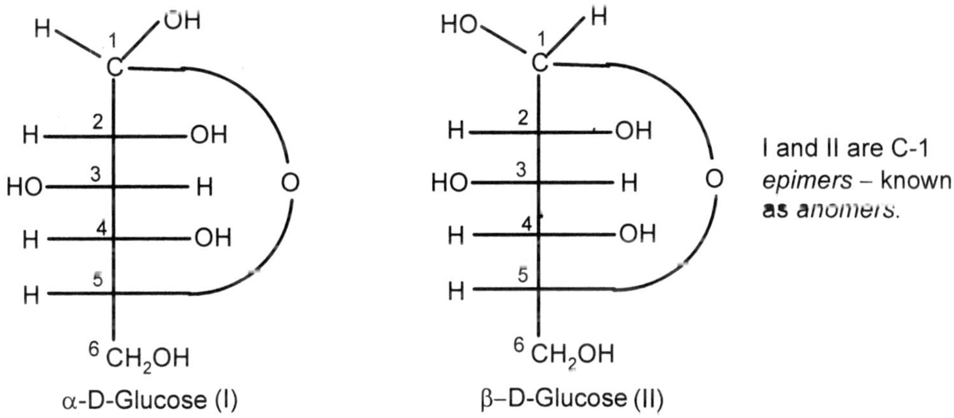

Problem 9.3 Draw projection formulas for D-glucose and D-fructose.

Solution.

D-Glucose

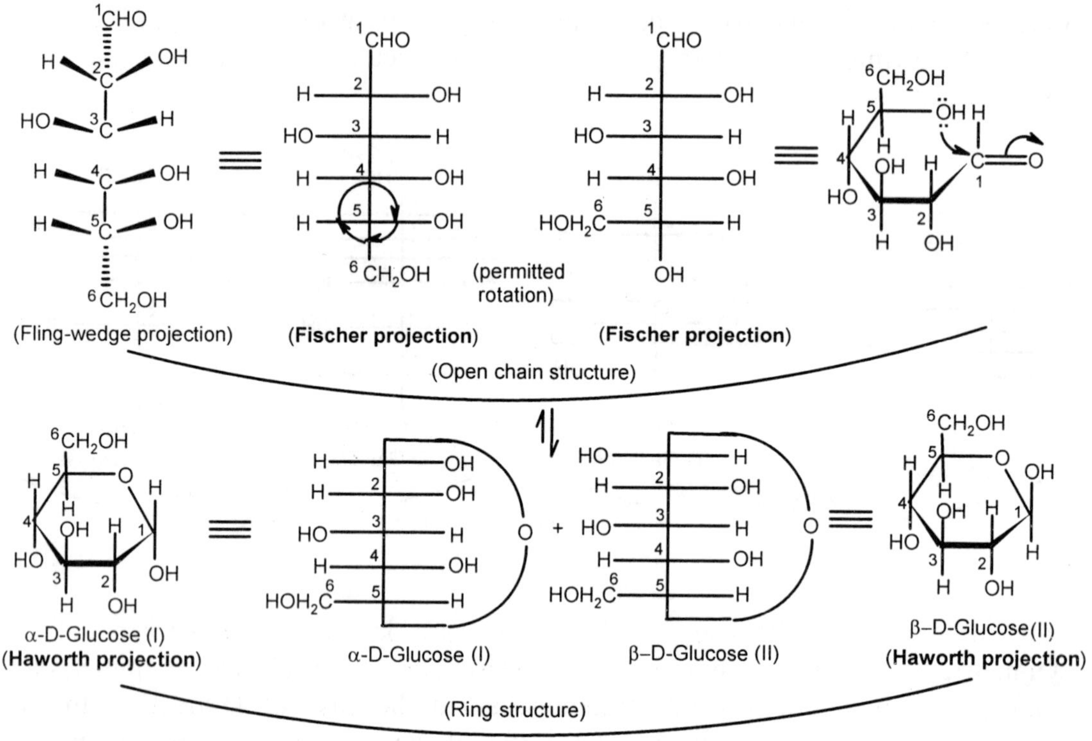

D-Fructose

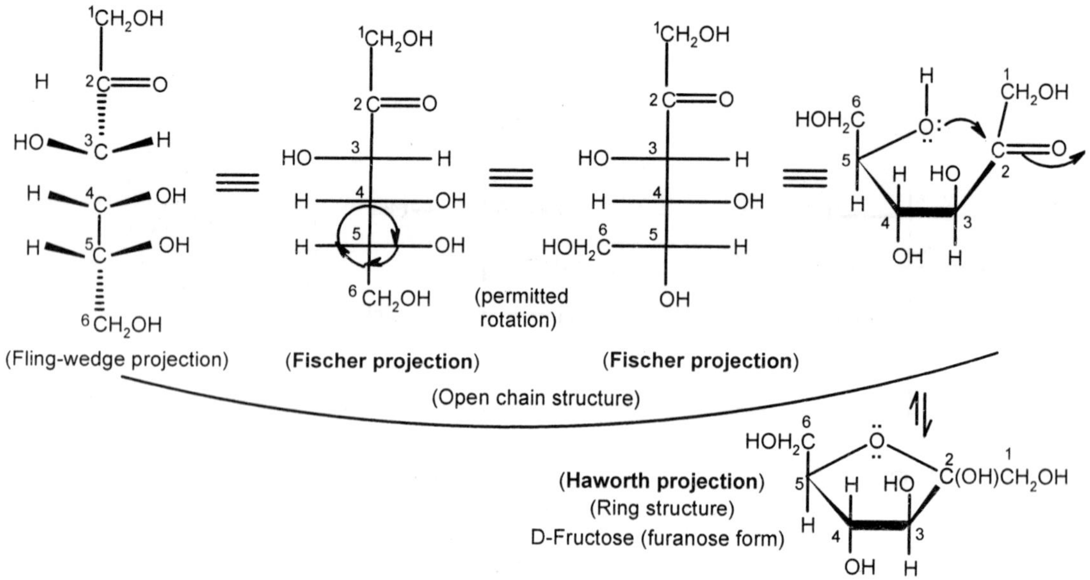

Problem 9.4 Show the conformation of D-glucose in pyranose form.

Solution. Like cyclohexanes, pyranose rings can also attain chair-like conformation with axial and equatorial substituents. Thus, the six-membered cyclic oxide structure of D-glucose is expected to exist preferentially in chair form; respective Haworth projections can easily be converted into chair representations as shown below:

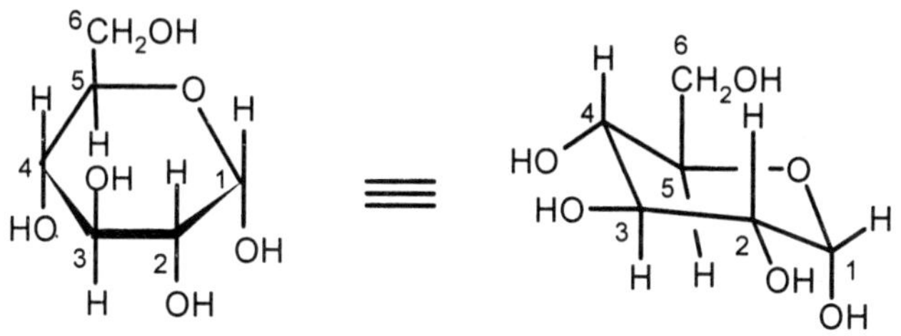

α-D-Glucose (**Haworth projection**) ≡ **Chair form** (I)

C_1–OH group is of axial orientation in α-D-glucose; all other substituents occupy equatorial positions.

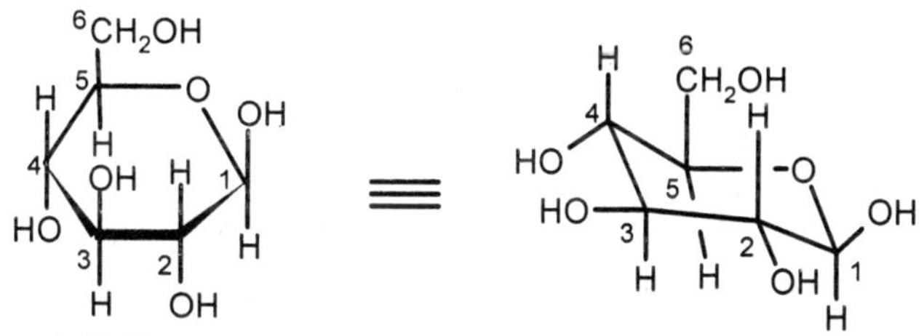

β-D-Glucose (**Haworth projection**) ≡ **Chair form** (II)

All the substituents in β-D-glucose including C_1–OH group take up equatorial positions.

From X-ray crystallographic studies it has been established that in crystalline state α-D-glucose exists in chair conformation (I), while the β-anomer remains in form (II). However, this conformational preference (chair form) is expected to be disturbed when these sugars go into solution because the oxide ring is relatively more susceptible to undergo interconversion between chair and boat forms than the simple cyclohexane ring does. But it has been observed that when the hydroxyl groups are acylated, as in α- and β-D-glucose-2,3,4,6-tetraacetates, the respective chair conformations are preferred having the bulky substituents at equatorial positions.

Problem 9.5 How do you suggest that D-glucose is an aldohexose?

Solution. D-Glucose (m.f. $C_6H_{12}O_6$) is a penta-hydroxyaldehyde, and can be written as:

$$OHC–(CHOH)_4–CH_2OH$$

The results of the following chemical reactions as carried out with D-glucose established that the molecule bears a six-carbon straight chain with five hydroxy groups and an aldehyde function. The chemical reactions are being cited below:

(i) Acetylation: On acetylation with acetic anhydride D-glucose givers a very stable pentaacetyl derivative suggesting thereby that it contains five hydroxyl functions on different carbons in its molecule.

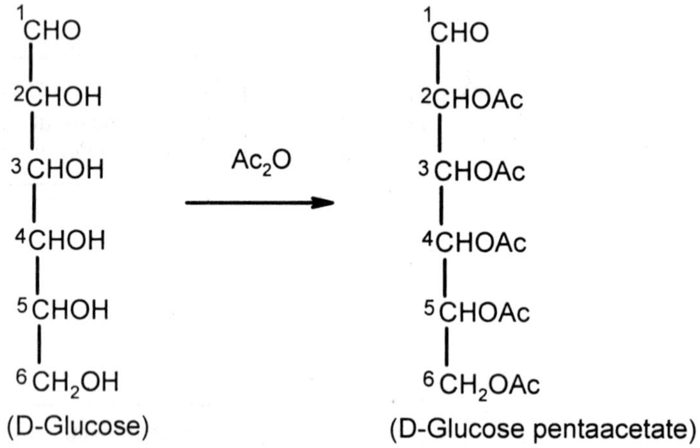

(ii) Reduction: D-Glucose on reduction with $NaBH_4$ forms D-sorbitol that results in the formation of *n*-hexane on treatment with hydroiodic acid in the presence of red phosphorus. Reduction may also be performed with $LiAlH_4$ or H_2/Ni. Thus, it can be suggested that the six carbons in glucose are in a straight chain.

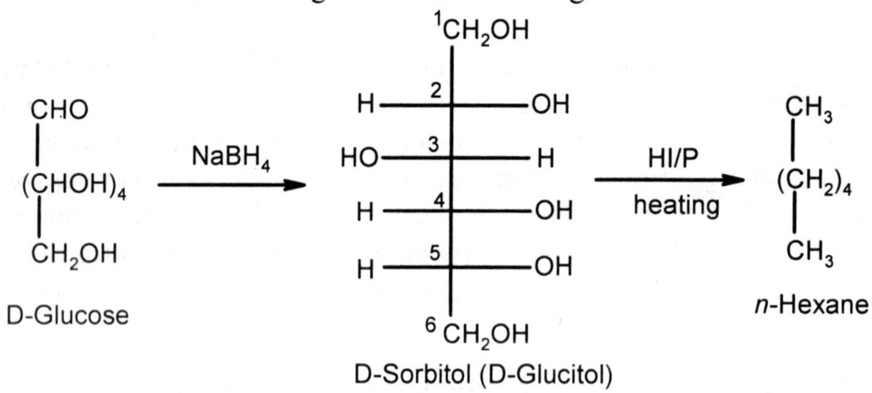

(iii) Reaction with carbonylic reagents: D-Glucose reacts with hydroxylamine to form the corresponding oxime derivative, and it also forms phenylhydrazone on reaction with one mole of the reagent; these chemical reactions are in accordance with the presence of a carbonyl group in the sugar molecule. The aldehydic nature of the carbonyl group as present can be ascertained from the positive responses of D-glucose toward Fehling's solution and Tollens' reagent.

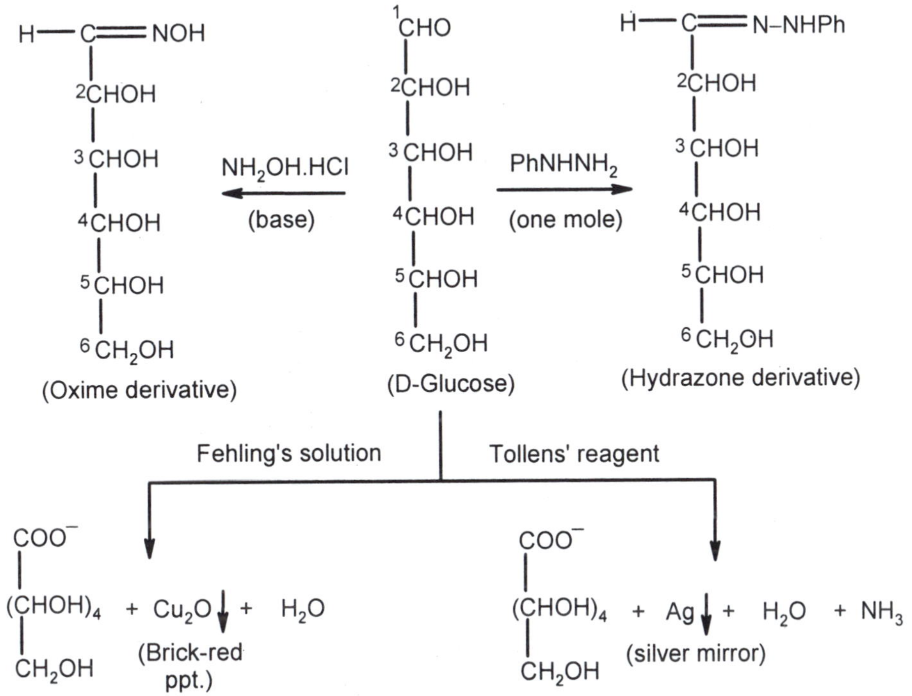

(iv) Reaction with HCN: D-Glucose reacts with HCN to form cyanohydrin that on hydrolysis followed by treating with HI/P results in the formation of *n*-heptanoic acid. Thus, the conversion of glucose (a six-carbon sugar) into a known straight chain heptanoic acid established that the carbonyl group to which HCN was added was at the end of a straight-chain of six-carbon atoms.

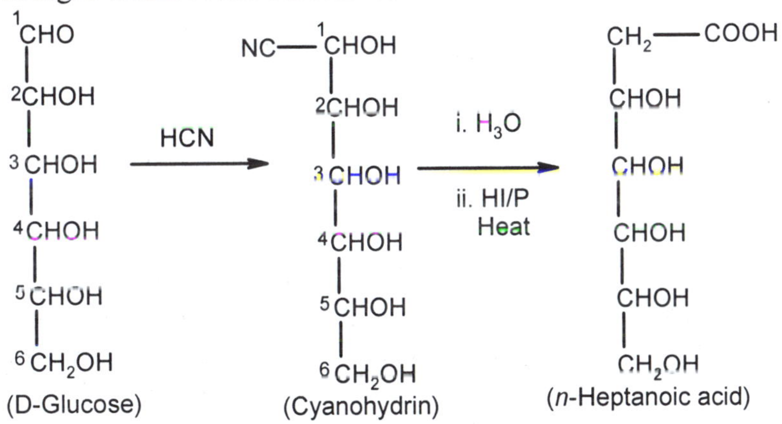

(v) Oxidation: Oxidation reactions have been carried out with three different oxidizing agents — bromine water, Conc. HNO_3, and HIO_4 (periodic acid). The results and conclusions there from are mentioned below:

(a) Oxidation with bromine water: Bromine water, being a mild oxidizing agent, oxidizes only the aldehyde group into carboxylic acid function. Bromine water can't oxidize the alcoholic functions of a sugar, and is also unable to oxidize keto group. Hence, it serves as a useful reagent to distinguish aldoses from ketoses.

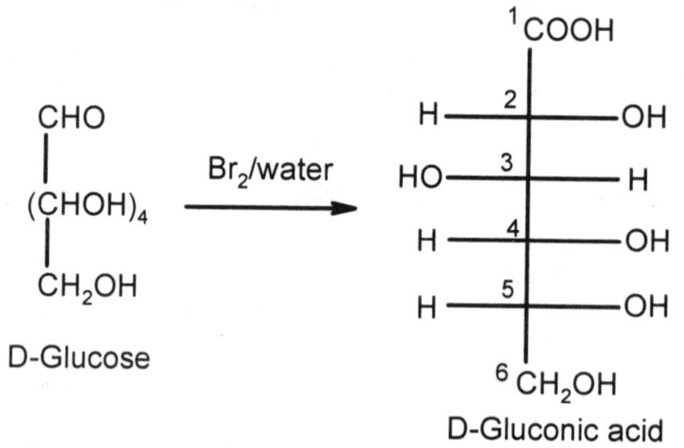

(b) Oxidation with Conc. HNO_3: Concentrated nitric acid, being more powerful oxidizing agent than bromine water, oxidizes both the aldehyde group and the terminal $-CH_2OH$ group of an aldose into carboxylic acid functions.

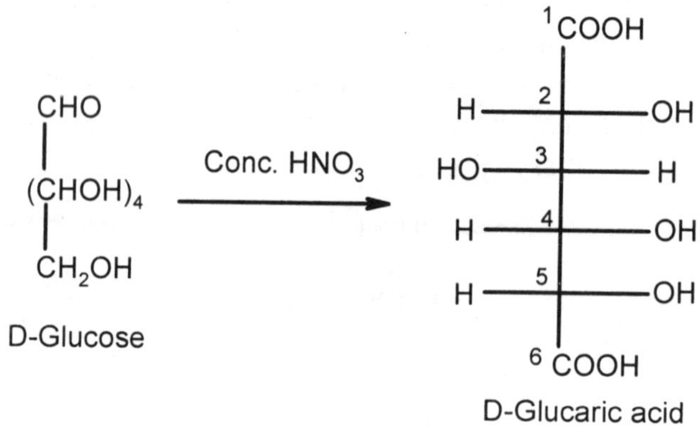

(c) Oxidation with HIO_4 (periodic acid): The highly reactive periodic acid cleaves carbon-carbon bond containing –OH groups. It has been observed that one mole of glucose consumes five moles of periodic acid to yield five moles of formic acid and one mole of formaldehyde. The reaction scheme can be shown as below:

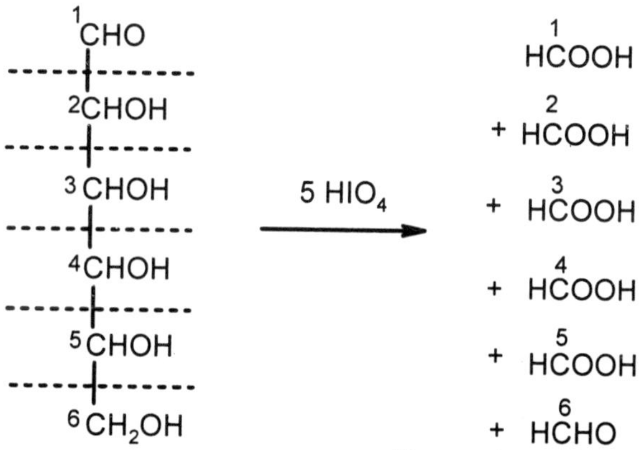

D-Glucose (five moles of formic acid, and one mole of formaldehyde)

In conclusion, in the light of the aforediscussed chemical observations the structure of D-glucose can unequivocally be suggested as 2,3,4,5,6-pentahydoxyhexanal.

Problem 9.6 Discuss Fischer's proposition to the configuration of glucose.

Solution.

$$OHC\text{—}\overset{*}{C}H(OH)\text{—}\overset{*}{C}H(OH)\text{—}\overset{*}{C}H(OH)\text{—}\overset{*}{C}H(OH)\text{—}CH_2OH$$

(Glucose)

(+)-Glucose is an aldohexose that has four chiral centers, and can therefore be any one of 2^4=16 possible stereoisomers (i.e. eight pairs of enantiomers). On the basis of chemical logic Emil Fischer solved this problem (for outstanding works on carbohydrate chemistry he was awarded the 1902 Nobel Prize in chemistry). How Fischer arrived his final proposition is being presented here step by step:

STEP-I: For simplify the matter, Fischer arbitrarily choosed only eight isomers of the aldohexose out of the possible sixteens, having the C_5–OH group on the right hand side in Fischer projections (D-series). He started his thinking considering the four possible isomers of a D-aldopentose, i.e. D-arabinose (I – IV).

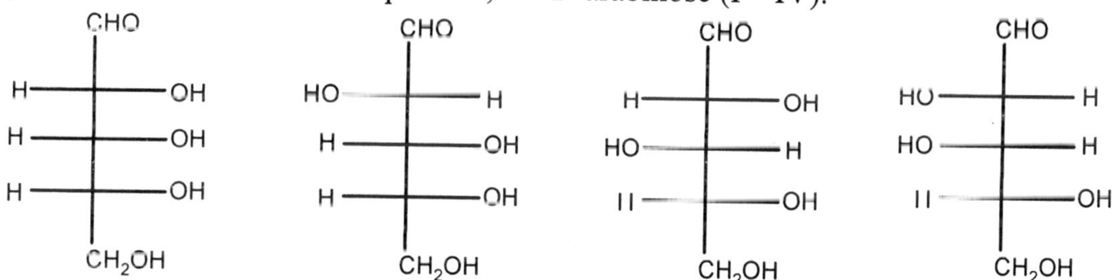

(I) (II) (III) (IV)

STEP-II: Eight possible isomers of D-aldohexoses (V – XII) are feasible from the four isomers of D-arabinose (I – IV) of D-arabinose on Kiliani-Fischer chain extension reaction; one of these eight D-aldohexoses stands surely for D-glucose, characterization of which is the task here.

D-Aldopentoses
(four possible stereoisomers)

D-Aldohextoses
(eight possible stereoisomers)

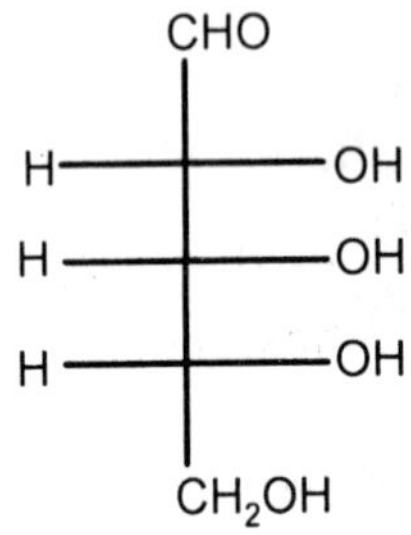

(I)

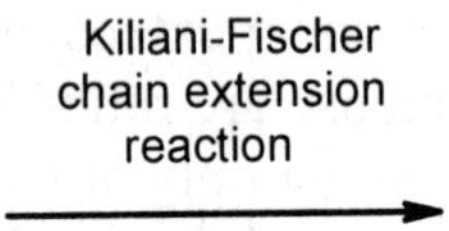

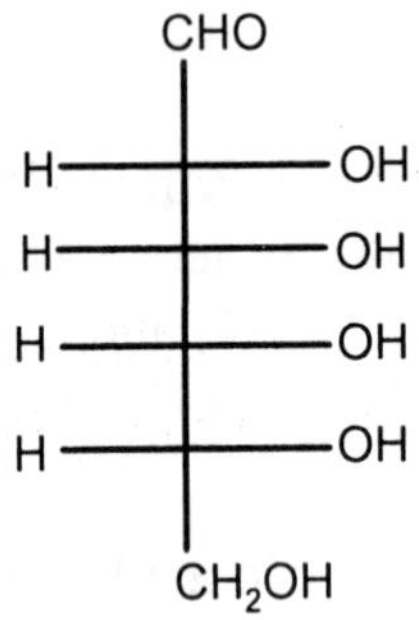

(V)

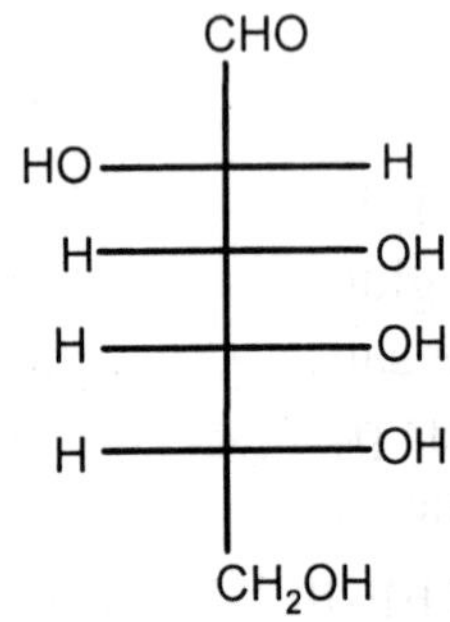

(VI)

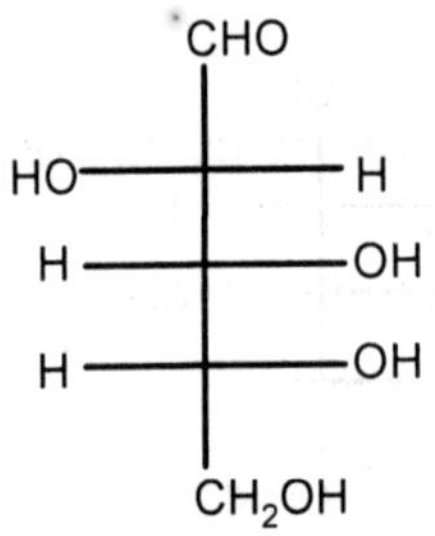

(II)

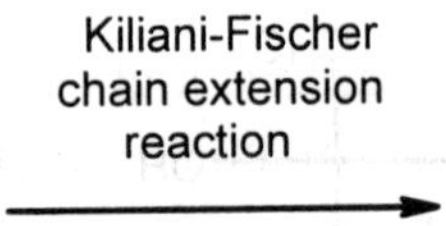

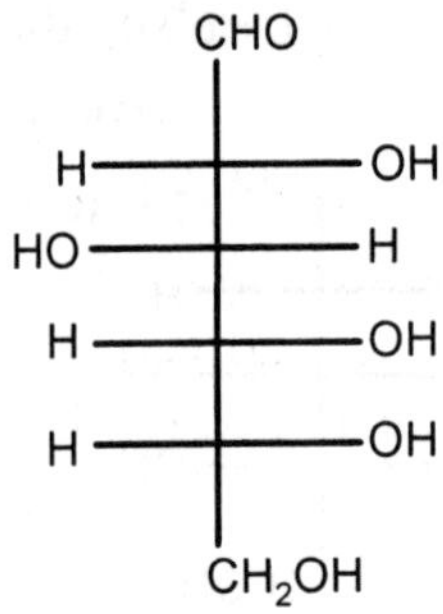

(VII)

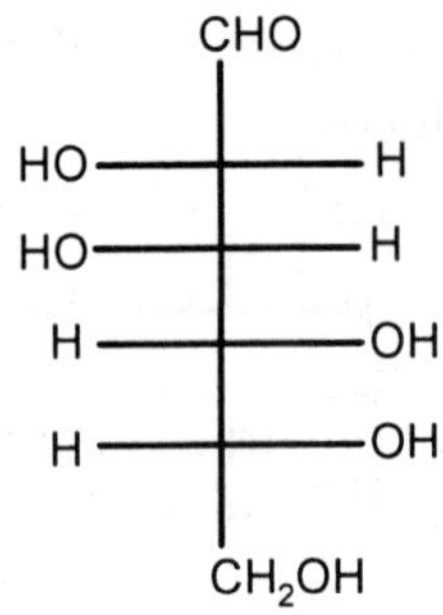

(VIII)

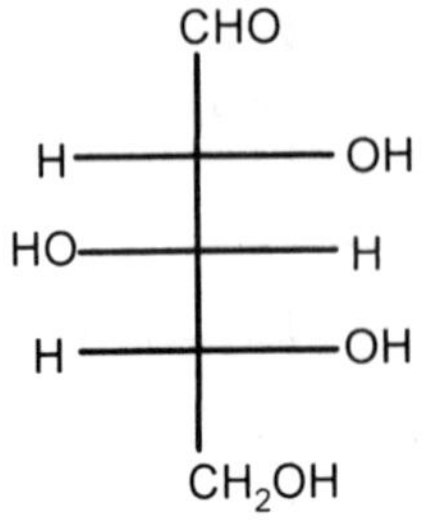

(III)

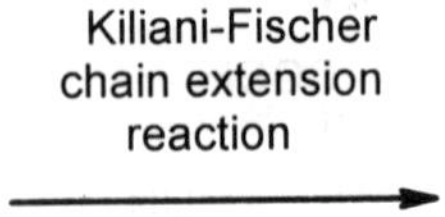

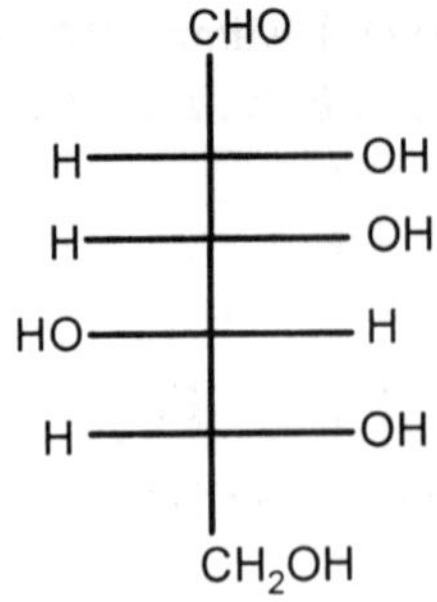

(IX)

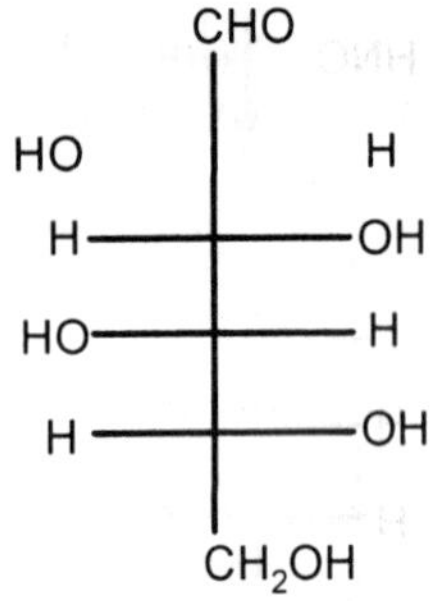

(X)

(see on next page)

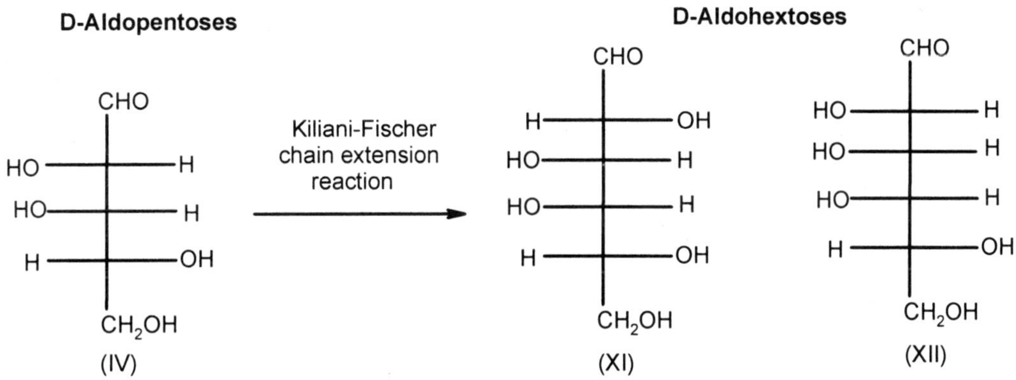

It is the fact that arabinose on Kiliani-Fischer chain extension furnishes a mixture of glucose and mannose. Thus, glucose and mannose are rationally supposed to have the same configuration at C-3, C-4, and C-5 differing only at C-2. Hence, among the four pairs of D-aldohexoses (V – XII), one pair would denote the stereochemistry of glucose and mannose.

STEP-III: On oxidation with nitric acid, arabinose results in the formation of optically active dicarboxylic acid. Scrutiny of the four structures of arabinose (I–IV) reveals that only (II) and (IV) will give optically active products. This rules out isomers (I) and (III) for D-arabinose, and also the isomers (V), (VI), (IX) and (X) for glucose/mannose.

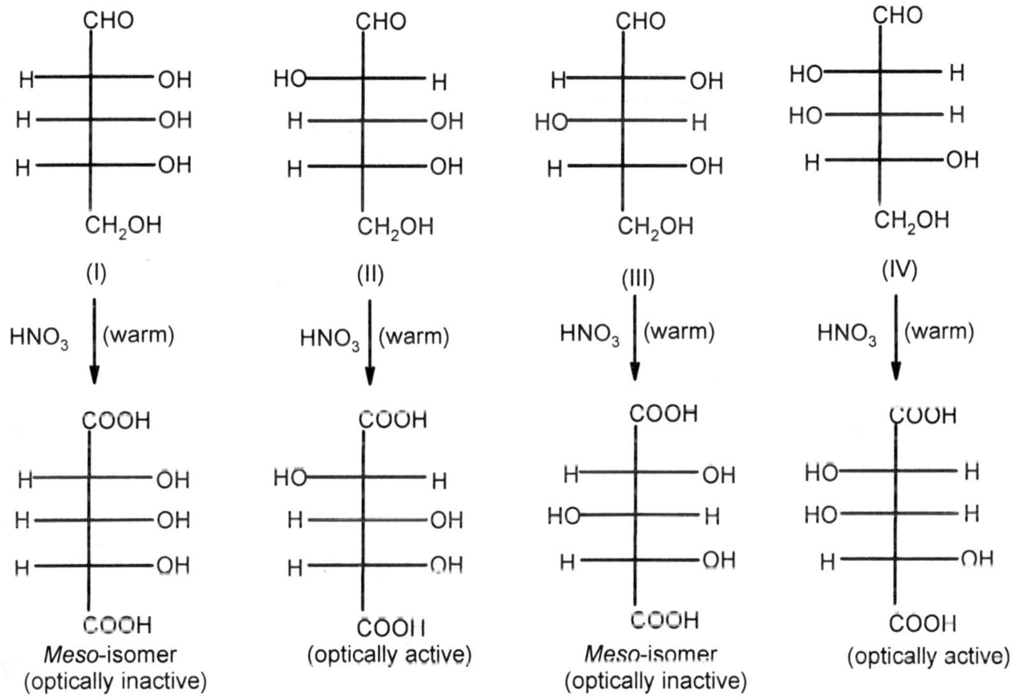

Thus, glucose and mannose may form epimeric pairs as (VII and VIII) or (XI and XII).

STEP-IV: Both the epimers, (+)-glucose and (+)-mannose, on oxidation with nitric acid yield optically active dicarboxylic acids. From the structural scrutiny it is evident that both VII and VIII would result in the formation of optically active dicarboxylic acids, but in the remaining pair (XI and XII) only structure XII would give active acid, not the XI — it would lead to give optically inactive *meso*-isomer.

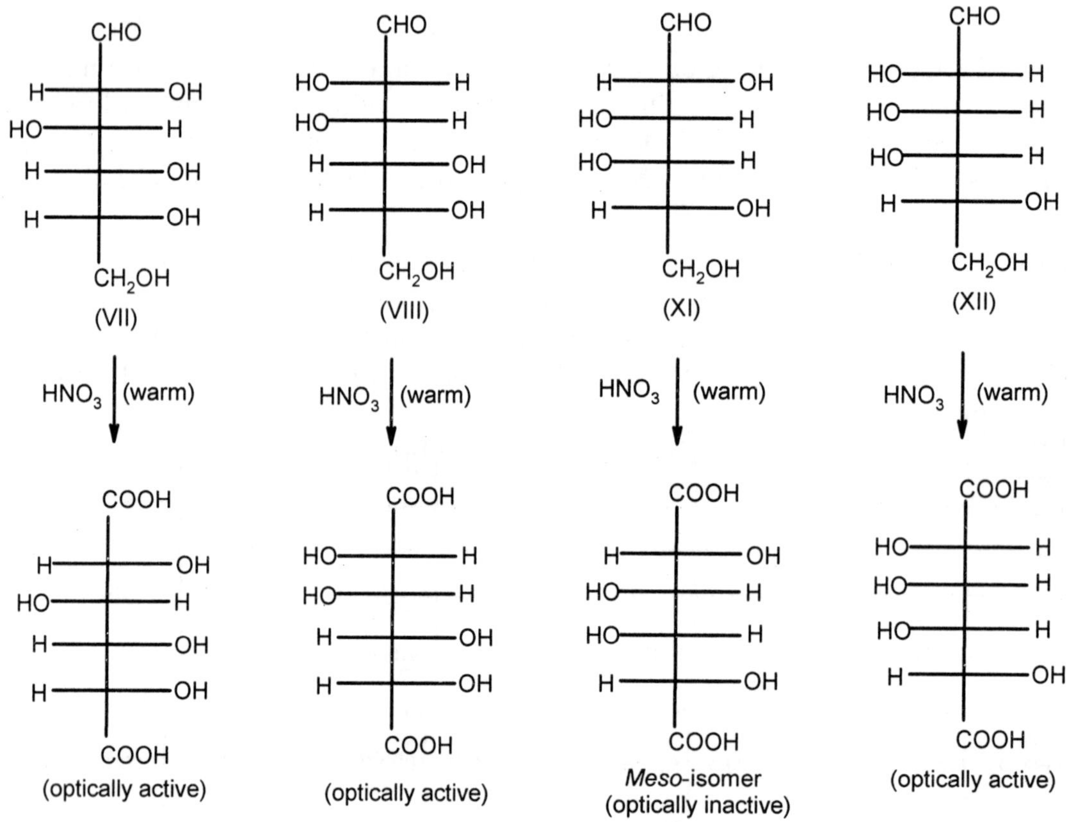

Thus, (VII) and (VIII) must denote the epimeric pair of glucose and mannose, and structure (II) stands for D-arabinose.

STEP-V: In the aldaric acids derived from VII and VIII, both the end carboxylic acid functions are generated from the oxidation of the end –CHO and –CH_2OH groups. If these two end functionalities in any of these respective structures (VII & VIII) were to be interchanged, then also the same aldaric acid would result. Thus, the aldaric acid (glucaric acid) obtained from compound (VII) could also be formed on oxidation of another aldohexose (i.e. L-glucose).

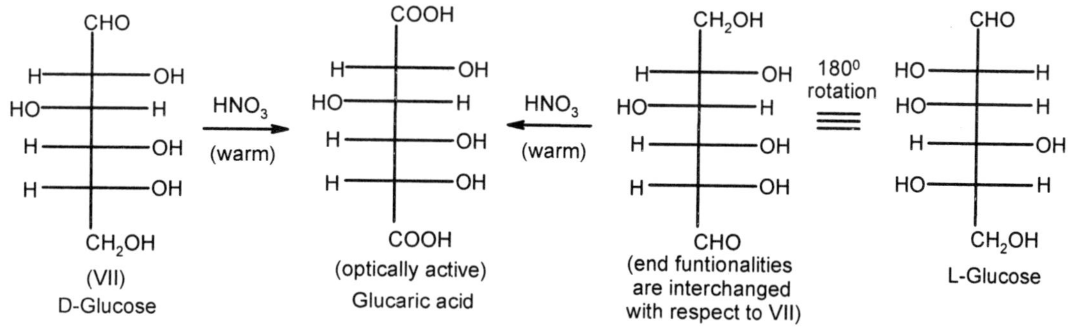

But the aldaric acid (i.e. mannaric acid) from VIII could not be obtained from the another one structure like that of VII, because the "other" aldohexose generated by interchanging the end groups becomes identical with VIII as evidenced from simple rotation through 180^0

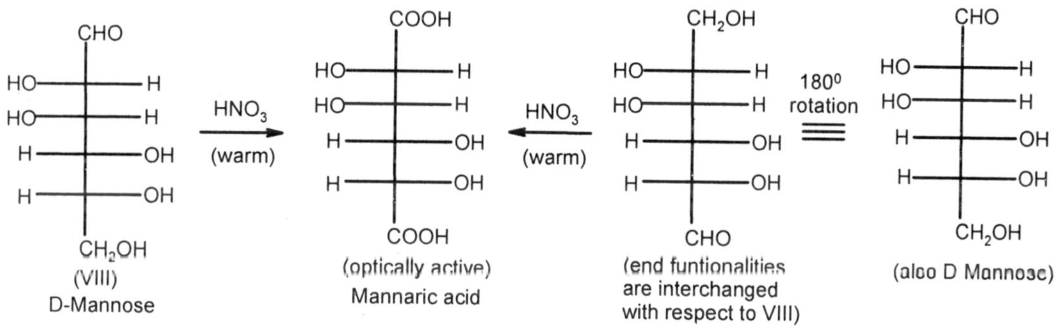

Thus, D-glucose must have the structure VII and D-mannose must have structure VIII.

Problem 9.7 Cite the limitations of open chain structure for D-glucose. Propose ring structure for the compound along with evidence.

Solution. *Limitations of open chain structure for D-glucose:*

The open chain structure of D-glucose is given as:

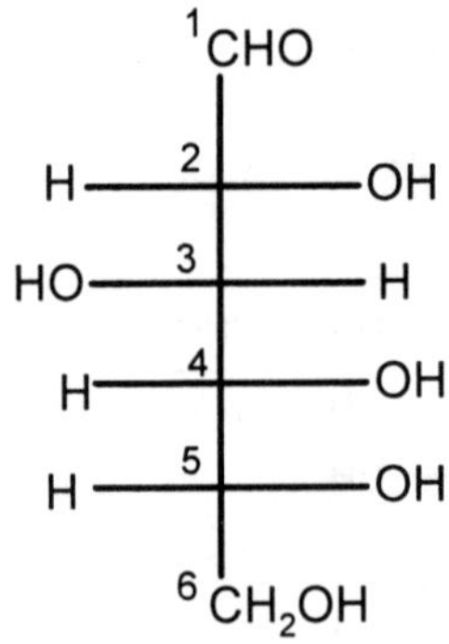

D-Glucose

Although this structure for D-glucose satisfies a number of chemical observations, yet it suffers from limitations. These limitations are cited below:

(i) As we know aldehydes give positive Schiff's test and form sodium bisulphate addition product. But both the tests are negative for glucose suggesting that the aldehydic functionality is not free enough.

(ii) Aldehydes react with two moles of alcohols to form acetals, but in contrast glucose reacts with one mole of alcohol to form methylglucoside, an acetal derivative.

(iii) D(+)-Glucose exists in two anomeric forms that undergo mutarotation; open chain structure can't solely explain the phenomenon.

(iv) More interestingly, the IR spectrum of glucose does not record absorption in the region ~1740–1720 cm^{-1} specified for carbonyl stretching frequency of an aldehyde group, and also its ^{1}H-NMR spectrum does not record any signal at ~δ9.7 for aldehydic proton.

All these facts are suggestive of the phenomenon that glucose predominantly exists not as an open chain aldehyde but as cyclic hemiacetal formed by the intramolecular nucleophilic addition of the hydroxy group at C-5 to the carbonyl carbon (C-1). The phenomenon is shown in the following scheme:

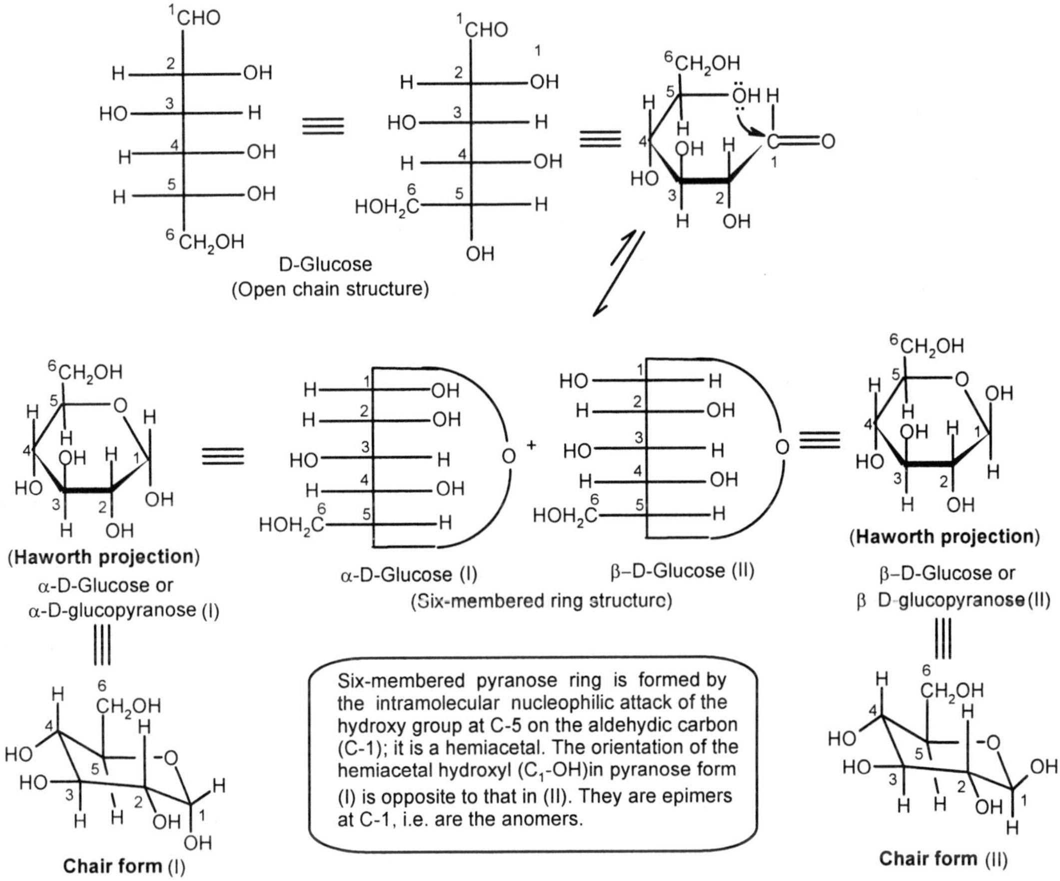

Evidence in favour of pyranose ring structure for D-glucose:

Haworth, Hirst and their co-workers offered a chemical evidence in favour of the six-membered pyranose structure for D-glucose. D-Glucose on drastic methylation either with dimethylsulphate/alkali or with methyl iodide/silver oxide gives pentamethyl ether derivative (II), which is converted to the tertramethyl glucose (III) on hydrolysis in aqueous acid. This tetramethyl ether derivative (III) yielded dimethoxytartaric acid (V) on oxidation with hot nitric acid, and in another way it gave a lactone (IV) in bromine/water, which in turn yielded xylotrimethoxyglutaric acid (VI; this can be directly obtained by the oxidation of methylated xylose) on oxidation with nitric acid. All these results are well-fitted with a six- and not a five-membered ring.

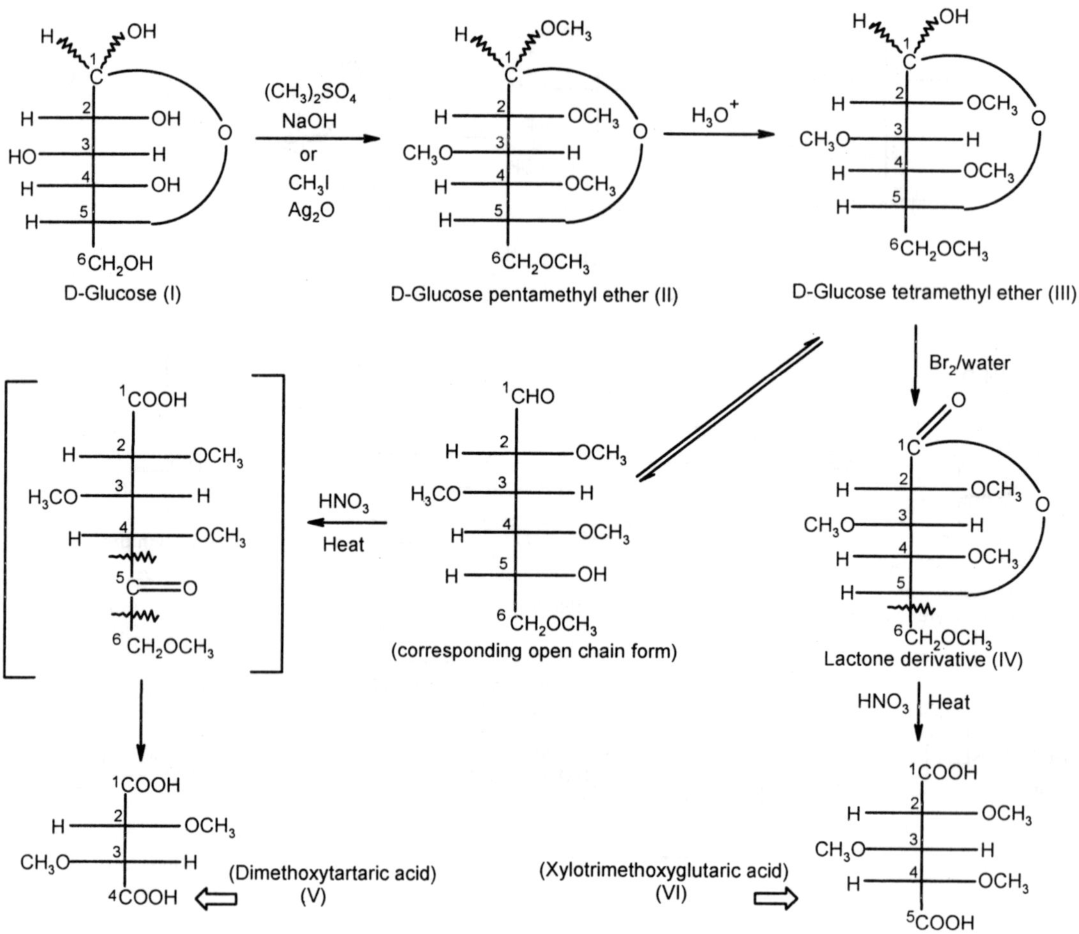

Hudson also offered evidence in favour of the six-membered pyranose structure; his method was to oxidize methyl glucoside (VII; α or β) with periodic acid. Consumption of two moles of the acid as well as formation of one mole of formic acid along with a dialdehyde (VIII) is possible only if glucose exists in pyranose ring structure. The dialdehyde (VIII) on oxidation with bromine water followed by acid hydrolysis furnishes glyoxalic acid and glyceric acid one mole each. The reaction scheme is given below:

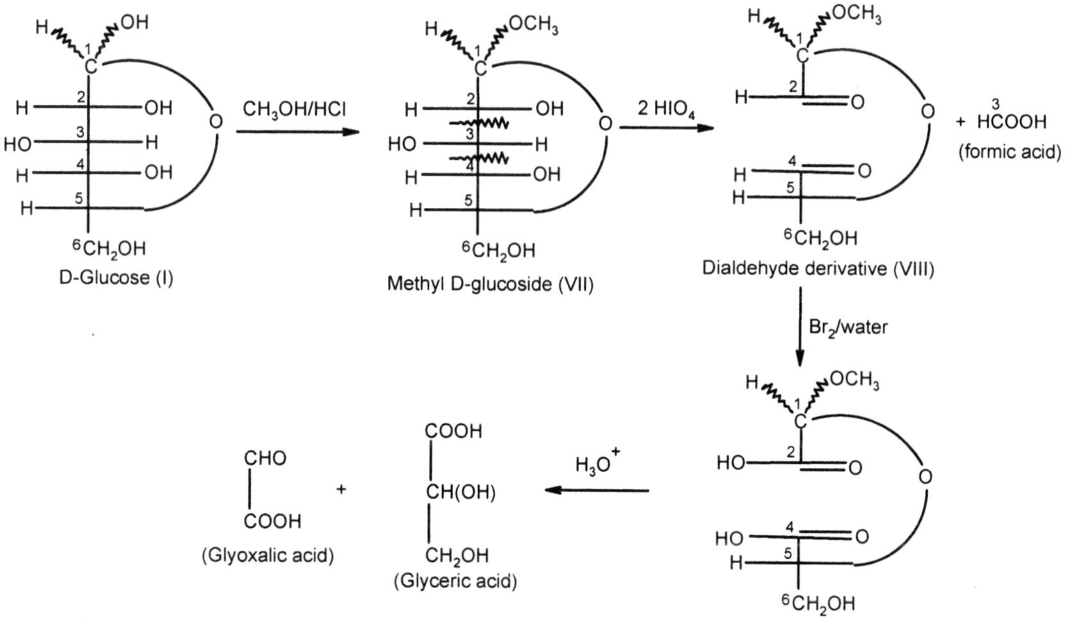

Problem 9.8 "Under certain conditions glucose may exist in furanose ring structure" — how far this statement is correct?

Solution. It has been established that depending upon the reactions conditions D-glucose furnishes methyl D-glucosides having two distinctly different ring structures — one is five-membered ring methyl D-glucofuranosides (α and β-isomers) and another is six-membered ring methyl D-glucopyranosides (α and β-isomers).

Under low temperatures and short reaction times, D-glucose furnishes primarily the five-membered ring methyl D-glucofuranosides on dissolving the substance in methanol in the presence of hydrochloric acid.

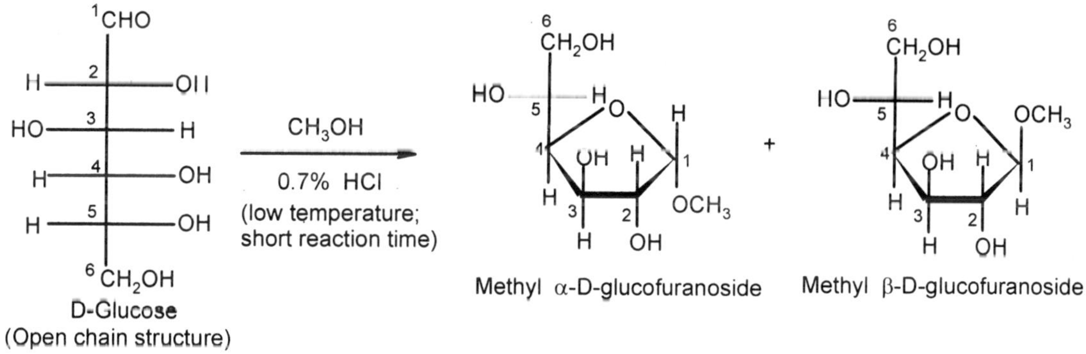

On the other hand, D-glucose on prolonged heating with methanol in the presence of hydrochloric acid gives mainly the six-membered ring methyl D-glucopyranosides.

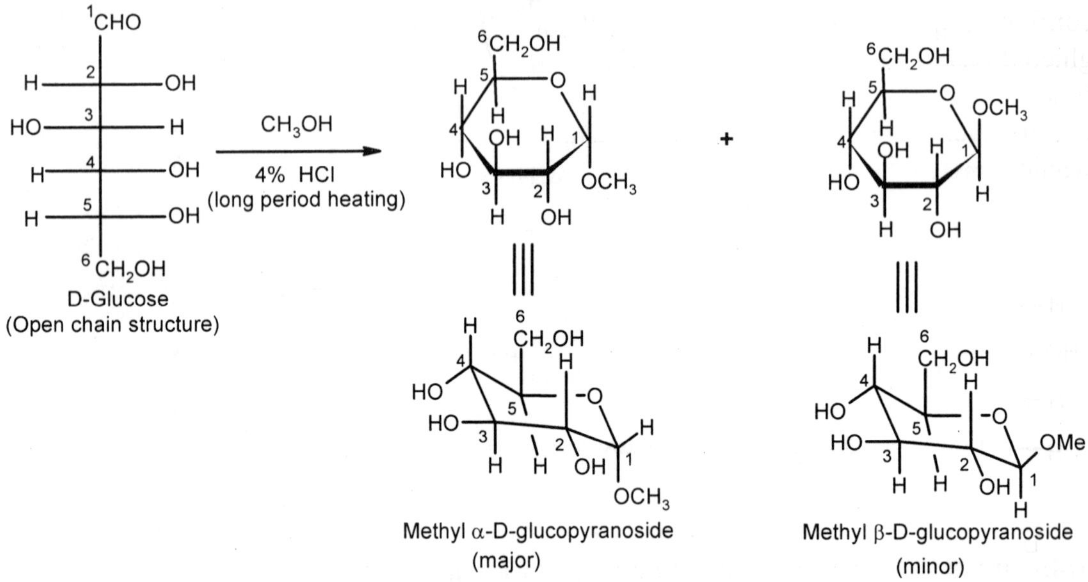

Five-membered rings are formed faster than six-membered rings; the methyl D-glucofuranosides as formed fast at low temperatures are the kinetically controlled products while the methyl D-glucopyranosides, formed at relatively higher temperatures and over longer period of times, are the thermodynamically controlled products. However, under reaction conditions the five-membered rings can reopen so as to develop equilibrium among all possible forms — at relatively higher temperature and longer period of time an equilibrium is established, where methyl D-glucopyranosides are appeared as predominant products with the α-isomer (having the methoxy group at axial position) being the thermodynamically most stable.

Problem 9.9 Comment on the structure of D(–)-fructose.

Solution. Fructose is a 2-ketohexose, and is levorotatory ($[\alpha]_D = -92^0$).

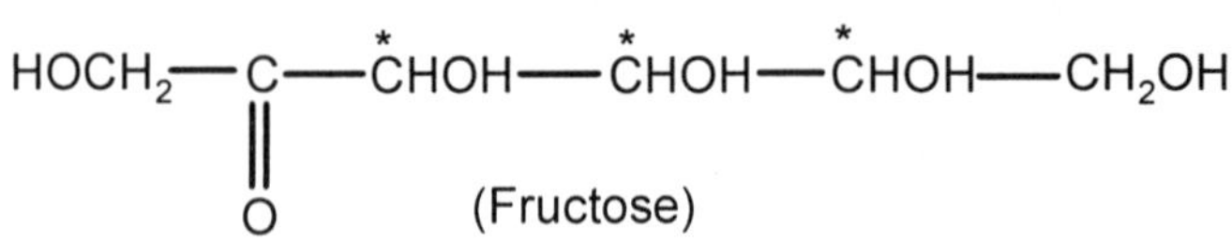

It has been observed that both glucose and fructose on reaction with excess of phenylhydrazine form the *same* osazone; since osazone formation involves only the first two carbons (C-1 & C-2) in a sugar molecule, it appears logically that the configurations of C-3, C-4 and C-5 in both the molecules are identical. Like D-glucose (I), D-fructose also bears five hydroxyl functions — the difference in functionality is that glucose is an aldohexose while fructose is a ketohexose. Hence, C-2 in fructose must be involved in forming the keto function. Therefore, D-fructose would be (II):

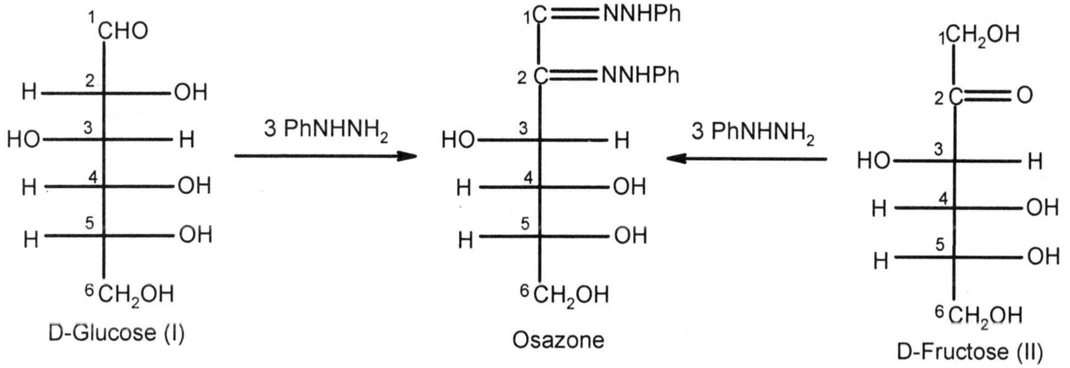

The following chemical evidences support the proposed structure (II) for D-fructose:

(i) On reduction with H_2/Ni D-fructose forms sorbitol and mannitol, which on heating with HI/P yield the completely reduced product *n*-hexane.

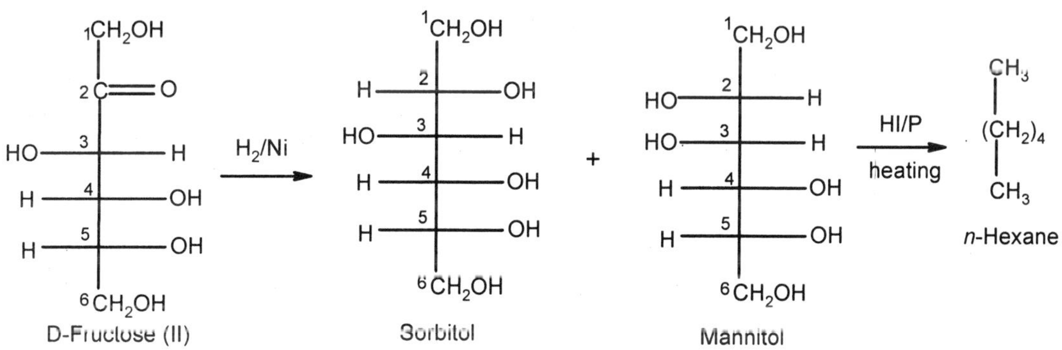

(ii) Fructose forms a very stable pentaacetyl derivative on treatment with acetic anhydride.

(iii) Fructose gives oxime as well as phenylhydrazone derivative as expected.

(iv) Bromine water cannot oxidize fructose.

(v) Fructose on addition with HCN yields a cyanohydrin that on hydrolysis followed by treatment with HI/P (heating condition) produces 2-methylhexanoic acid. This suggests the presence of the carbonyl group in fructose molecule as keto function at C-2.

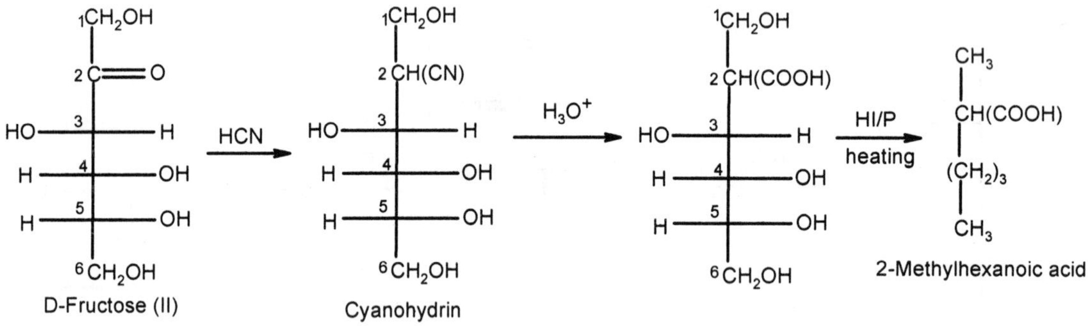

D-Fructose predominantly exists in ring structure — two ring structures are possible for D-fructose: one is five-membered furanose form, and another is six-membered pyranose form. But it is the furanose structure that primarily predominates for D-fructose in combined states (i.e. in disaccharides/polysaccharides).

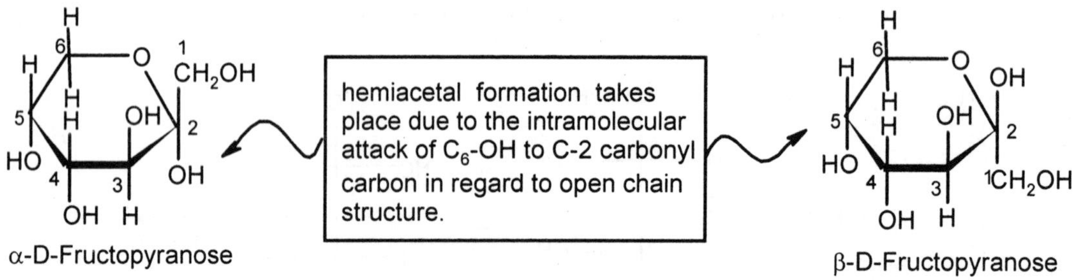

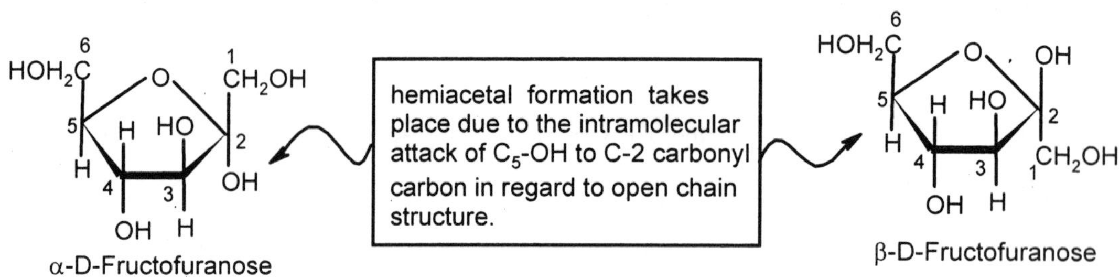

Problem 9.10 D-Glucose, D-Mannose and D-Fructose yield same osazone on reaction with excess of phenylhydrazine. Comment.

Solution. D-Glucose and D-mannose are C-2 epimers, and D-fructose is a ketohexose. It has been observed that all of these compounds on reaction with excess of phenylhydrazine produce the *same osazone derivative*. One mole of any of the three compounds reacts with three moles of phenylhydrazine to yield one mole of osazone along with one mole of aniline and one mole of ammonia. The formation of same osazone from these compounds is because of the fact that osazone formation involves only the first two carbons (*i.e.* C-1 & C-2) retaining the rest portion intact.

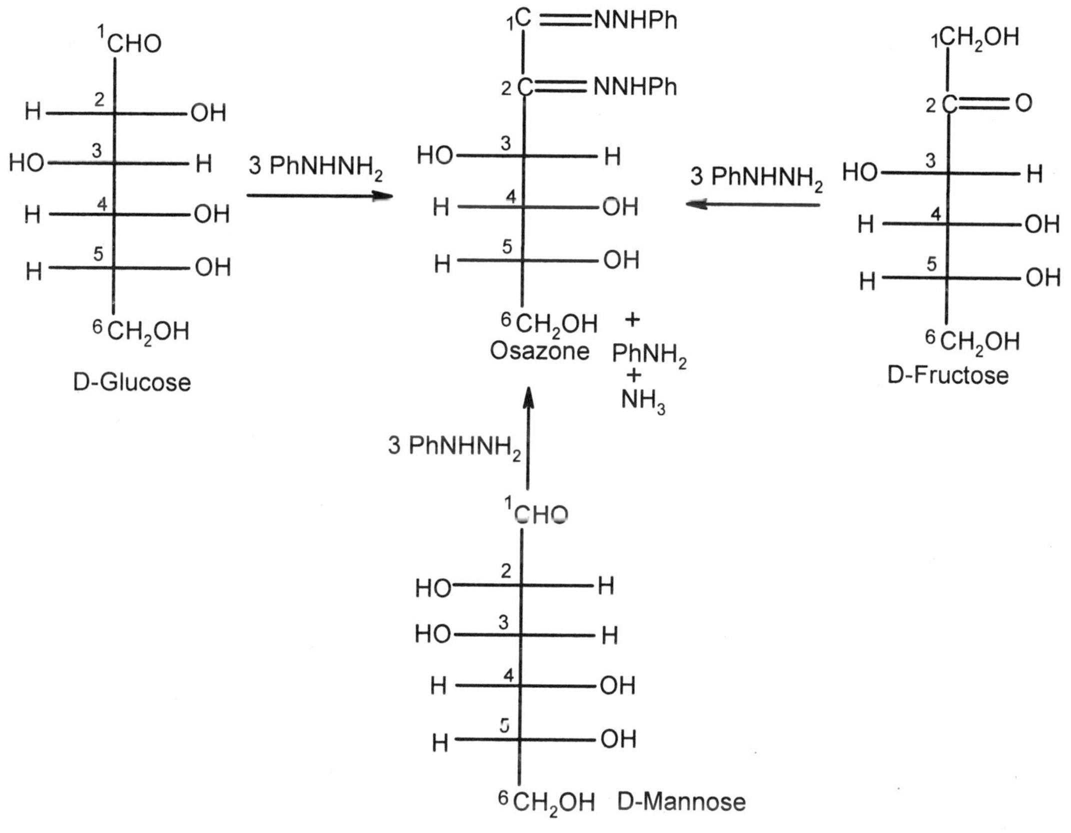

Osazone reaction is believed to proceed through the initial formation of phenylhydrazone derivative of the corresponding carbonyl group in the first step, which in turn undergoes *Amadori rearrangement* to form the final product. The mechanistic pathways for both aldohexoses and ketohexose are shown below:

Substrate: D-Glucose/D-Mannose

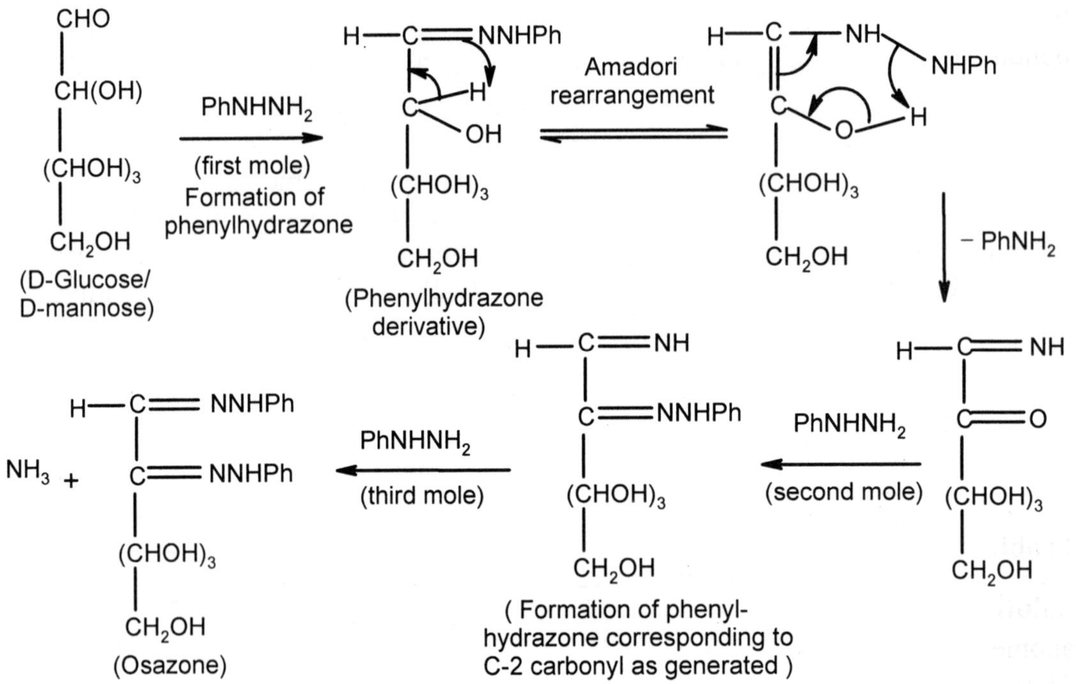

Substrate: D-Fructose

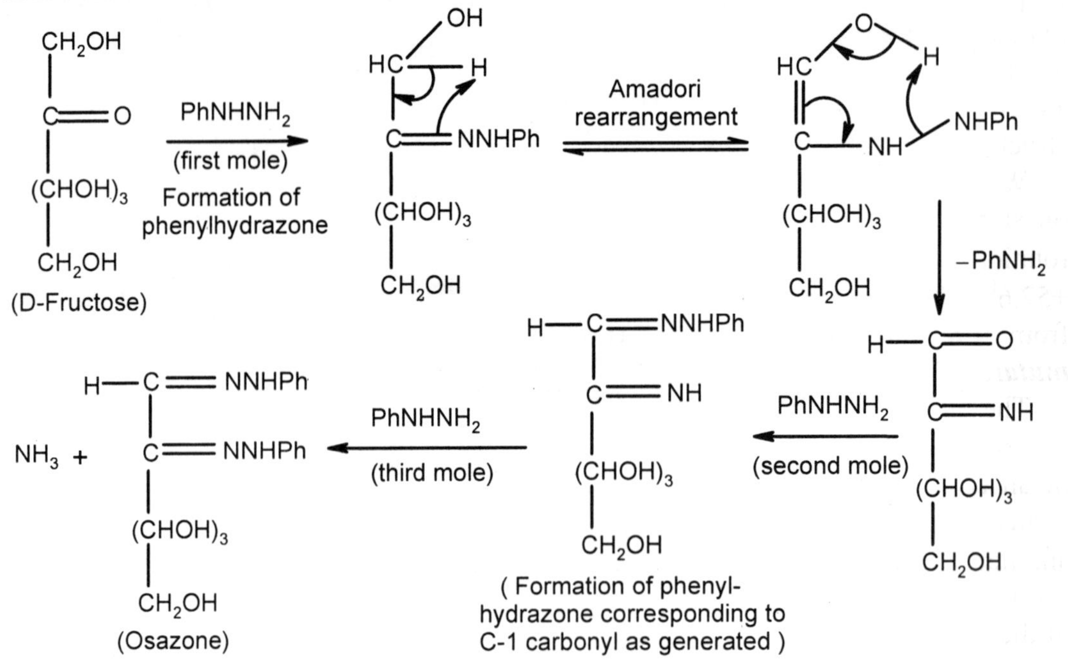

Osazone formation is restricted to the first two carbons — does not undergo further forward; this phenomenon was justified by Fieser and Fieser (1944) by assuming that the osazone as formed becomes stabilized by means of intramolecular hydrogen bonding (i.e. chelation) forming apparently a six-membered ring as shown below:

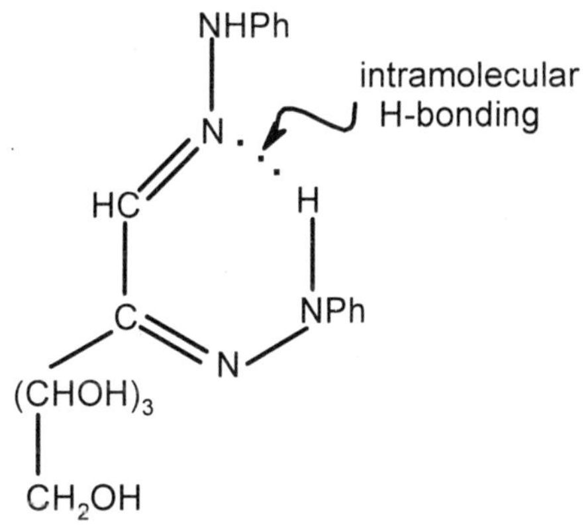

Problem 9.11 Discuss the phenomenon of mutarotaion of D(+)-glucose.

Solution. As we know D(+)-glucose predominantly exists in ring form as in two anomeric structures — one is α-D-glucopyranose and another is β-D-glucopyranose. These anomers are diastereoisomeric in relation, and generally are of different properties. They have been isolated in pure form as crystalline solids — α-D-glucopyranose has a melting point of 146^0C and a specific rotation of $+112.2^0$, while the β-anomer has a melting point ranging from 148^0–155^0C and a specific rotation of $+18.7^0$. In the solid state the two forms do not interconvert and are quite stable. The structures have been unambiguously settled by X-ray crystallographic studies.

When a sample of either of the pure anomers of D(+)-glucose is dissolved in water, on standing there occurs an interesting change in specific rotation — the specific rotation of the solution containing α-isomer decreases gradually from $+112.2^0$ to $+52.6^0$, while the rotation of the solution containing the β-isomer gradually increases from $+18.7^0$ to the same value of $+52.6^0$. This change in specific rotation is called ***mutarotation***.

The inherent cause of mutarotation of D(+)-glucose is the interconversion between the two anomers in solution through open-chain structure leading to the establishment of an equilibrium between the open-chain form of D(+)-glucose and the α and β-forms of the cyclic hemiacetals. A solution, initially containing either of the pure anomers, undergoes equilibration to the same mixture of α and β-pyranose forms — at equilibrium an intermediate rotation results from equilibrium concentrations of the anomers. Such conformational change occurring in solution phase is termed as

first order asymmetric transformation. For D(+)-glucose, the specific rotation at equilibrium is observed as $+52.6^0$.

It has been found that the process is catalyzed by either acid or base, and is particularly more effective if both are present in the same solution (water acts as an amphoteric solvent). The actual mechanism of mutaroration is still a matter of much discussion, however, the well-accepted path is shown below:

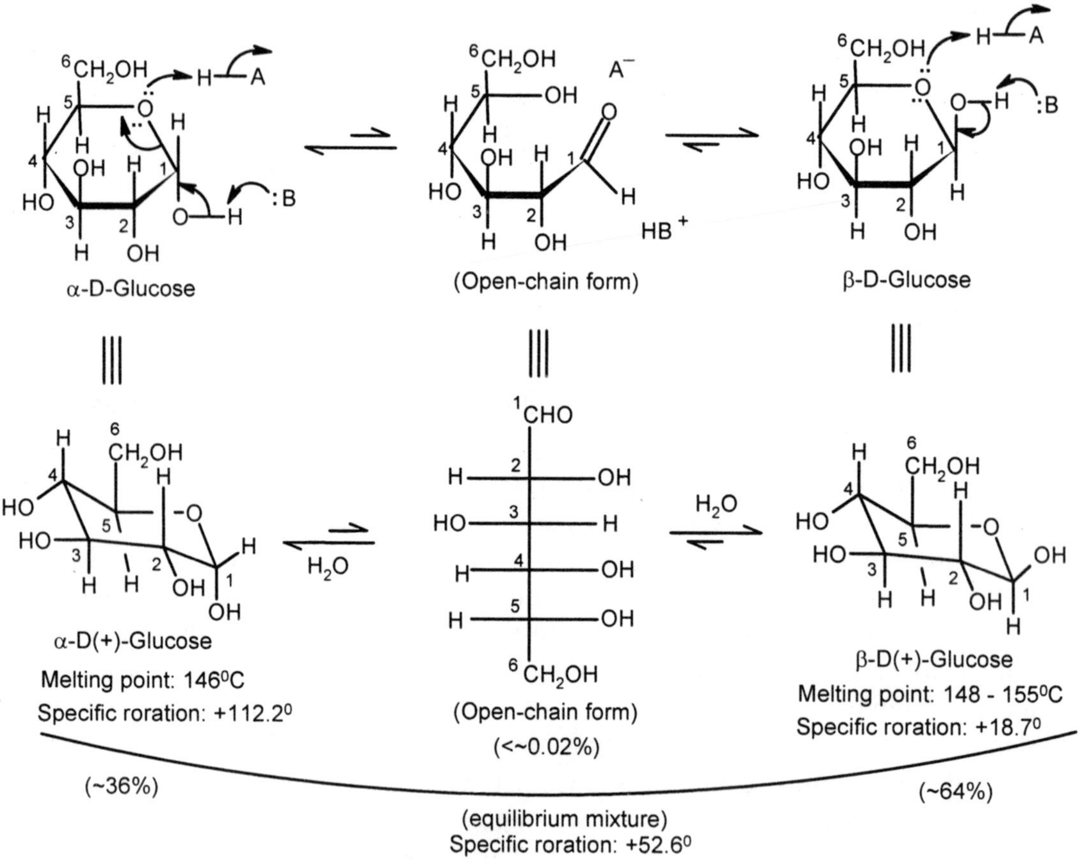

At equilibrium, the concentration of open-chain form is very very small (< ~0.02%); thus assuming this concentration negligible it is possible to calculate the equilibrium-concentration of α-anomer as 36% and β-anomer as 64%.

However, it is also to be kept in mind that at equilibrium all of the four possible cyclic forms (pyranose and furanose forms) of D(+)-glucose are present along with the open-chain structure,although pyranose forms predominate in water almost completely; furanose forms constitute only less than 0.5% at equilibrium. ^{13}C-NMR spectral study of D(+)-glucose in water has revealed the existence of all the possible forms with the composition as: α-glucopyranose (38.8%), β-glucopyranose (60.9%), α-glucofuranose (0.14%), β-glucofuranose (0.15%) and the open-chain form as hydrate (0.0045%).

Problem 9.12 "In aqueous medium, the α-anomer of D-glucopyranose predominates over the β-anomer at equilibrium, but the reverse situation is observed in case of D-mannose in the same medium" – comment critically.

Solution. Predominance of α- or β-pyranose form of a particular carbohydrate at equilibrium, when remains in solution, is dependent on a number of effects; however, two significant effects are the ***anomeric effect*** and the ***solvation effect***. These two effects act in different directions but are comparable in magnitude in aqueous solution; hence α-pyranose form becomes predominant for some carbohydrates while the other one for others.

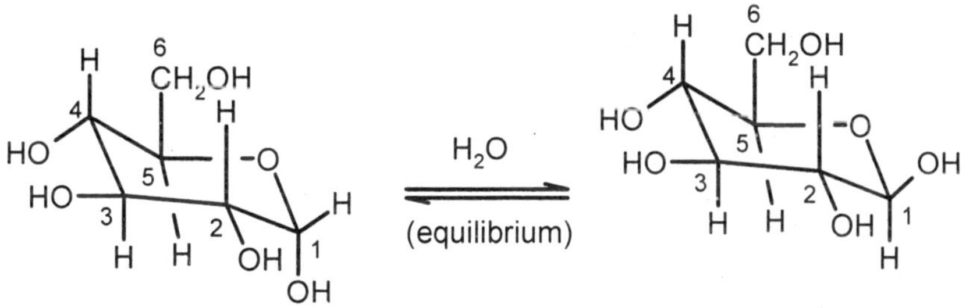

α-D(+)-Glucose (~36%) β-D(+)-Glucose (~64%)

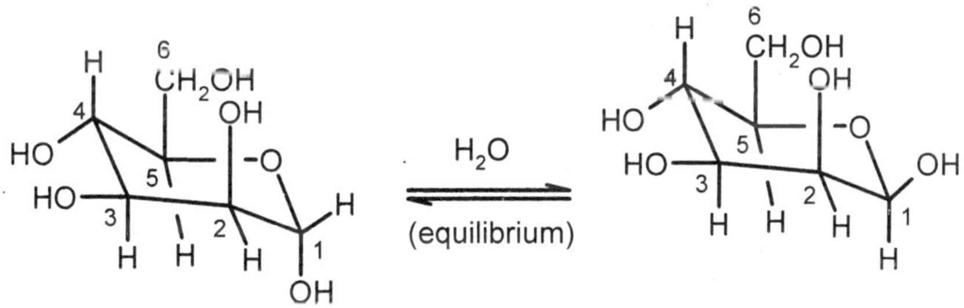

α-D-Mannose (~ 69%) β-D Mannose (~ 31%)

In both the carbohydrates anomeric effect plays — this is a stereoelectronic repulsive effect that becomes more effective when the hydroxy group at anomeric carbon (*i.e.* C-1) is oriented equatorially so that the nonbonding electrons on the ring oxygen atom can repel the nonbonding electrons on hydroxyl oxygen more effectively; such electronic repulsion becomes less effective when the hydroxyl group is oriented axially because in that situation the respective nonbonding electrons are farther apart.

Thus, anomeric effect preferentially stabilizes the α-pyranose form where the –OH group on anomeric carbon is axially oriented. On the other hand, an equatorial hydrŏxyl function being less crowded is solvated more effected by water molecules than an axial one; this solvation effect results in the stabilization of the β-pyranose form. These two effects compete each other, and the resultant directs the final situation as observed in the present cases.

Problem 9.13 "α-Anomer of methyl D-glucopyranoside is more stable than the β-anomer" – justify the statement.
Solution.

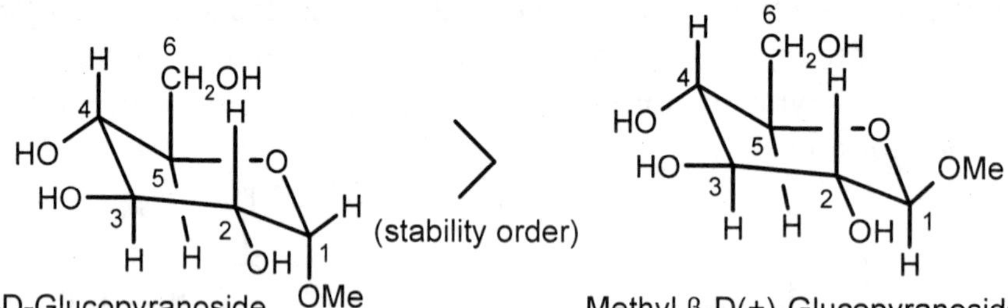

Methyl α-D-Glucopyranoside Methyl β-D(+)-Glucopyranoside

In the β-isomer, the nonbonding orbitals containing the nonbonding electrons on the ring oxygen atom are closer to the nonbonding orbitals containing the nonbonding electrons on the equatorially oriented methoxyl group than in the α-isomer. Therefore, electronic repulsive interaction is more pronounced in the former making the α-pyranose form more stable. This electronic interaction effect is called the *anomeric effect*.

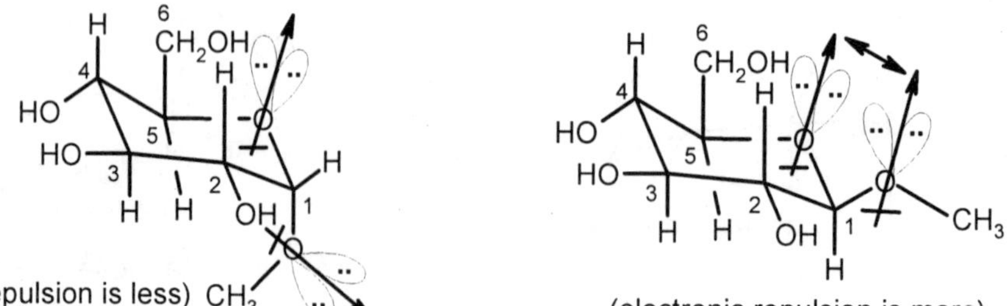

(electronic repulsion is less) α-D-Glucopyranoside

(electronic repulsion is more) β-D(+)-Glucopyranoside

Problem 9.14 Outline the Kiliani-Fischer synthesis in carbohydrate chemistry.
Solution. The Kiliani-Fischer synthesis is a way to lengthen an aldose carbon chain by introducing one more carbon atom; the newly introduced carbon atom becomes C-1, *i.e.* carbon atom of the aldehydic functionality in the lengthened aldose. A general reaction scheme is shown below:

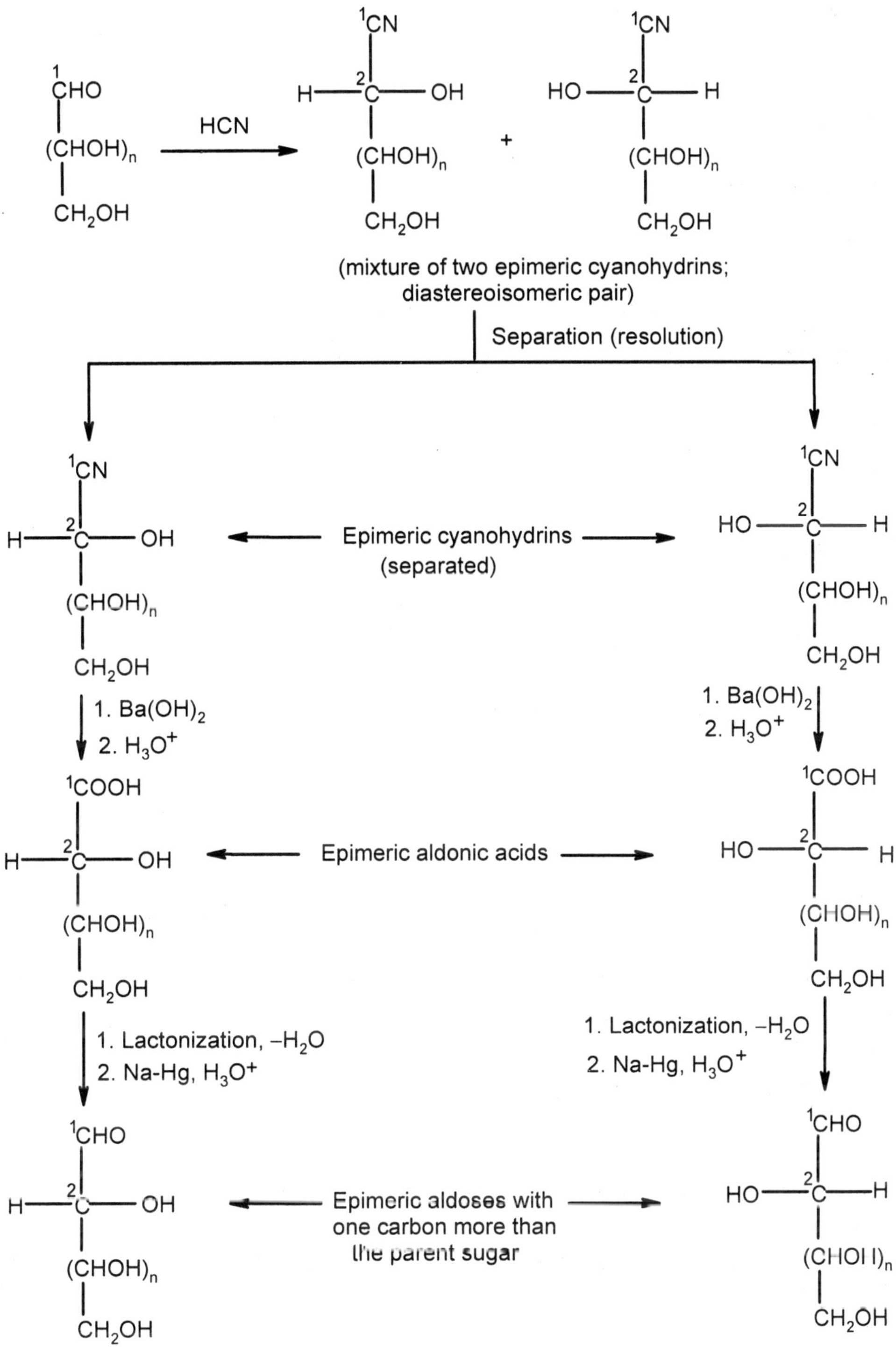
HCN
(mixture of two epimeric cyanohydrins; diastereoisomeric pair)
Separation (resolution)
Epimeric cyanohydrins (separated)
1. Ba(OH)$_2$
2. H$_3$O$^+$
Epimeric aldonic acids
1. Lactonization, –H$_2$O
2. Na-Hg, H$_3$O$^+$
Epimeric aldoses with one carbon more than the parent sugar

Problem 9.15 Mention carbon chain-shortening reactions in carbohydrate chemistry.

Solution. Two important methods of descending the sugar series are: Whol's method and Ruff's method.

(i) ***Wohl's method:***

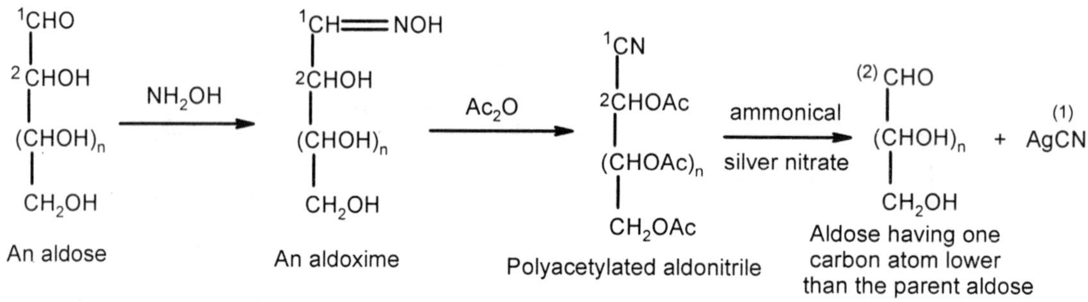

(ii) ***Ruff's method:***

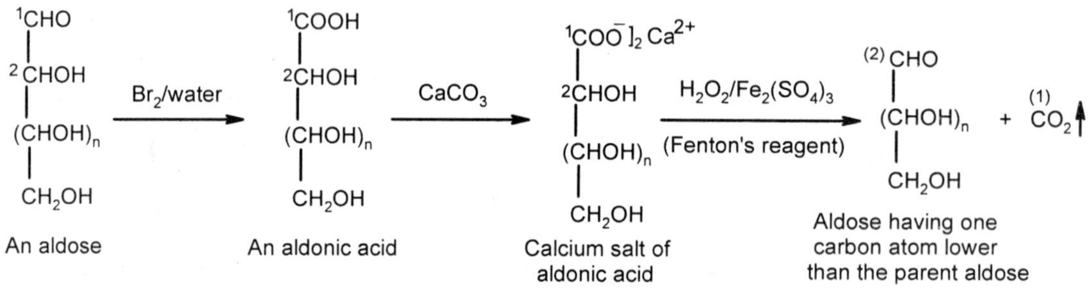

Problem 9.16 How do you convert an aldose to ketose, and *vice versa*?

Solution. An aldose on reaction with excess of phenylhydrazine forms corresponding osazone, which on hydrolysis with aqueous hydrochloric acid yields a α-keto aldehyde derivative (known as osone). On treatment with Zn in acetic acid medium, only the aldehyde functionality is reduced to primary alcohol leaving the keto group intact; thus a ketose is formed ultimately.

The keto group in a ketose is reduced to secondary alcohol by H_2/Ni; the reduced derivative, when treated with HNO_3, one of the terminal $-CH_2OH$ group is oxidized to carboxylic acid function. This polyhydroxy monocarboxylic acid derivative forms γ-lactone on hot condition, which ultimately is converted to an aldose on reduction with Na-Hg in mild acidic medium. The reaction schemes for both the conversions are shown below:

Aldose to ketose

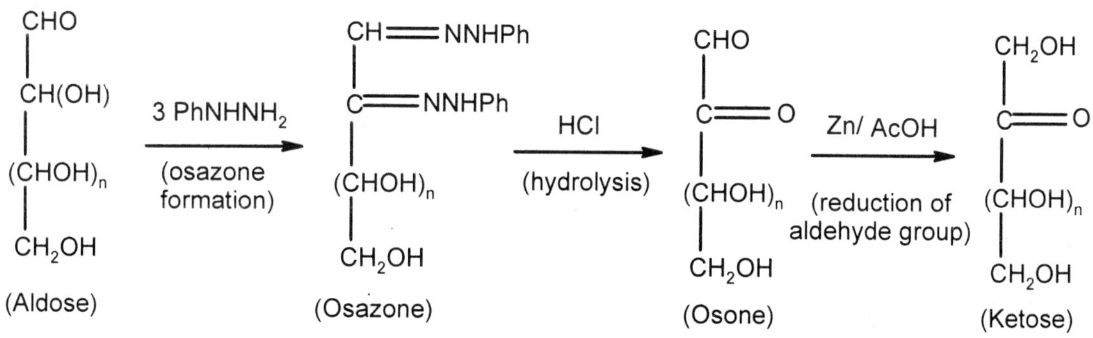

Ketose to aldose

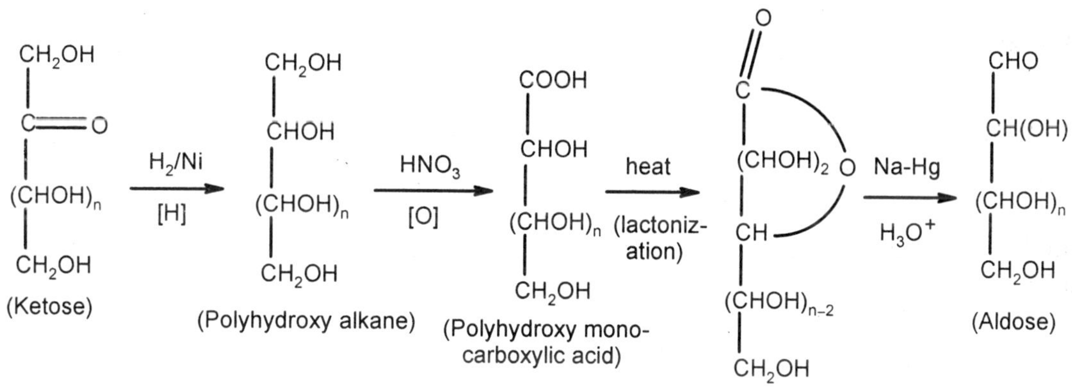

Problem 9.17 What are reducing and nonreducing sugars? Give examples.

Solution. Tollens' reagent and Benedict's solution (that contains cupric citrate complex ion) or the related Fehling's solution (that contains cupric tartrate complex ion) are reduced by aldehyde functionality forming respectively a silver mirror and brick-red precipitate of Cu_2O. Sugars that give positive tests toward these reagents are called as ***reducing sugars***; alternatively sugars incapable to reduce such specific reagents are called ***nonreducing sugars***. Both aldoses and ketoses give positive tests with Benedict's reagent (or with Tollens' reagent), although they exist primarily as cyclic hemiacetals. Ketoses under the reaction conditions equilibrate with aldoses by way of *enediol intermediates*, and the aldoses as formed are oxidized by the reagent.

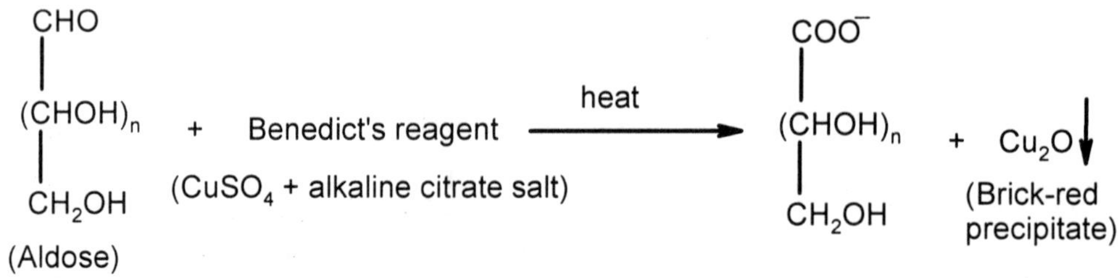

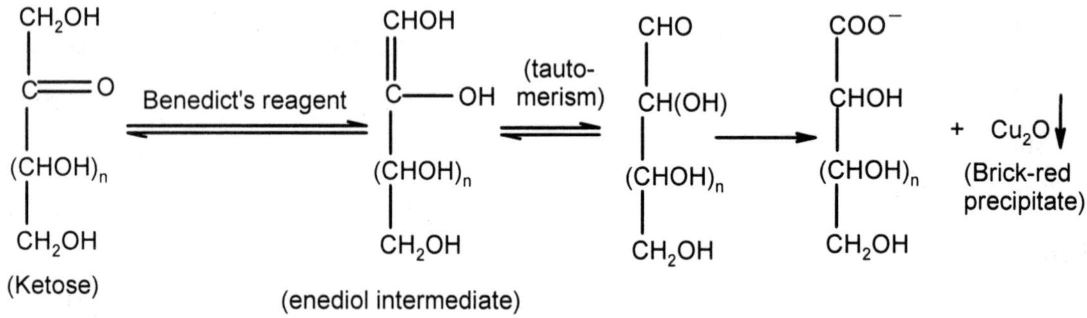

Actually, carbohydrates having hemiacetal linkage behave as reducing sugars. In aqueous solution these hemiacetals remain in equilibrium with open chain aldehydes or α-hydroxy ketones, although concentrations of these open chain forms are very small. These open chain forms take part in reducing the oxidizing reagents, thereby, shifting the equilibrium to produce more such forms until one of the reactants is completely exhausted.

Carbohydrates having acetal linkage behave as nonreducing sugars because they fail to give positive test with such specific reagents; due to the presence of acetal-like linkage they cannot exist in equilibrium with open-chain aldehydes or α-hydroxy ketones under the reaction conditions.

Examples:

Reducing sugars: Glucose, fructose, galactose, arabinose, lactose, mannose, maltose, xylose, ribose, etc.

Nonreducing sugars: Methyl D-glucopyranoside, sucrose, trehalose, etc.

Problem 9.18 Draw the structure of sucrose.

Solution. Sucrose is a disaccharide arising out of the combination of one molecule of D-glucose and one molecule of D-fructose. It is formed by [1,2]-linkage where α-D-glucopyranose is attached through its C-1 to the C-2 of β-D-fructofuranose, *i.e.* sucrose has been found to be α-D-glucopyranosyl-β-D-fructofuranoside. The structural forms are shown below:

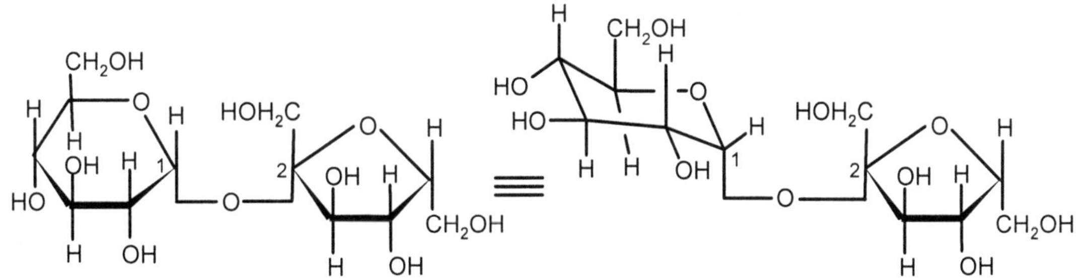

D(+)-Sucrose

Sucrose is a nonreducing sugar, and does not exhibit mutarotation.

Problem 9.19 What do you mean to say "invert sugar"?

Solution. As wc know sucrose is a disaccharide of D-glucose and D-fructose. Thus on hydrolysis sucrose will form a mixture of these two monosaccharides in equal amounts. Again, sucrose is *dextrorotatory* with a specific rotation of $+66.5^0$ while its constituent monosaccharides are of opposite optical rotations — D-glucose is *dextrorotatory* (specific rotation of $+52.6^0$) but D-fructose is *levorotatory* (specific rotation of -92^0). Since D-fructose has high specific rotation (with opposite sign) in comparison to D-glucose, net optical rotation of the resulting mixture as obtained on hydrolysis of sucrose would become *levo* with a magnitude of -39.4^0 $[(-)92^0 + (+)52.6^0]$. Thus the hydrolytic mixture has an optical rotation opposite in direction to that of sucrose; hence hydrolysis of sucrose (or cane-sugar) is known as the *inversion of cane-sugar*, and the hydrolytic mixture is said to be the "*invert sugar*".

Supplementary problems

SP-9.1 "All methyl pyranosides of the α-D-hexose series have the same configuration at C-1 and C-5" — offer evidence in favour of the statement.

SP-9.2 Why in the formation of glycosides, the hydroxyl group at the anomeric carbon is replaced by an alkoxy group under mild acidic condition while other hydroxyls remain unaffected?

SP-9.3 Write down the structure of a monosaccharide of your choice, and how would you experimentally determine whether it belongs to D or L series?

SP-9.4 Illustrate how does the isolation of 2,3,4-trimethoxyglutaric acid from the oxidation of pentamethyl glucose help to establish the ring structure of glucose.

SP-9.5 How would you prepare the oxime of an aldohexose? What would happen if this oxime is successively treated with acetic anhydride and then warmed with ammonical silver nitrate solution? What is the final product formed? Explain all the steps.

SP-9.6 Write down the conformations of α- and β-D(+)-glucopyranose. Which anomer predominates in an aqueous solution of D(+)-glucose and why?

SP-9.7 D-Glucose on successive methylation, hydrolysis and oxidation with nitric acid affords a *meso*-trimethoxyglutaric acid and (+)-dimethoxysuccinic acid. Write down the different steps and sum up the information you derive from these experiments regarding the structure and configuration of D-glucose.

SP-9.8 Draw the α- and β-pyranose and α- and β-furanose forms of D-glucose. How would you distinguish between methyl glucopyranoside and methyl glucofuranoside by periodic acid oxidation? Write down the products and amounts of periodic acid consumed in each case?

SP-9.9 How do epimers and anomers differ?

SP-9.10 Briefly discuss the anomeric effect in D-glucose and its derivatives.

SP-9.11 Offer explanation in each case:

(i) D-Fructose gives a positive Tollens' test?

(ii) The rate of oxidation of β-D-glucopyranose by bromine water is 250 times as fast as that of α-D-glucopyranose.

(iii) In an equilibrium mixture of methyl α-D-glucopyranoside and methyl β-D-glucopyranoside, the α-anomer predominates.

(iv) The specific rotation of β-D-glucose changes from $+18.7^0$ to a steady value of $+52.6^0$, when dissolved in water.

(v) The rate of acid-catalyzed hydrolysis of methyl α-D-fructofuranoside is some 10^5 times as fast as that of β-anomer.

(vi) The β-methylglucopyranoside of 2-deoxy-D-glucose undergoes hydrolysis several thousand times faster than that of D-glucose itself.

SP-9.12 State with reasoning how Fischer established the configuration of D-arabinose.

SP-9.13 Establish the ring size of D-glucose and draw its stable conformation.

SP-9.14 Write down the hexagonal structures of α-D-glucose and β-D-glucose and explain the phenomenon of mutarotation exhibited by glucose.

SP-9.15 How would you convert an aldose into its epimer?

SP-9.16 On reaction with acetone D-mannose yields a di-isopropylidene derivative that still bears the reducing property. Propose a probable structure of this derivative.

SP-9.17 Indicate stepwise the conversion of an aldopentose to an aldohexose. Explain the reactions involved.

SP-9.18 Give mechanism of the reaction of D-glucose with excess of phenylhydrazone.

SP-9.19 D-Glucose, D-fructose and D-mannose form identical osazone. Justify the phenomenon.

SP-9.20 Discuss the application of periodic acid reaction in the determination of ring size of aldohexoses.

SP-9.21 Propose a mechanism to account for the fact that D-gluconic acid and D-mannonic acid are interconverted when either is heated in pyridine catalyst.

SP-9.22 Reduction of L-glucose with sodium borohydride results the same alditol as obtained from the reduction of D-glucose.

SP-9.23 On oxidation with nitric acid D-glucose gives an optically active aldaric acid, but D-allose yields an optically inactive aldaric acid. Explain.

SP-9.24 Kiliani-Fischer chain lengthening of D-arabinose results in the formation of a mixture of D-glucose and D-mannose. Explain.

SP-9.25 How can you prove that α-D-glucose is linked to β-fructose in sucrose?

SP-9.25 Sucrose cannot reduce Tollens' reagent, but maltose can. Justify.

SP-9.26 What do you mean to say "*inversion of cane-sugar*" and "*invert sugar*"?

APPENDICES

Appendix 1

Applications of Some Useful Reagents (including Organometallics) in Organic Synthesis

1

Reagent: ***9-Borabicyclo[3.3.1]nonane (9-BBN)***

Structure:

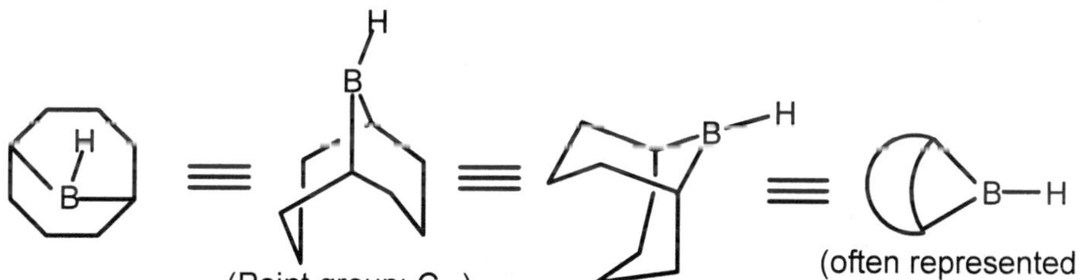

(Point group: C_{2v})

(often represented in this form)

Main objective(s): 9-Borabicyclo[3.3.1]nonane (9-BBN) is an organoborane compound and used in organic chemistry as a hydroboration reagent. 9-BBN is more stable than other boranes in the pure state due to the formation of hydride-bridged dimers in the solid state. The reagent hydroborates preferentially less-hindered bonds, and the reaction occurs with high regioselectivity. For more selective hydroboration, 9-BBN appears to be is an excellent reagent.

Representative illustrations:

(i)

(ii)

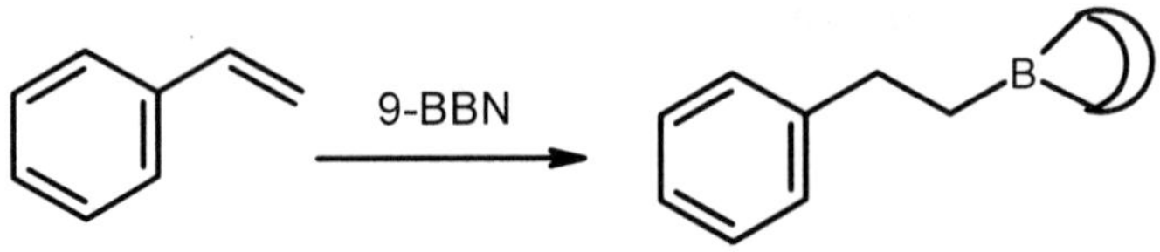

(iii)

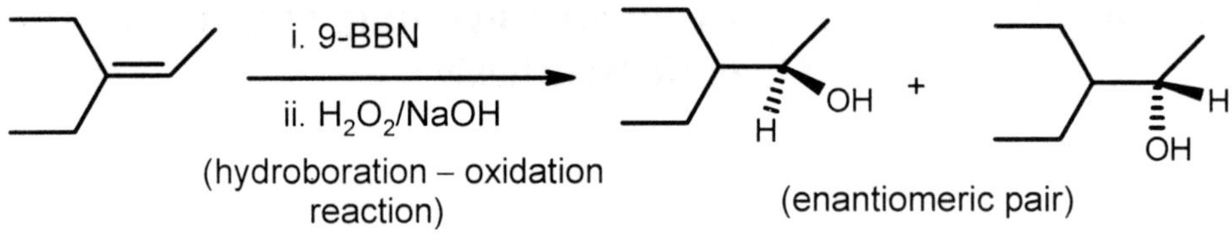

(iv)

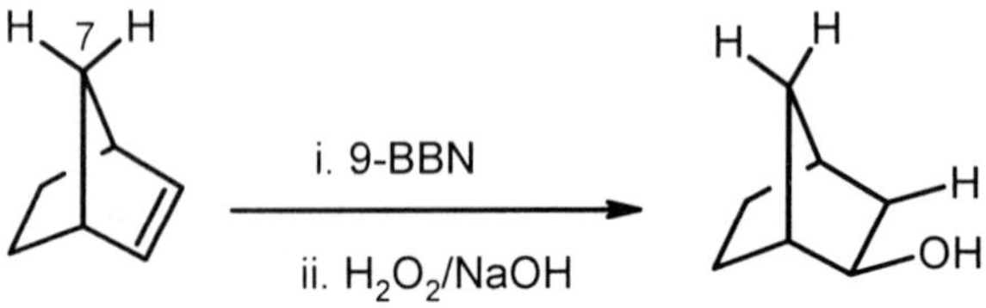

(*exo*-alcohol)

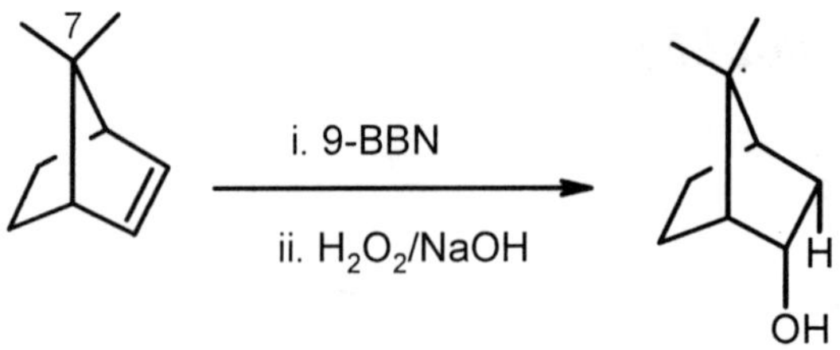

(*endo*-alcohol)
(7-dimethyls create a hindered-zone preventing the attack of the bulky 9-BBN from the *exo*-side)

(v)

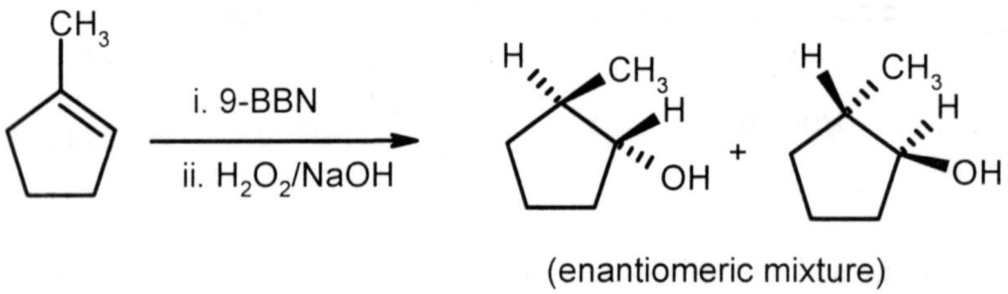

(vi)

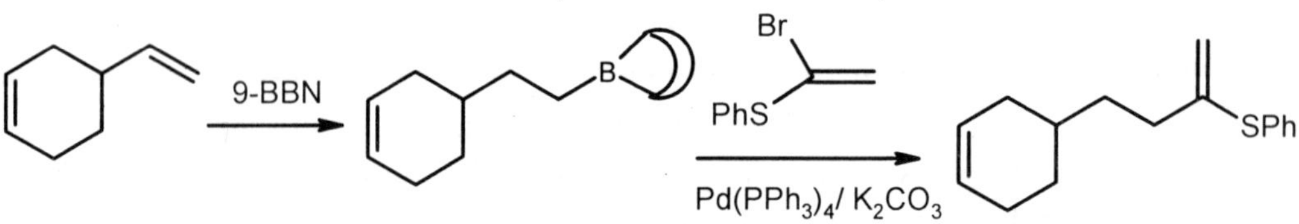

(*Org. Syntheses*, **Coll. Vol. 9**, 107(1998); *ibid*, **Vol. 71**, 89(1993)]

(vii) 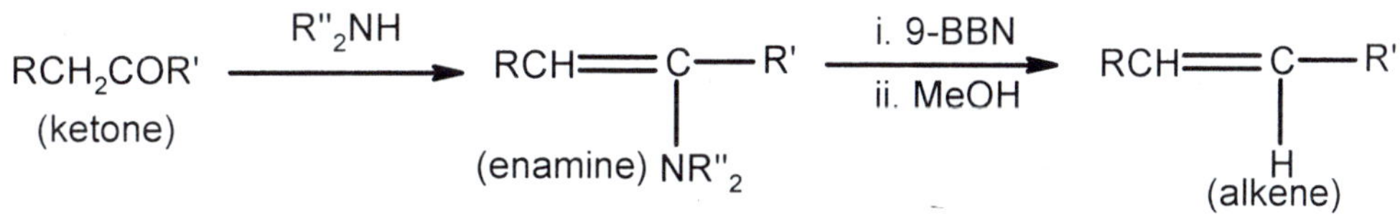

(indirect way to convert a ketone to an alkene) [*J. Org. Chem.*, **56**, 1543(1991)]

2

Reagent: *n-Butyl lithium*
Structure: $CH_3CH_2CH_2CH_2Li$ (written as: *n*-BuLi)
Main objective(s): Used as strong base and also for lithiations.
Representative illustrations:

(i) 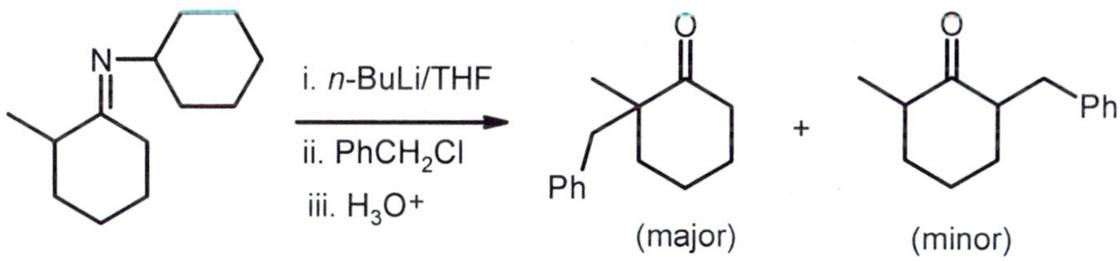

[*J. Am. Chem. Soc.*, **104**, 2081(1982)]

(ii) 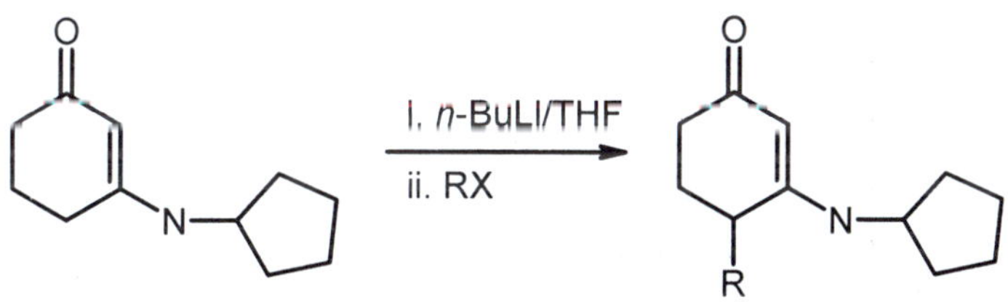

[*Tet. Lett.*, 3963(1974)]

(iii) 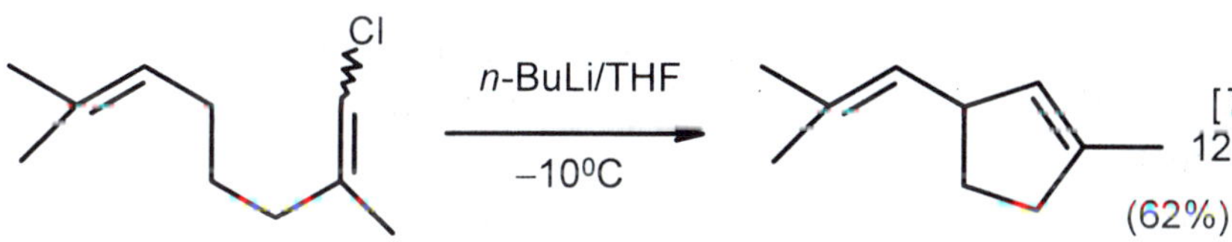

[*Tet. Lett.*, 1207(1974)]

(iv) 

(v) 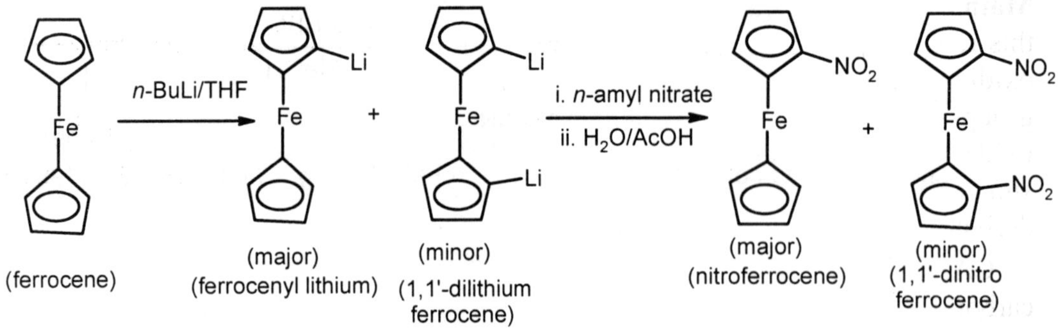

3

Reagent: ***Ceric ammonium nitrate (CAN)***[1]

Also named as ammonium cerium (IV) nitrate, ammonium hexanitrato-cerate(IV), or ammonium nitratocerate(IV).

Structure: CAN is a salt comprised of the anion $[Ce^{IV}(NO_3)_6]^{2-}$ and a pair of NH_4^+ counter ions. X-ray crystallographic studies of CAN reveals that cerium (IV) locates at the centre of the complex-anion where each nitrato group is linked to the cerium atom in a bidentate manner as shown below (one unit):

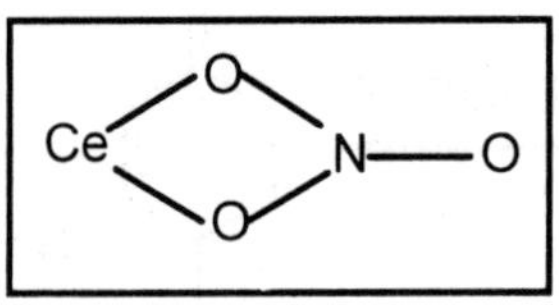

The CeO_{12} core formulates an icosahedron. Although the N–O bonds (on the metal side of the nitrato group) are not symmetrical, yet the anion $[Ce(NO_3)_6]^{2-}$ is regarded to possess T_h symmetry.

Thus the formula of CAN should be written as: $(NH_4)_2[Ce^{IV}(NO_3)_6]$.

Ref. 1: For detailed discussion see the Review Article by Hwu, J. R. and King, K-Y., Versatile reagent ceric ammonium nitrate in modern chemical synthesis, *Current Science*, **81**(18), 1043(2001).

Main objective(s): CAN is an efficient oxidizing agent. Being non-hygroscopic solid, this reagent is readily avialable in pure state and can be used easily. It is a stronger oxidizing agent than even Cl_2. CAN is useful as an oxidant for many functional groups, e.g. oxidation of C-H bonds; oxidation of alcohols, phenols, and ethers; oxidation of nitroalkanes. Oxidative halogenation can be promoted by CAN as an *in situ* oxidant, for benzylic bromination, the iodination of ketones and uracil derivatives.

CAN is used with versatile role in various chemical transformations that include carbon–carbon bond formation, oxidative carbon–carbon bond cleavage, nitration, and removal of protecting groups also.

Representative illustrations:

(i)

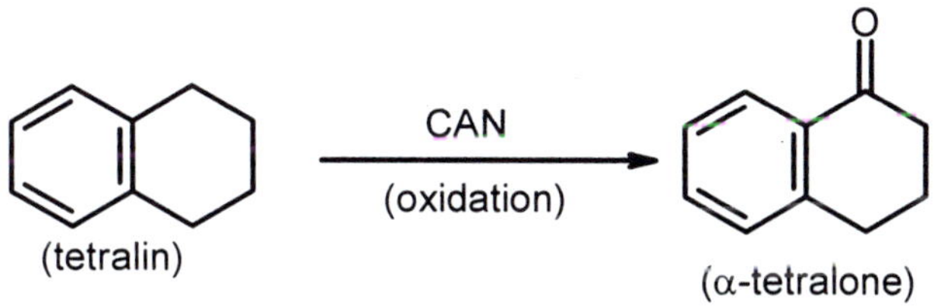

(ii)

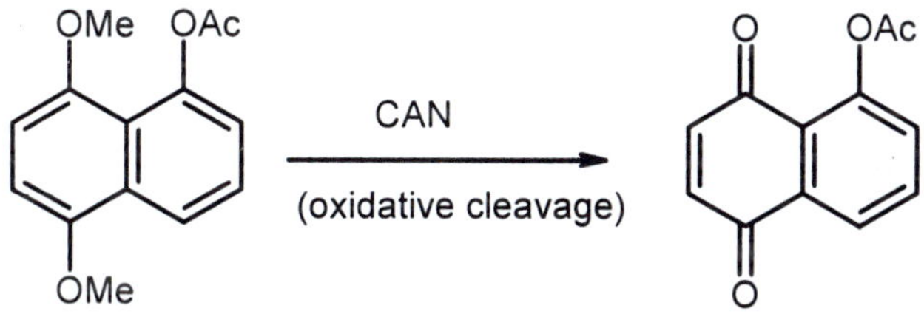

[*J. Org. Chem.*, **51**, 271(1986)]

(iii)

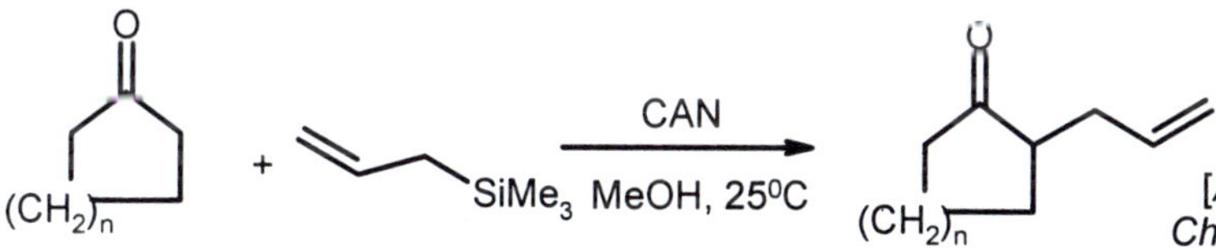

[*Appl. Organomet. Chem.*, **11**, 381(1997)]

[Yield: n =1(35%); n =2(39%); n =4(25%); n =(75%)]

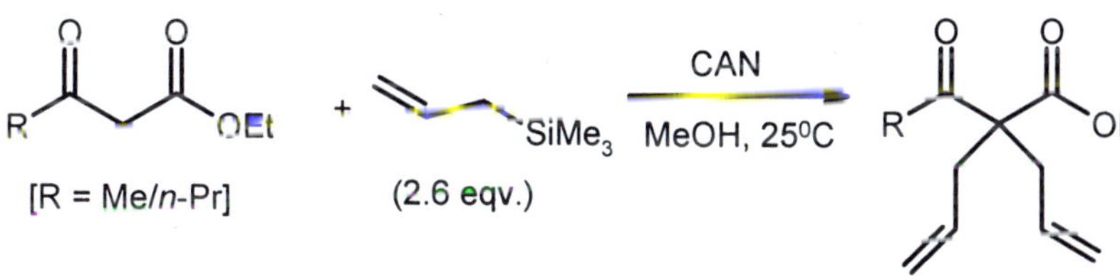

[R =Me(81%); R = *n*-Pr(76%)]

(iv)

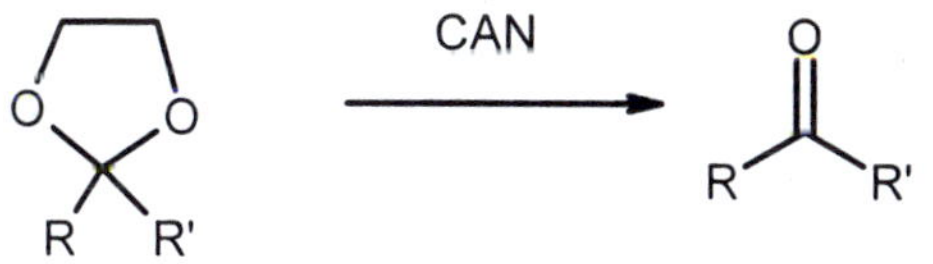

[*Tet. Lett.*, **40**, 1799(1999)]

(v) $RCONHNH_2 \xrightarrow{\text{CAN}} [RCON{=}NH] \xrightarrow[\text{25}^\circ\text{C}]{\text{R'OH}} RCOOR'$ [*Tet. Lett.*, **40**, 4429(1999)]

(vi)

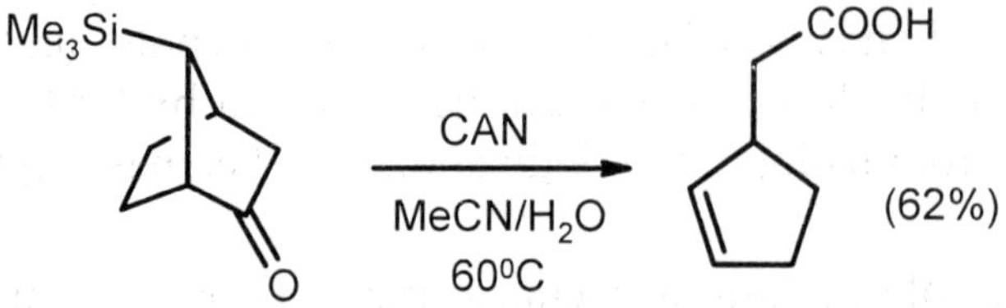

[*J. Am. Chem. Soc.*, **122**, 5899(2000)]

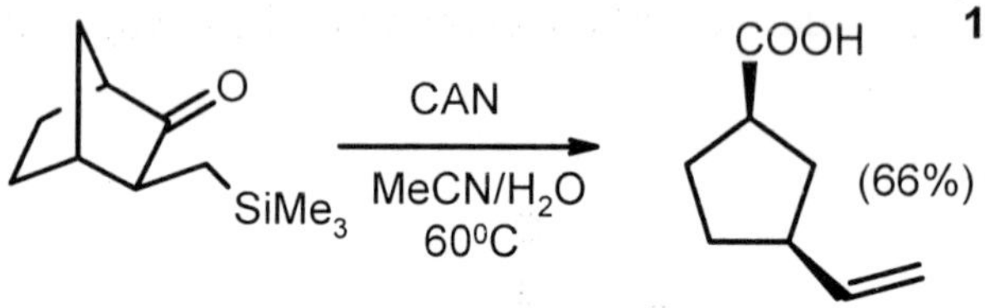

4

Reagent: ***2,3-Dichloro-5,6-dicyano-1,4-benzoquinone (DDQ)***

Structure:

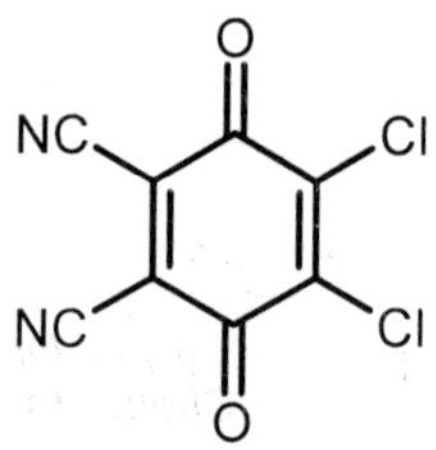

Main objective(s): Generally used as oxidizing agent leading to aromatization, dehydrogenation, oxidative coupling, and also for dehydration.

Representative illustrations:

(i)

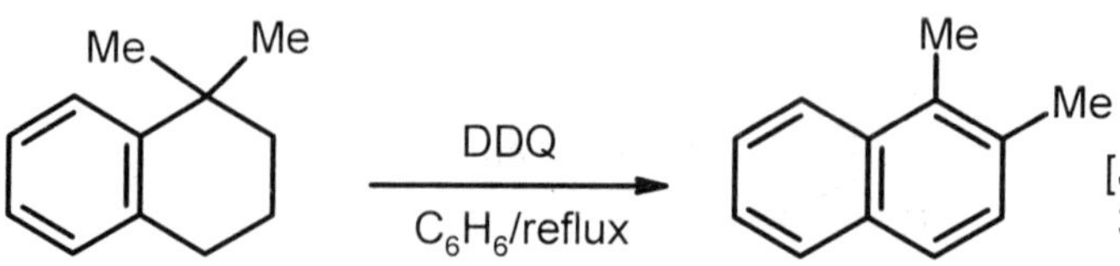

[*J. Chem. Soc.*, 3123, 3133(1960)]

(ii) 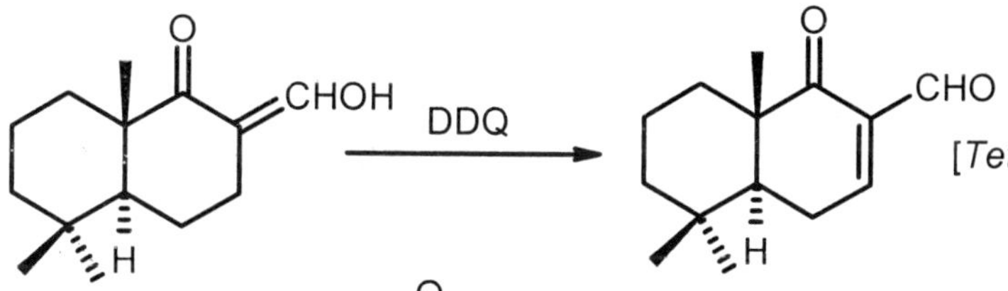

[*Tet. Lett.*, 3119(1985)]

(iii) 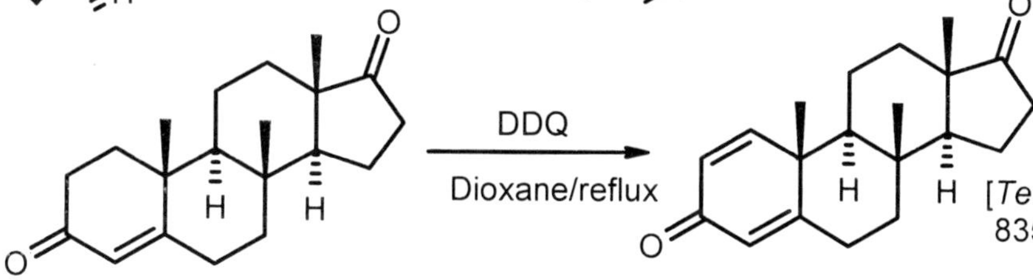

[*Tet. Lett.*, 835(1962)]

(iv) 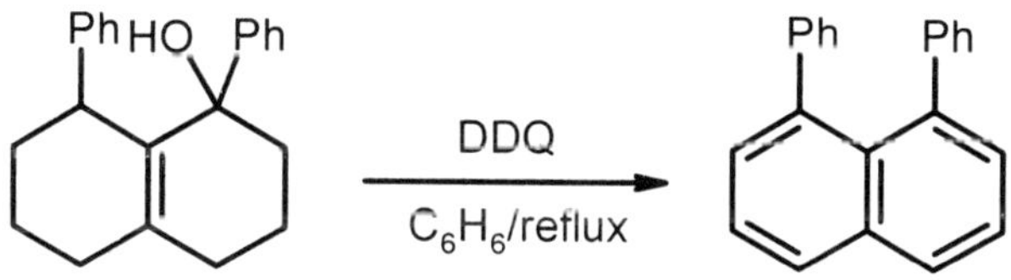

[*J. Org. Chem.*, **30**, 2942(1965)]

5

Reagent: ***1,3-Dicyclohexylcarbodiimide (DCC)***

Structure: 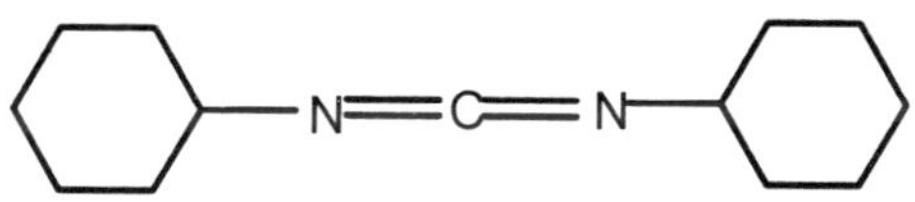

Main objective(s): Generally used as dehydrating agent under mild conditions in the synthesis of various compounds such as esters, amides, anhydrides, lactones, lactums, peptides, etc.

Representative illustrations:

(i) $RCOOH + R'OH \xrightarrow{DCC} RCOOR'$ (ester)

(ii) $RCOOH + R'NH_2 \xrightarrow{DCC} RCONHR'$ (amide)

(iii) 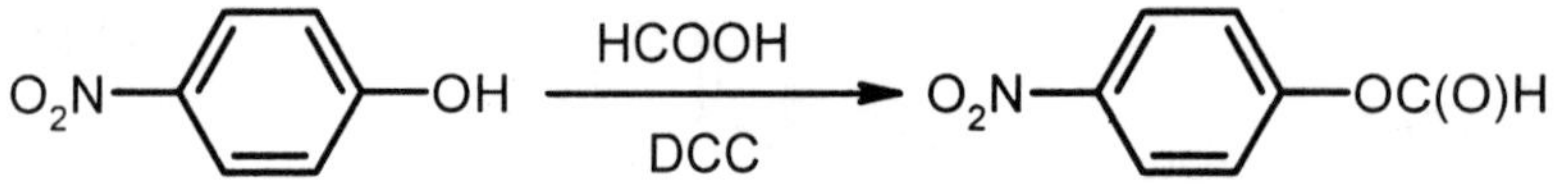

[*Synthesis*, 979(1982)]

(iv) 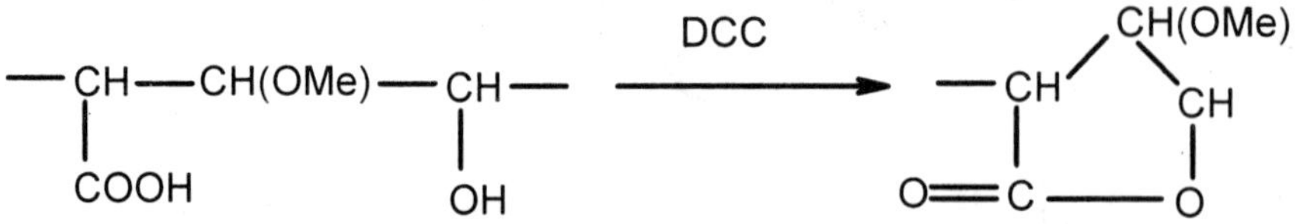

(γ-lactone)

(v) 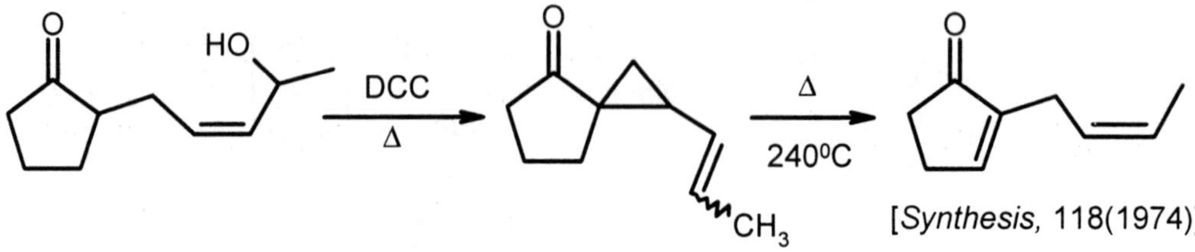

[*Synthesis*, 118(1974)]

6

Reagent: *Grignard Reagents*

Structure: Grignard reagents (RMgX) in solution (dry ether) are believed to maintain an equilibrium between a number of species (*vide* Schlenk equilibrium); amounts of various species is largely dependent upon different factors like nature of 'R' and of 'X', the solvent, temperature and concentration. Both monomeric and dimeric forms (bearing solvent molecules) of Grignard reagents are possible:

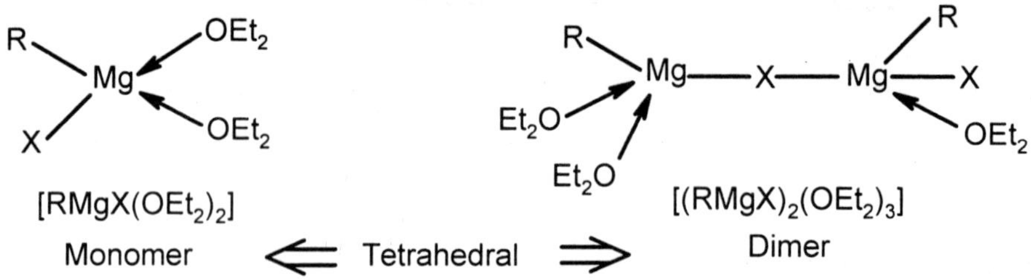

Besides the tetrahedral structural skeleton, penta-coordinated trigonal bipyramidial species are also detected. For convenience, Grignard reagents will be represented here as $R^{\delta-}$— $^{\delta+}MgX$, where R = aliphatic/aromatic group, and X = Cl, Br, or I.

Main objective(s): Multipurpose uses including nucleophilic addition to compounds bearing multiple-bond groups (*e.g.* >C=O, –CN, etc.).

Representative illustrations:

(i) Reaction with compounds having active hydrogen atoms

$$RMgX \xrightarrow{R'OH} RH + Mg(OR')X$$

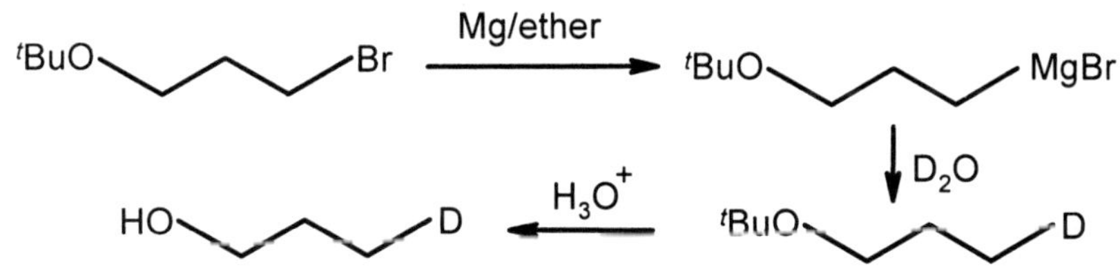

(ii) Formation of primary alcohols

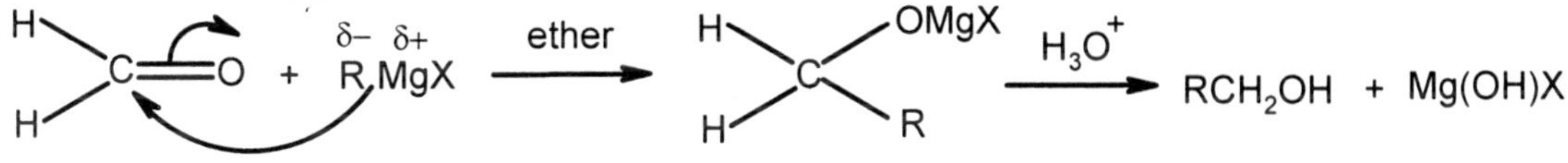

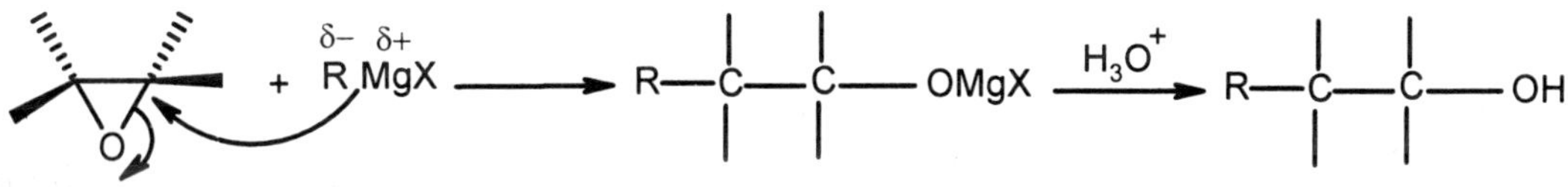

(iii) Formation of secondary alcohols

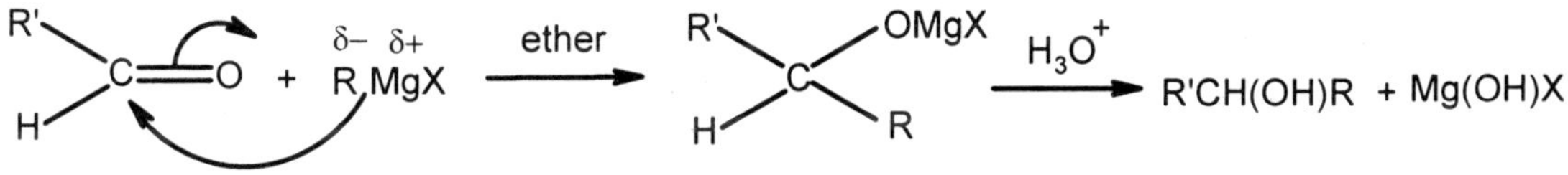

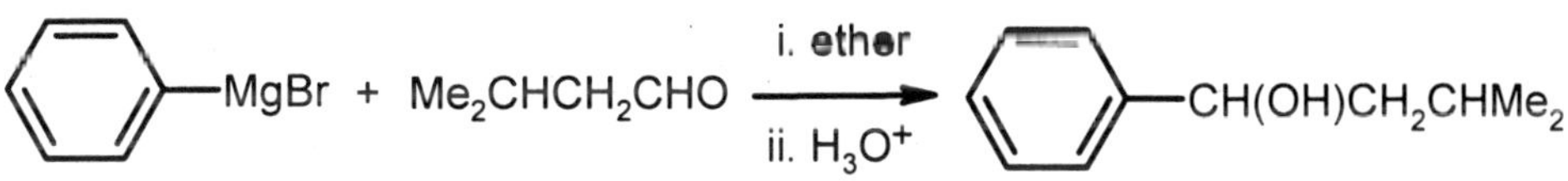

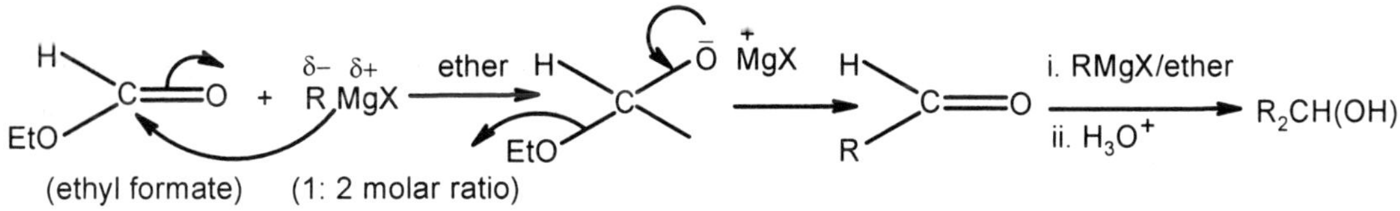

(iv) Formation of tertiary alcohols

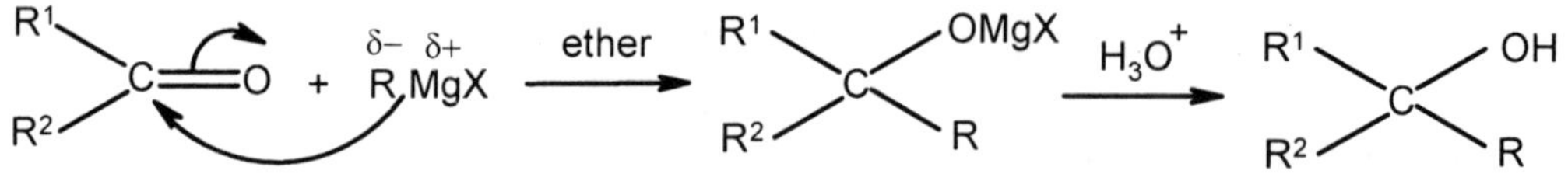

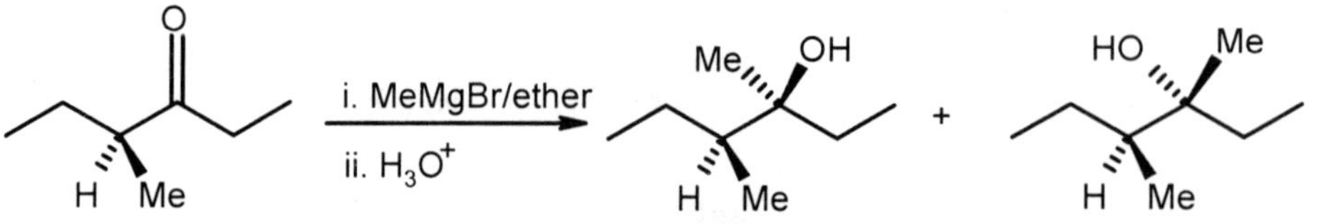

anti-product (major) *syn*-product (minor)

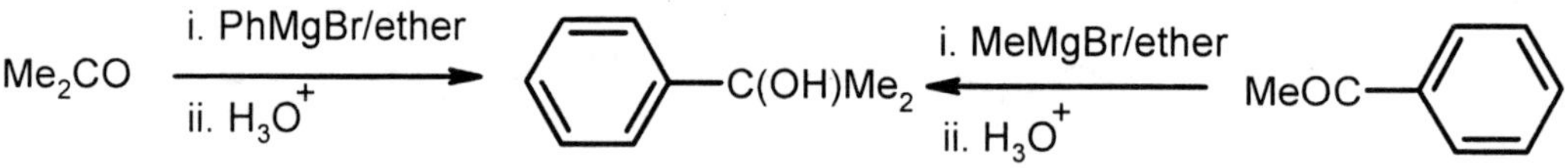

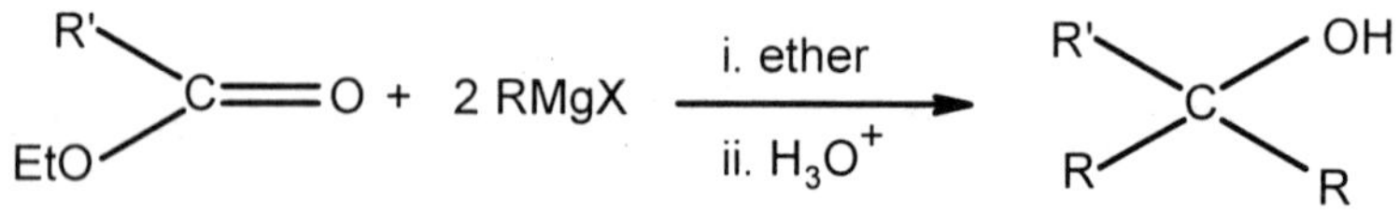

(v) Formation of aldehydes

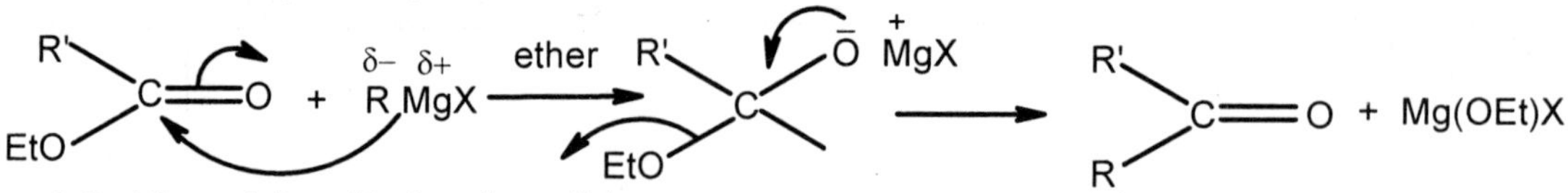

(ethyl formate) (1: 1 molar ratio)

(vi) Formation of ketones

$$R'CN + RMgX \xrightarrow{\text{ether}} [RR'C{=}NMgX] \xrightarrow{H_3O^+} \underset{\text{(ketimine)}}{[RR'C{=}NH]} \xrightarrow{\text{hydrolysis}} RCOR'$$

(vii) Formation of carboxylic acids

$$RMgX + CO_2\text{(dry ice)} \xrightarrow{\text{ether}} RCOOMgX \xrightarrow{H_3O^+} RCOOH$$

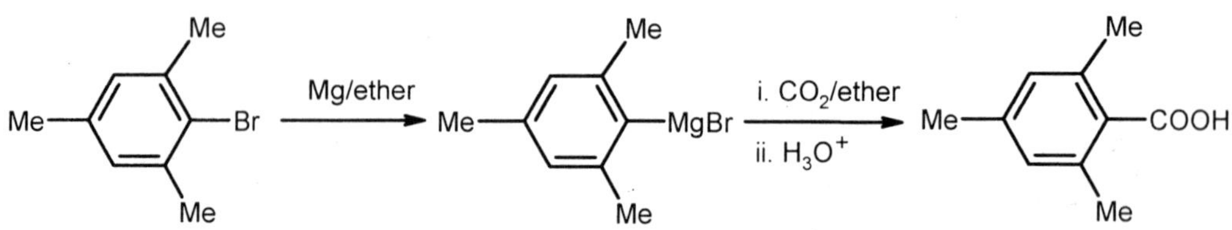

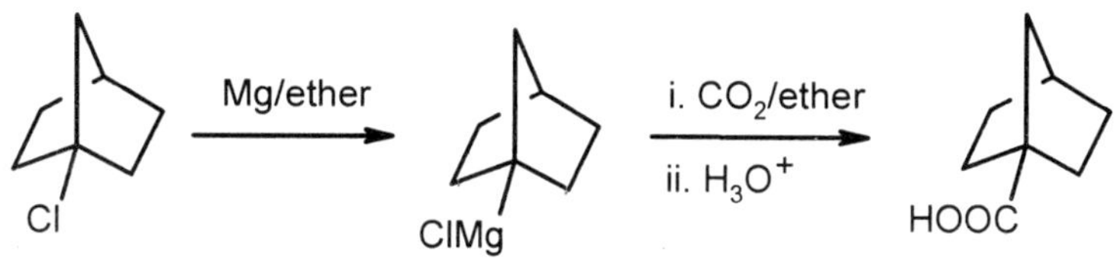

(viii) Formation of amines

$$RMgX + ClNH_2 \xrightarrow{ether} RNH_2 + Mg(X)Cl$$

$$RMgX + H_2N{-}OCH_2Ph \xrightarrow{ether} RNH_2 + Mg(OCH_2Ph)X$$

(ix) Formation of diazoamines

$$R'{-}N{=}\overset{+}{N}{=}\overset{-}{N} + RMgX \xrightarrow{ether} R'{-}N{=}N{-}N(R)MgX \xrightarrow{H_3O^+} R'{-}N{=}N{-}NHR$$

(x) Formation of N,N-dialkyl/diaryl hydroxylamines

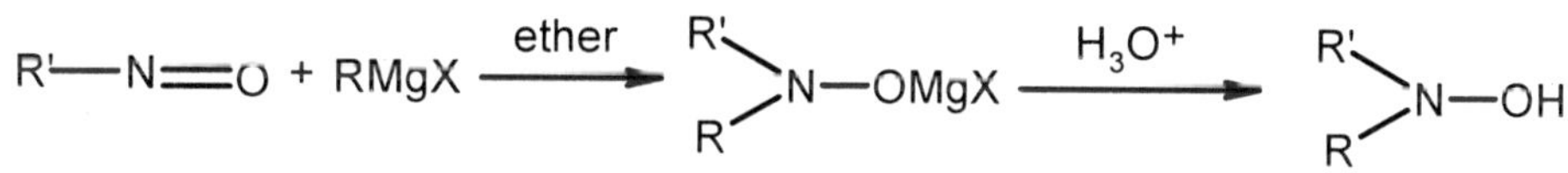

Few more reactions

(viii)

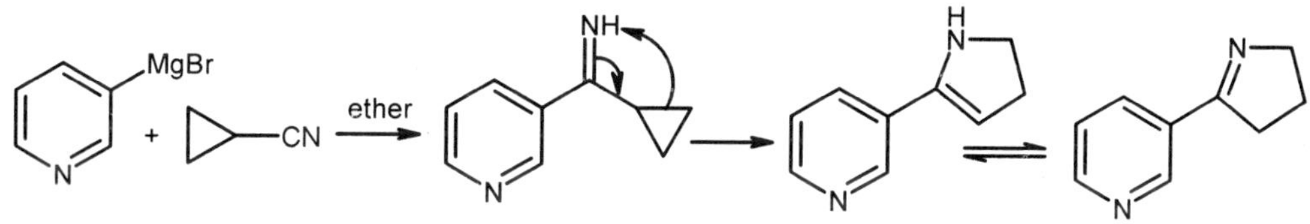

(ix) In the presence of copper salts, Grignard reagents undergo 1,4-addition to enone systems.

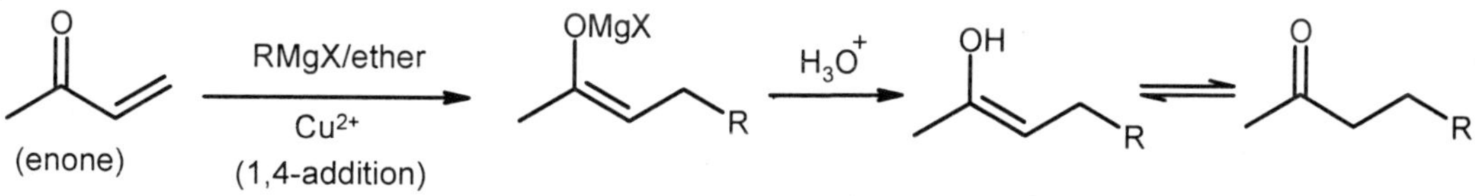

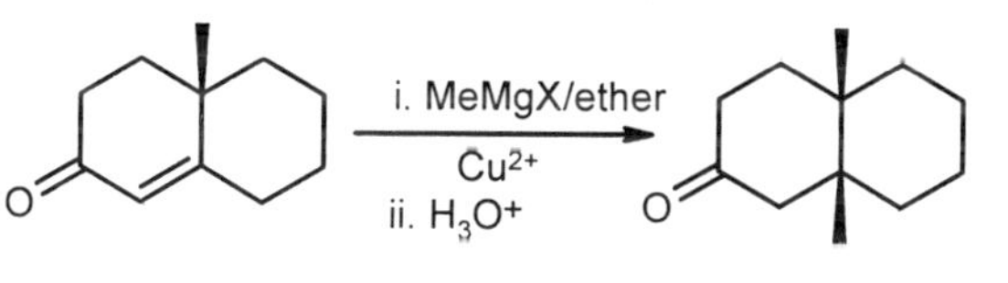

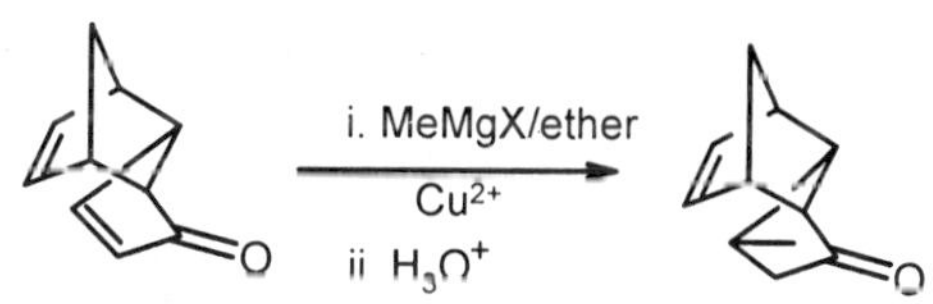

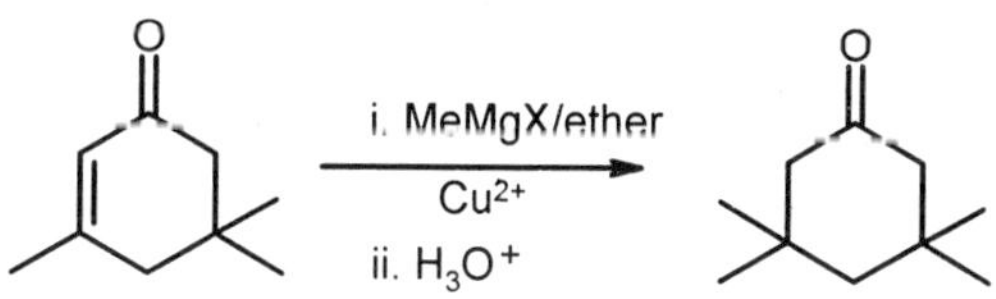

(x) Grignard reagents react with anhydrides in the presence of (–)-sparteine to yield highly enatioselective product [*Angew. Chem.*,**114**, 1099(2002)].

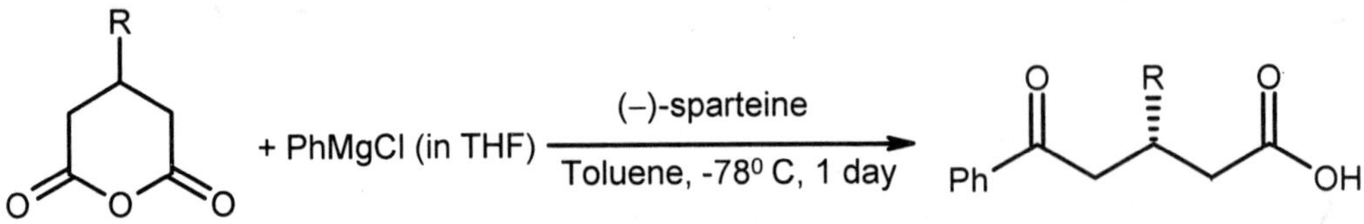

(xi) Though Grignard reagents undergo mostly nucleophilic addition reaction, single electron transfer (SET) mechanism may operate in case of sterically hindered substrates. As for example:

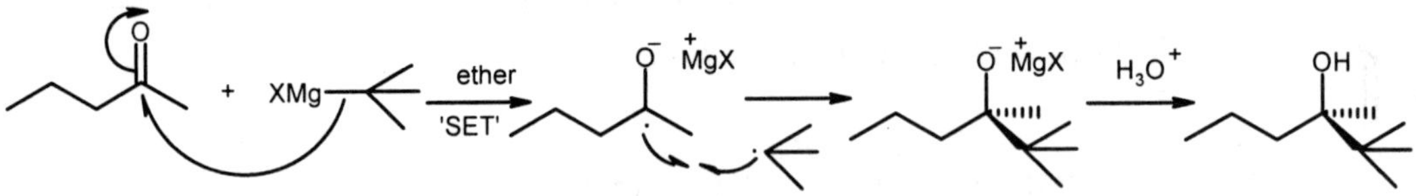

7

Reagent: ***Lithium di-isopropylamide (LDA)***

Structure:

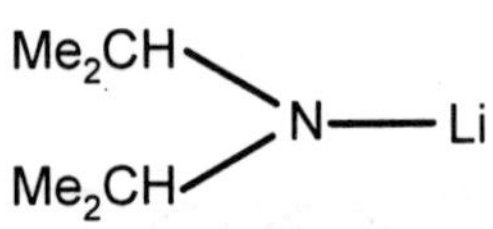

Main objective(s): LDA is a strong and hindered base that abstracts a proton from the less substituted α-carbon of compounds containing carbonyl function at low temperature (~ –78^{0}C) in a non-polar solvent (*e.g.* THF or ether) leading to an irreversible formation of *kinetic enolates*.

Representative illustrations:

(i)

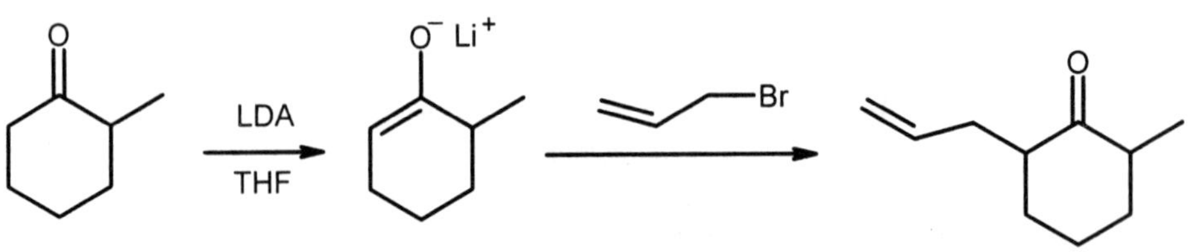

(ii)

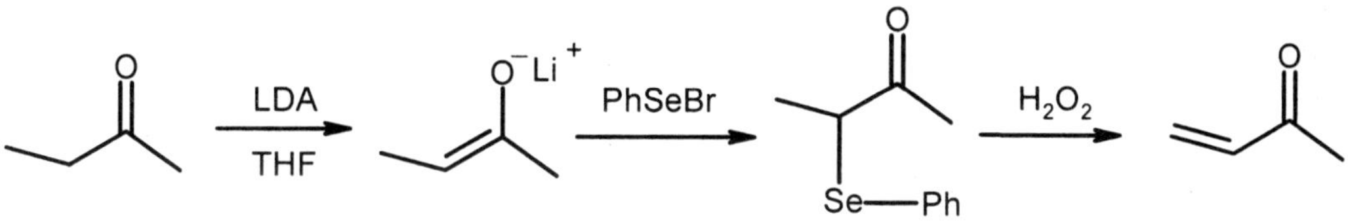

(iii)

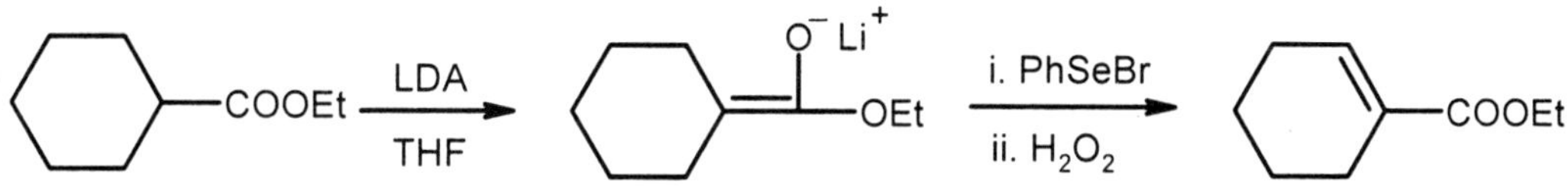

(iv)

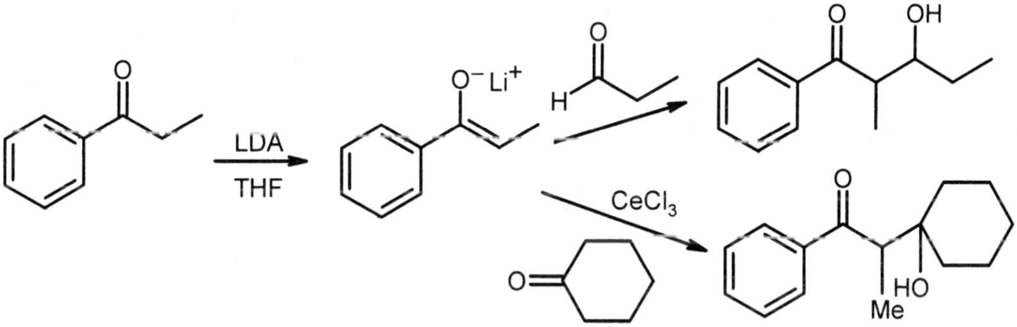

(v)

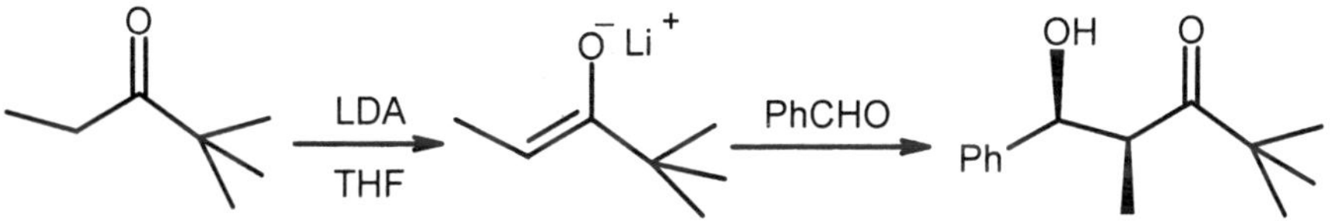

[*J. Org. Chem.*, **46**, 191(1981)]

(vi)

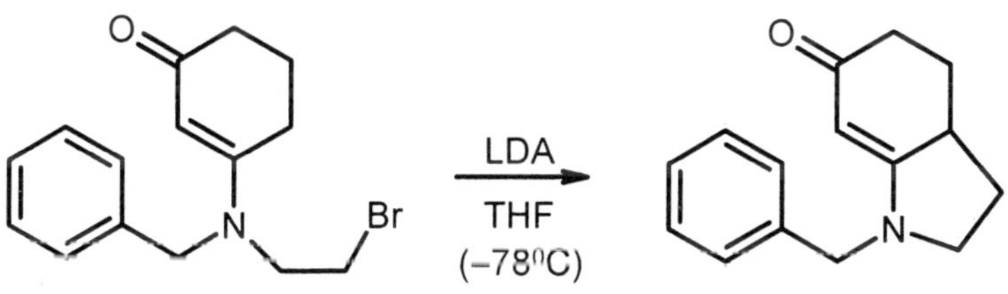

[*Tet. Lett.*, 3963(1974)]

(vii)

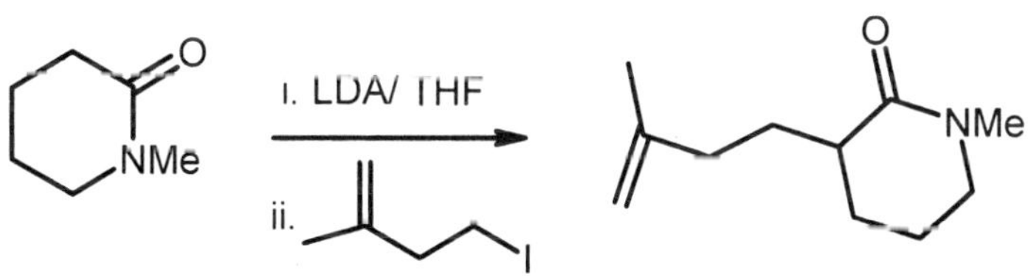

[*J. Org. Chem.*, **39**, 2475 (1974)]

8

Reagent: ***Lithium diorganocuprates (Gilman reagents)***

Structure: R_2CuLi

$$2\ RLi + CuI \xrightarrow{\text{ether}} R_2\overset{-}{Cu}\ \overset{+}{Li}$$

$$\underset{\text{(methyllithium)}}{2\ (CH_3)_2Li} + \underset{\text{(cuprous iodide)}}{CuI} \xrightarrow{\text{ether}} \underset{\text{(lithium dimethylcuprate) a Gilman reagent}}{(CH_3)_2CuLi}$$

Main objective(s): Gilman reagents undergo organometallic coupling reactions with alkyl/aryl halides (not the fluorides) where one of the organic moieties from the reagent replaces the halogen of the substrate molecule, resulting in the formation of a new carbon–carbon bond. These reagents also attack epoxides and α,β-unsatuared carbonyl compounds.

Representative illustrations:

(i) $CH_3(CH_2)_4CH_2I + (CH_3)_2CuLi \xrightarrow{\text{ether}} CH_3(CH_2)_4CH_2CH_3$

(ii) 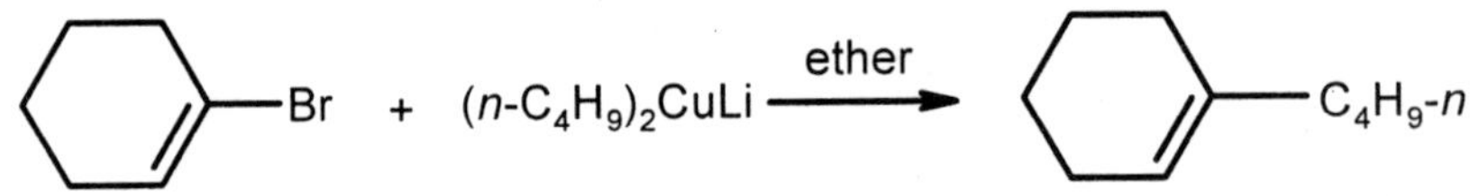

(iii) 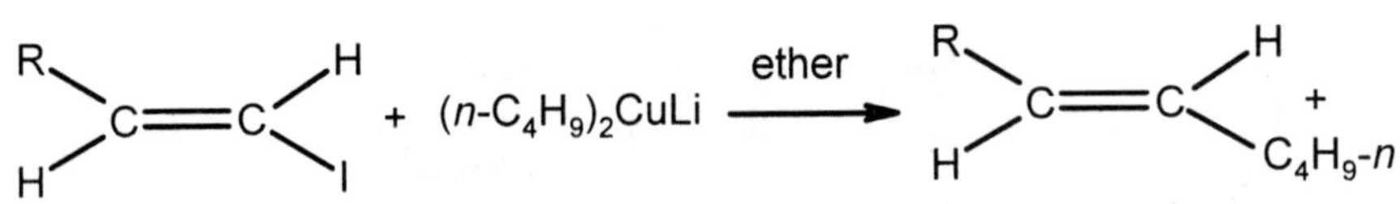

(iv) 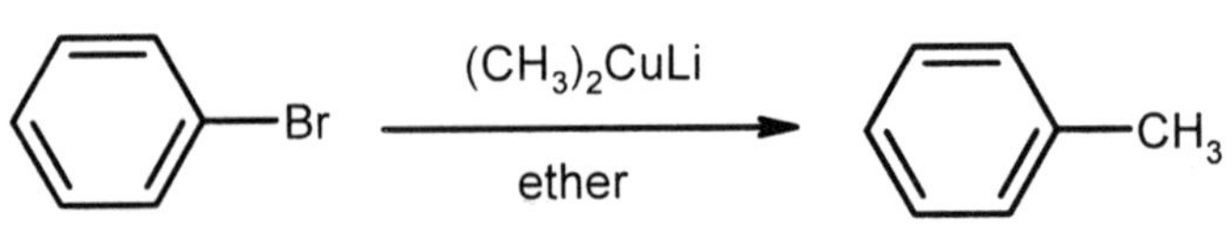

(v) 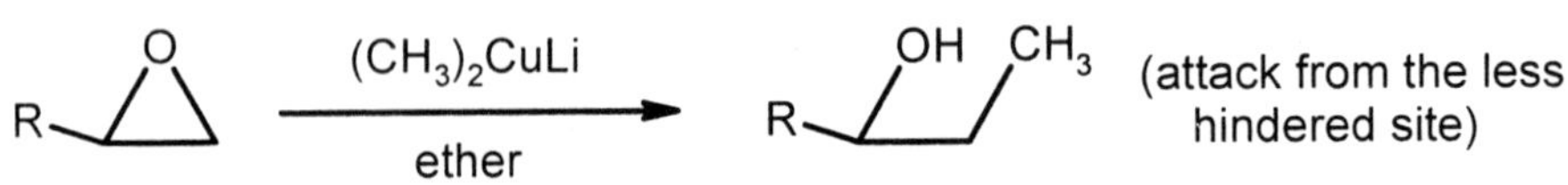

(vi)

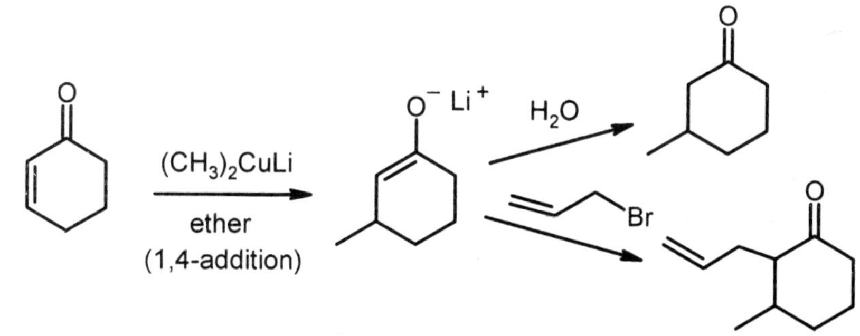

(vii)

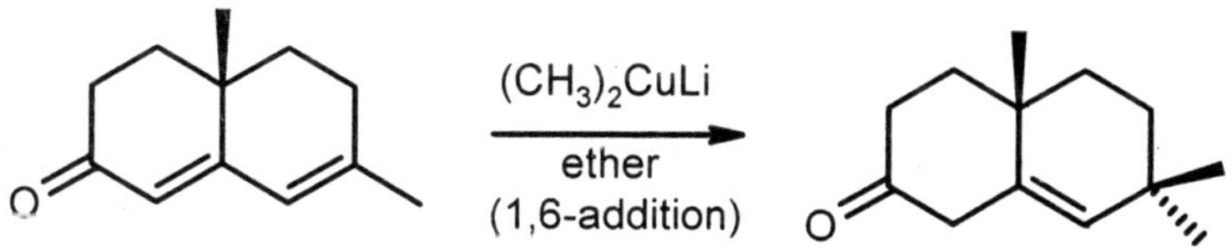

(viii)

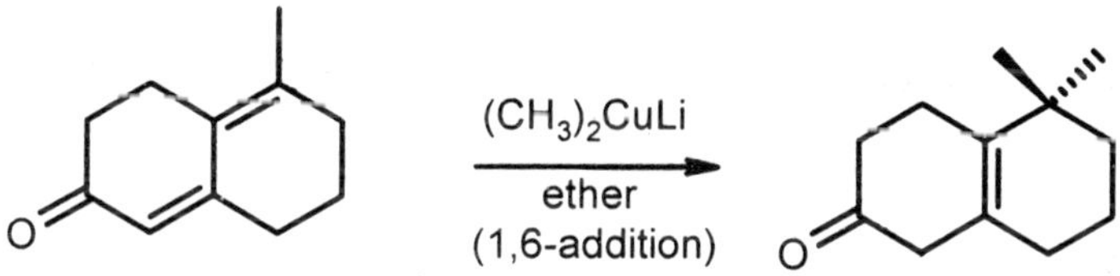

(ix)

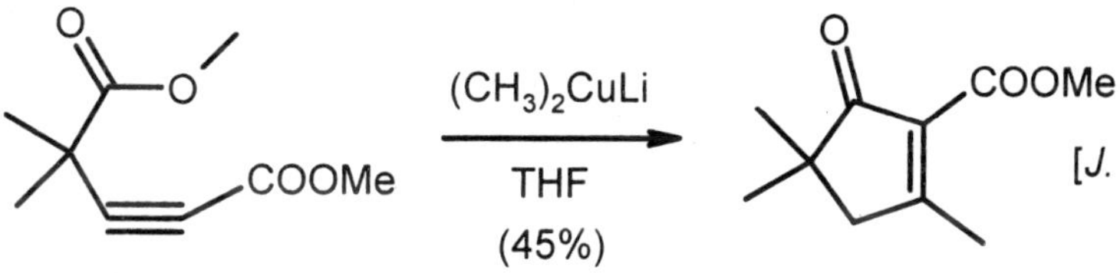

[*J. Org. Chem.*, **49**(16), 3033(1984)]

(x)

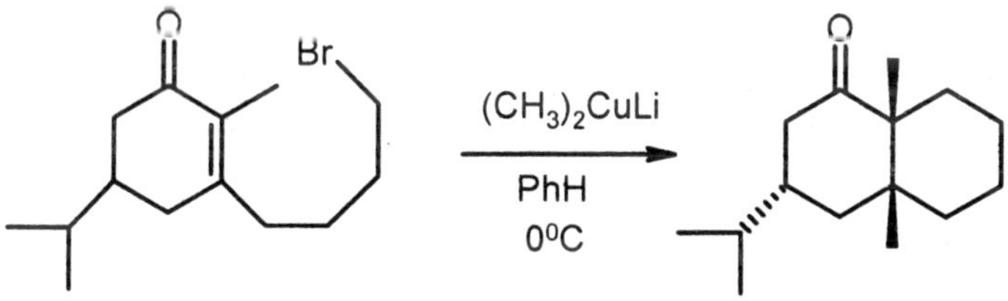

[*Tet. Lett.*, 2591(1974)]

9

Reagent: ***N-Bromosuccinimide (NBS)***

Structure:

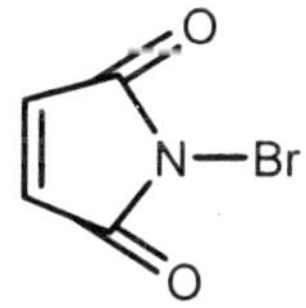

Main objective(s): Specific for allylic and benzylic brominations.

Representative illustrations:

(i)

(ii) $CH_3CH{=}CHCOOEt \xrightarrow{NBS/CCl_4} BrCH_2CH{=}CHCOOEt$

(iii) $(CH_3)_2CHCH{=}CHCOOEt \xrightarrow{NBS/CCl_4} (CH_3)_2C(Br)CH{=}CHCOOEt$ [*J. Org. Chem.*, **24**, 1577(1959)]

(iv) $CH_3CH{=}CHCH_2CH_3 \xrightarrow{NBS/CCl_4} CH_3CH{=}CHCH(Br)CH_3$ (major) + $BrCH_2CH{=}CHCH_2CH_3$ (minor)

(v)

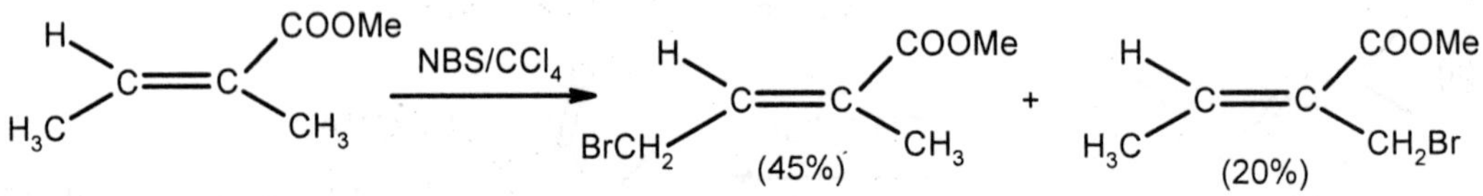

(vi)

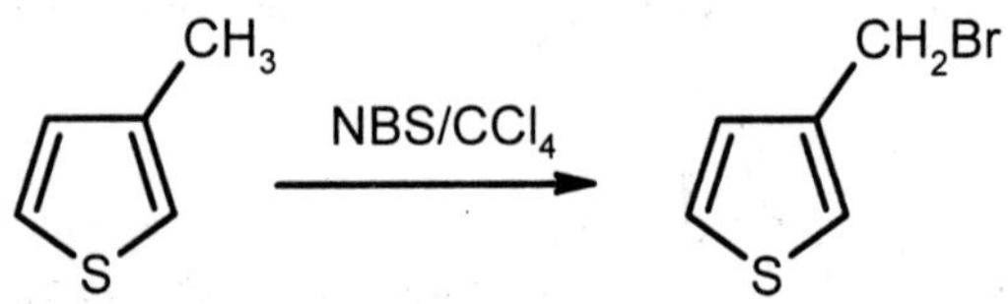

[*Org. Syn.*, **Coll. Vol. 4**, 921(1963)]

(vii)

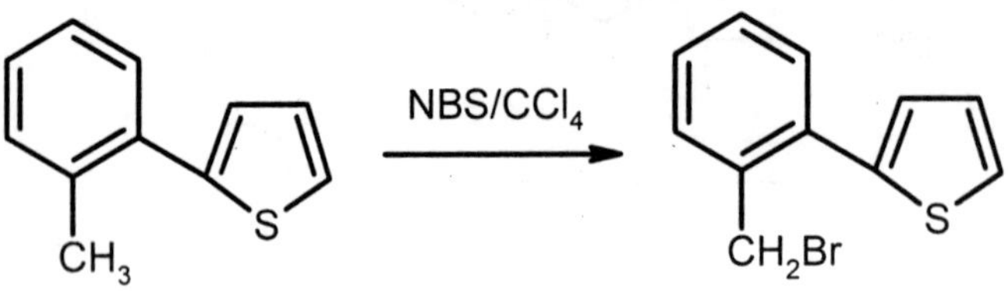

[*J. Org. Chem.*, **51**, 142(1986)]

(viii)

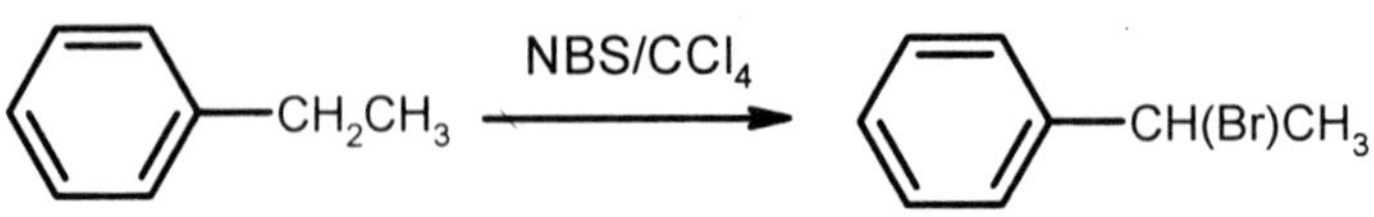

Appendix 3

Chemical shifts for some selected types of protons

Category 1: ***Approximate chemical shifts for different types of protons (ppm)***

[Reference peak: $(CH_3)_4Si$ 0 δ (ppm)]

Proton	δ	Proton	δ
CH_3–C	0.9-1.3	–C–CH_2–OR	3.4
CH_3–C–C=C	1.1	–C–CH_2–OH	3.6
CH_3–C–O	1.4	–C–CH_2–OAr	4.3
CH_3–C=C	1.6	–C=CH_2	5.3
CH_3–NR_2	2.2	R_3CH	1.5
CH_3–Ar	2.3	Alkynyl proton	2.5–2.7
CH_3–CO –R	2.2	–C–CH–C–O	2.0
CH_3–CO –Ar	2.6	–C–CH–Ar	3.0
CH_3–CO –OR	2.0	–C–CH–CO–R	2.7
CH_3–CO –OAr	2.4	–C–CH–O–R	3.3-4.0
CH_3–Br	2.7	–C–CH–O–H	3.4-4.0
CH_3 – Cl	3.1	–C–CH–NO_2	4.7
CH_3–OR	3.3	RCOO-CH	3.7–4.1
CH_3–OAr	3.8	–CH–COOR	2.0–2.2
CH_3–CO –NHR	2.0	–CH–COOH	2.0–2.6
CH_3 –O–CO –R	2.7	R_2CH–C–Cl	1.6
CH_3–S	2.1	HC–CO–	2.0–2.7
CH_3–C–NO_2	1.6	RCHO	9.7 – 10.0
–C=C(CH_3)–CO–	1.8	H–CO–O–	8.0
R_2CH_2	1.3	–C=CH–	4.9
R–CH_2–C–Cl	1.7	–CH–Br	2.7–4.1
–C–CH_2–C–C=C	1.7	–CH–Cl	3.1–4.1
–C–CH_2–C–O	1.9	–CH–I	2.0–4.0
–C–CH_2–C=C	2.3	Ar–H	6.5–8.0
–C–CH_2–Ar	2.7	Ar–OH	4.0–7.0
–C–CH_2–CO–R	2.4	CH=CH–OH	>15
–C–CH_2–CO–OR	2.2	RCOOH	11.0–12.0

Category 2: ***Approximate chemical shifts for some protons in alicyclic and heterocyclic molecules***

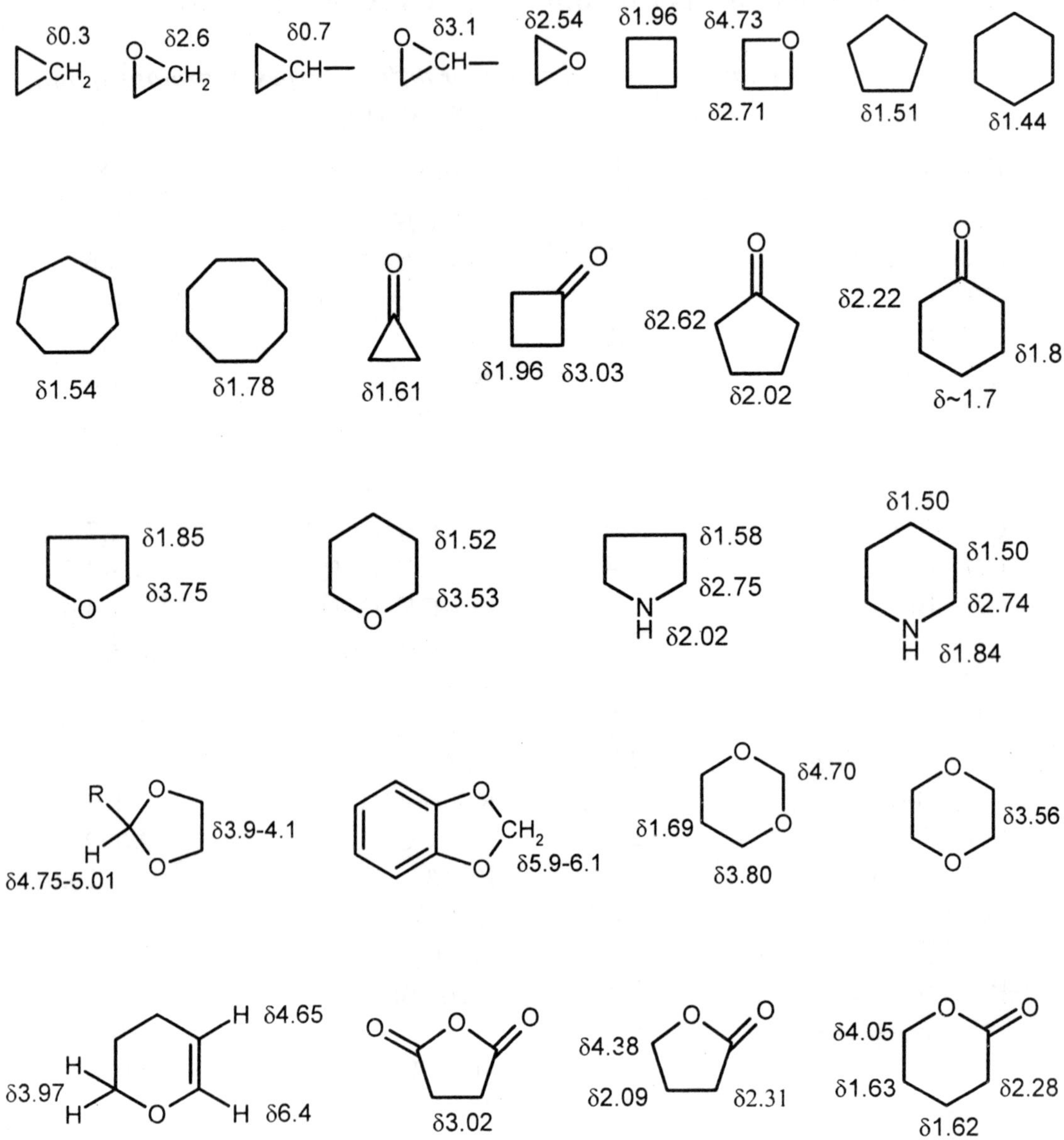

Appendix 4

Reference List

Author's acknowledgements

This book '***Organic Chemistry Through Solved Problems***' has been designed to present various aspects of Organic Chemistry in a problem solving approach, and in true sense it is a problem-solution based study material. During preparation of this book, a good number of problems have been adapted from the previous year question papers of a number of Indian Universities as well as previous year question papers of different competitive examinations involving Chemistry as a subject. For exact and specific clues to the problems, the author has consulted numerous original literature sources (a very few of which have been referred in the text in respective places) in combination with various standard reference books. The author is wholeheartedly acknowledging the help from all these sources, and he expresses his deep gratitude to the respective authors and publishers concerned. A list of standard reference books that have been deeply consulted by the author is cited below as a recommendation for further studies:

Books Consulted, and Recommended for Further Reading

1. March's Advanced Organic Chemistry (Reactions, Mechanisms, and Structure), 5th Edn., Michael B. Smith and Jerry March, John Wiley & Sons, Inc., 2001.
2. Organic photochemistry, J.M. Coxon and B.H. Halton, Cambridge University Press, 1974.
3. Organic Reactions and Orbital Symmetry (2nd Edn.), T.L. Gilchrist and R.C. Storr, Cambridge University Press, 1979.
4. Modern Molecular Photochemistry, Nicholas J. Turro, The Benjamin/Cummings Publishing Co., Inc., 1978.
5. The Conservation of Orbital Symmetry, R. B. Woodward and R. Hoffman, Academic Press, 1970.
6. Pericyclic Reactions – A Textbook, S. Sankararaman, Wiley-VCH Verlag, 2005.

7. Photochemistry and Pericyclic Reactions (2nd Edn.), Jagdamba Singh and Jaya Singh, New Age International Publishers, 2005.
8. Organic Chemistry (6th Edn.), Robert Thornton Morrison and Robert Neilson Boyd, Prentice-Hall of India Pvt. Ltd., New Delhi, 1988.
9. Organic Chemistry, Frederick G. Bordwell, The Macmillan Company, New York, 1963.
10. Mechanism and Structure in Organic Chemistry, Edwin S. Gould, Holt, Rinehart and Winston, Inc., New York, 1959.
11. Organic Chemistry (8th Edn.), T. W. Graham Solomons and Craig B. Fryhle, John Wiley & Sons, Inc., 2004.
12. Organic Chemistry (2nd Edn.), John McMurry, Brooks/Cole Publishing Company, Wadsworth, Inc., California, 1988.
13. Organic Chemistry (3rd Edn.), Francis A. Carey, Tata McGraw-Hill Edition, New Delhi, 1999.
14. Basic Principles of Organic Chemistry, John D. Roberts and Marjorie C. Caserio, W. A. Benjamin, Inc., USA, 1964.
15. Heterocyclic Chemistry (Synthses, Reactions and Mechanisms), Raj K. Bansal, Wiley Eastern Limited, New Delhi, 1999.
16. The Chemistry of Heterocyclic Compounds, A. A. Morton, McGraw-Hill Book Company, Inc., 1946.
17. Organic Synthesis, Michael B. Smith, McGraw-Hill, Inc., 1994.
18. Advanced Organic Chemistry (Reaction Mechanisms), Reinhard Bruckner, Harcourt/Academic Press, 2002.
19. Organic Chemistry (3rd Edn.), G. M. Loudon, The Benjamin/Cummings Publishing Co., Inc., 1988.
20. The Logic of Chemical Synthesis, E.J. Corey and Xue-Min Cheng, John Wiley & Sons, Inc., 1989.
21. Mechanism and Theory in Organic Chemistry (3rd Edn.), Thomas H. Lowry and K. S. Richardson, Addison-Wesley, 1998.
22. Organic Chemistry (5th Edn.), L. G. Wade, Jr., Pearson Education, Inc., 2004.
23. Annual Reports in Organic Synthsis-1974, L. S. Hegedus and S. R. Wilson (Editors), Academic Press, 1975.
24. A Guidebook to Mechanism in Organic Chemistry (6th Edn.), Peter Sykes, Orient Longman Ltd., 1970.
25. The Modern Structural Theory of Organic Chemistry, Lloyd N. Ferguson, Prentice-Hall of India Pvt. Ltd., New Delhi, 1963.
26. Heterocyclic Chemistry (4th Edn.), J. A. Joule and K. Mills, Blackwell Science, 2000.
27. Organic Chemistry (6th Edn.), Vol. 1, I. L. Finar, ELBS/Longman, 1985.

28. Organic Chemistry (5th Edn.), Vol. 2, I. L. Finar, ELBS/Longman, 1986.
29. Advanced Organic Synthesis (Methods and Techniques), Richard S. Monson, Academic Press, 1971.
30. A Guide to Understanding Basic Organic Reactions, R. C. Whitfield, Longman, 1966.
31. Handbook of Structural Analysis, Yasuhide Yukawa (Editor), W. A. Benjamin, Inc., 1965.
32. Advanced Organic Chemistry (Reactions and Mechanisms), 2nd Edn., B. Miller and R. Prasad, Pearson Education, Inc/Dorling Kindersley (India) Pvt. Ltd., 2006.
33. Organic Chemistry, J. Clayden, N. Greeves, S. Warren and P. Wothers, Oxford University Press, 2004.
34. Some Modern Methods of Organic Synthesis (3rd Edn.), W. Carruthers, Cambridge University Press, 1993.
35. Nomenclature of Organic Compounds, S. C. Pal, Narosa Publishing House Pvt. Ltd., New Delhi, 2006.
36. Organic Chemistry (3rd Edn.), M. A. Fox and J. K. Whitesell, Jones and Bartlett Publishers, Inc., 2004.
37. Organic Synthesis: The Disconnection Approach, Stuart Warren, John Wiley & Sons, Inc., 1982.
38. Organic Chemistry (3rd Edn.), Seyhan Ege, AITBS Pub. & Distr., Delhi, 1998.
39. Organic Name Reactions: A Unified Approach, G. Brahmachari, Narosa Publishing House Pvt. Ltd., New Delhi, 2006.
40. Reagents for Organic Synthesis, Vol. 1 and 2, L. F. Fieser and M. Fieser, Wiley/Interscience, 1967 and 1969.
41. Organic Synthesis, C. Willis and M. Wills, Oxford University Press, 2003.
42. Name Reactions and Reagents in Organic Synthesis, B. P. Mundy and M. G. Ellerd, John Wiley & Sons, 1988.
43. Stereochemistry of Organic Compounds, E. L. Eliel, S. H. Wilson, L. N. Mander, John Wiley & Sons, Inc., 2003.
44. Stereochemistry of Carbon Compounds, E. L. Eliel, Tata McGraw-Hill Edition, New Delhi, 1988.
45. Stereochemistry of Organic Compounds (Principles and Applications), 2nd Edn., D. Nasipuri, Wiley Eastern Limited, New Delhi, 1994.
46. Conformational Analysis, E.L. Eliel, N.L. Allinger, S. J. Angyal and G. A Morrison, John Wiley & Sons, Inc., 1967.
47. Stereochemistry at a Glance, J. Eames and J. Peach, Blackwell Science, 2003.
48. Introduction to Stereochemistry, K. Mislow, W. A. Benjamin, New York, 1965.

49. Principles of Organic Stereochemistry, B. Testa, Marcel Dekker, New York, 1979.
50. Stereochemistry & Conformational Analysis, E. Juaristi, John Wiley & Sons, Inc., 1991.
51. Stereochemistry: Concepts and Applications, M. Nogradi, Pergamon Press, New York, 1981.
52. Introduction to Spectroscopy (3rd Edn.), D. L. Pavia, G. M. Lampman and G.S. Kriz, Harcourt, Inc., 2001.
53. Spectroscopic Identification of Organic Compounds (5th Edn.), R. M. Silverstein, G. C. Bassler and T. C. Morrill, John Wiley & Sons, Inc., 1991.
54. Spectroscopic Methods in Organic Chemistry (5th Edn.), D. H. Williams and I. Fleming, Tata McGraw-Hill Edition, New Delhi, 2004.
55. Organic Spectroscopy (3rd Edn.), William Kemp, Macmillan Press Ltd., 1991.
56. Solving Problems with NMR Spectroscopy, Atta-ur-Rahman and Md. Iqbal Choudhary, Academic Press, Inc., 1996.
57. Spectroscopy of Organic Compounds (4th Edn.), P. S. Kalsi, New Age International Publishers, New Delhi, 2000.
58. Elementary Organic Spectroscopy (Principles and Chemical Applications), Y. R. Sharma, S. Chand & Company Ltd., New Delhi, 1992.
59. Organic Spectroscopy (Principles and Chemical Applications), 2nd Edn., Jag Mohan, Narosa Publishing House Pvt. Ltd., New Delhi, 2004.
60. Applications of NMR Spectroscopy in Organic Chemistry, Norman S. Bhacca and D. H. Williams, Holden-Day, Inc., 1964.
61. Advances in Carbohydrate Chemistry (Book Series, Vol. 1-17), W. W. Pigman, M. L. Wolform, C. S. Hudson (Editors), Academic Press, 1945-1962.
62. Heterocyclic Compounds (Vol. 6), R. C. Elderfield (Editor), John Wiley & Sons, Inc., 1957.
63. Reaction Mechanism in Organic Chemistry (3rd Edn.), S. M. Mukherji and S. P. Singh, Macmillan India Limited, New Delhi, 1996.
64. Reaction Mechanisms (3rd Edn.), Raj K. Bansal, Tata McGraw-Hill Edition, New Delhi, 2005.
65. Elements of Organic Chemistry, J. H. Richards, D. J. Cram and George S. Hammond, McGraw-Hill, Inc., 1967.

Index

Abbreviations

Ac	Acetyl
AIBN	2,2'-Azobisisobutyronitrile
Ar	Aryl
9-BBN	9-Borabicyclo[3.3.1]nonane
BHT	Butylated hydroxy toluene
Boc/BOC	*tert*-Butyloxycarbonyl
CAN	Ceric ammonium nitrate
Cbz	Carboxybenzyl
DABCO	1,4-Diazabicyclo[2.2.2]octane
DBE	Double bond equivalent
DCC	*N,N*-Dicyclohexylcarbodiimide
DDQ	2,3-Dichloro-5,6-dicyano-1,4-benzoquinone
DEAD	Diethyl azodicarboxylate
DIBAL	Diisobutylaluminium hydride
DIBAL/ DIBAH	Diisobutylaluminum hydride
DMAP	4-Dimethylaminopyridine
DME	1,2-Dimethoxyethane
DMF	*N,N*-Dimethylformamide
DMSO	Dimethyl sulphoxide.
DSS	Sodium 2,2-dimethyl-2-silapentane-5-sulphonate
EPR	Electron paramagnetic resonance
ESR	Electron spin resonance
$h\nu$	Irradiation with light
LAH	Lithium aluminium hydride
LDA	Lithium diisopropylamide
MCPBA	*meta*-Chloroperoxybenzoic acid
MOM	Methoxymethyl
MVK	Methyl vinyl ketone
NBS	*N*-Bromosuccinimide
NMR	Nuclear magnetic resonance
PCC	Pyridinium chlorochromate
PMB	*p*-methoxybenzoate
PMR	Proton magnetic resonance
PNBA	*p*-nitrobenzoic acid
NMR	Nuclear magnetic resonance
PCC	Pyridinium chlorochromate
PMB	*p*-methoxybenzoate
PMR	Proton magnetic resonance
PNBA	*p*-nitrobenzoic acid
PPA	Polyphosphoric acid
THF	Tetrahydrofuran
TMS	Tetramethylsilane
TS	Transition state
UV	Ultraviolet

About the Author

After graduating from Visva-Bharati University in 1990 Dr. Goutam Brahmachari completed his masters in 1992 and received his Ph.D. degree in 1997 from the same University. With about nine years of teaching experience, he has published more than thirty research papers in the field of organic chemistry (Natural Products) in national and international journals of repute, and has authored the book, *Organic Name Reactions: A Unified Approach* (Narosa Publishing House, New Delhi; Co-published by Alpha Science International, Oxford) and also edited the book, *Chemistry of Natural Products: Recent Trends & Developments* (Research Signpost, India) along with four invited book chapters. Dr. Brahmachari presently holds the position of Reader in Chemistry, Visva-Bharati University, Santiniketan, West Bengal.